Elliptic Partial Differential Equations and
Quasiconformal Mappings in the Plane

Princeton Mathematical Series

EDITORS: PHILLIP A. GRIFFITHS, JOHN N. MATHER, AND ELIAS M. STEIN

1. The Classical Groups *by Hermann Weyl*
8. Theory of Lie Groups: I *by C. Chevalley*
9. Mathematical Methods of Statistics *by Harald Cramér*
14. The Topology of Fibre Bundles *by Norman Steenrod*
17. Introduction to Mathematical Logic, Vol. I *by Alonzo Church*
19. Homological Algebra *by H. Cartan and S. Eilenberg*
28. Convex Analysis *by R. T. Rockafellar*
30. Singular Integrals and Differentiability Properties of Functions *by E. M. Stein*
32. Introduction to Fourier Analysis on Euclidean Spaces *by E. M. Stein and G. Weiss*
33. Étale Cohomology *by J. S. Milne*
35. Three-Dimensional Geometry and Topology, Volume 1 *by William P. Thurston. Edited by Silvio Levy*
36. Representation Theory of Semisimple Groups: An Overview Based on Examples *by Anthony W. Knapp*
38. Spin Geometry *by H. Blaine Lawson, Jr., and Marie-Louise Michelsohn*
43. Harmonic Analysis: Real Variable Methods, Orthogonality, and Oscillatory Integrals *by Elias M. Stein*
44. Topics in Ergodic Theory *by Ya. G. Sinai*
45. Cohomological Induction and Unitary Representations *by Anthony W. Knapp and David A. Vogan, Jr.*
46. Abelian Varieties with Complex Multiplication and Modular Functions *by Goro Shimura*
47. Real Submanifolds in Complex Space and Their Mappings *by M. Salah Baouendi, Peter Ebenfelt, and Linda Preiss Rothschild*
48. Elliptic Partial Differential Equations and Quasiconformal Mappings in the Plane *by Kari Astala, Tadeusz Iwaniec, and Gaven Martin*

Elliptic Partial Differential Equations and Quasiconformal Mappings in the Plane

Kari Astala, Tadeusz Iwaniec, and Gaven Martin

PRINCETON UNIVERSITY PRESS
PRINCETON AND OXFORD

Copyright © 2009 by Princeton University Press

Published by Princeton University Press
41 William Street, Princeton, New Jersey 08540

In the United Kingdom: Princeton University Press
6 Oxford Street, Woodstock, Oxfordshire OX20 1TW

All Rights Reserved

Library of Congress Cataloging-in-Publication Data

Astala, Kari, 1953-
   Elliptic partial differential equations and quasiconformal mappings in the plane / Kari Astala, Tadeusz Iwaniec, and Gaven Martin.
     p. cm. – (Princeton mathematical series ; 48)
   Includes bibliographical references and index.
   ISBN 978-0-691-13777-3 (hardcover : acid-free paper) 1. Differential equations, Elliptic. 2. Quasiconformal mappings. I. Iwaniec, Tadeusz. II. Martin, Gaven. III. Title.
   QA377.A836 2009
   515'.93–dc22
                                                2008037099

British Library Cataloging-in-Publication Data is available

This book has been composed in LaTeX

The publisher would like to acknowledge the authors of this volume for providing the camera-ready copy from which this book was printed.

Printed on acid-free paper. ∞

press.princeton.edu

Printed in the United States of America

10 9 8 7 6 5 4 3 2 1

To our families
Tuulikki, Eero & Eeva
Grażyna & Krystyna
Dianne, Jennifer & Amy

Contents

Preface   xv

1 Introduction   1
   1.1 Calculus of Variations, PDEs and Quasiconformal Mappings . . .   2
   1.2 Degeneracy   6
   1.3 Holomorphic Dynamical Systems   8
   1.4 Elliptic Operators and the Beurling Transform   9

2 A Background in Conformal Geometry   12
   2.1 Matrix Fields and Conformal Structures   12
   2.2 The Hyperbolic Metric   15
   2.3 The Space $\mathbf{S}(2)$ 17
   2.4 The Linear Distortion   21
   2.5 Quasiconformal Mappings   22
   2.6 Radial Stretchings   28
   2.7 Hausdorff Dimension   30
   2.8 Degree and Jacobian   32
   2.9 A Background in Complex Analysis   34
     2.9.1 Analysis with Complex Notation   34
     2.9.2 Riemann Mapping Theorem and Uniformization   36
     2.9.3 Schwarz-Pick Lemma of Ahlfors   37
     2.9.4 Normal Families and Montel's Theorem   39
     2.9.5 Hurwitz's Theorem   40
     2.9.6 Bloch's Theorem   40
     2.9.7 The Argument Principle   41
   2.10 Distortion by Conformal Mapping   41
     2.10.1 The Area Formula   41
     2.10.2 Koebe $\frac{1}{4}$-Theorem and Distortion Theorem   44

3 The Foundations of Quasiconformal Mappings   48
   3.1 Basic Properties   48
   3.2 Quasisymmetry   49
   3.3 The Gehring-Lehto Theorem   51
     3.3.1 The Differentiability of Open Mappings   52

3.4	Quasisymmetric Maps Are Quasiconformal	58
3.5	Global Quasiconformal Maps Are Quasisymmetric	64
3.6	Quasiconformality and Quasisymmetry: Local Equivalence	70
3.7	Lusin's Condition $\mathcal{N}$ and Positivity of the Jacobian	72
3.8	Change of Variables	76
3.9	Quasisymmetry and Equicontinuity	78
3.10	Hölder Regularity	80
3.11	Quasisymmetry and δ-Monotone Mappings	83

4 Complex Potentials — 92

4.1	The Fourier Transform	98
	4.1.1 The Fourier Transform in L^1 and L^2	99
	4.1.2 Fourier Transform on Measures	100
	4.1.3 Multipliers	100
	4.1.4 The Hecke Identities	101
4.2	The Complex Riesz Transforms R^k	102
	4.2.1 Potentials Associated with R^k	103
4.3	Quantitative Analysis of Complex Potentials	104
	4.3.1 The Logarithmic Potential	105
	4.3.2 The Cauchy Transform	109
4.4	Maximal Functions and Interpolation	117
	4.4.1 Interpolation	117
	4.4.2 Maximal Functions	120
4.5	Weak-Type Estimates and L^p-Bounds	124
	4.5.1 Weak-Type Estimates for Complex Riesz Transforms	124
	4.5.2 Estimates for the Beurling Transform $\mathcal{S}$	128
	4.5.3 Weighted L^p-Theory for $\mathcal{S}$	130
4.6	BMO and the Beurling Transform	131
	4.6.1 Global John-Nirenberg Inequalities	132
	4.6.2 Norm Bounds in BMO	134
	4.6.3 Orthogonality Properties of $\mathcal{S}$	135
	4.6.4 Proof of the Pointwise Estimates	137
	4.6.5 Commutators	143
	4.6.6 The Beurling Transform of Characteristic Functions	146
4.7	Hölder Estimates	147
	4.7.1 Hölder Bounds for the Beurling Transform	147
	4.7.2 The Inhomogeneous Cauchy-Riemann Equation	149
4.8	Beurling Transforms for Boundary Value Problems	150
	4.8.1 The Beurling Transform on Domains	151
	4.8.2 L^p-Theory	153
	4.8.3 Complex Potentials for the Dirichlet Problem	155
4.9	Complex Potentials in Multiply Connected Domains	158

5 The Measurable Riemann Mapping Theorem: The Existence Theory of Quasiconformal Mappings — 161
- 5.1 The Basic Beltrami Equation — 163
- 5.2 Quasiconformal Mappings with Smooth Beltrami Coefficient — 165
- 5.3 The Measurable Riemann Mapping Theorem — 168
- 5.4 L^p-Estimates and the Critical Interval — 172
  - 5.4.1 The Caccioppoli Inequalities — 174
  - 5.4.2 Weakly Quasiregular Mappings — 178
- 5.5 Stoilow Factorization — 178
- 5.6 Factoring with Small Distortion — 184
- 5.7 Analytic Dependence on Parameters — 185
- 5.8 Extension of Quasisymmetric Mappings of the Real Line — 189
  - 5.8.1 The Douady-Earle Extension — 191
  - 5.8.2 The Beurling-Ahlfors Extension — 192
- 5.9 Reflection — 192
- 5.10 Conformal Welding — 193

6 Parameterizing General Linear Elliptic Systems — 195
- 6.1 Stoilow Factorization for General Elliptic Systems — 196
- 6.2 Linear Families of Quasiconformal Mappings — 198
- 6.3 The Reduced Beltrami Equation — 202
- 6.4 Homeomorphic Solutions to Reduced Equations — 204
  - 6.4.1 Fabes-Stroock Theorem — 206

7 The Concept of Ellipticity — 210
- 7.1 The Algebraic Concept of Ellipticity — 211
- 7.2 Some Examples of First-Order Equations — 213
- 7.3 General Elliptic First-Order Operators in Two Variables — 214
  - 7.3.1 Complexification — 215
  - 7.3.2 Homotopy Classification — 217
  - 7.3.3 Classification; $n = 1$ — 217
- 7.4 Partial Differential Operators with Measurable Coefficients — 221
- 7.5 Quasilinear Operators — 222
- 7.6 Lusin Measurability — 223
- 7.7 Fully Nonlinear Equations — 226
- 7.8 Second-Order Elliptic Systems — 231
  - 7.8.1 Measurable Coefficients — 233

8 Solving General Nonlinear First-Order Elliptic Systems — 235
- 8.1 Equations Without Principal Solutions — 236
- 8.2 Existence of Solutions — 237
- 8.3 Proof of Theorem 8.2.1 — 239
  - 8.3.1 Step 1: H Continuous, Supported on an Annulus — 239
  - 8.3.2 Step 2: Good Smoothing of H — 242
  - 8.3.3 Step 3: Lusin-Egoroff Convergence — 244
  - 8.3.4 Step 4: Passing to the Limit — 246

8.4	Equations with Infinitely Many Principal Solutions	248
8.5	Liouville Theorems	249
8.6	Uniqueness	253
	8.6.1 Uniqueness for Normalized Solutions	255
8.7	Lipschitz $H(z,w,\zeta)$	256

9 Nonlinear Riemann Mapping Theorems   259
9.1 Ellipticity and Change of Variables 261
9.2 The Nonlinear Mapping Theorem: Simply Connected Domains . 263
   9.2.1 Existence 264
   9.2.2 Uniqueness 267
9.3 Mappings onto Multiply Connected Schottky Domains 269
   9.3.1 Some Preliminaries 271
   9.3.2 Proof of the Mapping Theorem 9.3.4 273

10 Conformal Deformations and Beltrami Systems   275
10.1 Quasilinearity of the Beltrami System 275
   10.1.1 The Complex Equation 276
10.2 Conformal Equivalence of Riemannian Structures 279
10.3 Group Properties of Solutions 280
   10.3.1 Semigroups 283
   10.3.2 Sullivan-Tukia Theorem 285
   10.3.3 Ellipticity Constants 287

11 A Quasilinear Cauchy Problem   289
11.1 The Nonlinear $\bar{\partial}$-Equation 289
11.2 A Fixed-Point Theorem 290
11.3 Existence and Uniqueness 291

12 Holomorphic Motions   293
12.1 The λ-Lemma 294
12.2 Two Compelling Examples 296
   12.2.1 Limit Sets of Kleinian Groups 296
   12.2.2 Julia Sets of Rational Maps 297
12.3 The Extended λ-Lemma 298
   12.3.1 Holomorphic Motions and the Cauchy Problem 299
   12.3.2 Holomorphic Axiom of Choice 300
12.4 Distortion of Dimension in Holomorphic Motions 306
12.5 Embedding Quasiconformal Mappings in Holomorphic Flows . . 309
12.6 Distortion Theorems 310
12.7 Deformations of Quasiconformal Mappings 313

13 Higher Integrability   316
13.1 Distortion of Area 317
   13.1.1 Initial Bounds for Distortion of Area 318
   13.1.2 Weighted Area Distortion 319

	13.1.3 An Example 323
	13.1.4 General Area Estimates 324
13.2	Higher Integrability 327
	13.2.1 Integrability at the Borderline........................... 330
	13.2.2 Distortion of Hausdorff Dimension 332
13.3	The Dimension of Quasicircles 333
	13.3.1 Symmetrization of Beltrami Coefficients 336
	13.3.2 Distortion of Dimension 338
13.4	Quasiconformal Mappings and BMO 343
	13.4.1 Quasiconformal Jacobians and A_p-Weights 345
13.5	Painlevé's Theorem: Removable Singularities 347
	13.5.1 Distortion of Hausdorff Measure 351
13.6	Examples of Nonremovable Sets 357

14 L^p-Theory of Beltrami Operators   362

14.1	Spectral Bounds and Linear Beltrami Operators 365
14.2	Invertibility of the Beltrami Operators 366
	14.2.1 Proof of Invertibility; Theorem 14.0.4 368
14.3	Determining the Critical Interval 369
14.4	Injectivity in the Borderline Cases 373
	14.4.1 Failure of Factorization in $W^{1,q}$ 376
	14.4.2 Injectivity and Liouville-Type Theorems 378
14.5	Beltrami Operators; Coefficients in VMO 382
14.6	Bounds for the Beurling Transform 385

15 Schauder Estimates for Beltrami Operators   389

15.1	Examples 390
15.2	The Beltrami Equation with Constant Coefficients 391
15.3	A Partition of Unity 392
15.4	An Interpolation 394
15.5	Hölder Regularity for Variable Coefficients 395
15.6	Hölder-Caccioppoli Estimates 398
15.7	Quasilinear Equations 400

16 Applications to Partial Differential Equations   403

16.1	The Hodge $*$ Method.. 404
	16.1.1 Equations of Divergence Type: The A-Harmonic Operator 405
	16.1.2 The Natural Domain of Definition 406
	16.1.3 The A-Harmonic Conjugate Function 408
	16.1.4 Regularity of Solutions 409
	16.1.5 General Linear Divergence Equations 411
	16.1.6 A-Harmonic Fields 414
16.2	Topological Properties of Solutions 418
16.3	The Hodographic Method 420
	16.3.1 The Continuity Equation 420

	16.3.2 The p-Harmonic Operator $\operatorname{div}|\nabla|^{p-2}\nabla$ 423
	16.3.3 Second-Order Derivatives 424
	16.3.4 The Complex Gradient 427
	16.3.5 Hodograph Transform for the p-Laplacian 430
	16.3.6 Sharp Hölder Regularity for p-Harmonic Functions 431
	16.3.7 Removing the Rough Regularity in the Gradient 432
16.4	The Nonlinear $\mathcal{A}$-Harmonic Equation 433
	16.4.1 δ-Monotonicity of the Structural Field 435
	16.4.2 The Dirichlet Problem 441
	16.4.3 Quasiregular Gradient Fields and $C^{1,\alpha}$-Regularity 445
16.5	Boundary Value Problems 449
	16.5.1 A Nonlinear Riemann-Hilbert Problem 452
16.6	G-Compactness of Beltrami Differential Operators 456
	16.6.1 G-Convergence of the Operators $\partial_{\bar{z}} - \mu_j \partial_z$ 457
	16.6.2 G-Limits and the Weak*-Topology 459
	16.6.3 The Jump from $\partial_{\bar{z}} - \nu\overline{\partial_z}$ to $\partial_{\bar{z}} - \mu\partial_z$ 461
	16.6.4 The Adjacent Operator's Two Primary Solutions 462
	16.6.5 The Independence of $\Phi_z(z)$ and $\Psi_z(z)$ 463
	16.6.6 Linear Families of Quasiregular Mappings 464
	16.6.7 G-Compactness for Beltrami Operators 467

17 PDEs Not of Divergence Type: Pucci's Conjecture   472
17.1 Reduction to a First-Order System 475
17.2 Second-Order Caccioppoli Estimates 476
17.3 The Maximum Principle and Pucci's Conjecture 478
17.4 Interior Regularity 481
17.5 Equations with Lower-Order Terms 483
      17.5.1 The Dirichlet Problem 486
17.6 Pucci's Example 488

18 Quasiconformal Methods in Impedance Tomography: Calderón's Problem   490
18.1 Complex Geometric Optics Solutions 493
18.2 The Hilbert Transform $\mathcal{H}_\sigma$ 495
18.3 Dependence on Parameters 497
18.4 Nonlinear Fourier Transform 499
18.5 Argument Principle 502
18.6 Subexponential Growth 504
18.7 The Solution to Calderón's Problem 510

19 Integral Estimates for the Jacobian   514
19.1 The Fundamental Inequality for the Jacobian 514
19.2 Rank-One Convexity and Quasiconvexity 518
      19.2.1 Burkholder's Theorem 521
19.3 L^1-Integrability of the Jacobian 523

20 Solving the Beltrami Equation: Degenerate Elliptic Case — 527

- 20.1 Mappings of Finite Distortion; Continuity 529
  - 20.1.1 Topological Monotonicity 530
  - 20.1.2 Proof of Continuity in $W^{1,2}$ 534
- 20.2 Integrable Distortion; $W^{1,2}$-Solutions and Their Properties . . . 534
- 20.3 A Critical Example 540
- 20.4 Distortion in the Exponential Class 543
  - 20.4.1 Example: Regularity in Exponential Distortion 543
  - 20.4.2 Beltrami Operators for Degenerate Equations 545
  - 20.4.3 Decay of the Neumann Series 549
  - 20.4.4 Existence Above the Critical Exponent 554
  - 20.4.5 Exponential Distortion: Existence of Solutions 557
  - 20.4.6 Optimal Regularity 560
  - 20.4.7 Uniqueness of Principal Solutions 563
  - 20.4.8 Stoilow Factorization 564
  - 20.4.9 Failure of Factorization in $W^{1,q}$ When $q < 2$ 567
- 20.5 Optimal Orlicz Conditions for the Distortion Function 570
- 20.6 Global Solutions 576
  - 20.6.1 Solutions on $\mathbb{C}$ 576
  - 20.6.2 Solutions on $\hat{\mathbb{C}}$ 578
- 20.7 A Liouville Theorem 579
- 20.8 Applications to Degenerate PDEs 580
- 20.9 Lehto's Condition 581

21 Aspects of the Calculus of Variations — 586

- 21.1 Minimizing Mean Distortion 586
  - 21.1.1 Formulation of the General Problem 588
  - 21.1.2 The L^1-Grötzsch Problem 588
  - 21.1.3 Sublinear Growth: Failure of Minimization 591
  - 21.1.4 Inverses of Homeomorphisms of Integrable Distortion . . . 592
  - 21.1.5 The Traces of Mappings with Integrable Distortion 595
- 21.2 Variational Equations 599
  - 21.2.1 The Lagrange-Euler Equations 601
  - 21.2.2 Equations for the Inverse Map 604
- 21.3 Mean Distortion, Annuli and the Nitsche Conjecture 606
  - 21.3.1 Polar Coordinates 611
  - 21.3.2 Free Lagrangians 612
  - 21.3.3 Lower Bounds by Free Lagrangians 613
  - 21.3.4 Weighted Mean Distortion 615
  - 21.3.5 Minimizers within the Nitsche Range 616
  - 21.3.6 Beyond the Nitsche Bound 618
  - 21.3.7 The Minimizing Sequence and Its BV-limit 619
  - 21.3.8 Correction Lemma 622

Appendix: Elements of Sobolev Theory and Function Spaces   **624**
   A.1   Schwartz Distributions 624
   A.2   Definitions of Sobolev Spaces 627
   A.3   Mollification 628
   A.4   Pointwise Coincidence of Sobolev Functions 630
   A.5   Alternate Characterizations 630
   A.6   Embedding Theorems 633
   A.7   Duals and Compact Embeddings 636
   A.8   Hardy Spaces and BMO 637
   A.9   Reverse Hölder Inequalities 640
   A.10 Variations of Sobolev Mappings 640

Basic Notation   **643**

Bibliography   **647**

Index   **671**

Preface

This book presents the most recent developments in the theory of planar quasiconformal mappings and their wide-ranging applications in partial differential equations and nonlinear analysis, conformal geometry, holomorphic dynamical systems, singular integral operators, inverse problems, the geometry of mappings and, more generally, the calculus of variations. It is a simply amazing fact that the mathematics that underpins the geometry, structure and dimension of such concepts as Julia sets and limit sets of Kleinian groups, the spaces of moduli of Riemann surfaces, conformal dynamical systems and so forth is the *very same* as that which underpins existence, regularity, singular set structure and so forth for precisely the most important class of differential equations one meets in physical applications, namely, second-order divergence-type equations. All these subjects are inextricably linked in two dimensions by the theory of quasiconformal mappings.

There have been profound developments in the three or four decades since the publication of Lars Ahlfors' beautiful little book [8] and the classical text of Olli Lehto and Kalle Virtanen [229]. Indeed, whole subjects have blossomed, conformal and holomorphic dynamics, holomorphic motions, nonlinear partial differential equations and connections with the calculus of variations, to name just a few.

This book gives a fairly comprehensive account of the modern theory, but for those planning to present a semester course in the theory of quasiconformal mappings and their applications in modern complex analysis, the contents of Chapters 3 and 5, with selected applications chosen from Chapters 12, 13 and some of the later chapters, should provide ample material at an easy pace. Further, the material in Chapter 4 presents a reasonable and self-contained introduction to harmonic analysis and the theory of singular integral operators in two dimensions.

The latter parts of the book present perhaps the most recent advances in the area. Indeed, more than a few results and proofs in this monograph are new. These chapters also serve to illustrate the wide applicability of the ideas and techniques developed in the earlier part of the book.

It is our pleasure to acknowledge the wide-ranging support we have had from a number of places that has made this book possible. First, we have all been partly supported by the Academy of Finland, the Marsden Fund of New

Zealand and the National Science Foundation of the United States at one time or another. We all shared Research in Peace fellowships at Institute Mittag-Leffler (Sweden) where the first real progress toward a book was made. Of course our home institutions—Massey University (New Zealand), Syracuse University (United States) and the University of Helsinki (Finland)—have all hosted and supported us as a group at various times.

There are many people to thank as well. In particular, Pekka Koskela let us use his notes on quasisymmetric functions (a good part of Chapter 3), Stanislaw Smirnov let us present his unpublished proof of the dimension bounds for quasicircles and Laszlo Lempert communicated Chirka's proof of the λ-lemma to us while we were at Oberwolfach. There are also the people who read various drafts of the book and made substantial and valuable comments. These include Tomasz Adamowicz, Samuel Dillon, Daniel Faraco, Peter Haïssinsky, Jarmo Jääskeläinen, Matti Lassas, Martti Nikunen, Jani Onninen, Lassi Päivärinta, Istvan Prause, Eero Saksman, Carlo Sbordone, Ignacio Uriarte-Tuero and Antti Vähäkangas. We would also like to thank the team at Princeton University Press—Kathleen Cioffi, Carol Dean, Lucy Day Hobor and Vickie Kearn—who skillfully guided us through the production process and whose considerable efforts improved this book.

Finally, during the writing of this book the "quasi-world" was saddened by the premature death of one of its leading figures, Juha Heinonen. We wish to record here the deep respect we have for Juha and the contributions he made. He was an inspiration to all of us.

Kari Astala, Tadeusz Iwaniec, Gaven Martin

Helsinki, Syracuse and Auckland, 2009

Elliptic Partial Differential Equations and Quasiconformal Mappings in the Plane

Chapter 1

Introduction

This book relates the most modern aspects and most recent developments in the theory of planar quasiconformal mappings and their application in conformal geometry, partial differential equations (PDEs) and nonlinear analysis. There are profound applications in such wide-ranging areas as holomorphic dynamical systems, singular integral operators, inverse problems, the geometry of mappings and, more generally, the calculus of variations—all of which are presented here. It is a simply amazing fact that the mathematics that underpins the geometry, structure and dimension of such concepts as Julia sets and limit sets of Kleinian groups, the spaces of moduli of Riemann surfaces, conformal dynamical systems and so forth is the *very same* as that which underpins existence, regularity, singular set structure and so forth for precisely the most important class of equations one meets in physical (and other) applications, namely, second-order divergence-type equations. All these theories are inextricably linked in two dimensions by the theory of quasiconformal mappings.

Because of these and other compelling applications, there has recently been considerable pressure to extend classical results from conformal geometry to more general settings, for instance, to obtain optimal bounds on the existence, regularity and geometric properties of solutions of quasilinear and general nonlinear systems in the plane both in the classical elliptic setting and now in the degenerate elliptic setting. Here one moves from the established theory of quasiconformal mappings, through the theory of weakly quasiregular mappings, and comes to the more general class of Sobolev mappings of finite distortion. This progression is natural as one seeks greater knowledge about the fine properties of these mappings for implementation. Even for such well-known problems as the nonlinear $\bar{\partial}$-problem, we find that precise L^2-bounds lead to a simple and beautiful proof of the extension theorem for holomorphic motions. In the same vein, we use optimal regularity to prove Pucci's conjecture, as well as related precise results to give a solution to Calderón's problem on impedance tomography and also to Painlevé's problem on the size and structure of removable singular sets for solutions to elliptic and degenerate elliptic equations.

These precise results are in a large part due to a new understanding of the relationship between quasiconformal mappings and holomorphic flows on the one hand, and, on the other, precise results on the L^p-invertibility of classes of singular integral operators called Beltrami operators. However, there have been other recent developments in the theory of quasiconformal mappings—notably in the field of analysis on metric spaces principally established by Heinonen and Koskela. These advances could not leave a book such as ours untouched, for they clarify many of the basic facts and the precise hypotheses necessary to prove them and often provide elementary and clear proofs. Thus the reader will find novelty and simplicity here even for the foundations of the theory, which now go back more than half a century.

Another novelty in the approach of this book is the use of many of the significant advances in harmonic analysis made over the last few decades; these include H^1-BMO duality, maximal function estimates, the theory of nonlinear commutators and integral estimates for Jacobian determinants both above and below their natural Sobolev domain of definition, all crucial for our studies of optimal regularity and nonlinear PDEs, as well as the Painlevé problem on removable singularities. The reader will have ample opportunity to see these powerful modern techniques in diverse applications.

1.1 Calculus of Variations, PDEs and Quasiconformal Mappings

The strong interplay among the calculus of variations, partial differential equations and the geometric theory of mappings (which is what this book is all about) has a long and distinguished history—going back at least to d'Alembert who in 1746 first related the derivatives of the real and imaginary part of a complex function in his work on hydrodynamics [51, p. 497]. These equations came to be known as the Cauchy-Riemann equations.

Conservation laws and equations of motion or state in physics and mathematics are described by divergence-type second-order differential equations. This is no accident. It is a fundamental precept of physics that a system acts so as to minimize some action functional—Hamilton's principle of least action. Hamilton's principle applies quite generally to classical fields such as the electromagnetic, gravitational and even quantum fields. We are therefore naturally led to study the minima of energy functionals, regularity of minimizers and other aspects of the calculus of variations. We give a classical problem a review in the next section. Loosely, minima satisfy an associated Euler-Lagrange equation that appears in divergence form as a result of integration by parts in the derivation of the equation. Similar examples appear in continuum mechanics and materials science.

On the other hand, general conservation laws are described as follows. Suppose the flux density of a scalar quantity e, such as density, concentration, temperature or energy, is $q = \mathcal{A}(z, \nabla e)$, a function of the gradient of e. A basic

1.1. CALCULUS OF VARIATIONS

assumption of continuum physics is that the gain of the physical quantity in a domain Ω corresponds to the loss of this quantity across the boundary $\partial\Omega$. Thus

$$\int_{\partial\Omega} q \cdot \nu = \int_{\Omega} f$$

Here f denotes the source density and ν denotes the outer normal. The above identity is called the conservation law with respect to the flux q and leads (we describe how in Section 16.3) to the differential equation

$$\dot{e} + \text{div}(q) = \dot{e} + \text{div}\mathcal{A}(z, \nabla e) = f$$

This is a conservation law for the physical quantity e. In the steady-state case we obtain a second-order equation in divergence form for q.

With so many compelling applications in hand, it is no wonder that there is considerable interest in the topological and analytic properties of the minimizers of various functionals and also in solutions of second-order equations in divergence form. These topological and analytic properties describe, for instance, the flow lines of the field and the structure and size of any singular set.

Let us explain using an elementary example from the calculus of variations how related first- and second-order equations might arise. Consider deforming the unit disk $\mathbb{D}$ to another domain Ω minimizing energy. This was in fact Riemann's approach to his mapping theorem, and which he called the Dirichlet principle. He obtained the desired conformal mapping as an absolute minimizer of the Dirichlet energy. Weierstrass showed Riemann's argument was not generally valid, however Hilbert later ironed out the details—ultimately requiring some regularity of $\partial\Omega$. As this discussion suggests, the example is quite classical, but it's solution contains many key ideas and provides us with some important lessons.

Problem. *Given a simply connected domain Ω,*
  (a) find the homeomorphism of minimal energy mapping the disk to Ω,
  (b) find the minimizer subject to prescribed boundary values.

The energy of a mapping is defined as the Dirichlet integral, so we are asked to find

$$\min_{f:\mathbb{D}\to\Omega} \left\{ \int_{\mathbb{D}} \|Df(z)\|^2 \, dz \right\}, \qquad \|Df\|^2 = |f_x|^2 + |f_y|^2,$$

over all homeomorphisms, with the possible restriction $f|\partial\mathbb{D} = g_o$. In order to solve this problem (if it is possible at all), we should consider the correct function space to start looking for a solution. For the minimum to be finite, we certainly need for there to be some mapping f_0 satisfying the hypotheses (the gradient of f_0 should be square-integrable with correct boundary values). Given this mapping, we can then assume the existence of a sequence tending to the minimum (a *minimizing sequence*). Then comes the difficult problem of proving this sequence has a convergent subsequence whose limit is sufficiently

regular to satisfy the hypotheses (thus the need for a priori estimates). For the problem in hand, Hadamard's inequality for matrices $A \in \mathbb{R}^{2\times 2}(\mathbb{C})$ states $\|A\|^2 = \operatorname{tr}(A^t A) \geqslant 2 \det A$ and therefore gives the pointwise almost everywhere estimate
$$\|Df(z)\|^2 \geqslant 2\, J(z,f) = 2 \det Df(z)$$
(we consider only orientation-preserving homeomorphisms, meaning that the Jacobian $J(z,f) \geqslant 0$ almost everywhere in $\mathbb{D}$.) Then for every homeomorphism of Sobolev class $W^{1,2}(\mathbb{D})$, we have
$$\int_{\mathbb{D}} \|Df\|^2 \geqslant 2 \int_{\mathbb{D}} J(z,f) = 2|f(\mathbb{D})| = 2|\Omega|,$$
providing a lower bound on the minimum. Consequently, if there is to be an absolute minimizer f achieving this lower bound we must have it solving the first(!)-order equation for an absolute minimizer
$$\|Df(z)\|^2 = 2\, J(z,f)$$
Some linear algebra (we have equality in Hadamard's estimate) shows this to be equivalent to
$$D^t f(z)\, Df(z) = J(z,f)\, \mathbf{I},$$
where $\mathbf{I}$ is the identity matrix. This is the equation for a conformal mapping, of course—in complex notation this system is the Cauchy-Riemann equations (which points to the virtue of complex notation). Back to our problem, if we prescribe the boundary values and they happen not to be those of a conformal mapping, then a minimizer cannot achieve our a priori lower bound. Another approach is to vary a supposed minimizer f by a parameterized family of homeomorphisms of $\mathbb{D}$ that are the identity near the boundary, say φ_t normalized so $\varphi_0(z) = z$. Since f is a minimizer we must have
$$\frac{d}{dt} \int_{\mathbb{D}} \|D(f \circ \varphi_t)\|^2 \bigg|_{t=0} = 0,$$
leading to the second-order Euler-Lagrange equation for f, $\operatorname{div} Df = \Delta f = 0$. Thus the minimum should be a harmonic mapping with the given boundary values, and the question boils down to whether our prescribed boundary values g_0 have a harmonic homeomorphic extension to $\mathbb{D}$. The Poisson formula gives a harmonic function, and we are left to discuss the topological properties of this solution. A way forward here is to show that the Jacobian is continuous and does not vanish (so local injectivity) and use the monodromy theorem, but the geometry of the domain and the boundary values must come into play. For instance, without some convexity assumption on Ω the mean value of g_0 may lie outside Ω. It is a classical theorem of Choquet, Kneser and Rado that as soon as Ω is convex, one can solve the posed problem with homeomorphic boundary data and the solution is a smooth diffeomorphism.

We may consider the above problem in more general circumstances. For instance, if $H : \Omega \to \mathbb{R}^{2\times 2}$, symmetric and positive definite, is some measurable

1.1. CALCULUS OF VARIATIONS

function describing some material property of Ω, we could seek to minimize the new energy functional

$$\int_{\mathbb{D}} \langle H(f(z))Df(z), Df(z) \rangle$$

We use Hadamard's inequality in the form

$$\langle \sqrt{H}Df, \sqrt{H}Df \rangle \geq 2\sqrt{\det H} \det(Df)$$

and, as before, an absolute minimizer must satisfy the nonlinear PDE

$$D^t f H(f) Df = \sqrt{\det H(f)}\, J(z,f)\, \mathbf{I}$$

It is only in two dimensions that such an equation is not overdetermined (this accounts for higher-dimensional rigidity), and we have the possibility of finding a solution in quite reasonable generality. For conformal geometry we are interested in the case $\det H \equiv 1$ yielding the nonlinear Beltrami equation

$$D^t f\, H(f)\, Df = J(z,f)\mathbf{I}$$

If in the above we consider a tensor field $G : \mathbb{D} \to \mathbb{R}^{2\times 2}$, $\det G \equiv 1$, we have

$$D^t f\, Df = J(z,f)\, G,$$

equivalent to a linear (over $\mathbb{C}$) equation called the complex Beltrami equation,

$$\frac{\partial f}{\partial \bar{z}} = \mu(z)\, \frac{\partial f}{\partial z},$$

which we will spend quite a bit of time discussing. Finally, if we consider a constrained problem and look for the Euler-Lagrange equation we are quickly led to second-order equations in divergence form (arising from the necessary integration by parts) for the real and imaginary parts of $f = u + iv$,

$$\operatorname{div} G^{-1} \nabla u = 0, \qquad \operatorname{div} G^{-1} \nabla v = 0,$$

and ultimately to more general second-order equations in divergence form.

There are a few important points we would like to draw from this discussion regarding minima of variational problems:

- Unconstrained or absolute minimizers of variational functionals are likely to satisfy first-order differential equations.

- Constrained or stationary mappings will likely satisfy a second-order differential equation.

- We may well find stationary solutions that are not minimizers. Indeed, there might not be a minimizer within the class of homeomorphisms.

Of course, in the most general setting of multiple connected domains, one would consider minimizers in a given homotopy class of maps between domains, or more generally, homotopy classes of maps between Riemann surfaces. Moreover we would seek to minimize more general functionals. Here we find clear connections with Teichmüller theory, surface topology and so forth.

A significant portion of this book is given over to the study of the equations like those we have discovered above where we will seek existence, uniqueness and optimal regularity and so forth for their solutions—and also for the counterparts to these equations in other settings. Later we shall discuss recent developments in the study of existence and uniqueness properties for mappings between planar domains whose boundary values are prescribed and have the smallest mean distortion—this will bring the relevance of the first example discussed above back into focus because of a surprising connection with harmonic mappings and other surprises as well. Indeed, the analogy here with Teichmüller theory is quite strong. This theory is partly concerned with extremal quasiconformal mappings in a homotopy class. These mappings minimize the L^∞-norm of the distortion. We investigate what happens when the L^1-norm of the distortion is minimized instead. Further, in these studies we will find many new and unexpected phenomena concerning existence, uniqueness and regularity for these extremal problems where the functionals are polyconvex but typically not convex. These seem to differ markedly from phenomena observed when studying multi-well functionals in the calculus of variations. The phenomena observed concerning mappings between annuli present a case in point.

In two dimensions, the methods of complex analysis, conformal geometry and quasiconformal mappings provide powerful techniques, not available in other dimensions, to solve highly nonlinear partial differential equations, especially those in divergence form. Of course the relevance of divergence-type equations to quasiconformal mappings is not new. It has been evident to researchers for at least 70 years, beginning with M.A. Lavrentiev [224], C.B. Morrey [271, 273, 272], R. Caccioppoli [83], L. Bers and L. Nirenberg [54, 56, 57], B. Bojarski [68], Finn [126, 127] and Serrin [325], among many others. In the literature one finds concrete applications in materials science, particularly, nonlinear elasticity, gas flow and fluid flow, and in the calculus of variations going back generations.

One of the primary aims of this book is to give a thorough account of this classical theory from a modern perspective and connect it with the most recent developments.

1.2 Degeneracy

As we have suggested, the equations we consider arise naturally in hydrodynamics, nonlinear elasticity, holomorphic dynamics and several other areas. A good part of this book is concerned with these equations at the extreme limits of regularity and related assumptions on the coefficients. A particular aim is

1.2. DEGENERACY

to develop tools to handle these situations where a system of equations might degenerate. Here is a natural example.

In two-dimensional hydrodynamics, the fluid velocity (the gradient of the potential function—see for instance (16.51)—satisfies a Beltrami equation that degenerates as the flow approaches a critical value, the local speed of sound; see (16.54).

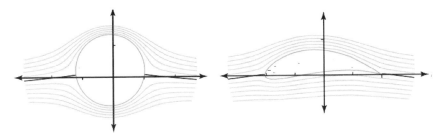

Subsonic fluid flow around a disk and Joukowski aerofoil

What happens as we break the speed of sound? In the 1950s when this was a problem of particular importance, the existence of a shock wave boundary was supposed, presupposing, albeit with good evidence, the particular structure of the singular set; see for instance [100, 144]. On one side of the shock boundary one had an elliptic equation, and on the other a hyperbolic equation. The very early approaches had to assume some degree of analyticity and used various schemes of successive approximations: the Rayleigh-Janzen expansion of the potential function in a power series in the stream Mach number or a modification of this method due to Prandtl and the solution of mixed (transonic) flows by means of power series in the space variables. Perturbative methods were also employed, most based on von Kármán's similarity law for transonic flow [207]. The state of the art as of the late 1950s is described in L. Bers' well-known book [54], and although there has been a great deal of literature on the subject since, most has focused on the study of shock waves in a similar sort of setup (and of course in higher dimensions). In this book we will describe the precise limits of existence and regularity and the structure of the singular set in the degenerate setting—but where there is no shock wave. This allows for isolated points (or even Cantor sets) where one might have degeneracies such as infinite density or pressure. The precise conditions are described in terms of bounded mean oscillation (BMO) bounds on the distortion function of the coefficient - leading to the theory of mappings with exponentially integrable distortion. This was first realized by G. David [103], and here we present substantial sharpening and refinement of these early results. When applied in the setting described above, this theory shows the topological properties of the streamlines and so forth to be the same as those for subsonic incompressible flows (really the Stoilow factorization theorem showing that these mappings are topologically equivalent to analytic mappings).

Thus a significant problem addressed in this book is to see how to relax the classical assumptions on the Beltrami equations making them uniformly elliptic, so as to study the nonuniformly elliptic (that is, degenerate elliptic) setting and yet save as much of the theory as possible.

1.3 Holomorphic Dynamical Systems

There are two basic examples of holomorphic dynamical systems. First, is the classical Fatou-Julia theory of iteration of rational mappings of the sphere, [90, 123, 203, 262]. Hardly anyone has not seen the beautiful pictures [293] of the Mandelbrot set and associated Julia sets of quadratic mappings.

The Julia set of a quadratic polynomial.

The theory of quasiconformal maps has played a key role in the study of these conformal dynamical systems ever since D. Sullivan, A. Douady and their coauthors introduced them to the theory [108, 111, 237, 257, 341]. Ideas such as quasiconformal surgery show how one conformal dynamical system can be constructed from another.

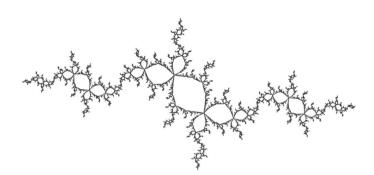

The Julia set after quasiconformal surgery: grafted with Douady's rabbit.

1.4. ELLIPTIC OPERATORS

A crucial discovery for us, which underpins a good deal of our approach in this book, is the concept of holomorphic motions intoduced by R. Mañé, P. Sad and D. Sullivan [237] and the subsequent conjectures on the extension of these motions by Sullivan and W.P. Thurston [342] and the solution by Z. Słodkowski [329]. This discovery really shows the notions of holomorphically parameterized flows and quasiconformal mappings to be inextricably linked.

In this book we will provide tools that allow one to study the structure, dimension and other properties of Julia sets. As far as the question of degeneracy goes, we will see that the Julia set of $\lambda z + z^2$ is a $\frac{1+|\lambda|}{1-|\lambda|}$-quasicircle if $|\lambda| < 1$ (equivalently for $z^2 + c$ when $c = \lambda/2 - \lambda^2/4$ lies in the primary component of the Mandelbrot set) and degeneracy occurs as $|\lambda| \to 1$. The dynamical systems obtained are quasiconformally equivalent on hyperbolic components of parameter space. The intriguing question of what happens as $|\lambda| \to 1$ (or generally moves to the boundary of a hyperbolic component) and the uniform bounds in the theory of quasiconformal mappings are lost is still to some measure unresolved. Haïssinsky has shown, using David's work, that for real $\lambda \nearrow 1$ ($c \nearrow \frac{1}{4}$) the sequence of Julia sets converges to a Jordan curve—the cauliflower Julia set—that is the image of the unit circle under a mapping of exponentially integrable distortion.

The second classical example of quasiconformal mappings being applied in conformal dynamical systems is the way they arise naturally in the study of Kleinian groups; through Teichmüller theory and moduli spaces. The modern approach goes back to Bers' seminal work on simultaneous uniformization [53] and Ahlfors' use of quasiconformal mappings in proving geometric finiteness [5]. The key idea again is that in moduli space the Kleinian groups in question are quasiconformally equivalent. What happens as one goes to the boundary and considers, for instance, degenerating sequences of quasi-fuchsian groups? Again we loose the uniform estimates needed in the classical theory of quasiconformal mappings and need to analyze a degenerate situation. We hope that mappings of finite distortion may play a future role in the analytic understanding of these questions.

1.4 Elliptic Operators and the Beurling Transform

The types of first-order equations $\mathcal{L}f = 0$ we have seen above have evolved from study of the Cauchy-Riemann operators,

$$\mathcal{L}_1 f = \frac{\partial}{\partial \bar{z}} f, \qquad \mathcal{L}_2 f = \frac{\partial}{\partial z} f$$

The solutions to $\mathcal{L}_i f = 0$, $i = 1, 2$, represent analytic and anti-analytic functions. A quantatative distinction between these two classes of mappings is that the former are orientation-preserving and the later orientation-reversing (or positive versus negative Jacobian determinant). In fact, this topological dichotomy of solutions applies to *all* first-order elliptic PDEs in the complex plane. The continuous deformation of a general elliptic system $\mathcal{L}f = 0$, perhaps by varying the coefficients, will never change the orientation of solutions unless ellipticity is violated at some moment.

This idea leads to the homotopy classification of all first-order elliptic systems and the corresponding differential operators into the two classes represented by the Cauchy-Riemann equations $\mathcal{L}_1 f = 0$ and its dual $\mathcal{L}_2 f = 0$. The fundamental connection between these classes is made via the Beurling transform, about which we will have much to say. It is a singular integral operator $\mathcal{S}$ of Calderón-Zygmund type bounded in $L^p(\mathbb{C})$, $1 < p < \infty$. It is the remarkable property

$$\mathcal{S} \circ \frac{\partial}{\partial \bar{z}} = \frac{\partial}{\partial z} : C_0^\infty(\mathbb{C}) \to C_0^\infty(\mathbb{C})$$

intertwining the Cauchy-Riemann operators that makes it so important in the L^p-theory of elliptic operators. There are six homotopy classes of second-order elliptic operators in the complex plane, or three equivalence classes of elliptic equations, comprising of combinations of the z- and $\bar{z}$-derivatives. The most important of these is the complex Laplace equation $\frac{\partial^2}{\partial z \partial \bar{z}} f = 0$. Notice that for this equation the "factors" $\frac{\partial}{\partial \bar{z}}$ and $\frac{\partial}{\partial z}$ come from different homotopy classes.

For the other two classes, the first-order factors come from the same homotopy class and this partly explains why the equations and their solutions have significantly different features. For instance, the Fredholm alternative fails; as an example, the equation $f_{\bar{z}\bar{z}} = 0$, $f|\partial\mathbb{D} = 0$, admits the uncountable family of solutions $f(z) = (1 - |z|^2)h(z)$, where h is holomorphic and continuous in the closed unit disk.

A major innovation in this book is the study of second-order PDEs of divergence form in the complex plane,

$$\text{div}\mathcal{A}(z, \nabla u) = 0 \tag{1.1}$$

when $\mathcal{A}$ is only supposed δ-monotone, with *no additional regularity assumption*. Here we are still able to obtain a reduction to a first-order system for the complex gradient $f = u_z$ of a solution and show that it is a quasiregular mapping, if $\mathcal{A}$ is spatially independent. In this way the significant results we obtain for such mappings apply to show that f has good regularity and nice topological properties which the solution u then inherits.

Another important approach is via the duality given by the Hodge $*$ operator. Here we reduce the $\mathcal{A}$-harmonic equation (1.1) to the first-order system

$$-\mathcal{A}(z, \nabla u) = *\nabla v$$

for a function v called the $\mathcal{A}$-harmonic conjugate of u. This approach is particularly useful when $\mathcal{A}(z, \nabla u) = A(z)\nabla u$ for some measurable $A : \Omega \to \mathbb{R}^{2\times 2}$. This leads us to the quasiregular mapping $f = u + iv$ in much the same way as an analytic function is composed of a harmonic function and its harmonic conjugate.

1.4. ELLIPTIC OPERATORS

There are also very interesting questions concerning the convergence of *sequences* of operators that we shall address in the book. Here we will meet the notion of G-convergence, which emerges in quite a natural way and exploits the normal family (equicontinuity) properties of quasiregular mappings.

While we give evidence of substantial progress in the theory of elliptic second-order equations in the complex plane, we are sure there remains many interesting phenomena to be discovered and interesting connections to other areas of mathematics to be found.

Chapter 2

A Background in Conformal Geometry

We have mentioned the strong connections between PDEs and geometry. In this chapter we give a gentle introduction to conformal geometry in the plane and describe some of the connections between conformal and Riemannian geometry with PDEs in two dimensions. We shall also try to give a clear account of how the PDEs we shall spend much of our time studying arise from a number of differing perspectives.

In this book the development of the theory of quasiconformal mappings really starts at the beginning of Chapter 3. The material in the present chapter will be quite familiar to those with some experience in geometric function theory. Such readers may therefore skip this chapter and proceed to Chapter 3, returning to the present chapter for background material only when necessary.

2.1 Matrix Fields and Conformal Structures

Let $\mathbf{S}(2)$ denote the space of 2×2 positive definite symmetric matrices with real entries and having determinant equal to 1. A positive definite matrix can be used to define an inner product using the standard (Euclidean) inner product between vectors by the rule

$$\langle \eta, \eta \rangle_G = \langle \eta, G\eta \rangle \qquad (2.1)$$

Recall that a Riemannian structure is, roughly, an inner product defined on the tangent bundle of a manifold. Such a structure gives rise to a metric distance defined as the length of the shortest curve between two points. The length of a curve is the sum (integral) of the lengths of its tangent vectors.

Typically Riemannian structures are assumed smooth and second-order invariants such as curvature determine the local isometry type. The equations determining an isometry are nonlinear and overdetermined as they essentially

2.1. MATRIX FIELDS AND CONFORMAL STRUCTURES

prescribe the Jacobian of a mapping, and this is why compatibility conditions such as curvature are necessary. However, in this book, motivated by applications in PDEs and dynamics, we shall largely consider measurable Riemannian structures. Despite the loss of such higher-order invariants, we shall see that there is a surprisingly rich theory.

For planar domains various differential-geometric constructions are somewhat easier to describe, as we have global coordinates and the tangent bundle trivializes. In particular, an orientable Riemannian metric on a planar domain Ω can simply be viewed as a map $A : \Omega \to \mathbb{R}_+ \cdot \mathbf{S}(2)$, the space of symmetric positive definite matrices. If

$$A = A(z) = \begin{bmatrix} a & b \\ b & c \end{bmatrix},$$

then the length element induced by A is

$$ds = \langle dz, A(z)dz \rangle^{1/2} = \sqrt{a\,dx^2 + 2b\,dxdy + c\,dy^2} \tag{2.2}$$

The most important special case we shall deal with is the hyperbolic metric studied in the next section. We shall henceforth restrict ourselves only to orientable stuctures without saying so every time. This is equivalent to the restriction $\det A(z) > 0$.

If $B = B(z)$ is another Riemannian metric on Ω', a map $f : \Omega \to \Omega'$ is an isometry if and only if it preserves the inner-products. That is, for all $z \in \Omega$, if ξ, ζ are tangent vectors at z, then

$$\langle f_*\xi, f_*\zeta \rangle_B = \langle \xi, \zeta \rangle_A \tag{2.3}$$

Of course, $f_*\xi = Df(z)\xi$, and in view of (2.1) the equation (2.3) is written as

$$\langle Df(z)\xi, B(f(z))Df(z)\zeta \rangle = \langle \xi, A(z)\zeta \rangle \tag{2.4}$$

From here we quickly find that

$$D^t f(z) B(f(z)) Df(z) = A(z) \tag{2.5}$$

for all $z \in \Omega$. This is the first-order partial differential equation for a map between the metric structures determined by A and B to be an isometry.

If we simply take determinants of both sides of (2.5) and write $a(z) = \det^{1/2} A$ and $b(z) = \det^{1/2} B$, we obtain the equation

$$J(z, f)b(f(z)) = a(z), \qquad z \in \Omega \tag{2.6}$$

for the unknown mapping f. Solving (2.6) is already a formidable task. For instance, even when $b \equiv 1$, identifying those functions a that are the Jacobians of mappings is an important outstanding problem in analysis. Some results are known once one assumes some smoothness; see for instance [275, 310, 313].

Two Riemannian structures $\langle \cdot, \cdot \rangle_A$ and $\langle \cdot, \cdot \rangle_B$ on a domain Ω are said to be *conformally equivalent* if there is a positive function $\phi : \Omega \to \mathbb{R}_+$ such that

$$\langle \cdot, \cdot \rangle_{A(z)} = \phi(z) \langle \cdot, \cdot \rangle_{B(z)},$$

or equivalently, $A(z) = \phi(z) B(z)$. Locally, ϕ effects a change of scale, but all angles are preserved, as

$$\frac{\langle \xi, \zeta \rangle_{\phi A}}{\sqrt{\langle \xi, \xi \rangle_{\phi A} \langle \zeta, \zeta \rangle_{\phi A}}} = \frac{\langle \xi, \phi A \zeta \rangle}{\sqrt{\langle \xi, \phi A \xi \rangle \langle \zeta, \phi A \zeta \rangle}} = \frac{\langle \xi, \zeta \rangle_A}{\sqrt{\langle \xi, \xi \rangle_A \langle \zeta, \zeta \rangle_A}}$$

Measurable Structures and Conformal Mappings

Every Riemannian structure A on Ω is conformally equivalent to another, say

$$G : \Omega \to \mathbf{S}(2), \qquad (2.7)$$

for which the determinant is identically equal to 1, namely, $G = (\det A)^{-1/2} A$. We say that G is bounded if the set of matrices $\{G(z) : z \in \Omega\}$ is a bounded subset of the 2×2 matrices or, equivalently, bounded as a subset of $\mathbb{R}^4$.

Definition 2.1.1. *If $G : \Omega \to \mathbf{S}(2)$ is bounded and measurable, we call G a measurable conformal structure on Ω. If $H : \Omega' \to \mathbf{S}(2)$ is a conformal structure on Ω', a homeomorphic mapping $f : \Omega \to \Omega'$ is said to be* conformal *from (Ω, G) to (Ω', H) if f preserves angles. This means that, for $z \in \Omega$ and unit vectors ξ, ζ*

$$\langle f_* \xi, f_* \zeta \rangle_H = \phi \langle \xi, \zeta \rangle_G \qquad (2.8)$$

Here ϕ is an unspecified real-valued function. As G and H are only assumed measurable, such an equation is only supposed to hold almost everywhere and accordingly, ϕ is only assumed to be measurable.

However, we shall often have cause to abuse terminology. A homeomorphism (of sufficient Sobolev regularity) $f : \Omega \to \Omega'$ is holomorphic, or complex analytic, if and only if it preserves the standard conformal structure, that is, f satisfies (2.8) with $G = H \equiv \mathbf{I}$. Thus, when there are no obvious measurable conformal structures in sight, we shall use the term "conformal", as one usually does in complex analysis, to mean a holomorphic injection, an injective mapping having a complex derivative at each point of its domain. There will be very little opportunity for confusion here.

As a differential equation, (2.8) reads as

$$D^t f(z) H(f(z)) Df(z) = \phi(z) G(z) \qquad (2.9)$$

Since $\det(G) = \det(H) = 1$, we must in fact have $\phi(z) = J(z, f)$, and the resulting equation, called the *Beltrami system*, becomes

$$D^t f(z) H(f(z)) Df(z) = J(z, f) G(z) \qquad (2.10)$$

2.2. THE HYPERBOLIC METRIC

and will be the focus of much of our study. A first glance suggests this equation is overdetermined and nonlinear, however, we shall soon see that it is well determined and, while not always linear, it satisfies the nice property of being quasilinear, as does the related nonlinear equation

$$D^t f(z) H(z, f) D f(z) = J(z, f) G(z, f) \qquad (2.11)$$

It is difficult at first to appreciate the importance of solving such equations as (2.10). Let us point out simply that if $G = H \equiv \mathbf{I}$, then a solution f satisfies the Cauchy-Riemann equations and therefore represents an analytic equivalence between domains. As a further example, if $\Omega = \Omega' = \mathbb{D}$, the unit disk, then the existence of solutions shows that all Riemannian metrics on $\mathbb{D}$ are equivalent by a conformal change of variables. However, solutions do not exist in complete generality; one needs to place restrictions on G and H, such as boundedness, to guarantee ellipticity. Further, there are topological restrictions on the domains Ω and Ω', and even when there are no topological obstructions, in multiply connected domains there are many "conformal invariants" represented by Teichmüller spaces or moduli spaces.

At this point the reader may ask why we have chosen to speak of measurable structures and not of smooth structures. There are a number of reasons for this. First, in many applications, such as those in elasticity theory, the matrices G and H describe the properties of various media and are seldom smooth. In other applications, such as in holomorphic dynamics, we must construct conformal structures G and H by various infinite processes that will cause smoothness to be lost. From the point of view of mapping theory or the calculus of variations, one seeks the minima of a certain problems. Seldom will these minima be smooth; therefore the equations these minima satisfy should not have smooth coefficients. Yet it is from these equations that we deduce properties, such as continuity or the existence of partial derivatives, of minima. In Teichmüller theory the distance between topologically equivalent surfaces is measured in terms of the mapping of smallest distortion. Such extrema are almost never smooth and yet have very nice structure reflecting the geometry and topology of the surfaces in question. Also, from other points of view, particularly those of PDEs, it is the families of equations and the properties of their solutions that are important. Compactness requirements, that these families of equations should be closed, lead naturally to consideration of those equations with measurable coefficients.

2.2 The Hyperbolic Metric

One of the more useful tools in complex analysis is the hyperbolic metric of a planar domain. We discuss here the hyperbolic metric of the unit disk, sometimes refered to as the Poincaré plane or disk. Later we shall use this metric of $\mathbb{D}$, together with the uniformization theorem, to define the hyperbolic metric of an arbitrary planar domain (other than the full plane or the punctured plane). This is quite a technical and deep fact from complex analysis, in fact, one of the most important results and themes of 19th century mathematics, and we shall not offer proofs here. Mostly, we shall need only the hyperbolic metric on the disk and the triply punctured sphere.

For each $a \in \mathbb{D}$ and $\theta \in \mathbb{R}$, the linear fractional transformation

$$\phi_a(z) = e^{i\theta} \frac{z-a}{1-\bar{a}z} \tag{2.12}$$

defines a holomorphic self-homeomorphism of the disk $\mathbb{D}$ to itself. We compute

$$\phi_a'(z) = e^{i\theta} \frac{1-|a|^2}{(1-\bar{a}z)^2},$$

from which we have the identity

$$\frac{|\phi_a'(z)|}{1-|\phi_a(z)|^2} = \frac{1}{1-|z|^2} \tag{2.13}$$

Equation (2.13) expresses the fact that each $\phi_a : \mathbb{D} \to \mathbb{D}$ is an isometry of the Riemannian metric

$$ds_{hyp}(z) = \frac{2|dz|}{1-|z|^2}, \quad z \in \mathbb{D}$$

This corresponds to the choice of matrix

$$A = \begin{bmatrix} \frac{4}{(1-|z|^2)^2} & 0 \\ 0 & \frac{4}{(1-|z|^2)^2} \end{bmatrix}$$

in (2.2). Integrating this metric provides the hyperbolic metric $\rho_{\mathbb{D}}$ of the unit disk $\mathbb{D}$,

$$\rho_{\mathbb{D}}(z,w) = \inf_\gamma \int_\gamma ds_{hyp} \tag{2.14}$$

where the infimum is over all rectifiable curves γ joining z to w in $\mathbb{D}$. From the definition at (2.14) the triangle inequality is clear, and so $\rho_{\mathbb{D}}(z,w)$ is a metric.

Next, symmetry considerations and an integration quickly reveal that

$$\rho_{\mathbb{D}}(0,z) = \log \frac{1+|z|}{1-|z|},$$

while using the transitivity of the group of linear fractional transformations of the disk gives us the more general formula

$$\rho_{\mathbb{D}}(z,w) = \log \frac{|1-\bar{z}w| + |z-w|}{|1-\bar{z}w| - |z-w|}$$

As a consequence, we can make the useful observation that, for all $z \in \mathbb{D}$,

$$|z| = \tanh \frac{1}{2} \rho_{\mathbb{D}}(0,z) \tag{2.15}$$

Actually the group of all linear fractional transformations of $\mathbb{D}$, described by (2.12), is isomorphic to the group $PSL(2,\mathbb{R})$, the projective group of 2×2 matrices with real entries and determinant 1. To see this, note that the map

$$\Phi : z \mapsto i\,\frac{1-z}{1+z}$$

is conformal from $\mathbb{D}$ onto the upper half-space $\mathbb{H} = \{z : \Im m(z) > 0\}$. In fact, Φ is an isometry from the hyperbolic metric of the disk to the metric

$$ds = \frac{|dz|}{\Im m(z)}, \qquad z \in \mathbb{H},$$

giving us another model for the hyperbolic plane. The reflection principle quickly identifies the conformal (and hence isometric!) transformations of $\mathbb{H}$ as the linear fractional transformations

$$z \mapsto \frac{az+b}{cz+d}, \qquad a,b,c,d \in \mathbb{R},\ \ ad-bc \neq 0$$

Since we may multiply the numerator and denominator of this fractional transformation by any nonzero constant without affecting the transformation, we may normalize so that $ad - bc = 1$. The reader may wish to verify that this estabishes a topological isomorphism between the group of orientation-preserving isometries of $\mathbb{H}$ and $PSL(2,\mathbb{R})$.

Discrete subgroups of linear fractional transformations are called Fuchsian groups, and it is a fact that every Riemann surface, other than the sphere, torus, plane and punctured plane, admits a complete hyperbolic metric. Further, each such surface can be identified with the orbit space (or quotient) of a Fuchsian group. Thus hyperbolic geometry plays a central role in the theory of surfaces. More of the basic facts concerning hyperbolic geometry, hyperbolic trigonometry, discrete groups and Riemann surfaces can be found in the well-known texts of Beardon [48], Ahlfors-Sario [12] and Farkas-Kra [122], among many others.

2.3 The Space S(2)

There is a natural differential geometric structure on the space $\mathbf{S}(2)$ of symmetric positive definite 2×2 matrices with determinant equal to 1. This structure is induced in turn on the space of measurable conformal structures on a domain, as for instance in (2.7). It is the purpose of this section to take time to recount this important geometric fact.

As we shall see later, the measurable conformal structure induced by a quasiconformal mapping may be described either from the point of view of real analysis and the space $\mathbf{S}(2)$ or in the terminology of complex analysis and the hyperbolic plane $\mathbb{D}$. Both aspects are important for our study, and this dichotomy of real and complex analysis is typical for many topics considered in this monograph.

Reflecting this phenomenon, we shall show that in fact the space $\mathbf{S}(2)$ is isometric to the hyperbolic plane $\mathbb{D}$. This isometry is effected by the correspondences $G \to \mu$, where

$$G = \begin{bmatrix} g_{11} & g_{12} \\ g_{12} & g_{22} \end{bmatrix} \mapsto \mu = \frac{g_{11} - g_{22} - 2ig_{12}}{g_{11} + g_{22} + 2} \tag{2.16}$$

In fact, this correspondence goes much deeper than just this one-to-one isomorphism, and as we shall see it partly reflects the correspondence between the real and complex distortion functions of quasiconformal mappings. We shall see it appearing in a number of different forms.

The space $\mathbf{S}(2)$ is clearly two-dimensional, and any $A \in \mathbf{S}(2)$ can be written in the form

$$A = O^t \begin{bmatrix} \lambda & 0 \\ 0 & 1/\lambda \end{bmatrix} O$$

for some $O \in SO(2, \mathbb{R})$ and $\lambda > 0$. The general linear group $GL(2, \mathbb{R})$ acts transitively on the right on $\mathbf{S}(2)$ via the rule

$$X[G] = |\det X|^{-1} X^t G X \qquad X \in GL(2, \mathbb{R});\ G \in \mathbf{S}(2) \tag{2.17}$$

The Riemannian metric

$$ds^2 = \frac{1}{2} \operatorname{tr}(Y^{-1} dY)^2 \tag{2.18}$$

on $\mathbf{S}(2)$ gives rise to a metric distance, which we denote by $\rho(G, H)$, for $G, H \in \mathbf{S}(2)$. This metric is invariant under the right action of $GL(2, \mathbb{R})$ and makes $\mathbf{S}(2)$ isometric to the hyperbolic plane $\mathbb{D}$. In particular, $\mathbf{S}(2)$ is simply connected and complete. See [162, p. 518] for this computation in all dimensions.

Here we sketch in two dimensions an elementary argument showing that $\mathbf{S}(2)$ and $\mathbb{D}$ are isometric. We have already indicated how a given matrix in $\mathbf{S}(2)$ should correspond to a point in $\mathbb{D}$, namely,

$$G = \begin{bmatrix} g_{11} & g_{12} \\ g_{12} & g_{22} \end{bmatrix} \mapsto \mu = \frac{g_{11} - g_{22} - 2ig_{12}}{g_{11} + g_{22} + 2} \tag{2.19}$$

Clearly this map is continuous. We next compute that

$$\begin{aligned} (\operatorname{tr}(G) + 2)^2 |\mu|^2 &= (g_{11} - g_{22})^2 + 4g_{12}^2 = (g_{11} + g_{22})^2 - 4 \\ &= \operatorname{tr}^2(G) - 4 \end{aligned}$$

Thus $|\mu| < 1$, and two matrices G and H have the same image only if

$$\frac{\operatorname{tr}(G) - 2}{\operatorname{tr}(G) + 2} = \frac{\operatorname{tr}(H) - 2}{\operatorname{tr}(H) + 2},$$

2.3. THE SPACE S(2)

which implies that $\operatorname{tr}(G) = \operatorname{tr}(H)$ as both $\operatorname{tr}(G), \operatorname{tr}(H) \geqslant 2$. It is now quite clear in view of (2.19) that the map $G \mapsto \mu$ is an injection. Next, the elementary observation that $\operatorname{tr}(G) = |G| + |G|^{-1}$ implies

$$|\mu| = \frac{|G|-1}{|G|+1}, \qquad |G| = \frac{1+|\mu|}{1-|\mu|}$$

where we continue to use $|\cdot|$ to denote the operator norm (that is, the largest singular value) of a matrix.

On the diagonal matrices

$$\Lambda = \begin{bmatrix} \lambda & 0 \\ 0 & 1/\lambda \end{bmatrix},$$

we can compute the metric (2.18) directly,

$$\rho(\mathbf{I}, \Lambda) = |\log \lambda|$$

Because any $G \in \mathbf{S}(2)$ can be diagonalized and because of the invariance of the metric under the action in (2.17) we have

$$\rho(\mathbf{I}, G) = \rho(\mathbf{I}, O^t \Lambda O) = \rho(\mathbf{I}, \Lambda) = \log |G| \qquad (2.20)$$

As the hyperbolic metric $\rho_\mathbb{D}$ on the unit disk $\mathbb{D}$ is given by the Riemannian metric $ds = 2|dz|/(1-|z|^2)$, we have the formula

$$\rho_\mathbb{D}(0, \mu) = \log \frac{1+|\mu|}{1-|\mu|} = \log |G| \qquad (2.21)$$

The two formulas (2.20) and (2.21) show that the map $G \mapsto \mu$ is an isometry on geodesic rays from the identity to geodesic lines passing through the origin and is therefore onto (it is not too difficult to invert this map, however, we shall do it later in a more general setting).

To show that (2.16) provides a global isometry between $\mathbf{S}(2)$ and $\mathbb{D}$, we need the following lemma.

Lemma 2.3.1. *Suppose that under the correspondence (2.16) we have $G \mapsto \mu$ and $H \mapsto \nu$, where $G, H \in \mathbf{S}(2)$. Then*

$$X \mapsto \frac{\mu - \nu}{1 - \overline{\mu}\nu},$$

where $X = \sqrt{G} H^{-1} \sqrt{G} = \sqrt{G}[H^{-1}]$.

Proof. First, we find the square root of the positive definite matrix G. This is a positive definite matrix $\sqrt{G}$ such that $\sqrt{G} \cdot \sqrt{G} = G$. By direct calculation

$$\sqrt{G} = \frac{1}{\sqrt{\operatorname{tr}(G)+2}} \begin{bmatrix} g_{11}+1 & g_{12} \\ g_{12} & g_{22}+1 \end{bmatrix} = \frac{1}{\sqrt{\operatorname{tr}(G)+2}} \left(\begin{bmatrix} g_{11} & g_{12} \\ g_{12} & g_{22} \end{bmatrix} + \mathbf{I} \right)$$

Next, given the matrices $G, H \in \mathbf{S}(2)$, we see that

$$X = \sqrt{G}H^{-1}\sqrt{G} = \frac{1}{\operatorname{tr}(G)+2}\left(GH^{-1}G + GH^{-1} + H^{-1}G + H^{-1}\right)$$

Therefore

$$\begin{aligned}
(\operatorname{tr}(G)+2)x_{11} &= g_{12}^2 h_{11} + (1+g_{11})(h_{22} + g_{11}h_{22} - 2g_{12}h_{12}) \\
(\operatorname{tr}(G)+2)x_{22} &= (1+g_{22})^2 h_{11} + g_{12}^2 h_{22} - 2(1+g_{22})g_{12}h_{12} \\
(\operatorname{tr}(G)+2)x_{12} &= g_{12}(h_{11} + g_{22}h_{11} + h_{22} + g_{11}h_{22}) - h_{12}(g_{11} + g_{22} + 2g_{11}g_{22})
\end{aligned}$$

Moreover,

$$\operatorname{tr}(X) + 2 = g_{11}h_{22} + g_{22}h_{11} - 2g_{12}h_{12} + 2$$

$$\frac{x_{11} - x_{22}}{x_{11} + x_{22} + 2} =$$

$$\frac{(2 + g_{11}(2 + g_{11} - g_{22}))h_{22} - (2 + g_{22}(2 - g_{11} + g_{22}))h_{11} + 2(g_{22} - g_{11})g_{12}h_{12}}{(g_{11} + g_{22} + 2)(g_{11}h_{22} + g_{22}h_{11} - 2g_{12}h_{12} + 2)}$$

and

$$\frac{2x_{12}}{x_{11} + x_{22} + 2} =$$

$$\frac{2g_{12}(h_{11} + g_{22}h_{11} + h_{22} + g_{11}h_{22}) - 2h_{12}(g_{11} + g_{22} + 2g_{11}g_{22})}{(g_{11} + g_{22} + 2)(g_{11}h_{22} + g_{22}h_{11} - 2g_{12}h_{12} + 2)}$$

We have identified G with μ and similarly H with ν. We may hence compute $\frac{\mu-\nu}{1-\bar{\mu}\nu}$ using (2.19). On the other hand, we calculated the corresponding expressions for the matrix $X = \sqrt{G}H^{-1}\sqrt{G}$ above. Comparing the results, we find that indeed

$$X \mapsto \frac{\mu - \nu}{1 - \bar{\mu}\nu}$$

under our proposed isometry. □

Of course, the map $z \mapsto (\mu - z)/(1 - \bar{\mu}z)$ is an isometry of the hyperbolic plane. Thus the previous lemma gives

$$\begin{aligned}
\rho_{\mathbb{D}}(\mu, \nu) &= \rho_{\mathbb{D}}\left(0, \frac{\mu - \nu}{1 - \bar{\mu}\nu}\right) \\
&= \rho(\mathbf{I}, X) = \rho(\mathbf{I}, \sqrt{G}[H^{-1}]) \\
&= \rho(\mathbf{I}, \sqrt{G}^{-1}[H]) \\
&= \rho(\sqrt{G}[\mathbf{I}], H) = \rho(G, H)
\end{aligned}$$

The second-to-last equality holds by virtue of the isometric $GL(2, \mathbb{R})$ action and the last equality as $\sqrt{G}[\mathbf{I}] = G$. We have proved the following theorem.

Theorem 2.3.2. *The map $G \mapsto \mu$ induces an isometry between the space $\mathbf{S}(2)$ with the metric $ds^2 = \frac{1}{2}(Y^{-1}dY)^2$ and the hyperbolic plane $(\mathbb{D}, \rho_{\mathbb{D}})$.*

2.4 The Linear Distortion

Given a homeomorphism $f : \Omega \to \Omega'$, we may introduce a quantity that measures the deviation from f to a conformal mapping.

Definition 2.4.1. *The* linear distortion *of f is the measurable function defined by*
$$H(z, f) = \limsup_{r \to 0} \frac{\max_{|\zeta|=r} |f(z + \zeta) - f(z)|}{\min_{|\zeta|=r} |f(z + \zeta) - f(z)|} \tag{2.22}$$

The first example to consider is that of the linear mapping, which we write conveniently in the complex notation
$$f(z) = az + b\bar{z}$$

Indeed, we have decomposed the $\mathbb{R}$-linear mapping f as the sum of a complex linear operator $f_+(z) = az$ and a complex antilinear operator $f_-(z) = b\bar{z}$. A moment's thought gives
$$H(z, f) = \frac{|a| + |b|}{|a| - |b|}$$
$$J(z, f) = |a|^2 - |b|^2$$

In the first identity we have implicitly assumed that f is orientation-preserving, for example, that $J(z, f) > 0$ or that $|a| > |b|$.

More generally then, if f has a derivative $Df(z)$ at z, we may use the complex differential operators
$$\frac{\partial f}{\partial z} = f_z = \frac{1}{2}(f_x - i f_y), \qquad \frac{\partial f}{\partial \bar{z}} = f_{\bar{z}} = \frac{1}{2}(f_x + i f_y)$$

and write the derivative as
$$Df(z) h = \frac{\partial f}{\partial z}(z) h + \frac{\partial f}{\partial \bar{z}}(z) \bar{h}, \qquad h \in \mathbb{C} = \mathbb{R}^2$$

In particular, $h \mapsto f_z(z) h$ is the $\mathbb{C}$-linear part of $Df(z)$, and $h \mapsto f_{\bar{z}}(z) \bar{h}$ is its antilinear part. Then the norm of the derivative
$$|Df(z)| = \sup\{|Df(z) h| : |h| = 1\} = \left|\frac{\partial f}{\partial z}(z)\right| + \left|\frac{\partial f}{\partial \bar{z}}(z)\right| = |f_z| + |f_{\bar{z}}| \tag{2.23}$$

and the Jacobian
$$J(z, f) = \left|\frac{\partial f}{\partial z}(z)\right|^2 - \left|\frac{\partial f}{\partial \bar{z}}(z)\right|^2 = |f_z|^2 - |f_{\bar{z}}|^2 \tag{2.24}$$

In terms of the complex derivatives,
$$H(z, f) = \frac{\max_{|\zeta|=1} |(D_\zeta f)(z)|}{\min_{|\zeta|=1} |(D_\zeta f)(z)|} = \frac{|f_z(z)| + |f_{\bar{z}}(z)|}{|f_z(z)| - |f_{\bar{z}}(z)|}$$
$$= |Df(z)|^2 J(z, f)^{-1} \tag{2.25}$$

at points where $J(z,f) > 0$. In particular, if f has a complex derivative at z, then $H(z,f) = 1$. Thus if f is *conformal* (that is, holomorphic and injective), we have $H(z,f) \equiv 1$. The reader may wish to attempt to establish the converse to this statement, assuming differentiability.

Suggested by (2.11), if f is differentiable almost everywhere, we may write

$$G_f(z) = J(z,f)^{-1} D^t f(z) Df(z)$$

whenever $J(z,f) > 0$. The quantity $G_f(z)$ is called the *distortion tensor* of the mapping f.

When $Df(z) = 0$, the mapping satisfies the condition (2.8), with $\phi = 0$, and thus at those points we may set $G_f(z) = \mathbf{I}$. However, where $Df(z)$ is nonzero but degenerate, that is, $J(z,f) = 0$ but $Df(z) \neq 0$, there is no meaningful definition for $G_f(z)$.

Assuming the set of all such degenerate points has measure zero, then $G_f : \Omega \to \mathbf{S}(2)$ is a measurable map and, in particular, f is a conformal mapping from (Ω, G_f) to $(\Omega', \mathbf{I})$. Moreover,

$$H(z,f) = |G_f(z)| = e^{\rho(G_f(z), \mathbf{I})} \qquad (2.26)$$

Thus the deviation (2.22) from f to a conformal map is simply another measure of the distance between the conformal structure induced by f on Ω (namely, G_f) and the Euclidean structure $\mathbf{I}$.

Under the isometry $\mathbf{S}(2) \to \mathbb{D}$, the mapping f induces a mapping $G_f(z) \mapsto \mu_f(z)$. Now $\mu_f : \Omega \to \mathbb{D}$ will be a measurable map, and evidently,

$$H(z,f) = \frac{1 + |\mu_f(z)|}{1 - |\mu_f(z)|} \qquad (2.27)$$

It is important to realize that equations (2.25)–(2.27) hold only where f has a derivative. However, the formula for the linear distortion makes sense in much more generality; for instance, one may compute that $z \mapsto z|z|^{-1/2}$ has linear distortion 1 at the origin where it certainly does not have a derivative. Nevertheless, one expects the boundedness of the linear distortion of a mapping to imply some differentiability and geometric properties of a mapping.

It is perhaps a little surprising that the theory of mappings of bounded linear distortion (or quasiconformal mappings) is so rich.

2.5 Quasiconformal Mappings

Quasiconformal mappings are principally mappings of "bounded distortion". However, there are in fact many ways to measure the distortion of a mapping, and we shall consider a number of them. We consider first the geometric definition through the linear distortion function in (2.22). For clarity of exposition we use here the term mapping of bounded distortion, even if later these turn out to be precisely the quasiconformal mappings.

2.5. QUASICONFORMAL MAPPINGS

Mappings of Bounded Distortion

Definition 2.5.1. *A homeomorphism* $f : \Omega \to \Omega'$ *is called a mapping of bounded distortion if it is orientation-preserving and if its linear distortion is uniformly bounded,*

$$\sup_{z \in \Omega} H(z, f) < \infty \qquad (2.28)$$

For a more detailed discussion on the concept of orientation-preserving mappings, see Section 2.8. It is important to note that in (2.28) uniform boundedness, and requiring this at every point instead of up to measure zero, is necessary to build a useful theory. For instance, requiring only that $H(z, f) < \infty$ for all $z \in \Omega$ does not give one the desirable compactness or regularity properties. It does motivate the notion of mappings of finite distortion, as discussed later in Chapter 20, when additional integrability conditions are imposed on $H(z, f)$, but without such extra properties not much can be said. Indeed, simply note here that for a diffeomorphism of a domain Ω we have $H(z, f) < \infty$ everywhere, while the space of diffeomorphisms of a domain has very few compactness properties.

It is a little more difficult to see that

$$\operatorname{ess\,sup}_{z \in \Omega} H(z, f) = \|H(z, f)\|_\infty \leqslant K < \infty \qquad (2.29)$$

alone also does not imply good regularity for homeomorphisms. The problem here is basically with the differentiability properties of the mappings. Equation (2.28) implies that the homeomorphism f is differentiable almost everywhere, and in fact it can be shown that if f is assumed a priori to be in the Sobolev space $W^{1,1}(\Omega)$ of integrable functions whose first distributional derivatives are integrable, then conversely (2.29) implies $H(z, f)$ is uniformly bounded.

To see the problems that arise when we assume only (2.29), consider the following example. Let $u(x)$ denote the Cantor function defined on the real line. Thus $u(x)$ is continuous and increasing with $u'(x) = 0$ almost everywhere. Set

$$f(z) = x + u(x) + iy, \qquad z = x + iy$$

Then f is a homeomorphism $\mathbb{C} \to \mathbb{C}$ and almost everywhere f is differentiable with

$$Df(z) = \begin{bmatrix} 1 & 0 \\ 0 & 1 \end{bmatrix}$$

In particular, the Jacobian determinant $J(z, f) = 1$ almost everywhere. We compute that $H(z, f) = 1$ almost everywhere, so

$$\operatorname{ess\,sup}_{z \in \mathbb{C}} H(z, f) = 1$$

Further, the distortion tensor $G_f(z) = \mathbf{I}$ almost everywhere. However, f is not conformal; it is not even of Sobolev class. The change-of-variables formula does not hold (f does not preserve the measure of sets even though $J(z, f) = 1$ almost everywhere). The reader may wish to discover other nasty properties of this homeomorphism. This example reinforces the distinction between functions that are merely differentiable almost everywhere and Sobolev functions. (The reader who is unfamiliar with Sobolev space theory will find the requisite definitions and so forth in the Appendix.)

While the definition (2.28) of mappings of bounded distortion is aesthetically pleasing, in practice it is very difficult to work with. Also, it requires the a priori knowledge that one is working with an injective mapping (at least locally). It is for such reasons that one turns to the analytic definition.

Here we return to the identities (2.25). At points z where f is differentiable with positive Jacobian, we see that the linear distortion $H(z, f)$ is precisely the ratio of the largest and the smallest directional derivatives at z. This interpretation leads to the following definition of a quasiconformal mapping.

Definition of a Quasiconformal Map

Definition 2.5.2. *A homeomorphism $f : \Omega \to \Omega'$ is called K-quasiconformal if it is orientation-preserving, if*

$$f \in W^{1,2}_{loc}(\Omega), \tag{2.30}$$

and if the directional derivatives satisfy

$$\max_\alpha |\partial_\alpha f(z)| \leqslant K \min_\alpha |\partial_\alpha f(z)| \tag{2.31}$$

for almost every $z \in \Omega$.

Often it is convenient to formulate (2.31) as the distortion inequality

$$|Df(z)|^2 \leqslant K J(z, f) \quad \text{for almost every } z \in \Omega$$

The smallest constant $K = K(f)$ for which (2.31) holds almost everywhere is called the *distortion* of the mapping f.

The above formulation of a quasiconformal mapping in terms of Sobolev spaces is the most useful and flexible, allowing estimates, in taking limits and so on. There are, however, a few subtleties we need to discuss. The requirement $f \in W^{1,2}_{loc}(\Omega)$ implies that f has partial derivatives f_x and f_y almost everywhere (see the Appendix), but being a Sobolev function is not enough for f to be differentiable almost everywhere. For (2.31) to be meaningful one need only set

$$\partial_\alpha f(z) = \cos(\alpha) f_x(z) + \sin(\alpha) f_y(z), \quad \alpha \in [0, 2\pi], \tag{2.32}$$

then for almost all $z \in \Omega$, condition (2.31) is well defined as written.

2.5. QUASICONFORMAL MAPPINGS

The condition ties together the partial derivatives of f and hence provides geometric information on the mapping. As such it is enough to start the development of the theory. The reader should note, though, that in the next chapter we will show that all *homeomorphic* Sobolev functions are differentiable almost everywhere; with this result the directional derivatives retain their usual meaning,

$$\partial_\alpha f(z) = \lim_{r \to 0} \frac{f(z + r e^{i\alpha}) - f(z)}{r}$$

Definition of a Quasiregular Map

Now we also want to give up the hypothesis that f is injective, so as to be able to define a wider and more flexible class of mappings particularly useful in the study of planar elliptic PDEs. This is quite analogous to moving to analytic functions from conformal mappings.

We shall require only that the mapping $f \in W^{1,2}_{loc}(\Omega)$, that it is orientation-preserving, so $J(z,f) = |f_z|^2 - |f_{\bar{z}}|^2 \geqslant 0$ almost everywhere, and that it satisfies the conditions (2.31) and (2.32) on partial derivatives. We shall call these mappings K-*quasiregular*.

The reader may wonder about the choice of the Sobolev regularity $W^{1,2}_{loc}$ in the definitions of quasiconformal and quasiregular mappings. For the homeomorphic $W^{1,1}_{loc}(\Omega)$-maps, this regularity follows automatically from the distortion inequality (2.31) since, by Corollary 3.3.6 from the next chapter, for any compact subset $A \subset \Omega$,

$$\int_A |Df|^2 \leqslant K(f) \int_A J(z,f) \leqslant K(f)|f(A)| < \infty \qquad (2.33)$$

However, the last estimate fails for general nonhomeomorphic Sobolev mappings, and it is for these that the precise regularity $f \in W^{1,2}_{loc}(\Omega)$ is necessary. This condition will guarantee the local integrability of the Jacobian, the key property in the geometric study of mappings.

Before dwelling further on these notions, let us make it clear that quasiconformal mappings are precisely the mappings of bounded distortion. In the literature the conditions (2.30) and (2.31) taken together are called the *analytic definition* of a quasiconformal mapping while (2.28) is termed the *geometric definition*.

For a diffeomorphism the equivalence of these two definitions is clear, as the reader may quickly verify. The general case lies much deeper. That every quasiconformal mapping satisfies (2.28) will be shown in Theorem 3.6.2. For the converse direction, in fact much less than (2.28) is required. The recent surprising result of Heinonen and Koskela [160] allows the lim sup condition in the definition of the linear distortion to be replaced by a lim inf condition. It

The lim inf theorem

Theorem 2.5.3. *Let $f : \Omega \to \Omega'$ be an orientation-preserving homeomorphism between planar domains. If there is $H < \infty$ such that for every $z \in \Omega$*

$$\liminf_{r \to 0} \frac{\max_{|\zeta|=r} |f(z+\zeta) - f(z)|}{\min_{|\zeta|=r} |f(z+\zeta) - f(z)|} \leqslant H, \tag{2.34}$$

then f is quasiconformal.

The proof of this result will take us rather too far from the themes of this book as it uses ideas from the theory of quasisymmetric mappings of metric spaces and the moduli of curve families. The theorem was further refined in [205] where exceptional sets of zero length are allowed. A key point here is that in applications, such as in conformal dynamics (see for instance [303]), at each point one need only find estimates or control the geometry on some sequence of radii tending to 0 as opposed to every sequence tending to 0. It would be very good to have an analytic proof for this result at hand.

Problem. Give an analytic proof for Theorem 2.5.3.

The Beltrami Equation

The previous section relates in (2.25) the linear distortion to the complex derivatives at points where the mappings are differentiable with positive Jacobian. A similar analysis can be made for the analytic definition (2.30) and (2.31) of a quasiconformal mapping. These complex analytic reductions have far-reaching consequences.

A local analysis again leads to consideration of the linear (over $\mathbb{R}$) mappings $f : z \mapsto az + b\bar{z}$. That f is orientation-preserving implies $|b| \leqslant |a|$. We leave it to the reader to verify that $\max_\alpha |\partial_\alpha f(z)| = |a| + |b| = |Df(z)|$ and $\min_\alpha |\partial_\alpha f(z)| = |a| - |b| = J(z,f)/|Df(z)|$ when $|b| \leqslant |a|$, as this calculation is more or less a repetition of that given to find (2.25). As a consequence, under the interpretation (2.32), the distortion inequality (2.31) achieves the form

$$\left|\frac{\partial f}{\partial z}\right| + \left|\frac{\partial f}{\partial \bar{z}}\right| \leqslant K \left(\left|\frac{\partial f}{\partial z}\right| - \left|\frac{\partial f}{\partial \bar{z}}\right| \right),$$

an inequality equivalent to

$$\left|\frac{\partial f}{\partial \bar{z}}\right| \leqslant \frac{K-1}{K+1} \left|\frac{\partial f}{\partial z}\right| \tag{2.35}$$

Furthermore, write $\mu(z) = f_{\bar{z}}(z)/f_z(z)$ when $f_z(z) \neq 0$ and, say, $\mu(z) = 0$ otherwise. This expresses the inequality (2.35) as a linear partial differential equation. We have thus shown the following theorem.

2.5. QUASICONFORMAL MAPPINGS

Theorem 2.5.4. *Suppose $f: \Omega \to \Omega'$ is a homeomorphic $W^{1,2}_{loc}$-mapping. Then f is K-quasiconformal if and only if*

$$\frac{\partial f}{\partial \bar{z}}(z) = \mu(z) \frac{\partial f}{\partial z}(z) \quad \text{for almost every } z \in \Omega, \tag{2.36}$$

where μ, called the Beltrami coefficient of f, is a bounded measurable function satisfying

$$\|\mu\|_\infty \leqslant \frac{K-1}{K+1} < 1$$

Often in the literature the term *complex dilatation of f* is used for the Beltrami coefficient μ. We shall later see for quasiconformal mappings that $f_z \neq 0$ almost everywhere. Thus the Beltrami coefficient is uniquely defined up to a set of measure zero. The reader should also note the relations

$$\|\mu\|_\infty = \frac{K(f)-1}{K(f)+1} \quad \text{and} \quad K(f) = \frac{1+\|\mu\|_\infty}{1-\|\mu\|_\infty}$$

From Theorem 2.5.4 and Weyl's Lemma A.6.10 we have a quick proof showing that for a quasiconformal mapping f the distortion $K(f) = 1$ if and only if f is conformal, that is, holomorphic and injective. If we need to give up the hypothesis that f is injective, then we see that f is 1-quasiregular if and only if it is an analytic function.

The differential equation (2.36) is called the *Beltrami equation*. It is this equation that provides the connections from the geometric theory of quasiconformal mappings to complex analysis and to elliptic PDEs. Understanding and utilizing these relations underlie many of the themes of this monograph.

It is in the setting described above that planar quasiconformal mappings were first studied around 1928 by Grötzsch. The term "quasiconformal" was coined by Ahlfors in 1935 [1, 2] when this class of mappings proved to be an integral tool in his geometric development of Nevanlinna theory based on the "length-area" method. Teichmüller found a fundamental connection between quasiconformal mappings and quadratic differentials in his studies on extremal mappings between Riemann surfaces [350] around 1939. Developments of the length-area method led to the definition of quasiconformal mappings in terms of the distortion of the modulus of curve families by Pfluger [295]. These were systematically studied in their own right by Ahlfors from 1953 [3].

The class of quasiconformal diffeomorphisms is not closed under uniform limits. Thus the generalization to Sobolev spaces is absolutely necessary if one is to solve various extremal problems by taking limits. After making this generalization we will find that the limit of a bounded sequence of quasiconformal mappings is either quasiconformal or constant. Therefore it is in this setting that the class of quasiconformal mappings becomes more flexible and has a greater range of applications.

The equivalence between the geometric definition and the analytic definition in the Sobolev setting was shown by Gehring and Lehto in 1959 [137]. This is a relatively deep fact that we shall explore to various extents in this book, giving new and quite different proofs from a modern perspective through the notions of quasisymmetry and also through the theory of holomorphic motions. The connection among quasiconformal mappings, Teichmüller theory and quadratic differentials has been intensively investigated by Ahlfors-Bers, Reich-Strebel and Lehto and others; see [339] and [228] and the references therein.

There are two further routes to the theory of planar quasiconformal mappings. These are via the conformal modulus, the approach taken in Lehto-Virtanen's classic text [229], and the modern approach via "holomorphic motions" due to Sullivan [237]. We shall discuss this last approach quite extensively later.

2.6 Radial Stretchings

There is a class of examples that it is important to have at hand as they typically provide extremal examples. These are the radial stretchings, mappings $f : \mathbb{D}(0, R) \to \mathbb{C}$ of the form

$$f(z) = \frac{z}{|z|} \rho(|z|), \qquad f(0) = 0 \qquad (2.37)$$

Here the function $t \mapsto \rho(t) > 0$, $0 \leqslant t < R$, is assumed to be continuous and strictly increasing. For $\rho(0) = 0$ the mapping f is continuous at the origin.

Basic examples include $\rho(t) = t^K$ and $\rho(t) = t^{1/K}$ giving rise to the mappings

$$f_1(z) = z|z|^{K-1}$$

and

$$f_2(z) = z|z|^{\frac{1}{K}-1},$$

respectively. These are the standard radial stretchings we shall see frequently, for they arise as extremals for the problems on Hölder continuity, and integrability of the differential, for quasiconformal mappings. For future reference note here that $f_1 = f_2^{-1}$.

We may calculate the differential and distortions of a radial stretching f directly from the definition at points where the derivative $\dot{\rho}$ exists. In complex notation, using the simple identity $\partial_{\bar{z}} |z| = z\,(2|z|)^{-1}$, we have

$$\frac{\partial f}{\partial z}(z) = \frac{1}{2}\left[\dot{\rho}(|z|) + \frac{\rho(|z|)}{|z|}\right] \qquad (2.38)$$

$$\frac{\partial f}{\partial \bar{z}}(z) = \frac{1}{2}\frac{z}{\bar{z}}\left[\dot{\rho}(|z|) - \frac{\rho(|z|)}{|z|}\right] \qquad (2.39)$$

Thus we obtain

$$|Df(z)| = \left|\frac{\partial f}{\partial z}\right| + \left|\frac{\partial f}{\partial \bar{z}}\right| = \max\left\{\dot{\rho}(|z|), \frac{\rho(|z|)}{|z|}\right\} \qquad (2.40)$$

2.6. RADIAL STRETCHINGS

$$J(z, f) = \left|\frac{\partial f}{\partial z}\right|^2 - \left|\frac{\partial f}{\partial \bar{z}}\right|^2 = \frac{\rho(|z|)\dot{\rho}(|z|)}{|z|} \tag{2.41}$$

It is easy to see that for any radial stretching $H(0, f) = 1$. Away from 0 we see that

$$H(z, f) = |Df(z)|^2 J(z, f)^{-1} = \max\left\{\frac{|z|\dot{\rho}(|z|)}{\rho(|z|)}, \frac{\rho(|z|)}{|z|\dot{\rho}(|z|)}\right\}$$

In particular, for $f(z) = z|z|^{K-1}$, $K > 0$, we have

$$H(z, f) = \max\{|K|, |K|^{-1}\}$$

Furthermore, we find that f satisfies the complex Beltrami equation

$$f_{\bar{z}} = \mu(z) f_z,$$

where the *Beltrami coefficient* μ given by

$$\mu(z) = \frac{z}{\bar{z}} \frac{|z|\dot{\rho}(|z|) - \rho(|z|)}{|z|\dot{\rho}(|z|) + \rho(|z|)} \tag{2.42}$$

Hence

$$K(z) = \frac{1 + |\mu(z)|}{1 - |\mu(z)|} = \max\left\{\frac{|z|\dot{\rho}(|z|)}{\rho(|z|)}, \frac{\rho(|z|)}{|z|\dot{\rho}(|z|)}\right\} \tag{2.43}$$

Again, if $f(z) = z|z|^{K-1}$, we have

$$J(z, f) = K|z|^{2(K-1)}, \qquad |Df(z)| = K|z|^{K-1} \tag{2.44}$$

and

$$\mu_f(z) = \frac{K-1}{K+1} \frac{z}{\bar{z}}$$

More generally, if f is a radial stretching, then $J(z, f)$ is always locally integrable. Indeed,

$$\int_{|z|\leqslant\lambda} J(z, f) = 2\pi \int_0^\lambda \rho(t)\dot{\rho}(t)dt = \pi\rho^2(\lambda),$$

the last term being the area of the disk of radius $\rho(\lambda) = f(\lambda)$.

Naturally, the above expressions can be found in real notation, too. For instance, the full Jacobian matrix has the form

$$Df(z) = \frac{\rho(|z|)}{|z|}\mathbf{I} + \left(\dot{\rho}(|z|) - \frac{\rho(|z|)}{|z|}\right)\frac{z \otimes z}{|z|^2},$$

where we have used (and will use elsewhere) the shorthand notation

$$z \otimes z = \begin{bmatrix} x^2 & xy \\ xy & y^2 \end{bmatrix}$$

for $z = x + iy$. For further details and identities in this notation see, e.g., [191].

2.7 Hausdorff Dimension

In this section we give the basic definition and properties of Hausdorff dimension. Hausdorff dimension and its close relatives play an important role in planar complex analysis, providing a way of measuring the size of quite general sets. A good reference is Mattila's book [251] on geometric measure theory. The basic definitions and foundational results for the theory are due to Carathéodory [88] and Hausdorff [158] in the second decade of the 20th century.

We shall often meet fractal sets with nonintegral dimension in the theory of quasiconformal mappings. Particularly, they often occur as dynamically defined invariant subsets of the plane (for instance, limit sets of Kleinian groups and Julia sets of rational maps). Self-similarity is often a natural property of dynamically defined sets. Quasiconformal mappings are one of the natural geometric tools to study sets of nonintegral dimension, as it is under quasiconformal deformations of conformal dynamical systems that the invariant sets change their dimensions. Thus a substantial part of our later studies will be aimed at trying to estimate just how much a quasiconformal mapping can change the dimension of a planar set. In fact, we shall be able to present optimal bounds later; see Theorem 13.2.10. It may come as a bit of a surprise to the reader unfamiliar with such things, but distortion of dimension is intimately connected with integrability properties of the derivatives of a mapping.

We begin with a fairly general, but not the most general, construction.

Let $\mathcal{F}$ be the family of all Borel subsets of $\mathbb{C}$ and $\eta : [0, \infty] \to [0, \infty]$ an increasing homeomorphism. For any $0 < \delta \leqslant \infty$ and $A \subset \mathbb{C}$ we set

$$\mathcal{H}_{\eta,\delta}(A) = \inf \left\{ \sum_{i=1}^{\infty} \eta(\mathrm{diam}(E_i)) \ : \ A \subset \bigcup_{i=1}^{\infty} E_i, \ E_i \in \mathcal{F}, \ \mathrm{diam}(E_i) < \delta \right\}$$

The function $\mathcal{H}_{\eta,\delta}$ is monotonic and subadditive. Notice that for $A \subset \mathbb{C}$ we have $\mathcal{H}_{\eta,\delta_1}(A) \leqslant \mathcal{H}_{\eta,\delta_2}(A)$ if $\delta_2 < \delta_1$. Hence we can define $\mathcal{H}_\eta$ on subsets A of $\mathbb{C}$ by

$$\mathcal{H}_\eta(A) = \lim_{\delta \searrow 0} \mathcal{H}_{\eta,\delta}(A)$$

The following is an exercise for the reader, see [251] for a proof.

Theorem 2.7.1. *The function $\mathcal{H}_\eta$ defines a Borel regular measure on $\mathbb{C}$.*

s-Dimensional Hausdorff Measure

If we put $\eta(t) = t^s$, the resulting Borel measure is called the s-dimensional Hausdorff measure, denoted $\mathcal{H}^s$. This measure is not locally finite on $\mathbb{C}$ unless $s \geqslant 2$ and is therefore not a Radon measure. However, if $A \subset \mathbb{C}$ is such that for some $s \leqslant 2$ $\mathcal{H}^s(A) < \infty$, then upon restricting to A we do obtain such a measure. That is, $\mathcal{H}^s|A$ is a Radon measure.

2.7. HAUSDORFF DIMENSION

Integral-dimensional Hausdorff measures of course play a central role. $\mathcal{H}^0(A)$ is just the cardinality of A. When $s = 1$, the measure $\mathcal{H}^1$ is a generalized length measure. For a rectifiable curve α we can show that $\mathcal{H}^1(\alpha)$ is simply the length of α, while for $s = 2$ we obtain (up to a constant scaling factor) the usual Lebesgue measure. Note that for any $s > 2$ and $A \subset \mathbb{C}$, $\mathcal{H}^s(A) \equiv 0$.

There is no great dependence of Hausdorff measure on the type of covering sets in $\mathcal{F}$. For instance, one may restrict oneself only to covers by convex sets without affecting the measure. If, however, one asks for optimal coverings by sets (say disks) of the *same* diameter, then one is led to the study of Minkowski measures and dimensions; see for instance [251].

The next result we want to quote, [251, Theorem 4.7], quickly leads to the central concept of Hausdorff dimension.

Theorem 2.7.2. *Let $0 \leqslant s < t < \infty$ and $A \subset \mathbb{C}$. Then*

- $\mathcal{H}^s(A) < \infty$ *implies* $\mathcal{H}^t(A) = 0$, *and*
- $\mathcal{H}^t(A) > 0$ *implies* $\mathcal{H}^s(A) = \infty$.

We may make use of Theorem 2.7.2 to define Hausdorff dimension,

Hausdorff Dimension

A Sierpiński gasket of dimension $\dim_\mathcal{H} = \frac{\log 3}{\log 2}$

The Hausdorff dimension of a set $A \subset \mathbb{C}$ is defined as

$$\dim_\mathcal{H}(A) = \sup\{s : \mathcal{H}^s(A) > 0\} = \inf\{t : \mathcal{H}^t(A) < \infty\}$$

Of course, the Hausdorff dimension of a finite set is 0, of a rectifiable curve it is 1 and for an open set it is 2. The topology of a set has implications for its dimension. For instance, any connected nondegenerate subset of $\mathbb{C}$ has dimension at least 1. However, totally disconnected closed sets, usually called Cantor sets, can have any dimension from 0 to 2. Readers may wish to construct such examples for themselves.

Next we record a few simple but useful properties:

- If $A \subset B \subset \mathbb{C}$, then $\dim_\mathcal{H}(A) \leqslant \dim_\mathcal{H}(B) \leqslant 2$
- For $A_1, A_2, \ldots \subset \mathbb{C}$, $\dim_\mathcal{H}(\cup_i A_i) = \sup_i \dim_\mathcal{H}(A_i)$.
- If $s < \dim_\mathcal{H}(A)$, then $\mathcal{H}^s(A) = \infty$.
- If $t > \dim_\mathcal{H}(A)$, then $\mathcal{H}^t(A) = 0$.

Note also that Lipschitz maps, those mappings $f : \Omega \to \mathbb{C}$ satisfying the estimate

$$|f(z) - f(w)| \leqslant L|z - w| \tag{2.45}$$

for some $L > 0$ and all $z, w \in \Omega$, do not increase the dimension. In particular, Hausdorff dimension will be preserved under bilipschitz mappings—those mappings for which both f and f^{-1} are Lipschitz.

2.8 Degree and Jacobian

There are two basic approaches to the notion of local degree for a mapping, the algebraic (see for instance Dold [107]) and the analytic (see for instance Lloyd [234]). Both these notions try to capture the idea of counting the number of solutions $z \in \Omega$ to the equation $f(z) = w$. The reader will be familiar with the winding number of a piecewise smooth closed curve $\gamma : [0, 1] \to \mathbb{C}$,

$$W_\gamma(a) = \frac{1}{2\pi i} \int_\gamma \frac{dz}{z - a}$$

If Ω is bounded by a piecewise smooth Jordan curve γ and $w \notin f(\partial\Omega)$, $a \in \Omega$, then the degree of f at w is simply

$$\deg_\Omega(f, w) = W_\gamma(a) \cdot W_{f \circ \gamma}(w),$$

where $W_\gamma(a) = \pm 1$ accounting for the two possible orientations of the boundary and is constant for $a \in \Omega$. In fact, the winding number is constant on components of $\mathbb{C} \setminus f(\partial\Omega)$. Analytically the degree can be defined as

$$\deg_\Omega(f, w) = \sum_{z \in f^{-1}(w)} \operatorname{sgn} J(z, f)$$

Remark. For a continuous map $f : \Omega \to \mathbb{C}$ and $p \in f(\Omega)$ with $f^{-1}(p)$ compact, the local topological degree is then generally defined as the unique integer representing the homomorphism $H_2(\hat{\mathbb{C}}) \to H_2(\hat{\mathbb{C}}) \cong \mathbb{Z}$ induced on the second homology by the excision isomorphism; see [107, Definition 5.1]. Any map $g : \hat{\mathbb{C}} \to \hat{\mathbb{C}}$ has a well-defined degree independent of a chosen point, so we simply write $\deg(g)$ for such maps.

2.8. DEGREE AND JACOBIAN

These two notions are equivalent for sufficiently regular maps and are more generally proven to be equivalent via approximation. The topological definition has the virtue of clearly defining a homotopy invariant notion, while the analytic definition is computable. We recall the basic facts concerning degree.

Theorem 2.8.1. *Let $f : \Omega \to \mathbb{C}$ be continuous, $p \in \mathbb{C} \setminus f(\partial\Omega)$ and $f^{-1}(p)$ compact.*

- *The degree $\deg_\Omega(f, p)$ is constant on components of $\mathbb{C} \setminus f(\partial\Omega)$. That is, if p_1 and p_2 lie in the same component of $\mathbb{C} \setminus f(\partial\Omega)$, then $\deg_\Omega(f, p_1) = \deg_\Omega(f, p_2)$ [234, Theorem 2.1.3].*

- *If $H : [0,1] \times \Omega \to \mathbb{C}$ is a homotopy, Ω a bounded domain, and $w_t = H(t, z_0) \notin H(t, \partial\Omega)$ for $0 \leqslant t \leqslant 1$, then $\deg_\Omega(H_t, w_t)$ is independent of t [234, Theorem 2.2.4].*

- *If f is a homeomorphism, then $\deg_\Omega(f, z) = \pm 1$ for all $z \in \Omega$ [107, IV.5].*

- *Degree is multiplicative. Suppose $g : \hat{\mathbb{C}} \to \hat{\mathbb{C}}$ is continuous, then [107, IV.5]*

$$\deg_{g^{-1}(\Omega)}(f \circ g, g(p)) = \deg_\Omega(f, p) \deg(g)$$

We say that a homeomorphism f is *orientation-preserving* if $\deg_\Omega(f, p) \equiv 1$.

For a linear mapping $A : \mathbb{C} \to \mathbb{C}$, it is easy to check (from any of the above definitions) that

$$\deg_\Omega(A, p) = \deg(A) = \operatorname{sign} J(p, A) \in \{\pm 1\}, \qquad p \in A(\Omega) \qquad (2.46)$$

We note the following corollary.

Corollary 2.8.2. *Let $f : \Omega \to \mathbb{C}$ and suppose that f is differentiable at z_0 with Jacobian determinant $J(z_0, f) \neq 0$. Put $w = f(z_0)$. Then there is $\varepsilon > 0$ so that*

$$\deg_{\mathbb{D}(z_0, r)}(f, w) = \operatorname{sgn} J(z_0, f), \qquad r < \varepsilon$$

Proof. We assume $z_0 = f(z_0) = 0$ and $A = Df(0)$ is a linear map of determinant $J(0, f)$. The Taylor approximation shows that $|f(z) - Az| < o(|z|)$, and so there is $\varepsilon > 0$ such that $|z| < \varepsilon$ implies $|f(z)| > c|z|$, where c depends on $\|A^{-1}\|$. Thus $0 \notin f(\partial\mathbb{D}(0, r))$ and $0 = f^{-1}(f(0))$ for $f|\mathbb{D}(0, r)$, $r < \varepsilon$. Now the homotopy $H(t, z) = \frac{1}{t} f(tz)$ connects the mapping f to the linear transformation A, and the Lipschitz estimate shows $0 \notin H(t, \partial\mathbb{D}(0, r))$. The result now follows by homotopy invariance and (2.46). □

In particular, we see that if a homeomorphism f is differentiable at z_0 with Jacobian determinant $J(z_0, f) \neq 0$, then f is orientation-preserving if and only if $J(z_0, f) > 0$.

2.9 A Background in Complex Analysis

Throughout this book we shall largely be using complex analytic methods. This section is intended to provide, without much proof, some of the basic facts we shall use concerning conformal mappings and so forth.

2.9.1 Analysis with Complex Notation

It will often be convenient to handle with the complex notation also the basic results needed from real analysis. For instance, we have already used the fact that if f is differentiable at a point z, the differential can be written as

$$Df(z)\,h = \frac{\partial f}{\partial z}(z)\,h + \frac{\partial f}{\partial \bar{z}}(z)\,\bar{h}$$

Furthermore, the operator norm of the derivative takes the form in (2.23) and the Jacobian determinant has the form in (2.24).

Similarly, we may identify the $\mathbb{C}$-linear and the antilinear parts of the derivative $D(f \circ g)(z)$ of a composed function. Consequently, the chain rule takes the form

$$\begin{aligned}
(f \circ g)_z(z) &= f_w(g(z))\,g_z(z) + f_{\bar{w}}(g(z))\,\overline{g_{\bar{z}}(z)} & (2.47) \\
(f \circ g)_{\bar{z}}(z) &= f_w(g(z))\,g_{\bar{z}}(z) + f_{\bar{w}}(g(z))\,\overline{g_z(z)} & (2.48)
\end{aligned}$$

whenever the appropriate pointwise derivatives exist. Moreover, observing that $\partial_z z = \partial_{\bar{z}} \bar{z} = 1$ and $\partial_{\bar{z}} z = \partial_z \bar{z} = 0$, we see that any partial differential operator on $\mathbb{C}$ can be written in terms of ∂_z and $\partial_{\bar{z}}$ only.

If f is a homeomorphism differentiable at a point z with $Df(z)$ invertible, many times it will be useful to calculate in the complex notation the derivative of the inverse $h = f^{-1}$ at the point $w = f(z)$. Here we start from

$$(h \circ f)(z) = z$$

and compute

$$\begin{aligned}
h_w(w)\,f_z(z) + h_{\bar{w}}(w)\,\overline{f_{\bar{z}}(z)} &= 1 \\
h_w(w)\,f_{\bar{z}}(z) + h_{\bar{w}}(w)\,\overline{f_z(z)} &= 0
\end{aligned}$$

From these two equations we may eliminate h_w to find $\left(|f_z|^2 - |f_{\bar{z}}|^2\right) h_{\bar{w}} = -f_{\bar{z}}$. Hence, wherever $J(z, f) = |f_z|^2 - |f_{\bar{z}}|^2 \neq 0$, we have at $w = f(z)$ the identities

$$\frac{\partial h}{\partial \bar{w}}(w) = -\frac{f_{\bar{z}}(z)}{|f_z(z)|^2 - |f_{\bar{z}}(z)|^2} = -\frac{f_{\bar{z}}(z)}{J(z, f)} \quad (2.49)$$

$$\overline{\frac{\partial h}{\partial w}}(w) = \frac{f_z(z)}{|f_z(z)|^2 - |f_{\bar{z}}(z)|^2} = \frac{f_z(z)}{J(z, f)} \quad (2.50)$$

In particular, if f satisfies the equation $f_{\bar{z}} = \mu(z) f_z$, then

$$-\frac{\partial h}{\partial \bar{w}}(w) = \mu(h(w))\,\overline{\frac{\partial h}{\partial w}}(w), \qquad w = f(z) \quad (2.51)$$

2.9. A BACKGROUND IN COMPLEX ANALYSIS

Also, the familiar integral formulas can be written in this spirit. For instance, we have the following theorem.

Theorem 2.9.1. (Green's Formula) *Let Ω be a bounded domain with boundary $\partial\Omega$ consisting of a finite number of disjoint and rectifiable Jordan curves. Suppose that $f, g \in W^{1,1}(\Omega) \cap C(\overline{\Omega})$. Then*

$$\int_\Omega \left(\frac{\partial f}{\partial z} + \frac{\partial g}{\partial \bar{z}} \right) = \frac{i}{2} \int_{\partial\Omega} f \, d\bar{z} - g \, dz \tag{2.52}$$

For smooth functions this is just the usual Green's formula dressed in complex notation. The general case follows by approximation.

If in Green's formula we choose the functions $g(\tau) = \frac{\phi(\tau)}{z-\tau}$, $f \equiv 0$ and as the domain take $\Omega_\varepsilon = \Omega \setminus \mathbb{D}(z, \varepsilon)$, then letting $\varepsilon \to 0$ gives the classical *generalized Cauchy formula*, the starting point of all complex analysis,

$$\phi(z) = \frac{1}{2\pi i} \int_{\partial\Omega} \frac{\phi(\tau)}{\tau - z} d\tau - \frac{1}{\pi} \int_\Omega \frac{\phi_{\bar{\tau}}(\tau)}{\tau - z}, \qquad z \in \Omega \tag{2.53}$$

Here we have assumed that Ω is bounded, with $\partial\Omega$ a finite union of rectifiable Jordan curves, and

$$\phi \in W^{1,1}(\Omega) \cap C(\overline{\Omega})$$

Another immediate consequence of Green's formula concerns the integral of the Jacobian derivative $J(z, f) = |f_z(z)|^2 - |f_{\bar{z}}(z)|^2$. The Jacobian has a number of curious properties of fundamental importance in the geometric study of mappings; as an example, the integral of the Jacobian depends only on the boundary values of f. This phenomenon leads naturally to the notion of *null Lagrangians*, to be discussed in more detail in Chapter 19.

Corollary 2.9.2. *Let $f, g \in W^{1,2}(\Omega)$. Suppose f and g have the same boundary values in the sense of Sobolev functions, that is, $f - g \in W_0^{1,2}(\Omega)$, the closure of $C_0^\infty(\Omega)$ in $W^{1,2}(\Omega)$.*
Then

$$\int_\Omega J(z, f) = \int_\Omega J(z, g)$$

Proof. Approximating $f - g$ by functions in $C_0^\infty(\Omega)$ and f by $C^\infty(\Omega)$, we may assume that f and $g = f - (f - g)$ are smooth up to the boundary with $f = g$ on $\partial\Omega$. Consider the algebraic identity

$$f_z \overline{f_z} - f_{\bar{z}} \overline{f_{\bar{z}}} = g_z \overline{f_z} - g_{\bar{z}} \overline{f_{\bar{z}}} + \left[(f-g) \overline{f_z} \right]_z - \left[(f-g) \overline{f_{\bar{z}}} \right]_{\bar{z}}$$

By Green's formula the integral of the second term on the right is

$$\int_\Omega \left[(f-g) \overline{f_z} \right]_z = \frac{i}{2} \int_{\partial\Omega} (f-g) \overline{f_z} \, d\bar{z} = 0$$

since $f = g$ on $\partial\Omega$. Similarly the integral of the last term on the right vanishes. Hence we have

$$\int_\Omega J(z,f) = \int_\Omega f_z \overline{f_z} - f_{\bar{z}} \overline{f_{\bar{z}}} = \int_\Omega g_z \overline{f_z} - g_{\bar{z}} \overline{f_{\bar{z}}} = \int_\Omega f_z \overline{g_z} - f_{\bar{z}} \overline{g_{\bar{z}}}$$

since the Jacobian is real-valued. Interchanging the roles of f and g proves the claim. $\square$

2.9.2 Riemann Mapping Theorem and Uniformization

In order to avoid developing any machinery here that we shan't use elsewhere, we take a rather advanced standpoint and collect together basic results concerning conformal mappings and hyperbolic metrics and show how many of them follow from the uniformization theorem. Of course classically many of these results are basic ingredients for the proof of this central result [7]. The reader familiar with the basics of complex analysis may easily skip this material.

We recall that a conformal mapping of a domain $\Omega \subset \mathbb{C}$ is a holomorphic homeomorphism $\phi : \Omega \to \phi(\Omega) = \Omega'$. The first result we wish to present is the Riemann mapping theorem.

Theorem 2.9.3. (Riemann mapping theorem) *Let $\Omega \subset \mathbb{C}$ be a simply connected proper subdomain. Then there is a conformal surjection $\varphi : \mathbb{D} \to \Omega$. Moreover, if ψ is another such mapping, then $\psi^{-1} \circ \varphi : \mathbb{D} \to \mathbb{D}$ is a linear fractional transformation. In particular, given $z_0 \in \Omega$, there exists a unique conformal mapping $\varphi : \mathbb{D} \to \Omega$ with $\varphi(0) = z_0$ and $\varphi'(0) > 0$.*

The uniformization theorem of Koebe and Klein asserts a more general result by identifying the simply connected Riemann surfaces. Recall that a Riemann surface Σ is a two-dimensional topological manifold equipped with a conformal atlas. That is, charts (homeomorphisms onto their images) $\varphi_i : U_i \to \Sigma$, U_i open in $\mathbb{C}$, such that

- $\Sigma = \bigcup_i \varphi_i(U_i)$ and
- $\varphi_i^{-1} \circ \varphi_j : (\varphi_j^{-1} \circ \varphi_i)(U_i) \hookrightarrow \mathbb{C}$ is conformal for all i, j.

The identity chart shows every planar domain to be a Riemann surface.

Theorem 2.9.4. (Uniformization Theorem) *Let Σ be a simply connected Riemann surface. Then Σ is conformally equivalent to exactly one of $\mathbb{C}$, $\hat{\mathbb{C}}$ or $\mathbb{D}$.*

Now it is a basic fact of topology, more precisely covering space theory [147], that any Riemann surface admits a universal cover $\tilde{\Sigma}$, a simply connected Riemann surface, for which the deck transformations (the fundamental group) will act as a discrete group Γ of conformal transformations without fixed points. The discrete groups of conformal transformations of $\mathbb{C}$ and $\hat{\mathbb{C}}$ are easily identified. Any homeomorphism of $\hat{\mathbb{C}}$ has a fixed point. A conformal transformation of $\mathbb{C}$ is a similarity. A similarity without fixed points is a translation. It follows with

2.9. A BACKGROUND IN COMPLEX ANALYSIS

a little work that $\mathbb{C}$ is the universal cover only for Tori $\approx \mathbb{S} \times \mathbb{S}$ and cylinders $\mathbb{C} \setminus \{0\} \approx \mathbb{S} \times \mathbb{R}$. Every other Riemann surface therefore admits the disk as its universal cover.

We have already identified the conformal transformations of $\mathbb{D}$ as the linear fractional transformations $z \mapsto e^{i\theta}(z-a)/(1-\bar{a}z)$, $|a| < 1$, $\theta \in \mathbb{R}$. These maps act as hyperbolic isometries, as discussed earlier, and so the quotient inherits a hyperbolic metric in a natural way from the covering projection $\mathbb{D} \to \mathbb{D}/\Gamma = \Sigma$. Of importance to us later will be the case of the triply punctured sphere $\hat{\mathbb{C}} \setminus \{0, 1, \infty\} = \mathbb{C} \setminus \{0, 1\}$. There, the universal covering projection $\mathbb{D} \to \mathbb{C} \setminus \{0, 1\}$ is the Weierstrass $\mathcal{P}$-function [6]. Another example is the planar domain $\mathbb{D} \setminus \{0\}$. There the universal covering projection is $\exp\left(\frac{z-1}{z+1}\right)$. If we recall the conformal equivalence between $\mathbb{D}$ and $\mathbb{H}$, one can easily identify the associated group of deck transformations acting on $\mathbb{H}$ as the group $\langle z \mapsto z + k : k \in \mathbb{Z}\rangle$.

In general, for a given domain $\Omega \subset \mathbb{C}$, with $\partial\Omega$ having more than one finite boundary point, there is a universal covering map $\varphi : \mathbb{D} \to \Omega$ given by Theorem 2.9.4. The element of arc for the hyperbolic metric of Ω is given at each $z \in \Omega$ by

$$ds_{hyp} = \frac{2|\psi'(z)||dz|}{1 - |\psi(z)|^2}, \tag{2.54}$$

where ψ is a local inverse of φ defined near z. As all local inverses differ by composition with a linear fractional transformation of $\mathbb{D}$, the identity (2.13) shows this metric is well defined. As an example, let the domain $\Omega = \mathbb{D} \setminus \{0\}$ and the covering map $\psi(z) = \exp\left(\frac{z-1}{z+1}\right)$. Using (2.54), a calculation shows that

$$ds_{hyp} = \frac{1}{|z| \log \frac{1}{|z|}}, \qquad z \in \mathbb{D} \setminus \{0\} \tag{2.55}$$

There are many other ways to construct these hyperbolic metrics—for instance through potential theoretic methods or as the solutions to the PDE $\Delta \log \lambda = \lambda^2$, whence the metric $\lambda(z)|dz|$ has constant negative curvature; see [7].

The metric defined in (2.54) is complete and of constant curvature. If Ω is a planar domain with at least two finite points in its boundary, we call Ω a *hyperbolic domain*. Then we define the hyperbolic metric ρ_Ω by integrating the Riemannian metric in (2.54),

$$\rho_\Omega(z, w) = \inf_\gamma \int_\gamma ds_{hyp},$$

where the infimum is over all rectifiable curves γ joining z to w. This infimum is actually attained by a *hyperbolic geodesic*.

2.9.3 Schwarz-Pick Lemma of Ahlfors

Here we shall meet one of the most useful facts concerning holomorphic mappings. The classical Schwarz lemma states that if $\varphi : \mathbb{D} \to \mathbb{D}$ is holomorphic with $\varphi(0) = 0$, then $|\varphi(z)| \leqslant |z|$ and so $|\varphi'(0)| \leqslant 1$. Equality holds only for rotations. The usual proof is to apply the maximum principle to the holomorphic (!) function $\varphi(z)/z$, which does not exceed 1 in modulus on the circle $\partial \mathbb{D}$; see [6]. If $\varphi : \mathbb{D} \to \mathbb{D}$ is holomorphic and $z \in \mathbb{D}$ and we put

$$\Phi_z(\zeta) = \frac{z - \zeta}{1 - \bar{\zeta}z},$$

then

$$\varphi_z(\zeta) = (\Phi_{\varphi(z)} \circ \varphi \circ \Phi_z)(\zeta)$$

has $\varphi_z(\mathbb{D}) \subset \mathbb{D}$ and $\varphi_z(0) = 0$. Therefore, applying the Schwarz lemma to this composition with linear fractional transformations together with a bit of calculation gives the following invariant formulation.

Theorem 2.9.5. *If $\varphi : \mathbb{D} \to \mathbb{D}$ is holomorphic, then*

$$\frac{|\varphi'(z)|}{1 - |\varphi(z)|^2} \leqslant \frac{1}{1 - |z|^2} \tag{2.56}$$

In view of our discussion of the hyperbolic metric of the disk, (2.56) implies that φ is a local (and therefore global) contraction in the hyperbolic metric. If we have a holomorphic map $\varphi : \Omega \to \Omega'$ from one planar domain into another, then this mapping is a local (and hence global) contraction of the hyperbolic metrics. To see this, we simply lift the map via the respective holomorphic coverings to a holomorphic map of the disk $\mathbb{D}$. Locally, the holomorphic covering is an isometry, while the lifted map is a contraction, as we have observed. Thus the original map is a local (and hence global) contraction in view of the definition of the metric in (2.56). We record this as the following theorem.

Theorem 2.9.6. *Let $\varphi : \Omega \to \Omega'$ be a holomorphic mapping between hyperbolic domains. Then φ is a contraction of the hyperbolic metrics,*

$$\rho_{\Omega'}(\varphi(z), \varphi(w)) \leqslant \rho_\Omega(z, w)$$

for all $z, w \in \Omega$.

As a simple application, the identity map is holomorphic and therefore we have the following.

Corollary 2.9.7. *If $\Omega \subset \Omega'$, then for every $z, w \in \Omega$ we have*

$$\rho_{\Omega'}(z, w) \leqslant \rho_\Omega(z, w)$$

Then for $z \in \Omega'$ and $\Omega = \mathbb{D}(z, \text{dist}(z, \partial \Omega'))$ we apply Corollary 2.9.7 to get the following.

Corollary 2.9.8. *For $z \in \Omega'$*

$$\frac{1}{2 \text{dist}(z, \partial \Omega')} \leqslant \delta_{\Omega'}(z) \leqslant \frac{2}{\text{dist}(z, \partial \Omega')},$$

where $ds_{hyp} = \delta_{\Omega'}(z)|dz|$ is the length element of the hyperbolic metric of Ω'.

2.9. A BACKGROUND IN COMPLEX ANALYSIS

Here we have found it convenient to include the lower bound, which is a consequence of the Koebe $\frac{1}{4}$-theorem and in particular the estimate of (2.70) that we will come to in a moment.

2.9.4 Normal Families and Montel's Theorem

A family of continuous functions $\mathcal{F} = \{\varphi : \Omega \to \mathbb{C}\}$ is called *normal* if every sequence $\{\varphi_i\}_{i=1}^{\infty} \subset \mathcal{F}$ contains a subsequence converging locally uniformly to a limit function φ.

Two ideas from topology suffice to establish that a family is normal. These are the notion of equicontinuity and the Ascoli theorem.

Equicontinuity

Let (X, d) and (Y, σ) be metric spaces and let $\mathcal{F} = \{\varphi : X \to Y\}$ be a family of functions. We call $\mathcal{F}$ *equicontinuous* at $x_0 \in X$ if for every $\varepsilon > 0$ there is $\delta > 0$ such that $d(x, x_0) < \delta$ implies $\sigma(\varphi(x), \varphi(x_0)) < \varepsilon$ for all $\varphi \in \mathcal{F}$. We say that $\mathcal{F}$ is equicontinuous if it is equicontinuous at each $x_0 \in X$.

Next, the central result linking normal families and equicontinuity is the Arzela-Ascoli theorem.

Theorem 2.9.9. *If (X, d) is a separable metric space and (Y, σ) a compact metric space, then every equicontinuous family of mappings $\mathcal{F} = \{\varphi : X \to Y\}$ is a normal family.*

There are a number of variations of this theorem. One of the most common is to replace the hypothesis that (Y, σ) is a compact metric space by the hypothesis that the family $\mathcal{F}$ is locally bounded and (Y, σ) is complete.

In our applications we are going to take $Y = \hat{\mathbb{C}}$ and σ as the spherical metric arising from the Riemannian metric $ds_{sph} = |dz|/\sqrt{1+|z|^2}$. Then

$$\sigma(z, w) = \frac{|z-w|}{\sqrt{1+|z|^2}\sqrt{1+|w|^2}}$$

with the obvious interpretation if z or w is equal to ∞.

To use the Arzela-Ascoli theorem we need to relate the hyperbolic and spherical metrics. Notice that both these metrics are conformally equivalent to the Euclidean metric. The density of the hyperbolic metric is nowhere vanishing and tends uniformly to ∞ near $\partial \Omega$. Hence it has a positive minimum in Ω. As σ is bounded, we have the following theorem.

Theorem 2.9.10. *Let $\Omega \subset \mathbb{C}$ be a hyperbolic domain. Then there is a constant C_Ω such that*

$$\sigma(z, w) \leqslant C_\Omega \rho_\Omega(z, w)$$

for all $z, w \in \Omega$.

There is no uniform bound independent of domain. If $\Omega = \mathbb{D}(0,R)$, $R > 1$, then $\sigma(0,1) = \frac{1}{\sqrt{2}}$ and $\rho_\Omega(0,1) = \log \frac{R+1}{R-1} \to 0$ as $R \to \infty$.

Now as a consequence, we have Montel's theorem.

Theorem 2.9.11. *Let $z_0, z_1 \in \mathbb{C}$ and let Ω be a planar domain. Suppose that $\mathcal{F} \subset \{\varphi : \Omega \to \mathbb{C} \setminus \{z_0, z_1\}\}$ is a family of holomorphic maps omitting these two values. Then $\mathcal{F}$ is a normal family.*

Proof. $\mathbb{C} \setminus \{z_0, z_1\}$ is a hyperbolic domain, and therefore each element of $\mathcal{F}$ is a contraction of the hyperbolic metrics if Ω is also a hyperbolic domain. In that case to see that the family $\mathcal{F}$ is equicontinuous when viewed as mappings into $\hat{\mathbb{C}}$ we use Theorem 2.9.6 which tells us that for each $z, w \in \Omega$,

$$\sigma(\varphi(z), \varphi(w)) \leqslant C\, \rho_{\mathbb{C} \setminus \{z_0, z_1\}}(\varphi(z), \varphi(w)) \leqslant C\, \rho_\Omega(z, w)$$

with constant C depending only on z_0 and z_1. If Ω is not a hyperbolic domain we simply restrict to a hyperbolic subdomain and the result follows easily. $\square$

2.9.5 Hurwitz's Theorem

Hurwitz's theorem asserts that the limits of nonvanishing holomorphic functions are either nonvanishing or identically 0. It is usually proved as an application of Rouché's Theorem and the maximum principle.

Theorem 2.9.12. *Let $\Omega \subset \mathbb{C}$ be a domain and $\{\varphi_i\}_{i=1}^\infty$ be a sequence of nonvanishing holomorphic functions converging locally uniformly to $\varphi : \Omega \to \mathbb{C}$. Then either $\varphi \equiv 0$ or φ does not vanish in Ω*

Using this theorem the reader may easily prove the following important consequence.

Corollary 2.9.13. *Let $\Omega \subset \mathbb{C}$ be a domain and $\{\varphi_i\}_{i=1}^\infty$ be a sequence of conformal mappings converging locally uniformly to $\varphi : \Omega \to \mathbb{C}$. Then either φ is constant, or φ is conformal.*

2.9.6 Bloch's Theorem

Bloch's constant $\mathcal{B}_0$ is defined to be the supremum of all numbers δ such that each holomorphic function f in the unit disk $\mathbb{D}$ with $|f'(0)| = 1$ possesses a continuous inverse in some disk of radius δ; see [63]. We have the following theorem of Ahlfors; see for instance [7].

Theorem 2.9.14. $\mathcal{B}_0 \geqslant \sqrt{3}/4$.

For locally univalent mappings (so with the further assumption $f'(z) \neq 0$, $z \in \mathbb{D}$), the similarly defined constant is denoted $\mathcal{B}_\infty$, and it is known that $\mathcal{B}_\infty > \frac{1}{2}$. The exact determination of $\mathcal{B}_0$ remains a famously difficult open problem. The prevailing conjecture, due to Ahlfors and Grunsky (1937), asserts that $\mathcal{B}_0 = \Gamma(\frac{1}{3})\Gamma(\frac{11}{12})/\sqrt{1+\sqrt{3}}\,\Gamma(\frac{1}{4}) = 0.4719\ldots$.

2.9.7 The Argument Principle

We will later need the following formulation of the argument principle, which is essentially a corollary of Theorem 2.8.1. An elementary proof using classical complex analysis can be found in [6].

Theorem 2.9.15. *If f is holomorphic in $\mathbb{D}$, if it is continuous in $\overline{\mathbb{D}}$ and if $f|_{\partial \mathbb{D}}$ is homotopic to the identity relative to $\mathbb{C} \setminus \{0\}$, then $f(z_0) = 0$ at precisely one point $z_0 \in \mathbb{D}$.*

2.10 Distortion by Conformal Mapping

From a historical perspective, the notion of a quasiconformal mapping has its roots in the study of the geometric properties of conformal mappings. The philosophy here is that locally a conformal mapping is close to its linearization $z \mapsto f(z_0) + f'(z_0)(z - z_0)$—a similarity mapping of $\mathbb{C}$. Such mappings form a finite-dimensional family and because of this are quite "rigid." One expects that conformal mappings should inherit some of this geometric rigidity. This is a property that we will state in an explicit and quantitative manner in Theorem 2.10.9. In fact, this view leads naturally to the notion of a quasisymmetric mapping, to be studied in the next chapter. For the reader's convenience we describe the basic distortion theorems of conformal mappings.

2.10.1 The Area Formula

We shall first formulate and prove a classical result of Gronwall [148], soon thereafter rediscovered by Bieberbach [60], that is known as the area formula. It will have important applications later. It also leads us to the important Koebe distortion theorems.

Theorem 2.10.1. *Suppose that $f \in W^{1,2}_{loc}(\mathbb{C})$ is analytic outside the disk $\mathbb{D}(0,r)$ and has the expansion*
$$z + b_1 z^{-1} + b_2 z^{-2} + \cdots \tag{2.57}$$
near ∞. Then
$$\int_{\mathbb{D}(0,r)} J(z,f) = \pi \left(r^2 - \sum_{n=1}^{\infty} n|b_n|^2 r^{-2n} \right) \tag{2.58}$$
In particular, if f is orientation-preserving (so $J(z,f) \geqslant 0$ almost everywhere), then
$$\sum_{n=1}^{\infty} n|b_n|^2 r^{-2n} \leqslant r^2 \tag{2.59}$$

Proof. As f is analytic outside $\mathbb{D}(0,r)$, the power series in (2.57), converges uniformly on $\{z : |z| \geqslant s\}$ for any fixed $s > r$. Then we use Green's formula

(2.52) to write

$$\begin{aligned}
\int_{\mathbb{D}(0,s)} J(z,f) &= \int_{\mathbb{D}(0,s)} |f_z|^2 - |f_{\bar z}|^2 = \frac{1}{2i}\int_{|z|=s} \bar f\, f_z\, dz \\
&= \frac{1}{2i}\int_{|z|=s} (\bar z + \overline{b_1}\,\bar z^{-1} + \cdots)(1 - b_1 z^{-2} - 2b_2 z^{-3} + \cdots)dz \\
&= \frac{1}{2i}\int_{|z|=s} (s^2 - |b_1|^2 s^{-2} - 2|b_2|^2 s^{-4} - \cdots)\frac{dz}{z} \\
&= \pi(s^2 - |b_1|^2 s^{-2} - 2|b_2|^2 s^{-4} - \cdots)
\end{aligned}$$

Now we may let $s \to r$ to obtain (2.58) and of course (2.59) when $J(z,f) \geqslant 0$ almost everywhere. $\square$

We can use the area inequality in (2.58) to estimate certain functions as follows. Suppose that $f \in W^{1,2}_{loc}(\mathbb{C})$ is analytic outside the disk $\mathbb{D}(0,r)$ and has the expansion

$$z + b_1 z^{-1} + b_2 z^{-2} + \cdots \tag{2.60}$$

near ∞. We obtain from the Cauchy-Schwarz inequality the easy estimate

$$\begin{aligned}
|f(z) - z| &\leqslant \frac{|b_1|}{|z|} + \frac{|b_2|}{|z|^2} + \cdots \\
&\leqslant \left(\frac{|b_1|^2}{r^2} + \frac{2|b_2|^2}{r^4} + \frac{3|b_3|^2}{r^6} + \cdots\right)^{1/2}\left(\frac{r^2}{|z|^2} + \frac{r^4}{2|z|^4} + \frac{r^6}{3|z|^6} + \cdots\right)^{1/2} \\
&\leqslant r\left(\log \frac{|z|^2}{|z|^2 - r^2}\right)^{1/2}
\end{aligned}$$

Therefore, in particular, we obtain for $|z| > 1.3\,r$ the estimate

$$|f(z) - z| \leqslant r \tag{2.61}$$

If f is a homeomorphism of $\mathbb{C}$, then (2.59) holds even without assuming the $W^{1,2}_{loc}$-regularity. We therefore have the following corollary.

Corollary 2.10.2. *If $f \in W^{1,1}_{loc}(\mathbb{C})$ is a homeomorphism analytic outside the disk $\mathbb{D}(0,r)$ with $|f(z) - z| = o(1)$ at ∞, then*

$$|f(z)| < |z| + 3r, \quad \text{for all } z \in \mathbb{C} \tag{2.62}$$

Proof. The only things to observe are that, first, the assumption $|f(z) - z| = o(1)$ at ∞ together with analyticity implies a series expansion as in (2.59) and thus we have (2.61) for $|z| > 1.3r$, and second, as f is supposed a homeomorphism, the estimate clearly follows for $|z| < 1.3r$ as f is an open mapping. $\square$

2.10. DISTORTION BY CONFORMAL MAPPING

Another interesting application concerns the solutions to the Beltrami system with the normalization at ∞ as above (we shall later call them principal solutions). Suppose f is a $W^{1,2}_{loc}(\mathbb{C})$-solution to the equation

$$\frac{\partial f}{\partial \bar{z}} = \mu(z) \frac{\partial f}{\partial z}, \tag{2.63}$$

where $|\mu(z)| \leqslant k \chi_D$ for $D = \mathbb{D}(0, r)$ and $0 \leqslant k < 1$. Then f is analytic outside the disk D. Suppose we have the further normalisation that

$$|f(z) - z| = o(1), \quad z \text{ near } \infty \tag{2.64}$$

As we have noticed already, this gives us a series expansion as in (2.58) converging locally uniformly on $|z| > r$. We have the formula (see (2.53))

$$f(z) = z - \frac{1}{\pi} \int_D \frac{f_{\bar{z}}(\tau)}{\tau - z} d\tau$$

and hence

$$b_1 = \lim_{z \to \infty} z(f(z) - z) = \frac{1}{\pi} \int_D f_{\bar{z}}(\tau) \, d\tau,$$

which gives us the estimate

$$|b_1|^2 \leqslant \left(\frac{1}{\pi} \int_D |f_{\bar{z}}|\right)^2 \leqslant \frac{r^2}{\pi} \int_D |f_{\bar{z}}|^2$$

$$\leqslant \frac{k^2 r^2}{\pi(1-k^2)} \int_D (|f_z|^2 - |f_{\bar{z}}|^2)$$

$$= \frac{k^2 r^2}{(1-k^2)} \left(r^2 - \sum_{n=1}^{\infty} n|b_n|^2 r^{-2n}\right)$$

and hence

$$|b_1|^2 \left(1 + \frac{k^2}{1-k^2}\right) \leqslant \frac{k^2 r^2}{(1-k^2)} \left(r^2 - \sum_{n=2}^{\infty} n|b_n|^2 r^{-2n}\right)$$

$$\leqslant \frac{k^2 r^4}{(1-k^2)}$$

This establishes the following corollary.

Corollary 2.10.3. *For a solution of (2.63) normalized by (2.64), we have the estimate*

$$|b_1| \leqslant kr^2 \tag{2.65}$$

We have equality if and only if

$$f(z) = \begin{cases} z + \lambda r^2/z, & |z| \geqslant r \\ z + \lambda \bar{z}, & |z| \leqslant r \end{cases} \tag{2.66}$$

where λ is any complex number with $|\lambda| = k$.

The only thing not clear in the corollary is the claim (2.66) regarding equality. However, this follows by consideration of the fact that, in order to have equality in (2.65), we must have had equality in every step of our calculation.

2.10.2 Koebe $\frac{1}{4}$-Theorem and Distortion Theorem

The Koebe $\frac{1}{4}$-Theorem is one of the first and also one of the most powerful distortion theorems one meets in complex analysis. With the correct interpretation this result implies universal distortion estimates in hyperbolic disks, satisfied by *all* conformal mappings.

To find these we first consider the analytic function $g(z) = z+b_0+b_1z^{-1}+\cdots$, which we suppose is conformal in the exterior of the unit disk and further that $g(z) \neq w$ for $|z| > 1$. Then the branch

$$h(z) = \sqrt{g(z^2) - w} = z + \frac{1}{2}(b_0 - w)z^{-1} + \cdots$$

is well defined and conformal in the exterior disk. Furthermore, for any $r > 1$ its restriction to $\{z : |z| > r\}$ extends to a global mapping $h \in W^{1,2}_{loc}(\mathbb{C})$. Thus Theorem 2.10.1 gives

$$|w - b_0| \leqslant 2 \tag{2.67}$$

Often it is convenient to use this result in the following form.

Theorem 2.10.4. *Suppose $g : \mathbb{C} \to \mathbb{C}$ is a homeomorphism, which is conformal in the exterior of the unit disk. If g has the development $g(z) = z+b_0+b_1z^{-1}+\cdots$ for $|z| > 1$, then*

$$g(\mathbb{D}) \subset \mathbb{D}(b_0, 2)$$

The famous Koebe $\frac{1}{4}$-theorem is a quick consequence.

Theorem 2.10.5. *Suppose that $\varphi : \mathbb{D} \to \mathbb{C}$ is conformal and normalized by $\varphi(0) = 0$ and $\varphi'(0) = 1$. Then*

$$\mathbb{D}(0, \frac{1}{4}) \subset \varphi(\mathbb{D})$$

Proof. With our assumptions $\varphi(z) = z + a_2 z^2 + \cdots$ for $z \in \mathbb{D}$. The conjugate

$$g(z) = \frac{1}{\varphi(z^{-1})} = z - a_2 + \mathcal{O}(\frac{1}{z}), \qquad |z| > 1,$$

never vanishes, and thus (2.67) implies the classical bound of Bieberbach,

$$|a_2| \leqslant 2 \tag{2.68}$$

Also if $w \notin \varphi(\mathbb{D})$, the function

$$\varphi_1(z) = \frac{w\,\varphi(z)}{w - \varphi(z)} = z + (a_2 + \frac{1}{w})z^2 + \mathcal{O}(z^2)$$

satisfies the assumptions of the theorem, and hence we have additionally

$$|a_2 + \frac{1}{w}| \leqslant 2 \tag{2.69}$$

2.10. DISTORTION BY CONFORMAL MAPPING

Combining the bounds shows that $|w| \geq 1/4$ whenever $w \notin \varphi(\mathbb{D})$. □

We may view Theorem 2.10.4 as the counterpart to Koebe's result at ∞. In bounded domains the following form of Koebe's $\frac{1}{4}$-theorem applies in fact to all conformal mappings, independently of their normalization.

Theorem 2.10.6. *Suppose that f is conformal in a domain Ω with $f(\Omega) = \Omega' \subset \mathbb{C}$. Let $z_0 \in \Omega$. Then*

$$\frac{1}{4}|f'(z_0)|\mathrm{dist}(z_0, \partial\Omega) \leq \mathrm{dist}\left(f(z_0), \partial\Omega'\right) \leq |f'(z_0)|\mathrm{dist}(z_0, \partial\Omega) \tag{2.70}$$

The first inequality in (2.70) follows from Koebe's theorem, applied to

$$\varphi(z) = \frac{f(z_0 + zd) - f(z_0)}{d\, f'(z_0)}, \qquad d = \mathrm{dist}(z_0, \partial\Omega),$$

while the latter inequality is a consequence of the Schwarz lemma, applied to $f^{-1}: \mathbb{D}\left(f(z_0), d'\right) \to \mathbb{D}(z_0, d)$, where $d' = \mathrm{dist}\left(f(z_0), \partial\Omega'\right)$.

The Bieberbach bound (2.68) also provides us with uniform distortion estimates as soon as we are able to express it in an invariant form. To reveal this we introduce, for each mapping f conformal in $\mathbb{D}$, the *Koebe transform*

$$\varphi(z) = \frac{f\left(\frac{z+w}{1+\overline{w}z}\right) - f(w)}{(1-|w|^2)f'(w)}, \qquad z \in \mathbb{D} \tag{2.71}$$

Here $w \in \mathbb{D}$ is arbitrary.

An elementary calculation gives

$$\varphi''(0) = (1-|w|^2)\frac{f''(w)}{f'(w)} - 2\overline{w} \tag{2.72}$$

Since φ is conformal in $\mathbb{D}$ with $\varphi(0) = 0$ and $\varphi'(0) = 1$, Bieberbach's coefficient estimate yields the following theorem.

Theorem 2.10.7. *If f is conformal in the unit disk $\mathbb{D}$, then*

$$(1-|w|^2)\frac{|f''(w)|}{|f'(w)|} \leq 6, \qquad w \in \mathbb{D}$$

We are now in a position to prove the first of the Koebe distortion theorems. For our purposes an invariant formulation, such as the following, is the most prefered.

Theorem 2.10.8. *Suppose that f is conformal in the unit disk $\mathbb{D}$ and $z, w \in \mathbb{D}$. Then*

$$e^{-3\rho_\mathbb{D}(z,w)} \leq \frac{|f'(z)|}{|f'(w)|} \leq e^{3\rho_\mathbb{D}(z,w)}$$

Proof. Since f is conformal, the function $g(z) = \log f'(z)$ is analytic in $\mathbb{D}$. Theorem 2.10.7 tells us that $|g'(z)| \leq 3\, ds_{hyp}(z)$, and the claim follows via an integration,

$$\left|\log \frac{|f'(z)|}{|f'(w)|}\right| \leq |g(z) - g(w)| \leq 3\rho_{\mathbb{D}}(z,w) \qquad \square$$

It is remarkable that each of Theorems 2.10.5 – 2.10.8 is sharp, as the reader may verify using the Koebe function $f_0(z) = \frac{z}{(1-z)^2} = \frac{1}{4}\left(\frac{1+z}{1-z}\right)^2 - \frac{1}{4}$. The function f_0 maps the unit disk $\mathbb{D}$ conformally onto $\mathbb{C} \setminus (-\infty, -\frac{1}{4}]$.

According to Theorem 2.10.8, one may consider the derivative of a conformal mapping as almost constant on hyperbolic disks! It is to be expected that then, with suitable interpretation, the mapping itself should almost be a similarity when restricted to a subdomain bounded in the hyperbolic metric.

This fact turns out to be true and is perhaps most conveniently expressed in the notation of the next theorem. Here note that a homeomorphism f in a domain Ω is a similarity if and only if

$$\frac{|f(z) - f(w)|}{|f(\zeta) - f(w)|} = \frac{|z-w|}{|\zeta - w|} \qquad \text{for all } z, w, \zeta \in \Omega \tag{2.73}$$

Theorem 2.10.9. *Suppose that f is conformal in the unit disk $\mathbb{D}$. Let z_1, z_2 and $w \in \mathbb{D}$ with*

$$\rho_{\mathbb{D}}(z_1, w) + \rho_{\mathbb{D}}(z_2, w) \leq M < \infty$$

Then

$$\frac{|f(z_1) - f(w)|}{|f(z_2) - f(w)|} \leq e^{4M} \frac{|z_1 - w|}{|z_2 - w|} \tag{2.74}$$

Proof. We will use the Koebe transform (2.71) and evaluate $\varphi(\zeta_j)$, where $z_j = (\zeta_j + w)/(1 + \overline{w}\zeta_j)$ and $j = 1, 2$. Then $\zeta_j = (z_j - w)/(1 - \overline{w}z_j)$. Hence

$$\frac{f(z_1) - f(w)}{f(z_2) - f(w)} = \frac{\varphi(\zeta_1)}{\varphi(\zeta_2)} = \frac{z_1 - w}{z_2 - w} \frac{\varphi(\zeta_1)}{\zeta_1} \frac{\zeta_2}{\varphi(\zeta_2)} \frac{1 - \overline{w}z_2}{1 - \overline{w}z_1}$$

To estimate the last expression we note that

$$\log\left|\frac{1 - \overline{w}z_2}{1 - \overline{w}z_1}\right| \leq \log\left(\frac{1 + \left|\frac{z_1 - w}{1 - \overline{w}z_1}\right|}{1 - \left|\frac{z_2 - w}{1 - \overline{w}z_2}\right|}\right)$$

$$\leq \rho_{\mathbb{D}}(z_1, w) + \rho_{\mathbb{D}}(z_2, w)$$

Since $\rho_{\mathbb{D}}(\zeta_j, 0) = \rho_{\mathbb{D}}(z_j, w)$, it remains to show that

$$e^{-3\rho_{\mathbb{D}}(\zeta,0)} \leq \frac{|\varphi(\zeta)|}{|\zeta|} \leq e^{3\rho_{\mathbb{D}}(\zeta,0)}, \qquad \zeta \in \mathbb{D} \tag{2.75}$$

2.10. DISTORTION BY CONFORMAL MAPPING

In fact, by Theorem 2.10.8

$$|\varphi(\zeta)| = \left|\int_0^\zeta \frac{\varphi'(z)}{\varphi'(0)}dz\right| \leq \int_0^{|\zeta|} e^{3\rho_{\mathbb{D}}(z,0)}|dz| \leq |\zeta|e^{3\rho_{\mathbb{D}}(\zeta,0)}$$

For the former of the inequalities in (2.75), note that this is clear if $|\varphi(\zeta)| \geq 1/4$. Otherwise, by Koebe $\frac{1}{4}$-theorem, the interval $[\varphi(\zeta), 0] \subset \varphi(\mathbb{D})$. As $\varphi'(z)dz$ has a constant argument on $\varphi^{-1}[\varphi(\zeta), 0]$, we have

$$|\varphi(\zeta)| = \int_0^{|\zeta|} |\varphi'(z)|\,|dz| \geq \int_0^{|\zeta|} e^{-3\rho_{\mathbb{D}}(z,0)}|dz| \geq |\zeta|e^{-3\rho_{\mathbb{D}}(\zeta,0)}$$

Combining these estimates gives the inequality (2.74). □

The above theorem is an invariant version of the second Koebe distortion theorem, expressing in a compact and quantitative manner the fact that locally every conformal mapping is close to a similarity. Here, though, no claim is made on the sharpness of (2.74) in terms of the exponent $4M$. On the other hand, an important fact in Theorem 2.10.9 is the conformal invariance; via a change of variables it applies immediately to all mappings f conformal in a simply connected domain Ω.

We note also the following immediate consequence.

Corollary 2.10.10. *Conformal mappings of the plane are similarities.*

Proof. With a scaling, the estimate (2.74) holds in any disk $\mathbb{D}(0, r) = r\mathbb{D}$. If we denote $M_r = \rho_{r\mathbb{D}}(z_1, w) + \rho_{r\mathbb{D}}(z_1, w)$, then (2.74) attains the form

$$\frac{|z_1 - w|}{|z_2 - w|}e^{-4M_r} \leq \frac{|f(z_1) - f(w)|}{|f(z_2) - f(w)|} \leq \frac{|z_1 - w|}{|z_2 - w|}e^{4M_r} \quad (2.76)$$

Fixing the points z_1, z_2 and w but letting $r \to \infty$ gives $M_r = \rho_{r\mathbb{D}}(z_1, w) + \rho_{r\mathbb{D}}(z_1, w) = \rho_{\mathbb{D}}(z_1/r, w/r) + \rho_{\mathbb{D}}(z_1/r, w/r) \to 0$. Hence f satisfies (2.73). □

Bounds on the distortion of ratios such as in (2.74) quickly yield a large spectrum of various geometric properties. Indeed, the geometric study of mappings requires general notions that allow such conclusions, and for much larger classes than just (the very rigid) conformal mappings. These considerations will lead in a natural manner to the concept of quasisymmetry, which is studied and utilized in the next section. In this terminology, Theorem 2.10.9 tells us that all conformal mappings are uniformly quasisymmetric in subdomains with bounded hyperbolic diameter.

Chapter 3

The Foundations of Quasiconformal Mappings

One of the more interesting and important recent developments in the theory of quasiconformal mappings has been the development of the study of these mappings on more general spaces than those that are locally Euclidean. In particular, the work of Heinonen and Koskela [161] develops the theory in certain metric spaces where some reasonable measure theoretic and geometric properties hold. Part of the motivation behind this work is that often while studying classical problems in geometry one is naturally led to singular spaces as quotients or, for instance, to the boundaries of geometric objects such as groups. In these situations one has no a priori knowledge that the space in question is a manifold. While we do not discuss these developments in this book, this work does have important ramifications. It has led to a new understanding, and in particular new and simplified proofs, of many of the foundational properties of quasiconformal mappings. Here we will present some of this material, partly making use of the lecture notes of P. Koskela [215].

3.1 Basic Properties

Here is our starting point.

Definition 3.1.1. *An orientation-preserving homeomorphism $f : \Omega \to \Omega'$ is K-quasiconformal, $1 \leqslant K < \infty$, if*

$$f \in W^{1,2}_{loc}(\Omega)$$

and

$$\max_\alpha |\partial_\alpha f(z)| \leqslant K \min_\alpha |\partial_\alpha f(z)| \qquad (3.1)$$

for almost every $z \in \Omega$.

3.2. QUASISYMMETRY

In particular, a mapping f is 1-quasiconformal if and only if it is conformal, that is, a univalent holomorphic mapping. The reader should not forget the precise meaning or background of (3.1) nor the complex interpretation through the Beltrami equation, both discussed in Section 2.5. However, the complex analytic approach will not show its real virtues much before Section 5.

Beginning with the above definition, we will prove in this chapter the first and most fundamental properties of quasiconformal mappings as embodied in the following two theorems.

Theorem 3.1.2. *Let $f : \Omega \to \Omega'$ be a K-quasiconformal mapping from the domain $\Omega \subset \mathbb{C}$ onto $\Omega' \subset \mathbb{C}$ and let $g : \Omega' \to \mathbb{C}$ be a K'-quasiconformal mapping. Then*

- *$f^{-1} : \Omega' \to \Omega$ is K-quasiconformal.*
- *$g \circ f : \Omega \to \mathbb{C}$ is KK'-quasiconformal.*
- *For all measurable sets $E \subset \Omega$, $|E| = 0$ if and only if $|f(E)| = 0$.*
- *The Jacobian determinant $J(z, f) > 0$ almost everywhere in Ω.*

Thus the family of quasiconformal mappings forms a pseudogroup with respect to composition (actually a group if the domain and range are the same). The next theorem addresses local compactness.

Theorem 3.1.3. *Let $f_\nu : \Omega \to \mathbb{C}$, $\nu = 1, 2, \ldots$, be a bounded sequence of K-quasiconformal mappings defined on the domain $\Omega \subset \mathbb{C}$. Then there is a subsequence converging locally uniformly on Ω to a mapping f,*

$$f_{\nu_k} \to f,$$

and f is either a K-quasiconformal mapping or a constant.

The key concept we shall use in this chapter to establish these results is that of quasisymmetry, which we now look at.

3.2 Quasisymmetry

The very definition of a quasiconformal map supposes the function to be defined on an open set. However, from time to time in this book we will be concerned with the restrictions of quasiconformal mappings to smaller sets or mappings perhaps defined on fractal-like sets that might have a quasiconformal extension. This suggests that we seek a notion of (quasi)conformality applicable in general subsets of the plane (or even in arbitrary metric spaces).

In fact, we have already met such a notion, in the invariant formulation of the Koebe distortion estimates in Theorem 2.10.9. This point of view leads us naturally to the very useful concept of quasisymmetry. The notion was introduced by Ahlfors and Beurling [11] on the real line and formulated for general metric spaces by Tukia and Väisälä [359].

Definition of Quasisymmetry

Definition 3.2.1. *Let $\eta : [0, \infty) \to [0, \infty)$ be an increasing homeomorphism, $A \subset \mathbb{C}$ and*
$$f : A \to \mathbb{C}$$
a mapping. We say f is η-quasisymmetric if for each triple $z_0, z_1, z_2 \in A$ we have
$$\frac{|f(z_0) - f(z_1)|}{|f(z_0) - f(z_2)|} \leqslant \eta\left(\frac{|z_0 - z_1|}{|z_0 - z_2|}\right) \tag{3.2}$$
Should f be defined on an open set, we will assume that it is orientation-preserving and further, we say f is quasisymmetric if there is some η as above for which f is η-quasisymmetric.

Notice that by definition $\eta(0) = 0$ and $\eta(t) \to \infty$ as $t \to \infty$. A typical example of such a homeomorphism η is $\eta(t) = C t^\alpha$, for constants $C \geqslant 1$ and $\alpha > 0$. In fact, as a consequence of what is proved later, η can always be chosen in this form (see Section 3.10). Note also that the condition at (3.2) can be made to make sense even for mappings of metric spaces.

It is an immediate consequence of the definition that a quasisymmetric mapping is continuous and injective, and by Lemma 3.2.2 below we see that it is a homeomorphism onto its image. Furthermore, when A is an open set, $z_0 \in A$ and $r < \text{dist}(z_0, \partial A)$; then, with a suitable choice of z_1 and z_2 on the circle $\{|z - z_0| = r\}$, we have
$$\frac{\max_{|z-z_0|=r} |f(z) - f(z_0)|}{\min_{|z-z_0|=r} |f(z) - f(z_0)|} = \frac{|f(z_1) - f(z_0)|}{|f(z_2) - f(z_0)|} \leqslant \eta(1) \tag{3.3}$$

The right hand side of (3.3) is independent of r. We may take the limit as $r \to 0$ on the left to see that quasisymmetric mappings defined on open subsets are, in particular, mappings of bounded distortion. This suggests a strong relationship between quasiconformality and quasisymmetry, which we shall also explore in this chapter.

It is not at all obvious that the inverse of a quasiconformal homeomorphism is quasiconformal. However, for quasisymmetric mapping this is an elementary observation.

Lemma 3.2.2. *Let $f : \Omega \to \Omega'$ be an η-quasisymmetric mapping onto. Then $f^{-1} : \Omega' \to \Omega$ is σ-quasisymmetric with*
$$\sigma(t) = \frac{1}{\eta^{-1}(1/t)}$$

Proof. As f is injective and onto, given a triple of points a_i, we may put $z_i = f^{-1}(a_i)$, $i = 0, 1, 2$. Then we have, from the fact that f is η-quasisymmetric,

$$\frac{|a_0 - a_2|}{|a_0 - a_1|} \leq \eta\left(\frac{|z_0 - z_2|}{|z_0 - z_1|}\right),$$

and hence

$$\frac{|f^{-1}(a_0) - f^{-1}(a_1)|}{|f^{-1}(a_0) - f^{-1}(a_2)|} \leq 1/\eta^{-1}\left(\frac{|a_0 - a_2|}{|a_0 - a_1|}\right),$$

which is what we wanted. □

There are no bounded conformal mappings of the plane—similarly, for quasisymmetric mappings.

Lemma 3.2.3. *Every entire quasisymmetric mapping f (that is defined in the whole complex plane $\mathbb{C}$) is a surjection, $f(\mathbb{C}) = \mathbb{C}$.*

Proof. The condition (3.2) implies $|f(z)| \to \infty$ when $|z| \to \infty$ since we may fix z_0, z_1 and let $z_2 \to \infty$. Thus f extends continuously to $\hat{\mathbb{C}}$ with $f(\infty) = \infty$. Thus $f(\hat{\mathbb{C}})$ is open and closed in $\hat{\mathbb{C}}$; that is $f(\hat{\mathbb{C}}) = \hat{\mathbb{C}}$. □

These observations suggest the route we will take to establishing the basic properties of quasiconformal mappings. After auxiliary results we first show that quasisymmetric mappings are quasiconformal. The principal problem here is in proving the function lies in the correct Sobolev class, $W^{1,2}_{loc}(\Omega)$. Next we show that quasiconformal mappings of $\mathbb{C}$ are quasisymmetric. The principal problem in that case is establishing control on global distortion functions. It will then follow, for instance, that the inverse of a K-quasiconformal mapping is quasiconformal. That the inverse is in fact K-quasiconformal will not be too difficult to establish.

Classically, the equivalence of quasiconformality and quasisymmetry for maps of $\mathbb{C}$ was a consequence of the modulus definition of quasiconformality (see for instance [229]) once one has some knowledge of extremal rings and the like. Here we shall offer a couple of alternate routes to this result directly in what follows and somewhat later as a consequence of the theory of holomorphic motions.

3.3 The Gehring-Lehto Theorem

Throughout the theory of quasiconformal mappings we often need quite subtle methods to control the geometry of mappings. In fact, in our later studies we will be basically on the borderline of where a geometric study of mappings is even possible. However, conditions (2.30) and (2.31) for the partial derivatives, or the bound (2.28) for the linear distortion, are rather weak by themselves. To improve the situation we need a few additional basic facts from real analysis to carry the geometric information up from the infinitesimal level.

First we establish a fairly well-known and general theorem of Gehring and Lehto [137] which asserts that an open mapping with finite partial derivatives at almost every point is differentiable at almost every point. For homeomorphisms the result was earlier established by Menchoff [258]. The proof must use properties of the plane, as it is false in higher dimensions, although certain analogs exist; see [180]. Second, with the Gehring-Lehto result we are able to connect the volume derivatives and the pointwise Jacobians $J(z, f)$ and thereby to obtain a first version of the change-of-variables formula.

3.3.1 The Differentiability of Open Mappings

We recall that a mapping $f : \Omega \subset \mathbb{C} \to \mathbb{C}$ is *open* if $f(U)$ is open for every open $U \subset \Omega$.

To begin with we will need the following refinements of the concept of density. Let E be a measurable subset of $\mathbb{C}$. A point $z_0 = x_0 + iy_0 \in E$ is called a point of x-density if x_0 is a point of linear density of the set $\{x \in \mathbb{R} : x + iy_0 \in E\}$. Similarly we have the notion of y-density. A point $z_0 \in E$ that is both a point of x-density and of y-density will be called a point of xy-density. Of course, as soon as we are able to establish the measurability of the set of points of xy-density, then Fubini's theorem implies that such points have full measure in E. Indeed, the set E_1 consisting of points of x-density has full measure in E for otherwise $|E \setminus E_1| > 0$ and by Fubini's theorem we could find y_0 such that the set $\{x \in \mathbb{R} : x + iy_0 \in E \setminus E_1\}$ has positive linear measure. But this would contradict the Lebesgue density theorem on the real line. Analogously, the set E_2 of points of y-density is measurable with full measure and therefore the intersection of these two sets has full measure. This is of course the set of all points of xy-density.

To show the measurability of E_1, it is enough to consider closed sets E. We denote by $E_{n,k}$ the set of points $x + iy \in E$ such that

$$\mathcal{H}^1(\{t \in [a,b] : t + iy \in E\}) \geq \left(1 - \frac{1}{n}\right)(b-a)$$

whenever

$$a < x < b \quad \text{and} \quad 0 < b - a \leq \frac{1}{k}$$

Then clearly

$$E_1 = \bigcap_{n=1}^{\infty} \bigcup_{k}^{\infty} E_{n,k},$$

and it suffices to show that the sets $E_{n,k}$ are closed. Here, let $z_j = x_j + iy_j \in E_{n,k}$ with $z_j \to z_0 = x_0 + iy_0$. If $a < x_0 < b$ with $b - a < 1/k$, then $a < x_j < b$ for all j large enough, and since E is closed,

$$\mathcal{H}^1(\{t \in [a,b] : t + iy_0 \in E\}) \geq \mathcal{H}^1\left(\bigcap_{\ell=1}^{\infty} \bigcup_{j=\ell}^{\infty} \{t \in [a,b] : t + iy_j \in E\}\right)$$
$$\geq \left(1 - \frac{1}{n}\right)(b-a)$$

3.3. THE GEHRING-LEHTO THEOREM

Therefore $z_0 \in E_{n,k}$ which proves that $E_{n,k}$ is closed and hence that E_1 is measurable. We argue in the same manner to see E_2 is measurable. We have thus established that the set of all points of xy-density in E is measurable with full measure.

We next have the following lemma whose proof is an elementary argument in linear density and measure theory.

Lemma 3.3.1. *Let $\varepsilon > 0$ and $z_0 = x_0 + iy_0 \in E$ be a point of xy-density of E. Then, for all $z = a + ib$ sufficiently close to z_0, there is a rectangle*

$$R = [x_1, x_2] \times [y_1, y_2]$$

containing z and such that

$$(x_2 - x_1) < 2\varepsilon |a - x_0|, \quad (y_2 - y_1) < 2\varepsilon |b - y_0|$$

and such that the points $x_1 + iy_0$, $x_2 + iy_0$, $x_0 + iy_1$ and $x_0 + iy_2$ all lie in E.

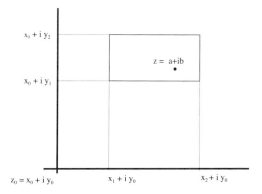

Choosing the rectangle $R = [x_1, x_2] \times [y_1, y_2]$

Proof. We may assume that $x_0 = y_0 = 0$, and so z_0 is the point at the origin. We may also assume $a, b > 0$. Let

$$E_x = \{x \in \mathbb{R} : x + i0 \in E\}, \quad E_y = \{y \in \mathbb{R} : 0 + iy \in E\}$$

Since $x = 0$ is a point of density of E_x, each of the intervals $(a - \varepsilon a, a)$ and $(a, a + \varepsilon a)$ contains points of E_x provided that a is sufficiently small, say $a < \delta$. We pick $x_1 \in (a - \varepsilon a, a)$ and $x_2 \in (a, a + \varepsilon a)$ such that both $x_1 + i0$ and $x_2 + i0$ lie in E. Similarly, we find $y_1 \in (b - \varepsilon b, b)$ and $y_2 \in (b, b + \varepsilon b)$ with both $0 + iy_1$ and $0 + iy_2 \in E$, again provided b is sufficiently small. Now we have $x_2 - x_1 < 2\varepsilon a$ and $y_2 - y_1 < 2\varepsilon b$, as desired. □

We shall now prove the following theorem.

Theorem 3.3.2. *Let $f : \Omega \to \mathbb{C}$ be a continuous open mapping. Then f is differentiable almost everywhere in Ω if and only if f has finite first partials almost everywhere.*

Proof. If f is differentiable almost everywhere, then f has finite first partials almost everywhere. It is the converse that we need to establish. Thus we assume that the partials f_x and f_y exist and are finite at almost every point of Ω. It will be enough to prove that f is in fact differentiable at almost every point of a given compact subset $X \subset \Omega$. Let t be a real number, $0 < |t| < \mathrm{dist}(X, \partial\Omega)$. We define, for those $z = x + iy$ at which both partials exist and are finite, the function

$$F_t(z) = \left| \frac{f(x+t,y) - f(x,y)}{t} - f_x(x,y) \right| + \left| \frac{f(x,y+t) - f(x,y)}{t} - f_y(x,y) \right| \quad (3.4)$$

It is easy to see that the set where F_t is defined is a Borel set [316, p. 70]. Thus F_t is a Borel function defined almost everywhere on X and it follows that the functions

$$g_n(z) = \sup_{0 < |t| < 1/n} F_t(z)$$

are also Borel for sufficiently large n (as it suffices to let t run through only rational values). From our assumption on the partial derivatives of f, we see that

$$g_n(z) \to 0, \quad \text{almost everywhere as } n \to \infty$$

Now by the theorems of Egoroff and Lusin, there is an increasing sequence of compact subsets $X_1 \subset X_2 \subset \cdots \subset X$ with

$$\left| X - \bigcup_{\nu=1}^{\infty} X_\nu \right| = 0$$

for which we have for each ν that the functions f_x, f_y are continuous in X_ν and

$$F_t(z) \to 0, \quad \text{uniformly on } X_\nu$$

Of course, it will now suffice to fix a ν, put $E = X_\nu$ and prove that f is differentiable at any $z_0 \in E$ that is an xy-density point of E.

Let $0 < \varepsilon < 1$. We aim to prove the estimate

$$|f(z) - f(z_0) - f_x(z_0)(x - x_0) - f_y(z_0)(y - y_0)|$$
$$\leqslant \varepsilon(4 + |f_x(z_0)| + |f_y(z_0)|)(|x - x_0| + |y - y_0|) \quad (3.5)$$

whenever $z \in \Omega$ is sufficiently close to z_0. This is enough to ensure the differentiability of f at z_0.

Now, for all $z \in E$ sufficiently close to z_0, we have

$$|f_x(z) - f_x(z_0)| < \varepsilon, \qquad |f_y(z) - f_y(z_0)| < \varepsilon, \quad (3.6)$$

3.3. THE GEHRING-LEHTO THEOREM

while $|F_t(z)| < \varepsilon$ if t is small. Next

$$\begin{aligned}|f(z) - f(z_0) &- f_x(z_0)(x - x_0) - f_y(z_0)(y - y_0)| \\ &\leq |f(z) - f(x + iy_0) - f_y(x + iy_0)(y - y_0)| \\ &+ |f(x + iy_0) - f(z_0) - f_x(z_0)(x - x_0)| \\ &+ |f_y(x + iy_0) - f_y(z_0)||y - y_0|\end{aligned}$$

We now assume that z is sufficiently close to z_0 so as to be able to apply (3.6). Accordingly, we arrive at the estimate

$$\begin{aligned}|f(z) - f(z_0) &- f_x(z_0)(x - x_0) - f_y(z_0)(y - y_0)| \\ &\leq F_{y-y_0}(x + iy_0)|y - y_0| + F_{x-x_0}(z_0)|x - x_0| + \varepsilon|y - y_0| \\ &\leq 2\varepsilon(|x - x_0| + |y - y_0|) \end{aligned} \quad (3.7)$$

whenever $z = x + iy \in \Omega$ is sufficiently close to z_0 and in addition $x + iy_0 \in E$. Similarly, we have this same estimate whenever $z = x + iy \in \Omega$, $x_0 + iy \in E$ and z is sufficiently close to z_0.

Up to this point we have not used the fact that f is open, and we do so now. That f is open implies that f satisfies the maximum principle—maxima occur on the boundary. In particular, for each point $z \in \Omega$ close to z_0, let R be the rectangle given by Lemma 3.3.1. Using the maximum principle, we find that the expression

$$|f(\zeta) - f(z_0) - f_x(z_0)(u - x_0) - f_y(z_0)(v - y_0)|$$

considered as a function of $\zeta = u + iv$ takes its maximum value on the boundary of R. Hence at the maximum point $\zeta \in \partial R$,

$$\begin{aligned}|f(z) - f(z_0) &- f_x(z_0)(x - x_0) - f_y(z_0)(x - y_0)| \\ &\leq |f(\zeta) - f(z_0) - f_x(z_0)(u - x_0) - f_y(z_0)(v - y_0)| \\ &+ |f_x(z_0)||u - x| + |f_y(z_0)||v - y|\end{aligned}$$

Furthermore, for each boundary point $\zeta = u + iv \in \partial R$, either $u + iy_0 \in E$ or $x_0 + iv \in E$. In view of the estimate in (3.7),

$$|f(\zeta) - f(z_0) - f_x(z_0)(u - x_0) - f_y(z_0)(v - y_0)| \leq 2\varepsilon(|u - x_0| + |v - y_0|)$$

As $|u - x| \leq \varepsilon|x - x_0|$ and $|v - y| \leq \varepsilon|y - y_0|$, the above estimates prove (3.5), establishing the theorem. □

We note that in the proof we used only the maximum principle on rectangles, an apparently weaker condition than assuming f is open.

The above theorem applies, in particular, to Sobolev homeomorphisms.

Corollary 3.3.3. *Every homeomorphism $f \in W^{1,1}_{loc}(\Omega)$ is differentiable almost everywhere.*

We now observe the following consequence of the Gehring-Lehto theorem and Corollary 2.8.2 concerning local degree.

Theorem 3.3.4. *Suppose $f : \Omega \to \Omega'$ is a homeomorphism between planar domains of Sobolev class $W^{1,1}_{loc}(\Omega)$. The Jacobian determinant, $J(z,f) = \det Df(z)$, does not change sign; that is, either*

- *$J(z,f) \geqslant 0$ almost everywhere in Ω or*

- *$J(z,f) \leqslant 0$ almost everywhere in Ω.*

Proof. Since f is a homeomorphism, the local degree $\deg_\Omega(f,p)$ is independent of the point $p \in f(\Omega)$. Next, Corollary 2.8.2 implies that if a is a point of differentiability for f, then this local degree is the sign of the Jacobian (should it not be zero). The Gehring-Lehto theorem implies that f is differentiable almost everywhere so that if $J(a,f) > 0$ on a set of positive measure, then $\deg_\Omega(f,p) \equiv 1$ and the Jacobian cannot be negative. □

In particular, we see that if the homeomorphism f is *orientation-preserving*, that is, $\deg_\Omega(f,p) \equiv 1$, then $J(z,f) \geqslant 0$ almost everywhere. The converse holds if $J(a,f) > 0$ for at least one point a.

Actually, the above result holds in greater generality. The assumption that f is a homeomorphism is too strong. A mapping $f : \Omega \to f(\Omega)$ is *discrete* if the inverse image of a point $p \in f(\Omega)$ is discrete in Ω. Then there is the well-known Stoilow factorization theorem that states that every discrete open mapping f has the form $f = \varphi \circ F$, where F is a homeomorphism and φ is holomorphic. We will later give a fairly detailed exposition of this factorization theorem in the Sobolev setting, for mappings of bounded distortion in Section 5.5 and for mappings of finite distortion in Chapter 20. The following result is a simple consequence of Corollary 3.3.4 and this factorization.

Theorem 3.3.5. *Suppose $f : \Omega \to \mathbb{C}$ is a discrete open mapping of Sobolev class $W^{1,1}_{loc}(\Omega)$. Then the Jacobian determinant, $J(z,f) = \det Df(z)$, does not change sign.*

The interesting question here of course is whether the Jacobian determinant can be identically 0 for such a mapping. There are simple examples based on the Cantor function $(x,y) \mapsto (c(x),c(y))$, $0 \leqslant x,y, \leqslant 1$ and $c : [0,1] \to [0,1]$, strictly increasing with $c'(x) = 0$ for all but a countable set of points. These are homeomorphisms for which the Jacobian is identically 0, but such examples are not of Sobolev class. On the other hand, there are Sobolev homeomorphisms for which the Jacobian vanishes on a set of positive measure.

Next, let us connect the Gehring-Lehto theorem to the notion of volume derivative. Every homeomorphism $f : \Omega \to \Omega'$ induces a natural Borel measure on Ω, the pullback of the Lebesgue measure

$$\nu(E) = |f(E)|$$

3.3. THE GEHRING-LEHTO THEOREM

The *volume derivative* of the homeomorphism f is, by definition, the derivative ν with respect to Lebesgue measure,

$$D\nu(z) = \lim_{r \to 0} \frac{|f\mathbb{D}(z,r)|}{|\mathbb{D}(z,r)|}, \tag{3.8}$$

where the limit exists for almost every $z \in \Omega$ and belongs to $L^1_{loc}(\Omega)$ [251]. In fact,

$$\nu(E) = \int_E D\nu(z)\,dz + \nu_s(E) \qquad E \subset \Omega \text{ a Borel subset}, \tag{3.9}$$

where ν_s is the singular part of ν.

If a homeomorphism $f \in W^{1,1}_{loc}(\Omega)$, then from Corollary 2.8.2 f is differentiable almost everywhere. It follows that $D\nu(z) = |\det Df(z)| \equiv |J(z,f)|$ at almost every $z \in \Omega$. In particular, we have the following version of the change-of-variables formula.

Corollary 3.3.6. *If* $f \in W^{1,1}_{loc}(\Omega, \Omega')$ *is an orientation-preserving homeomorphism, then the Jacobian determinant* $J(z,f)$ *is locally integrable and*

$$\int_E J(z,f)\,dz \leqslant |f(E)|, \qquad E \subset \Omega \text{ a Borel set.} \tag{3.10}$$

Moreover, if $0 \leqslant v \in C(\Omega')$,

$$\int_\Omega v(f(z))\,J(z,f)\,dz \leqslant \int_{\Omega'} v(z)\,dz$$

Note that here we need to restrict ourselves to Borel sets E only since the homeomorphic image of a measurable set need not remain measurable. (For instance, the Cantor function described above maps a set of zero measure to a set of positive measure. Every set of positive measure contains a nonmeasurable subset. The preimage of such a subset lies in a set of measure 0 and so is measurable.) For f to preserve Lebesgue-measurable sets, it must satisfy Lusin's condition $\mathcal{N}$, which we will consider more extensively for quasiconformal mappings in Section 3.7. Next, however, we present an elementary proof of the fact that $W^{1,2}_{loc}$-homeomorphisms preserve sets of measure 0.

Theorem 3.3.7. *Let* Ω *and* Ω' *be planar domains and* $f : \Omega \to \Omega'$ *be a* $W^{1,2}_{loc}(\Omega, \Omega')$-*homeomorphism. If* $E \subset \Omega$ *has measure* 0, *then the measure of* $f(E)$ *is also* 0,

$$|E| = 0 \Rightarrow |f(E)| = 0 \tag{3.11}$$

Proof. We may assume that f is orientation-preserving and hence that $J(z,f) \geqslant 0$ almost everywhere, see Section 2.8. If $E \subset \Omega$ is compact, choose a subdomain U of Ω containing E such that the boundary ∂U consists of finitely many line segments, each of which is parallel to the real or imaginary axis and for which

$f|\partial U$ is absolutely continuous. Since f is a homeomorphism that is absolutely continuous on almost all lines it is possible to find such U. We claim that

$$|f(U)| = \int_U J(z, f)\, dz \qquad (3.12)$$

This is enough to imply (3.11) since if $|E| = 0$ and $|f(E)| > 0$, then

$$\begin{aligned}\int_U J(z,f) &= \int_{U\setminus E} J(z,f) \leqslant |f(U\setminus E)| = |f(U)| \setminus f(E)| \\ &= |f(U)| - |f(E)| < |f(U)|\end{aligned}$$

which contradicts (3.12) as $|f(U)| < \infty$.

We put $V = f(U)$ and compute

$$\begin{aligned}|f(U)| &= |V| = \int_V 1 = \frac{1}{2i}\int_{\partial V} \bar z\, dz \qquad \text{(Green's formula (2.52))} \\ &= \frac{1}{2i}\int_{\partial U} \bar f\, df \qquad \text{(change of variables)}\end{aligned}$$

In the last inequality the change-of-variables in the one-dimensional integral is enough since f is an absolutely continuous homeomorphism on ∂U. Note that, for more general Sobolev mappings, the index or winding number of the mapping f would have to be taken into account.

Next,

$$\frac{1}{2i}\int_{\partial U} \bar f\, df = \frac{1}{2i}\int_{\partial U} (\bar f\, f_z\, dz + \bar f\, f_{\bar z}\, d\bar z) = \int_U |f_z|^2 - |f_{\bar z}|^2$$

The last equality holds here by virtue of Green's formula (2.52) for smooth mappings and by an approximation for continuous $W^{1,2}$-mappings, keeping uniform bounds in L^2 on the derivatives. □

The local integrability of the Jacobian of a Sobolev mapping is of fundamental importance in much of the analysis we will perform. For a homeomorphism this property is automatic, by Corollary 3.3.6. Otherwise, it is the minimum prerequisite for obtaining any reasonable results. As we will see, this will become particularly evident in the theory of degenerate elliptic equations and mappings of finite distortion in Chapter 20.

3.4 Quasisymmetric Maps Are Quasiconformal

Here we present a first step in the circle of results that relate quasiconformality and quasisymmetry.

Theorem 3.4.1. *Suppose that $f : \Omega \to \Omega'$ is an η-quasisymmetric mapping. Then f is quasiconformal. In particular, $f \in W^{1,2}_{loc}(\Omega)$.*

3.4. QUASISYMMETRIC MAPS ARE QUASICONFORMAL

This section will cover the proof of Theorem 3.4.1, with the several auxiliary results required. Our main obstacle here is that the definition of a quasisymmetric mapping gives no indications as to the possible Sobolev regularity or absolutely continuity properties. In fact, a one-dimensional quasisymmetric mapping need not be absolutely continuous [11], and hence we are discussing genuinely higher-dimensional phenomena.

In obtaining Sobolev regularity for quasisymmetric mappings, the following notion from [161] will prove particularly valuable. For a given $z \in \Omega$ and $\varepsilon > 0$, we define the Borel function

$$L_f^\varepsilon(z) = \sup \left\{ \frac{|f(z+h) - f(z)|}{|h|} : 0 < |h| < \min\{\operatorname{dist}(z, \partial\Omega), \varepsilon\} \right\} \quad (3.13)$$

If we let $\varepsilon \to 0$, then

$$L_f^\varepsilon(z) \to L_f(z), \quad (3.14)$$

the *maximal derivative*. It is possible for this maximal derivative, and hence $L_f^\varepsilon(z)$, to be infinite. However, when f is quasisymmetric, $L_f^\varepsilon(z) > 0$ for almost every z.

We now need the following lemma.

Lemma 3.4.2. *If $f : \Omega \to \Omega'$ is quasisymmetric and $z_0 \in \Omega$, then for each $\varepsilon > 0$ the function $2L_f^\varepsilon(z)$ is an upper gradient for the function $|f(z) - f(z_0)|$. That is, by definition,*

$$|f(z) - f(z_0)| \leq 2 \int_\gamma L_f^\varepsilon(\zeta) |d\zeta| \quad (3.15)$$

for each rectifiable curve $\gamma \subset \Omega$ joining z to z_0.

Proof. Fix $z, z_0 \in \Omega$ and γ a rectifiable curve in Ω joining them. Let $\ell(\gamma)$ denote the arc length of this curve. Since $L_f^\varepsilon(z)$ is nondecreasing in ε, we may assume $\varepsilon < \operatorname{dist}(\gamma, \partial\Omega)$. We consider two separate cases.

First, suppose that $\operatorname{diam}(\gamma) \leq \varepsilon$. Choose a $\zeta \in \gamma$. Then from the triangle inequality we have

$$\begin{aligned} |f(z) - f(z_0)| &\leq \frac{|f(z) - f(\zeta)|}{|z - \zeta|} |z - \zeta| + \frac{|f(z_0) - f(\zeta)|}{|z_0 - \zeta|} |z_0 - \zeta| \\ &\leq 2\ell(\gamma) L_f^\varepsilon(\zeta) \end{aligned}$$

Hence

$$\int_\gamma L_f^\varepsilon(\zeta) |d\zeta| \geq \frac{1}{2\ell(\gamma)} |f(z) - f(z_0)| \, \ell(\gamma) = \frac{1}{2} |f(z) - f(z_0)|,$$

which completes the first case.

Now suppose that $\operatorname{diam}(\gamma) > \varepsilon$. We reparameterise $\gamma : [0,1] \to \Omega$ with $\gamma(0) = z_0$ and $\gamma(1) = z$ and then choose $0 = t_0 < t_1 < \cdots < t_n = 1$ so that
$$\operatorname{diam}(\gamma([t_j, t_{j+1}])) < \varepsilon$$

Then, using the first case,
$$\begin{aligned}|f(z) - f(z_0)| &\leq \sum_{j=0}^{n-1} |f(\gamma(t_{j+1})) - f(\gamma(t_j))| \\ &\leq 2\sum_{j=0}^{n-1} \int_{\gamma([t_j,t_{j+1}])} L_f^\varepsilon(\zeta)|d\zeta| \\ &\leq 2\int_\gamma L_f^\varepsilon(\zeta)|d\zeta|,\end{aligned}$$

which establishes the remaining case. $\square$

Next, we want to point out a few easy but useful consequences of the quasisymmetric condition. These relate the diameter and measure of subsets of the plane under quasisymmetric (or, using the as yet unproven Theorem 3.5.3, of quasiconformal) mappings.

Lemma 3.4.3. *Suppose $f : \Omega \to \Omega'$ is η-quasisymmetric and $B \subset \Omega$ is a disk. Then*
$$\operatorname{diam}(f(B))^2 \leq C_0 |f(B)|,$$
where the constant C_0 depends only on $\eta(1)$.

Proof. Suppose $B = \mathbb{D}(z_0, s)$ and let r be the radius of an inscribed disk for $f(B)$, $r = \min_{|z-z_0|=s} |f(z) - f(z_0)|$. Clearly, $|f(B)| \geq \pi r^2$. Then (3.3) gives
$$\operatorname{diam}(f(B)) \leq 2 \max_{|z-z_0|=s} |f(z) - f(z_0)| \leq 2\eta(1) r,$$
from which the result follows with $C_0 = \frac{4}{\pi} \eta(1)^2$. $\square$

Corollary 3.4.4. *Suppose $f : \Omega \to \Omega'$ is η-quasisymmetric. Then, for all points $z \in \mathbb{D}(z_0, s) \subset \Omega$,*
$$|f(z) - f(z_0)|^2 \leq \frac{4}{\pi} \eta(1)^2 |f(\mathbb{D}(z_0, s))|$$

We see that quasisymmetric images of disks are roundish objects. To view this from another angle, look at the image $f(\mathbb{D}(z_0, r))$. With the notation
$$R := \max_{|\zeta-z_0|=r} |f(\zeta) - f(z_0)| \leq \eta(1) \min_{|\zeta-z_0|=r} |f(\zeta) - f(z_0)|, \qquad (3.16)$$
we have the following lemma.

3.4. QUASISYMMETRIC MAPS ARE QUASICONFORMAL

Lemma 3.4.5. *If $\Omega = f(B)$ is the image of a disk $B = \mathbb{D}(z_0, r)$ under a quasisymmetric mapping, then*

$$\mathbb{D}(w_0, \delta R) \subset \Omega \subset \mathbb{D}(w_0, R), \qquad \delta = \frac{1}{\eta(1)},$$

where $w_0 = f(z_0)$ and the radius R is given in (3.16).

In other words, the image $\Omega = f(B)$ has roughly the size of a disk of radius $R \simeq \operatorname{diam}((f(B))$.

Next, we have the following lemma establishing the integrability properties of the upper gradient via the maximal function.

Lemma 3.4.6. *If $f : \Omega \to \Omega'$ is η-quasisymmetric and $2\mathbb{D}(z_0, r) \subset \Omega$, then for each $\varepsilon < r$ we have*

$$|\{\zeta \in \mathbb{D}(z_0, r) : L_f^\varepsilon(\zeta) > t\}| \leqslant C(\eta) t^{-2} |f(2\mathbb{D}(z_0, r))|, \qquad (3.17)$$

where $C(\eta) = (10\eta(1))^2$ depends only on the quasisymmetric distortion.

An estimate of the form

$$|\{z \in E : |h(z)| > t\}| \leqslant C t^{-p}$$

such as in (3.17) with $p = 2$ is commonly refered to as a *weak-L^p-type* estimate. The connection is via the Chebychev inequality for L^p functions. Thus (3.17) says that $L_f^\varepsilon \in \text{weak-}L_{loc}^2(\Omega)$.

Proof. We begin with the standard approach to studying maximal functions. If $t > 0$ and

$$E_t = \{\zeta \in \mathbb{D}(z_0, r) : L_f^\varepsilon(\zeta) > t\},$$

then for all $x \in E_t$ we can find $r_x \in [0, r]$ such that for some real θ_x

$$|f(x + r_x e^{i\theta_x}) - f(x)| > t r_x \qquad (3.18)$$

Using the well-known $\frac{1}{5}$-covering theorem [251] we can find a countable family of disjoint disks $\mathbb{D}_j = \mathbb{D}(x_j, r_{x_j})$ such that

$$\mathbb{D}_j \subset \mathbb{D}(z_0, 2r) \subset \Omega, \qquad E_t \subset \bigcup_{j=1}^\infty 5\mathbb{D}_j$$

Then, using (3.18) and quasisymmetry via Corollary 3.4.4,

$$\begin{aligned}
|E_t| &\leqslant 25 \sum_{j=1}^\infty |\mathbb{D}_j| = 25\pi \sum_{j=1}^\infty r_{x_j}^2 \\
&\leqslant 25\pi \sum_{j=1}^\infty \frac{|f(x + r_x e^{i\theta_x}) - f(x)|^2}{t^2} \\
&\leqslant 100\, t^{-2}\, \eta(1)^2 \sum_{j=1}^\infty |f(\mathbb{D}_j)| \\
&\leqslant 100\, t^{-2}\, \eta(1)^2 |f(2\mathbb{D}(z_0, r))|
\end{aligned}$$

The penultimate inequality follows, as the disks $\mathbb{D}_j$ are disjoint and so therefore are their images under the homeomorphism f. Thus we have established the result with
$$C(\eta) = (10\, \eta(1))^2,$$
as desired. □

If we integrate the distribution function for which we have the estimate at (3.17), we quickly obtain the following corollary.

Corollary 3.4.7. *Suppose that $f : \Omega \to \Omega'$ is η-quasisymmetric, that $B = \mathbb{D}(z_0, r)$ is a disk with $2B \subset \Omega$ and that $\varepsilon < r$. Then for all $1 \leq p < 2$ we have*

$$\frac{1}{|B|} \int_B [L_f^\varepsilon]^p \leq C(p, \eta) \left(\frac{|f(2B)|}{|B|} \right)^{p/2} \tag{3.19}$$

Proof. Using (3.17), an integration gives, for any $t_0 > 0$,

$$\begin{aligned}
\frac{1}{|B|} \int_B [L_f^\varepsilon]^p &= \frac{p}{|B|} \int_0^\infty t^{p-1} |\{z \in B : L_f^\varepsilon > t\}| dt \\
&\leq \frac{p}{|B|} \int_0^{t_0} t^{p-1} |B| dt + \frac{p}{|B|} C(\eta) \int_{t_0}^\infty t^{p-3} |f(2B)| dt \\
&\leq t_0^p + C(\eta) \frac{p}{2-p} \frac{|f(2B)|}{|B|} t_0^{p-2}
\end{aligned}$$

Choosing $t_0 = \sqrt{|f(2B)|/|B|}$ implies the required bound (3.19) with $C(p, \eta) = 1 + pC(\eta)/(2-p)$. □

Next, we establish the absolute continuity of quasisymmetric mappings as a prelude to the more general result we seek. The reader will see that the key ideas are in fact here, and the role of L_f^ε becomes evident: On the one hand, because of Lemma 3.4.2 this quantity is large enough to majorize the variation of f, while on the other, for a quasisymmetric mapping it resembles the actual derivative with integrability properties similar to those of (3.10) or (2.33).

Theorem 3.4.8. *Let $f : \Omega \to \Omega'$ be η-quasisymmetric. Then $f \in W^{1,1}_{loc}(\Omega)$.*

Proof. Let $Q = [a, b] \times [c, d]$ be a rectangle compactly contained in Ω. For the moment we will use real variable notation with $x \in [a, b]$, $y \in [c, d]$ and write $f = u + iv$. First, by Fubini's theorem and Corollary 3.4.7, for $\mathcal{H}^1$-almost every $y \in [c, d]$ and all $1 \leq p < 2$, we know that the function

$$x \mapsto L_f^\varepsilon(x + iy) \in L^p([a, b] \times \{y\})$$

Next, by Lemma 3.4.2 we have

$$|u(a, y) - u(b, y)| \leq |f(a + iy) - f(b + iy)| \leq 2 \int_a^b L_f^\varepsilon(x + iy) dx < \infty,$$

3.4. QUASISYMMETRIC MAPS ARE QUASICONFORMAL

and similarly for any sequence of disjoint subintervals $(a_j, b_j) \subset (a, b)$, we have

$$\sum_{j=1}^{n} |u(b_j, y) - u(a_j, y)| \leq 2 \sum_{j=1}^{n} \int_{a_j}^{b_j} L_f^\varepsilon(x + iy) dx$$
$$\leq 2 \int_E L_f^\varepsilon(x + iy) dx,$$

where $E = \cup_{j=1}^n (a_j, b_j)$. This shows us that the function $x \mapsto u(x, y)$ is absolutely continuous for almost every $y \in [c, d]$. Of course, entirely analogous arguments show that $y \mapsto u(x, y)$ is absolutely continuous for almost every $x \in [a, b]$ and that v has the same properties. This gives us more than just the existence of the partial derivatives $\frac{\partial u}{\partial y}$ almost everywhere and the like; we also know that $x \mapsto L_f^\varepsilon(x + iy) \in L^1([a, b] \times \{y\})$ and that the partials

$$\left|\frac{\partial u}{\partial x}\right|, \left|\frac{\partial u}{\partial y}\right| \leq 2L_f^\varepsilon \in L^1(Q),$$

and this is enough to guarantee that $f \in W^{1,1}(Q)$. As $Q \subset \Omega$ was arbitrary, the result follows. $\square$

Now it is a relatively easy matter to establish Theorem 3.4.1, which claims that if $f : \Omega \to \Omega'$ is η-quasisymmetric, then f is quasiconformal. The only real issue is to establish that $f \in W^{1,2}_{loc}(\Omega)$. We recall Corollary 3.4.4, which gives for all z with $|z - z_0| = r$ the estimate

$$|f(z) - f(z_0)| \leq \frac{2}{\sqrt{\pi}} \eta(1) |f(\mathbb{D}(z_0, r))|^{1/2},$$

and hence

$$\frac{1}{2\eta(1)} \frac{|f(z) - f(z_0)|}{|z - z_0|} \leq \left(\frac{|f(\mathbb{D}(z_0, r))|}{|\mathbb{D}(z_0, r)|}\right)^{1/2}.$$

We also have

$$|f(\mathbb{D}(z_0, r))| \leq \pi \max_{|z - z_0| = r} |f(z) - f(z_0)|^2$$
$$= \pi \eta(1)^2 \min_{|z - z_0| = r} |f(z) - f(z_0)|^2,$$

from which we deduce that, for all z with $|z - z_0| = r$,

$$\left(\frac{|f(\mathbb{D}(z_0, r))|}{|\mathbb{D}(z_0, r)|}\right)^{1/2} \leq \eta(1) \frac{|f(z) - f(z_0)|}{|z - z_0|},$$

and hence we have the inequality

$$\frac{1}{2\eta(1)} \frac{|f(z) - f(z_0)|}{|z - z_0|} \leq \left(\frac{|f(\mathbb{D}(z_0, r))|}{|\mathbb{D}(z_0, r)|}\right)^{1/2} \leq \eta(1)^2 \frac{|f(z) - f(z_0)|}{|z - z_0|} \qquad (3.20)$$

These inequalities take us back to (3.8) and the volume derivative f, that is the derivative $D\nu(z)$ of the pullback measure $\nu(E) = |f(E)|$. Indeed, we see that $D\nu(z)$ is comparable to $L_f(z)^2$, the square of the maximal derivative. In Theorem 3.4.8 we have already established the existence almost everywhere of the partial derivatives, and now we obtain

$$|\partial_x f|^2, \ |\partial_y f|^2 \leqslant L_f(z)^2 \leqslant 4\eta(1)\, D\nu \in L^1_{loc}(\Omega) \tag{3.21}$$

As f is continuous, we have $f \in L^2_{loc}(\Omega)$, which together with what we have proven is enough to put $f \in W^{1,2}_{loc}(\Omega)$.

Since quasisymmetric mappings are homeomorphisms, it remains to show that they satisfy the distortion estimate (3.1). However, the quasisymmetry condition immediately yields, for all z, w with $|z - z_0| = |w - z_0| = r$, that

$$\frac{|f(z) - f(z_0)|}{|z - z_0|} \leqslant \eta(1) \frac{|f(w) - f(z_0)|}{|w - w_0|}$$

To complete the argument we need the Gehring-Lehto theorem, and Corollary 3.3.3 in particular, which say that f is differentiable almost everywhere. At points of differentiability,

$$\lim_{r \to 0} \frac{f(z_0 + re^{i\alpha}) - f(z_0)}{r} = Df(z_0)e^{i\alpha} = f_x(z_0)\cos(\alpha) + f_y(z_0)\sin(\alpha)$$

It follows that for almost every $z \in \Omega$, $|\partial_\alpha f(z)| \leqslant \eta(1)|\partial_\beta f(z)|$ for all directions $\alpha, \beta \in [0, 2\pi]$. Hence f is $K = \eta(1)$ quasiconformal. The proof of Theorem 3.4.1 is complete. □

3.5 Global Quasiconformal Maps Are Quasisymmetric

In this section we shall prove that quasiconformal mappings $f : \mathbb{C} \to \mathbb{C}$ are quasisymmetric. Thus we must connect the analytic properties of these mappings with the geometric properties. As mentioned earlier, the classical route to establish this connection is via the modulus of curve families and the subsequent identification of certain extremal curve families. One then turns geometric data into analytic data by estimating the modulus of these curve families.

Here we shall take an alternate, but quite evidently equivalent, approach. As we do not as yet care for best constants fairly elementary geometric arguments will suffice. Good estimates for the associated constants will appear later for instance as a consequence of the theory of holomorphic motions. We connect geometry and analysis through the L^2-capacity and the important oscillation Lemma 3.5.1. We are going to have to deal with globally defined and surjective quasiconformal maps $f : \mathbb{C} \to \mathbb{C}$. It is quite clear that the conformal map of the

3.5. GLOBAL QUASICONFORMAL MAPS

disk $\mathbb{D}$ onto the infinite strip $\{z \in \mathbb{C} : -1 < \Re e(z) < 1\}$ is not quasisymmetric. Thus in obtaining local results, that is, for quasiconformal mappings $f : \Omega \to \Omega'$, we will somehow need to incorporate the geometry of the image $\Omega' = f(\Omega)$. We hope the reader will see the necessary modifications of our arguments to develop these results. However, we shall circumvent the problem by reducing the local case to the global one via conformal mappings and an extension property. For details see Lemma 3.6.1.

Let us begin by defining the oscillation of a real-valued continuous function u on a compact set Y as

$$\operatorname{osc}(u, Y) = \max_{z, w \in Y} |u(z) - u(w)|$$

We have the following lemma.

Lemma 3.5.1. *Let $u \in W^{1,2}(\mathbb{D}(0,b))$ be continuous on $\overline{\mathbb{D}(0,b)}$ and real-valued. Then for all $a < b$ we have*

$$\int_a^b \operatorname{osc}(u, \mathbb{S}(t))^2 \, \frac{dt}{t} \leqslant 2\pi \int_{a < |z| < b} |\nabla u|^2 \qquad (3.22)$$

Proof. Since $u \in W^{1,2}(\mathbb{D}(0,b))$, u is absolutely continuous on almost all of the circles $\mathbb{S}(t)$, $0 < t < b$. This is simply because the function $u \circ \exp \in W^{1,2}$ is absolutely continuous on lines; see Lemma A.5.2. Note that smooth bilipschitz maps preserve all Sobolev spaces, e.g., by Lemma A.5.4.

It follows that, for almost every $t \in [0, b]$,

$$\operatorname{osc}(u, \mathbb{S}(t)) \leqslant \int_0^{2\pi} |\nabla u|(te^{i\theta}) \, t \, d\theta \leqslant t \left[2\pi \int_0^{2\pi} |\nabla u|^2 (te^{i\theta}) d\theta \right]^{1/2}$$

We square this and integrate with respect to t to find

$$\int_a^b \operatorname{osc}(u, \mathbb{S}(t))^2 \, \frac{dt}{t} \leqslant 2\pi \int_a^b t \int_0^{2\pi} |\nabla u|^2 (te^{i\theta}) d\theta \, dt$$

$$= 2\pi \int_{a < |z| < b} |\nabla u|^2,$$

which completes the proof. $\square$

To apply the oscillation lemma to quasiconformal mappings we need to be able compose them with suitable test functions. Here we are content with a preliminary version of the chain rule and leave the more general cases to Theorem 3.8.2.

Lemma 3.5.2. *Let $f \in W^{1,2}_{loc}(\mathbb{C})$ be a K-quasiconformal homeomorphism of $\mathbb{C}$. If the function v is Lipschitz in a domain Ω, then*

$$v \circ f \in W^{1,2}_{loc}(\Omega'),$$

where $\Omega' = f^{-1}\Omega$. Moreover,

$$\int_{\Omega'} |\nabla(v \circ f)|^2 \leqslant K \int_{\Omega} |\nabla v|^2 \qquad (3.23)$$

Proof. That $v \circ f \in W^{1,2}_{loc}(\Omega')$ follows from Lemma A.5.5 or the characterization of Sobolev functions via absolute continuity on lines (ACL, Lemma A.5.2). To prove (3.23) first let v be C^∞-smooth. At points of differentiability of f, hence almost everywhere,

$$|\nabla(v \circ f)|^2(z) = |D^t f(z)|^2 |\nabla v|^2(f(z)) \leqslant K |\nabla v|^2(f(z)) J(z, f)$$

In view of Corollary 3.3.6, we have

$$\int_{\Omega'} |\nabla(v \circ f)|^2 \leqslant K \int_{\Omega'} |\nabla v|^2(f(z)) J(z, f) \leqslant K \int_{\Omega} |\nabla v|^2,$$

thus proving the claim for smooth v. For the general case, assuming $\nabla v \in L^2(\Omega)$, we have a sequence $v_\varepsilon \in C^\infty(\Omega)$ converging locally uniformly to v with $\|\nabla v - \nabla v_\varepsilon\|_2 \to 0$. If $\phi \in C_0^\infty(\Omega')$, then

$$\int_{\Omega'} \phi \nabla(v \circ f) = -\int_{\Omega'} (v \circ f) \nabla \phi = -\lim_{\varepsilon \to 0} \int_{\Omega'} (v_\varepsilon \circ f) \nabla \phi = \lim_{\varepsilon \to 0} \int_{\Omega'} \phi \nabla(v_\varepsilon \circ f)$$

The last term is less than

$$\lim_{\varepsilon \to 0} K^{1/2} \|\phi\|_2 \|\nabla v_\varepsilon\|_2 = K^{1/2} \|\phi\|_2 \|\nabla v\|_2$$

Taking the supremum over $\phi \in C_0^\infty(\Omega')$ with $\|\phi\|_2 = 1$ proves the claim for a general Lipschitz function v. $\square$

We are now ready to showing that global quasiconformal mappings are quasisymmetric.

Theorem 3.5.3. *Suppose $f : \mathbb{C} \to \mathbb{C}$ is a K-quasiconformal homeomorphism. Then f is η-quasisymmetric, where η depends only on K.*

Proof. By assumption, f is onto, $f(\mathbb{C}) = \mathbb{C}$. Let $x, y, z \in \mathbb{C}$ and begin with the case $|x - y| = r|x - z|$, where $r \geqslant 1$. Our aim is to show that $|f(x) - f(y)| \leqslant \eta(r)|f(x) - f(z)|$ for some function η depending only on K.

We first normalize our mapping. If $\phi, \psi : \mathbb{C} \to \mathbb{C}$ are similarities, then both

$$\frac{f(x) - f(y)}{f(x) - f(z)} = \frac{(\phi \circ f)(x) - (\phi \circ f)(y)}{(\phi \circ f)(x) - (\phi \circ f)(z)}, \qquad \frac{x - y}{x - z} = \frac{\psi(x) - \psi(y)}{\psi(x) - \psi(z)},$$

so after appropriate substitutions (preserving K-quasiconformality) we may assume that in fact $x = f(x) = 0$, $z = f(z) = 1 = \inf_{|\zeta|=1} |f(\zeta)|$ and $|y| = r \geqslant 1$. Further, we may as well assume that

$$L = |f(y)| = \sup_{\zeta \in \mathbb{D}(0,r)} |f(\zeta)| > r,$$

3.5. GLOBAL QUASICONFORMAL MAPS

for otherwise any homeomorphism η dominating the identity will suffice.

Now define the function

$$v(z) = \begin{cases} 0, & |z| \geq L \\ 1, & |z| \leq 1 \\ \log(L/|z|)/\log L, & 1 \leq |z| \leq L \end{cases}$$

Then v is Lipschitz, and we compute

$$\int_{\mathbb{C}} |\nabla v|^2 = 2\pi \int_1^L \frac{1}{\log^2 L} t \frac{1}{t^2} \, dt = \frac{2\pi}{\log L}$$

In fact, it is not difficult to show that v minimizes the Dirichlet energy among functions with $v(z) = 0$ for $|z| \geq L$ and $v(z) = 1$ for $|z| \leq 1$.

Since f is quasiconformal, Lemma 3.5.2 provides the bounds

$$\int_{\mathbb{C}} |\nabla(v \circ f)|^2 \leq K \int_{\mathbb{C}} |\nabla v|^2 = \frac{2\pi K}{\log L} \qquad (3.24)$$

On the other hand, the function $u = v \circ f$ has the following properties. If we put

$$E = f^{-1}(\overline{\mathbb{D}}), \quad \text{and} \quad F = f^{-1}(\mathbb{C} \setminus \mathbb{D}(0, L)),$$

then

1. $u \in W^{1,2}_{loc}(\mathbb{C})$ is continuous,

2. E is compact and connected with $u|E = 1$,

3. F is closed, connected and unbounded with $u|F = 0$,

4. $0, 1 \in E$ and $F \cap \overline{\mathbb{D}}(0, r) \neq \emptyset$, where $r \geq 1$.

We next claim that in fact, for any function u with sets E and F satisfying properties 1.–4. above, we have

$$\frac{1}{4\pi} \log(1 + 2r^{-2}) \leq \int_{\mathbb{C}} |\nabla u|^2 \qquad (3.25)$$

To establish this we need a geometric lemma.

Lemma 3.5.4. *Let $z_0 \in \mathbb{C}$ with $|z_0| = r \geq 1$. Then there is $w_0 \in \mathbb{C}$ and an interval $[t_0, t_1]$ with*

$$\frac{t_1}{t_0} \geq \sqrt{1 + 2r^{-2}}$$

such that for all $t \in (t_0, t_1)$ we have

- $\mathbb{S}(w_0, t)$ *separates 0 and 1.*
- $\mathbb{D}(w_0, t)$ *contains z_0.*

Proof. We consider two cases. Suppose z_0 is closer to 0 than to 1. We put $w_0 = z_0/2$ and compute that $|w_0 - 1| \geqslant \sqrt{r^2/4 + 1/2}$. This gives a lower bound for the ratio
$$|w_0 - 1|/|w_0| \geqslant \sqrt{1 + 2r^{-2}}, \tag{3.26}$$
which is attained when z_0 is equidistant from 0 and 1. If z_0 is closer to 1 than to 0, we put $w_0 = (1 + z_0)/2$. We then compute $|w_0| = \frac{1}{2}\sqrt{1 + r^2 + 2r\cos\theta}$ and $|w_0 - 1| = \frac{1}{2}\sqrt{1 + r^2 - 2r\cos\theta}$, where $z_0 = re^{i\theta}$. The ratio $|w_0|/|w_0 - 1|$ is smallest when $\cos\theta = 1/2r$ where z_0 is equidistant from 0 and 1 again. Therefore in this case the difference has the same lower bound
$$|w_0|/|w_0 - 1| = \sqrt{1 + 2r^{-2}} \tag{3.27}$$

The bounds in (3.26) and (3.27) now easily imply that we have circles $\mathbb{S}(w_0, t)$ with the desired properties. □

We now return to our proof of Theorem 3.5.3. We wish to establish the estimate in (3.25) for an arbitrary function u subject to the constraints itemized in properties 1. – 4. If we choose a point $z_0 \in F \cap \mathbb{S}(r)$, then Lemma 3.5.4 provides us with a point w_0 and an interval $I = [t_0, t_1]$ for which, given the fact that E and F are connected, we have $\mathbb{S}(w_0, t) \cap E \neq \emptyset$ and $\mathbb{S}(w_0, t) \cap F \neq \emptyset$ for all $t \in I$. On such a circle we have u assuming both the values 0 and 1, and hence by the oscillation Lemma 3.5.1,

$$\begin{aligned} 1 &\leqslant \operatorname{osc}^2(u, \mathbb{S}(w_0, t)), \\ \int_{t_0}^{t_1} \frac{dt}{t} &\leqslant 2\pi \int_{t_0 < |z - w_0| < t_1} |\nabla u|^2 \\ &\leqslant 2\pi \int_{\mathbb{C}} |\nabla u|^2 \end{aligned}$$

This establishes the inequality
$$\log \frac{t_1}{t_0} \leqslant 2\pi \int_{\mathbb{C}} |\nabla u|^2$$

Now we use our lower bound on the ratio t_1/t_0 to obtain
$$\log \sqrt{1 + 2r^{-2}} \leqslant 2\pi \int_{\mathbb{C}} |\nabla u|^2,$$
which establishes (3.25). Next, we combine this inequality with that of (3.24) to find
$$\frac{1}{4\pi} \log(1 + 2r^{-2}) \leqslant \int_{\mathbb{C}} |\nabla u|^2 \leqslant \frac{2\pi K}{\log L},$$
which we rearrange to read
$$\log L \leqslant \frac{8\pi^2 K}{\log(1 + 2r^{-2})},$$

3.5. GLOBAL QUASICONFORMAL MAPS

and after recalling our normalizations we have

$$\frac{|f(x) - f(y)|}{|f(x) - f(z)|} \leq \eta\left(\frac{|x-y|}{|x-z|}\right),$$

where

$$\eta(r) = \exp\left(\frac{8\pi^2 K}{\log(1 + 2r^{-2})}\right), \qquad r \geq 1 \qquad (3.28)$$

Before studying the remaining case $|x - y| = r|x - z|$ with $r \leq 1$, we note that the above argument can be turned into a systematic method, the moduli of ring domains, leading to yet another possible approach to quasiconformality.

Let us then settle the remaining case in Theorem 3.5.3, where we fix x, y and z satisfying $|x - y| = r|x - z|$ with $r < 1$. Notice that by what we have already shown, the quasiconformal homeomorphism f is *weakly quasisymmetric*. This means that we have a constant $H = \eta(1) < \infty$ such that for every triple $z_1, z_2, w \in \mathbb{C}$,

$$|z_1 - w| \leq |z_2 - w| \quad \text{implies} \quad |f(z_1) - f(w)| \leq H|f(z_2) - f(w)| \qquad (3.29)$$

Using this condition, we first see that whenever

$$|w - x| = 2R \geq 2|x - y|, \qquad (3.30)$$

then $f(\mathbb{D}(w, R))$ contains a disk of radius $H^{-3}|f(x) - f(y)|$. Namely, if w_1 is the midpoint of the segment $[x, w]$, then

$$|x - y| \leq |x - w_1| = |w - w_1| = R,$$

and using (3.29) thrice gives $|f(x) - f(y)| \leq H^3|f(w) - f(\zeta)|$ for every $\zeta \in \mathbb{S}^1(w, R)$.

Next, assume $|x - z| = \rho = a^n|x - y|$ for some number $3 \leq a \leq 9$, $n \in \mathbb{N}$. Then one finds at least n disjoint subdisks

$$B_j = \mathbb{D}(w_j, R_j) \subset \mathbb{D}(x, \rho), \qquad \text{with } |w_j - x| = 2R_j \geq 2|x - y|, \; j = 1, \ldots, n$$

Since the disks have disjoint images,

$$n\pi|f(x) - f(y)|^2 \leq H^6 \sum_{j=1}^n |f(B_j)| \leq H^6 |f(\mathbb{D}(x, \rho))|$$

$$\leq \pi H^6 \max_{|\zeta - x| = \rho} |f(\zeta) - f(x)|^2 \leq \pi H^8 |f(z) - f(x)|^2$$

We see that

$$\frac{|f(x) - f(y)|}{|f(x) - f(z)|} \leq \eta\left(\frac{|x - y|}{|x - z|}\right) \qquad (3.31)$$

where now

$$\eta(t) = H^4 (2 \log 3)^{1/2} \left(\log(1/t)\right)^{-1/2} \to 0 \qquad \text{as } t \to 0 \qquad (3.32)$$

Finally, we combine our formulas for $\eta|_{[0,1]}$ and $\eta|_{[1,\infty]}$ to get a function defined on $[0,\infty]$ for which (3.31) holds. We then dominate this function by a strictly increasing, continuous bijection of $[0,\infty]$ and thus retain (3.31) for all triples x, y and z. This completes the proof of Theorem 3.5.3. $\square$

Notice that for r large we have $\log(1 + 2r^{-2}) \approx 2/r^2$. Thus for η as in (3.32), we have $\eta(r) \approx e^{4\pi^2 K r^2}$. For the radial stretching $f(z) = z|z|^{K-1}$, we have

$$\left| \frac{f(z) - f(0)}{f(1) - f(0)} \right| = |z|^K,$$

which shows that any η satisfying the conclusions of Theorem 3.5.3 must have $\eta(r) \geq r^K$ for $r > 1$. One expects the estimate $\eta(r) \approx r^K$ to be the correct growth. We shall later give two proofs for this fact, one based on the isoperimetric inequality in Section 3.10 and the other on the holomorphic motions, in Section 12.6.

Despite the poor estimate of η above, Theorem 3.5.3 is fundamental in establishing uniform global properties of quasiconformal mappings.

3.6 Quasiconformality and Quasisymmetry: Local Equivalence

We have seen above that a quasiconformal mapping $f : \mathbb{C} \to \mathbb{C}$ is a quasisymmetric homeomorphism, while a quasisymmetric mapping $g : \Omega \to \Omega'$ is quasiconformal. We have already pointed out that there are conformal, and so 1-quasiconformal, mappings that are not quasisymmetric. Yet we still have to establish the basic properties of quasiconformal mappings given in Theorem 3.1.2. The following theorem, not too far from the well-known Stoilow factorization, will be enough to easily deduce the local results we need to complete this proof.

Lemma 3.6.1. Let $f : \Omega \to \Omega'$ be K-quasiconformal. Let $\overline{\mathbb{D}}(z_0, r) \subset \Omega$. Then there is a K-quasiconformal map (hence quasisymmetric) $g : \mathbb{C} \to \mathbb{C}$ onto, and a conformal map $\varphi : \mathbb{D}(z_0, r) \to f(\mathbb{D}(z_0, r))$ such that

$$f|\mathbb{D}(z_0, r) = (\varphi \circ g)|\mathbb{D}(z_0, r)$$

Proof. The domain $f(\mathbb{D}(z_0, r))$ is a simply connected Jordan domain. There is a conformal map

$$\varphi : \mathbb{D}(z_0, r) \to f(\mathbb{D}(z_0, r)) \qquad \text{with } \varphi(z_0) = f(z_0)$$

The Carathéodory theorem [89] asserts that φ extends to a homeomorphism from $\overline{\mathbb{D}}(z_0, r)$ onto $\overline{f(\mathbb{D}(z_0, r))}$. Then

$$h = \varphi^{-1} \circ f : \overline{\mathbb{D}}(z_0, r) \to \overline{\mathbb{D}}(z_0, r) \tag{3.33}$$

3.6. LOCAL EQUIVALENCE

is a homeomorphism. From Lemma 3.5.2 we have $h \in W^{1,2}_{loc}(\mathbb{D}(z_0, r))$, and as φ is conformal, h satisfies the condition (3.1) whenever f does. Thus $h|\mathbb{D}(z_0, r)$ is quasiconformal. Let

$$\Phi(z) = r^2(\bar{z} - \bar{z}_0)^{-1}$$

denote the anticonformal inversion in the circle $\mathbb{S}(z_0, r)$. We define the function g by the rule

$$g(z) = \begin{cases} h(z), & z \in \mathbb{D}(z_0, r) \\ (\Phi \circ h \circ \Phi)(z), & z \in \mathbb{C} \setminus \mathbb{D}(z_0, r) \end{cases} \qquad (3.34)$$

Clearly, $g : \mathbb{C} \to \mathbb{C}$ is a homeomorphism onto. Also, it is clear that $g|\mathbb{D}(z_0, r)$ and $g|\mathbb{C} \setminus \mathbb{D}(z_0, r)$ are K-quasiconformal. We leave it to the reader to verify the fact that g is K-quasiconformal. The only issue is to establish that $g \in W^{1,2}_{loc}$ near points of $\mathbb{S}(z_0, r)$. This is a fairly general fact concerning continuous functions that lie in a Sobolev space on either side of a sufficiently regular curve, and it can be proved in a rather straightforward manner using the characterization of Sobolev functions via absolute continuity; see Lemma A.5.2 in the Appendix.

Finally, observe that on $\mathbb{D}(z_0, r)$ we have $\varphi \circ g = \varphi \circ h = f$. $\square$

It is often useful to formulate the above local result in the following uniform manner: A homeomorphism of Ω is quasiconformal if and only if it is quasisymmetric in hyperbolic disks of Ω. Stating this principle in a quantitative manner gives the following result, a generalization of Theorem 2.10.9.

Theorem 3.6.2. *Suppose $f : \Omega \to \Omega'$ is a homeomorphism and suppose $z_0 \in \Omega$ with $\mathbb{D}(z_0, 2r) \subset \Omega$. Let $B = \mathbb{D}(z_0, r)$. If f is K-quasiconformal in Ω, then the restriction $f|B$ is η-quasisymmetric, where $\eta = \eta_K$ depends only on K. Conversely, if f is η-quasisymmetric, then $f|B$ is K-quasiconformal, where $K = K(\eta)$ only.*

Proof. Since the converse direction holds even globally, by Theorem 3.4.1, we need to prove only the first assertion. We may assume $\mathbb{D}(z_0, 2r) = \mathbb{D} \in \Omega$, and use the factorization $f = \varphi \circ g$ in Lemma 3.6.1. Here $g : \mathbb{C} \to \mathbb{C}$ is quasisymmetric with $\eta = \eta_g$ depending only on K, $g(\mathbb{D}) = \mathbb{D}$ and $g(0) = 0$, and φ is conformal on $\mathbb{D}$. If $|z| \leq 1/2$ and $|\zeta| = 1$,

$$\frac{|g(0) - g(\zeta)|}{|g(z) - g(\zeta)|} \leq \eta(2)$$

Therefore $\mathrm{dist}(g(z), \mathbb{S}) \geq 1/\eta(2)$ and the hyperbolic distance

$$\rho_{\mathbb{D}}(g(z), 0) \leq \log\left(\frac{\eta(2) + 1}{\eta(2) - 1}\right) = M, \qquad |z| \leq \frac{1}{2}$$

Lemma 2.10.9 now implies

$$\frac{|f(z_1) - f(z_0)|}{|f(z_2) - f(z_0)|} = \frac{|\varphi(g(z_1)) - \varphi(g(z_0))|}{|\varphi(g(z_2)) - \varphi(g(z_0))|} \leq e^{16M} \frac{|g(z_1) - g(z_0)|}{|g(z_2) - g(z_0)|}$$

$$\leq e^{16M} \eta\left(\frac{|z_1 - z_0|}{|z_2 - z_0|}\right), \qquad |z_j| \leq \frac{1}{2}, \, j = 0, 1, 2,$$

proving the quasisymmetry in $\mathbb{D}(0, 1/2)$. □

With Lemma 3.2.3 we now have for any quasiconformal mapping on the entire plane, that $f(\mathbb{C}) = \mathbb{C}$. Indeed, we have the next immediate corollary to the theorem.

Theorem 3.6.3. *If f is a quasiconformal map defined on $\mathbb{C}$, then f is quasisymmetric and, in particular, setting $f(\infty) = \infty$ defines a continuous homeomorphism of the Riemann sphere $\hat{\mathbb{C}}$.*

As another corollary of the above local equivalence, we also see that the analytic definition of quasiconformality (2.30) implies the geometric definition (2.28).

3.7 Lusin's Condition $\mathcal{N}$ and Positivity of the Jacobian

The previous theorems already imply that the inverse of a quasiconformal mapping f is also quasiconformal. However, the results have given rather poor bounds on the distortion of f^{-1}. A moment's consideration of the differentials leads one to expect $K(f^{-1}) = K(f)$. To prove this precise relation, we need to understand the absolute continuity properties of quasiconformal mappings. As a simple consequence, we also get the rather important fact that the Jacobian of a quasiconformal mapping is positive almost everywhere.

Classically, the fact that sets of Lebesgue measure zero are preserved by quasiconformal mappings involves a fairly subtle investigation of the properties of their Jacobians. Here we again show the utility of the notion of quasisymmetry.

Given measurable functions $f : \Omega \to \Omega'$ and $g : \Omega' \to \Omega''$, in general their composition is not a measurable function. However, in the geometric study of mappings it is necessary to avoid all unnessary constraints on such natural operations as the composition. It is for this reason, among many others, that the following Lusin condition arises.

Definition 3.7.1. *Let $f : \Omega \to \Omega'$ be a measurable mapping. We say that f satisfies Lusin's condition $\mathcal{N}$ if for all $E \subset \Omega$,*

$$|E| = 0 \quad \Rightarrow \quad |f(E)| = 0$$

We leave it for the reader to verify that, by Lusin's theorem, if a measurable function f satisfies Lusin's condition $\mathcal{N}$, then (and only then) f takes measurable sets to measurable sets.

Naturally, one frequently needs to study mappings that preserve measurability under inverse images. This leads us to the following condition.

Definition 3.7.2. *Let $f : \Omega \to \Omega'$ be a measurable mapping. We say that f satisfies Lusin's condition $\mathcal{N}^{-1}$ if*

$$|f^{-1}(E)| = 0 \quad \text{whenever} \quad E \subset \Omega' \quad \text{and} \quad |E| = 0$$

3.7. POSITIVITY OF THE JACOBIAN

Note that if a measurable $f : \Omega \to \Omega'$ satisfies Lusin's condition $\mathcal{N}^{-1}$, then the inverse image $f^{-1}(E)$ is Lebesgue-measurable whenever E is. In particular, the composition $u \circ f$ with any measurable function on Ω' is measurable. Without condition $\mathcal{N}^{-1}$ this need not be true.

Note also that if $f : \Omega \to \Omega'$ satisfies $\mathcal{N}^{-1}$ and $|\Omega| < \infty$, then by the Radon-Nikodym theorem, for any $\varepsilon > 0$ we have $\delta > 0$ such that

$$|E| < \delta \quad \Rightarrow \quad |f^{-1}(E)| < \varepsilon$$

We saw in Theorem 3.3.7 that $W^{1,2}_{loc}$-homeomorphisms have the Lusin property $\mathcal{N}$. In particular, by Theorem 3.4.1, this holds for all quasisymmetric mappings. Since the inverse is quasisymmetric as well, we have the fundamental properties $\mathcal{N}$ and $\mathcal{N}^{-1}$ for all quasisymmetric mapping.

For quasisymmetric mappings there are also other ways to approach the Lusin properties. We will next give an alternative argument that also works in higher dimensions. We begin with a form of the Poincaré inequality in terms of the maximal derivative (3.14).

Lemma 3.7.3. Let $f : \Omega \to \Omega'$ be an η-quasisymmetric mapping. Let $z, z_0 \in \Omega$ with $|z - z_0| = r < d(z_0, \partial\Omega)$. Then

$$|f(z) - f(z_0)|^2 \leqslant C_\eta \int_{\mathbb{D}(z_0, r)} L_f^2,$$

where $C_\eta = \frac{4}{\pi} \eta^2(1) \eta^2(2)$.

Proof. We know from (3.21) that $L_f \in L^2_{loc}(\Omega)$. Without loss of generality we may assume that $z_0 = 0$. Let $w \in \mathbb{C}$, $|w| = r$. Together with quasisymmetry, Lemma 3.4.2 gives us, for all suitably small $\varepsilon > 0$,

$$|f(z) - f(0)| \leqslant \eta(1)|f(w) - f(0)| \leqslant \eta(1)\eta(2)|f(w) - f(w/2)|$$
$$\leqslant 2\eta(1)\eta(2) \int_{w/2}^{w} L_f^\varepsilon$$

Integrating over $|w| = r$ implies

$$|f(z) - f(0)| \leqslant \frac{2\eta(1)\eta(2)}{2\pi} \frac{2}{r} \int_{r/2 < |z| < r} L_f^\varepsilon$$
$$\leqslant \frac{2\eta(1)\eta(2)}{\pi r} \int_{\mathbb{D}(0,r)} L_f^\varepsilon \qquad (3.35)$$

We now let $\varepsilon \to 0$ and use the monotone convergence theorem, as $\varepsilon \mapsto L_f^\varepsilon$ is increasing and $L_f^\varepsilon \in L^1$ according to Corollary 3.4.7. We obtain

$$|f(z) - f(z_0)| \leqslant \frac{2\eta(1)\eta(2)}{\pi r} \int_{\mathbb{D}(0,r)} L_f$$

Finally, we can apply Hölder's inequality to obtain

$$|f(z) - f(z_0)|^2 \leq \left(\frac{2\eta(1)\eta(2)}{\pi r}\right)^2 \pi r^2 \int_{\mathbb{D}(0,r)} L_f^2,$$

which is what we wanted. $\square$

Our main concern is the absolutely continuity of the pullback

$$m(E) = |f(E)|$$

of the Lebesque measure under a quasisymmetric mapping f. But we also need other, more geometric, properties of the measure. For instance, m satisfies the doubling property.

Lemma 3.7.4. *Let $f : \Omega \to \Omega'$ be η-quasisymmmetric, $s > 0$ and $B \subset \Omega$ a disk for which $sB \subset \Omega$. Then*

$$|f(B)| \leq \eta^2(1/s)|f(sB)|$$

Proof. Let $B = \mathbb{D}(z_0, r)$. We make the following calculation

$$\begin{aligned}
|f(B)| &\leq \pi \max_{|z-z_0|=r} |f(z) - f(z_0)|^2 \\
&= \pi \frac{\max_{|z-z_0|=r} |f(z) - f(z_0)|^2}{\min_{|z-z_0|=sr} |f(z) - f(z_0)|^2} \min_{|z-z_0|=sr} |f(z) - f(z_0)|^2 \\
&\leq \pi \, \eta^2(1/s) \min_{|z-z_0|=sr} |f(z) - f(z_0)|^2 \leq \eta^2(1/s)|f(sB)|,
\end{aligned}$$

which is what we wanted to prove. $\square$

We can now establish the result we seek in this section.

Theorem 3.7.5. *Let $f : \Omega \to \Omega'$ be quasisymmetric. Then $|E| = 0$ if and only if $|f(E)| = 0$.*

Proof. As $|E| = 0$, we can choose a countable cover by disks whose union has arbitrarily small measure,

$$E \subset \bigcup_j \mathbb{D}_j, \quad \sum_{j=1}^\infty \operatorname{diam}(\mathbb{D}_j)^2 \leq \varepsilon$$

Using a Vitali-type covering lemma, we find from this cover a sequence of pairwise disjoint disks $\{\tilde{\mathbb{D}}_k\}$ such that

$$E \subset \bigcup_k (5\tilde{\mathbb{D}}_k)$$

3.7. POSITIVITY OF THE JACOBIAN

Lemma 3.7.4 gives the uniform bound $|f(5\tilde{\mathbb{D}}_k)| \leqslant \eta(5)^2|f(\tilde{\mathbb{D}}_k)|$, so that

$$|f(E)| \leqslant \sum_k \eta(5)^2|f(\tilde{\mathbb{D}}_k)| \leqslant C_\eta \sum_k \int_{\tilde{\mathbb{D}}_k} L_f^2$$
$$= C_\eta \int_{\cup_k \tilde{\mathbb{D}}_k} L_f^2 \leqslant C_\eta \int_{\cup_j \mathbb{D}_j} L_f^2, \qquad (3.36)$$

where the second inequality uses Lemma 3.7.3. We have already seen in (3.21) that quasisymmetry implies

$$L_f^2 \leqslant 4\eta^2(1) D\nu,$$

where $D\nu$ is the volume derivative of the measure $\nu(E) = |f(E)|$, defined in (3.8). Therefore $L_f \in L^2_{loc}(\Omega)$. In particular, if we have sets with $|E_j| \to 0$ as $j \to \infty$, then $\int_{E_j} L_f^2 \to 0$ as $j \to \infty$. But with this observation we deduce from (3.36) that $|f(E)| = 0$ when we let $\varepsilon \to 0$. We have already observed that f^{-1} is quasisymmetric, so that if $|f(E)| = 0$, then $|E| = |f^{-1}(f(E))| = 0$ also. The proof of the theorem is complete. $\square$

Corollary 3.7.6. *Let $f : \Omega \to \Omega'$ be K-quasiconformal. Then for a set $E \subset \Omega$ we have $|E| = 0$ if and only if $|f(E)| = 0$. In other words, quasiconformal mappings satisfy both conditions $\mathcal{N}$ and $\mathcal{N}^{-1}$.*

In particular, it follows that f preserves the class of Lebesgue-measurable sets, $J(z, f) > 0$ almost everywhere, and

$$|f(E)| = \int_E J(z, f) \qquad (3.37)$$

for all Lebesgue-measurable subsets $E \subset \Omega$.

Proof. Via Theorem 3.6.2, the previous result applies to quasiconformal mappings so that f preserves the class of Lebesgue-measurable sets and the pullback measure $\nu(E) = |f(E)|$ is absolutely continuous, $\nu_s(E) \equiv 0$ (see (3.9)). Since f is an orientation-preserving Sobolev homeomorphism, $D\nu = J(z, f)$ and hence (3.9) proves the identity (3.37) for E Borel. The extension to Lebesgue-measurable sets is immediate. Put $E = \{z \in \Omega : J(z, f) = 0\}$. Then

$$0 = \int_E J(z, f) = |f(E)|,$$

so $|f(E)| = 0$ and hence by Theorem 3.7.5 we have $|E| = 0$. $\square$

As another important consequence, we see that for a quasiconformal mapping $f_z \neq 0$ almost everywhere, so that the Beltrami equation

$$\frac{\partial f}{\partial \bar{z}} = \mu(z) \frac{\partial f}{\partial z}$$

determines Beltrami coefficient $\mu = \mu_f$ uniquely up to a set of measure zero.

We are now also able to complete the proof of the first fundamental Theorem 3.1.2. It remains to establish the following group properties of quasiconformal mappings.

Theorem 3.7.7. *Let $f : \Omega \to \Omega'$ be K-quasiconformal. Then $f^{-1} : \Omega' \to \Omega$ is K-quasiconformal. If $g : \Omega' \to \Omega''$ is K'-quasiconformal, then the composition $g \circ f : \Omega \to \Omega''$ is KK'-quasiconformal.*

Proof. It follows from Theorem 3.6.2 that $f^{-1} : \Omega' \to \Omega$ and $g \circ f : \Omega \to \Omega''$ are quasiconformal mappings. The only issue left to resolve is the precise estimate for $K(f^{-1})$ and for $K(g \circ f)$. To see these, take $w = f^{-1}(z)$, a point where f is differentiable with $J(w, f) > 0$, and observe that $\mathbf{I} = Df(w)Df^{-1}(z)$ and $1 = J(w, f)J(z, f^{-1})$. Then

$$\begin{aligned} |Df^{-1}(z)|^2 &= |Df(w)^{-1}|^2 \\ &= |Df(w)|^2/J(w,f)^2 \\ &\leqslant K/J(w,f) = KJ(z, f^{-1}) \end{aligned}$$

The points of differentiability have full measure by the Gehring-Lehto theorem, but to justify this calculation for almost every z, we need to see that f does not compress a set of positive measure to a set of zero-measure. This is in fact Lusin's condition $\mathcal{N}$, which we have just established in Corollary 3.7.6.

Similarly for the composition, let f be differentiable at a and g at $f(a)$, both with positive Jacobian. We know that such points a have full measure. Moreover,

$$|D(g \circ f)(a)|^2 \leqslant |Dg(f(a))|^2|Df(a)|^2 \leqslant KK'J(f(a), g)J(a, f) = KK'J(a, g \circ f)$$

We have thus also completed the proof of Theorem 3.1.2. □

3.8 Change of Variables

Here we use a few well-known facts from measure theory to establish the change-of-variables formula and the chain rule for quasiconformal mappings.

Theorem 3.8.1. *Let $f : \Omega \to \Omega'$ be a quasiconformal mapping and $u \in L^1(\Omega')$. Then*

$$u(f(z))J(z, f) \in L^1(\Omega),$$

and we have

$$\int_\Omega u(f(z))J(z,f)\, dz = \int_{\Omega'} u(w)\, dw \qquad (3.38)$$

Proof. Corollary 3.7.6 tells us that (3.38) holds for all characteristic functions $u = \chi_E$. We now have to extend this equality to $u \in L^1(\Omega)$ and can take u non-negative. Since f satisfies Lusin's condition $\mathcal{N}^{-1}$, $u(f(z))$ is measurable.

3.8. CHANGE OF VARIABLES

Certainly, we may find simple functions $u_\nu \leqslant u$ converging almost everywhere to u, and for these u_ν we have

$$\int_\Omega u_\nu(f(z))J(z,f) = \int u_\nu(w) \to \int u(w)$$

The only problem we have to overcome then is to show that $u_\nu(f(z)) \to u(f(z))$ almost everywhere. This is clear from Theorem 3.7.5. □

The most convenient way to present the chain rule is in terms of the Royden algebra $R(\Omega)$ of a domain Ω. Recall that the elements of this algebra are the continuous and bounded functions with distributional derivatives in $L^2(\Omega)$,

$$R(\Omega) = C(\Omega) \cap L^\infty(\Omega) \cap \mathbb{W}^{1,2}(\Omega)$$

Here, and throughout this monograph, we use the notation

$$\mathbb{W}^{1,p}(\Omega) = \{v : \nabla v \in L^p(\Omega, \mathbb{C}^2)\}$$

for the homogeneous Sobolev space, that is, for the space of (locally integrable) functions with L^p-integrable gradient. No assumptions are made here on the L^p-integrability of the function itself.

For a bounded domain, however, $R(\Omega) \subset W^{1,2}(\Omega)$. The Royden algebra is a Banach algebra with the norm

$$\|v\|_* = \|v\|_\infty + \|\nabla v\|_2$$

According to the next result, quasiconformal mappings of Ω preserve the Royden algebra $R(\Omega)$. In fact, the reader may easily verify that a quasiconformal $f : \Omega \to \Omega'$ induces an algebra isomorphism $T_f : R(\Omega') \to R(\Omega)$, $T_f(v) = v \circ f$. Somewhat deeper lies the fact that every algebra isomorphism of the Royden algebra is of this type [230].

Theorem 3.8.2. *Let $f : \Omega \to \Omega'$ be a K-quasiconformal mapping and let $u \in R(\Omega')$. Then the composition $v = u \circ f$ lies in the Royden algebra $R(\Omega)$ with derivative*

$$\nabla v(z) = D^t f(z) \nabla u(f(z)), \quad \text{almost everywhere in } \Omega \tag{3.39}$$

Further, we have the estimate

$$\int_\Omega |\nabla v(z)|^2 \, dz \leqslant K \int_{\Omega'} |\nabla u(z)|^2 \, dz \tag{3.40}$$

Proof. The main point is showing that $u \circ f$ has square-integrable distributional derivatives, and the argument is a variation of Lemma 3.5.2. We first assume that $u_\beta \in C^\infty(\overline{\Omega'})$ converge to u locally uniformly and in $\mathbb{W}^{1,2}(\Omega)$. From Lemma 3.5.2 we have $u_\beta \circ f \in \mathbb{W}^{1,2}(\Omega)$ and at the points of differentiability of f, hence almost everywhere,

$$\nabla(u_\beta \circ f)(z) = D^t f(z) \nabla u_\beta(f(z)),$$

which we may write equivalently in the integral form

$$-\int_\Omega \nabla\varphi(z)\, u_\beta(f(z)) = \int_\Omega \varphi(z) D^t f(z) \nabla u_\beta(f(z)) \qquad (3.41)$$

for every test funtion $\varphi \in C_0^\infty(\Omega)$.

Here passing to the limit as $\beta \to \infty$ on the left hand side poses no difficulty whatsoever as $u_\beta(f(z)) \to u(f(z))$ locally uniformly. However, we must justify passing to the limit on the right hand side. We do this as follows, using the fact that f is K-quasiconformal to give $|D^t f|^2 \leqslant KJ(z, f)$.

$$\int_\Omega |D^t f(z) \nabla u_\beta(f(z)) - D^t f(z) \nabla u(f(z))|^2$$
$$= \int_\Omega |D^t f(z)[\nabla u_\beta - \nabla u](f(z))|^2$$
$$\leqslant K \int_\Omega J(z,f) |[\nabla u_\beta - \nabla u](f(z))|^2$$
$$= K \int_{\Omega'} |\nabla u_\beta - \nabla u|^2$$

by the change-of-variable formula (3.38). Now,

$$\int_{\Omega'} |\nabla u_\beta - \nabla u|^2 \to 0 \quad \text{as } \beta \to \infty$$

We end up with the identity

$$-\int_\Omega \nabla\varphi(z)\, u(f(z)) = \int_\Omega \varphi(z) D^t f(z) \nabla u(f(z)) \qquad (3.42)$$

for every $C_0^\infty(\Omega)$-test function. Directly from the definition of the Sobolev space we see that $v(z) = u(f(z))$ lies in the algebra $W^{1,2}(\Omega) \cap C(\Omega) \cap L^\infty(\Omega)$, and its gradient is as given at (3.39). The integral inequality is immediate from the change-of-variable formula of Theorem 3.8.1. $\square$

3.9 Quasisymmetry and Equicontinuity

Next, we set about proving Theorem 3.1.3. We first establish the result for quasisymmetric mappings, which will in turn establish the desired result. First, let us prove that a family of normalized quasisymmetric mappings is equicontinuous. We shall use this fact a bit later on in the book. We draw the reader's attention to Section 2.9.4 where the definition of equicontinuity is given.

Theorem 3.9.1. *Let $A \subset \mathbb{C}$ with $0, 1 \in A$. Then the family of all η-quasisymmetric maps $f : A \to \mathbb{C}$ with $f(0) = 0$ and $f(1) = 1$ is equicontinuous.*

Proof. Let $a_0 \in A$. Then for any $a \in A$,

$$\frac{|f(a_0) - f(a)|}{|f(a_0) - f(1)|} \leqslant \eta\left(\frac{|a_0 - a|}{|a_0 - 1|}\right) \quad \text{and} \quad \frac{|f(a_0) - f(a)|}{|f(a_0) - f(0)|} \leqslant \eta\left(\frac{|a_0 - a|}{|a_0|}\right)$$

3.9. QUASISYMMETRY AND EQUICONTINUITY

If $a_0 \in \{0,1\}$, we then have $|f(a_0) - f(a)| \leq \eta(|a_0 - a|)$, independent of f, and so the family is equicontinuous at 0 and 1. More generally, equicontinuity follows directly provided we can show that one of $|f(a_0) - f(1)| = |f(a_0) - 1| \leq 1 + |f(a_0)|$ or $|f(a_0) - f(0)| = |f(a_0)|$ is bounded independent of f. That is, we want a bound on $|f(a_0)|$ independent of f. But this follows as

$$|f(a_0)| = \frac{|f(a_0) - f(0)|}{|f(1) - f(0)|} \leq \eta(|a_0|),$$

and so the family of maps is equicontinuous. $\square$

Of course, we now know from the Arzela-Ascoli theorem that a normalized family of η-quasisymmetric maps is a normal family.

The following corollary is an easy consequence of Theorem 3.9.1.

Corollary 3.9.2. *Let $A \subset \mathbb{C}$ and $a, b \in A$, $a \neq b$. Then the family of all η-quasisymmetric maps $f : A \to \mathbb{C}$ with $f(a) = a$ and $f(b) = b$ is normal.*

Using the compactness properties of the family of similarities of $\mathbb{C}$, it is easy to modify Corollary 3.9.2 so as to relax the condition $f(a) = a$ and $f(b) = b$ by pre- and post-composing with similarity mappings. One only needs a uniform bound on $|f(a)|$ and $|f(b)|$. In particular then, Corollary 3.9.2 implies the following corollary.

Corollary 3.9.3. *Let $A \subset \mathbb{C}$. Then a locally bounded family of η-quasisymmetric mappings $f : A \to \mathbb{C}$ is normal. In particular, the limit of a sequence of η-quasisymmetric mappings is either η-quasisymmetric or a constant.*

It is worth pointing out here that in effect a family of quasisymmetric mappings has an a priori additional normalization, namely, that $f(\infty) = \infty$. When discussing families of quasiconformal mappings of $\hat{\mathbb{C}}$, this additional normalization at infinity may not hold. However, normalization using the family of all linear fractional transformations of $\hat{\mathbb{C}}$ quickly leads to the idea that a family of quasiconformal mappings of $\hat{\mathbb{C}}$ normalized by their values at three points is a normal family.

Of course, equicontinuity is a local property, and therefore all the above statements are valid for families of K-quasiconformal mappings should A be an open set. We record this in the following theorem.

Theorem 3.9.4. *Let $\{f_\nu\}_{\nu=0}^{\infty}$ be a family of K-quasiconformal mappings $f_\nu : \mathbb{C} \to \mathbb{C}$ normalized by the conditions $f_\nu(0) = 0$ and $f_\nu(1) = 1$. Then $\{f_\nu\}_{\nu=0}^{\infty}$ is a normal family. Moreover, every limit mapping is a nonconstant K-quasiconformal homeomorphism of $\hat{\mathbb{C}}$.*

Proof. Since every limit mapping f of a sequence $\{f_\nu\}$ of normalized K-quasiconformal mappings is η-quasisymmetric, where η depends only on K,

80 CHAPTER 3. QUASICONFORMAL MAPPINGS

it remains to show the exact distortion bound $K(f) \leqslant K$. For this, note first that for every disk B,

$$\int_B |Df_\nu(z)|^2 \leqslant K \int_B J(z, f_\nu) = K|f_\nu(B)| \leqslant C(K, B)$$

As $f_\nu(z)$ converges locally uniformly to $f(z)$, the derivatives $Df_\nu \rightharpoonup Df$ converge weakly in $L^2(B)$ for every disk B. Hence we may apply the weak lower semicontinuity of the L^2-norm,

$$\int_B |Df(z)|^2 \leqslant \liminf_{\nu \to \infty} \int_B |Df_\nu(z)|^2$$
$$\leqslant K \liminf_{\nu \to \infty} \int_B J(z, f_\nu) = K \lim_{\nu \to \infty} |f_\nu(B)| = K|f(B)|$$

Since the inequality holds for every disk, Lebesgue's Density theorem tells us that $|Df(z)|^2 \leqslant KJ(z, f)$ almost everywhere. □

3.10 Hölder Regularity

In this section we will determine the optimal Hölder exponent of K-quasiconformal mappings. Identifying these mappings through the notion of quasisymmetry naturally gave them equicontinuity properties. However, the modulus of continuity provided by Theorem 3.5.3 is far from the optimal one.

It turns out that each K-quasiconformal mapping is locally $\frac{1}{K}$-Hölder continuous. In the literature there are several different proofs for this fact. We have chosen here an elegant argument, due to Morrey [271], that is based on the isoperimetric inequality. Later, in Section 12.6, we will give another approach using the theory of holomorphic motions.

Theorem 3.10.1. (Isoperimetric Inequality) *Suppose $\Omega \subset \mathbb{C}$ is a bounded Jordan domain with rectifiable boundary $\partial \Omega$. Then*

$$|\Omega| \leqslant \frac{1}{4\pi} \left[\mathcal{H}^1(\partial \Omega) \right]^2, \tag{3.43}$$

where $|\Omega|$ is the area of Ω and $\mathcal{H}^1(\partial \Omega)$ is the length of $\partial \Omega$.

Proof. After scaling we may assume that the length of $\partial \Omega$ is precisely 2π. We will use the arc length parameterization $\gamma : [0, 2\pi] \to \partial \Omega$, which makes γ a 2π-periodic absolutely continuous function with $|\gamma'(t)| = 1$ for almost every $t \in [0, 2\pi]$. Thus

$$2\pi = \int_0^{2\pi} |\gamma'(t)|^2 = 2\pi \sum_{-\infty}^{\infty} n^2 |c_n|^2,$$

where

$$\gamma(t) = \sum_{-\infty}^{\infty} c_n e^{int}$$

3.10. HÖLDER REGULARITY

is the series representation of γ.

On the other hand, by Green's formula (2.52), with $f(z) = z$ and $g(z) = \bar{z}$, the area of Ω attains the form

$$2|\Omega| = \frac{i}{2}\int_{\partial\Omega} z\,d\bar{z} - \bar{z}\,dz = \int_0^{2\pi} \Im m\left[\overline{\gamma(t)}\gamma'(t)\right]dt = 2\pi \sum_{-\infty}^{\infty} n|c_n|^2$$

Since $n \leq n^2$ for every n, we see that $|\Omega| \leq \pi$. Taking the normalization $\mathcal{H}^1(\partial\Omega) = 2\pi$ into account, we arrive precisely at (3.43). Notice that, by the above proof, circles are the only boundaries for which we have equality. □

The images of a circle under K-quasiconformal mappings of $\mathbb{C}$ have a uniformly bounded "excentricity," and the circular distortion function $\lambda(K)$ is a natural quantity measuring this deviation from a round circle.

If we write

$$\mathcal{F} = \{f : \mathbb{C} \to \mathbb{C},\ K - \text{quasiconformal}, f(0) = 0 \text{ and } f(1) = 1\},$$

then the circular distortion can be defined by

$$\lambda(K) = \sup\{|f(e^{i\theta})| : f \in \mathcal{F},\ 0 \leq \theta \leq 2\pi\} \tag{3.44}$$

Explicit estimates [229] show that

$$1 < \lambda(K) \leq \frac{1}{16} e^{\pi K} \quad \text{for } 1 < K < \infty;\ \lambda(K) \to 1 \text{ as } K \to 1$$

For an approach to these bounds via holomorphic motions and hyperbolic metrics, see Section 12.6.

Since K-quasiconformality is preserved under pre- and post-composition with similarities, we have the well-known result of Mori [268] that

$$|f(z_0) - f(z_1)| \leq \lambda(K)|f(z_0) - f(z_2)| \tag{3.45}$$

whenever $|z_0 - z_1| = |z_0 - z_2|$ and $f : \mathbb{C} \to \mathbb{C}$ is K-quasiconformal.

With these tools we are ready for the Hölder bounds.

Theorem 3.10.2. *Suppose $f : \mathbb{C} \to \mathbb{C}$ is K-quasiconformal with $f(0) = 0$ and $f(1) = 1$. Then*

$$|f(z)| \leq \lambda(K)^2 |z|^{1/K}, \qquad 0 \leq |z| \leq 1 \tag{3.46}$$

The K-quasiconformal mapping $f(z) = z|z|^{1/K-1}$ shows that the Hölder exponent $\frac{1}{K}$ is optimal.

Proof. Let $t > 0$ and write $B = \mathbb{D}(0, t)$. Abbreviate $J = J(z, f)$. By the isoperimetric inequality and Hölder's inequality, we obtain for almost all $t \geq 0$,

$$\int_B J \leq \frac{1}{4\pi}\left(\int_{\partial B} |Df|\right)^2 \leq \frac{K|\partial B|}{4\pi}\int_{\partial B} \frac{|Df|^2}{K}$$

$$\leq \frac{Kt}{2}\int_{\partial B} J$$

Next, if we define the increasing function
$$\phi(t) = \int_{\mathbb{D}(0,t)} J(z,f),$$
then the previous estimate may be written in the form $\phi(t) \leqslant \frac{Kt}{2}\phi'(t)$, or equivalently,
$$\frac{d}{dt}\left(t^{-2/K}\phi(t)\right) \geqslant 0 \qquad \text{for almost every } t > 0 \tag{3.47}$$

Integrating this over the interval $[t,1]$, we arrive at
$$\phi(t) \leqslant \phi(1) t^{2/K} \qquad 0 \leqslant t \leqslant 1 \tag{3.48}$$

Suppose now that $|z| = r < 1$. We use (3.45) and (3.48) to get
$$\begin{aligned}|f(z)| &\leqslant \lambda(K) \inf_{|w|=r} |f(w)| \leqslant \lambda(K)\left(\pi^{-1}|f\mathbb{D}(0,r)|\right)^{1/2} \\ &\leqslant \lambda(K)\left(\pi^{-1}|f\mathbb{D}|\right)^{1/2} r^{1/K} \leqslant \lambda(K)^2\, r^{1/K}\end{aligned}$$

This completes the proof. $\square$

Applying Theorem 3.10.2 and arguing as in the proof of Theorem 3.6.2 and (3.33) or (3.34), we obtain the precise degree of Hölder regularity valid for general quasiconformal mappings.

Corollary 3.10.3. *Every K-quasiconformal mapping $f : \Omega \to \Omega'$ is locally $\frac{1}{K}$-Hölder continuous. More precisely, if a disk $B \subset 2B \subset \Omega$, then*
$$|f(z) - f(w)| \leqslant C(K)\operatorname{diam}(fB)\frac{|z-w|^{1/K}}{(\operatorname{diam} B)^{1/K}}, \qquad z, w \in B,$$
where the constant $C(K)$ depends only on K.

Naturally, the inequality (3.46) cannot hold for large values of $|z|$. To have a global version of Theorem 3.10.2, we need to interpret it in terms of η-quasisymmetric mappings. As a result, we have an explicit estimate for the function η in (3.2).

Corollary 3.10.4. *If $f : \mathbb{C} \to \mathbb{C}$ is K-quasiconformal and $z_0, z_1, z_2 \in \mathbb{C}$, then*
$$\frac{|f(z_0) - f(z_1)|}{|f(z_0) - f(z_2)|} \leqslant \eta\left(\frac{|z_0 - z_1|}{|z_0 - z_2|}\right), \tag{3.49}$$

where
$$\eta(t) = \eta_K(t) = \lambda(K)^{2K} \max\{t^K, t^{1/K}\}, \qquad t \in [0,\infty) \tag{3.50}$$

3.11. δ-MONOTONE MAPPINGS

Proof. If $F : \mathbb{C} \to \mathbb{C}$ is a K-quasiconformal mapping fixing 0 and 1, then

$$\lambda(K)^{-2K} |z|^K \leqslant |F(z)| \leqslant \lambda(K)^2 |z|^{1/K}, \qquad z \in \overline{\mathbb{D}} \tag{3.51}$$

Here the left inequality follows from Theorem 3.10.2 applied to the inverse $F^{-1} : \mathbb{C} \to \mathbb{C}$. Note that the left estimate holds trivially if $|F(z)| > 1$.

Suppose first that $|z_0 - z_1| \leqslant |z_0 - z_2|$. We make use of the auxiliary K-quasiconformal mapping

$$\phi(z) = \frac{f(z_0 + z(z_2 - z_0)) - f(z_0)}{f(z_2) - f(z_0)}$$

Clearly, $\phi(0) = 0$, $\phi(1) = 1$ and

$$\phi(w) = \frac{f(z_1) - f(z_0)}{f(z_2) - f(z_0)} \quad \text{for } w = \frac{z_1 - z_0}{z_2 - z_0} \in \overline{\mathbb{D}}$$

Hence in this case (3.49) follows from the second inequality in (3.51).

If $|z_0 - z_1| \geqslant |z_0 - z_2|$ a similar reasoning with $F(z) = 1/\phi(1/z)$ and the first inequality in (3.51) prove the claim. □

Surprisingly, this shows that if $f : \mathbb{C} \to \mathbb{C}$ is η-quasisymmetric for *some* η, then in fact η can be assumed to have the form in (3.50).

3.11 Quasisymmetry and δ-Monotone Mappings

Quasiconformal mappings can be constructed in many different ways: as solutions to PDEs, as holomorphic motions or deformations in complex dynamics; as isothermal coordinate changes in differential geometry and so on. In this section we describe yet another interesting construction for quasiconformal mappings. We show the remarkable fact, proven by L.V. Kovalev [218], that a δ-monotone mapping is quasiconformal. Monotone mappings arise naturally in operator theory and nonlinear functional analysis, and they have found wide application in partial differential equations for decades, particularly second-order PDEs of divergence-type.

On the other hand, from the point of view of quasiconformal mappings, monotone mappings have additional special properties not shared by general quasiconformal mappings. For example, any convex combination of monotone mappings is monotone. On the geometric side, the image of a line by any δ-monotone mapping is (up to a rotation) the graph of a Lipschitz function.

The proofs of the results mentioned above will allow us the opportunity to establish other important results regarding quasisymmetry. For instance, we shall show the key fact that an entire weakly quasisymmetric mapping is quasisymmetric, Lemma 3.11.3 below.

Definition 3.11.1. *A function* $h : \mathbb{R}^2 \to \mathbb{R}^2$ *is called* δ-*monotone*, $0 < \delta \leqslant 1$, *if for every* $z, w \in \mathbb{R}^2$,

$$\langle h(z) - h(w), z - w \rangle \geqslant \delta |h(z) - h(w)||z - w| \qquad (3.52)$$

There is no supposition of continuity here. It is clear that the notion generalizes to subdomains of $\mathbb{R}^n$ or more generally to Hilbert spaces, but we shall consider only the planar case.

To get a feel for the notion of monotonicity, let us decide when a linear map $h(z) = \alpha z + \beta \bar{z}$ of the complex plane $\mathbb{C}$ is δ-monotone. As monotonicity is invariant under adding a constant, we need an estimate at 0, where the condition $\langle h(z), z \rangle \geqslant \delta |h(z)||z|$ can be written as

$$\Re[(\alpha z + \beta \bar{z}) \bar{z}] \geqslant \delta |\alpha z + \beta \bar{z}|, \qquad |z| = 1 \qquad (3.53)$$

Assuming $\beta \neq 0$, we ask that $\Re(\frac{\alpha}{|\beta|} + \lambda) \geqslant \delta |\frac{\alpha}{|\beta|} + \lambda|$ for every $|\lambda| = 1$, or that the disk with center $\alpha/|\beta|$ and radius 1 is contained in the cone

$$C(\delta) = \{ z = x + iy : \delta |y| \leqslant \sqrt{1 - \delta^2}\, x \}$$

The set of such possible center points forms another cone with the same opening and direction as $C(\delta)$ but with vertex $z_0 = \frac{1}{\sqrt{1-\delta^2}}$. Hence the requirement of δ-monotonicity takes the form

$$\delta |\Im m(\alpha)| = \delta \left|\Im m\left(\alpha - \frac{|\beta|}{\sqrt{1-\delta^2}}\right)\right| \leqslant \sqrt{1-\delta^2}\, \Re\left(\alpha - \frac{|\beta|}{\sqrt{1-\delta^2}}\right)$$

Multiplying and reorganizing, we have that the linear map $h(z) = \alpha z + \beta \bar{z}$ is δ-monotone if and only if

$$|\beta| + \delta |\Im m\, \alpha| \leqslant \sqrt{1-\delta^2}\, \Re\, \alpha \qquad (3.54)$$

As a particular consequence, under δ-monotonicity we have $|\beta| \leqslant \sqrt{1-\delta^2}\, |\alpha|$, so that the linear distortion of h,

$$K(h) = \frac{|\alpha| + |\beta|}{|\alpha| - |\beta|} \leqslant \frac{1 + \sqrt{1-\delta^2}}{1 - \sqrt{1-\delta^2}} \qquad (3.55)$$

Equality occurs here for $h(z) = z + k\bar{z}$, where $k = \sqrt{1 - \delta^2} \in [0, 1)$. Thus linear δ-monotone mappings are quasiconformal. Kovalev's theorem shows that this remains true in general for nonlinear δ-monotone mappings. We will actually show the result with the estimate (3.55) on K, and so the linear examples show that the bound is sharp for every $\delta > 0$.

For general monotone mappings we go about proving weak quasisymmetry, the notion briefly introduced in Section 3.4 in the proof of Theorem 3.4.1. We now need to show that in fact, that this notion is just another equivalent route to quasisymmetry.

3.11. δ-MONOTONE MAPPINGS

Recall that a mapping f in a domain Ω was defined as weakly quasisymmetric if there is a constant $H < \infty$ such that for all $z_1, z_2, w \in \Omega$,

$$|z_1 - w| \leqslant |z_2 - w| \quad \text{implies} \quad |f(z_1) - f(w)| \leqslant H|f(z_2) - f(w)| \quad (3.56)$$

Note that this a priori does not require f to be continuous. However, that readily follows.

Lemma 3.11.2. *Let f be a weakly quasisymmetric function in a domain $\Omega \subset \mathbb{C}$. Then f is either a homeomorphism or a constant.*

Proof. If f is not constant, to see that the mapping is a homeomorphism onto its image, it is clearly enough just to establish continuity. Suppose that f is not continuous at $z_0 \in \Omega$. Then there is a sequence of points $z_j \to z_0$ such that for some $\varepsilon > 0$ we have

$$|f(z_j) - f(z_0)| \geqslant \varepsilon$$

Passing to a subsequence, we may assume that $|z_{j+1} - z_0| < \frac{1}{2}|z_j - z_0|$. This in turn implies

$$|z_{j+1} - z_0| \leqslant |z_{j+1} - z_j| \quad (3.57)$$

Now weak quasisymmetry implies the image sequence is bounded,

$$|f(z_j) - f(z_0)| \leqslant H|f(z_1) - f(z_0)|$$

for all j. We may again pass to a subsequence so as to be able to assume that $f(z_j) \to a \in \mathbb{C}$, $a \neq f(z_0)$. But now of course we have from (3.57)

$$|f(z_{j+1}) - f(z_0)| \leqslant |f(z_{j+1}) - f(z_j)|,$$

which is a clear contradiction as the left hand side is bounded below by ε and the right hand side tends to 0.

Thus f is continuous and hence a homeomorphism. □

Indeed, for global mappings the lemma is quickly improved, and the following lemma will have a key role to play.

Entire Weakly Quasisymmetric Maps are Quasisymmetric

Lemma 3.11.3. *Every weakly H-quasisymmetric mapping $f : \mathbb{C} \to \mathbb{C}$ is η-quasisymmetric, where η depends only on the constant H.*

Proof. Quasisymmetry asks for a continuous strictly increasing bijection $\eta : [0, \infty] \to [0, \infty]$ such that

$$\frac{|f(x) - f(y)|}{|f(x) - f(z)|} \leqslant \eta\left(\frac{|x - y|}{|x - z|}\right), \quad x, y, z \in \mathbb{C} \quad (3.58)$$

We actually have already seen in (3.29) – (3.32) that weak quasisymmetry gives the required $\eta(t)$ for all $0 \leqslant t \leqslant 1$, with $\eta(t) \to 0$ as $t \to 0$. Hence the condition (3.58) holds for all triples with $|x - y| \leqslant |x - z|$.

For triples with $|x-y| = r|x-z|$, assume $n \leq r < n+1$, $n \in \mathbb{N}$, and set

$$y_t = x + t\frac{|z-x|}{|y-x|}(y-x), \qquad t > 0$$

Using (3.56), we have for $1 \leq j \leq n$,

$$|f(y_{j+1}) - f(y_j)| \leq H|f(y_j) - f(y_{j-1})| \leq H^j|f(y_1) - f(x)| \leq H^{j+1}|f(z) - f(x)|$$

Therefore

$$|f(y) - f(x)| \leq H|f(y_{n+1}) - f(y_n)| + \sum_{j=1}^{n}|f(y_j) - f(y_{j-1})| \leq nH^{n+2}|f(z) - f(x)|$$

Hence we have (3.58) as soon as $\eta(t) \geq tH^{t+2}$ for $t \geq 1$. □

The reader may have noticed that for the Sobolev properties in Theorem 3.4.1 we actually used weak quasisymmetry (3.56) with $H = \eta(1)$ instead of the full power of η-quasisymmetric mappings. Hence that already implies that weakly quasisymmetric mappings are quasiconformal, thus quasisymmetric by Theorem 3.5.3 or 3.6.2.

We can now return to prove Kovalev's theorem.

Theorem 3.11.4. *A δ-monotone mapping $h : \mathbb{C} \to \mathbb{C}$ is a quasisymmetric homeomorphism onto $\mathbb{C}$. In particular, h is quasiconformal and $h \in W_{loc}^{1,2}(\mathbb{C})$.*

Proof. We adopt the following notation for cones in the plane,

$$C_w^\delta(z) = \left\{\zeta \in \mathbb{R}^n : \left|\frac{\zeta-w}{|\zeta-w|} - \frac{z-w}{|z-w|}\right| < \frac{\delta}{2}\right\} \qquad (3.59)$$

The cone $C_w^\delta(z)$ has w as its vertex and opens up in the direction $z - w$. It is the union of all rays starting at w and making an angle less than $2\arcsin(\delta/4)$ with the ray in direction $z - w$.

Next, as h is δ-monotone, we see that if $\zeta \in C_w^\delta(z)$, then

$$|h(\zeta) - h(w)| \leq \frac{2}{\delta}\langle h(\zeta) - h(w), \frac{z-w}{|z-w|}\rangle \qquad (3.60)$$

This is because

$$\frac{|\zeta-w|}{|z-w|}\langle h(\zeta) - h(w), z - w\rangle$$

$$= \langle h(\zeta) - h(w), \zeta - w\rangle - \langle h(\zeta) - h(w), \zeta - w - \frac{|\zeta-w|}{|z-w|}(z-w)\rangle$$

$$\geq (\delta - \frac{\delta}{2})|h(\zeta) - h(w)||\zeta - w|,$$

and rearranging the nonzero terms gives (3.60).

3.11. δ-MONOTONE MAPPINGS

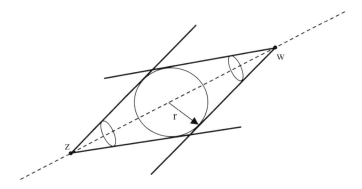

Intersection of cones $C_z^\delta(w)$ and $C_w^\delta(z)$

From this we deduce the following basic estimate for h simply by adding the relevant estimates obtained by swapping z and w: If h is δ-monotone and

$$\zeta \in C_z^\delta(w) \cap C_w^\delta(z) = Q_{z,w}^\delta, \tag{3.61}$$

then

$$|h(\zeta)-h(w)|+|h(\zeta)-h(z)| \leqslant \frac{2}{\delta} \langle h(z)-h(w), \frac{z-w}{|z-w|} \rangle \leqslant \frac{2}{\delta} |h(z)-h(w)| \tag{3.62}$$

The intersection of the cones $Q_{z,w}^\delta$ in (3.61) is a rectangular parallelipiped with z and w as opposite vertices.

The following easy lemma concerning convex sets will be useful.

Lemma 3.11.5. *Let $L = \{t\zeta : t \geqslant 0\} \cup \{t\eta : t > 0\}$ with $\zeta, \eta \in \mathbb{S}^1$ not equal or antipodal. Let Ω be a proper convex subset of $\mathbb{C}$. Then there is a Euclidean motion ψ of the plane so that $\psi(0) \notin \Omega$, yet for some s, t we have both $\psi(t\zeta), \psi(s\eta) \in \Omega$.*

Proof. Choose point $x \in \partial\Omega$ with a uniquely defined support line and an inward normal α. Rotate and translate so that x is the image of 0 while the image of $\zeta + \eta$ is parallel to α. Now move the image of 0 in the direction $-\alpha$, away from Ω. For a sufficiently small move, the image of L will then have the desired properties. □

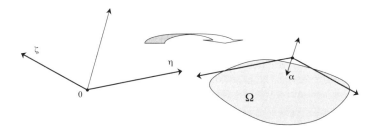

To complete the proof of Theorem 3.11.4, define for each $|\alpha| = 1$

$$\mathcal{F}_\alpha = \{h : \mathbb{C} \to \mathbb{C} : h \text{ is } \delta\text{-monotone}, h(0) = 0 \text{ and } |h(\alpha)| = 1\}$$

and consider the set

$$X = \{z \in \mathbb{C} : \sup_{h \in \mathcal{F}_\alpha} |h(z)| < \infty\} \tag{3.63}$$

Then X is nonempty, as $\{0, \alpha\} \subset X$. Furthermore, the set is convex: Given $z, w \in X$, apply (3.62) with the choice $\zeta \in [z, w] \cap \Omega$. If $X \neq \mathbb{C}$, using Lemma 3.11.5 we can find $z_0 \in \mathbb{C} \setminus X$ and two points $u, v \in X$ such that $z_0 \in Q^\delta_{u,v}$. Again, (3.62) implies that $\sup_{h \in \mathcal{F}_\alpha} |h(z_0)| < \infty$ and hence $z_0 \in X$, contradicting the choice of the point. Thus $X = \mathbb{C}$.

We need uniformity in the estimates (3.63). For this simply choose two points z, w so that $\mathbb{D} \subset Q^\delta_{z,w}$. With (3.62) we obtain

$$\sup_{|\zeta| \leq 1} |h(\zeta)| \leq H = \left(\frac{1}{2} + \frac{1}{\delta}\right) \left(\sup_{h \in \mathcal{F}_\alpha} |h(z)| + \sup_{h \in \mathcal{F}_\alpha} |h(w)|\right) < \infty \tag{3.64}$$

whenever $h \in \mathcal{F}_\alpha$. In fact, in (3.64) we can choose the bound $H = H(\delta) < \infty$ to be independent of the choice of $\alpha \in \mathbb{S}^1$. This is possible because given any mapping h, we may set $h_\alpha(z) = \alpha h(\bar{\alpha} z)$. Then h_α is δ-monotone whenever h is, and similarly, $h_\alpha(z) \in \mathcal{F}_\alpha$ if and only if $h \in \mathcal{F}_1$.

Given now any δ-monotone mapping $h : \mathbb{C} \to \mathbb{C}$, $z_0 \in \mathbb{C}$ and $r > 0$, choose $\eta \in \mathbb{S}^1$, such that

$$\min_{|\zeta| = r} |h(z_0 + \zeta) - h(z_0)| = |h(z_0 + r\eta) - h(z_0)|$$

Then define

$$g(z) = \frac{h(z_0 + rz) - h(z_0)}{|h(z_0 + r\eta) - h(z_0)|}$$

and note that g is a δ-monotone mapping, $g(0) = 0$ and $|g(\eta)| = 1$. Hence (3.64) gives

$$\frac{\max_{|\zeta|=r} |h(z_0 + \zeta) - h(z_0)|}{\min_{|\zeta|=r} |h(z_0 + \zeta) - h(z_0)|} = \max_{|\xi|=1} |g(\xi)| < H,$$

and thus h is weakly quasisymmetric. With Lemma 3.11.3 we have now completed the proof of Theorem 3.11. □

For convenience we formulated Theorem 3.11.4 for mappings only of $\mathbb{C}$, but it is not difficult to adopt the proof to a local version; see [30, 218].

Note that Theorem 3.11.4 does not hold for mappings that are merely monotone, that is, satisfy

$$\langle h(z) - h(w), z - w \rangle \geq 0, \qquad z, w \in \mathbb{C} \tag{3.65}$$

3.11. δ-MONOTONE MAPPINGS

Namely, take any singular increasing homeomorphism $\phi : \mathbb{R} \to \mathbb{R}$, such as the Cantor function, and set $h(x,y) = (\phi(x), \phi(y))$. This satisfies (3.65), but $h \notin W^{1,1}_{loc}(\mathbb{C})$.

To estimate the distortion of δ-monotone mappings, we shall go about deriving a first-order quasilinear equation for them. In fact, the equation characterizes this class of mappings.

Theorem 3.11.6. *Let $0 < \delta \leqslant 1$. Then a mapping $f \in W^{1,2}_{loc}(\mathbb{C})$ is δ-monotone if and only if*

$$\left|\frac{\partial f}{\partial \bar{z}}\right| + \delta \left|\Im m \frac{\partial f}{\partial z}\right| \leqslant \sqrt{1-\delta^2}\, \Re e\, \frac{\partial f}{\partial z} \quad \text{almost everywhere} \tag{3.66}$$

In particular, every δ-monotone mapping $f : \mathbb{C} \to \mathbb{C}$ is K-quasiconformal with

$$K = \frac{1+\sqrt{1-\delta^2}}{1-\sqrt{1-\delta^2}} \tag{3.67}$$

This bound on distortion is sharp for every δ.

Proof. Given a δ-monotone mapping, it is quasiconformal and hence differentiable almost everywhere. At the points of differentiability we consider the first-order Taylor expansion

$$f(z+\omega) - f(z) = f_\zeta(z)\omega + f_{\bar{\zeta}}(z)\bar{\omega} + o(|\omega|)$$

and rewrite our formulation (3.52) for monotonicity as

$$\Re e((f_\zeta \omega + f_{\bar{\zeta}} \bar{\omega})\bar{\omega}) \geqslant \delta |f_\zeta \omega + f_{\bar{\zeta}} \bar{\omega}||\omega| + o(|\omega|)|\omega|$$

We divide this inequality by $|\omega|^2$ and let this tend to zero to obtain

$$\Re e(f_\zeta + \lambda f_{\bar{\zeta}}) \geqslant \delta |f_\zeta + \lambda f_{\bar{\zeta}}|, \qquad |\lambda| = 1 \tag{3.68}$$

Arriving now at the condition (3.53), we may repeat the analysis there and see that the complex derivatives of f satisfy pointwise almost everywhere the inequality (3.66).

Conversely, if $f \in W^{1,2}_{loc}(\mathbb{C})$ satisfies (3.66), then

$$\left|\frac{\partial f}{\partial \bar{z}}\right| \leqslant \sqrt{1-\delta^2}\, \Re e\, \frac{\partial f}{\partial z} \leqslant \sqrt{1-\delta^2}\, \left|\frac{\partial f}{\partial z}\right|, \tag{3.69}$$

where $\sqrt{1-\delta^2} < 1$. Thus f is a quasiregular mapping, and in particular continuous and differentiable almost everywhere; see, e.g., Theorem 5.5.2 below. For a linear mapping $h(z) = \alpha z + \beta \bar{z}$, we showed earlier that the conditions (3.52) and (3.54) are equivalent. Hence, if f satisfies the inequality (3.66), then at almost every $z \in \mathbb{C}$ the derivative $Df(z)$ is a δ-monotone linear mapping,

$$\langle Df(z)\omega, \omega\rangle \geqslant \delta |Df(z)\omega|\,|\omega|, \qquad \omega \in \mathbb{C}, \tag{3.70}$$

with the same fixed $\delta \in (0,1]$. Given two points $z, w \in \mathbb{C}$, assume that f is absolutely continuous on the line segment $[z, w]$. Then (3.70) gives

$$\begin{aligned}\langle f(z) - f(w), z - w\rangle &= \int_0^1 \langle Df(w + t(z-w))(z-w), z-w\rangle\, dt \\ &\geq \delta|z-w| \int_0^1 |Df(w+t(z-w))(z-w)|\, dt \\ &\geq \delta|z-w||f(z) - f(w)| \end{aligned} \qquad (3.71)$$

As Sobolev mappings are absolutely continuous on almost every line, by the continuity of f we reach the required inequality, namely,

$$\langle f(z) - f(w), z - w\rangle \geq \delta|z-w||f(z) - f(w)|, \qquad z, w \in \mathbb{C}$$

It remains only to prove the bound (3.67) on the distortion. But this is easy; the bounds follow already from (3.69) (see Theorem 2.5.4), and they are attained by the δ-monotone mappings $h(z) = z + \sqrt{1-\delta^2}\,\bar{z}$. This completes the proof of the theorem. $\square$

In particular, it follows from the theorem that any $W^{1,2}$-solution to the quasilinear Beltrami equation (3.66) is injective!

Monotone mappings have surprising geometric properties. On the one hand they have as bad singularities as any quasiconformal mapping and, for instance, distort the Hausdorff dimension in the worst possible manner. For this, see for instance Theorem 13.6.1. On the other hand, they exhibit strong rectifiability properties not shared by general quasiconformal mappings, as demonstrated by the next theorem.

Theorem 3.11.7. *Let $h : \mathbb{C} \to \mathbb{C}$ be a δ-monotone mapping, $\zeta \in \mathbb{C}$, $|\zeta| = 1$ and $\omega \in \mathbb{C}$. Let $\mathcal{L}$ denote the line*

$$\mathcal{L} = \{z = \omega + t\zeta : t \in \mathbb{R}\}$$

Then $h(\mathcal{L})$ is the graph of a $\frac{1}{\delta}$-Lipschitz function, in particular rectifiable.

More generally, the image of any C^∞-smooth curve under a δ-monotone mapping is rectifiable.

Proof. By means of translation and conjugation by complex multiplication ζ, we can assume $\mathcal{L}$ is the real line $\mathbb{R}$.

We will then show that whenever $\varepsilon(t)$ is a $\mathbb{R}$-valued Lipschitz function on $\mathbb{R}$, with Lipschitz constant $\leq \sqrt{\delta/2}$, then any δ-monotone mapping takes its graph

$$\Gamma(\varepsilon) = \{(t, \varepsilon(t)) : t \in \mathbb{R}\} \subset \mathbb{C}$$

to another Lipschitz graph. From this we also have the rectifiability of the images of smooth curves under δ-monotone mappings.

3.11. δ-MONOTONE MAPPINGS

Let us use the notation $w_t = t + i\varepsilon(t)$ for the points of the graph of ε. We will apply (3.60), that

$$\left| \frac{\zeta - w}{|\zeta - w|} - \frac{z - w}{|z - w|} \right| \leqslant \frac{\delta}{2} \tag{3.72}$$

implies

$$|h(\zeta) - h(w)| \leqslant \frac{2}{\delta} \left\langle h(\zeta) - h(w), \frac{z - w}{|z - w|} \right\rangle \tag{3.73}$$

Here, choose $\zeta = w_t$, $w = w_s$ and $z - w = t - s$, where $t > s$. Then

$$\left| \frac{w_t - w_s}{|w_t - w_s|} - \frac{t - s}{|t - s|} \right| \leqslant \frac{|\varepsilon(t) - \varepsilon(s)|^2}{|t - s|^2} \leqslant \frac{\delta}{2}$$

by our assumption on the Lipschitz norm. This gives

$$(t - s)\langle h(w_t) - h(w_s), 1 \rangle = \langle h(w_t) - h(w_s), t - s \rangle \geqslant \frac{\delta}{2} |h(w_t) - h(w_s)| \, |t - s| > 0$$

The above argument shows that $t \mapsto \langle h(w_t), 1 \rangle$ is strictly increasing. Therefore there is a strictly increasing homeomorphism $\gamma : \mathbb{R} \to \mathbb{R}$ such that

$$\Re\, h(w_{\gamma(t)}) = \langle h(w_{\gamma(t)}), 1 \rangle \equiv t$$

The function γ gives us the Lipschitz parameterisation. We set

$$\phi(t) = h(w_{\gamma(t)}) \in h(\Gamma(\varepsilon)), \qquad t \in \mathbb{R}$$

Then ϕ is $\frac{2}{\delta}$-Lipschitz mapping since for $t > s$ we have

$$\begin{aligned}
(\gamma(t) - \gamma(s))|\phi(t) - \phi(s)| &= |\gamma(t) - \gamma(s)| \, |h(w_{\gamma(t)}) - h(w_{\gamma(s)})| \\
&\leqslant \frac{2}{\delta} \langle h(w_{\gamma(t)}) - h(w_{\gamma(s)}), \gamma(t) - \gamma(s) \rangle \\
&= \frac{2}{\delta} (\gamma(t) - \gamma(s))(t - s)
\end{aligned}$$

The Lipschitz bound follows by clearing the (nonzero) common term. Thus $\phi(t) = t + iu(t)$, where

$$u(t) = \Im \phi(t) = \Im\, h(w_{\gamma(t)})$$

is a $2/\delta$-Lipschitz function.

In the case where $\varepsilon(t) \equiv 0$, the above argument slightly simplifies, and we see that actually then $u(t)$ has Lipschitz constant $\leqslant 1/\delta$. $\square$

Chapter 4

Complex Potentials

The basic tools for solving general linear elliptic systems are singular integral operators. Even for nonlinear systems, the methods from the linear theory often provide the estimates needed to establish existence and regularity. In the setting of quasilinear elliptic equations, the fact that each solution satisfies some linear elliptic equation makes the appproach via singular integrals particularly successful.

In our approach we shall meet three fundamental operators and many variations of them. It is the purpose of this chapter to develop the theory of these operators. This chapter may also be viewed as an introduction to harmonic analysis in the complex plane.

The Logarithmic Potential

Definition 4.0.8. *Given a smooth density function $\phi \in C_0^\infty(\mathbb{C})$, we define its logarithmic potential by the formula*

$$(\mathcal{L}\phi)(z) := \frac{2}{\pi} \int_\mathbb{C} \phi(\tau) \log|z - \tau|\, d\tau \tag{4.1}$$

As the logarithmic potential commutes with translations, we have the identity

$$\mathcal{L}(D\phi) = D(\mathcal{L}\phi) \tag{4.2}$$

for $\phi \in C_0^\infty(\mathbb{C})$, where D is any constant coefficient linear (over the field $\mathbb{C}$) differential operator, say

$$D\phi = \sum_{0 \leqslant k, \ell \leqslant N} a_{k\ell}\, \frac{\partial^{k+\ell}\phi}{\partial z^k \partial \bar{z}^\ell} \tag{4.3}$$

The function $\mathcal{L}\phi$ is an infinitely smooth solution to the Poisson equation

$$\frac{\partial^2}{\partial z \partial \bar{z}}\, \mathcal{L}\phi = \phi \tag{4.4}$$

Directly from the integral formula in (4.1), we may deduce the following asymptotic behavior at ∞:

$$(\mathcal{L}\phi)(z) = \left(\frac{2}{\pi}\int_C \phi(\tau)d\tau\right)\log|z| + o(1) \qquad \text{as } z \to \infty \qquad (4.5)$$

It then follows that $\mathcal{L}\phi$ is uniformly continuous in the complex plane $\mathbb{C}$.

The Cauchy Transform

As with the logarithmic potential, we consider functions $\phi \in C_0^\infty(\mathbb{C})$ and define the Cauchy transform as follows.

Definition 4.0.9. *The Cauchy transform is defined by the rule*

$$(\mathcal{C}\phi)(z) := \frac{\partial \mathcal{L}\phi}{\partial z} = \frac{1}{\pi}\int_C \frac{\phi(\tau)}{z-\tau}d\tau \qquad (4.6)$$

Here we have applied (4.2) with $D = \partial_z$ and then integration by parts, that is, Green's formula (2.52), for

$$\mathcal{L}(\phi_z) = \lim_{\varepsilon \to 0}\frac{1}{\pi}\int_{\mathbb{C}\setminus\mathbb{D}(z,\varepsilon)} \phi_z(\tau)\log|z-\tau|^2\,d\tau = \frac{1}{\pi}\int_C \frac{\phi(\tau)}{z-\tau}d\tau \qquad (4.7)$$

Similarly, with (2.53) we obtain $\mathcal{L}(\phi_{\bar{z}z}) = \phi$, that is, (4.4).

The kernel of the Cauchy transform is the function $\frac{1}{\pi\tau}$, which is a fundamental solution for the Cauchy-Riemann operator $\frac{\partial}{\partial \bar{z}}$; we may write (2.53) and the above identities (4.4) and (4.6) in the compact form

$$\mathcal{C} \circ \frac{\partial}{\partial \bar{z}} = \frac{\partial}{\partial \bar{z}} \circ \mathcal{C} = \mathbf{I}, \qquad (4.8)$$

with $\mathbf{I}$ the identity operator. This means that the inhomogeneous Cauchy-Riemann system

$$\frac{\partial}{\partial \bar{z}} f = \phi \qquad (4.9)$$

can be solved uniquely up to a holomorphic function $h(z)$ by the rule

$$f(z) = h(z) + (\mathcal{C}\phi)(z) \qquad (4.10)$$

Applying the differential operator $\frac{\partial}{\partial z}$ to both sides of (4.2), the formula in (4.10) becomes the commutation rule for the Cauchy transform,

$$\mathcal{C}(D\phi) = D(\mathcal{C}\phi), \qquad (4.11)$$

$\phi \in C_0^\infty(\mathbb{C})$. As $\mathcal{C}$ is a convolution-type operator, we see that for every smooth test function $\eta \in C_0^\infty(\mathbb{C})$,

$$\int_C \bar{\eta}\,\mathcal{C}\phi = -\int \phi\,\mathcal{C}\bar{\eta} = -\int_C \phi\,\overline{\mathcal{C}^*\eta},$$

where the operator $\mathcal{C}^*$, called the *formal adjoint* of the Cauchy transform, is defined by $\mathcal{C}^*\eta = \overline{\mathcal{C}\bar\eta}$,

$$(\mathcal{C}^*\eta)(z) = \frac{1}{\pi}\int_\mathbb{C} \frac{\eta(\tau)}{\bar z - \bar\tau}\, d\tau \tag{4.12}$$

The formal adjoint $\mathcal{C}^*$ can also be defined by the rule

$$\mathcal{C}^*\eta = \frac{\partial \mathcal{L}\eta}{\partial \bar z} = \mathcal{L}(\eta_{\bar z}) \tag{4.13}$$

for $\eta \in C_0^\infty(\mathbb{C})$. Once again, the solutions to the inhomogeneous anti-Cauchy-Riemann equations

$$\frac{\partial}{\partial z} f = \eta \tag{4.14}$$

can be found as

$$f(z) = g(z) + (\mathcal{C}^*\eta)(z), \tag{4.15}$$

where this time g is any anti-holomorphic function.

The Beurling Transform

If we again differentiate the logarithmic potential, we obtain a further singular integral operator, the Beurling transform

$$(\mathcal{S}\phi)(z) := \frac{\partial^2}{\partial z^2} \mathcal{L}\phi = \frac{\partial}{\partial z} \mathcal{C}\phi = -\frac{1}{\pi}\int_\mathbb{C} \frac{\phi(\tau)}{(z-\tau)^2}\, d\tau \tag{4.16}$$

Here the last identity is only a formal one, as the kernel is not integrable. A rigorous interpretation requires the notion of Cauchy principal value. To get this, let $\phi \in C_0^\infty(\mathbb{C})$ and again apply Green's formula (2.52),

$$\begin{aligned}(\mathcal{S}\phi)(z) &= \mathcal{C}(\phi_z)(z) = \lim_{\varepsilon\to 0} \frac{1}{\pi}\int_{\mathbb{C}\setminus \mathbb{D}(z,\varepsilon)} \frac{\phi_z(\tau)}{z-\tau}\, d\tau \\ &= \lim_{\varepsilon\to 0} \frac{-1}{\pi}\int_{\mathbb{C}\setminus \mathbb{D}(z,\varepsilon)} \frac{\phi(\tau)}{(z-\tau)^2}\, d\tau - \lim_{\varepsilon\to 0}\frac{i}{2\pi}\int_{\partial\mathbb{D}(z,\varepsilon)} \frac{\phi(\tau)}{z-\tau}\, d\bar\tau\end{aligned}$$

The last integral vanishes in the limit as $\varepsilon \to 0$. Hence, for $\phi \in C_0^\infty(\mathbb{C})$, an equivalent definition for the Beurling transform is given in terms of its principal-value integral

$$(\mathcal{S}\phi)(z) = -\frac{1}{\pi} \lim_{\varepsilon\to 0} \int_{|z-\tau|>\varepsilon} \frac{\phi(\tau)}{(z-\tau)^2}\, d\tau \tag{4.17}$$

Put briefly, the Beurling transform is a singular convolution operator of Calderón-Zygmund type with the nonintegrable kernel $-\frac{1}{\pi}\tau^{-2}$. A very good source for the basics of the theory of Calderón-Zygmund operators continues to be Stein's book [336].

For functions $\phi \in C_0^\infty(\mathbb{C})$ the limit at (4.17) exists at every point $z \in \mathbb{C}$. If $\phi \in L^p(\mathbb{C})$, the principal value (4.17) still exists at *almost every* $z \in \mathbb{C}$. This is, however, more subtle, and we will return to this question a little later, in Theorem 4.0.10.

The fundamental importance of the Beurling transform in complex analysis is furnished by the identity

$$\mathcal{S} \circ \frac{\partial}{\partial \bar{z}} = \frac{\partial}{\partial z}, \qquad (4.18)$$

initially valid for functions contained in the space $C_0^\infty(\mathbb{C})$. In other words, $\mathcal{S}$ intertwines the Cauchy-Riemann operators $\frac{\partial}{\partial \bar{z}}$ and $\frac{\partial}{\partial z}$, a fact that explains the importance of the operator in complex analysis. One easily obtains (4.18) from (4.6) and (4.8), for we may write

$$\mathcal{S} \circ \frac{\partial}{\partial \bar{z}} = \frac{\partial}{\partial z} \circ \mathcal{C} \circ \frac{\partial}{\partial \bar{z}} = \frac{\partial}{\partial z} \circ \mathbf{I} = \frac{\partial}{\partial z}$$

We also record the commutation rule for linear differential operators as in (4.3) and $\phi \in C_0^\infty(\mathbb{C})$,

$$D(\mathcal{S}\phi) = \mathcal{S}(D\phi)$$

Next observe that $\mathcal{S}$ is symmetric,

$$\int_\mathbb{C} \eta(z)(\mathcal{S}\phi)(z) = -\frac{1}{\pi} \lim_{\varepsilon \to 0} \int_{|z-\tau|>\varepsilon} \frac{\eta(z)\phi(\tau)}{(z-\tau)^2}\, d\tau\, dz = \int_\mathbb{C} \phi(\tau)(\mathcal{S}\eta)(\tau) \qquad (4.19)$$

for $\eta, \phi \in C_0^\infty(\mathbb{C})$.

For $f \in C_0^\infty(\mathbb{C})$ we may use integration by parts to produce the identity

$$\int_\mathbb{C} |f_z|^2 = \int_\mathbb{C} f_z(\bar{f})_{\bar{z}} = \int_\mathbb{C} f_{\bar{z}}(\bar{f})_z = \int_\mathbb{C} |f_{\bar{z}}|^2 \qquad (4.20)$$

and this formula expresses the fact that $\mathcal{S}$ acts as an isometry in $L^2(\mathbb{C})$,

$$\|\mathcal{S}\phi\|_2 = \|\phi\|_2, \qquad (4.21)$$

or at least for those ϕ of the form $\phi = f_{\bar{z}}$ with $f \in C_0^\infty(\mathbb{C})$. It is easy to see that these functions are dense in $L^2(\mathbb{C})$ since an $h \in L^2(\mathbb{C})$ is orthogonal to all $f_{\bar{z}}$, $f \in C_0^\infty(\mathbb{C})$, precisely when $h_{\bar{z}} = 0$ in the sense of distributions. In view of Weyl's Lemma A.6.10 such an h is a holomorphic function contained in $L^2(\mathbb{C})$, thus $h \equiv 0$. Note that the corresponding density argument works even for $L^p(\mathbb{C})$, $1 < p < \infty$, but not for $p = 1$ or $p = \infty$, where in fact density fails.

From these considerations we see in particular that $\mathcal{S}$ extends to a continuous and surjective operator, in fact to a unitary operator acting on $L^2(\mathbb{C})$. The Hilbert adjoint of the Beurling transform $\mathcal{S} : L^2(\mathbb{C}) \to L^2(\mathbb{C})$ coincides with the inverse of $\mathcal{S}$, $\mathcal{S}^{-1} : L^2(\mathbb{C}) \to L^2(\mathbb{C})$ defined by

$$\mathcal{S}^{-1}\phi(z) = \overline{\mathcal{S}\bar{\phi}}(z) = \frac{1}{\pi} \int_\mathbb{C} \frac{\phi(\tau)}{(\bar{z}-\bar{\tau})^2}\, d\tau \qquad (4.22)$$

Therefore all the bounds we obtain below for $\mathcal{S}$ will also hold for the inverse operator $\mathcal{S}^{-1} = \overline{\mathcal{S}}$.

This, when combined with a straightforward approximation, generalizes (4.18) to any locally integrable function f with square-integrable distributional derivatives,

$$\mathcal{S}\frac{\partial f}{\partial \bar{z}} = \frac{\partial f}{\partial z} \qquad (4.23)$$

As an illustration, let us calculate the Beurling transform of the characteristic function of the unit disk $\mathbb{D} = \{z : |z| < 1\}$. Searching for a function with $f_{\bar{z}} = \chi_{\mathbb{D}}$ leads to the choice

$$f(z) = \begin{cases} \bar{z}, & |z| \leqslant 1 \\ 1/z, & |z| > 1 \end{cases}$$

Since the derivatives of the function are square-integrable, we have from (4.23)

$$\mathcal{S}\chi_{\mathbb{D}} = -\frac{1}{z^2}\chi_{\mathbb{C}\setminus\mathbb{D}} \qquad (4.24)$$

With a change of variables, or modifying the argument, we identify the transform of the characteristic function

$$\chi = \chi_{\mathbb{D}(a,r)}(z)$$

of a general disk $\mathbb{D}(a,r)$ as

$$\mathcal{S}\chi = \begin{cases} 0, & |z-a| \leqslant r \\ -r^2/(z-a)^2, & |z-a| > r \end{cases} \qquad (4.25)$$

The fact that $\mathcal{S}$ is bounded also on $L^p(\mathbb{C})$, $1 < p < \infty$, lies somewhat deeper. But once boundedness has been established, the above identities generalize to functions in $L^p(\mathbb{C})$ as well. We will return to these topics in Section 4.5.

As the Beurling transform commutes with translations, it is a convolution operator, although strictly speaking $1/z^2 \notin L^1(\mathbb{C})$. Nevertheless, straightforward arguments such as differential identities like (4.23) or Fourier methods provide a variety of different approaches for analyzing $\mathcal{S}$.

Convergence of the Integrals

It is important to note that the integrals defining the logarithmic potential and the Cauchy transform actually converge for almost every $z \in \mathbb{C}$ whenever the function ϕ is compactly supported and integrable. That this is so follows directly from Fubini's theorem. Given a compact set F, we perform the double integral to find that both

$$\int_F \int_\mathbb{C} |\phi(\tau) \log|z-\tau|| \, dz \, d\tau < \infty$$

and

$$\int_F \int_\mathbb{C} \frac{|\phi(\tau)|}{|z-\tau|} \, dz \, d\tau < \infty$$

Then Fubini's theorem tells us not only that the integral is defined for almost all $z \in \mathbb{C}$, but also that the functions

$$(\mathcal{L}\phi)(z) = \frac{2}{\pi} \int_{\mathbb{C}} \phi(\tau) \log |z - \tau| d\tau$$

and

$$(\mathcal{C}\phi)(z) = \frac{1}{\pi} \int_{\mathbb{C}} \frac{\phi(\tau)}{z - \tau} d\tau$$

are locally integrable.

As far as the Beurling transform is concerned, the above argument fails since the local integrability of the kernel is now lost. Stronger techniques are thus required. In the theory of singular integrals the pointwise behavior is usually controlled by the *maximal transforms* such as

$$\mathcal{S}_* f(z) = \frac{1}{\pi} \sup_{\varepsilon > 0} \left| \int_{|z-\tau|>\varepsilon} \frac{f(\tau)}{(z-\tau)^2} d\tau \right| \qquad (4.26)$$

Note here that the integrals $\int_{|z-\tau|>\varepsilon} \frac{f(\tau)}{(z-\tau)^2} d\tau$ with $\varepsilon > 0$ are all absolutely convergent for any $f \in L^p(\mathbb{C})$, $1 < p < \infty$, by Hölder's inequality. Hence the maximal Beurling transform is well defined for such f.

However, in the special case of the Beurling transform there, is an alternate route, avoiding maximal transforms. We describe a wonderful elementary argument observed recently by Mateu and Verdera [250] for proving the almost everywhere convergence of the principal-value integral defining $\mathcal{S}f$.

To start with, note that by continuity of the operator, given $f \in L^2(\mathbb{C})$, the image $\mathcal{S}f$ is a well-defined function in $L^2(\mathbb{C})$, unique up to a set of measure zero. Of course, the identity in (4.19) extends by continuity to all $\phi, \eta \in L^2(\mathbb{C})$. Then, using (4.25) and the symmetry of $\mathcal{S}$, we calculate

$$-\frac{1}{\pi} \int_{|z-\tau|>\varepsilon} \frac{f(\tau)}{(z-\tau)^2} d\tau = \frac{1}{\pi\varepsilon^2} \int_{\mathbb{C}} \mathcal{S}\left(\chi_{\mathbb{D}(z,\varepsilon)}\right) f = \frac{1}{\pi\varepsilon^2} \int_{\mathbb{C}} \chi_{\mathbb{D}(z,\varepsilon)} \mathcal{S}f$$

In other words, for every disk $\mathbb{D}(z,\varepsilon)$ we have the identity

$$-\frac{1}{\pi} \int_{|z-\tau|>\varepsilon} \frac{f(\tau)}{(z-\tau)^2} d\tau = \frac{1}{|\mathbb{D}(z,\varepsilon)|} \int_{\mathbb{D}(z,\varepsilon)} \mathcal{S}f \qquad (4.27)$$

But by Lebesgue's theorem the last integral converges almost everywhere with limit $\mathcal{S}f(z)$. In particular, the principal value converges almost everywhere. We assumed here that $f \in L^2(\mathbb{C})$, but as soon as $\mathcal{S}$ has been extended to $L^p(\mathbb{C})$, the argument works there as well.

We state the conclusion as a separate theorem.

Theorem 4.0.10. *Suppose $f \in L^p(\mathbb{C})$, $1 < p < \infty$. Then the limit*

$$(\mathcal{S}f)(z) = -\frac{1}{\pi} \lim_{\varepsilon \to 0} \int_{|z-\tau|>\varepsilon} \frac{f(\tau)}{(z-\tau)^2}$$

exists at almost every $z \in \mathbb{C}$, precisely at the Lebesgue points of the function $\mathcal{S}f \in L^p(\mathbb{C})$.

Suitable versions of the result hold also for L^∞ and L^1. As another consequence of (4.27), note that we obtain a pointwise bound for the maximal Beurling transform,
$$\mathcal{S}_* f \leq \mathcal{M}(\mathcal{S}f),$$
where $\mathcal{M}$ is the Hardy-Littlewood maximal function, defined at a given point $a \in \mathbb{C}$ in terms of disks centered at a. We discuss this maximal function in a moment in Section 4.4.2.

4.1 The Fourier Transform

We shall now take a little time to present and discuss aspects of the basic theory of the Fourier transform and the concepts of multipliers for singular integral operators of convolution type.

We begin with the Schwartz class $\mathscr{S}(\mathbb{C})$ of rapidly decreasing smooth functions $f \in C^\infty(\mathbb{C})$ such that
$$\lim_{z \to \infty} |z|^m \frac{\partial^{k+\ell} f(z)}{\partial z^k \, \partial \bar{z}^\ell} = 0$$
for all k, ℓ and m in $\mathbb{N} = \{0, 1, 2, \ldots\}$. The *Fourier* transform $\mathcal{F}$ acts on the Schwartz class by the rule
$$(\mathcal{F}f)(\tau) = \hat{f}(\tau) = \int_\mathbb{C} e^{i\pi(z\bar{\tau} + \bar{z}\tau)} f(z) dz$$
from which it is not difficult to see that $\mathcal{F} : \mathscr{S}(\mathbb{C}) \to \mathscr{S}(\mathbb{C})$. In fact, the Fourier transform has an inverse,
$$\mathcal{F}^{-1} : \mathscr{S}(\mathbb{C}) \to \mathscr{S}(\mathbb{C}),$$
given by the formula
$$(\mathcal{F}^{-1}\phi)(z) = \int_\mathbb{C} e^{-i\pi(z\bar{\tau} + \bar{z}\tau)} \phi(\tau) d\tau$$

We identify a few important relations between these operators,
$$\begin{aligned}
\mathcal{F} \circ \mathcal{F}^{-1} &= \mathcal{F}^{-1} \circ \mathcal{F} = \mathbf{I} : \mathscr{S}(\mathbb{C}) \to \mathscr{S}(\mathbb{C}) \\
(\mathcal{F}f)(-\tau) &= (\mathcal{F}^{-1}f)(\tau) \\
(\mathcal{F}\hat{f})(\tau) &= f(-\tau) \\
\overline{(\mathcal{F}f)(\tau)} &= (\mathcal{F}^{-1}\bar{f})(\tau),
\end{aligned}$$

4.1. THE FOURIER TRANSFORM

and the following differentiation formulas,

$$(\mathcal{F}f_{\bar{z}})(\tau) = -i\pi\tau \hat{f}(\tau) \tag{4.28}$$
$$(\mathcal{F}f_z)(\tau) = -i\pi\bar{\tau}\hat{f}(\tau) \tag{4.29}$$

Conversely, we have the formulas

$$\hat{f}_\tau(\tau) = i\pi\widehat{\bar{z}f}(\tau)$$
$$\hat{f}_{\bar{\tau}}(\tau) = i\pi\widehat{zf}(\tau)$$

With these identities we can interpret any constant coefficient differential operator as a Fourier multiplier.

4.1.1 The Fourier Transform in L^1 and L^2

For applications we want to extend the domain of definition of the Fourier transform. To do this let us introduce the class of continuous functions vanishing at ∞, denoted $C_0(\hat{\mathbb{C}})$. Since $\mathscr{S}(\mathbb{C}) \subset L^1(\mathbb{C})$ is dense, it follows that $\mathcal{F} : L^1(\mathbb{C}) \to C_0(\hat{\mathbb{C}})$ with the norm estimate

$$\|\mathcal{F}\|_\infty \leqslant \|f\|_1$$

We note that the space $L^1(\mathbb{C})$ is an algebra with respect to the convolution

$$(f * g)(z) = \int_\mathbb{C} f(z-w)g(w)dw$$

as

$$\|f * g\|_1 \leqslant \|f\|_1 \|g\|_1$$

On the other hand, $C_0(\hat{\mathbb{C}})$ is an algebra with respect to pointwise multiplication

$$(\phi\psi)(\tau) = \phi(\tau)\psi(\tau)$$

as

$$\|\phi\psi\|_\infty \leqslant \|\phi\|_\infty \|\psi\|_\infty$$

The Fourier transform is actually a homomorphism between these algebras,

$$\widehat{f*g}(z) = \hat{f}(z)\hat{g}(z) \tag{4.30}$$

Fubini's theorem gives us the important Parseval identity,

$$\int_\mathbb{C} \phi(\tau)\overline{\hat{f}(\tau)}d\tau = \int_\mathbb{C} \hat{\phi}(z)\overline{f(z)}\,dz, \tag{4.31}$$

so that $\mathcal{F}$ extends as a unitary operator on $L^2(\mathbb{C})$,

$$\|\mathcal{F}f\|_2 = \|f\|_2$$

By interpolation we can then identify the Fourier transform as a continuous operator

$$\mathcal{F}: L^q(\mathbb{C}) \to L^p(\mathbb{C}); \qquad 1 \leqslant q \leqslant 2 \leqslant p \leqslant \infty, \quad \frac{1}{p} + \frac{1}{q} = 1$$

The uniform bound here is called the Hausdorff-Young inequality

$$\|\mathcal{F}f\|_p \leqslant \|f\|_q$$

In fact, the sharp constant in this inequality was identified by W. Beckner [50] as

$$p^{1/p} \|\mathcal{F}f\|_p \leqslant q^{1/q} \|f\|_q$$

4.1.2 Fourier Transform on Measures

Recall that $C_0(\hat{\mathbb{C}})$ is the Banach space of continuous functions that vanish at ∞ equipped with the supremum norm. Its dual, $C_0(\hat{\mathbb{C}})^*$, is the Banach space of all finite complex-valued Borel measures on $\mathbb{C}$, denoted

$$C_0(\hat{\mathbb{C}})^* = \mathscr{M}(\mathbb{C}) = \left\{ \nu : \|\nu\| = \int_{\mathbb{C}} |d\nu| < \infty \right\}$$

The subspace of absolutely continuous measures (with repect to Lebesgue measure on $\mathbb{C}$) is usually identified with $L^1(\mathbb{C})$ by the isometry

$$L^1(\mathbb{C}) \hookrightarrow \mathscr{M}(\mathbb{C}), \qquad \phi \mapsto \phi(x,y)\, dx\, dy$$

The Fourier transform can be defined on measures

$$\mathcal{F}: \mathscr{M}(\mathbb{C}) \to L^\infty(\mathbb{C})$$

by the rule

$$(\mathcal{F}\nu)(\tau) = \hat{\nu}(\tau) = \int_{\mathbb{C}} e^{i\pi(z\bar{\tau}+\bar{z}\tau)}\, d\nu(z)$$

The convolution of Borel measures $\nu_1 * \nu_2$, $\nu_1, \nu_2 \in \mathscr{M}(\mathbb{C})$, is again a Borel measure. We can identify this convolution as a linear functional defined on $C_0(\hat{\mathbb{C}})$ via the following action on test functions $\eta \in C_0(\hat{\mathbb{C}})$

$$(\nu_1 * \nu_2)[\eta] = \int_{\mathbb{C}} \int_{\mathbb{C}} \eta(z+\tau)\, d\nu_1(z)\, d\nu_2(\tau)$$

4.1.3 Multipliers

Let $\mathbf{m} = \mathbf{m}(\tau)$ be a bounded measurable function on $\mathbb{C}$. Via the Fourier transform and multiplication, this defines a bounded operator

$$T_{\mathbf{m}} : L^2(\mathbb{C}) \to L^2(\mathbb{C})$$

4.1. THE FOURIER TRANSFORM

by setting
$$T_{\mathbf{m}} f = \mathcal{F}^{-1}(\mathbf{m}\hat{f})$$
We may compute the norm of this operator as
$$\|T_{\mathbf{m}} f\|_2 = \|\mathbf{m}\hat{f}\|_2 \leqslant \|\mathbf{m}\|_\infty \|f\|_2$$
And in fact,
$$\|T_{\mathbf{m}}\|_2 = \|\mathbf{m}\|_\infty$$
We call $\mathbf{m}$ the L^2-multiplier or the Fourier multiplier of the operator. Our principal interest is going to be in multipliers that are homogeneous, are of degree 0 and are smooth away from the origin. In this case the operator $T_{\mathbf{m}}$ will extend continuously to all of the $L^p(\mathbb{C})$-spaces, $1 < p < \infty$, and
$$\|T_{\mathbf{m}} f\|_p \leqslant \|T_{\mathbf{m}}\|_p \|f\|_p$$
When we have an estimate such as that above, we say that $\mathbf{m}$ is an L^p-multiplier. The boundedness of $T_{\mathbf{m}}$ on $L^p(\mathbb{C})$ is a fundamental fact known as the Marcinkiewicz multiplier theorem, see [335, Chapter VI]. A little later we will present a proof for a special class of multiplier operators, the complex Riesz transforms. With some modifications the proof can be adapted to the general case.

4.1.4 The Hecke Identities

Let $P_k(z) = P_k(x,y)$, $z = x + iy$, be a complex-valued homogeneous harmonic polynomial of degree k. Having in mind the convolution formula in (4.30), we state the Hecke identities in the following elegant (and purely symbolic) manner:
$$\left(\frac{P_k(z)}{|z|^{k+2}}\right)^\wedge = \frac{2\pi i^k}{k} \frac{P_k(\tau)}{|\tau|^k}$$
The true meaning of this identity is that the operator $T_{\mathbf{m}} : L^p(\mathbb{C}) \to L^p(\mathbb{C})$, whose multiplier
$$\mathbf{m} = \frac{2\pi i^k}{k} \frac{P_k(z)}{|z|^k}$$
takes the form
$$(T_{\mathbf{m}})(z) = \lim_{\varepsilon \to 0} \int_{|z-\tau|>\varepsilon} \frac{P_k(z-\tau)}{|z-\tau|^{k+2}} \phi(\tau) \, d\tau$$
at least for functions $\phi \in C_0^\infty(\mathbb{C})$.

Here are a few basic examples.
$$\frac{\tau^k}{|\tau|^k} = \frac{k}{2\pi i^k} \left(\frac{z^k}{|z|^{k+2}}\right)^\wedge, \quad k = 1, 2, 3, \ldots$$
$$\frac{\bar{\tau}^k}{|\tau|^k} = \frac{k}{2\pi i^k} \left(\frac{\bar{z}^k}{|z|^{k+2}}\right)^\wedge, \quad k = 1, 2, 3, \ldots,$$

so that for every $k = 1, 2, \ldots$ there is a singular integral operator $R^k : L^p(\mathbb{C}) \to L^p(\mathbb{C})$ defined by

$$(R^k \phi)(z) = \frac{k}{2\pi i^k} \lim_{\varepsilon \to 0} \int_{|z-\tau|>\varepsilon} \frac{(\bar{z}-\bar{\tau})^k}{|z-\tau|^{k+2}} \phi(\tau) \, d\tau$$

whose multiplier is

$$\mathbf{m}_k(\tau) = \left(\frac{\bar{\tau}}{|\tau|}\right)^k, \qquad \tau \neq 0$$

Of course, for $k = 0$ we set R^0 as the identity operator, $(R^0 \phi)(z) = \phi(z)$. We shall see in a moment that this notation is no accident. The operators R^k will turn out to be the kth power of another operator (the Riesz transform).

If we return to the differentiation formulas in (4.28) and (4.29), we find that for smooth $f \in C_0^\infty(\mathbb{C})$,

$$\tau(\widehat{f_z})(\tau) = \bar{\tau}(\widehat{f_{\bar{z}}})(\tau),$$

or in other words

$$(\mathcal{F} \circ \mathcal{S})\phi = \mathbf{m}_2 \, \mathcal{F}\phi,$$

again at least for ϕ of the form $\phi = f_{\bar{z}}$, with $f \in C_0^\infty(\mathbb{C})$. On such functions we therefore see that $R^2 = \mathcal{S}$, the Beurling transform.

Corollary 4.1.1. *The Beurling transform $\mathcal{S}$, represented by the singular integral operator in (4.17), is a Fourier multiplier operator with multiplier*

$$\mathbf{m}_2(\tau) = \frac{\bar{\tau}}{\tau} = \left(\frac{\bar{\tau}}{|\tau|}\right)^2$$

4.2 The Complex Riesz Transforms R^k

Parseval's identity immediately implies that the $L^2(\mathbb{C})$ adjoint of an operator with multiplier $\mathbf{m}(\tau)$ has multiplier $\overline{\mathbf{m}}(\tau)$. Since for $k = 1, 2, \ldots$,

$$\overline{\mathbf{m}}_k(\tau) = \left(\frac{\tau}{|\tau|}\right)^k = \left(\frac{\bar{\tau}}{|\tau|}\right)^{-k},$$

it is natural to use the symbol R^{-k} for the operators that are the $L^2(\mathbb{C})$-adjoints of R^k. In particular, then

$$(R^{-k}\phi)(z) = \frac{k}{2\pi i^k} \lim_{\varepsilon \to 0} \int_{|z-\tau|>\varepsilon} \frac{(z-\tau)^k}{|z-\tau|^{k+2}} \phi(\tau) \, d\tau$$

In fact, the operators R^k and R^{-k} are mutual inverses of each other, as the product of the multipliers is identically equal to 1. The general rule for composition of these operators can be seen to be

$$R^k \circ R^\ell = R^{k+\ell}, \qquad k, \ell \in \mathbb{Z}$$

We summarize the above discussion in the following theorem.

4.2. THE COMPLEX RIESZ TRANSFORMS R^k

Theorem 4.2.1. *We have for each $k \in \mathbb{Z}$,*

$$(R^k \phi)^{\wedge}(\tau) = \left(\frac{|\tau|}{\tau}\right)^k \hat{\phi}(\tau)$$

and

$$(R^k \phi)(z) = \frac{|k|}{2\pi i^{|k|}} \int_{\mathbb{C}} \frac{(\bar{z} - \bar{\tau})^k}{|z - \tau|^{k+2}} \phi(\tau) \, d\tau,$$

where the integral is to be understood in terms of the Cauchy principal value.

The generator of the group $\{R^k\}_{k=-\infty}^{k=\infty}$ is the *complex Riesz transform* $(R\phi)(z) = (R^1\phi)(z)$ with multiplier $\mathbf{m}_1 = \bar{\tau}/|\tau|$.

In fact, given any function F analytic in the annulus $\{\zeta : \rho < |\zeta| < 1/\rho\}$, we can define a singular integral operator via the Laurent series expansion of F. We have

$$F(\zeta) = \sum_{k=-\infty}^{\infty} a_k \zeta^k$$

and so we put

$$F(R) = \sum_{k=-\infty}^{\infty} a_k R^k : L^2(\mathbb{C}) \to L^2(\mathbb{C})$$

The multiplier of this operator is

$$\mathbf{m}_F(\tau) = F(\bar{\tau}/|\tau|)$$

The range of exponents $p \neq 2$ for which the operator $F(R)$ remains bounded in $L^p(\mathbb{C})$ depends on the radius ρ (or more precisely the radius of convergence of the Laurent series). In the case of a polynomial, say $P(z)$, we can calculate that

$$P(R) = a_0 \mathbf{I} + a_1 R + a_2 R^2 + \cdots + a_n R^n$$

and

$$P(R)\phi = a_0 \phi + \frac{1}{2\pi i} \int_{\mathbb{C}} \frac{P'\left(\frac{|z-\tau|}{iz - i\tau}\right)}{(z-\tau)|z-\tau|} \phi(\tau) \, d\tau,$$

the integral being understood, as usual, in terms of the Cauchy principal value.

4.2.1 Potentials Associated with R^k

For given $k \in \mathbb{Z}$, $k \neq 0$, we may define

$$(\mathcal{L}_k \phi)(z) = \frac{-2}{\pi |k| i^{|k|}} \int_{\mathbb{C}} \frac{(z - \tau)^k}{|z - \tau|^k} \phi(\tau) \, d\tau,$$

while we set $(\mathcal{L}_0 \phi)(z)$ to be the logarithmic potential

$$(\mathcal{L}_0 \phi)(z) = \frac{2}{\pi} \int_{\mathbb{C}} \log|z - \tau| \, \phi(\tau) \, d\tau$$

For $k \neq 0$ the Cauchy transforms are then defined by

$$\left(\frac{\partial}{\partial z}\mathcal{L}_k\right)\phi = \frac{-k}{\pi|k|i^{|k|}}\int_{\mathbb{C}}\frac{(z-\tau)^{k-1}}{|z-\tau|^k}\phi(\tau)\,d\tau$$

$$\left(\frac{\partial}{\partial \bar{z}}\mathcal{L}_{-k}\right)\phi = \frac{-k}{\pi|k|i^{|k|}}\int_{\mathbb{C}}\frac{(\bar{z}-\bar{\tau})^{k-1}}{|z-\tau|^k}\phi(\tau)\,d\tau,$$

which in the case $k=1$ reduces to the usual Newtonian or Riesz potential

$$\left(\frac{\partial}{\partial z}\mathcal{L}_1\right)\phi = \frac{i}{\pi}\int_{\mathbb{C}}\frac{\phi(\tau)}{|z-\tau|}\,d\tau$$

Upon further differentiation we find that for $k \neq 0$

$$\begin{aligned}\frac{\partial^2}{\partial z\bar{z}}(\mathcal{L}_k\phi) &= \frac{\partial^2}{\partial zz}(\mathcal{L}_{k+2}\phi)\\ &= \frac{\partial^2}{\partial \bar{z}\bar{z}}(\mathcal{L}_{k-2}\phi)\\ &= \frac{|k|}{2\pi|k|i^{|k|}}\int_{\mathbb{C}}\frac{(z-\tau)^k}{|z-\tau|^{k+2}}\phi(\tau)\,d\tau\\ &= R^{-k}\phi\end{aligned}$$

4.3 Quantitative Analysis of Complex Potentials

We now begin the study of the $L^p(\mathbb{C})$-theory of the three fundamental operators, the logarithmic, the Cauchy and the Beurling transforms, in greater detail. We will also be dealing with a number of other function spaces familiar to harmonic analysis such as BMO, those functions of bounded mean oscillation.

Definition 4.3.1. *For a domain $\Omega \subset \mathbb{C}$ the space $BMO(\Omega)$ consists of the locally integrable functions u such that*

$$\|u\|_* = \sup_B \frac{1}{|B|}\int_B |u - u_B| < \infty$$

Here the supremum is taken over all disks $B \subset \Omega$, and the notation u_B stands for the average of u over the disk B

$$u_B = \frac{1}{|B|}\int_B u$$

In this section we start with the logarithmic and the Cauchy operators. The discussion on the Beurling transform in the following sections is more extensive and often appears together with a discussion of its close associates, the complex Riesz transforms.

4.3.1 The Logarithmic Potential

The operator $\mathcal{L}$ extends to the function space $L^1(\mathbb{C})$ with its range in the space $BMO(\mathbb{C})$. Indeed, for $\phi \in C_0^\infty(\mathbb{C})$ we have

$$\|\mathcal{L}\phi\|_* \leq \frac{2}{\pi} \, \|\log|\cdot|\,\|_* \, \|\phi\|_1 \tag{4.32}$$

Since $C_0^\infty(\mathbb{C})$ is dense in $L^1(\mathbb{C})$ and $\mathcal{L}\phi \in C^\infty(\mathbb{C})$, whenever $\phi \in C_0^\infty(\mathbb{C})$ we observe that, in fact,

$$\mathcal{L}: L^1(\mathbb{C}) \to VMO_*(\mathbb{C}),$$

where the space $VMO_*(\mathbb{C})$ is taken to mean the completion of the uniformly continuous functions on $\mathbb{C}$ in the $BMO(\mathbb{C})$ norm.

If, however, we restrict ourselves to those functions with mean value 0, say

$$L_\bullet^1(\mathbb{C}) = \left\{ \phi \in L^1(\mathbb{C}) : \int_\mathbb{C} \phi = 0 \right\},$$

then the asymptotic estimate in (4.5) shows that

$$\mathcal{L}: L_\bullet^1(\mathbb{C}) \to VMO(\mathbb{C}),$$

where $VMO(\mathbb{C})$ is the completion of $C_0^\infty(\mathbb{C})$ in $BMO(\mathbb{C})$.

In order to familiarize ourselves with the logarithmic transform, we shall make a calculation in the case of the characteristic function χ of a disk $\mathbb{D}(a,r)$. Here the complex notation is particularly helpful. First, for the unit disk $\mathbb{D}$, the function

$$\psi(z) = \begin{cases} |z|^2 - 1, & z \in \mathbb{D} \\ 2\log|z|, & z \notin \mathbb{D} \end{cases} \tag{4.33}$$

satisfies $\psi_{z\bar{z}} = \chi_\mathbb{D}$ and has the right asymptotics (4.5). Hence $\mathcal{L}\chi_\mathbb{D} = \psi$. For the characteristic function χ of an arbitrary disk $\mathbb{D}(a,r)$, we use a change of variables (or similar reasoning as in (4.33)) to get

$$(\mathcal{L}\chi)(z) = \begin{cases} |z-a|^2 + r^2 \log r^2 - r^2, & z \in \mathbb{D}(a,r) \\ 2r^2 \log|z-a|, & z \notin \mathbb{D}(a,r) \end{cases} \tag{4.34}$$

This formula shows that the estimate in (4.32) is actually sharp and that $\mathcal{L}\chi$ lies in $BMO(\mathbb{C})$ and not in $VMO(\mathbb{C})$. We in fact have the following theorem.

Theorem 4.3.2. *The logarithmic potential operator $\mathcal{L}$ acts continuously from $L^1(\mathbb{C})$ into $BMO(\mathbb{C})$ and*

$$\|\mathcal{L}\|_{L^1(\mathbb{C}) \to BMO(\mathbb{C})} = \frac{2}{\pi} \, \|\log|\cdot|\,\|_*, \tag{4.35}$$

while $\mathcal{L}: L_\bullet^1(\mathbb{C}) \to VMO(\mathbb{C})$.

Unfortunately, the integral formula at (4.1) for the extension of the logarithmic potential to the space $L^1(\mathbb{C})$ can no longer be valid. We must consider $\mathcal{L}\varphi$ as an equivalence class of functions modulo the constants. For example, a representative of this class can be chosen by the formula

$$(\mathcal{L}_*\phi)(z) = \frac{2}{\pi} \int_{\mathbb{C}} \phi(\tau) \log \frac{|z-\tau|}{1+|\tau|} \, d\tau$$

This integral converges for every z because of the cancellation of the logarithmic singularity of the integral at $\tau = \infty$. For $\phi \in C_0^\infty$ the formula $(\mathcal{L}_*\phi)$ agrees with $\mathcal{L}\phi$ modulo a constant. This constant, in general, depends on the function ϕ in question.

The identity in (4.4) is fundamental to our studies, and therefore we want to extend it to more general classes of functions. In particular, we would like this identity for $\phi \in L^1(\mathbb{C})$. The following lemma deals with this. It shows that up to a harmonic function (4.4) applies.

Lemma 4.3.3. *Let $f \in L^1_{loc}(\mathbb{C})$. Suppose that*

$$\phi = \frac{\partial^2 f}{\partial z \partial \bar{z}} \in L^1(\mathbb{C})$$

(or $L^1_\bullet(\mathbb{C})$, respectively) in the sense of distributions. Then there exists a harmonic function h defined in $\mathbb{C}$ and unique up to a constant such that $f - h \in BMO(\mathbb{C})$. Therefore

$$\mathcal{L}\phi = \mathcal{L}\frac{\partial^2}{\partial z \partial \bar{z}}(f-h) = f - h$$

lies in $VMO_*(\mathbb{C})$ (or $VMO(\mathbb{C})$, respectively).

Proof. That $\phi = f_{z\bar{z}} \in L^1(\mathbb{C})$ in the sense of distributions means that, for all test functions $\eta \in C_0^\infty(\mathbb{C})$, we have

$$\int_{\mathbb{C}} \eta \, \phi = \int_{\mathbb{C}} f \, \eta_{z\bar{z}}$$

To find a harmonic function h we set

$$h(z) = (f - \mathcal{L}\phi)(z) \in L^1_{loc}(\mathbb{C})$$

Since $\mathcal{L}$ is symmetric, we have, in the sense of distributions, $h_{z\bar{z}} = 0$. Consequently, by Weyl's Lemma A.6.10 we conclude that h is harmonic in the usual sense. Then from Theorem 4.3.2 we have $f - h = \mathcal{L}\phi \in BMO(\mathbb{C})$. For uniqueness we argue as follows. If $f - h_1$ and $f - h_2$ both lie in $BMO(\mathbb{C})$, then $h = h_1 - h_2 \in BMO(\mathbb{C})$, and so we may appeal to the following Lemma 4.3.4 which is basically a Liouville-type result. $\square$

4.3. QUANTITATIVE ANALYSIS

Lemma 4.3.4. *Let $h \in BMO(\mathbb{C})$ be a harmonic function. Then h is constant.*

Proof. We first show that h, as above, is bounded, say
$$|h(z)| \leq |h(0)| + 4\|h\|_*$$
To establish this we consider two large disks $\mathbb{D}(z, R) \subset \mathbb{D}(0, 2R) = B$. The mean value property of harmonic functions tells us that $h(0) = h_B$, independent of the radius R. Then

$$\begin{aligned}|h(z) - h(0)| &= \frac{1}{\pi R^2}\left|\int_{\mathbb{D}(z,R)} (h - h_B)\right| \\ &\leq \frac{4}{4\pi R^2}\int_B |h - h_B| \\ &\leq 4\|h\|_*,\end{aligned}$$

as desired. That bounded harmonic functions are constant is the classical Liouville theorem and can be proved as follows. Consider $z_1, z_2 \in \mathbb{C}$ and for large R set $\mathbb{D}_i = \mathbb{D}(z_i, R)$, $i = 1, 2$. Then

$$\begin{aligned}|h(z_1) - h(z_2)| &= \left|\frac{1}{\pi R^2}\left(\int_{\mathbb{D}_1} h - \int_{\mathbb{D}_2} h\right)\right| \\ &= \frac{1}{\pi R^2}\left|\int_{(\mathbb{D}_1 \cup \mathbb{D}_2)\setminus(\mathbb{D}_1 \cap \mathbb{D}_2)} h\right| \\ &\leq \frac{\|h\|_\infty}{\pi R^2}|(\mathbb{D}_1 \cup \mathbb{D}_2) \setminus (\mathbb{D}_1 \cap \mathbb{D}_2)| \to 0\end{aligned}$$

as $R \to \infty$. Thus h is constant. $\square$

We demonstrate the utility of Lemma 4.3.3 in the computation of the logarithmic potential in the following example. Consider the function

$$f(z) = \log(1 + |z|^2), \quad \frac{\partial^2 f}{\partial z \partial \bar{z}}(z) = \frac{1}{(1 + |z|^2)^2}$$

Since $f_{z\bar{z}} \in L^1(\mathbb{C})$ and as $f \in BMO(\mathbb{C})$, Lemma 4.3.3 tells us that in fact

$$\mathcal{L}\left(\frac{1}{(1 + |z|^2)^2}\right) = \log(1 + |z|^2)$$

up to a constant. Note that $f \notin VMO(\mathbb{C})$.

As an exercise to further the reader's understanding of the matters at hand, we suggest that they try to extend the logarithmic potential $\mathcal{L}$ to the space of Borel measures with range in $BMO(\mathbb{C})$ and with the uniform bound

$$\|\mathcal{L}\nu\|_* \leq \frac{2}{\pi}\|\log|\cdot|\,\|_* \int_\mathbb{C} |d\nu|$$

In this particular case the mean value or moment condition

$$\int_{\mathbb{C}} d\nu = 0$$

will be insufficient to ensure that a function in the range of the operator lies in the space $VMO(\mathbb{C})$. For instance, take Dirac masses at the points $\tau = \pm 1$,

$$\nu = \delta_1 + \delta_{-1},$$

and then make a calculation to show that

$$(\mathcal{L}\nu)(z) = \frac{2}{\pi} \log \frac{|z-1|}{|z+1|},$$

which fails to have vanishing mean oscillation at the two points $z = \pm 1$. The logarithmic potential $\mathcal{L} : \mathcal{M}(\mathbb{C}) \to BMO(\mathbb{C})$ also fails to be continuous with respect to the weak* convergence of measures. Indeed, the sequence $\nu_n = \delta_{1/n}$ converges weak* to $\nu_0 = \delta_0$,

$$\nu[\eta] = \eta(1/n) \to \eta(0) = \nu_0[\eta]$$

for every test function $\eta \in C_0(\hat{\mathbb{C}})$. However,

$$(\mathcal{L}\nu_n)(z) = \frac{2}{\pi} \log \left| z - \frac{1}{n} \right|$$

does not converge to $\frac{2}{\pi} \log |z|$ in the space $BMO(\mathbb{C})$.

The next natural question is to determine for which density functions $\phi \in L^1(\mathbb{C})$ the logarithmic potential $\mathcal{L}\phi$ is bounded. A necessary condition is provided by (4.5), which tells us that at least $\int_{\mathbb{C}} \phi = 0$ must hold. This moment condition therefore suggests that we consider the Hardy space $H^1(\mathbb{C})$.

Theorem 4.3.5. *The logarithmic potential operator $\mathcal{L}$ acts continuously from the Hardy space $H^1(\mathbb{C})$ into $C_0(\hat{\mathbb{C}})$ and*

$$\|\mathcal{L}\phi\|_\infty \leqslant C \|\phi\|_{H^1(\mathbb{C})}, \tag{4.36}$$

where C is an absolute constant.

Proof. We need only show the estimate in (4.36) for functions in $C_0^\infty(\mathbb{C})$ with integral $\int \phi = 0$. We use the BMO-H^1 duality [125] to write

$$|\mathcal{L}\phi(z)| \leqslant C \|\phi\|_{H^1} \|\log |\cdot|\|_* \tag{4.37}$$

for an absolute constant C. The remainder of the proof is immediate, as those functions in $C_0^\infty(\mathbb{C})$ with integral zero are dense in $H^1(\mathbb{C})$. □

The reader who is familiar with Hardy spaces may wish to observe that, using the atomic decomposition [335, p. 101] of functions $\phi \in H^1(\mathbb{C})$,

$$\phi(z) = \sum_{k=1}^\infty \lambda_k a_k(z), \qquad \sum_{k=1}^\infty |\lambda_k| \leqslant C \|\phi\|_{H^1(\mathbb{C})},$$

4.3. QUANTITATIVE ANALYSIS

we can compute $\mathcal{L}\phi$ in a straightforward manner by identifying the logarithmic potential applied to each atom $a_k \in L^1_\bullet(\mathbb{C})$. The sequence $\|\mathcal{L}a_k\|_\infty$ is bounded, and therefore the series

$$\mathcal{L}\phi(z) = \sum_{k=1}^\infty \lambda_k \mathcal{L}a_k(z)$$

converges uniformly in $\hat{\mathbb{C}}$,

$$\|\mathcal{L}\phi\|_\infty \leqslant C \sum_{k=1}^\infty |\lambda_k| \leqslant C\|\phi\|_{H^1(\mathbb{C})}$$

Further regularity and integrability properties of the logarithmic potential acting on $L^p(\mathbb{C})$-spaces can be formulated in terms of its derivatives,

$$\mathcal{C}\phi = \frac{\partial}{\partial z}\mathcal{L}\phi, \qquad \mathcal{C}^*\phi = \frac{\partial}{\partial \bar{z}}\mathcal{L}\phi$$

These are the Cauchy transform and its adjoint. We observe that both of these transforms are dominated by the Riesz potential,

$$|(\mathcal{C}\phi)(z)| + |(\mathcal{C}^*\phi)(z)| \leqslant \frac{2}{\pi}\int_\mathbb{C} \frac{|\phi(\tau)|}{|z-\tau|}\, d\tau$$

This observation will aid our discussion of the $L^p(\mathbb{C})$-theory of the Cauchy transform next.

4.3.2 The Cauchy Transform

As with the logarithmic potential, we begin our discussion in $L^1(\mathbb{C})$. Before attacking the general theory, let us calculate the Cauchy transform of the characteristic function of the disk $\mathbb{D}(a,r)$. Here, note that (4.34) with the definition $\mathcal{C}\phi = \partial_z \mathcal{L}\phi$ gives

$$(\mathcal{C}\chi_{\mathbb{D}(a,r)})(z) = \begin{cases} \overline{z-a}, & |z-a| \leqslant r \\ r^2/(z-a), & |z-a| > r \end{cases}$$

As a particular instance,

$$(\mathcal{C}\chi_\mathbb{D})(z) = \begin{cases} \bar{z}, & |z| \leqslant 1 \\ 1/z, & |z| > 1, \end{cases} \tag{4.38}$$

as we should have already guessed, for according to (4.10) we have $(\mathcal{C}\chi_\mathbb{D})_{\bar{z}} = \chi_\mathbb{D}$ up to a holomorphic function. Notice that $\mathcal{C}$ has mapped our discontinuous function $\chi_{\mathbb{D}(a,r)}$ into a quite nice function, a function at least Lipschitz and bounded.

Of course, $\chi_\mathbb{D} \in L^1$ while $f = \mathcal{C}\chi_\mathbb{D} \notin L^2(\mathbb{C})$. There is, however, a natural function space to which f belongs. This is the Marcinkiewicz class $weak\text{-}L^2(\mathbb{C})$. The precise formulation takes the following form.

Theorem 4.3.6. *The Cauchy transform extends to a bounded linear operator from $L^1(\mathbb{C})$ into the space weak-$L^2(\mathbb{C})$. That is, $\mathcal{C}$ is an operator of weak-type $(1,2)$. Precisely, we have the estimate*

$$|\{z \in \mathbb{C} : |\mathcal{C}\phi(z)| > t\}| \leq \frac{C}{t^2} \|\phi\|_1^2 \qquad (4.39)$$

for every $t > 0$ and $\phi \in L^1(\mathbb{C})$.

This result is the weak-type part of the celebrated Hardy-Littlewood-Sobolev theorem on Riesz potentials, see [335, p. 120].

In particular, the identities $\mathcal{C}f_{\bar{z}} = (\mathcal{C}f)_{\bar{z}} = f$ extend to $f \in W^{1,1}(\mathbb{C})$. For a compactly supported function ϕ, the Cauchy transform of ϕ is going to be analytic near ∞ where its Laurent series expansion takes the form

$$(\mathcal{C}\phi)(z) = \left(\frac{1}{\pi}\int_{\mathbb{C}} \phi(\tau)d\tau\right)\frac{1}{z} + \text{higher powers of } \frac{1}{z}$$

To ensure that $\mathcal{C}\phi$ is going to be square-integrable near ∞, we are forced to assume that $\int_{\mathbb{C}} \phi(\tau)d\tau = 0$. The Hardy space $H^1(\mathbb{C})$ enjoys this moment condition, and it is here that we shall find our $L^2(\mathbb{C})$-bounds.

Theorem 4.3.7. *The Cauchy transform extends continuously to $H^1(\mathbb{C})$ with range in $L^2(\mathbb{C})$, $\mathcal{C} : H^1(\mathbb{C}) \to L^2(\mathbb{C})$, with uniform bounds*

$$\|\mathcal{C}\phi\|_{L^2(\mathbb{C})} \leq C\|\phi\|_{H^1(\mathbb{C})} \qquad (4.40)$$

Proof. Consider $\phi \in C_0^\infty(\mathbb{C})$ with mean value zero, $\int_{\mathbb{C}} \phi(\tau)d\tau = 0$. This class of functions is dense in $H^1(\mathbb{C})$. Given a test function $\eta \in C_0^\infty(\mathbb{C})$, we can write

$$\int_{\mathbb{C}} \eta\, \mathcal{C}\phi = -\int_{\mathbb{C}} \phi\, \mathcal{C}\eta$$
$$\leq C\|\phi\|_{H^1(\mathbb{C})} \cdot \|\mathcal{C}\eta\|_*$$

using the BMO-H^1 duality theorem. It is the term $\|\mathcal{C}\eta\|_*$ that we want to estimate. We shall soon see that $\mathcal{C} : L^2(\mathbb{C}) \to BMO(\mathbb{C})$ is continuous, in Theorem 4.3.9 below. Thus $\|\mathcal{C}\eta\|_* \leq C\|\eta\|_{L^2(\mathbb{C})}$. Since η was an arbitrary test function, Theorem 4.3.7 is proved. □

We now want to look at the case $1 < p < 2$ as well. To do this we need to recall the Hardy-Littlewood-Sobolev theorem on Riesz potentials [335, p. 120]. Accordingly, there is an absolute constant C such that

$$\left\| \int_{\mathbb{C}} \frac{|\phi(\tau)|}{|z-\tau|} d\tau \right\|_{\frac{2p}{2-p}} \leq \frac{C}{(p-1)(2-p)} \|\phi\|_p$$

Then we have the following theorem.

4.3. QUANTITATIVE ANALYSIS

Theorem 4.3.8. *For all $1 < p < 2$, the Cauchy transform is continuous from $L^p(\mathbb{C})$ into $L^{2p/(2-p)}(\mathbb{C})$. Moreover, we have the estimate*

$$\|\mathcal{C}\phi\|_{\frac{2p}{2-p}} \leqslant \frac{C}{(p-1)(2-p)} \|\phi\|_p$$

for some absolute constant C.

Now we examine the Cauchy transform in $L^2(\mathbb{C})$ and obtain the following two very useful results.

Theorem 4.3.9. *The Cauchy transform extends continuously to $L^2(\mathbb{C})$ with range in the space $VMO(\mathbb{C})$, and*

$$\|\mathcal{C}\phi\|_* \leqslant C\|\phi\|_2 \tag{4.41}$$

for some absolute constant $C \leqslant 2$.

Proof. First, we give a meaning to $\mathcal{C}\phi$ as an equivalence class of functions modulo constants. Namely, for $\phi \in L^2(\mathbb{C})$, the following integral converges for almost every $z \in \mathbb{C}$,

$$(\mathcal{C}\phi)(z) = \frac{1}{\pi} \int_\mathbb{C} \left[\frac{1}{z-\tau} + \frac{\chi_{\mathbb{C}\setminus\mathbb{D}}}{\tau}\right] \phi(\tau) d\tau \tag{4.42}$$

We then need to establish the inequality in (4.41) for $\phi \in C_0^\infty(\mathbb{C})$, as these functions are dense in $L^2(\mathbb{C})$. By rescaling the variables z and τ we need only establish the following inequality on the unit disk, as opposed to an arbitrary disk,

$$\frac{1}{\pi} \int_\mathbb{D} |(\mathcal{C}\phi)(z) - (\mathcal{C}\phi)_\mathbb{D}| dz \leqslant 2 \|\phi\|_{L^2(\mathbb{C})}$$

For $\phi \in C_0^\infty(\mathbb{C})$ the integral average of the Cauchy transform $\mathcal{C}\phi$ over the disk, by Fubini's theorem, is equal to

$$\begin{aligned}(\mathcal{C}\phi)_\mathbb{D} &= \frac{1}{\pi} \int_\mathbb{C} \phi(\tau) \left(\frac{1}{\pi} \int_\mathbb{D} \frac{dz}{z-\tau}\right) d\tau \\ &= -\frac{1}{\pi} \int_\mathbb{C} \phi(\tau)(\mathcal{C}\chi_\mathbb{D})(\tau) d\tau.\end{aligned}$$

We now apply the formula in (4.38) for the Cauchy transform of the characteristic function of the disk to see that

$$(\mathcal{C}\phi)_\mathbb{D} = -\frac{1}{\pi} \int_\mathbb{D} \bar{\tau}\phi(\tau) d\tau - \frac{1}{\pi} \int_{\mathbb{C}\setminus\mathbb{D}} \frac{\phi(\tau)}{\tau} d\tau$$

Hence

$$(\mathcal{C}\phi)(z) - (\mathcal{C}\phi)_\mathbb{D} = \frac{1}{\pi} \int_\mathbb{D} \left(\frac{1}{z-\tau} + \bar{\tau}\right) \phi(\tau) d\tau + \frac{1}{\pi} \int_{\mathbb{C}\setminus\mathbb{D}} \left(\frac{1}{z-\tau} + \frac{1}{\tau}\right) \phi(\tau) d\tau$$

The mean oscillation of $\mathcal{C}\phi$ over the unit disk is estimated as

$$\frac{1}{\pi}\int_\mathbb{D} |(\mathcal{C}\phi)(z) - (\mathcal{C}\phi)_\mathbb{D}| \leq \frac{1}{\pi}\int_\mathbb{C} |\phi(\tau)|\,|\psi(\tau)|d\tau$$
$$\leq \frac{1}{\pi}\|\phi\|_{L^2(\mathbb{C})}\|\psi\|_{L^2(\mathbb{C})} \leq 2\|\phi\|_{L^2(\mathbb{C})},$$

where we have defined $\psi(\tau)$ by the following rule: For $|\tau| \leq 1$ we have

$$\psi(\tau) = \frac{1}{\pi}\int_\mathbb{D}\left|\frac{1}{z-\tau}+\bar{\tau}\right|dz \leq |\tau| + \frac{1}{\pi}\int_\mathbb{D}\frac{dz}{|z|} = |\tau| + 2$$

Then for $|\tau| > 1$ we have

$$\psi(\tau) = \frac{1}{\pi}\int_\mathbb{D}\left|\frac{1}{z-\tau}+\frac{1}{\tau}\right|dz = \frac{1}{\pi|\tau|^2}\int_\mathbb{D}\left|\frac{z\tau}{z-\tau}\right|dz \leq \frac{2}{|\tau|^2}$$

We then find that $\|\psi\|^2_{L^2(\mathbb{C})} < 4\pi^2$, as desired. $\square$

With the norm bounds proven, we can now use in quite general settings the basic identities that relate the operators $\mathcal{L}$, $\mathcal{C}$ and $\mathcal{S}$.

Theorem 4.3.10. *Suppose $\phi \in L^2(\mathbb{C})$. Then the weak derivatives*

$$(\mathcal{C}\phi)_{\bar{z}} = \phi, \qquad (\mathcal{C}\phi)_z = \mathcal{S}\phi \qquad (4.43)$$

Proof. Let $\eta \in C_0^\infty(\mathbb{C})$. The element $\mathcal{S}\phi \in L^2(\mathbb{C})$ is uniquely determined via the extension (4.21), and by (4.11) and (4.19) the distributional derivatives satisfy

$$\int_\mathbb{C} \eta\,(\mathcal{C}\phi)_z = -\int_\mathbb{C} \eta_z \mathcal{C}\phi = \int_\mathbb{C} (\mathcal{C}\eta_z)\,\phi = \int_\mathbb{C} (\mathcal{S}\eta)\phi = \int_\mathbb{C} \eta\mathcal{S}\phi,$$

and similarly for the $\partial_{\bar{z}}$-derivative. $\square$

It is also clear that the identities (4.43) hold for functions in $L^p(\mathbb{C})$ once we have extended $\mathcal{S}$ to this space.

In our applications of $\mathcal{C}$ to the problems of existence and uniqueness of quasilinear elliptic systems in the plane, we shall be very much concerned with the continuity of $\mathcal{C}\phi$, for certain given ϕ, on the Riemann sphere $\hat{\mathbb{C}}$. Precisely, we will be looking for those densities ϕ whose Cauchy transform is continuous and vanishes at ∞, symbolically

$$\mathcal{C}\phi \in C_0(\hat{\mathbb{C}}) \qquad (4.44)$$

To ensure the continuity of $\mathcal{C}\phi$ we need at least $\phi \in L^p(\mathbb{C})$ for some $p > 2$. We extend $\mathcal{C}$ to $L^p(\mathbb{C})$ by (4.42). When $2 < p < \infty$, the integral still converges and we have $(\mathcal{C}\phi)_{\bar{z}} = \phi$, modulo an additive constant, whenever $\phi \in L^p(\mathbb{C})$. On

4.3. QUANTITATIVE ANALYSIS

the other hand, this is not enough to guarantee the boundedness of $(\mathcal{C}\phi)(z)$ as $z \to \infty$. Here is an example to illustrate this. Consider the function

$$f(z) = [1 - \chi_{\mathbb{D}}(z)] \log |z|^2$$

and set

$$\phi = f_{\bar{z}} = \overline{f_z} = \frac{1 - \chi_{\mathbb{D}}(z)}{\bar{z}} \in L^p(\mathbb{C}), \qquad p > 2$$

Hence

$$\mathcal{C}\phi(z) = f(z) \notin L^\infty(\mathbb{C}), \quad \text{modulo an additive constant}$$

The failure of the boundedness at ∞ may also be seen directly from the integral formula

$$(\mathcal{C}\phi)(z) = \frac{1}{\pi} \int_{\mathbb{C}} \frac{\phi(\tau)}{z - \tau} \, d\tau \tag{4.45}$$

The point here is that the function $\tau \mapsto (z-\tau)^{-1}$ defined near ∞ fails to belong to the dual space $L^q(\mathbb{C})$ as $q = p/(p-1) < 2$. The way out of this apparently vicious cycle is to assume that $\phi \in L^p(\mathbb{C}) \cap L^q(\mathbb{C})$ for a Hölder conjugate pair (p,q), with $1 < q < 2 < p < \infty$. We shall see later as the theory develops that this is a natural assumption and is satisfied by the solutions to certain elliptic equations of interest to us. The reader may wish to verify that $L^p(\mathbb{C}) \cap L^q(\mathbb{C})$, with (p,q) a Hölder conjugate pair, is a Banach space when equipped with the norm

$$\|\phi\|_p + \|\phi\|_q \tag{4.46}$$

In any case, we can now formulate the result we want.

Theorem 4.3.11. *Let $1 < q < 2 < p < \infty$ be a Hölder conjugate pair. Then the Cauchy transform is a continuous linear mapping from $L^p(\mathbb{C}) \cap L^q(\mathbb{C})$ into the space $C_0(\hat{\mathbb{C}})$. Moreover, we have the estimate*

$$\|\mathcal{C}\phi\|_\infty \leqslant \frac{2}{\sqrt{2-q}} \sqrt{\|\phi\|_p \|\phi\|_q} \leqslant \frac{1}{\sqrt{2-q}} (\|\phi\|_p + \|\phi\|_q)$$

for every $\phi \in L^p(\mathbb{C}) \cap L^q(\mathbb{C})$.

Proof. We split the integral in (4.45) as

$$\mathcal{C}\phi(z) = \frac{1}{\pi} \int_{|z-\tau| \leqslant R} \frac{\phi(\tau)}{z - \tau} \, d\tau + \frac{1}{\pi} \int_{|z-\tau| \geqslant R} \frac{\phi(\tau)}{z - \tau} \, d\tau,$$

where $R > 0$ is to be chosen. Next an application of Hölder's inequality gives

$$|(\mathcal{C}\phi)(z)| \leqslant \frac{1}{\pi} \|\phi\|_p \left(\int_{|z-\tau| \leqslant R} \frac{d\tau}{|z-\tau|^q} \right)^{1/q} + \frac{1}{\pi} \|\phi\|_q \left(\int_{|z-\tau| \geqslant R} \frac{d\tau}{|z-\tau|^p} \right)^{1/p}$$

$$= \frac{1}{\pi} \|\phi\|_p \left(\frac{2\pi R^{2-q}}{2-q} \right)^{1/q} + \frac{1}{\pi} \|\phi\|_q \left(\frac{2\pi R^{2-p}}{p-2} \right)^{1/p}$$

We then choose R so that these two terms are equal, so

$$R^2 = \frac{\|\phi\|_q}{\|\phi\|_p} \left(\frac{2\pi}{p-2}\right)^{1/p} \left(\frac{2-q}{2\pi}\right)^{1/q}$$

This then gives

$$|\mathcal{C}\phi(z)| \leqslant \frac{1}{\pi} \left(\frac{2\pi}{p-2}\right)^{1/2p} \left(\frac{2\pi}{2-q}\right)^{1/2q} \sqrt{\|\phi\|_q \|\phi\|_p}$$

Next, assuming $\frac{1}{p} + \frac{1}{q} = 1$, an elementary computation shows that in fact

$$|\mathcal{C}\phi(z)|^2 \leqslant \frac{4}{2-q} \|\phi\|_p \|\phi\|_q, \tag{4.47}$$

as claimed. Finally, it is clear that $\mathcal{C}\phi(z) = o(1)$ at ∞ for every $\phi \in C_0^\infty(\mathbb{C})$. That this property remains valid for all those $\phi \in L^p(\mathbb{C}) \cap L^q(\mathbb{C})$ is easily seen by an approximation argument. $\square$

When considering functions vanishing near ∞, one may further refine the estimates of the Cauchy transform.

Theorem 4.3.12. *Let $1 < p < \infty$. If $\phi \in L^p(\mathbb{C})$ and $\phi(\tau) = 0$ for $|\tau| \geqslant R$, then*

- $\|\mathcal{C}\phi\|_{L^p(D_{2R})} \leqslant 6R \|\phi\|_p$
- $\|\mathcal{C}\phi(z) - \frac{1}{\pi z}\int \phi\|_{L^p(\mathbb{C} \setminus D_{2R})} \leqslant \frac{2R}{(p-1)^{1/p}} \|\phi\|_p$

Thus, in particular, for $1 < p \leqslant 2$,

$$\|\mathcal{C}\phi\|_{L^p(\mathbb{C})} \leqslant \frac{8R}{(p-1)^{1/p}} \|\phi\|_p \quad \text{provided} \quad \int \phi = 0$$

For $p > 2$ this vanishing condition is not needed, and we have

$$\|\mathcal{C}\phi\|_{L^p(\mathbb{C})} \leqslant \left(6 + 3(p-2)^{-1/p}\right) R \|\phi\|_p, \qquad p > 2$$

Proof. For the first integral we use a truncated Cauchy kernel $K(\xi) = \frac{1}{\pi \xi}\chi_{\mathbb{D}(0,3R)}$, where for $|z| \leqslant 2R$ we have $\mathcal{C}\phi(z) = \int_{\mathbb{C}} K(z-\xi)\phi(\xi) := \mathcal{C}_R\phi(z)$. By Young's inequality for convolutions,

$$\|\mathcal{C}\phi\|_{L^p(D_{2R})} \leqslant \|\mathcal{C}_R\phi\|_{L^p(\mathbb{C})} \leqslant \|K\|_1 \|\phi\|_p = 6R \|\phi\|_p$$

For the second inequality we write, for $|z| > 2R$,

$$\left|\mathcal{C}\phi(z) - \frac{1}{\pi z}\int\phi\right| = \frac{1}{\pi}\left|\int_{\mathbb{C}}\left(\frac{1}{z-\xi} - \frac{1}{z}\right)\phi(\xi)\right|$$
$$\leqslant \frac{2R}{\pi|z|^2}\|\phi\|_1 \leqslant \frac{2R(\pi R^2)^{1/q}}{\pi|z|^2}\|\phi\|_p,$$

4.3. QUANTITATIVE ANALYSIS

from which the result follows by integration. The remaining estimates are straightforward consequences. □

To see how $\mathcal{C}$ operates on the spaces $L^p(\mathbb{C})$, with $2 < p < \infty$, we look at Hölder continuity estimates. To this effect we introduce the Hölder spaces $C^\alpha(\mathbb{C})$, $0 < \alpha \leqslant 1$, of continuous functions $\eta : \mathbb{C} \to \mathbb{C}$ that satisfy the Hölder condition

$$\|\eta\|_{C^\alpha(\mathbb{C})} = \sup_{z \neq w} \frac{|\eta(z) - \eta(w)|}{|z - w|^\alpha} < \infty \qquad (4.48)$$

We should emphasize here that $C^\alpha(\mathbb{C})$ is, modulo constant functions, a Banach space. We also point out that smooth functions are not dense in this space, however, given $F \in C^\alpha(\mathbb{C})$, our standard mollification procedure (see the Appendix) gives a sequence of smooth functions that remain bounded in $C^\alpha(\mathbb{C})$ and converge to F as continuous functions. Precisely, we have

$$\begin{aligned} F_\varepsilon(z) &= (F * \Phi_\varepsilon)(z) = \int_\mathbb{C} F(\tau)\Phi_\varepsilon(z - \tau)\, d\tau \\ &= \int_\mathbb{C} F(z - \tau)\Phi_\varepsilon(\tau)\, d\tau, \end{aligned}$$

which shows us that in fact

$$\|F_\varepsilon\|_{C^\alpha} \leqslant \|F\|_{C^\alpha}, \quad \text{for all } \varepsilon > 0, \qquad (4.49)$$

and moreover,

$$\begin{aligned} |F_\varepsilon(z) - F(z)| &\leqslant \|F\|_{C^\alpha} \int_\mathbb{C} |\tau|^\alpha \Phi_\varepsilon(z - \tau)\, d\tau \\ &\leqslant \varepsilon^\alpha \|F\|_{C^\alpha}, \end{aligned}$$

which establishes the uniform convergence $F_\varepsilon \to F$ as $\varepsilon \to 0$.

A straightforward application of Hölder's inequality leads us to the following result.

Theorem 4.3.13. *The Cauchy transform acts continuously from the space $L^p(\mathbb{C})$, $2 < p < \infty$, into the space $C^\alpha(\mathbb{C})$, $\alpha = 1 - \frac{2}{p}$. We have the estimate*

$$\|\mathcal{C}\phi\|_{C^\alpha(\mathbb{C})} \leqslant C_p \|\phi\|_p, \qquad (4.50)$$

where the constant $C_p \leqslant \frac{12\, p^2}{p-2}$.

Proof. At points $z \neq w$ we calculate by Hölder's inequality

$$\begin{aligned} \frac{|\mathcal{C}\phi(z) - \mathcal{C}\phi(w)|}{|z - w|^{1-2/p}} &= \frac{1}{\pi} \frac{1}{|z - w|^{1-2/p}} \left| \int_\mathbb{C} \left(\frac{\phi(\tau)}{z - \tau} - \frac{\phi(\tau)}{w - \tau} \right) d\tau \right| \\ &= \frac{1}{\pi} |z - w|^{2/p} \left| \int_\mathbb{C} \frac{\phi(\tau)\, d\tau}{(z - \tau)(w - \tau)} \right| \end{aligned}$$

$$\leqslant \frac{1}{\pi}\|\phi\|_p \left| \int_{\mathbb{C}} \frac{|z-w|^{2/(p-1)} d\tau}{|z-\tau|^{p/(p-1)}|w-\tau|^{p/(p-1)}} \right|^{(p-1)/p}$$

$$= \frac{1}{\pi}\|\phi\|_p \left| \int_{\mathbb{C}} \frac{d\tau}{|\tau|^{p/(p-1)}|1-\tau|^{p/(p-1)}} \right|^{(p-1)/p}$$

As $p/(p-1) < 2$, the integral converges at the singularities 0 and 1, while near ∞ the integrand is $\mathcal{O}(1/\tau^{2p/(p-1)})$ and so again converges. Thus it is clear that the sharp constant

$$C_p = \frac{1}{\pi} \left(\int_{\mathbb{C}} \frac{d\tau}{|\tau|^{p/(p-1)}|\tau-1|^{p/(p-1)}} \right)^{(p-1)/p} \leqslant \frac{12\, p^2}{p-2} \qquad (4.51)$$

is finite. We leave it to the reader to confirm the upper bound of (4.51), which we shall not use again. □

Here is an example that illustrates what happens when $p = \infty$. Let us consider the image of $L^\infty(\mathbb{C})$ under the Cauchy transform. Put

$$f(z) = z \log |z|^2 \chi_{\mathbb{D}} \qquad (4.52)$$

Then $\omega = f_{\bar{z}} = \frac{z}{\bar{z}} \chi_{\mathbb{D}} \in L^\infty(\mathbb{C})$ and $\mathcal{S}\omega = f_z = (1 + \log|z|^2)\chi_{\mathbb{D}} \in BMO(\mathbb{C})$. But f_z is not bounded. Therefore $f = \mathcal{C}\omega$ is not Lipschitz. Hence the Cauchy transform $\mathcal{C}$ does not take $L^\infty(\mathbb{C})$ to the space $C^1(\mathbb{C})$ or even to the space of Lipschitz continuous functions. We mention here only that the precise characterization of those functions that lie in the image of $L^\infty(\mathbb{C})$ under the Cauchy transform was given by Reimann [309].

In finding solutions to nonlinear differential equations, fixed-point theorems are particularly useful. In infinite dimensions these results, such as Schauder's fixed-point theorem, often require compactness of the mappings in question. If X and Y are Banach spaces and $U \subset X$, recall that a (not necessarily linear) mapping $T : U \to Y$ is called compact if the image $T(A)$ is relatively compact for any bounded $A \subset U$.

In describing the compactness properties of the Cauchy transform in the theorem below, we shall view the characteristic function χ_Ω of a set $\Omega \subset \mathbb{C}$ both as a function and as a multiplier operator $f \mapsto \chi_\Omega f$.

Theorem 4.3.14. *Let Ω be a bounded measurable subset of $\mathbb{C}$. Then the following operators are compact.*

- *For $2 < p \leqslant \infty$,*

$$\chi_\Omega \circ \mathcal{C} : L^p(\mathbb{C}) \to C^\alpha(\Omega), \qquad 0 \leqslant \alpha < 1 - \frac{2}{p}$$

- *For $1 \leqslant p \leqslant 2$,*

$$\chi_\Omega \circ \mathcal{C} : L^p(\mathbb{C}) \to L^s(\mathbb{C}), \qquad 1 \leqslant s < \frac{2p}{2-p}$$

4.4 Maximal Functions and Interpolation

The extensions of the logarithmic potential and the Cauchy transform to the L^p-setting was achieved by relatively straightforward estimates. The L^p-bounds for the singular integrals such as the Beurling transform require, however, sharper tools. The methods of interpolation and maximal functions are indispensable. Therefore in this section we take a quick look at these ideas.

4.4.1 Interpolation

We start with interpolation and give an elementary proof for the Riesz-Thorin interpolation theorem. Usually the result is based on complex analytic arguments. Here we avoid these and present a proof based on two simple well-known facts. The first is the density of simple functions, and the second is Hölder's inequality. There is no gain in simplicity by restricting ourselves to the setting in the complex plane, and therefore we work in complete generality.

Thus let (Ω, μ) be a sigma-finite measure space. We denote by $L^p(\Omega, \mu)$, $1 \leqslant p < \infty$, the Banach space of functions valued in a finite-dimensional inner-product space $(\mathbb{V}, \langle, \rangle)$ such that

$$\|f\|_p = \left(\int_\Omega |f|^p d\mu \right)^{1/p} < \infty$$

We have now the fundamental Riesz-Thorin theorem.

Theorem 4.4.1. *Fix exponents p_1 and p_2 with $1 \leqslant p_1 \leqslant p_2 < \infty$. Suppose that*

$$\mathcal{T} : L^{p_1}(\Omega, \mu) \cap L^{p_2}(\Omega, \mu) \to L^{p_1}(\Omega, \mu) \cap L^{p_2}(\Omega, \mu)$$

is a linear operator such that

$$\|\mathcal{T}\phi\|_{p_1} \leqslant \|\mathcal{T}\|_{p_1} \|\phi\|_{p_1} \tag{4.53}$$

$$\|\mathcal{T}\phi\|_{p_2} \leqslant \|\mathcal{T}\|_{p_2} \|\phi\|_{p_2} \tag{4.54}$$

for every $\phi \in L^{p_1}(\Omega, \mu) \cap L^{p_2}(\Omega, \mu)$. Then for every p, $p_1 \leqslant p \leqslant p_2$, $\mathcal{T}$ extends as a bounded linear operator from $L^p(\Omega, \mu)$ into itself. Moreover, for every $f \in L^p(\Omega, \mu)$, we have the uniform estimate

$$\|\mathcal{T}f\|_p \leqslant \|\mathcal{T}\|_{p_1}^\alpha \|\mathcal{T}\|_{p_2}^\beta \|f\|_p, \tag{4.55}$$

where α and β are determined from the relations

$$\frac{1}{p} = \frac{\alpha}{p_1} + \frac{\beta}{p_2}, \qquad \alpha + \beta = 1 \tag{4.56}$$

Proof. In order to avoid the rather delicate question concerning the existence of extremal functions for the p-norm of $\mathcal{T}$, we shall proceed as follows. We decompose Ω into a finite number N of mutually disjoint μ-measurable sets, say $\Omega = \bigcup_{k=1}^{N} \Omega_k$. For each such decomposition, there is a finite-dimensional subspace $\mathcal{L}(\Omega, \mu) \subset L^{p_1}(\Omega, \mu) \cap L^{p_2}(\Omega, \mu) \subset L^p(\Omega, \mu)$ of simple functions $\phi = \sum_{k=1}^{N} \alpha_k \chi_{\Omega_k}$ where $\alpha_k \in \mathbb{V}$ and assumed equal to 0 if $\mu(\Omega_k) = \infty$. The union of all such finite dimensional subspaces is dense in $L^p(\Omega, \mu)$. As a function uniformly continuous on a dense subset extends continuously, we need only establish the uniform norm bound at (4.55) for such functions $\phi \in \mathcal{L}(\Omega, \mu)$. Of course, Hölder's inequality shows that

$$\|\phi\|_p \leqslant \|\phi\|_{p_1}^{\alpha} \|\phi\|_{p_2}^{\beta} \tag{4.57}$$

with α and β as above. Our proof will exploit this inequality.

For $p \in [p_1, p_2]$ let us set

$$A_p = \max \left\{ \frac{\|\mathcal{T}\phi\|_p}{\|\phi\|_p} : \phi \in \mathcal{L}(\Omega, \mu) \right\} \tag{4.58}$$

The maximum A_p is achieved, since by homogeneity we may restrict ourselves to the coefficients normalized by $\sum_{k=1}^{N} |\alpha_k|^2 = 1$ and as $\|\mathcal{T}\phi\|_p^p = \int_{\Omega} |\sum_{k=1}^{N} \alpha_k g_k|^p$ for the finite collection of functions $g_k = \mathcal{T}\chi_{\Omega_k} \in L^{p_1}(\Omega, \mu) \cap L^{p_2}(\Omega, \mu)$. In fact, A_p is a well-defined continuous function of $p \in [p_1, p_2]$.

For the given p assume the maximum in (4.58) is attained for, say, $f \in \mathcal{L}(\Omega, \mu)$. Then

$$\|\mathcal{T}f\|_p = A_p \|f\|_p, \tag{4.59}$$

while

$$A_p^p \int_{\Omega} |f + th|^p d\mu - \int_{\Omega} |\mathcal{T}f + t\mathcal{T}h|^p d\mu \geqslant 0 \tag{4.60}$$

for every $h \in \mathcal{L}(\Omega, \mu)$ and $t \in \mathbb{R}$. The left hand side of (4.60), for $p > 1$, is a continuously differentiable function of the real variable t which assumes its minimum at $t = 0$. Its derivative must vanish at $t = 0$, and therefore

$$A_p^p \int_{\Omega} \langle |f|^{p-2} f, h \rangle d\mu = \int_{\Omega} \langle |\mathcal{T}f|^{p-2} \mathcal{T}f, \mathcal{T}h \rangle d\mu$$

We apply this identity to the function

$$h = |f|^{\frac{p-r}{r-1}} f \in \mathcal{L}(\Omega, \mu),$$

where we treat $r \in [p_1, p]$ with $r(p-1)/(r-1) \leqslant p_2$ as a free parameter. At this point we apply Hölder's inequality,

$$\begin{aligned} A_p^p \int_{\Omega} |f|^{\frac{pr-r}{r-1}} &\leqslant \|\mathcal{T}h\|_r \big\| |\mathcal{T}f|^{p-1} \big\|_{\frac{r}{r-1}} = \|\mathcal{T}h\|_r \|\mathcal{T}f\|_{\frac{pr-r}{r-1}}^{p-1} \\ &\leqslant A_r \|h\|_r A_{\frac{pr-r}{r-1}}^{p-1} \|f\|_{\frac{pr-r}{r-1}}^{p-1} \\ &= A_r A_{\frac{pr-r}{r-1}}^{p-1} \|f\|_{\frac{pr-r}{r-1}}^{\frac{pr-r}{r-1}}, \end{aligned}$$

4.4. MAXIMAL FUNCTIONS AND INTERPOLATION

from which we deduce

$$A_p \leqslant A_r^{1/p} A_{\frac{pr-r}{r-1}}^{1-1/p} \qquad (4.61)$$

This is already the inequality (4.55) in the special case $\frac{1}{p} = \frac{\alpha}{r} + \frac{\beta(r-1)}{pr-r}$ with $\alpha = 1/p$ and $\beta = 1 - 1/p$. The remainder of the proof is a simple matter of analysis of the estimate (4.61). Let us consider the set $I \subset [p_1, p_2]$ consisting of those p such that

$$A_p \leqslant A_{p_1}^\alpha A_{p_2}^\beta; \qquad \frac{\alpha}{p_1} + \frac{\beta}{p_2} = \frac{1}{p}, \text{ and } \alpha + \beta = 1$$

Of course, $p_1, p_2 \in I$, while inequality (4.61) tells us that if $a, b \in I$, then $c = b\left(1 - \frac{1}{a}\right) + 1 \in I$ as well. To see this we write $b = (c-1)a/(a-1)$ and use (4.61) to estimate

$$A_c \leqslant A_a^{1/c} A_b^{1-1/c} \leqslant \left(A_{p_1}^{\alpha_a} A_{p_2}^{\beta_a}\right)^{1/c} \left(A_{p_1}^{\alpha_b} A_{p_2}^{\beta_b}\right)^{1-1/c} = A_{p_1}^{\alpha_c} A_{p_2}^{\beta_c},$$

where $\alpha_c + \beta_c = 1$ and $\frac{1}{c} = \frac{\alpha_c}{p_1} + \frac{\beta_c}{p_2}$. This property of the set I implies that it is dense in the interval $[p_1, p_2]$, and finally the continuity of $p \mapsto A_p$ implies $I = [p_1, p_2]$. $\square$

The estimate $\|T\|_p \leqslant \|T\|_{p_1}^\alpha \|T\|_{p_2}^\beta$ of Theorem 4.4.1 says that $t \mapsto \log \|T\|_{1/t}$ is convex on the interval $t \in [\frac{1}{p_2}, \frac{1}{p_1}]$. In particular, we have the following corollary.

Corollary 4.4.2. *Under the assumptions given in Theorem 4.4.1, the function $t \mapsto \|T\|_t$ is (locally) Lipschitz-continuous on (p_1, p_2).*

Basic bounds for singular integral operators are typically found from and based on weak-type estimates. Therefore we need to complement the Riesz-Thorin theorem with Marcinkiewicz interpolation.

If $1 \leqslant p_1 < p_2 < \infty$, we denote by $L^{p_1}(\Omega, \mu) + L^{p_2}(\Omega, \mu)$ the space of functions which can be split as $f = f_1 + f_2$, where $f_j \in L^{p_j}(\Omega, \mu)$, $j = 1, 2$. An operator T from $L^{p_1}(\Omega, \mu) + L^{p_2}(\Omega, \mu)$ to the space of measurable functions on Ω is said to be sub-additive if $|T(f+g)| \leqslant |Tf| + |Tg|$ pointwise.

Theorem 4.4.3. *Let T be a subadditive operator on $L^{p_1}(\Omega, \mu) + L^{p_2}(\Omega, \mu)$ such that*

$$\mu(\{z \in \Omega : |Tg(z)| > t\}) \leqslant \left(\frac{A_j}{t} \|g\|_{L^{p_j}(\Omega, \mu)}\right)^{p_j} \qquad (4.62)$$

holds for every $g \in L^{p_j}(\Omega, \mu)$, $j = 1, 2$. Then for all $p_1 < p < p_2$,

$$\|Tf\|_{L^p(\Omega,\mu)} \leqslant A_p \|f\|_{L^p(\Omega,\mu)}, \qquad f \in L^p(\Omega, \mu), \qquad (4.63)$$

where the constant A_p depends only on p, A_j and p_j, $j = 1, 2$.

Of course, in general, the constant A_p blows up at the endpoints p_1 and p_2. The proof below gives the following bound for the p-norm,

$$\|\mathcal{T}\|_p \leqslant A_p \leqslant \frac{pA_1^{p_1}}{\alpha_1^{p_1}(p-p_1)} + \frac{pA_2^{p_2}}{\alpha_2^{p_2}(p_2-p)}, \tag{4.64}$$

where α_j are any positive numbers with $\alpha_1 + \alpha_2 = 1$.

Proof. Fix $f \in L^p(\Omega, \mu)$ and let $\lambda(t) = \mu(\{z \in \Omega : |\mathcal{T}f(z)| > t\})$ be the distribution function of $\mathcal{T}f$. Fix positive numbers α_1, α_2 such that $\alpha_1 + \alpha_2 = 1$. For each $0 < t < \infty$, we split $f = f_1 + f_2$, where

$$f_1(z) = \begin{cases} f(z) & \text{if } |f(z)| > t \\ 0 & \text{if } |f(z)| < t \end{cases} \quad \text{and} \quad f_2(z) = \begin{cases} 0 & \text{if } |f(z)| > t \\ f(z) & \text{if } |f(z)| < t \end{cases}$$

Since $|\mathcal{T}f| \leqslant |\mathcal{T}f_1| + |\mathcal{T}f_2|$, the assumption (4.62) gives

$$\begin{aligned}
\lambda(t) &\leqslant \mu(\{z \in \Omega : |\mathcal{T}f_1(z)| > \alpha_1 t\}) + \mu(\{z \in \Omega : |\mathcal{T}f_2(z)| > \alpha_2 t\}) \\
&\leqslant \frac{A_1^{p_1}}{\alpha_1^{p_1} t^{p_1}} \int |f_1|^{p_1} \, d\mu + \frac{A_2^{p_2}}{\alpha_2^{p_2} t^{p_2}} \int |f_2|^{p_2} \, d\mu \\
&= \frac{A_1^{p_1}}{\alpha_1^{p_1} t^{p_1}} \int_{\{|f|>t\}} |f|^{p_1} \, d\mu + \frac{A_2^{p_2}}{\alpha_2^{p_2} t^{p_2}} \int_{\{|f|<t\}} |f|^{p_2} \, d\mu
\end{aligned}$$

We can now integrate the distribution function and apply Fubini's theorem,

$$\begin{aligned}
\int_\Omega |\mathcal{T}f|^p \, d\mu &= p \int_0^\infty t^{p-1} \lambda(t) \, dt \\
&\leqslant \frac{pA_1^{p_1}}{\alpha_1^{p_1}} \int_0^\infty t^{p-p_1-1} \int_{\{|f|>t\}} |f|^{p_1} + \frac{pA_2^{p_2}}{\alpha_2^{p_2}} \int_0^\infty t^{p-p_2-1} \int_{\{|f|<t\}} |f|^{p_2} \\
&= \frac{pA_1^{p_1}}{\alpha_1^{p_1}} \int_\Omega |f|^{p_1} \int_0^{|f|} t^{p-p_1-1} + \frac{pA_2^{p_2}}{\alpha_2^{p_2}} \int_\Omega |f|^{p_2} \int_{|f|}^\infty t^{p-p_2-1}
\end{aligned}$$

Unwinding the integral gives (4.63), with the bound (4.64). $\square$

4.4.2 Maximal Functions

Maximal functions have found an important role in the analysis of singular integral operators. Indeed the theories of these two analytic objects are deeply intertwined. Various different maximal functions have been developed to study singular operators defined on different spaces.

4.4. MAXIMAL FUNCTIONS AND INTERPOLATION

We collect in this subsection a few basic notions and results from the theory of maximal functions and begin by introducing the following definition.

The Hardy-Littlewood Maximal Function

Definition 4.4.4. *For $f \in L^1_{loc}(\mathbb{C})$ the Hardy-Littlewood maximal function is defined by the rule*

$$(\mathcal{M}f)(z) = \sup\left\{\frac{1}{|Q|}\int_Q |f| : z \in Q\right\}$$

where the supremum is to be considered over all squares Q containing the point z. More generally, we have for $f \in L^r_{loc}$ the r-maximal functions

$$(\mathcal{M}_r f)(z) = \sup\left\{\left(\frac{1}{|Q|}\int_Q |f|^r\right)^{1/r} : z \in Q\right\}, \quad 0 < r < \infty \quad (4.65)$$

Sometimes it is convenient to use disks instead of squares in the definition of the maximal operator. Of course, this gives an equivalent notion, up to a multiplicative constant.

Similarly, we define the Hardy-Littlewood maximal function for every measure $\mu \in \mathcal{M}(\mathbb{C})$,

$$(\mathcal{M}\mu)(z) = \sup\left\{\frac{1}{|Q|}\int_Q d\mu : z \in Q\right\}$$

The most important fact we shall need concerning the Hardy-Littlewood maximal function is the following basic weak-type estimate; see [335].

Theorem 4.4.5. *Let $\mu \in \mathcal{M}(\mathbb{C})$. Then*

$$|\{z : (\mathcal{M}\mu)(z) > \alpha\}| \leq \frac{9}{\alpha}\|\mu\|$$

In particular, the above theorem holds for the measures $d\mu = f dz$, where $f \in L^1(\mathbb{C})$. Since the maximal function is clearly bounded for every $f \in L^\infty(\mathbb{C})$, we may modify the proof of Theorem 4.4.3 to obtain the following theorem.

Theorem 4.4.6. *For each $1 < p \leq \infty$, there is a constant C_p, depending only on p, such that*

$$\|\mathcal{M}f\|_{L^p(\mathbb{C})} \leq C_p \|f\|_{L^p(\mathbb{C})}, \quad f \in L^p(\mathbb{C}) \quad (4.66)$$

We now introduce another useful maximal function;

The Spherical Maximal Function

$$\phi^\flat(z) = \sup_{r>0}\left\{\left(\frac{1}{2\pi r}\int_{\partial B}|\phi(\tau)|^2 d\tau\right)^{1/2} : B = \mathbb{D}(z,r)\right\}$$

According to Bourgain's theorem [73], the spherical maximal operator $\phi \mapsto \phi^\flat$ acts continuously from $L^p(\mathbb{C}) \to L^p(\mathbb{C})$ for all $p > 2$. For $p = 2$, we have $\phi^\flat \equiv \infty$ for some function $\phi \in L^2(\mathbb{C})$.

The reader should be aware that replacing disks by squares in this definition of the spherical maximal function would, surprisingly, destroy all the boundedness properties that we will need.

Fefferman-Stein Sharp Maximal Function

The Fefferman-Stein *sharp* maximal function is defined for $\phi \in L^2_{loc}(\mathbb{C})$ by the rule

$$\phi^\#(z) = \sup_{r>0} \left\{ \left(\frac{1}{|B|} \int_B |\phi(\tau) - \phi_B|^2 d\tau \right)^{1/2} : B = \mathbb{D}(z,r) \right\}$$

We will later need the fact that the p-norm of the sharp maximal function dominates the p-norm of the Hardy-Littlewood maximal function, [335, p. 148]. More precisely, we have the following theorem.

Theorem 4.4.7. *Suppose that $\phi \in L^r(\mathbb{C})$ for some $1 < r < \infty$ and that $\phi^\# \in L^p(\mathbb{C})$, $1 < p < \infty$. Then $\phi \in L^p(\mathbb{C})$ and*

$$\|\phi\|_p \leqslant C_p \|\phi^\#\|_p$$

Associated with the sharp maximal function is the BMO_2 norm,

$$\|\phi\|_{BMO_2} = \|\phi^\#\|_\infty \qquad (4.67)$$

This is a norm in the space $BMO(\mathbb{C})$, equivalent to the standard BMO-norm $\|u\|_*$. The equivalence follows from the John-Nirenberg inequality and (4.94).

The sharp maximal function makes transparent many arguments related to BMO. For instance, we have the following simple but useful principle which we will need later.

Lemma 4.4.8. *Suppose F is a Lipschitz function on $\mathbb{R}$, with Lipschitz constant L. Then for $\phi \in L^1_{loc}(\mathbb{C})$,*

$$[F(\phi)]^\#(z) \leqslant L \phi^\#(z)$$

In particular, if $\phi \in BMO(\mathbb{C})$, then $F(\phi) \in BMO(\mathbb{C})$ with

$$\|F(\phi)\|_{BMO_2} \leqslant L \|\phi\|_{BMO_2}$$

Proof. For a disk $B = \mathbb{D}(z,r)$,

$$\frac{1}{|\mathbb{D}|} \int_\mathbb{D} |F(\phi) - [F \circ \phi]_\mathbb{D}|^2 d\tau \leqslant \frac{1}{|\mathbb{D}|} \int_\mathbb{D} |F(\phi) - F(\phi_\mathbb{D})|^2 d\tau \leqslant \left[L \phi^\#(z) \right]^2,$$

4.4. MAXIMAL FUNCTIONS AND INTERPOLATION

from which the claim follows. $\square$

Maximal operators have delicate smoothing properties. As an example, consider the Hardy-Littlewood maximal function of an L^1-function or even of a measure μ. There we have the well-known Coifman-Rochberg theorem [96].

Theorem 4.4.9. *Let μ be a Borel measure in $\mathbb{C}$ such that its Hardy-Littlewood maximal function $\mathcal{M}(\mu)$ is finite at a single point (and therefore at every point). Then $\log \mathcal{M}(\mu) \in BMO(\mathbb{C})$, and its norm is bounded by an absolute constant,*

$$\|\log \mathcal{M}(\mu)\|_* \leqslant C \tag{4.68}$$

Proof. We will actually show that the function $\omega = \sqrt{\mathcal{M}(\mu)}$ belongs to the Muckenhoupt class of A_1 weights. More precisely, for an absolute constant $C < \infty$, we have

$$\frac{1}{|Q|} \int_Q \omega \leqslant C \inf_Q \omega \quad \text{for every square } Q \subset \mathbb{C} \tag{4.69}$$

From this the claim readily follows. To prove the A_1 estimate, split μ into two measures

$$\mu_1 = \chi_{2Q}\mu \quad \text{and} \quad \mu_2 = (1 - \chi_{2Q})\mu$$

Thus

$$\int_Q \sqrt{\mathcal{M}(\mu)} \leqslant \int_Q \sqrt{\mathcal{M}(\mu_1)} + \int_Q \sqrt{\mathcal{M}(\mu_2)} \tag{4.70}$$

For the first integral we make use of Kolmogorov's inequality, showing that $\mathcal{M}(\mu_1) \in L^p(\mathbb{C})$ for every $0 < p < 1$. This is obtained simply by integrating the distribution bound (4.66),

$$\int_Q \sqrt{\mathcal{M}(\mu_1)} = \int_0^\infty \frac{1}{2} t^{-1/2} |\{z \in Q : |\mathcal{M}(\mu_1)| > t\}| dt$$

$$\leqslant \frac{1}{2}|Q| \int_0^A t^{-1/2} dt + 9\|\mu_1\| \int_A^\infty \frac{1}{2} t^{-1/2-1} \leqslant 6\sqrt{\|\mu_1\| |Q|}$$

when we choose $A = 9\|\mu_1\|/|Q|$. Therefore

$$\frac{1}{|Q|} \int_Q \sqrt{\mathcal{M}(\mu_1)} \leqslant 6\sqrt{\mathcal{M}(\mu)(\xi)}, \qquad \xi \in 2Q \tag{4.71}$$

For the second integral in (4.70) we use the following observation,

$$\sup_Q \mathcal{M}(\mu_2)(z) \leqslant 32 \inf_Q \mathcal{M}(\mu) \tag{4.72}$$

Indeed, given $z \in Q$, choose a square $Q = Q_z$ containing z, so that $\mathcal{M}(\mu_2)(z) \leqslant 2\frac{\mu_2(Q_z)}{|Q_z|}$. Then Q_z must intersect the complement of $2Q$, as otherwise $\mu_2(Q_z) = 0$. But since also $Q \cap Q_z$ is nonempty, the square is so large that $2Q \subset 4Q_z$. In particular,

$$\mathcal{M}(\mu_2)(z) \leqslant 32 \frac{\mu_2(4Q_z)}{|4Q_z|} \leqslant 32 \inf_{4Q_z} \mathcal{M}(\mu_2) \leqslant 32 \inf_Q \mathcal{M}(\mu_2),$$

proving (4.72).

Now substituting (4.71) and (4.72) into (4.70) gives

$$\frac{1}{|Q|}\int_Q \sqrt{\mathcal{M}(\mu)} \leqslant (6+4\sqrt{2})\inf_Q \sqrt{\mathcal{M}(\mu)}$$

Having established that $\omega = \sqrt{\mathcal{M}(\mu)} \in A_1$, the theorem follows quickly. Given a square $Q \subset \mathbb{C}$, we apply Jensen's inequality to write

$$\frac{1}{|Q|}\int_Q |2\log\omega - (2\log\omega)_Q| \leqslant \frac{4}{|Q|}\int_Q |\log\omega - \inf_Q \log\omega| = \frac{4}{|Q|}\int_Q [\log\omega - \inf_Q \log\omega]$$

$$\leqslant 4\log\left(\frac{1}{|Q|}\int_Q \omega\right) - 4\inf_Q \log\omega \leqslant 4\log(6+4\sqrt{2})$$

The claim (4.68) is now proved, with the constant $C = 4\log(6+4\sqrt{2})$. □

4.5 Weak-Type Estimates and L^p-Bounds

We have already calculated in (4.25) the action of the Beurling transform on the characteristic function $\chi_{\mathbb{D}(a,r)}(z)$ of the disk $\mathbb{D}(a,r)$, obtaining

$$\mathcal{S}\chi_{\mathbb{D}(a,r)} = \begin{cases} 0, & |z-a| \leqslant r \\ -r^2/(z-a)^2, & |z-a| > r \end{cases}$$

The formula shows that in fact $\mathcal{S}\chi_\mathbb{D}$ is not an element of $L^1(\mathbb{C})$. However, $\mathcal{S}\chi_\mathbb{D}$ does lie in the Marcinkiewicz space weak-$L^1(\mathbb{C})$. This is an example of a general phenomenon, that the Beurling transform takes functions in $L^1(\mathbb{C})$ to the space weak-$L^1(\mathbb{C})$.

In the sequel it will be important to establish that the $L^p(\mathbb{C})$ norms of the kth iterate of the complex Riesz transform do not grow as a power but rather that the growth is a polynomial in k. It is not at all clear what the precise asymptotics here are. But we prove bounds for $\|R^k\|_p$ that are linear in k^2. This will be most efficiently achieved in the context of weak-$L^1(\mathbb{C})$. Therefore we now seek to give a unified treatment of these operators.

4.5.1 Weak-Type Estimates for $\mathcal{S}$ and Complex Riesz Transforms

Let $f \in L^1(\mathbb{C})$ and $\alpha > 0$. Let us put

$$\Omega = \{z : (\mathcal{M}f)(z) > \alpha\}$$

Then Theorem 4.4.5 implies that Ω is an open set of measure

$$|\Omega| \leqslant \frac{9}{\alpha}\|f\|_1 \qquad (4.73)$$

4.5. L^p-BOUNDS

The complement of Ω is a closed set $F = \{z : (\mathcal{M}f)(z) \leqslant \alpha\}$. In particular,
$$|f(z)| \leqslant \alpha, \qquad z \in F \tag{4.74}$$

Following Stein [335], we consider a Whitney-type decomposition of Ω,
$$\Omega = \bigcup_j Q_j,$$

where Q_j are mutually disjoint cubes such that
$$\operatorname{diam}(Q_j) \leqslant \operatorname{dist}(F, Q_j) \leqslant 4\operatorname{diam}(Q_j) \tag{4.75}$$

Note that as $Q_j \subset \Omega$ directly from the definition of the maximal function, we have for all the squares Q_j,
$$\frac{1}{|Q_j|} \int_{Q_j} |f| \leqslant \alpha \tag{4.76}$$

L^1-Estimate of the Kernel of R^k

For convenience we recall the definition of the complex Riesz transforms R^k, $k = \pm 1, \pm 2, \ldots$,
$$(R^k f)(z) = \int_{\mathbb{C}} K(z - \tau) f(\tau) \, d\tau, \tag{4.77}$$

where the kernel K is given by
$$K(z) = \frac{|k|}{2\pi i^{|k|}} \frac{\bar{z}^k}{|z|^{k+2}} \tag{4.78}$$

We compute that
$$\begin{aligned}
|K_z| + |K_{\bar{z}}| &= \frac{|k|}{2\pi}(|1 - \frac{k}{2}| + |1 + \frac{k}{2}|)|z|^{-3} \\
&= \frac{1}{2\pi} \max\{2, k^2\} |z|^{-3}
\end{aligned} \tag{4.79}$$

Now, for $|z| > |\tau|$, we estimate
$$\begin{aligned}
|K(z - \tau) - K(z)| &= \left| \int_0^1 \frac{d}{dt} K(z - t\tau) dt \right| \\
&\leqslant |\tau| \int_0^1 (|K_z(z - t\tau)| + |K_{\bar{z}}(z - t\tau)|) \, dt \\
&= \frac{\max\{2, k^2\}}{2\pi} |\tau| \int_0^1 \frac{dt}{|z - t\tau|^3} \\
&\leqslant \max\{2, k^2\} \frac{1}{4\pi} \left[\frac{1}{(|z| - |\tau|)^2} - \frac{1}{|z|^2} \right]
\end{aligned}$$

If we now integrate this over the set $|z| > 2|\tau|$, we find

$$\int_{|z|>2|\tau|} |K(z-\tau) - K(z)|dz \leqslant \frac{\max\{2, k^2\}}{4\pi} \int_{|z|>2|\tau|} \left[\frac{1}{(|z|-|\tau|)^2} - \frac{1}{|z|^2}\right] dz$$
$$\leqslant \max\{2, k^2\},$$

which, after a simple substitution yields

$$\int_{|z-\tau_0|>2|\tau-\tau_0|} |K(z-\tau) - K(z-\tau_0)|dz \leqslant \max\{2, k^2\} \qquad (4.80)$$

We now return to the decomposition of Ω into squares. If τ_j is the center of the square Q_j and $\tau \in Q_j$, then for every $z \in F$, the complement of Ω, we find from (4.75) that $|z - \tau_j| > 2|\tau - \tau_j|$. Hence for $\tau \in Q_j$, we have

$$\int_F |K(z-\tau) - K(z-\tau_j)|dz \leqslant \max\{2, k^2\} \qquad (4.81)$$

The "Good" and "Bad" Parts of f

We set f_{Q_j} as the integral mean $f_{Q_j} = \frac{1}{|Q_j|} \int_{Q_j} f$ and decompose f as the sum of two functions,

$$f = g + b,$$

where the "good" part

$$g(z) = \begin{cases} f(z) & \text{if } z \in F \\ f_{Q_j} & \text{if } z \in Q_j \end{cases}$$

so that in particular $|g(z)| \leqslant \alpha$. The "bad" part

$$b(z) = \sum_j [f(z) - f_{Q_j}] \chi_{Q_j}(z)$$

Note that the bad part $b \in L^1(\mathbb{C})$ and has the redeeming feature that its integral mean on any square Q_j is zero. Indeed,

$$\|b\|_{L^1(\mathbb{C})} = \sum_j \|f(z) - f_{Q_j}\|_{L^1(Q_j)} \leqslant 2 \int_\Omega |f|$$

On the other hand, the good part lies in $L^2(\mathbb{C})$ and we have

$$\int_\mathbb{C} |g|^2 \leqslant \alpha \int_\mathbb{C} |g| = \alpha \int_F |f| + \alpha \sum_j \int_{Q_j} |f_{Q_j}|$$
$$\leqslant \alpha \int_F |f| + \alpha \int_\Omega |f| = \alpha \|f\|_{L^1(\mathbb{C})}$$

4.5. L^p-BOUNDS

We first want a weak-type estimate for the good part of f. Since R^k is an isometry in $L^2(\mathbb{C})$, we have

$$|\{z : |R^k g(z)| > \alpha/2\}| \leq \frac{4}{\alpha^2} \int_\mathbb{C} |R^k g|^2 = \frac{4}{\alpha^2} \int_\mathbb{C} |g|^2$$

$$\leq \frac{4}{\alpha} \|f\|_{L^1(\mathbb{C})}$$

Now the weak-type estimate for the bad part is

$$|\{z \in F : |R^k b(z)| > \alpha/2\}| \leq \frac{2}{\alpha} \int_F |R^k b| \leq \frac{2}{\alpha} \sum_j \int_F \left| R^k([f(\cdot) - f_{Q_j}]\chi_{Q_j}) \right|$$

$$= \frac{2}{\alpha} \sum_j \int_F \left| \int_{Q_j} [K(z - \tau) - K(z - \tau_j)][f(\tau) - f_{Q_j}] d\tau \right|$$

$$\leq \frac{2}{\alpha} \sum_j \int_{Q_j} |f(\tau) - f_{Q_j}| \left(\int_F |[K(z - \tau) - K(z - \tau_j)]| dz \right) d\tau$$

$$\leq \frac{2}{\alpha} \sum_j \int_{Q_j} |f(\tau) - f_{Q_j}| \max\{2, k^2\} d\tau \leq \frac{\max\{8, 4k^2\}}{\alpha} \int_\Omega |f|$$

Next, by the Hardy-Littlewood weak-type estimate, Theorem 4.4.5, we have

$$\left|\left\{z \in \Omega : |R^k b(z)| > \frac{\alpha}{2}\right\}\right| \leq |\Omega| \leq \frac{9}{\alpha} \|f\|_{L^1(\mathbb{C})}$$

Thus

$$\left|\left\{z \in \mathbb{C} : |R^k b(z)| > \frac{\alpha}{2}\right\}\right| \leq (9 + 4\max\{2, k^2\}) \frac{\|f\|_{L^1(\mathbb{C})}}{\alpha}$$

Finally, we may add these weak estimates for the good and bad parts of f to obtain

$$|\{z \in \mathbb{C} : |R^k f(z)| > \alpha\}| \leq \frac{4k^2 + 17}{\alpha} \|f\|_{L^1(\mathbb{C})} \tag{4.82}$$

Since the complex Riesz transforms are Fourier multiplier operators, with a unimodular multiplier, they are isometries in $L^2(\mathbb{C})$. In particular, they satisfy the weak-type condition (4.62) with $p_2 = 2$. Marcinkiewicz interpolation (Theorem 4.4.3) shows now that R^k, $k \in \mathbb{Z}$, are bounded on $L^p(\mathbb{C})$ for $1 < p \leq 2$. Since the adjoint of the complex Riesz transform R^k is R^{-k}, we have the boundedness for all $1 < p < \infty$. Last, using the Riesz-Thorin and Marcinkiewicz interpolation theorems with the weak-type bounds in (4.82), we obtain the next basic result we have been aiming at.

Corollary 4.5.1. *For all $1 < p < \infty$ and $k \in \mathbb{Z}$,*

$$\|\mathcal{R}^k\|_p \leq C_p(1 + k^2),$$

where the constant C_p depends only on p. In particular, the Beurling operator $\mathcal{S}$ extends continuously to $L^p(\mathbb{C})$, $1 < p < \infty$.

In the case of the Beurling transform, the weak-L^1-bounds take the following form.

Theorem 4.5.2. *The Beurling transform $\mathcal{S}$ extends to $L^1(\mathbb{C})$ with values in the space weak-$L^1(\mathbb{C})$, with the estimate*

$$|\{z \in \mathbb{C} : |(\mathcal{S}\phi)(z)| > t\}| \leq \frac{A_\mathcal{S}}{t} \|\phi\|_{L^1(\mathbb{C})}, \qquad A_\mathcal{S} \leq 30, \qquad (4.83)$$

for all $t > 0$ and $\phi \in L^1(\mathbb{C})$.

The upper bound for the constant $A_\mathcal{S}$ in the above weak-type estimate is not the best possible. It is not even clear what the best constant might be. However, the inequality does not hold with $A_\mathcal{S} = 1$, as the following example shows. Consider

$$f(z) = \bar{z} \log |z|^2 \chi_\mathbb{D}(z)$$

and its complex derivatives $\phi = f_{\bar{z}} = (1 + \log|z|^2)\chi_\mathbb{D}$ and $\mathcal{S}\phi = f_z = \frac{\bar{z}}{z}\chi_\mathbb{D}$. Thus by easy computation

$$|\{z \in \mathbb{C} : |\mathcal{S}\phi(z)| \geq 1\}| = \pi, \qquad \|\phi\|_{L^1(\mathbb{C})} = \frac{2\pi}{e}$$

We see that for this function the constant in (4.83) with $t = 1$ equals $\frac{e}{2} > 1.35$. Thus

$$A_\mathcal{S} > 1.35$$

It is worth noticing that the Marcinkiewicz interpolation (Theorem 4.4.3) gives the following asymptotic estimate, connecting the weak-type estimate and the L^p-norms $\mathbf{S}_p$ of the Beurling operator,

$$\lim_{p \to 1} (p-1) \, \mathbf{S}_p \leq A_\mathcal{S}$$

There is a well-known conjecture, discussed in more detail in the next section, that actually $(p-1)\mathbf{S}_p = 1$ for all $1 < p \leq 2$. From the above discussion, even if true, this result cannot be achieved from the weak-type bounds.

4.5.2 L^p-Estimates for the Beurling Transform

Because of its importance, we shall spend some time discussing the fine properties of the Beurling transform $\mathcal{S}$ on the spaces $L^p(\mathbb{C})$, $1 < p < \infty$. Consider first, as we did for the Cauchy transform, the function $\varphi(z) = |z|^\beta \chi_\mathbb{D}$, $\beta > -1$. Then

$$\begin{aligned}(\mathcal{S}\varphi)(z) &= \frac{\partial}{\partial z}(\mathcal{C}\varphi)(z) = \frac{2}{\beta+2}\frac{\partial}{\partial z}\left(|z|^\beta \bar{z}\chi_\mathbb{D} + z^{-1}\chi_{\mathbb{C}\setminus\mathbb{D}}\right) \\ &= \frac{1}{\beta+2}\left(\beta|z|^\beta \bar{z}/z\,\chi_\mathbb{D} - 2z^{-2}\chi_{\mathbb{C}\setminus\mathbb{D}}\right)\end{aligned}$$

4.5. L^p-BOUNDS

Then for $1 < p \leqslant 2$ and $-2/p < \beta < 0$, an elementary computation shows that when $\beta \searrow -2/p$, we have

$$\frac{\|\mathcal{S}\varphi\|_p}{\|\varphi\|_p} \to \frac{1}{p-1}, \quad 1 < p \leqslant 2,$$

establishing a lower bound for the operator norm $\|\mathcal{S} : L^p(\mathbb{C}) \to L^p(\mathbb{C})\|$.

Summing up, we have the fundamental L^p-boundedness of the Beurling operator $\mathcal{S}$.

Theorem 4.5.3. *The Beurling transform $\mathcal{S}$ extends to a continuous operator from $L^p(\mathbb{C})$ into $L^p(\mathbb{C})$ for all $1 < p < \infty$.*

$$\|\mathcal{S}\phi\|_p \leqslant \mathbf{S}_p \|\phi\|_p \tag{4.84}$$

The norm of the operator $\mathcal{S} : L^p(\mathbb{C}) \to L^p(\mathbb{C})$, denoted by $\mathbf{S}_p$, is a locally Lipschitz function of p with the property that

$$\mathbf{S}_2 = 1 \tag{4.85}$$

$$\mathbf{S}_p = \mathbf{S}_q \tag{4.86}$$

for every Hölder conjugate pair $1 < q \leqslant 2 \leqslant p < \infty$. For $p \geqslant 2$,

$$(q-1)\mathbf{S}_q = \frac{1}{p-1}\mathbf{S}_p \geqslant 1 \tag{4.87}$$

We recall here the well-known conjecture [172], mentioned in the previous section, that equality actually holds in (4.87).

Conjecture 4.5.4.

$$\mathbf{S}_p = \begin{cases} p - 1, & p \geqslant 2 \\ 1/(p-1), & p \leqslant 2 \end{cases} \tag{4.88}$$

There are many reasons for believing this conjecture, as it is intimately related to the regularity properties of quasiconformal mappings. It will also arise naturally at several different contexts in this monograph. Currently, the best-known bounds are $\mathbf{S}_p \leqslant A(p-1)$ for $p > 2$, where $A < 1.575$ [44]. Using $\mathbf{S}_2 = 1$ and interpolation, this gives the estimate

$$\mathbf{S}_p \leqslant 1 + 3(p-2), \quad p \geqslant 2 \tag{4.89}$$

Corollary 4.5.5. *For all $0 \leqslant \varepsilon < 1$,*

$$1 \leqslant \|\mathcal{S}\|_{2\pm\varepsilon} \leqslant 1 + \frac{3\varepsilon}{1-\varepsilon}$$

4.5.3 Weighted L^p-Theory for $\mathcal{S}$

In this brief subsection we wish to discuss the weighted L^p-theory of $\mathcal{S}$, although not in any great generality, since we shall need these results later. For these purposes we need to discuss A_p-weights introduced by Muckenhoupt [276].

A positive weight function ω defined on $\mathbb{C}$ belongs to the class $A_p = A_p(\mathbb{C})$ and is called an A_p-*weight* if

$$|\omega|_{A_p} = \sup_B \left(\frac{1}{|B|}\int_B \omega\right)\left(\frac{1}{|B|}\int_B \omega^{-1/(p-1)}\right)^{p-1} < \infty \qquad (4.90)$$

Here $1 < p < \infty$ and the supremum is taken over all disks $B \subset \mathbb{C}$, while the quantity $|\omega|_{A_p}$ will be referred to as the A_p-norm of the weight ω.

For a discussion of the basic properties of A_p-weights we refer the reader to [336]. We do note here that it is simply Hölder's inequality which shows that

$$A_p \subset A_q \qquad \text{whenever } p \leqslant q$$

The fact that

$$A_\infty = \bigcup_{p<\infty} A_p$$

is quite a bit deeper. Here the weights $\omega \in A_\infty$ are defined as follows: We require a pair of constants $0 < \gamma, \delta < 1$ such that for all disks B and all subsets $E \subset B$,

$$|E| \leqslant \gamma |B| \quad \Rightarrow \quad |\omega(E)| \leqslant \delta |\omega(B)|$$

Later in Chapter 14, when comparing the derivatives u_z and $u_{\bar{z}}$ for functions with gradient contained in a weighted space L_w^p, we will make use of the following classical theorem due to R. Coifman and C. Fefferman [94].

Theorem 4.5.6. *Let $1 < p < \infty$. Regular Calderón-Zygmund operators, and the Beurling transform $\mathcal{S}$ in particular, are bounded on the weighted spaces*

$$L_\omega^p(\mathbb{C}) = \{f : \|f\|_{L_\omega^p} = \left(\int_\mathbb{C} |f|^p \omega\right)^{1/p} < \infty\}$$

if and only if the weight ω belongs to A_p.

It is clear that in Theorem 4.5.6 the operator norm $\|\mathcal{S}\|_{L_\omega^p(\mathbb{C})}$ is a function of $|\omega|_{A_p}$, but the Coifman-Fefferman theorem leaves the dependence rather vague. Recently, S. Petermichl and A. Volberg [294], applying heat extensions and Bellman function methods, were able to describe the correct dependence on the A_p-norm $|\omega|_{A_p}$ here. Their improved version of the Coifman-Fefferman theorem shows that the dependence is linear. The following Petermichl-Volberg inequality will be crucial in proving the optimal Stoilow factorization and Sobolev regularity results for weakly quasiregular mappings.

4.6. BMO AND THE BEURLING TRANSFORM

Theorem 4.5.7. *For each $1 < p < \infty$ there is a constant $C(p) < \infty$, such that the Beurling transform satisfies the inequality*

$$\|\mathcal{S}\phi\|_{L^p_\omega} \leqslant C(p)\|\omega\|_{A_p}\|\phi\|_{L^p_\omega}, \qquad 1 < p < \infty,$$

for all weights $\omega \in A_p$ and all functions $\phi \in L^p_\omega$.

4.6 BMO and the Beurling Transform

In our subsequent studies of the regularity of solutions to elliptic PDEs, we will be concerned with the action of the Beurling transform on the space $BMO(\mathbb{C})$ and its subspace $L^\infty(\mathbb{C})$. It is quite clear that we are going to have to modify the integral formula defining the Beurling transform in order to achieve anything. In fact, our definition of $\mathcal{S}$ on $BMO(\mathbb{C})$ is going to depend on an initial modification of the Cauchy transform. Thus, for $\phi \in BMO(\mathbb{C})$, we define the modified Cauchy transform as

$$(\mathcal{C}_*\phi)(z) = \frac{1}{\pi}\int_\mathbb{C}\left[\frac{1}{z-\tau} + \frac{1}{\tau} + \frac{z}{\tau(\tau-1)}\right]\phi(\tau)d\tau \qquad (4.91)$$

The kernel

$$\frac{1}{z-\tau} + \frac{1}{\tau} + \frac{z}{\tau(\tau-1)} = \frac{z(z-1)}{\tau(\tau-1)(z-\tau)},$$

and so we see from Corollary 4.6.3 below that the integral converges for every $z \in \mathbb{C}$ when $\phi \in BMO(\mathbb{C})$. Note that if ϕ has compact support, then $\mathcal{C}_*\phi$ and $\mathcal{C}\phi$ differ by a linear function,

$$(\mathcal{C}_*\phi)(z) = (\mathcal{C}\phi)(z) + \frac{1}{\pi}\int_\mathbb{C}\frac{\phi(\tau)d\tau}{\tau} + z\left(\frac{1}{\pi}\int_\mathbb{C}\frac{\phi(\tau)d\tau}{\tau(\tau-1)}\right)$$

In this case we compute the complex derivatives to obtain

$$\frac{\partial}{\partial z}\mathcal{C}_*\phi = \mathcal{S}\phi + \frac{1}{\pi}\int_\mathbb{C}\frac{\phi(\tau)d\tau}{\tau(\tau-1)}$$

A BMO function is defined only up to a constant, hence this suggests that for $\phi \in BMO(\mathbb{C})$ we can define the Beurling transform by the rule

$$(\mathcal{S}\phi)(z) = \frac{-1}{\pi}\int_\mathbb{C}\left[\frac{1}{(z-\tau)^2} - \frac{1}{\tau(\tau-1)}\right]\phi(\tau)d\tau \qquad (4.92)$$

Again, the principal value of this integral exists for almost every $z \in \mathbb{C}$. This is most easily seen from Theorem 4.0.10, since $BMO(\mathbb{C}) \subset L^2_{loc}(\mathbb{C})$ and we may write $\phi = \phi\chi_B + \phi\chi_{\mathbb{C}\setminus B}$ to have the convergence of the principal value at almost every point of any disk $B \subset \mathbb{C}$.

4.6.1 Global John-Nirenberg Inequalities

The celebrated John-Nirenberg inequality establishes the existence of a positive δ such that for every disk $B \subset \mathbb{C}$,

$$\frac{1}{|B|} \int_B \exp\left(2\delta \frac{|u - u_B|}{\|u\|_*}\right) \leq 2 \tag{4.93}$$

In particular, since $x^p \leq p^p e^x$, we find that

$$\left(\frac{1}{|B|} \int_B |u - u_B|^p\right)^{1/p} \leq \frac{p}{\delta} \|u\|_* \quad \text{for } p \geq 1 \tag{4.94}$$

We want to amalgamate these local estimates into one global estimate. One of the corollaries of this global estimate is going to be that near every point $a \in \mathbb{C}$ a BMO function behaves at worst like $\log|z - a|$ in some average sense. This sort of behavior also occurs at ∞. To make such results precise we introduce the *truncated logarithm function*, $\log_*(t) = \max\{1, |\log(t)|\}$.

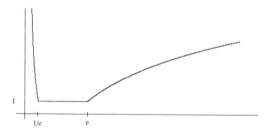

The graph of $\log_*(t)$

Theorem 4.6.1. *There is an $\varepsilon > 0$ such that, for every nonconstant $u \in BMO(\mathbb{C})$ and disk $B = \mathbb{D}(a, R)$, we have*

$$\int_{\mathbb{C}} \exp\left(\frac{\varepsilon |u(z) - u_B|}{\|u\|_* \log_* \left|\frac{z-a}{R}\right|}\right) \lambda(|z - a|) \, dz \leq 2, \tag{4.95}$$

where $\lambda : (0, \infty) \to [0, \infty)$ can be any decreasing function, not necessarily bounded, such that

$$\int_{\mathbb{C}} \lambda(|z|) \, dz = 1 \tag{4.96}$$

Proof. Because of the homogeneity with respect to u, as well as the scale and translation invariance with respect to z, we may assume

- $a = 0$, $R = 1$ and $B = \mathbb{D}(0, 1)$, the unit disk.
- $\|u\|_* = 1$ and $u_{\mathbb{D}} = 0$.

4.6. BMO AND THE BEURLING TRANSFORM

This reduces the problem to showing

$$\int_{\mathbb{C}} \exp\left(\frac{\varepsilon |u(z)|}{\log_* |z|}\right) \lambda(|z|)\, dz \leqslant 2 \tag{4.97}$$

Let $D_m = \mathbb{D}(0, 2^m)$, $m \in \mathbb{Z}$ and $u_m = \frac{1}{|D_m|}\int_{D_m} u$. Directly from the definition of the BMO-norm we know that

$$|u_m - u_{m-1}| = \frac{1}{|D_{m-1}|}\left|\int_{D_{m-1}} u - u_m\right| \leqslant \frac{1}{|D_{m-1}|}\int_{D_{m-1}} |u - u_m|$$

$$\leqslant \frac{4}{|D_m|}\int_{D_m} |u - u_m| \leqslant 4\|u\|_* = 4$$

We can then iterate these estimates to obtain

$$|u_m| \leqslant 4|m| \tag{4.98}$$

Now let δ be as in (4.93) and we use it to compute

$$\int_{D_m \setminus D_{m-1}} \exp\left(\frac{\delta|u(z)|}{\log_* |z|}\right) \lambda(|z|)$$

$$\leqslant \lambda(2^{m-1}) \int_{D_m \setminus D_{m-1}} \exp\left(\frac{\delta(|u(z) - u_m| + 4|m|)}{\frac{1}{2} + \frac{1}{4}|m|}\right)$$

$$\leqslant e^{16\delta} \lambda(2^{m-1}) \int_{D_m} \exp\left(2\delta |u(z) - u_m|\right)$$

$$\leqslant e^{16\delta} \lambda(2^{m-1}) 2\pi 4^m \leqslant 11 e^{16\delta} \int_{D_{m-1} \setminus D_{m-2}} \lambda(|z|)$$

We then sum these inequalities over $m \in \mathbb{Z}$ to find

$$\int_{\mathbb{C}} \exp\left(\frac{\delta|u(z)|}{\log_* |z|}\right) \lambda(|z|)\, dz \leqslant 11 e^{16\delta}$$

As the final step, we raise this inequality to the power $\frac{\varepsilon}{\delta} < 1$, where

$$\varepsilon = \frac{\delta \log 2}{16\delta + \log 11}$$

In view of the normalization in (4.96) showing λ to be a probability measure, an application of Hölder's inequality then gives

$$\int_{\mathbb{C}} \exp\left(\frac{\varepsilon |u(z)|}{\log_* |z|}\right) \lambda(|z|) \leqslant (11 e^{16\delta})^{\varepsilon/\delta} \leqslant 2$$

This completes the proof. $\square$

The comparison $x^p \leqslant p^p e^x$, for $x \geqslant 0$ and $p > 0$, gives the following corollary.

Corollary 4.6.2. *With λ as in (4.96), $u \in BMO$ and $B = \mathbb{D}(a, R)$, we have*

$$\int_{\mathbb{C}} \left(\frac{|u(z) - u_B|}{\|u\|_* \log_* \left|\frac{z-a}{R}\right|} \right)^p \lambda(|z - a|) \, dz \leqslant C_p, \qquad 0 < p < \infty$$

The inequality in (4.95) admits an interesting interpretation for the local behavior of a BMO function u near any point $a \in \mathbb{C}$. We simply let λ tend to the Dirac mass at the origin. This quickly shows that, as z approaches the point a, $|u - u_B|$ is controlled in an average sense by $\|u\|_* \log \frac{R}{|z-a|}$. This describes the worst possible behaviour of u at any point.

A less precise, though still useful, global estimate is given in the following corollary.

Corollary 4.6.3. *For each $1 \leqslant s < \infty$ and $u \in BMO(\mathbb{C})$, we have*

$$\left(\int_{\mathbb{C}} \frac{|u(z) - u_{\mathbb{D}}|^s}{1 + |z|^3} \right)^{1/s} \leqslant C_s \|u\|_*$$

4.6.2 Norm Bounds in BMO

The Beurling transform defines a bounded operator also on the space $BMO(\mathbb{C})$, and the operation is most conveniently understood in terms of the pointwise estimates for the different maximal functions we shall next establish in Theorem 4.6.4. As important consequences, we will identify the norm of the Beurling transform $\mathcal{S} : L^\infty(\mathbb{C}) \to BMO(\mathbb{C})$ and also get a good bound for its operator norm on $BMO(\mathbb{C})$. Actually, these results can also be used to give an alternative proof for the $L^p(\mathbb{C})$ boundedness of $\mathcal{S}$.

Theorem 4.6.4. *For every $z \in \mathbb{C}$ we have*

$$(\mathcal{S}\phi)^\#(z) \leqslant 3\,\phi^\flat(z) \qquad \text{if } \phi \in L^\infty(\mathbb{C}) \tag{4.99}$$

$$(\mathcal{S}\phi)^\#(z) \leqslant 2\sqrt{3}\,\phi^\#(z) \qquad \text{if } \phi \in BMO(\mathbb{C}) \tag{4.100}$$

The reader may notice that the latter inequality is better than the usual such estimate for singular integral operators giving bounds in terms of the Hardy-Littlewood maximal function; see [336, Proposition 2, p. 157].

We shall then consider the operator norms of the Beurling transform acting on L^∞ and BMO, following [174]. It is important to note that actually there is no canonical choice for the norm in the space BMO even if in most cases one uses Definition 4.3.1, that is, the norm $\|u\|_*$. In most situations the particular choice of the BMO-norm plays no role, but in cases such as we will encounter, when looking for precise bounds for operator norms, a specific choice must be made. In the next two theorems we will consider $BMO(\mathbb{C})$ equipped with the norm introduced earlier in (4.67),

$$\|\phi\|_{BMO_2} = \|\phi^\#\|_\infty = \sup \left(\frac{1}{|B|} \int_B |\phi(\tau) - \phi_B|^2 d\tau \right)^{1/2} \tag{4.101}$$

4.6. BMO AND THE BEURLING TRANSFORM

where the supremum is taken over all disks $B \subset \mathbb{C}$.

To emphasize the choice made in (4.101), we use the notation $BMO_2(\mathbb{C})$ for the space $BMO(\mathbb{C})$ equipped with the norm $\|\cdot\|_{BMO_2}$.

We then have the following theorem.

Theorem 4.6.5. *The Beurling transform acts from $L^\infty(\mathbb{C}) \to BMO_2(\mathbb{C})$ with the sharp norm bound*

$$\|\mathcal{S} : L^\infty(\mathbb{C}) \to BMO_2(\mathbb{C})\| = 3$$

For the norm of the Beurling transform on $BMO_2(\mathbb{C})$, its explicit value remains an open problem even if we have the following closely related estimate,

Theorem 4.6.6. *The Beurling transform acts from $BMO_2(\mathbb{C})$ into $BMO_2(\mathbb{C})$ with the norm bound*

$$\|\mathcal{S} : BMO_2(\mathbb{C}) \to BMO_2(\mathbb{C})\| \leqslant 2\sqrt{3}$$

From these results one can deduce for the operator $\mathcal{S}$ the L^p-bounds, $p \geqslant 2$, using Theorem 4.4.7 and the $L^p(\mathbb{C})$-bounds for the maximal functions. Namely, for $\phi \in C_0^\infty(\mathbb{C})$ we have the uniform estimate

$$\begin{aligned}\|\mathcal{S}\phi\|_p &\leqslant C_p \|(\mathcal{S}\phi)^\#\|_p \leqslant C_p 2\sqrt{3} \|\phi^\#\|_p \\ &\leqslant C_p \|\mathcal{M}_2 \phi\|_p \leqslant C_p \|\phi\|_p,\end{aligned}$$

where $\mathcal{M}$ is the Hardy-Littlewood maximal function.

We need to develop a few techniques in order to establish Theorem 4.6.4.

4.6.3 Orthogonality Properties of $\mathcal{S}$

Our proofs of sharp bounds for the Beurling transform are going to depend on certain orthogonality properties of this operator. For these we shall need to develop a little theory and discuss a few examples first. For every $k = 1, 2, \ldots$ let

$$\phi_k(z) = \bar{z}^k \chi_{\mathbb{D}}(z)$$

Then

$$(\mathcal{S}\phi_k)(z) = (\chi_{\mathbb{D}}(z) - 1)z^{-k-2} \tag{4.102}$$

Namely, as with the example in (4.52), we use the identity $\mathcal{S}f_{\bar{z}} = f_z$, where now

$$f_k(z) = \frac{1}{k+1}\left[\bar{z}^{k+1}\chi_{\mathbb{D}}(z) + (1 - \chi_{\mathbb{D}}(z))z^{-k-1}\right]$$

From linearity and the power series expansion, we deduce the following corollary of (4.102).

Corollary 4.6.7. Let $\mathcal{A}^2(\mathbb{D})$ denote the space of functions which are square-integrable and analytic in $\mathbb{D}$. Then for every $h \in \mathcal{A}^2(\mathbb{D})$ we have

$$\mathcal{S}(\chi_{\mathbb{D}} \bar{h})(z) = 0, \qquad z \in \mathbb{D}$$

The two sets $\mathbb{D}$ and $\mathbb{C} \setminus \mathbb{D}$ naturally induce an orthogonal decomposition of $L^2(\mathbb{C})$ as

$$L^2(\mathbb{C}) = L^2(\mathbb{D}) \oplus L^2(\mathbb{C} \setminus \mathbb{D}),$$

obtained by decomposing any $\phi \in L^2(\mathbb{C})$ as

$$\phi = b + \omega; \qquad b = \phi \chi_{\mathbb{D}}, \ \omega = \phi \chi_{\mathbb{C} \setminus \mathbb{D}}$$

Lemma 4.6.8. Let $b \in L^2(\mathbb{D})$ and $\omega \in L^2(\mathbb{C} \setminus \mathbb{D})$. Then the functions $\mathcal{S}b$ and $\mathcal{S}\omega$ are orthogonal in $L^2(\mathbb{D})$ and $L^2(\mathbb{C} \setminus \mathbb{D})$. That is,

$$\int_{\mathbb{D}} (\mathcal{S}b)(z) \overline{(\mathcal{S}\omega)(z)} dz = 0 \qquad (4.103)$$

$$\int_{\mathbb{C} \setminus \mathbb{D}} (\mathcal{S}b)(z) \overline{(\mathcal{S}\omega)(z)} dz = 0 \qquad (4.104)$$

Moreover,

$$(\mathcal{S}b)_{\mathbb{D}} = \frac{1}{\pi} \int_{\mathbb{D}} (\mathcal{S}b)(z) dz = 0 \qquad (4.105)$$

Proof. Since $\mathcal{S}$ is symmetric, the first identity we wish to establish takes the form

$$\int_{\mathbb{C}} (\mathcal{S}b) \chi_{\mathbb{D}} \overline{\mathcal{S}\omega} = \int_{\mathbb{C}} b \, \mathcal{S}(\chi_{\mathbb{D}} \overline{\mathcal{S}\omega}) = \int_{\mathbb{D}} b \, \mathcal{S}(\chi_{\mathbb{D}} \overline{\mathcal{S}\omega})$$

Observe that $\mathcal{S}\omega \in \mathcal{A}^2(\mathbb{D})$, and so by Corollary 4.6.7 the function $\mathcal{S}(\chi_{\mathbb{D}} \overline{\mathcal{S}\omega})$ vanishes in $\mathbb{D}$. The second identity follows directly from the first once we note that $\mathcal{S}$ is a unitary operator on $L^2(\mathbb{C})$ and that the supports of b and ω are disjoint. Finally, if we replace $\mathcal{S}\omega$ in the first identity by $\chi_{\mathbb{D}} \in \mathcal{A}^2(\mathbb{D})$ and again argue as above, we obtain (4.105). $\square$

From this lemma we obtain the following key identity.

Corollary 4.6.9. Let $\phi \in L^2(\mathbb{C})$. Then

$$\int_{\mathbb{D}} |\mathcal{S}\phi - (\mathcal{S}\phi)_{\mathbb{D}}|^2 = \int_{\mathbb{D}} |\mathcal{S}b|^2 + \int_{\mathbb{D}} |\mathcal{S}\omega - (\mathcal{S}\omega)_{\mathbb{D}}|^2, \qquad (4.106)$$

where $\phi = b + \omega$, $b \in L^2(\mathbb{D})$ and $\omega \in L^2(\mathbb{C} \setminus \mathbb{D})$.

4.6. BMO AND THE BEURLING TRANSFORM

4.6.4 Proof of the Pointwise Estimates

Having the above tools in hand, we now set about proving Theorem 4.6.4, basically following the arguments of [174].

Let $\phi \in BMO(\mathbb{C})$. For every disk $B = \mathbb{D}(a, R) \subset \mathbb{C}$, we aim to establish the two inequalities

$$\frac{1}{\pi R^2} \int_B |\mathcal{S}\phi(z) - (\mathcal{S}\phi)_B|^2 \, dz \leqslant 12 \, |\phi^\#(a)|^2 \qquad (4.107)$$

and

$$\frac{1}{\pi R^2} \int_B |\mathcal{S}\phi(z) - (\mathcal{S}\phi)_B|^2 \, dz \leqslant 9 \, |\phi^\flat(a)|^2 \qquad (4.108)$$

Once Theorem 4.6.4 has been established with these inequalities, as a consequence we obtain norm bounds for the Beurling transform,

$$\|\mathcal{S} : BMO_2(\mathbb{C}) \to BMO_2(\mathbb{C})\| \leqslant 2\sqrt{3}$$

and

$$\|\mathcal{S} : L^\infty(\mathbb{C}) \to BMO_2(\mathbb{C})\| \leqslant 3$$

The first norm estimate gives Theorem 4.6.6 and shows in particular that the Beurling transform preserves the space $BMO(\mathbb{C})$. We will then show that the second norm bound is sharp. In fact, the norm is attained at some function f with $\|f\|_\infty = 1$.

There is no loss of generality in assuming that $B = \mathbb{D}$, the unit disk.

Lemma 4.6.10. *Let $\phi \in BMO(\mathbb{C})$. Then*

$$\frac{1}{\pi} \int_\mathbb{D} |\mathcal{S}\phi - (\mathcal{S}\phi)_\mathbb{D}|^2 \leqslant \frac{1}{\pi} \int_\mathbb{D} |\phi - \phi_\mathbb{D}|^2 + \sum_{k=1}^\infty (k+1)|c_k|^2, \qquad (4.109)$$

where

$$c_k = -\frac{1}{\pi} \int_{|\tau| \geqslant 1} \frac{\phi(\tau) d\tau}{\tau^{k+2}} \qquad (4.110)$$

Proof. We will establish the inequality by proving a stronger result, the following identity:

$$\frac{1}{\pi} \int_\mathbb{D} |\mathcal{S}\phi - (\mathcal{S}\phi)_\mathbb{D}|^2 = \frac{1}{\pi} \int_\mathbb{D} |\phi - \phi_\mathbb{D}|^2 + \sum_{k=1}^\infty (k+1) \left| \frac{1}{\pi} \int_{|\tau| \geqslant 1} \frac{\phi(\tau) d\tau}{\tau^{k+2}} \right|^2$$

$$- \sum_{k=1}^\infty (k+1) \left| \frac{1}{\pi} \int_{|\tau| \leqslant 1} \tau^k \phi(\tau) d\tau \right|^2 \qquad (4.111)$$

The first term on the right hand side of this identity is clearly bounded. The last term is finite, as it is dominated by $\|\phi\|_{L^2(\mathbb{D})}$ as Bessel's inequality shows us when applied to the orthonormal basis $\sqrt{\frac{k+2}{2\pi}}\tau^k$ for $L^2(\mathbb{D})$. Since $k \geq 1$, by Corollary 4.6.3 each summand of the middle series is a finite number. Therefore there is no problem in giving a meaning to the right hand side, and once (4.111) is proven, the lemma follows.

We first prove the identity (4.111) for functions in $L^2(\mathbb{C})$. Given $\phi \in L^2(\mathbb{C})$, we write $\phi = b + \omega$, $b = \phi\chi_{\mathbb{D}} \in L^2(\mathbb{D})$ and $\omega = \phi\chi_{\mathbb{C}\setminus\mathbb{D}} \in L^2(\mathbb{C} \setminus \mathbb{D})$. We use Taylor's expansion to write, for $z \in \mathbb{D}$,

$$\begin{aligned}(\mathcal{S}\omega)(z) &= -\frac{1}{\pi}\int_{\mathbb{C}\setminus\mathbb{D}}\frac{\omega(\tau)d\tau}{(z-\tau)^2} \\ &= -\frac{1}{\pi}\sum_{k=0}^{\infty}(k+1)z^k\int_{|\tau|>1}\frac{\omega(\tau)d\tau}{\tau^{k+2}}\end{aligned}$$

As ω is supported outside $\mathbb{D}$, $\mathcal{S}\omega$ is holomorphic in $\mathbb{D}$ and therefore its integral mean over $\mathbb{D}$ is simply its value at 0. Thus

$$(\mathcal{S}\omega)(z) - (\mathcal{S}\omega)_{\mathbb{D}} = \sum_{k=1}^{\infty}(k+1)c_k z^k, \quad z \in \mathbb{D},$$

where the constants c_k are as in (4.110). As we have already pointed out, the system $\{z^k\}_{k=0}^{\infty}$ is orthogonal and Bessel's identity then shows us that

$$\frac{1}{\pi}\int_{\mathbb{D}}|(\mathcal{S}\omega)(z) - (\mathcal{S}\omega)_{\mathbb{D}}|^2\,dz = \sum_{k=1}^{\infty}(k+1)|c_k|^2$$

This is the first infinite sum in (4.111).

We now need to compute the terms involved in $\mathcal{S}b$. For this we have

$$\begin{aligned}\frac{1}{\pi}\int_{\mathbb{D}}|\mathcal{S}b|^2 &= \frac{1}{\pi}\int_{\mathbb{C}}|\mathcal{S}b|^2 - \frac{1}{\pi}\int_{\mathbb{C}\setminus\mathbb{D}}|\mathcal{S}b|^2 = \frac{1}{\pi}\int_{\mathbb{C}}|b|^2 - \frac{1}{\pi}\int_{\mathbb{C}\setminus\mathbb{D}}\left|\frac{1}{\pi}\int_{\mathbb{D}}\frac{\phi(\tau)d\tau}{(z-\tau)^2}\right|^2 \\ &= \frac{1}{\pi}\int_{\mathbb{D}}|\phi|^2 - \frac{1}{\pi^3}\sum_{k=0}^{\infty}\int_{|z|\geq 1}\frac{(k+1)^2}{|z|^{2k+4}}\left|\int_{\mathbb{D}}\tau^k\phi(\tau)d\tau\right|^2 \\ &= \frac{1}{\pi}\int_{\mathbb{D}}|\phi|^2 - \left|\frac{1}{\pi}\int_{\mathbb{D}}\phi\right|^2 - \frac{1}{\pi^2}\sum_{k=1}^{\infty}(k+1)\left|\int_{\mathbb{D}}\tau^k\phi(\tau)d\tau\right|^2 \\ &= \frac{1}{\pi}\int_{\mathbb{D}}|\phi - \phi_{\mathbb{D}}|^2 - \frac{1}{\pi^2}\sum_{k=1}^{\infty}(k+1)\left|\int_{\mathbb{D}}\tau^k\phi(\tau)d\tau\right|^2\end{aligned}$$

We can then use Corollary 4.6.9 to complete the proof of (4.111) under the assumption $\phi \in L^2(\mathbb{C})$.

However, once the identity (4.111) has been established for all functions in $L^2(\mathbb{C})$, we readily see that it actually holds for $\phi \in BMO(\mathbb{C})$, namely, by

4.6. BMO AND THE BEURLING TRANSFORM

considering truncations of ϕ by multiplying with the characteristic functions of ever larger disks. According to Corollary 4.6.3 , the coefficients c_k are all bounded independently of the truncation, hence one can take the limit, proving the identity for all BMO functions. $\square$

At this point our goal is to estimate the series (4.109), assuming that $\phi \in BMO(\mathbb{C})$. Variational considerations suggest that we consider the holomorphic function (which appears in the corresponding Euler-Lagrange equation) defined in $\mathbb{C} \setminus \mathbb{D}$ by

$$a(\tau) = -\frac{1}{\pi}\sum_{k=1}^{\infty}(k+1)\bar{c}_k \tau^{-k-2} \qquad (4.112)$$

Directly from the definition of the numbers c_k, it follows that

$$\sum_{k=1}^{\infty}(k+1)|c_k|^2 = \sum_{k=1}^{\infty}(k+1)\bar{c}_k \left[-\frac{1}{\pi}\int_{|\tau|\geq 1}\tau^{-k-2}\phi(\tau)d\tau\right]$$
$$= \int_{|\tau|\geq 1}\phi(\tau)a(\tau)d\tau$$

In this identity we may subtract from the function ϕ its mean value over any sphere of radius $|\tau|$, thus this identity holds for Φ defined by

$$\Phi(\tau) = \phi(\tau) - \frac{1}{2\pi|\tau|}\int_{|z|=|\tau|}\phi(z)|dz|$$

The reason for this is simply that the coefficients $c_k = -\frac{1}{\pi}\int_{|\tau|\geq 1}\phi(\tau)\tau^{-k-2}d\tau$ are not affected by subtracting from $\phi(\tau)$ any function of $|\tau|$. We may now apply Hölder's inequality in the following fashion:

$$\sum_{k=1}^{\infty}(k+1)|c_k|^2 \leq \left(\int_{|\tau|\geq 1}|\tau|^3 |a(\tau)|^2\right)^{1/2}\left(\int_{|\tau|\geq 1}|\Phi(\tau)|^2|\tau|^{-3}\right)^{1/2} \qquad (4.113)$$

The functions $\tau^{-k-2}|\tau|^{3/2}$, $k=1,2,3\ldots$, are mutually orthogonal in $L^2(\mathbb{C}\setminus\mathbb{D})$, and we may calculate that

$$\int_{|\tau|\geq 1}|\tau|^{-2k-1}d\tau = 2\pi \int_1^\infty \frac{dr}{r^{2k}} = \frac{2\pi}{2k-1}$$

for $k=1,2,\ldots$. This orthogonality allows the following computation.

$$\int_{|\tau|\geq 1}|\tau|^3|a(\tau)|^2 d\tau = \frac{1}{\pi^2}\int_{|\tau|\geq 1}\left|\sum_{k=1}^{\infty}(k+1)\bar{c}_k\tau^{-k-2}|\tau|^{3/2}\right|^2 d\tau$$
$$= \frac{1}{\pi^2}\sum_{k=1}^{\infty}(k+1)^2|c_k|^2 \int_{|\tau|\geq 1}|\tau|^{-2k-1}d\tau$$

$$= \frac{2}{\pi} \sum_{k=1}^{\infty} \frac{(k+1)^2}{2k-1} |c_k|^2$$

$$\leqslant \frac{4}{\pi} \sum_{k=1}^{\infty} (k+1) |c_k|^2$$

Now the inequalities (4.113) and (4.114) together give

$$\sum_{k=1}^{\infty} (k+1)|c_k|^2 \leqslant \frac{4}{\pi} \int_{|\tau| \geqslant 1} |\Phi(\tau)|^2 |\tau|^{-3} \, d\tau \qquad (4.114)$$

$$= \frac{4}{\pi} \int_1^{\infty} \frac{1}{r^3} \left(\int_{|\tau|=r} |\Phi(\tau)|^2 |d\tau| \right) dr \qquad (4.115)$$

Directly from the definition of the spherical maximal function, we infer that

$$\sum_{k=1}^{\infty} (k+1)|c_k|^2 \leqslant 8 \sup_{r \geqslant 1} \frac{1}{2\pi r} \int_{|\tau|=r} |\Phi(\tau)|^2 |d\tau|$$

$$\leqslant 8 \sup_{r \geqslant 1} \frac{1}{2\pi r} \int_{|\tau|=r} |\phi(\tau)|^2 |d\tau|$$

$$\leqslant 8 |\phi^{\flat}(0)|^2$$

We also have

$$\frac{1}{\pi} \int_{\mathbb{D}} |\phi - \phi_{\mathbb{D}}|^2 \leqslant \frac{1}{\pi} \int_{\mathbb{D}} |\phi|^2 = \frac{1}{\pi} \int_0^1 \left(\int_{|\tau|=r} |\phi(\tau)|^2 |d\tau| \right) dr$$

$$\leqslant \frac{1}{\pi} \int_0^1 2\pi r \, |\phi^{\flat}(0)|^2 dr = |\phi^{\flat}(0)|^2 \qquad (4.116)$$

We now combine this with (4.109) to obtain

$$\frac{1}{\pi} \int_{\mathbb{D}} |\mathcal{S}\phi - (\mathcal{S}\phi)_{\mathbb{D}}|^2 \leqslant |\phi^{\flat}(0)|^2 + 8|\phi^{\flat}(0)|^2 = 9|\phi^{\flat}(0)|^2 \leqslant 9\|\phi\|_{\infty}^2 \qquad (4.117)$$

In other words, we have established (4.108).

To complete the proof of Theorem 4.6.4, we need to establish the inequality (4.107) for the unit disk. Precisely, we need to show that

$$\frac{1}{\pi} \int_{\mathbb{D}} |\mathcal{S}\phi - (\mathcal{S}\phi)_{\mathbb{D}}|^2 \leqslant 12 \, |\phi^{\#}(0)|^2 \leqslant 12 \, \|\phi\|_{BMO_2(\mathbb{C})}^2$$

In view of the identity in (4.111) and the inequality in (4.114), we have

$$\frac{1}{\pi} \int_{\mathbb{D}} |\mathcal{S}\phi - (\mathcal{S}\phi)_{\mathbb{D}}|^2$$

4.6. BMO AND THE BEURLING TRANSFORM

$$\leqslant \frac{1}{\pi}\int_{\mathbb{D}}|\phi - \phi_{\mathbb{D}}|^2 + \sum_{k=1}^{\infty}(k+1)|c_k|^2$$

$$\leqslant \frac{1}{\pi}\int_{\mathbb{D}}|\phi - \phi_{\mathbb{D}}|^2 + \frac{4}{\pi}\int_1^{\infty}\frac{1}{r^3}\left(\int_{|\tau|=r}|\Phi(\tau)|^2|d\tau|\right)dr \quad (4.118)$$

Here the first integral on the right hand side is already dominated by $|\phi^{\#}(0)|^2$. In the proof of (4.117) we used the estimate $\int_{|\tau|=r}|\Phi|^2 \leqslant 2\pi r\|\phi\|_{\infty}^2$, but in order to proceed we now need to estimate the last integral by terms involving the L^2-mean oscillations of ϕ.

Lemma 4.6.11. *For almost every disk $\mathbb{D}_r = \mathbb{D}(0,r)$, $r > 0$, we have*

$$\int_{|\tau|=r}|\Phi(\tau)|^2|d\tau| \leqslant \frac{d}{dr}\int_{\mathbb{D}_r}|\phi(\tau) - \phi_{\mathbb{D}_r}|^2\,d\tau \quad (4.119)$$

Proof. The integral mean of a function $\phi \in L^2$ is the orthogonal projection to the closed subspace of constant functions. In particular, the L^2-distance of ϕ to the constant functions is attained precisely by the integral mean. To make use of this fact choose an arbitrary $R > r$ and write

$$\int_{\mathbb{D}_R \setminus \mathbb{D}_r}|\Phi|^2 = \int_r^R\left(\int_{|\tau|=t}|\Phi(\tau)|^2|d\tau|\right)dt$$

$$= \int_r^R\left(\int_{|\tau|=t}|\phi(\tau) - \phi_{\partial\mathbb{D}_t}|^2|d\tau|\right)dt$$

$$\leqslant \int_r^R\left(\int_{|\tau|=t}|\phi(\tau) - \phi_{\mathbb{D}_R}|^2|d\tau|\right)dt$$

$$= \int_{\mathbb{D}_R}|\phi(\tau) - \phi_{\mathbb{D}_R}|^2\,d\tau - \int_{\mathbb{D}_r}|\phi(\tau) - \phi_{\mathbb{D}_R}|^2\,d\tau$$

$$\leqslant \int_{\mathbb{D}_R}|\phi(\tau) - \phi_{\mathbb{D}_R}|^2\,d\tau - \int_{\mathbb{D}_r}|\phi(\tau) - \phi_{\mathbb{D}_r}|^2\,d\tau$$

Thus

$$\frac{1}{R-r}\int_r^R\left(\int_{|\tau|=t}|\Phi|^2\right) \leqslant \frac{1}{R-r}\left[\int_{\mathbb{D}_R}|\phi - \phi_{\mathbb{D}_R}|^2 - \int_{\mathbb{D}_r}|\phi - \phi_{\mathbb{D}_r}|^2\right]$$

Taking the limit $R \to r$ establishes the lemma. $\square$

We now return to the last integral in (4.118). We have

$$\frac{4}{\pi}\int_1^{\infty}\frac{1}{r^3}\left(\int_{|\tau|=r}|\Phi(\tau)|^2|d\tau|\right)dr$$

$$\leqslant \frac{4}{\pi} \int_1^\infty \frac{1}{r^3} \left(\frac{d}{dr} \int_{\mathbb{D}_r} |\phi - \phi_{\mathbb{D}_r}|^2 \right) dr$$

$$= -\frac{4}{\pi} \int_{\mathbb{D}} |\phi - \phi_{\mathbb{D}}|^2 + \frac{12}{\pi} \int_1^\infty \frac{1}{r^4} \left(\int_{\mathbb{D}_r} |\phi - \phi_{\mathbb{D}_r}|^2 \right) dr$$

$$\leqslant -\frac{4}{\pi} \int_{\mathbb{D}} |\phi - \phi_{\mathbb{D}}|^2 + 12 \int_1^\infty \frac{1}{r^2} |\phi^\#(0)|^2 \, dr$$

We now put this back into (4.118) to obtain

$$\frac{1}{\pi} \int_{\mathbb{D}} |\mathcal{S}\phi - (\mathcal{S}\phi)_{\mathbb{D}}|^2 \leqslant -3\frac{1}{\pi} \int_{\mathbb{D}} |\phi - \phi_{\mathbb{D}}|^2 + 12|\phi^\#(0)|^2$$

$$\leqslant 12|\phi^\#(0)|^2 \leqslant 12 \, \|\phi\|^2_{BMO_2(\mathbb{C})}$$

This last estimate completes the proof of Theorem 4.6.4. It also gives Theorem 4.6.6. $\square$

From Theorem 4.6.6 we see in particular that $\mathcal{S}$ maps $BMO(\mathbb{C})$ into itself. This fact provides an effective tool also in calculating the Beurling transform of fairly general functions in $L^\infty(\mathbb{C})$ or even in the more general class of $BMO(\mathbb{C})$ functions.

Lemma 4.6.12. *Let f be a Schwartz distribution whose $\bar{z}$-derivative $\phi = f_{\bar{z}}$ lies in $BMO(\mathbb{C})$. Then there is a function $h = h(z)$, holomorphic in $\mathbb{C}$ and unique up to a linear function, such that $(f - h)_z = f_z - h_z \in BMO(\mathbb{C})$.*

In particular, if both $f_{\bar{z}}$ and f_z lie in $BMO(\mathbb{C})$, then

$$\mathcal{S} \frac{\partial f}{\partial \bar{z}} = \frac{\partial f}{\partial z}$$

up to a constant.

Proof. The uniqueness of h up to a complex linear function already follows from Lemma 4.3.4. For the existence of the function h, we argue as before in Lemma 4.3.4 by considering the Schwartz distribution

$$h = f - \mathcal{C}_* \frac{\partial f}{\partial \bar{z}},$$

where the operator $\mathcal{C}_*$ is defined in (4.91). Of course, the $\bar{z}$-derivative of h vanishes in the distributional sense, and so by Weyl's Lemma A.6.10 we identify h as an entire function. As $f_{\bar{z}} \in BMO(\mathbb{C})$, we have already shown that the function $f_z - h_z = \mathcal{S} f_{\bar{z}}$ lies in $BMO(\mathbb{C})$. $\square$

Proof of Theorem 4.6.5

We need to show that the estimate (4.117) is the best possible. In order for this bound to be sharp, we need to have equality at each of the estimates we have made for the function ϕ, where we may assume $|\phi| \leqslant 1$. From general principles

4.6. BMO AND THE BEURLING TRANSFORM

in functional analysis, we expect that ϕ is an extreme point of the unit ball in $L^\infty(\mathbb{C})$; that is, ϕ should be a unimodular function, $|\phi| \equiv 1$. We first verify that the function
$$\phi(\tau) = \frac{\tau^2}{|\tau|^2}\chi_\mathbb{D}(\tau) + \frac{\tau^3}{|\tau|^3}\chi_{\mathbb{C}\setminus\mathbb{D}}(\tau)$$
satisfies these requirements. To calculate $(\mathcal{S}\phi)(z)$ we use Lemma 4.6.12 and get
$$(\mathcal{S}\phi)(z) = (1 + \log|z|^2)\chi_\mathbb{D}(z) - 4z\chi_\mathbb{D}(z) - 3\frac{z}{|z|}\chi_{\mathbb{C}\setminus\mathbb{D}}(z)$$
This is immediate from the formula $\mathcal{S}f_{\bar{z}} = f_z$ applied to
$$f(z) = (z\log|z|^2)\chi_\mathbb{D}(z) - 2z^2\chi_\mathbb{D}(z) - 2\frac{z^2}{|z|}\chi_{\mathbb{C}\setminus\mathbb{D}}(z)$$
Next, the mean value $(\mathcal{S}\phi)_\mathbb{D} = 0$. Then $\|\phi\|_\infty = 1$ with
$$\frac{1}{\pi}\int_\mathbb{D}|\mathcal{S}\phi - (\mathcal{S}\phi)_\mathbb{D}|^2 = \frac{1}{\pi}\int_\mathbb{D}|\mathcal{S}\phi|^2$$
$$= \frac{1}{\pi}\int_\mathbb{D}|1 + \log|z|^2 - 4z|^2 = 9$$

In other words, we have shown that
$$\|\mathcal{S} : L^\infty(\mathbb{C}) \to BMO_2(\mathbb{C})\| = 3$$
precisely, proving Theorem 4.6.5. $\square$

4.6.5 Commutators

For any $b \in BMO(\mathbb{C})$, let $b\mathcal{S} - \mathcal{S}b$ denote the commutator of the Beurling transform $\mathcal{S}$ and the multiplication operator $\phi \mapsto b \cdot \phi$. The commutator acts on $C_0^\infty(\mathbb{C})$ by the rule
$$(b\mathcal{S} - \mathcal{S}b)\phi = b \cdot (\mathcal{S}\phi) - \mathcal{S}(b \cdot \phi)$$
Pointwise multiplication by a BMO function need of course not preserve $L^p(\mathbb{C})$-spaces. However, we have the powerful theorem of Coifman, Rochberg and Weiss [97] that through the commutator, multiplication induces a bounded L^p-operation.

Theorem 4.6.13. *Suppose $b \in BMO(\mathbb{C})$ and $1 < p < \infty$. Then the commutator $b\mathcal{S} - \mathcal{S}b$ extends to a bounded operator on $L^p(\mathbb{C})$, with the uniform bound*
$$\|(\mathcal{S}b - b\mathcal{S})\phi\|_{L^p(\mathbb{C})} \leqslant C_p\|b\|_*\|\phi\|_{L^p(\mathbb{C})},$$
where C_p depends only on p.

The theorem also extends to more general translation-invariant singular integrals such as the Riesz-transforms and their iterates.

Proof. Let $F = (b\mathcal{S} - \mathcal{S}b)\phi$, where $\phi \in C_0^\infty(\mathbb{C})$. For a given $p > 1$ choose and fix an auxiliary exponent $1 < r < p$. We will prove for the sharp maximal function

$$F^\#(z) = \sup\Big\{\frac{1}{|B|}\int_B |F(w) - F_B|, \; z \in B \subset \mathbb{C} \text{ a disk}\Big\},$$

the following pointwise inequality:

$$F^\#(z) \leqslant C_r \|b\|_* (\mathcal{M}_r \phi + \mathcal{M}_r \mathcal{S}\phi) \qquad (4.120)$$

Here $\mathcal{M}_r$ is the Hardy-Littlewood maximal function from (4.65). Once this has been established, the theorem follows immediately from the estimates

$$\|F\|_p \leqslant C_p \|F^\#\|_p \leqslant C_p C_r \|b\|_* (\|\mathcal{M}_r \phi\|_p + \|\mathcal{M}_r \mathcal{S}\phi\|_p),$$

where the first inequality comes from Theorem 4.4.7.

For (4.120) choose a point z_0 and a disk B containing it. Because of translation and scaling invariance, we may assume that $B = \mathbb{D}$, the unit disk. Moreover, the identity $b\mathcal{S} - \mathcal{S}b = (b - b_\mathbb{D})\mathcal{S} - \mathcal{S}(b - b_\mathbb{D})$ shows that we may assume $b_\mathbb{D} = 0$. We then split the function F into four parts,

$$F = b\mathcal{S}\phi - \mathcal{S}(b\chi_{2\mathbb{D}}\phi) - \big[\mathcal{S}(b\chi_{\mathbb{C}\setminus 2\mathbb{D}}\phi) - C_0\big] - C_0, \qquad (4.121)$$

where $C_0 = \mathcal{S}(b\chi_{\mathbb{C}\setminus 2\mathbb{D}}\phi)(0)$ is a constant in $\mathbb{D}$. For the first term we apply Hölder's inequality,

$$\frac{1}{|\mathbb{D}|}\int_\mathbb{D} |b\mathcal{S}\phi| \leqslant \Big(\frac{1}{|\mathbb{D}|}\int_\mathbb{D} |b|^{r'}\Big)^{1/r'} \Big(\frac{1}{|\mathbb{D}|}\int_\mathbb{D} |\mathcal{S}\phi|^r\Big)^{1/r} \leqslant C_r \|b\|_* (\mathcal{M}_r \mathcal{S}\phi)(z_0)$$

by (4.94). For the second term of (4.121) we use Hölder's inequality and the L^p-boundedness of the Beurling transform,

$$\frac{1}{|\mathbb{D}|}\int_\mathbb{D} |\mathcal{S}(b\chi_{2\mathbb{D}}\phi)| \leqslant C_q \|b\chi_{2\mathbb{D}}\phi\|_q \leqslant C_q \Big(\int_{2\mathbb{D}} |b|^{\frac{qr}{r-q}}\Big)^{\frac{r-q}{qr}} \Big(\int_{2\mathbb{D}} |\phi|^r\Big)^{1/r}$$
$$\leqslant C_{q,r} \|b\|_* (\mathcal{M}_r \phi)(z_0)$$

It remains to estimate the third term in (4.121). Here, for $z \in \mathbb{D}$,

$$|\mathcal{S}(b\chi_{\mathbb{C}\setminus 2\mathbb{D}}\phi)(z) - C_0| = \Big|\frac{1}{\pi}\int_{|\xi|\geqslant 2}\Big[\frac{1}{(z-\xi)^2} - \frac{1}{\xi^2}\Big] b(\xi)\phi(\xi)\Big|$$

$$\leqslant \frac{3}{\pi}\int_{|\xi|\geqslant 2} \frac{|b(\xi)||\phi(\xi)|}{|\xi|^3} \leqslant \frac{3}{\pi}\Big(\int_{|\xi|\geqslant 2} \frac{|b(\xi)|^{r'}}{|\xi|^3}\Big)^{1/r'} \Big(\int_{|\xi|\geqslant 2} \frac{|\phi(\xi)|^r}{|\xi|^3}\Big)^{1/r}$$

$$\leqslant \|b\|_* \Big(\int_{|\xi|\geqslant 2} \frac{|\phi(\xi)|^r}{|\xi|^3}\Big)^{1/r}$$

4.6. BMO AND THE BEURLING TRANSFORM

where the last bound again is found using (4.94). To tidy up the remaining integral term, we divide the complement of $2\mathbb{D}$ into a union of annuli,

$$\int_{|\xi| \geqslant 2} \frac{|\phi(\xi)|^r}{|\xi|^3} \leqslant \sum_{j=1}^{\infty} 2^{-3j} \int_{2^j \leqslant |\xi| \leqslant 2^{j+1}} |\phi(\xi)|^r$$

$$\leqslant \sum_{j=1}^{\infty} 4\pi \, 2^{-j} \frac{1}{|2^{j+1}\mathbb{D}|} \int_{2^{j+1}\mathbb{D}} |\phi(\xi)|^r \leqslant 4\pi \bigl(\mathcal{M}_r \, \phi(z_0)\bigr)^r$$

Returning now to (4.121), we have

$$\frac{1}{|\mathbb{D}|} \int_{\mathbb{D}} |F(w) - F_{\mathbb{D}}| \leqslant \frac{2}{|\mathbb{D}|} \int_{\mathbb{D}} |F(w) + C_0| \leqslant C_{r,p} \|b\|_* \bigl[(\mathcal{M}_r \, \phi)(z_0) + (\mathcal{M}_r \, \mathcal{S}\phi)(z_0)\bigr],$$

proving (4.120). □

Compactness of the commutator operators requires VMO symbols. Since the definition of VMO seems to vary slightly in the literature, we emphasize that throughout this monograph $VMO(\mathbb{C})$ stands for the closure of $C_0^\infty(\mathbb{C})$ in $BMO(\mathbb{C})$. This is equivalent to taking the closure of $C(\hat{\mathbb{C}})$ in $BMO(\mathbb{C})$.

We have the following result due to A. Uchiyama [361]. Note that $VMO(\mathbb{C})$ agrees, modulo constants, with the space CMO employed by him, see [361, Lemma 3].

Theorem 4.6.14. *For each $b \in VMO(\mathbb{C})$ the linear operator*

$$\phi \mapsto (Sb - bS)\phi$$

is compact and bounded on $L^p(\mathbb{C})$, $p \in (1, \infty)$.

Proof. First let $b \in C_0^\infty(\mathbb{C})$. Consider an arbitrary function $f = \mathcal{C}\omega$ with $\omega \in L^p(\mathbb{C})$. As $(bf)_z = \mathcal{S}\bigl((bf)_{\bar{z}}\bigr)$,

$$\begin{aligned}
(b\mathcal{S} - \mathcal{S}b)\omega &= bf_z - \mathcal{S}(bf_{\bar{z}}) \\
&= \mathcal{S}(b_{\bar{z}}f) - b_z f \\
&= \mathcal{S}(b_{\bar{z}}\mathcal{C}\omega) - b_z \mathcal{C}\omega
\end{aligned}$$

From this representation and Theorem 4.3.14, we see that $b\mathcal{S} - \mathcal{S}b$ is a compact operator in all spaces $L^p(\mathbb{C})$, $1 < p < \infty$.

For a general $b \in VMO(\mathbb{C})$ we approximate and choose $b_j \in C_0^\infty(\mathbb{C})$ converging to b in $BMO(\mathbb{C})$. By Theorem 4.6.13 we know the commutator $b\mathcal{S} - \mathcal{S}b$ is going to be the limit in the operator norm topology of compact operators, namely,

$$\begin{aligned}
\|(b\mathcal{S} - \mathcal{S}b) - (b_j\mathcal{S} - \mathcal{S}b_j)\| &= \|(b - b_j)\mathcal{S} - \mathcal{S}(b - b_j)\| \\
&\leqslant C_p \|b - b_j\|_* \to 0
\end{aligned}$$

as $j \to \infty$. From this the compactness follows. □

4.6.6 The Beurling Transform of Characteristic Functions

Because of the importance of the Beurling transform to the general theory of quasiconformal mappings, in this section we shall give some explicit calculations of the transform acting on characteristic functions of planar domains. We have already calculated this for the case of a disk in (4.25).

Let Ω be a bounded open set in $\mathbb{C}$ with piecewise smooth boundary $\Gamma = \partial\Omega$. Then for $z \notin \partial\Omega$ we have from Green's formula (2.52),

$$(\mathcal{S}\chi_\Omega)(z) = -\frac{1}{\pi}\lim_{\varepsilon \to 0}\int_{\Omega\setminus\mathbb{D}(z,\varepsilon)}\frac{1}{(z-\tau)^2} = -\frac{i}{2\pi}\int_{\partial\Omega}\frac{d\bar{\tau}}{z-\tau} + \frac{i}{2\pi}\lim_{\varepsilon\to 0}\int_{\partial\mathbb{D}(z,\varepsilon)}\frac{d\bar{\tau}}{z-\tau}$$

We therefore deduce that

$$(\mathcal{S}\chi_\Omega)(z) = \frac{1}{2\pi i}\int_\Gamma \frac{d\bar{\tau}}{z-\tau}$$

Then for domains with piecewise smooth boundary, we see that $(\mathcal{S}\chi_\Omega)(z)$ is analytic off the boundary curve. This is true more generally.

Corollary 4.6.15. *The Beurling transform $\mathcal{S}\chi$ of the characteristic function χ of a measurable set E is analytic away from the boundary of E.*

Proof. For $z_0 \notin \partial E$ there exists a disk $B = \mathbb{D}(z_0, r)$ that does not intersect the boundary ∂E. Let $F = E \setminus B$. Then $\mathcal{S}\chi_F = \mathcal{S}\chi_E$ inside B by (4.25). From the integral representation we see that $\mathcal{S}\chi_F$ is complex analytic outside F, hence in a neighborhood of the point z_0. □

We now calculate the Beurling transform applied to the characteristic function of a polygon. Let Ω be a polygon bounded by a finite number of segments,

$$\Gamma = [z_1, z_2] \cup [z_2, z_3] \cup \cdots \cup [z_n, z_{n+1}], \qquad z_{n+1} = z_1$$

We first need to evaluate the line integrals $\int_{z_k}^{z_{k+1}}(z-\tau)^{-1}d\bar{\tau}$ for $z \notin [z_k, z_{k+1}]$. For $z \notin [0, R]$ we have

$$\int_0^R \frac{dt}{z-t} = -\mathrm{Log}\left(1 - \frac{R}{z}\right),$$

where Log is the principal branch of logarithm (defined in $\mathbb{C}\setminus(-\infty, 0]$). Changing variables gives

$$\int_{z_k}^{z_{k+1}} \frac{d\bar{\tau}}{z-\tau} = -\frac{\overline{z_{k+1}} - \overline{z_k}}{z_{k+1} - z_k}\mathrm{Log}\left(\frac{z - z_{k+1}}{z - z_k}\right)$$

Hence, for every point z not lying on the boundary $\partial\Omega$ of the polygon,

$$(\mathcal{S}\chi_\Omega)(z) = -\frac{1}{2\pi i}\sum_{k=1}^n \frac{\overline{z_{k+1}} - \overline{z_k}}{z_{k+1} - z_k}\mathrm{Log}\left(\frac{z - z_{k+1}}{z - z_k}\right) \qquad (4.122)$$

4.7. HÖLDER ESTIMATES

We note here the structure of the logarithmic singularities at the vertices of the polygon Ω. In particular, if we let Q denote the rectangle with vertices $\omega, -\omega, \bar{\omega}$ and $-\bar{\omega}$, with ω in the first quadrant of the complex plane, then formula (4.122) reads as

$$S\chi_Q(z) = \frac{1}{2\pi i}\left(\text{Log}\frac{z-\omega}{z-\bar{\omega}} - \text{Log}\frac{z+\bar{\omega}}{z-\omega} + \text{Log}\frac{z+\omega}{z+\bar{\omega}} - \text{Log}\frac{z-\bar{\omega}}{z+\omega}\right)$$

Therefore

$$S\chi_Q(z) = \frac{i}{\pi}\text{Log}\frac{z^2-\bar{\omega}^2}{z^2-\omega^2}, \qquad \text{for } z \notin Q,$$

with the well-defined principal branch of the logarithm.

4.7 Hölder Estimates

Hölder estimates for singular integral operators play an important role in uncovering properties of solutions to elliptic equations in the complex plane. We now discuss such estimates, particularly as they pertain to the Beurling transform.

4.7.1 Hölder Bounds for the Beurling Transform

As we will see, the L^p-theory of the operator S is an essential tool to study the theory of weak solutions of elliptic systems in the complex plane. The C^α-estimates for S play an equal role in the Hölder setting—that is, in obtaining Hölder continuity estimates for solutions to these sorts of equations and similarly for the space BMO. To take care of some (nonintegrable) constant terms, we need to adjust the definition of the Beurling transform as in (4.92) and set

$$(S\phi)(z) = \frac{-1}{\pi}\int_{\mathbb{C}}\left[\frac{1}{(z-\tau)^2} - \frac{1}{\tau(\tau-1)}\right]\phi(\tau)d\tau$$

Here the integral, computed by means of principal values near the singularity $\tau = z$, converges for all $z \in \mathbb{C}$ whenever $\varphi \in C^\alpha(\mathbb{C})$, $0 < \alpha < 1$. The integral also converges near ∞ since $|\varphi(\tau)| = \mathcal{O}(|\tau|^\alpha)$ as $\tau \to \infty$, and thus the integrand has integrable decay at ∞.

Theorem 4.7.1. *The Beurling operator S is a bounded linear operator*

$$S : C^\alpha(\mathbb{C}) \to C^\alpha(\mathbb{C}), \qquad 0 < \alpha < 1$$

Precisely, we have for every $\varphi \in C^\alpha(\mathbb{C})$ the estimate

$$\|S\varphi\|_{C^\alpha} \leqslant \frac{5}{\alpha(1-\alpha)}\|\varphi\|_{C^\alpha} \tag{4.123}$$

Proof. Since elements in C^α are defined only modulo a constant, for the proof it is convenient to consider the operator

$$(S_0\varphi)(z) = (S\varphi)(z) - (S\varphi)(0) = -\frac{1}{\pi}\int_C \left[\frac{1}{(z-\tau)^2} - \frac{1}{\tau^2}\right]\varphi(\tau)\,d\tau \quad (4.124)$$

We have defined the operator S_0 in precisely the above way so that it vanishes on constant functions. This can be seen if we compute $(S_0 1)(z)$ via the definition, but perhaps it is easier to observe this from the explicit formula (4.25) for the Beurling transform of the characteristic function of a disk,

$$(S_0 1)(z) = \lim_{R\to\infty}[(S\chi_R)(z) - (S\chi_R)(0)] = 0,$$

where $\chi_R = \chi_{\mathbb{D}(0,R)}$ and for sufficiently large R both z and 0 lying in the disk. In particular, we can speak of S_0 as an operator defined on equivalence classes of functions modulo the constants.

We must now establish the inequality

$$\begin{aligned}|(S_0\varphi)(z) - (S_0\varphi)(w)| &= \frac{1}{\pi}\left|\int_C \left[\frac{1}{(z-\tau)^2} - \frac{1}{(w-\tau)^2}\right]\varphi(\tau)\,d\tau\right| \\ &\leq \frac{5}{\alpha(1-\alpha)}\|\varphi\|_{C^\alpha}|z-w|^\alpha\end{aligned}$$

We consider an affine change of variable, $2\tau = (z-w)\zeta + z + w$. The points z and w correspond to $\zeta = 1$ and $\zeta = -1$, respectively. Thus putting

$$\phi(\zeta) = \varphi\left(\frac{z-w}{2}\zeta + \frac{z+w}{2}\right),$$

we reduce to the case $z = 1$ and $w = -1$ and therefore to showing that

$$\frac{1}{\pi}\left|\int_C \left[\frac{1}{(\zeta-1)^2} - \frac{1}{(\zeta+1)^2}\right]\phi(\zeta)\,d\zeta\right| \leq \frac{2^\alpha 5}{\alpha(1-\alpha)}\|\phi\|_{C^\alpha} \quad (4.125)$$

At this point it is clear that we may further assume $\|\phi\|_{C^\alpha} = 1$ and $\phi(0) = 0$. This gives us $|\phi(\zeta)| \leq |\zeta|^\alpha$ and $|\phi(\zeta) - \phi(\pm 1)| \leq |\zeta \pm 1|^\alpha$. We now split the integral on the left hand side of (4.125) as

$$= \int_{|\zeta|\leq 2}\left[\frac{1}{(\zeta-1)^2} - \frac{1}{(\zeta+1)^2}\right]\phi(\zeta)\,d\zeta + \int_{|\zeta|>2}\frac{4\zeta}{(\zeta^2-1)^2}\phi(\zeta)\,d\zeta$$

$$= \int_{|\zeta|\leq 2}\frac{\phi(\zeta)-\phi(1)}{(\zeta-1)^2}\,d\zeta - \int_{|\zeta|\leq 2}\frac{\phi(\zeta)-\phi(-1)}{(\zeta+1)^2}\,d\zeta + \int_{|\zeta|>2}\frac{4\zeta}{(\zeta^2-1)^2}\phi(\zeta)\,d\zeta$$

where we have used (4.25) in the form

$$\frac{1}{\pi}\int_{|\zeta|\leq 2}\frac{d\zeta}{(\zeta\pm 1)^2} = -(S\chi)(\pm 1) = 0, \qquad \chi = \chi_{\mathbb{D}(0,2)}$$

4.7. HÖLDER ESTIMATES

We may estimate the first two integrals as follows. Note that

$$\frac{|\phi(\zeta) - \phi(\pm 1)|}{|\zeta \pm 1|^2} \leq \frac{\|\phi\|_{C^\alpha}}{|\zeta \pm 1|^{2-\alpha}} = \frac{1}{|\zeta \pm 1|^{2-\alpha}},$$

so that

$$\left| \int_{|\zeta| \leq 2} \frac{\phi(\zeta) - \phi(1)}{(\zeta - 1)^2} \, d\zeta \right| \leq \int_{|\zeta| \leq 2} \frac{1}{|\zeta - 1|^{2-\alpha}} \, d\zeta$$

$$\leq \left| \int_{|\zeta| \leq 2} \frac{d\zeta}{|\zeta|^{2-\alpha}} \right| = \frac{2^{\alpha+1} \pi}{\alpha}$$

Here we have used a little symmetrization argument to get the second inequality. Of course, this estimate also holds for the -1 case.

Next, we wish to consider the last integral, which can in fact be explicitly computed using the L^2-orthogonality of the power functions $|\zeta|^{-p} \zeta^{-k}$, for $k = 0, 1, 2, \ldots$, and $p > 2$. In particular,

$$\left| \frac{1}{\pi} \int_{|\zeta| > 2} \frac{4\zeta}{(\zeta^2 - 1)^2} \phi(\zeta) \, d\zeta \right| \leq \frac{4}{\pi} \int_{|\zeta| > 2} \frac{1}{|\zeta|^{3-\alpha}} \left| 1 + \frac{1}{\zeta^2} + \frac{1}{\zeta^4} + \cdots \right|^2 d\zeta$$

$$= \frac{4}{\pi} \int_{|\zeta| > 2} \left(\frac{1}{|\zeta|^{3-\alpha}} + \frac{1}{|\zeta|^{7-\alpha}} + \cdots \right) d\zeta$$

$$= 2^{\alpha+2} \left(\frac{1}{1-\alpha} + \frac{1}{(5-\alpha)2^4} + \frac{1}{(9-\alpha)2^8} + \cdots \right)$$

$$\leq \frac{2^{2+\alpha}}{1-\alpha} + \frac{2^\alpha}{15}$$

Finally, putting these estimates together, we see that

$$\|\mathcal{S}\varphi\|_{C^\alpha} \leq \left(\frac{4}{\alpha(1-\alpha)} + \frac{1}{15} \right) \|\varphi\|_{C^\alpha} \tag{4.126}$$

This proves the theorem. $\square$

4.7.2 The Inhomogeneous Cauchy-Riemann Equation

The action of the Cauchy operator on Hölder spaces is most conveniently described in terms of $C^{1,\alpha}(\mathbb{C})$, the space of functions $f : \mathbb{C} \to \mathbb{C}$ whose first partial derivatives f_z and $f_{\bar{z}}$ belong to the space of Hölder continuous functions $C^\alpha(\mathbb{C})$.

We define

$$\|f\|_{C^{1,\alpha}} = \|Df\|_{C^\alpha} := \left\| \frac{\partial f}{\partial z} \right\|_{C^\alpha} + \left\| \frac{\partial f}{\partial \bar{z}} \right\|_{C^\alpha} \tag{4.127}$$

We now set about establishing the following useful theorem, showing that $\mathcal{C} : C^\alpha(\mathbb{C}) \to C^{1,\alpha}(\mathbb{C})$ is a bounded linear operator.

Theorem 4.7.2. *Given $h \in C^\alpha(\mathbb{C})$, $0 < \alpha < 1$, the equation*

$$\frac{\partial f}{\partial \bar{z}} = h \qquad (4.128)$$

has a solution in $C^{1,\alpha}(\mathbb{C})$. Such a solution is unique up to an additive affine function $\varphi(z) = az + b$, and we have the estimate

$$\|Df\|_{C^\alpha} \leqslant \frac{5}{\alpha(1-\alpha)} \|h\|_{C^\alpha}$$

Proof. The Cauchy transform associated with the operator $\mathcal{S}_0$ from (4.124) is given by the rule

$$(\mathcal{C}_0 \varphi)(z) = \frac{1}{\pi} \int_{\mathbb{C}} \left[\frac{1}{z-\tau} + \frac{1}{\tau} + \frac{z}{\tau^2} \right] \varphi(\tau)\, d\tau = \frac{1}{\pi} \int_{\mathbb{C}} \frac{z^2}{\tau^2} \frac{\varphi(\tau)}{z-\tau}\, d\tau \qquad (4.129)$$

The result follows immediately from Theorem 4.7.1 and the estimate (4.126) since $f = \mathcal{C}_0 h$ satisfies

$$\|Df\|_{C^\alpha} = \|f_{\bar{z}}\|_{C^\alpha} + \|\mathcal{S}f_{\bar{z}}\|_{C^\alpha} \leqslant \left(1 + \frac{4}{\alpha(1-\alpha)} + \frac{1}{15}\right) \|h\|_{C^\alpha} \leqslant \frac{5\|h\|_{C^\alpha}}{\alpha(1-\alpha)}$$

Of course, as with our earlier cases, the existence of some function f solving (4.128) is not really the issue—it is in identifying the space where it lies. □

4.8 Beurling Transforms for Boundary Value Problems

The inhomogeneous Cauchy-Riemann system

$$\frac{\partial f}{\partial \bar{z}} = \phi \qquad (4.130)$$

can also be studied in domains $\Omega \subset \mathbb{C}$. To ensure uniqueness we must of course impose some boundary conditions on the solution f. This can actually be done in a number of natural and different ways. It always depends on the problem under study how the choice of the boundary condition should be made. Once the choice has been made, this also determines the appropriate Cauchy and Beurling transforms for the problem in question.

We will not present a systematic development of all possible operators arising from different boundary normalizations but rather concentrate in this section on the few most important cases, to be applied later in this monograph. However, the boundary operators presented here also give a good illustration of the general methods.

4.8.1 The Beurling Transform on Domains

Solutions to the equation $f_{\bar{z}} = h$ in a domain Ω are determined only up to an additive analytic function. If Ω is bounded, a first natural choice for a condition to ensure uniqueness is to ask that one of the components of $f = u + iv$, say u, vanishes on the boundary $\partial\Omega$. Hence for sufficiently regular domains Ω we require
$$\Re e(f) = 0 \quad \text{on } \partial\Omega$$
We will denote by $W_{\Re}^{1,p}(\Omega)$ the space of complex-valued Sobolev functions $f = u + iv$ whose real part vanishes on the boundary of Ω. That is, $f = u + iv$ and
$$u \in W_0^{1,p}(\Omega, \mathbb{R}), \qquad v \in W^{1,p}(\Omega, \mathbb{R})$$

Let us start with the case $\phi \in L^2(\Omega)$. When Ω is simply connected with a sufficiently regular boundary, finding a solution to (4.130) in the class $W_{\Re}^{1,2}(\Omega)$ reduces to the basic Dirichlet problem: If some $f_0 \in W^{1,2}(\Omega)$ solves the equation, let u be harmonic in Ω with $u = \Re e\, f_0$ on $\partial\Omega$. Then $f = f_0 - (u+iv)$ with v the harmonic conjugate is a solution in the class $W_{\Re}^{1,2}(\Omega)$. For uniqueness, it only requires an integration by parts to show that, for $f \in W_{\Re}^{1,2}(\Omega)$, the integral of the Jacobian of $J(z, f) = |f_z|^2 - |f_{\bar{z}}|^2$ vanishes. Indeed, a simple manipulation gives
$$J(z, f) = -\frac{i}{2}(u_z v_{\bar{z}} - u_{\bar{z}} v_z), \qquad f = u + iv, \tag{4.131}$$
and therefore from Green's formula (2.52),
$$\int_\Omega J(z,f) = -\frac{i}{2} \int_\Omega (u_z v_{\bar{z}} - u_{\bar{z}} v_z) = \frac{1}{4} \int_{\partial\Omega} u(v_z + v_{\bar{z}}) = 0 \tag{4.132}$$
as $u|\partial\Omega = 0$. Thus, in particular, the $W_{\Re}^{1,2}(\Omega)$ solution to (4.130) with $\phi \in L^2$ is unique up to an imaginary constant. In any case, the z-derivative of f is unique, and so we may define an operator by the rule
$$\frac{\partial f}{\partial z} = \mathcal{S}_\Omega(\phi) \tag{4.133}$$

The operator $\mathcal{S}_\Omega : L^2(\Omega) \to L^2(\Omega)$ is an isometry since the integral of the Jacobian $J(z, f)$ vanishes. We call this operator the *Beurling transform in* Ω. To simplify the following discussion here we shall restrict ourselves to the case
$$\Omega = \mathbb{D}$$
The solution to (4.130) is then given explicitly by the formula
$$\begin{aligned} f(z) &= \frac{1}{\pi} \int_\mathbb{D} \left(\frac{\phi(\tau)}{z-\tau} - \frac{z\,\overline{\phi(\tau)}}{1-z\bar{\tau}} \right) d\tau \\ &= (\mathcal{C}_\mathbb{D}\,\phi)(z) \end{aligned} \tag{4.134}$$

The integral above defines the operator $\mathcal{C}_\mathbb{D}$, which is linear over the real numbers and will be called the Cauchy-transform on $\mathbb{D}$.

It is clear that $f = \mathcal{C}_\mathbb{D}\phi$ has a vanishing real part on the unit circle, as the integrand is purely imaginary there. The second term in the integrand represents an analytic function in $\mathbb{D}$ and therefore does not affect the $\bar{z}$-derivative of f. Consequently, we have

$$\frac{\partial}{\partial \bar{z}}(\mathcal{C}_\mathbb{D}\phi) = \phi \tag{4.135}$$

Now, of course, the Beurling transform on $\mathbb{D}$ is the z-derivative of $\mathcal{C}_\mathbb{D}$ and therefore can be found by the integral formula

$$(\mathcal{S}_\mathbb{D}\phi)(z) = \frac{\partial}{\partial z}(\mathcal{C}_\mathbb{D}\phi) = -\frac{1}{\pi}\int_\mathbb{D}\left(\frac{\omega(\tau)}{(z-\tau)^2} + \frac{\overline{\omega(\tau)}}{(1-z\bar{\tau})^2}\right)d\tau, \tag{4.136}$$

and as $\mathcal{S}_\mathbb{D}$ is an isometry in $L^2(\mathbb{D})$, we have

$$\Re\int_\mathbb{D}(\mathcal{S}_\mathbb{D}\phi)(\overline{\mathcal{S}_\mathbb{D}\psi}) = \Re\int\phi\bar\psi \tag{4.137}$$

for every $\phi, \psi \in L^2(\mathbb{D})$. For showing that $\mathcal{S}_\mathbb{D}$ is onto and to calculate its inverse, it is convenient to use another interpretation of the operator. To this end let us apply (4.134) to solve the inhomogeneous equation

$$\frac{\partial f}{\partial \bar{z}} = \bar\phi, \qquad f \in W^{1,2}_\Re(\mathbb{D}),$$

where ϕ is an arbitrary function in $L^2(\mathbb{D})$. Thus

$$\frac{\partial f}{\partial z} = \mathcal{S}_\mathbb{D}\bar\phi$$

We conjugate these two equations to obtain

$$\frac{\partial \bar f}{\partial z} = \phi, \qquad \frac{\partial \bar f}{\partial \bar z} = \overline{\mathcal{S}_\mathbb{D}\bar\phi}$$

As $\bar f \in W^{1,2}_\Re(\mathbb{D})$, we find the identity

$$\phi = \mathcal{S}_\mathbb{D}(\overline{\mathcal{S}_\mathbb{D}\bar\phi}),$$

which simply means that

$$\mathcal{S}_\mathbb{D}^{-1}\phi = \overline{\mathcal{S}_\mathbb{D}\bar\phi}$$

A further useful identity follows from (4.137) if we apply it to $\overline{\mathcal{S}_\mathbb{D}\psi}$ in place of ψ. This substitution yields

$$\Re\int_\mathbb{D}\psi(\mathcal{S}_\mathbb{D}\phi) = \Re\int\phi(\mathcal{S}_\mathbb{D}\psi) \tag{4.138}$$

again for all $\phi, \psi \in L^2(\mathbb{D})$. The reader may wish to check that the L^2-identities (4.137) and (4.138) are actually valid for any domain Ω and not just the unit disk, the derivation being essentially the same.

4.8.2 L^p-Theory

Now we briefly discuss the L^p-theory of the operator $\mathcal{S}_\mathbb{D}$, for $1 < p < \infty$. We begin with the following theorem.

Theorem 4.8.1. *The operator $\mathcal{S}_\mathbb{D} : L^2(\mathbb{D}) \to L^2(\mathbb{D})$ extends continuously to an operator $\mathcal{S}_\mathbb{D} : L^p(\mathbb{D}) \to L^p(\mathbb{D})$ for all $1 < p < \infty$.*

Proof. We first consider the case $p \geq 2$. Given $\phi \in L^p(\mathbb{D}) \subset L^2(\mathbb{D})$, we solve the equation
$$\frac{\partial f}{\partial \bar{z}} = \phi, \qquad f \in W^{1,2}_\Re(\mathbb{D}), \tag{4.139}$$
and extend the solution f to the entire complex plane by the rule
$$\tilde{f}(z) = \begin{cases} f(z), & |z| \leq 1 \\ -\overline{f(1/\bar{z})}, & |z| \geq 1 \end{cases}$$
As $\Re\{f\} = 0$ for $|z| = 1$, we see that in fact $\tilde{f} \in W^{1,2}(\mathbb{C})$. The complex derivatives of $\tilde{f}$ are easily computed to be
$$\frac{\partial \tilde{f}}{\partial \bar{z}}(z) = \begin{cases} f_{\bar{z}}(z), & |z| \leq 1 \\ -\overline{f_z(1/\bar{z})}/\bar{z}^2, & |z| \geq 1 \end{cases}$$
$$\frac{\partial \tilde{f}}{\partial z}(z) = \begin{cases} f_z(z), & |z| \leq 1 \\ -\overline{f_{\bar{z}}(1/\bar{z})}/z^2, & |z| \geq 1 \end{cases}$$
Notice that in fact $\tilde{f}_{\bar{z}} \in L^p(\mathbb{C})$. Indeed, we have the estimate
$$\int_\mathbb{C} |\tilde{f}_{\bar{z}}(z)|^p \, dz = \int_\mathbb{D} |\phi(z)|^p \, dz$$
$$+ \int_{|z| \geq 1} \frac{1}{|z|^{2p}} |f_{\bar{z}}(1/z)|^p \, dz$$
$$= \int_\mathbb{D} (1 + |z|^{2p-4}) |f_{\bar{z}}(z)|^p \, dz \leq 2 \int |\phi|^p$$
using an obvious substitution in the second integral.

A similar computation can now be performed for the z-derivative of $\tilde{f}$ if in fact we know that $f_z \in L^p(\mathbb{D})$, for we would then obtain the identity
$$\int_\mathbb{C} |\tilde{f}_z(z)|^p \, dz = \int_\mathbb{D} (1 + |z|^{2p-4}) |f_z(z)|^p \, dz \tag{4.140}$$

The identity (4.140) is certainly true for $p = 2$, hence $\tilde{f}_z \in L^2(\mathbb{C})$. We may therefore apply the usual Beurling transform $\mathcal{S} : L^2(\mathbb{C}) \to L^2(\mathbb{C})$, which shows us that
$$\frac{\partial \tilde{f}}{\partial z}(z) = \left(\mathcal{S} \frac{\partial \tilde{f}}{\partial \bar{z}}\right)(z)$$

But $\mathcal{S}$ is bounded in the spaces $L^p(\mathbb{C})$, $p \geqslant 2$, and we have already put $\tilde{f}_{\bar{z}} \in L^p(\mathbb{C})$. Thus (4.140) is in fact valid. Now, combining the identities above, we find that we have established

$$\int_{\mathbb{D}} (1+|z|^{2p-4})|\mathcal{S}_{\mathbb{D}}\phi|^p = \int_{\mathbb{C}} |\tilde{f}_z|^p = \int_{\mathbb{C}} |\mathcal{S}\tilde{f}_{\bar{z}}|^p$$
$$\leqslant \mathbf{S}_p^p \int_{\mathbb{C}} |\tilde{f}_{\bar{z}}|^p = \mathbf{S}_p^p \int_{\mathbb{D}} (1+|z|^{2p-4})|f_{\bar{z}}|^p$$

That is,

$$\int_{\mathbb{D}} (1+|z|^{2p-4})|\mathcal{S}_{\mathbb{D}}\phi|^p \leqslant \mathbf{S}_p^p \int_{\mathbb{D}} (1+|z|^{2p-4})|\phi|^p \qquad (4.141)$$

The estimate in (4.141) reduces to an equality when $p = 2$, as $\mathcal{S}$ is an isometry there. The fact that $1 \leqslant 1 + |z|^{2p-4} \leqslant 2$ leads us directly to the estimate

$$\int_{\mathbb{D}} |\mathcal{S}_{\mathbb{D}}\phi|^p \leqslant 2\mathbf{S}_p^p \int_{\mathbb{D}} |\phi|^p,$$

so that $\mathcal{S}_{\mathbb{D}}$ is in fact a bounded operator from $L^p(\mathbb{D})$ into $L^p(\mathbb{D})$; moreover, we have

$$\|\mathcal{S}_{\mathbb{D}}\|_p \leqslant 2^{1/p} \|\mathcal{S}\|_p \qquad (4.142)$$

We now need to examine that boundedness of the operator $\mathcal{S}_{\mathbb{D}}$ in the range $1 < q \leqslant 2$. Given such a q, let

$$\phi \in L^q(\mathbb{D}), \qquad \psi \in L^p(\mathbb{D}); \qquad \frac{1}{p} + \frac{1}{q} = 1$$

We use the identity in (4.138) to find that

$$\left| \Re \mathrm{e} \int_{\mathbb{D}} \psi \mathcal{S}_{\mathbb{D}} \phi \right| \leqslant \int |\phi| |\mathcal{S}_{\mathbb{D}} \psi|$$
$$\leqslant \|\phi\|_q \|\mathcal{S}_{\mathbb{D}} \psi\|_p \leqslant \|\mathcal{S}_{\mathbb{D}}\|_p \|\psi\|_p \|\phi\|_q$$

As the function ψ is arbitrary, we find that

$$\|\mathcal{S}_{\mathbb{D}} \phi\|_q \leqslant \|\mathcal{S}_{\mathbb{D}}\|_p \|\phi\|_q,$$

which shows that $\mathcal{S}_{\mathbb{D}} : L^q(\mathbb{D}) \to L^q(\mathbb{D})$ is bounded, and moreover, $\|\mathcal{S}_{\mathbb{D}}\|_q \leqslant \|\mathcal{S}_{\mathbb{D}}\|_p$ when p and q are Hölder conjugate indices. However, using the identity in (4.138) again in the obvious manner gives us the additional identity

$$\|\mathcal{S}_{\mathbb{D}}\|_q = \|\mathcal{S}_{\mathbb{D}}\|_p$$

Finally, these calculations suggest that it is possible that $\|\mathcal{S}_{\mathbb{D}}\|_p = \|\mathcal{S}\|_p = \mathbf{S}_p$.

4.8.3 Complex Potentials for the Dirichlet Problem

In this section we want to look at applications of complex potentials to the solution of the Dirichlet problem

$$\left.\begin{array}{rl} u_{z\bar{z}} &= \varphi, \quad \varphi \in L^2(\mathbb{D}) \\ u &\in W_0^{1,2}(\mathbb{D}) \cap W^{2,2}(\mathbb{D}), \end{array}\right\} \qquad (4.143)$$

where $4u_{z\bar{z}} = \Delta u$ is the Laplacian. This problem has a unique solution given explicitly by the formula

$$u(z) = \frac{2}{\pi} \int_{\mathbb{D}} \log \frac{|z-\tau|}{|1-z\bar{\tau}|} \varphi(\tau)\, d\tau, \qquad (4.144)$$

where the integral kernel is the Green's function in the unit disk $\mathbb{D}$. Given a function $\varphi \in L^p(\mathbb{D})$, $1 < p < \infty$, and u a solution to (4.143), we define the *Cauchy transform for the Dirichlet problem* as

$$(\mathcal{C}_\Delta \varphi)(z) = \frac{\partial u}{\partial z} = \frac{1}{\pi} \int_{\mathbb{D}} \left(\frac{1}{z-\tau} + \frac{\bar{\tau}}{1-z\bar{\tau}} \right) \varphi(\tau)\, d\tau \qquad (4.145)$$

The operator $\mathcal{C}_\Delta$ is hence induced by the z-derivative of the Green's function. It therefore depends also on the domain considered even if this dependence is not denoted explicitly. Note also that $\partial_{\bar{z}}(\mathcal{C}_\Delta \varphi) = \varphi$.

It is fairly clear that if $p > 2$, then $g = \mathcal{C}_\Delta \varphi$ is a continuous function on the closed disk $\bar{\mathbb{D}}$. Also, it is possible to see that for a real-valued function φ the quantity $zg(z)$ is real-valued on the unit circle, as for $|z| = 1$, we have

$$z\left(\frac{1}{z-\tau} + \frac{\bar{\tau}}{1-z\bar{\tau}}\right) = \frac{z}{z-\tau} + \frac{\bar{\tau}}{\bar{z}-\bar{\tau}} = \frac{1-|\tau|^2}{|z-\tau|^2}$$

Now, of course, we should define the corresponding Beurling transform as the z-derivative of $\mathcal{C}_\Delta \varphi$. We will call this new operator the *Beurling transform for the Dirichlet problem*. In the disk $\mathbb{D}$ this transform is expressed by the following singular integral formula:

$$(\mathcal{S}_\Delta \varphi)(z) = -\frac{1}{\pi} \int_{\mathbb{D}} \left(\frac{1}{(z-\tau)^2} + \frac{\bar{\tau}^2}{(1-z\bar{\tau})^2} \right) \varphi(\tau)\, d\tau \qquad (4.146)$$

We next propose to establish the following lemma.

Lemma 4.8.2. *For every real-valued function $\varphi \in L^2(\mathbb{D})$, we have*

$$\|\mathcal{S}_\Delta \phi\|_2 \leq \|\varphi\|_2 \qquad (4.147)$$

In fact, $\|\mathcal{S}_\Delta : L^2(\mathbb{D}) \to L^2(\mathbb{D})\| = 1$.

Proof. We may assume that φ is $C^\infty(\mathbb{D})$-smooth on $\bar{\mathbb{D}}$, so that its Cauchy transform $g = \mathcal{C}_\Delta \varphi$ is also smooth on $\bar{\mathbb{D}}$, and we have

$$zg(z) = \bar{z}\bar{g}(z), \qquad z \in \partial \mathbb{D}$$

Let us calculate the integral of the Jacobian determinant of g. Using Green's formula (2.52), we have

$$\int_\mathbb{D} J(z,g) = \int_\mathbb{D} (g\,\overline{g_{\bar z}})_z - (g\,\overline{g_z})_{\bar z} = \frac{i}{2}\int_{\partial\mathbb{D}} g\,\overline{g_z}\,d\bar z + g\,\overline{g_{\bar z}}\,dz \qquad (4.148)$$

$$= \frac{i}{2}\int_{\partial\mathbb{D}} z\bar z\, g\, d\bar g = \frac{i}{2}\int_{\partial\mathbb{D}} \bar z^2 \bar g\, d\bar g = \frac{i}{4}\int_{\partial\mathbb{D}} \bar z^2\, d\bar g^2$$

$$= -\frac{i}{4}\int_{\partial\mathbb{D}} \bar g^2\, d\bar z^2 = -\frac{i}{2}\int_{\partial\mathbb{D}} \bar z\bar g\, \bar g\, d\bar z = -\frac{i}{2}\int_{\partial\mathbb{D}} z|g|^2\, d\bar z$$

$$= -\frac{1}{2}\int_{\partial\mathbb{D}} |g|^2\, |dz| \qquad (4.149)$$

This then provides us with the identity

$$\int_\mathbb{D} |g_z|^2 = \int_\mathbb{D} |g_{\bar z}|^2 - \frac{1}{2}\int_{\partial\mathbb{D}} |g|^2 |dz| \qquad (4.150)$$

whenever $g \in W^{1,2}(\mathbb{D})$ and $zg - \bar z \bar g \in W_0^{1,2}(\mathbb{D})$. We see that $\|\mathcal{S}_\Delta \varphi\|_2 \leqslant \|\varphi\|_2$ and that equality is attained by taking $\varphi = g_{\bar z}$, where g vanishes on $\partial\mathbb{D}$. $\square$

Cauchy and Beurling Transforms for the Dirichlet Problem

We next make analogous computations for a smooth convex Jordan domain in $\mathbb{C}$. Let us suppose that the boundary of Ω is given by the defining equation $\Gamma(z) = 0$,

$$\bar\Omega = \{z : \Gamma(z) \leqslant 0,\ \Gamma_z \neq 0\},$$

where $\Gamma \in C^\infty(\bar\Omega)$ is a real-valued function defined in a neighborhood of $\partial\Omega$ and has a nonvanishing gradient. If we parameterize the boundary curve by arclength,

$$z(t) = (x(t), y(t)),$$

then upon differentiating the equation $\Gamma(z(t)) \equiv 0$ we obtain $\dot z \Gamma_z + \bar{\dot z}\Gamma_{\bar z} = 0$. This shows us that

$$\dot z = \rho(z) := \frac{i\overline{\Gamma_z}}{|\Gamma_z|}$$

is the boundary tangent vector of unit length.

Let $G(z,\tau)$, $z,\tau \in \bar\Omega$, $z \neq \tau$, denote the Green's function of Ω. Thus for any real-valued function $\varphi \in L^2(\Omega)$ we put

$$v = \mathcal{L}_\Delta(\varphi) = \int_\Omega G(z,\tau)\,\varphi(\tau)\,d\tau \in W_0^{1,2}(\Omega) \cap W^{2,2}(\Omega), \qquad (4.151)$$

which vanishes on $\partial\Omega$ in the sense of distributions and moreover satisfies the equation

$$\frac{\partial^2}{\partial z \partial \bar z} v = \varphi, \qquad \Re\left(\rho\frac{\partial v}{\partial z}\right) \in W_0^{1,2}(\Omega) \qquad (4.152)$$

4.8. BOUNDARY VALUE PROBLEMS

Given this, and in the spirit of the previous subsection, we introduce an associated Cauchy transform

$$\mathcal{C}_\Delta(\varphi)(z) = v_z = \int_\Omega G_z(z,\tau)\varphi(\tau)\,d\tau \tag{4.153}$$

and the associated Beurling transform

$$\mathcal{S}_\Delta(\varphi)(z) = v_{zz} = \int_\Omega G_{zz}(z,\tau)\varphi(\tau)\,d\tau \tag{4.154}$$

Explicit formulas for each of these operators in the case $\Omega = \mathbb{D}$, the unit disk, are given in (4.145) and (4.146). Arguing as in Theorem 4.8.1, we see that for the disk, $\mathcal{S}_\Delta : L^p(\Omega) \to L^p(\Omega)$ is a continuous linear operator. This is actually true for all bounded smooth domains Ω although we will not give the details here. What we will do is establish the following theorem.

Theorem 4.8.3. *Let Ω be a smoothly bounded convex domain. The operator norm*

$$\|\mathcal{S}_\Delta : L^2(\Omega) \to L^2(\Omega)\| = 1$$

Proof. We need to show that for all real-valued functions $v \in W^{2,2}(\Omega) \cap W^{1,2}_0(\Omega)$, the following inequality holds:

$$\int_\Omega |v_{zz}|^2 \leq \int_\Omega |v_{z\bar{z}}|^2 \tag{4.155}$$

Naturally, the inequality in (4.155) holds with equality for every $v \in C_0^\infty(\Omega)$. In general, we can of course assume that $v \in C^\infty(\bar{\Omega})$ and $v = 0$ on $\partial\Omega$, as these functions are dense in $W^{2,2}(\Omega) \cap W^{1,2}_0(\Omega)$. Now the inequality in (4.155) is an immediate consequence of the following lemma, which seems to be of independent interest. □

Lemma 4.8.4. *Let Ω be a smoothly bounded convex domain and $v \in C^\infty(\bar{\Omega})$ with $v = 0$ on $\partial\Omega$. Then*

$$2\int_\Omega \mathcal{H}v = \int_{\partial\Omega} |\nabla v|^2 \kappa \geq 0, \tag{4.156}$$

where $\mathcal{H}$ is the Hessian operator

$$\mathcal{H}v = v_{xx}v_{yy} - v_{xy}^2$$

and κ is the curvature of $\partial\Omega$.

Proof. If equal, both terms in (4.156) are positive as Ω is convex, so the curvature is positive. We observe that if $g = v_z$, then

$$\mathcal{H}v = -4(|g_z|^2 - |g_{\bar{z}}|^2) = -4J(z,g),$$

so that (4.156) has the alternate formulation in terms of the complex gradient of v. That is,

$$2\int_\Omega J(z,g) = -\int_{\partial\Omega} |g|^2 \kappa \leq 0 \qquad (4.157)$$

Let $z(s) = x(s)+iy(s)$ be the arc-length parameterisation of $\partial\Omega$. Then $\dot{x}^2+\dot{y}^2 \equiv 1$ and the curvature is given by $\kappa = \dot{x}\ddot{y} - \dot{y}\ddot{x}$. Since v vanishes on the boundary of Ω we have

$$\dot{x}\, v_x + \dot{y}\, v_y = 0, \qquad \text{on } \partial\Omega \qquad (4.158)$$

We differentiate (4.158) with respect to the arc length parameter s to obtain the following relation between the partials of v:

$$\dot{x}^2\, v_{xx} + \dot{y}^2\, v_{yy} + 2\dot{x}\,\dot{y}v_{xy} + \ddot{x}\, v_x + \ddot{y}\, v_y = 0 \qquad (4.159)$$

We also note that the Hessian volume form is exact,

$$2(v_{xx}v_{yy} - v_{xy}^2)\, dx\, dy = d[(v_x v_{xy} - v_y v_{xx})\, dx + (v_x v_{yy} - v_y v_{xy})\, dy]$$

We are in a situation where we may apply Stokes' or Green's formula to obtain

$$2\int_\Omega \mathcal{H}v = \int_{\partial\Omega} (v_x v_{xy} - v_y v_{xx})\, dx + (v_x v_{yy} - v_y v_{xy})\, dy$$

On the other hand, (4.158) shows us that there is a function $\lambda = \lambda(x,y)$ such that $v_x = \lambda \dot{y}$ and $v_y = -\lambda \dot{x}$ and where $\lambda^2 = |\nabla v|^2$. Of course, $dx = \dot{x}\, ds$ and $dy = \dot{y}\, ds$, and so have the equation

$$\begin{aligned}
2\int_\Omega \mathcal{H}v &= \int_{\partial\Omega} \lambda\left(\dot{x}^2 v_{xx} + \dot{y}^2 v_{yy} + 2\dot{x}\dot{y}v_{xy}\right) ds = -\int_{\partial\Omega} \lambda\left(\ddot{x}v_x + \ddot{y}v_y\right) ds \\
&= \int_{\partial\Omega} \lambda^2(\dot{x}\ddot{y} - \dot{y}\ddot{x})\, ds = \int_{\partial\Omega} |\nabla v|^2 \kappa(s)\, ds,
\end{aligned}$$

as required. $\square$

As a consequence of Theorem 4.8.3 and the Riesz-Thorin interpolation theorem, we must have

$$\|\mathcal{S}_\Delta : L^p(\Omega) \to L^p(\Omega)\| \to 1$$

as $p \to 2$.

4.9 Complex Potentials in Multiply Connected Domains

In what follows we shall encounter multiply connected domains and fairly general Riemann mapping theorems for problems associated with nonlinear elliptic equations posed on such domains. We will need the integral operators discussed above in this more general setting, and it is the purpose of this last section

4.9. MULTIPLY CONNECTED DOMAINS

to introduce them. In particular, we will need the Cauchy transform and the Beurling transform defined on multiply connected domains.

Thus let Ω be a smoothly bounded multiply connected domain, say of connectivity n. Thus $\hat{\mathbb{C}} \setminus \Omega$ has n components and $\partial\Omega$ consists of n smooth Jordan curves,

$$\partial\Omega = \Gamma_1 \cup \cdots \cup \Gamma_n$$

The Cauchy transform is the linear (over the real numbers) operator

$$\mathcal{C}_\Omega : L^p(\Omega) \to W^{1,p}(\Omega), \qquad 1 < p < \infty,$$

with the following properties:

1. $\frac{\partial}{\partial \bar{z}} \circ \mathcal{C}_\Omega = \mathbf{I} : L^p(\Omega) \to L^p(\Omega)$

2. $\frac{\partial}{\partial z} \circ \mathcal{C}_\Omega = \mathcal{S}_\Omega : L^p(\Omega) \to L^p(\Omega)$, where $\mathcal{S}_\Omega$ is the Beurling transform in Ω

3. To every boundary component Γ_i, $i = 1, 2, \ldots n$, there is a corresponding bounded linear functional

$$\mathcal{F}_i : L^p(\Omega) \to \mathbb{R}$$

such that

$$\Re\left\{(\mathcal{C}_\Omega \omega)(z)\right\} = \mathcal{F}_i \omega, \qquad z \in \Gamma_i$$

For such smoothly bounded domains, the Cauchy transform always exists and is unique up to an additive functional $\mathcal{F} : L^p(\Omega) \to \mathbb{C}$. In this way the Beurling transform, the z-derivative of $\mathcal{C}_\Omega$, is uniquely determined by the Cauchy transform and hence by Ω. Indeed, the construction of $\mathcal{C}_\Omega$ and $\mathcal{S}_\Omega$ reduces, via a conformal change of variables, to the case of Schottky's circular domain, that is, the plane $\mathbb{C}$ with n disjoint disks removed. An effective construction of Green's function for Schottky domains is given in [173]. Based on this construction we have explicit formulas for the kernel of the operators $\mathcal{C}_\Omega$ and $\mathcal{S}_\Omega$, describing their behavior at the boundary circles.

It is not too difficult to see, using integration by parts, that $\mathcal{S}_\Omega : L^2(\Omega) \to L^2(\Omega)$ is an isometry, however, we pose the following problem.

Problem. Is it true that $\|\mathcal{S}_\Omega\|_p = \|\mathcal{S}_\mathbb{C}\|_p = \mathbf{S}_p$ independently of the domain ?

To give an idea of the complexity involved in finding explicit formulas for these transforms even in relatively simple cases, we compute the Cauchy transform on the annulus $A = \{z : r < |z| < 1\}$. There we have

$$\begin{aligned}
(\mathcal{C}_A \omega)(z) &= \frac{1}{\pi} \int_A \left(\frac{\omega(\tau)}{z - \tau} - \frac{z\overline{\omega(\tau)}}{1 - z\bar{\tau}} \right) d\tau \\
&+ \frac{1}{2\pi} \int_A \frac{\omega(\tau)}{\tau} \sum_{k=1}^{\infty} \left(\frac{z + r^{2k}\tau}{z - r^{2k}\tau} + \frac{zr^{2k} + \tau}{zr^{2k} - \tau} \right) d\tau \\
&- \frac{1}{2\pi} \int_A \frac{\overline{\omega(\tau)}}{\tau} \sum_{k=1}^{\infty} \left(\frac{r^{2k} + z\bar{\tau}}{r^{2k} - z\bar{\tau}} + \frac{1 + r^{2k}z\bar{\tau}}{1 - r^{2k}z\bar{\tau}} \right) d\tau
\end{aligned}$$

The boundary functionals $\mathcal{F}_0, \mathcal{F}_1 : L^p(A) \to \mathbb{R}$ corresponding to the boundary circles $\Gamma_0 = \{|z| = 1\}$ and $\Gamma_1 = \{|z| = r\}$ are in fact given by

$$\begin{aligned} \mathcal{F}_0 \omega &= 0 \\ \mathcal{F}_1 \omega &= -\frac{1}{\pi} \mathfrak{Re} \int_A \frac{\omega(\tau)}{\tau} d\tau \end{aligned}$$

We should emphasize that the values of $(\mathcal{C}_A \omega)(z)$ on the boundary curves have constant real parts. However, these constants depend on the particular density function $\omega \in L^p(A)$. The general Riemann mapping problem for nonlinear elliptic PDEs will provide us with a particular ω associated with a normalized solution for each given system and these constants have a particular geometric meaning. They determine the moduli space of solutions to these nonlinear elliptic PDEs.

Chapter 5

The Measurable Riemann Mapping Theorem: The Existence Theory of Quasiconformal Mappings

It is the aim of this section to present a complete proof, in as elementary fashion as possible, of the classical existence theory for quasiconformal mappings, or what has come to be known as the measurable Riemann mapping theorem, originally due to Morrey in 1938 [271]. This theorem is a cornerstone in the various interactions among PDEs, complex analysis and geometry, the basic themes of this monograph. Roughly, it states that any bounded measurable conformal structure—or any measurable ellipse field—is equivalent to the usual (round) structure via a quasiconformal change of coordinates. The result underpins the relationship between quasiconformal mappings and Teichmüller theory, holomorphic dynamics, Kleinian groups, as well as the interactions between quasiconformal mappings and nonlinear elliptic PDEs in two-dimensions. For instance, a powerful technique in the study of holomorphic dynamical systems is called *quasiconformal surgery*. It plays a central role in renormalization and in the theory of polynomial-like mappings developed by A. Douady and J. Hubbard [111] and provides a method for constructing new rational maps with certain dynamical properties from existing ones. One modifies a rational map in some part of $\mathbb{C}$ (perhaps infinitely often in a sequence of preimages of a given Fatou component). This modification cannot be analytic, but if one is careful enough to make it quasiregular, then the measurable Riemann mapping theorem tells us that after a global quasiconformal change of coordinates, we may recover analyticity; see M. Shishikura's expository article [327] for a more detailed account (in the introduction there is an illustration of this technique).

A measurable conformal structure transported to the usual round structure by a quasiconformal mapping

Ultimately, the measurable Riemann mapping theorem asserts the existence of homeomorphic solutions to the Beltrami equation

$$\frac{\partial f}{\partial \bar{z}} = \mu(z) \frac{\partial f}{\partial z}$$

for arbitrary measurable μ subject to $\|\mu\|_\infty \leqslant k < 1$, and it was in this form that the result was first proven by Morrey [271]. There are alternate approaches to the solution such as that given in Lehto-Virtanen [229]. However, the theme of this book is analytic, and so we give a proof along the ideas of Bojarski [64] and the presentation of Ahlfors in his famous lecture notes [6]. This is not only for ease of presentation, as this approach ultimately gives far stronger results and is more amenable to generalization. In subsequent chapters we will present a variety of generalizations of this result that differ in regard to their assumptions. For instance, we will relax the uniform ellipticity bounds on the Beltrami equations or study more general types of equations both quasilinear and nonlinear.

For our proof of the classical result we will need only basic facts from harmonic analysis in the complex plane and these can all be found in the text, see Chapter 4.

First, we recall that for the Cauchy transform

$$\mathcal{C} : L_0^p(\mathbb{C}) \to W^{1,p}(\mathbb{C}), \ p > 2, \quad \text{with} \quad \mathcal{C} : C_0^\infty(\mathbb{C}) \to C^\infty(\mathbb{C})$$

Second, for $g \in L^2(\mathbb{C})$, or in $L^p(\mathbb{C})$ where $1 < p < \infty$, the derivatives

$$(\mathcal{C}g)_{\bar{z}} = g, \qquad (\mathcal{C}g)_z = \mathcal{S}g$$

In particular, $\mathcal{S}f_{\bar{z}} = f_z$ for functions $f \in W^{1,p}(\mathbb{C})$. Thus the main results we need are the boundedness of the Beurling transform $\mathcal{S} : L^p(\mathbb{C}) \to L^p(\mathbb{C})$ and the continuity of its p-norms. Both of these are general facts about Calderón-Zygmund-type operators (of which the Beurling transform is but one) and the Riesz-Thorin interpolation theorem. That $\mathbf{S}_2 = \|\mathcal{S}\|_{L^2(\mathbb{C}) \to L^2(\mathbb{C})} = 1$ was easy to see, while the Riesz-Thorin theorem implies that the map $p \mapsto \mathbf{S}_p$ is continuous in p. Therefore, for each $k < 1$, there is a Hölder conjugate pair of numbers

$$1 < Q(k) < 2 < P(k) < \infty$$

such that

$$k\|\mathcal{S}\|_{L^p(\mathbb{C}) \to L^p(\mathbb{C})} < 1, \qquad \text{for all } p \text{ with } Q(k) < p < P(k) \tag{5.1}$$

This is the deepest fact we shall use in the proof. Recall that it is believed that $\mathbf{S}_p = \max\{p-1, (p-1)^{-1}\}$. If this is the case, then the extremal exponents are $Q(k) = 1+k$ and $P(k) = 1 + 1/k$.

The reader should be aware that we will in fact get to these extremal exponents later in the text via completely different methods, as a consequence of the area distortion Theorem 13.2.3, but in a slightly weaker context of invertible Beltrami operators; see Sections 5.4 and 14.3. We will also discuss these p-norm estimates in a different setting through the concepts of quasiconvexity and rank-one convexity introduced by Morrey; see Chapter 19.

5.1 The Basic Beltrami Equation

As a first step toward the measurable Riemann mapping theorem, we assume that the Beltrami coefficient is compactly supported in some disk $\mathbb{D}_r = \mathbb{D}(0, r)$ and study the inhomogeneous problem.

Theorem 5.1.1. *Suppose that $0 \leqslant k < 1$ and $Q(k) < p < P(k)$, as defined in (5.1). If*

- $|\mu(z)| \leqslant k\chi_{\mathbb{D}_r}(z)$, $z \in \mathbb{C}$, *and*
- $\varphi \in L^p(\mathbb{C})$ *with compact support,*

then the equation

$$\frac{\partial \sigma}{\partial \bar{z}} = \mu \frac{\partial \sigma}{\partial z} + \varphi \qquad \text{for almost every } z \in \mathbb{C}, \tag{5.2}$$

has a unique solution σ with derivative $D\sigma \in L^p(\mathbb{C})$ and decay $\sigma(z) = \mathcal{O}(1/z)$ at infinity. In particular, for $p > 2$ we have $\sigma \in W^{1,p}(\mathbb{C})$.

Proof. We first consider the existence. As $\|\mu\|_\infty \mathbf{S}_p < 1$, the operator $(\mathbf{I} - \mu\mathcal{S})^{-1}$ defined by the Neumann series

$$(\mathbf{I} - \mu\mathcal{S})^{-1} = \mathbf{I} + \mu\mathcal{S} + \mu\mathcal{S}\mu\mathcal{S} + \mu\mathcal{S}\mu\mathcal{S}\mu\mathcal{S} + \cdots \tag{5.3}$$

is bounded from $L^p(\mathbb{C}) \to L^p(\mathbb{C})$. To see this simply note that the n-fold products

$$\|\mu\mathcal{S}\mu\mathcal{S}\cdots\mu\mathcal{S}\phi\|_{L^p(\mathbb{C})} \leqslant k\mathbf{S}_p \|\mu\mathcal{S}\cdots\mu\mathcal{S}\phi\|_{L^p(\mathbb{C})} \leqslant \cdots \leqslant (k\mathbf{S}_p)^n \|\phi\|_{L^p(\mathbb{C})},$$

so the series at (5.3) will converge in norm, in the space of bounded operators on $L^p(\mathbb{C})$, if $k\mathbf{S}_p < 1$.

The formula (5.3) shows that $(\mathbf{I} - \mu\mathcal{S})^{-1}\varphi \in L^p(\mathbb{C})$ is compactly supported. Hence

$$\sigma = \mathcal{C}\left((\mathbf{I} - \mu\mathcal{S})^{-1}\varphi\right) \tag{5.4}$$

has derivatives in $L^p(\mathbb{C})$ and the required decay at infinity. We need to check that σ defined above is a solution to (5.2). This is easy. We put $\omega = (\mathbf{I} - \mu \mathcal{S})^{-1}\varphi$,

$$\frac{\partial \sigma}{\partial \bar{z}} = \frac{\partial}{\partial \bar{z}} \mathcal{C}\omega = \omega$$

$$\frac{\partial \sigma}{\partial z} = \frac{\partial}{\partial z} \mathcal{C}\omega = \mathcal{S}\omega,$$

so that

$$\frac{\partial \sigma}{\partial \bar{z}} = \omega = \omega - \mu \mathcal{S}\omega + \mu \mathcal{S}\omega = (\mathbf{I} - \mu \mathcal{S})\omega + \mu \mathcal{S}\omega = \varphi + \mu \frac{\partial \sigma}{\partial z}$$

So σ is indeed a solution.

To see uniqueness we suppose that the two functions σ^1, σ^2 are solutions with L^p-integrable derivatives. The operator $(\mathbf{I} - \mu \mathcal{S})$ is invertible and so has trivial kernel. The difference $h = \sigma^1 - \sigma^2$ satisfies the homogeneous equation

$$(\mathbf{I} - \mu \mathcal{S})\frac{\partial h}{\partial \bar{z}} = \frac{\partial h}{\partial \bar{z}} - \mu \frac{\partial h}{\partial z} = 0$$

That is, $h_{\bar{z}} \equiv 0$. We take the Beurling transform of this, $h_z = \mathcal{S}h_{\bar{z}} = 0$, so that $\nabla h = 0$ and h is a constant. As h decays at infinity, this constant is 0. Thus $\sigma^1 = \sigma^2$, and we have uniqueness. $\square$

We are now able to present the first version of the measurable Riemann mapping theorem for compactly supported μ.

Theorem 5.1.2. *Suppose that $0 \leqslant k < 1$ and that $|\mu(z)| \leqslant k\chi_{\mathbb{D}_r}(z)$, $z \in \mathbb{C}$. Then there is a unique $f \in W^{1,2}_{loc}(\mathbb{C})$ such that*

$$\frac{\partial f}{\partial \bar{z}} = \mu(z) \frac{\partial f}{\partial z} \quad \text{for almost every } z \in \mathbb{C} \text{ and} \qquad (5.5)$$

$$f(z) = z + \mathcal{O}(1/z) \quad \text{as } z \to \infty. \qquad (5.6)$$

Moreover, $f \in W^{1,p}_{loc}(\mathbb{C})$ for every $2 \leqslant p < P(k)$.

Proof. We apply Theorem 5.1.1 with $\varphi = \mu$ to find a $\sigma \in W^{1,p}(\mathbb{C})$ solving the equation

$$\frac{\partial \sigma}{\partial \bar{z}} = \mu \frac{\partial \sigma}{\partial z} + \mu \qquad (5.7)$$

We put $f(z) = z + \sigma(z) \in W^{1,p}_{loc}(\mathbb{C})$, $2 \leqslant p < P(k)$. It is clear that f satisfies the Beltrami equation (5.5). We have identified σ as the Cauchy transform of a compactly supported function, and so directly from the definition we have $\sigma(z) = \mathcal{O}(1/z)$ near ∞. To see uniqueness, note that for any other solution $g \in W^{1,2}_{loc}(\mathbb{C})$ we have, $\nu(z) = g(z) - z$ satisfies (5.7). Since both f and g are holomorphic near ∞ with development as at (5.6), both σ and ν have derivatives in $L^2(\mathbb{C})$. As before, we compute

$$(\mathbf{I} - \mu \mathcal{S})(\sigma - \nu)_{\bar{z}} = 0$$

5.2. SMOOTH BELTRAMI COEFFICIENT

Since $\mathcal{S}$ is an L^2-isometry and $\|\mu\|_\infty \leqslant k < 1$, necessarily $(\sigma - \nu)_z = \mathcal{S}(\sigma - \nu)_{\bar{z}} = 0$ which shows us that $\sigma - \nu$ is a constant function in $L^2(\mathbb{C})$. That is $\sigma = \nu$. □

When μ is compactly supported with $|\mu(z)| \leqslant k < 1$ for $z \in \mathbb{C}$, the solution f of Theorem 5.1.2 is holomorphic in a neighborhood of ∞. There it has a Laurent series expansion. The normalization at (5.6) is quite natural in that it implies f is injective near ∞ and that as a function valued in $\hat{\mathbb{C}}$, f is continuous at ∞.

Principal Solutions

With μ as above, we will call the $W^{1,2}_{loc}(\mathbb{C})$ solutions to the Beltrami equation $f_{\bar{z}} = \mu f_z$ normalized by the condition $f(z) = z + \mathcal{O}(1/z)$ near ∞ the *principal solutions*.

In fact, the proof of Theorem 5.1.2 gives for principal solutions the representations

$$\begin{aligned}\frac{\partial f}{\partial \bar{z}} &= \frac{\partial \sigma}{\partial \bar{z}} = \mu + \mu \frac{\partial \sigma}{\partial z} = \mu + \mu \mathcal{S}\left(\frac{\partial \sigma}{\partial \bar{z}}\right) \\ &= \mu + \mu \mathcal{S}\left(\frac{\partial f}{\partial \bar{z}}\right) = \mu + \mu \mathcal{S}\mu + \mu \mathcal{S}\mu \mathcal{S}\mu + \cdots, \end{aligned} \quad (5.8)$$

where $f_{\bar{z}} \in L^p(\mathbb{C})$. The series converges in $L^p(\mathbb{C})$ whenever $Q(k) < p < P(k)$, in particular for $p = 2$. Moreover, we have the identities

$$\frac{\partial f}{\partial z} = 1 + \mathcal{S}\frac{\partial f}{\partial \bar{z}} \quad (5.9)$$

and

$$f(z) = z + \mathcal{C}\left(\frac{\partial f}{\partial \bar{z}}\right)(z), \quad z \in \mathbb{C} \quad (5.10)$$

5.2 Quasiconformal Mappings with Smooth Beltrami Coefficient

The existence of principal solutions was easy to establish. It requires more effort to prove they are homeomorphisms of the complex plane. We begin with the smooth case.

As a first step we establish a higher degree of regularity for the solution when the Beltrami coefficient is smooth.

Lemma 5.2.1. *Let $\mu \in C_0^\infty(\mathbb{C})$ with $|\mu(z)| \leqslant k < 1$ for $z \in \mathbb{C}$. Suppose that $\sigma_{\bar{z}} = \mu \sigma_z + \varphi$ and that $\varphi \in W_0^{1,p}(\mathbb{C})$ for some $2 < p < P(k)$. Then $\sigma \in W^{2,p}(\mathbb{C})$.*

Proof. From (5.4) and the uniqueness in Theorem 5.1.1, we have $\sigma_{\bar{z}} = \omega$, where

$$\begin{aligned}\omega &= \varphi + \mu \mathcal{S}\varphi + \mu \mathcal{S}\mu \mathcal{S}\varphi + \cdots \\ &= \omega_0 + \omega_1 + \omega_2 + \cdots\end{aligned}$$

We use the fact that $\mathcal{S} : W^{1,p}(\mathbb{C}) \to W^{1,p}(\mathbb{C})$ (see Chapter 4) to quickly see that $\omega_i \in W^{1,p}(\mathbb{C})$, $i = 1, 2, \ldots$. As $\mathcal{S}$ commutes with the z- and $\bar{z}$-derivatives,

$$\mathcal{S}(f_{\bar{z}}) = (\mathcal{S}f)_{\bar{z}}, \qquad \mathcal{S}(f_z) = (\mathcal{S}f)_z,$$

for $f \in W^{1,p}(\mathbb{C})$, $p > 1$, we have for any first-order constant coefficient differential operator D,

$$D = \alpha \frac{\partial}{\partial z} + \beta \frac{\partial}{\partial \bar{z}}, \qquad \alpha, \beta \in \mathbb{C} \tag{5.11}$$

the commutator relation $\mathcal{S} \circ D = D \circ \mathcal{S}$. Then we compute

$$D\omega_i = D\mu(\mathcal{S}\mu(\mathcal{S}\cdots\mu\mathcal{S}\varphi) + \mu\mathcal{S}D\mu(\mathcal{S}\cdots\mu\mathcal{S}\varphi) + \cdots + \mu\mathcal{S}\mu\mathcal{S}\cdots\mu\mathcal{S}D\varphi,$$

and thus

$$\|D\omega_i\|_{L^p(\mathbb{C})} \leqslant (i+1)\mathbf{S}_p(k\mathbf{S}_p)^{i-1}(\|\varphi\|_p\|D\mu\|_\infty + \|D\varphi\|_{L^p(\mathbb{C})})$$

As $k\mathbf{S}_p < 1$, the series $\sum_i D\omega_i$ converges uniformly in $L^p(\mathbb{C})$ to $D\omega \in L^p(\mathbb{C})$. It follows that $\omega = \sigma_{\bar{z}} \in W^{1,p}(\mathbb{C})$. Of course, then $\sigma_z = \mathcal{S}\sigma_{\bar{z}} \in W^{1,p}(\mathbb{C})$ as well. Consequently, $\sigma \in W^{2,p}(\mathbb{C})$. □

As a simple consequence, we obtain the following theorem.

Theorem 5.2.2. *Let $\mu, \varphi \in C_0^\infty(\mathbb{C})$ with $\|\mu\|_\infty < 1$ and let $2 < p < P(k)$. Then the $W^{1,p}$-solution σ to the equation $\sigma_{\bar{z}} = \mu\sigma_z + \varphi$ in fact satisfies $\sigma \in C^\infty(\mathbb{C})$.*

Proof. Of course, Lemma 5.2.1 immediately gives $\sigma \in W^{2,p}(\mathbb{C})$ and, moreover, for any differential operator D as in (5.11), we have

$$(D\sigma)_{\bar{z}} = D\sigma_{\bar{z}} = \mu(D\sigma)_z + (D\mu)\sigma_z + D\varphi$$

Then $(D\mu)\sigma_z + D\varphi \in W_0^{1,p}(\mathbb{C})$, and we again apply Lemma 5.2.1 to deduce $D\sigma \in W^{2,p}(\mathbb{C})$, from which it follows that $\sigma \in W^{3,p}(\Omega)$. Iterating this argument quickly gives $\sigma \in C^\infty(\mathbb{C})$ with all derivatives in $L^p(\mathbb{C})$. □

We are now ready to establish the topological properties of the principal solutions when the coefficients are smooth. We start with local injectivity.

Theorem 5.2.3. *Let $\mu \in C_0^\infty(\mathbb{C})$ with $|\mu(z)| \leqslant k < 1$ for $z \in \mathbb{C}$. Let $f \in W_{loc}^{1,2}(\mathbb{C})$ be the principal solution to $f_{\bar{z}} = \mu f_z$. Then $f \in C^\infty(\mathbb{C})$ and $J(z, f) > 0$ for all $z \in \mathbb{C}$.*

Proof. As smoothness follows from the previous theorem, the main point is to prove the positivity of the Jacobian $J(z, f)$ at every $z \in \mathbb{C}$. Let

$$k = \max_{\mathbb{C}} |\mu(z)| < 1$$

From Theorem 5.2.2, for $2 < p < P(k)$, there is a solution $\sigma \in W^{1,p}(\mathbb{C}) \cap C^\infty(\mathbb{C})$ to the auxiliary equation $\sigma_{\bar{z}} = \mu\sigma_z + \mu_z$. We put

$$F(z) = z + \mathcal{C}(\mu e^\sigma)(z)$$

5.2. SMOOTH BELTRAMI COEFFICIENT

for $z \in \mathbb{C}$. As $\mu \in C_0^\infty(\mathbb{C})$, we certainly have $F \in C^\infty(\mathbb{C})$. Moreover,
$$(\mu e^\sigma)_z = (\mu_z + \mu \sigma_z)e^\sigma = \sigma_{\bar z} \, e^\sigma = (e^\sigma)_{\bar z}$$
As $\sigma(z) = \mathcal{O}(1/z)$ near ∞ and $\sigma_{\bar z} \in L^p(\mathbb{C})$ is compactly supported,
$$e^\sigma - 1 \in W^{1,p}(\mathbb{C})$$
Hence
$$e^\sigma - 1 = \mathcal{C}((e^\sigma - 1)_{\bar z}) = \mathcal{C}((\mu e^\sigma)_z) = \mathcal{S}(\mu e^\sigma), \tag{5.12}$$
where we have used Theorem 5.2.2 to establish the last identity.

Returning to the definition of F, we have from (5.12) the identities
$$\frac{\partial F}{\partial \bar z} = \mu e^\sigma, \qquad \frac{\partial F}{\partial z} = 1 + \mathcal{S}(\mu e^\sigma) = e^\sigma$$
Together these imply that $F_{\bar z} = \mu F_z$ and the uniqueness of principal solutions, given by Theorem 5.1.2, shows that $F = f$. In particular, $f \in C^\infty(\mathbb{C})$. Furthermore,
$$J(z, F) = \left|\frac{\partial F}{\partial z}\right|^2 - \left|\frac{\partial F}{\partial \bar z}\right|^2 = |e^{2\sigma}|(1 - |\mu|^2) > 0$$
for every $z \in \mathbb{C}$. $\square$

We have proved the Jacobian is strictly positive in Theorem 5.2.3, and this has important topological consequences.

Theorem 5.2.4. *Suppose $\mu \in C_0^\infty(\mathbb{C})$ and $k = \|\mu\|_\infty < 1$. Then the principal solution $f \in W^{1,2}_{loc}(\mathbb{C})$ to the Beltrami equation*
$$\frac{\partial f}{\partial \bar z} = \mu(z) \frac{\partial f}{\partial z}$$
is a C^∞ diffeomorphism of the Riemann sphere $\hat{\mathbb{C}}$. Moreover, f is K-quasiconformal with
$$K = \frac{1+k}{1-k} < \infty$$

Proof. We know that f is $C^\infty(\mathbb{C})$ and $J(z, f) > 0$. Here $J(z, f) > 0$ implies, via the inverse function theorem, that f is a local homeomorphism. Once we have f a global homeomorphism, K-quasiconformality with $K = \frac{1+k}{1-k}$ is immediate from Theorem 2.5.4.

Since we can extend f to $\hat{\mathbb{C}}$ by setting $f(\infty) = \infty$, we find that f is a local homeomorphism $\hat{\mathbb{C}} \to \hat{\mathbb{C}}$ (recall that as a principal solution f is holomorphic near ∞ with Laurent series $z + \mathcal{O}(1/z)$ and therefore this extension is conformal, that is, a local homeomorphism, at ∞). The monodromy theorem then promotes f to a global homeomorphism. Actually, in our situation the topology is somewhat simpler. As f is a local homeomorphism and $\hat{\mathbb{C}}$ is compact, there is a cover of $\hat{\mathbb{C}}$ by a finite number of disks on each of which f is injective. Then it is easy to see that the set of points with a given fixed number of preimages is both open and closed in $\hat{\mathbb{C}}$. Therefore this set is empty or equal to $\hat{\mathbb{C}}$. However, as $f \in C^\infty(\mathbb{C})$, we see that ∞ has exactly one preimage. $\square$

5.3 Measurable Riemann Mapping Theorem

We want to relax the assumption on the smoothness of μ and still conclude the injectivity of the principal solution f to the Beltrami equation $f_{\bar z} = \mu f_z$. For this we need the next estimate.

Lemma 5.3.1. *Suppose* $|\mu|, |\nu| \leqslant k\chi_{\mathbb{D}_r}$, *where* $0 \leqslant k < 1$. *Let* $f, g \in W^{1,2}_{loc}(\mathbb{C})$ *be the principal solutions to the equations*

$$\frac{\partial f}{\partial \bar z} = \mu(z)\frac{\partial f}{\partial z}, \qquad \frac{\partial g}{\partial \bar z} = \nu(z)\frac{\partial g}{\partial z}$$

If for a number s we have $2 \leqslant p < ps < P(k)$, *then*

$$\|f_{\bar z} - g_{\bar z}\|_{L^p(\mathbb{C})} \leqslant C(p, s, k)\, r^{2/ps}\, \|\mu - \nu\|_{L^{ps/(s-1)}(\mathbb{C})}$$

Proof. We recall the representations for f and g from (5.8), $f_{\bar z} = \mu + \mu \mathcal{S} f_{\bar z}$ and $g_{\bar z} = \nu + \nu \mathcal{S} g_{\bar z}$. So $f_{\bar z} = (\mathbf{I} - \mu\mathcal{S})^{-1}\mu$. Then

$$\begin{aligned}
f_{\bar z} - g_{\bar z} &= \mu + \mu\mathcal{S} f_{\bar z} - \nu - \nu\mathcal{S} g_{\bar z} \\
&= (\mu - \nu) + \mu\mathcal{S} f_{\bar z} - \nu\mathcal{S} f_{\bar z} + \nu\mathcal{S} f_{\bar z} - \nu\mathcal{S} g_{\bar z} \\
&= (\mu - \nu)\chi_{\mathbb{D}_r} + (\mu - \nu)\mathcal{S} f_{\bar z} + \nu\mathcal{S}(f_{\bar z} - g_{\bar z}),
\end{aligned}$$

hence

$$(f_{\bar z} - g_{\bar z}) = (\mathbf{I} - \nu\mathcal{S})^{-1}[(\mu - \nu)(\chi_{\mathbb{D}_r} + \mathcal{S} f_{\bar z})]$$

Thus

$$\|f_{\bar z} - g_{\bar z}\|^p_{L^p(\mathbb{C})} \leqslant \|(\mathbf{I} - \nu\mathcal{S})^{-1}\|^p_{L^p(\mathbb{C})\to L^p(\mathbb{C})} \|(\mu - \nu)(\chi_{\mathbb{D}_r} + \mathcal{S} f_{\bar z})\|^p_{L^p(\mathbb{C})}$$

By Hölder's inequality,

$$\begin{aligned}
\|f_{\bar z} - g_{\bar z}\|^p_{L^p(\mathbb{C})} &\leqslant C(k, p)\left(\int |\mu - \nu|^{ps'}\right)^{1/s'}\left(\int |\chi_{\mathbb{D}_r} + \mathcal{S} f_{\bar z}|^{ps}\right)^{1/s} \\
&\leqslant C(k, p)\|\mu - \nu\|^p_{L^{ps'}(\mathbb{C})}\left(\int |\chi_{\mathbb{D}_r} + \mathcal{S}(\mathbf{I} - \mu\mathcal{S})^{-1}\mu|^{ps}\right)^{1/s} \\
&\leqslant C(k, p, s)r^{2/s}\|\mu - \nu\|^p_{L^{ps'}(\mathbb{C})}
\end{aligned}$$

This proves the result. $\square$

In order to complete the measurable Riemann mapping theorem in the compactly supported case, we shall use Lemma 5.3.1 together with a smoothing approximation via convolution. To this end let

$$\Phi(z) = \begin{cases} C\exp\left(\frac{1}{|z|^2 - 1}\right) & \text{if } |z| < 1 \\ 0 & \text{otherwise} \end{cases}$$

5.3. MEASURABLE MAPPING THEOREM

Here the constant C is chosen so that $\int_{\mathbb{C}} \Phi = 1$. Write

$$\phi_\varepsilon(z) = \varepsilon^2 \Phi\left(\frac{z}{\varepsilon}\right)$$

If μ is a measurable function with $|\mu| \leqslant k\chi_{\mathbb{D}_r}$, then

$$\mu_\varepsilon(z) = \phi_\varepsilon * \mu(z) = \int_{\mathbb{C}} \phi_\varepsilon(z-\tau)\mu(\tau)d\tau \qquad (5.13)$$

is a $C^\infty(\mathbb{C})$ function (differentiate under the integral sign) and

$$|\mu_\varepsilon(z)| \leqslant \int_{\mathbb{C}} \phi_\varepsilon(z-\tau)|\mu(\tau)| \leqslant k \int_{\mathbb{C}} \phi_\varepsilon(z-\tau)d\tau = k$$

Lebesgue's theorem gives $\mu_\varepsilon \to \mu$ in all $L^p(\mathbb{C})$-spaces, $1 \leqslant p < \infty$, as $\varepsilon \to 0$.

After these preparations we have the measurable Riemann mapping theorem for a compactly supported complex dilatation μ.

Theorem 5.3.2. *Let $|\mu| \leqslant k < 1$ be compactly supported and defined on $\mathbb{C}$. Then there is a unique principal solution to the Beltrami equation*

$$\frac{\partial f}{\partial \bar{z}} = \mu(z)\frac{\partial f}{\partial z} \quad \text{for almost every } z \in \mathbb{C},$$

and the solution $f \in W^{1,2}_{\text{loc}}(\mathbb{C})$ is a K-quasiconformal homeomorphism of $\mathbb{C}$.

Proof. Fix p and s with $2 < p < ps < P(k)$. Then approximate by convolution, $\mu_\varepsilon \to \mu$ in $L^{sp/(s-1)}(\mathbb{C})$. Let f_ε be the unique principal solution to the Beltrami equation

$$(f_\varepsilon)_{\bar{z}} = \mu_\varepsilon(z)(f_\varepsilon)_z$$

Theorem 5.2.4 implies that each f_ε is a K-quasiconformal diffeomorphism of $\mathbb{C}$, $K = \frac{1+k}{1-k}$. Thus each f_ε is quasisymmetric and holomorphic in the exterior disk $\mathbb{C} \setminus \mathbb{D}_{r+\varepsilon}$. The family of conformal maps

$$\psi(z) = \frac{1}{f_\varepsilon(\frac{1}{z})} \quad \text{for } |z| \leqslant \frac{1}{r+\varepsilon},$$

has $\psi(0) = 0$ and $\psi'(0) = 1$ and is thus a normal family. Hence f_ε converges locally uniformly to a nonconstant limit map in a neighbourhood of ∞. From the compactness properties of quasisymmetric mappings (see Section 3.9) it easily follows that for some subsequence $f_\varepsilon \to g$ a quasisymmetric mapping of $\mathbb{C}$. Of course, g is holomorphic in $\mathbb{C} \setminus \mathbb{D}_r$, and as holomorphic maps converge with their derivatives, $g(z) = z + \mathcal{O}(1/z)$ as $z \to \infty$.

Let f be the principal solution to the equation $f_{\bar{z}} = \mu f_z$. From Lemma 5.3.1 we see that $(f_\varepsilon)_{\bar{z}} \to f_{\bar{z}}$ in $L^p(\mathbb{C})$,

$$\|(f_\varepsilon)_{\bar{z}} - f_{\bar{z}}\|_{L^p(\mathbb{C})} \leqslant C(p,s,k) (2r)^{2/ps} \|\mu_\varepsilon - \mu\|_{L^{ps/(s-1)}(\mathbb{C})} \to 0$$

as $\varepsilon \to 0$. But now the identity (5.10) gives

$$f_\varepsilon(z) - f(z) = \mathcal{C}\big((f_\varepsilon)_{\bar{z}} - f_{\bar{z}}\big)(z) \to 0$$

where, according to Theorem 4.3.11, the convergence is uniform on $\mathbb{C}$. Thus $f = g$, and we conclude that f is a K-quasiconformal homeomorphism. $\square$

Finally, we want to remove the restriction that the distortion μ is compactly supported and thus prove the measurable Riemann mapping theorem in its full generality. Of course, once we remove the assumption on the support, the solutions f will no longer necessarily be holomorphic or even Lipschitz near ∞ and we can no longer speak of principal solutions. In this case it is more natural to distinguish mappings by their values at given points.

Definition 5.3.3. *A quasiconformal homeomorphism $f : \mathbb{C} \to \mathbb{C}$ is said to be normalized if*

$$f(0) = 0 \quad \text{and} \quad f(1) = 1$$

Sometimes we will write $f(\infty) = \infty$ as well, with the obvious interpretation. Similarly, if $\mu : \mathbb{C} \to \mathbb{C}$ is a measurable function and $|\mu(z)| \leq k < 1$, then a homeomorphic $W^{1,2}_{loc}(\mathbb{C})$-solution to the Beltrami equation $f_{\bar{z}} = \mu(z)f_z$ with $f(0) = 0$ and $f(1) = 1$ will be called a *normalized solution*. Such a solution is obviously a normalized quasiconformal mapping.

Theorem 5.3.4. *Let $|\mu(z)| \leq k < 1$ for almost every $z \in \mathbb{C}$. Then there is a solution $f : \hat{\mathbb{C}} \to \hat{\mathbb{C}}$ to the Beltrami equation*

$$\frac{\partial f}{\partial \bar{z}} = \mu(z) \frac{\partial f}{\partial z} \quad \text{for almost every } z \in \mathbb{C}, \qquad (5.14)$$

which is a K-quasiconformal homeomorphism normalized by the three conditions

$$f(0) = 0; \qquad f(1) = 1; \qquad f(\infty) = \infty$$

Furthermore, the normalized solution f is unique.

Proof. Let $\mu_r = \mu \chi_{\mathbb{D}(0,r)}$ and let $h_r : \hat{\mathbb{C}} \to \hat{\mathbb{C}}$ be the quasiconformal homeomorphism that is the principal solution to the Beltrami equation

$$h_{\bar{z}} = \mu_r(z) h_z, \quad \text{for almost every } z \in \mathbb{C}, \qquad (5.15)$$

as given by Theorem 5.3.2. We choose constants $a_r, b_r \in \mathbb{C}$ such that

$$f_r(z) = a_r h(z) + b_r$$

satisfies $f_r(0) = 0$ and $f_r(1) = 1$. Notice that f_r also satisfies the Beltrami equation (5.15). Now the family of normalized K-quasiconformal mappings $\{f_r\}_{r=1}^\infty$ is normal by Theorem 3.9.4, and thus we may extract a subsequence converging locally uniformly to a quasiconformal mapping $f : \mathbb{C} \to \mathbb{C}$. The question we have now is whether or not this limit map is actually a solution to (5.14). It is

5.3. MEASURABLE MAPPING THEOREM

useful to give the (positive) answer in a more general setup, hence we formulate the argument separately in the next lemma. Then to complete the proof of the theorem, we need to prove the uniqueness of the normalized solution. We shall leave this to Corollary 5.5.5 below. □

Good Approximation Lemma

Lemma 5.3.5. *Suppose we have a sequence of Beltrami coefficients* $\{\mu_n\}_{n \in \mathbb{N}}$ *such that*

$$\|\mu_n\|_{L^\infty(\mathbb{C})} \leqslant k < 1, \quad \text{for all } n \in \mathbb{N},$$

and such that the pointwise limit

$$\mu(z) := \lim_{n \to \infty} \mu_n(z)$$

exists almost everywhere. Let f_n be the normalized solutions to

$$\frac{\partial f}{\partial \bar{z}} = \mu_n(z) \frac{\partial f}{\partial z}, \quad n \in \mathbb{N}$$

Then the limit $f(z) = \lim_{n \to \infty} f_n(z)$ exists, the convergence is uniform on compact subsets of $\mathbb{C}$ and f solves the limiting Beltrami equation,

$$\frac{\partial f}{\partial \bar{z}} = \mu(z) \frac{\partial f}{\partial z} \quad \text{almost everywhere}$$

Proof. The mappings f_n are η-quasisymmetric, where η is independent of n by Theorem 3.5.3. For each fixed disk $\mathbb{D}_R$ we apply (3.37) and estimate

$$\int_{\mathbb{D}_R} |Df_n|^2 dz \leqslant K \int_{\mathbb{D}_R} J(z, f_n) dz = K |f_n(\mathbb{D}_R)| \leqslant \pi K \eta(R)^2$$

Hence the L^2-norms of the derivatives of f_n are locally uniformly bounded. We can extract a subsequence $\{f_{n_k}\}_{k=1}^\infty$ converging weakly in $W^{1,2}_{loc}(\mathbb{C})$ and locally uniformly on $\mathbb{C}$ to a mapping f.

Next suppose $\phi \in C_0^\infty(\mathbb{C})$. Then, since f_{n_k} satisfies the Beltrami equation with coefficient μ_{n_k}, we have

$$\int \phi \left(\frac{\partial f_{n_k}}{\partial \bar{z}} - \mu \frac{\partial f_{n_k}}{\partial z} \right) = \int \phi (\mu_{n_k} - \mu) \frac{\partial f_{n_k}}{\partial z} \tag{5.16}$$

When $k \to \infty$, the left expression in (5.16) tends to $\int \phi(f_{\bar{z}} - \mu f_z)$ by the weak convergence. The expression on the right is no more than

$$\|\phi(\mu_{n_k} - \mu)\|_{L^2} \|Df_{n_k}\|_{L^2(\mathbb{D}_R)} \leqslant \sqrt{\pi K} \eta(R) \|\phi(\mu_{n_k} - \mu)\|_{L^2}$$

if the disk $\mathbb{D}_R$ is so large that it contains the support of ϕ. By Lebesgue's Dominated convergence theorem this term tends to zero as $k \to \infty$. Thus f is a solution to (5.14). However, this same argument applies to any other convergent subsequence of the normal family $\{f_n\}$. Namely, the limit mapping is a normalized solution of the limiting equation. Since Corollary 5.5.5 tells us the normalized solution is unique, all convergent subsequences have the same limit and this proves the lemma. □

5.4 L^p-Estimates and the Critical Interval

Good L^p-estimates of the derivatives of quasiregular mappings become quite powerful tools in the geometric study of mappings. To some extent such estimates rely on the invertibility properties of the Beltrami operator

$$\mathbf{I} - \mu(z)\mathcal{S} : L^p(\mathbb{C}) \to L^p(\mathbb{C}), \qquad (5.17)$$

where $|\mu(z)| \leqslant k < 1$ for almost every $z \in \mathbb{C}$ and $\mathcal{S}$ is the Beurling transform. We made effective use of these operators in the previous sections, in the proof of the measurable Riemann mapping theorem, but the usefulness of the operators (5.17) goes well beyond the quasiconformal case. In this section we will gather together few of their properties as well as some immediate consequences. Chapter 14 will be devoted to a deeper study of these operators and their generalizations.

The proof of the measurable Riemann mapping theorem was based on the fundamental fact that, for every $0 \leqslant k < 1$, there are exponents

$$1 < q_k < 2 < p_k < \infty \qquad (5.18)$$

such that the inverse operator

$$(\mathbf{I} - \mu(z)\mathcal{S})^{-1} : L^p(\mathbb{C}) \to L^p(\mathbb{C}) \qquad (5.19)$$

exists whenever $p \in (q_k, p_k)$ and $\|\mu\|_\infty \leqslant k$. We may also require uniform norm bounds

$$\|(\mathbf{I} - \mu(z)\mathcal{S})^{-1}\|_p \leqslant C_0(k,p) < \infty \qquad (5.20)$$

depending only on k, p.

For instance, as described in the beginning of this chapter, the intervals of exponents $(Q(k), P(k))$ determined by

$$\|\mathcal{S}\|_{Q(k)} = \|\mathcal{S}\|_{P(k)} = \frac{1}{k}, \qquad Q(k) < 2 < P(k) \qquad (5.21)$$

satisfy these requirements and give the bounds $C_0(k,p) \leqslant \frac{1}{1-k\|\mathcal{S}\|_p}$.

The conjectural values of the p-norms of $\mathcal{S}$ suggest that the maximal interval where (5.18)–(5.20) will hold is described by

$$p_k = 1 + 1/k \quad \text{and} \quad q_k = 1 + k \qquad (5.22)$$

Somewhat curiously, this conjecture for the maximal range of invertibility of the Beltrami operators $\mathbf{I} - \mu(z)\mathcal{S}$ can in fact be shown to be true even if to this day we do not know the precise values of the p-norms $\|\mathcal{S}\|_p$. This result will be established in Chapter 14 after we have developed considerably more theory.

The Critical Interval

Since the L^p-invertibility properties of the operators $\mathbf{I} - \mu(z)\mathcal{S}$ have such a crucial role in the sequel, we introduce the notion of the critical interval.

5.4. THE CRITICAL INTERVAL

Definition 5.4.1. *Given $0 \leqslant k < 1$, the critical interval (q_k, p_k), where $1 \leqslant q_k < 2 < p_k \leqslant \infty$, is the largest open interval such that, for every $p \in (q_k, p_k)$ and every measurable μ with $|\mu(z)| \leqslant k$, the operator $\mathbf{I} - \mu(z)\mathcal{S}$ is invertible in $L^p(\mathbb{C})$ with*
$$\| (\mathbf{I} - \mu(z)\mathcal{S})^{-1} \|_p \leqslant C_0(k, p),$$
where the constant $C_0(k, p)$ depends only on p and k.

The role of the critical interval will become clear over the next few sections, but let us first consider a few simple consequences of the results in Section 5.1 and express them in terms of the critical interval. In the inequalities (5.23)–(5.25) below we assume that f is a principal solution to $f_{\bar{z}} = \mu f_z$, where $|\mu| \leqslant k\chi_{\mathbb{D}_r}$ and $0 \leqslant k < 1$.

First, by (5.8), $f_{\bar{z}} = (\mathbf{I} - \mu\mathcal{S})^{-1}\mu$. In particular,
$$\|f_{\bar{z}}\|_p \leqslant \pi^{1/p} C_0(k, p) \, r^{2/p}, \qquad p \in (q_k, p_k) \tag{5.23}$$

Since $f(z) = z + (\mathcal{C}f_{\bar{z}})(z)$, Theorem 4.3.13 gives, for all $2 < p < p_k$,
$$|f(z) - f(w)| \leqslant |z - w| + \pi^{1/p} C_p C_0(k, p) r^{2/p} |z - w|^{1-2/p}, \qquad z, w \in \mathbb{C} \tag{5.24}$$

Here the estimate is global and uniform. However, note from Theorem 3.10.2 that the correct Hölder continuity exponent for f is $1/K = 1 - 2/(1 + 1/k)$. Thus the above method cannot quite establish the optimal Hölder regularity, even if we have identified the critical interval.

More importantly, besides distortion of distance, the estimate (5.23) also yields bounds for the distortion of Lebesgue measure, or the area of sets $E \subset \mathbb{C}$. In fact, we have the uniform bounds
$$|f(E)| \leqslant 2|E| + C_1(p, k) \, r^{4/p} |E|^{1-2/p}, \qquad 2 \leqslant p < p_k, \tag{5.25}$$

for every measurable subset $E \subset \mathbb{C}$. Namely,
$$|f(E)| = \int_E J(z, f) \, dz \leqslant \int_E |f_z|^2 \, dz \leqslant 2 \int_E |f_z - 1|^2 \, dz + 2|E|$$
$$\leqslant 2|E| + 2\left(\int_{\mathbb{C}} |f_z - 1|^p \, dz\right)^{2/p} |E|^{1-2/p}$$

by Hölder's inequality. Here $\|f_z - 1\|_p \leqslant \mathbf{S}_p \|f_{\bar{z}}\|_p$, so that (5.25) holds with $C_1(k, p) = 2\pi^{2/p}(\mathbf{S}_p)^2 C_0(k, p)^2$.

The inequality (5.25) is a notable generalization of Corollary 3.7.6, giving quantitative and explicit measure distortion estimates.

The inequality (5.25) also raises the question of optimal bounds in the distortion of area. The predicted optimal bounds were conjectured by Gehring and Reich. Following [21] and [34], we will settle this in Chapter 13. This will lead to many important consequences.

5.4.1 The Cacioppoli Inequalities

We next wish to understand general, that is, not necessarily homeomorphic, solutions to the Beltrami equation. With this we are led to study the K-quasiregular mappings $g : \Omega \to \mathbb{C}$. From Section 2.5 recall that these are defined as orientation-preserving mappings

$$g \in W^{1,2}_{loc}(\Omega),$$

for which

$$\max_{\alpha} |\partial_\alpha g(z)| \leqslant K \min_{\alpha} |\partial_\alpha g(z)| \qquad \text{for almost every } z \in \Omega \qquad (5.26)$$

An equivalent, and for the Beltrami equation perhaps more transparent, formulation of (5.26) is

$$\left|\frac{\partial g}{\partial \bar{z}}\right| \leqslant k \left|\frac{\partial g}{\partial z}\right| \qquad \text{for almost every } z \in \Omega,$$

where $k = \frac{K-1}{K+1} < 1$. The bound may also be written in the equivalent form of the distortion inequality,

$$|Dg|^2 \leqslant K J(z,g)$$

The first elementary, although very useful, estimate for general quasiregular mappings is the Cacioppoli inequality, which controls the derivatives of a quasiregular mapping locally in terms of the function. The most basic form of the Cacioppoli estimate is

$$\int \eta^2 |Df|^2 \leqslant K^2 \int |f|^2 |\nabla \eta|^2, \qquad (5.27)$$

where $f \in W^{1,2}_{loc}(\Omega)$ is an arbitrary K-quasiregular mapping and η is a real-valued Lipschitz test function with compact support in Ω. The estimate (5.27) follows from (5.26) and the general inequality

$$\left|\int_\Omega \eta^2 J(z,f)\right| \leqslant \left(\int |f \nabla \eta|^2\right)^{1/2} \left(\int |\eta\, Df|^2\right)^{1/2} \qquad (5.28)$$

This in turn can be proven for all $f \in W^{1,2}_{loc}(\Omega)$ via an integration-by-parts argument, where the right hand side of (5.28) is finite by the Poincaré inequality.

We will show that the Beltrami operators, in fact, yield more general versions of the Cacioppoli estimates. More generally, we shall see the theme throughout this monograph that bounds for the Beurling transform yield improved versions of the integral estimates of Jacobians, and the Cacioppoli inequality is no exception.

Our next result establishes the higher integrability of quasiregular mappings, a result first obtained by B. Bojarski [64]. In terms of the associated linear elliptic Beltrami equation, Bojarski's theorem shows that an a priori $W^{1,2}_{loc}(\Omega)$-solution is in fact in $W^{1,p}_{loc}(\Omega)$, where $p > 2$ depends only on the ellipticity constant k. In particular, it will follow that all quasiregular mappings are continuous.

5.4. THE CRITICAL INTERVAL

Theorem 5.4.2. *Let $f \in W^{1,q}_{loc}(\Omega)$, for some $q \in (q_k, p_k)$, satisfy the distortion inequality*

$$\left|\frac{\partial f}{\partial \bar{z}}(z)\right| \leq k \left|\frac{\partial f}{\partial z}(z)\right|, \quad \text{for almost every } z \in \Omega \tag{5.29}$$

Then $f \in W^{1,p}_{loc}(\Omega)$ for every $p \in (q_k, p_k)$. In particular, f is continuous, and for every $s \in (q_k, p_k)$, the critical interval, we have the Caccioppoli estimate

$$\|\eta\, Df\|_s \leq C_s(k) \|f\nabla\eta\|_s \tag{5.30}$$

whenever η is a compactly supported Lipschitz function in Ω.

Proof. We write (5.29) as $f_{\bar{z}} = \mu(z) f_z$, $|\mu(z)| \leq k < 1$. Consider for a moment the compactly supported test function $F = \eta f \in W^{1,q}(\mathbb{C})$. The function F satisfies the homogeneous linear equation

$$F_{\bar{z}} = \mu F_z + (\eta_{\bar{z}} - \mu\eta_z)f \tag{5.31}$$

As F vanishes for large values of z, F is the Cauchy transform of its $\bar{z}$-derivative,

$$F(z) = (\mathcal{C}\omega)(z) = \frac{1}{\pi}\int_\mathbb{C} \frac{\omega(\tau)}{z-\tau}\,d\tau, \quad \omega = F_{\bar{z}}$$

Substituting this into (5.31) and simplifying yields the equation for ω,

$$\omega - \mu \mathcal{S}\omega = \phi, \tag{5.32}$$

where we have put $\phi = (\eta_{\bar{z}} - \mu\eta_z)f$ for simplicity. Now, of course,

$$F_{\bar{z}} = \omega = (\mathbf{I} - \mu\mathcal{S})^{-1}\phi$$

and

$$F_z = \mathcal{S}\omega = \mathcal{S}\circ(\mathbf{I} - \mu\mathcal{S})^{-1}\phi$$

Together these give the estimate

$$|DF| \leq |(\mathbf{I}-\mu\mathcal{S})^{-1}\phi| + |\mathcal{S}\circ(\mathbf{I}-\mu\mathcal{S})^{-1}\phi| \tag{5.33}$$

Now as $F \in W^{1,q}(\mathbb{C})$ by hypothesis on f, the Sobolev embedding theorem puts $F \in L^{q'}(\mathbb{C})$ for some $q' > q$, and of course similarly for ϕ. It follows from (5.33) that $F \in W^{1,q'}(\mathbb{C})$, provided that $q' < p_k$. Now we have set up a continuous induction with respect to the Sobolev exponent, based on the inequality (5.33). Evidently, the set of exponents p less than p_k for which we have $F \in W^{1,p}(\mathbb{C})$ has no upper bound smaller than p_k. Therefore we find that $F \in W^{1,p}(\mathbb{C})$ for every $p < p_k$. In fact, the Caccioppoli estimate follows from (5.33) as well, for

$$\|DF\|_s \leq C_s \|\phi\|_s \leq C_s \|f\nabla\eta\|_s$$

Recalling that $DF = D(\eta f) = \eta\, Df + f\nabla\eta$ then gives (5.30).

Finally, the continuity of f follows either from the Sobolev embedding, or directly from Theorem 4.3.11. □

A notable feature of the Caccioppoli inequality (5.30) is that it also holds for some exponents below the dimension, that is, for those exponents s with $q_k < s < 2$. This will be the key, for instance, in proving the removability of singularities.

The following efficient and convenient version of the Caccioppoli inequality actually formulates the result for a general Sobolev function. Here we have gradient bounds in terms of the function and the cancellation properties of its derivatives.

Theorem 5.4.3. *Suppose $\mu \in L^\infty(\Omega)$ with $\|\mu\|_\infty \leqslant k < 1$. Let $s \in (q_k, p_k)$. Then there is a constant $C = C_s(k) < \infty$ such that*

$$\|\eta\, Df\|_s \leqslant C\|fD\eta\|_s + C\|\eta\,(f_{\bar{z}} - \mu f_z)\|_s, \qquad (5.34)$$

for every function $f \in W^{1,s}_{loc}(\Omega)$ and for every compactly supported Lipschitz function η in Ω.

Proof. The argument is a repetition of Theorem 5.4.2. Indeed, differentiating the function $F = \eta f \in W^{1,s}(\mathbb{C})$ gives

$$\begin{aligned} F_{\bar{z}} &= \mu F_z + f\,(\eta_{\bar{z}} - \mu\eta_z) + \eta\,(f_{\bar{z}} - \mu f_z) \\ &= \mu F_z + \phi_1 + \phi_2, \end{aligned}$$

and arguing as in Theorem 5.4.2, we obtain

$$\|DF\|_s \leqslant C\,(\|\phi_1\|_s + \|\phi_2\|_s) \leqslant C\|f\nabla\eta\|_s + C\,\|\eta\,(f_{\bar{z}} - \mu f_z)\|_s$$

The proof follows. □

The utility of the Caccioppoli estimate is perhaps best illustrated in the following Liouville theorem. Refined variants of this result will be formulated later and used to establish uniqueness theorems for nonlinear equations in different settings. Recall that the classical Liouville theorem asserts that bounded entire functions are constant. We will establish, with the Caccioppoli inequality, a quasiregular counterpart; all bounded K-quasiregular mappings of $\mathbb{C}$ are constants.

More generally, any holomorphic entire function with sublinear growth must be a constant. For a K-quasiregular mapping $f : \mathbb{C} \to \mathbb{C}$ the analogous condition reads

$$|f(z)| = \mathcal{O}(|z|^\alpha), \qquad \alpha < 1 - \frac{2}{p_k}, \qquad (5.35)$$

where

$$K = \frac{1+k}{1-k}.$$

The following theorem proves this fact. Note that as soon as Chapter 14 has identified the critical interval, proving $p_k = 1 + \frac{1}{k}$ so that $1 - \frac{2}{p_k} = \frac{1}{K}$, the K-quasiconformal mapping

$$f(z) = z|z|^{1/K-1}, \qquad z \in \mathbb{C}$$

will show that, in fact, the following theorem is sharp.

5.4. THE CRITICAL INTERVAL

Theorem 5.4.4. *Suppose that $f \in W^{1,2}_{loc}(\mathbb{C})$ satisfies the distortion inequality in (5.29) and suppose that $2 \leq s < p_k$. If*

$$\int_{\mathbb{D}_R} |f|^s \leq C(1 + R^s), \qquad R > 0, \tag{5.36}$$

for some constant $C > 0$, then f is constant.

In particular, if such an f is bounded, then it is constant.

Proof. We begin with the the Caccioppoli estimate (5.30),

$$\|\eta\, Df\|_s \leq C_s(k) \|f \nabla \eta\|_s,$$

where we choose the test function

$$\eta(z) = \begin{cases} 1, & |z| \leq R \\ 2 - \frac{\log |z|}{\log R}, & R \leq |z| \leq R^2 \\ 0, & |z| \geq R^2 \end{cases} \tag{5.37}$$

We compute the gradient of η to be

$$|\nabla \eta(z)| = \begin{cases} \frac{1}{|z| \log R}, & R \leq |z| \leq R^2 \\ 0, & \text{elsewhere} \end{cases}$$

For given $R > 2$ choose an integer N so that

$$2^{N-1} < R < 2^N$$

and then compute

$$\begin{aligned}
\int_{\mathbb{C}} |f \nabla \eta|^s &= \int_{R < |z| < R^2} \left(\frac{1}{|z| \log R}\right)^s |f(z)|^s \\
&\leq \left(\frac{1}{\log R}\right)^s \sum_{j=1}^{N} \int_{2^{j-1} R < |z| < 2^j R} \frac{|f(z)|^s}{|z|^s} \\
&\leq \left(\frac{1}{\log R}\right)^s \sum_{j=1}^{N} 2^s (2^j R)^{-s} C(1 + 2^{js} R^s) \\
&\leq \frac{2^{s+1} C N}{(\log R)^s} \leq \frac{2^{s+2} C}{\log 2} (\log R)^{1-s},
\end{aligned}$$

where the last line follows from the choice of the integer N.

Thus for each $R > 2$ the Caccioppoli estimate gives

$$\int_{|z| \leq R} |Df|^s \leq C(K, s)(\log R)^{1-s}$$

As we let $R \to \infty$, we must conclude that in fact $|Df| \equiv 0$, hence f is constant as claimed. $\square$

5.4.2 Weakly Quasiregular Mappings

The self improving regularity properties of quasiregular mappings as described in Theorem 5.4.2 will frequently be useful in the sequel and motivate the following notion.

Definition of Weakly Quasiregular Mappings

Definition 5.4.5. *A function* $f \in W_{loc}^{1,q}(\Omega)$, *where* $1 \leqslant q$, *is said to be* weakly K-quasiregular *if f is orientation-preserving and satisfies the distortion inequality*

$$\max_\alpha |\partial_\alpha f(z)| \leqslant K \min_\alpha |\partial_\alpha f(z)| \quad \text{for almost every } z \in \Omega$$

Evidently Theorem 5.4.2 implies the following theorem.

Theorem 5.4.6. *Let* $q > q_k$ *and* $f \in W_{loc}^{1,q}(\Omega)$ *be weakly K-quasiregular with*

$$K \leqslant \frac{1+k}{1-k}$$

Then, $f \in W_{loc}^{1,2}(\Omega)$, *and hence f is in fact K-quasiregular.*

But weakly quasiregular mappings f with a small regularity exponent q need no longer be continuous, hence certainly not quasiregular. For example, a simple calculation shows that

$$f_0(z) = \frac{1}{z}|z|^{1-1/K} \tag{5.38}$$

is weakly K-quasiregular with $f_0 \in W_{loc}^{1,q}(\mathbb{C})$ for all $1 \leqslant q < \frac{2K}{K+1}$. However, f is not even continuous at $z = 0$. Note also that the Caccioppoli inequality (5.30) fails for f_0 as soon as the exponent $1 \leqslant s \leqslant \frac{2K}{K+1} = 1 + k$.

5.5 Stoilow Factorization

The Stoilow factorization theorem will be used to classify all solutions to the Beltrami equation within reasonable regularity classes. This theorem has its roots in two dimensional topology and states that any discrete open mapping h is topologically equivalent to an analytic function, $h = \varphi \circ f$, where f a homeomorphism and φ holomorphic (a mapping $h : \Omega \to \mathbb{C}$ is called discrete if $h^{-1}\{w\}$ is a discrete subset of Ω for every $w \in \mathbb{C}$). We will need to formulate and establish versions of this result from the point of view of PDEs and analysis. Instead of assuming open and discrete to prove a factorization, we seek to establish a factorization assuming quasiregularity. Once this is done, as a consequence, we obtain a classification of all quasiregular mappings, as it will follow that topologically these mappings are like holomorphic functions, in particular, open and discrete. With the factorization we thus obtain a clear understanding of quasiregular mappings in two dimensions.

5.5. STOILOW FACTORIZATION

The natural choice for regularity is $W^{1,2}_{loc}$, and we will present the factorization in this setup. However, we note that with Theorem 5.4.2 the regularity can immediately be reduced to $W^{1,p}_{loc}$ for any p in the critical interval (q_k, p_k).

In the sequel we will prove several generalizations of the Stoilow factorization theorem. For instance, in Chapter 20 we consider a variety of degenerate Beltrami equations. These cases will require a quite different approach than that presented here. Naturally though, the proofs given for the degenerate equations will apply here for quasiregular mappings. Thus we will eventually have in hand several alternative approaches to Theorem 5.5.1 and Stoilow factorization. The reader may wish to see how to adapt the proofs from our later Theorems 20.2.1 and 20.4.19 to the present case.

Theorem 5.5.1. (Stoilow Factorization) *Let $f : \Omega \to \Omega'$ be a homeomorphic solution to the Beltrami equation*

$$\frac{\partial f}{\partial \bar{z}} = \mu(z) \frac{\partial f}{\partial z}, \qquad f \in W^{1,1}_{loc}(\Omega), \tag{5.39}$$

with $|\mu(z)| \leqslant k < 1$ almost everywhere in Ω.

Suppose $g \in W^{1,2}_{loc}(\Omega)$ is any other solution to (5.39) on Ω. Then there exists a holomorphic function $\Phi : \Omega' \to \mathbb{C}$ such that

$$g(z) = \Phi(f(z)), \qquad z \in \Omega \tag{5.40}$$

Conversely, if Φ is holomorphic on Ω', then the composition $\Phi \circ f$ is a $W^{1,2}_{loc}$-solution to (5.39) in the domain Ω.

Note that given a Beltrami coefficient defined on Ω, we may always find such a homeomorphic solution f even if Ω is a complicated domain. For instance, simply extend μ by 0 to the whole plane, and then restrict the normalized solution defined on $\mathbb{C}$ to Ω.

Further note that, by Corollary 3.3.6, every homeomorphism of Sobolev class $W^{1,1}_{loc}(\Omega)$ has locally integrable Jacobian. In particular, as

$$|Df(z)|^2 \leqslant KJ(z,f),$$

where $K = \frac{1+k}{1-k}$, it follows that any homeomorphic solution $f \in W^{1,1}_{loc}(\Omega)$ actually belongs to $W^{1,2}_{loc}(\Omega)$.

Proof. If $g \in W^{1,2}_{loc}(\Omega)$ is a solution to (5.39), define

$$\Phi(z) = (g \circ f^{-1})(z), \qquad z \in f(\Omega)$$

This is a well-defined function that is continuous by Theorem 5.4.2. We are to show that Φ is holomorphic.

The function g lies in the Royden algebra of any relatively compact subdomain $\tilde{\Omega} \subset \Omega$. Theorem 3.8.2 tells that now $\Phi \in R(f\tilde{\Omega})$, and moreover, the chain rule holds for $\Phi = g \circ f^{-1}$. Writing this in complex notation yields

$$\Phi_{\bar{w}} = (g_z \circ f^{-1})(f^{-1})_{\bar{w}} + (g_{\bar{z}} \circ f^{-1}) \overline{(f^{-1})_w}, \qquad w = f(z),$$

for almost every $w \in f(\Omega)$.

Therefore we will need to find the derivatives of f^{-1} in terms of the Beltrami coefficient μ in (5.39). Indeed, denoting by h the inverse mapping f^{-1} and starting from
$$(h \circ f)(z) = z,$$
we may compute as in the identities (2.49) and (2.50) and obtain

$$\frac{\partial h}{\partial \overline{w}} = -\frac{f_{\bar{z}}}{|f_z|^2 - |f_{\bar{z}}|^2} \qquad (5.41)$$

$$\overline{\frac{\partial h}{\partial w}} = \frac{f_z}{|f_z|^2 - |f_{\bar{z}}|^2} \qquad (5.42)$$

wherever $J(z, f) = |f_z|^2 - |f_{\bar{z}}|^2 \neq 0$. Thus if f satisfies the equation $f_{\bar{z}} = \mu(z) f_z$, then
$$-h_{\overline{w}}(w) = \mu(h(w)) \, \overline{h_w(w)}, \qquad w = f(z) \qquad (5.43)$$
Consequently, for almost every $w \in f(\Omega)$,
$$\Phi_{\overline{w}}(w) = g_z(z) \, h_{\overline{w}}(w) + g_{\bar{z}}(z) \, \overline{h_w(w)} = \overline{h_w(w)} \big[-g_z \, \mu(z) + g_{\bar{z}} \big] = 0$$
Since $\Phi \in W^{1,2}_{loc}(f\Omega)$, we see from Weyl's Lemma A.6.10 that Φ is analytic.

The converse direction follows easily from the chain rule and is left for the reader to verify. □

In particular, we now see that if $f : \Omega \to \mathbb{C}$ is quasiregular, so that $f \in W^{1,2}_{loc}(\Omega)$ and $|f_{\bar{z}}| \leqslant k|f_z|$ almost everywhere in Ω with $0 \leqslant k < 1$, then

$$\frac{\partial f}{\partial \bar{z}} = \mu(z) \frac{\partial f}{\partial z}, \qquad \mu := \frac{f_{\bar{z}}}{f_z} \quad \text{with } \|\mu\|_\infty \leqslant k \qquad (5.44)$$

Corollaries 3.7.6 and 3.10.3 together with Theorem 5.5.1 now give the following topological properties of quasiregular mappings.

Corollary 5.5.2. *Suppose f is a quasiregular mapping defined on a subdomain $\Omega \subset \mathbb{C}$. Then*

- *f is open and discrete.*
- *f is locally Hölder-continuous, with exponent $\alpha = \frac{1}{K}$, where $K = K(f)$.*
- *f satisfies both conditions $\mathcal{N}$ and $\mathcal{N}^{-1}$.*
- *f is differentiable with nonvanishing Jacobian almost everywhere.*

We remind the reader of the local formulation of Stoilow factorization, which reduces to Theorem 5.5.1 via the Riemann mapping $\psi : f(\Omega) \to \Omega$.

Corollary 5.5.3. *Suppose g is quasiregular and defined in a simply connected domain Ω. Then $g = h \circ F$, where $F : \Omega \to \Omega$ is quasiconformal and h is holomorphic in Ω.*

5.5. STOILOW FACTORIZATION

We also note the the quasiconformal factor can be taken to be entire.

Corollary 5.5.4. *Suppose $g : \Omega \to \mathbb{C}$ is quasiregular. Then $g = h \circ F$, where $F : \mathbb{C} \to \mathbb{C}$ is quasiconformal and h is holomorphic in $F(\Omega)$.*

Proof. Let μ be the Beltrami coefficient of g. Obtain $\tilde{\mu} : \mathbb{C} \to \mathbb{D}$ by extending μ by 0 outside Ω. Then $\|\tilde{\mu}\|_\infty = \|\mu\|_\infty = k < 1$. Solve the Beltrami equation $F_{\bar{z}} = \tilde{\mu}(z) F_z$ for a normalized quasiconformal homeomorphism F. The first part of Theorem 5.5.1 assures us that $g = h \circ F$, as $F|\Omega$ is certainly a homeomorphic solution to $F_{\bar{z}} = \mu(z) F_z$ on Ω. □

As in the previous sections, and with the Stoilow factorization as well, we face the question of what is the weakest Sobolev regularity allowing the result to hold. Among nondegenerate equations the complete answer is given in Theorem 14.4.5, where we show that the factorization is still valid for weakly K-quasiregular mappings with

$$f \in W^{1,q}_{loc}(\Omega), \qquad q = \frac{2K}{K+1}$$

This is in fact the precise limit of factorization: The function f_0 from (5.38) is weakly K-quasiregular with $Df_0 \in weak\text{-}L^{2K/(K+1)}(\mathbb{C})$, but we cannot factor f_0 via a holomorphic function and a homeomorphism as it is not continuous at the origin.

On the other hand,

$$f_0(z) = \phi(f(z)),$$

where $f(z) = z|z|^{1/K-1}$ is K-quasiconformal and $\phi(z) = \frac{1}{z}$ is meromorphic. Thus one may wonder if there might exist a more general factorization in terms of meromorphic functions. However, in Theorem 14.4.6 we will construct for every $q < \frac{2K}{K+1}$ a weakly K-quasiregular mapping that is neither open nor discrete and is also not continuous. Similar examples are presented in [191, Theorem 6.5.1]. No kind of factorization is possible for these sorts of a functions.

Next, we complete the proof of the measurable Riemann mapping Theorem 5.3.4, obtaining the uniqueness from Stoilow factorization.

Corollary 5.5.5. *If μ is measurable with $|\mu(z)| \leqslant k < 1$ at almost every $z \in \mathbb{C}$, then the normalized $W^{1,2}_{loc}$-solution to the Beltrami equation $f_{\bar{z}} = \mu(z) f_z$ is unique.*

Proof. If f_1 and f_2 are normalized solutions to the same Beltrami equation, then Stoilow factorization implies $f_1 = \Phi \circ f_2$, where Φ is holomorphic in $\mathbb{C}$. Since f_1, f_2 are homeomorphisms, $\Phi : \mathbb{C} \to \mathbb{C}$ is conformal and by Corollary 2.10.10 in fact a similarity. Since Φ fixes 0, 1 and ∞, Φ is the identity. □

Theorem 5.5.1 and its proof have interesting ramifications: First, it follows that for every quasiregular mapping the Jacobian determinant $J(z, f) > 0$ for almost every z. Therefore its Beltrami coefficient is unique up to a set of measure zero. Indeed, as $J(z, f) \leqslant |f_z|^2$, the ratio

$$\mu(z) = \mu_f(z) = \frac{f_{\bar{z}}}{f_z}$$

is well-defined almost everywhere. This observation, with the chain rule and a direct computation, leads to an elegant formula for the Beltrami coefficient for the composition mappings. The reader should be aware of the relationship between this composition formula and the Möbius transformations of the disk. These connections will play an important role a bit later, see for instance Theorem 5.6.1 for a first elementary application.

Composition Formula for Beltrami Coefficients

Theorem 5.5.6. *Suppose $f : \Omega \to \Omega'$ is quasiconformal and $g : \Omega \to \mathbb{C}$ is quasiregular, with Beltrami coefficients μ_f and μ_g, respectively. Then the composition $g \circ f^{-1}$ is quasiregular in Ω', with Beltrami coefficient*

$$\mu_{g \circ f^{-1}}(w) = \frac{\mu_g(z) - \mu_f(z)}{1 - \mu_g(z)\overline{\mu_f(z)}} \left(\frac{f_z(z)}{|f_z(z)|} \right)^2, \qquad w = f(z) \qquad (5.45)$$

The identity (5.45) gives another proof for the fact that if two quasiconformal mappings f and g, defined on Ω, have the same Beltrami coefficient, then the map $g \circ f^{-1} : f(\Omega) \to g(\Omega)$ is conformal.

We establish other versions of the composition formula latter in (10.3.2).

In the study of quasilinear or general nonlinear elliptic PDEs, there is in fact no reason to limit oneself to those cases where the Beltrami coefficient depends only on the variable z. For instance, the next chapter gives natural examples related to the p-harmonic equation where the Beltrami coefficients will depend both on z and the function f. Systematic analysis of Beltrami-type equations with coefficients of the form $\mu(z, f)$, as well as other related equations, will be considered more closely in the sequel.

It takes little or no effort to find further applications of the Stoilow factorization theorem. We first present a quasiregular version of Montel's theorem.

Corollary 5.5.7. *Suppose $z_0, z_1 \in \mathbb{C}$ and let $g_\nu : \Omega \to \mathbb{C}$, $\nu = 1, 2, \ldots$, be a sequence of K-quasiregular mappings defined on a domain $\Omega \subset \mathbb{C}$, each omitting these two values. Then there is a subsequence converging locally uniformly on Ω to a mapping g,*

$$g_{\nu_k} \to g,$$

and g is either a K-quasiregular mapping or a constant.

5.5. STOILOW FACTORIZATION

Proof. There is no loss of generality in assuming $\Omega = \mathbb{D}$. There are several natural ways to modify the Stoilow factorization. Here we factor the quasiregular mappings g_ν as

$$g_\nu = \phi_\nu \circ F_\nu,$$

where the F_ν are K-quasiconformal with $F_\nu(\mathbb{D}) = \mathbb{D}$, $F_\nu(0) = 0$ and $F_\nu(1) = 1$, while the ϕ_ν are holomorphic in $\mathbb{D}$. These representations are obtained from the normalized solutions f_ν (as obtained in Corollary 5.5.4). Then we use the factorization of Theorem 5.5.1 by writing

$$g_\nu = \Phi_\nu \circ f_\nu = (\Phi_\nu \circ \psi_\nu^{-1}) \circ (\psi_\nu \circ f_\nu)$$

for suitable Riemann mappings $\psi_\nu : f_\nu(\mathbb{D}) \to \mathbb{D}$.

Both of the families $\{\phi_\nu\}$ and $\{F_\nu\}$ are normal, by Theorems 2.9.11 and 3.9.4, respectively. Hence the claim follows. $\square$

As a last example, the methods of the previous sections lead to a general approximation result.

Corollary 5.5.8. *Let $f : \Omega \to \mathbb{C}$ be a quasiregular mapping. Thus $f \in W^{1,2}_{loc}(\Omega)$ satisfies the Beltrami equation*

$$\frac{\partial f}{\partial \bar{z}} = \mu(z) \frac{\partial f}{\partial z}, \qquad |\mu(z)| \leqslant k < 1, \quad \text{for almost every } z \in \Omega$$

Then there are sequence of $C^\infty(\Omega)$-functions $\{f^\nu\}_{\nu=1}^\infty$ and $\{\mu_\nu\}_{\nu=1}^\infty$ such that each f^ν is a solution to the equation

$$\frac{\partial f^\nu}{\partial \bar{z}} = \mu_\nu(z) \frac{\partial f^\nu}{\partial z}, \qquad |\mu_\nu(z)| \leqslant k,$$

and where $f^\nu \to f$ locally uniformly in Ω and $\mu_\nu \to \mu$ almost everywhere. Furthermore, if f is quasiconformal, that is, a homeomorphism, then we can choose each f^ν to be a C^∞-diffeomorphism.

Proof. We first extend μ as zero outside Ω and then use Stoilow factorization to find $f = h \circ F$. Here h is holomorphic in the domain $\Omega' = F(\Omega)$ and $F : \mathbb{C} \to \mathbb{C}$ is the normalized quasiconformal mapping with the Beltrami coefficient μ. As in (5.13), we look at the smoothing approximations and set

$$\mu_\nu = \phi_{1/\nu} * \left[\mu \chi_{\mathbb{D}(0,\nu)} \right]$$

Clearly, $\mu_\nu \to \mu$ almost everywhere.

Let $F_\nu : \mathbb{C} \to \mathbb{C}$ be the normalized quasiconformal mapping with the Beltrami coefficient $\mu_\nu(z)$. According to Theorem 5.2.4, as each μ_ν is smooth and compactly supported, each F_ν is a C^∞-diffeomorphism of the extended plane, and similarly as in the proof of Theorem 5.3.4, we obtain $F_\nu \to F$ locally uniformly in $\mathbb{C}$. In particular,

$$f_\nu(z) := h \circ F_\nu(z) \to h \circ F = f(z)$$

locally uniformly in Ω. □

It is not difficult to see that above the convergence $f^\nu \to f$ occurs in $W^{1,s}_{loc}(\Omega)$, for all $s \in (q_k, p_k)$, the critical interval. In fact, for a quasiconformal mapping with a compactly supported Beltrami coefficient, this is a direct consequence of the proof given in Theorem 5.3.2. With Stoilow factorization and the convergence properties of holomorphic functions, the general case easily follows.

5.6 Factoring with Small Distortion

In this section we prove the following quite nice result. It is an elementary application of the existence theorem. It shows that a K-quasiconformal mapping of a planar domain can be factorized as the finite composition of n quasiconformal maps each with distortion $K^{1/n}$. It appears unlikely that this result has any counterpart in higher dimensions—but this is not known. The following analogous problem in two dimensions is as yet unsolved.

Problem. If $f : \mathbb{C} \to \mathbb{C}$ is L-biLipschitz, can we factor f into a composition with factors of strictly smaller bilischitz constant?

If so, one would further ask to bound the number of terms in the factorization in the obvious manner. We first prove the quasiconformal factorization result for mappings of $\mathbb{C}$.

Theorem 5.6.1. *Let $f : \mathbb{C} \to \mathbb{C}$ be K-quasiconformal and let $n \in \mathbb{N}$. Then we can write*
$$f = f_1 \circ f_2 \circ \cdots \circ f_n,$$
where each $f_i : \mathbb{C} \to \mathbb{C}$ is $K^{1/n}$-quasiconformal.

Proof. An elementary induction shows that is suffices to decompose f as $f = f_1 \circ g$ where f_1 is $K^{1/n}$-quasiconformal and g is $K^{1-1/n}$-quasiconformal. We may also assume that f is normalized so that $f(0) = 0$ and $f(1) = 1$. Let $\mu(z)$ be the complex Beltrami coefficient of f, so that $\mu(z) \in \mathbb{D}$ for $z \in \mathbb{C}$. Define another coefficent ν by setting $\nu(z) = t(z)\mu(z)$, where $0 \leqslant t(z) < 1$ is chosen so that the distance $\rho_\mathbb{D}(0, \nu(z)) = \frac{n-1}{n} \rho_\mathbb{D}(0, \mu(z))$. Here as usual, $\rho_\mathbb{D}$ denotes the hyperbolic metric of the unit disk.

Let g be the unique normalized solution to the Beltrami equation $g_{\bar{z}} = \nu(z) g_z$. Then g is quasiconformal with

$$\log(K_g) = \|\rho_\mathbb{D}(0, \nu(z))\|_\infty = \frac{n-1}{n} \|\rho_\mathbb{D}(0, \mu(z))\|_\infty = \frac{n-1}{n} \log(K)$$

We next set $f_1 = f \circ g^{-1}$ and then use the composition formula Theorem 5.5.6 to show that f_1 is quasiconformal with

$$\log(K_{f_1}) = \|\rho_\mathbb{D}(\nu(z), \mu(z))\|_\infty \leqslant \frac{1}{n} \|\rho_\mathbb{D}(0, \mu(z))\|_\infty = \frac{1}{n} \log(K),$$

and the result follows. □

In the case where f is defined on a domain Ω, we can write $f = \varphi \circ g$, where $g : \mathbb{C} \to \mathbb{C}$ is quasiconformal and φ is conformal by Corollary 5.5.4. Then factorising g and absorbing φ into the first factor gives the following corollary.

Theorem 5.6.2. *Let* $f : \Omega \to \Omega'$ *be* K*-quasiconformal and let* $n \in \mathbb{N}$. *Then we can write*
$$f = f_1 \circ f_2 \circ \cdots \circ f_n,$$
where each $f_i : f_{i+1} \circ \cdots \circ f_n(\Omega) \to \mathbb{C}$ *is* $K^{1/n}$*-quasiconformal.*

5.7 Analytic Dependence on Parameters

In this section we investigate how the solutions to a parameterized family of Beltrami equations depend on the parameters. We will see that under a proper normalization one in fact has holomorphic dependence, a result first obtained as a consequence of Bojarski's approach to the existence theory [64] and observed by Ahlfors-Bers [9]. This will be especially important when we later study the theory of holomorphic motions.

Recall that a Banach space-valued function of a complex variable λ, say $\lambda \mapsto g(\cdot, \lambda) \in L^p(\Omega)$ with $1 \leqslant p \leqslant \infty$, is holomorphic if there is a derivative $\partial_\lambda g \in L^p(\Omega)$ such that, for all $\alpha \in L^q(\Omega)$,

$$\langle \partial_\lambda g, \alpha \rangle := \int_\Omega \partial_\lambda g(z) \alpha(z) = \lim_{\eta \to 0} \int_\Omega \frac{g(z, \lambda + \eta) - g(z, \lambda)}{\eta} \alpha(z) \qquad (5.46)$$

Applying the Cauchy integral formula to each $\langle \partial_\lambda g, \alpha \rangle$, we obtain a vector-valued Cauchy representation for $\partial_\lambda g$, and from this one readily sees that (5.46) is equivalent to

$$\left\| \partial_\lambda g(z) - \frac{g(z, \lambda + \eta) - g(z, \lambda)}{\eta} \right\|_p \to 0, \quad \text{as } \eta \to 0$$

Moreover, one has pointwise analyticity almost everywhere.

Lemma 5.7.1. *Let* $1 \leqslant p \leqslant \infty$. *Suppose that the* L^p*-valued function* $\lambda \mapsto g(\cdot, \lambda) \in L^p(\Omega)$ *is holomorphic in the unit disk* $\mathbb{D}$.
Then, for almost all $z \in \Omega$, *the scalar functions*

$$\lambda \mapsto g(z, \lambda)$$

are holomorphic in $\mathbb{D}$.

Proof. From the vector-valued Cauchy integral formula, we also have the power series representation,

$$g(\cdot, \lambda) = \sum_{n=0}^{\infty} a_n(\cdot) \lambda^n$$

with convergence in L^p-norm. Here $a_n \in L^p(\Omega)$, $n \in \mathbb{N}$, with

$$\limsup_{n \to \infty} \|a_n\|_{L^p}^{1/n} \leqslant 1$$

Consider the sets $E_n = \{z \in \Omega : |a_n(z)| > n^{2/p} \|a_n\|_p\}$. Via the Chebychev inequality,

$$|E_n| \leqslant \left(\frac{1}{n^{2/p}\|a_n\|_p}\right)^p \|a_n\|_p^p = \frac{1}{n^2}$$

In particular, $|\cup_{n \geqslant j} E_n| \to 0$ as $j \to 0$, showing that the set $E = \cap_{j=1}^{\infty} \cup_{n \geqslant j} E_n$ has zero measure.

On the other hand, for any point $z \in \Omega \setminus E$,

$$\limsup_{n \to \infty} |a_n(z)|^{1/n} \leqslant \limsup_{n \to \infty} n^{2/np} \limsup_{n \to \infty} \|a_n\|_{L^p}^{1/n} \leqslant 1$$

Therefore for every $z \in \Omega \setminus E$ the power series $\sum_{n=0}^{\infty} a_n(z)\lambda^n$ converges in the unit disk $\mathbb{D}$, proving the claim. $\square$

Let us apply the above to the Beltrami equation

$$f_{\bar{z}} = \mu(z) f_z, \qquad |\mu(z)| \leqslant k\chi_G(z), \tag{5.47}$$

where G is a measurable set of finite area and $0 < k < 1$. We first look at the simplest case where the coefficient μ is multiplied by the complex parameter $\lambda \in \mathbb{D}(0, 1/k)$, that is, the equation

$$f_{\bar{z}}^\lambda = \lambda \mu(z) f_z^\lambda \tag{5.48}$$

The assumption $\|\lambda\mu(z)\|_\infty < 1$ guarantees that we may apply the existence theory of the previous section and write the principal solution in the form

$$f^\lambda(z) = z + \mathcal{C}\omega^\lambda, \qquad \omega \in L^2(\mathbb{C}) \tag{5.49}$$

The density function ω^λ is given by the Neumann series

$$\omega^\lambda = \lambda\mu + \lambda^2 \mu \mathcal{S}\mu + \cdots = f_{\bar{z}}^\lambda \tag{5.50}$$

and hence

$$1 + \mathcal{S}\omega^\lambda = 1 + \lambda \mathcal{S}\mu + \lambda^2 \mathcal{S}\mu\mathcal{S}\mu \cdots = f_z^\lambda \tag{5.51}$$

Since the Beurling transform is an isometry and $\|\mu\|_\infty \leqslant k$, the series converges in $L^2(\mathbb{C})$, representing a vector-valued analytic function. Thus $f_{\bar{z}}^\lambda$, f_z^λ depend, as an element of $L^2(\mathbb{C})$, analytically on λ.

The argument of Lemma 5.7.1 shows that pointwise analyticity is controlled by the function

$$\Gamma(z) := \limsup_{n \to \infty} \left|\frac{(\mathcal{S}\mu \cdots \mathcal{S}\mu)(z)}{nk^n}\right|$$

5.7. ANALYTIC DEPENDENCE

In particular, $\Gamma(z) < \infty$ for almost every $z \in \mathbb{C}$. Note also that as $\|\mathcal{S}\mu \cdots \mathcal{S}\mu\|_2 \leqslant k^n$, for the n times iterated operation, a simple use of Hölder's inequality shows

$$\Gamma(z) = 0 \quad \text{whenever } \operatorname{dist}(z, \operatorname{supp}(\mu)) \geqslant \varepsilon > 0 \tag{5.52}$$

We have the following theorem.

Theorem 5.7.2. *Let $z \in \mathbb{C}$ be a point for which $\Gamma(z) < \infty$ and assume f^λ is the principal solution to (5.48). Then for all $\lambda \in \mathbb{D}(0, 1/k)$ the complex derivatives $\partial_{\bar z} f^\lambda(z)$ and $\partial_z f^\lambda(z)$ exist. Moreover, the functions*

$$\lambda \mapsto \frac{\partial f^\lambda}{\partial \bar z}(z) \tag{5.53}$$

$$\lambda \mapsto \frac{\partial f^\lambda}{\partial z}(z) \tag{5.54}$$

are holomorphic in the disk $\mathbb{D}(0, 1/k)$.

Finally, we need to consider the analyticity of the function $\lambda \mapsto f^\lambda(z)$. For this we require some L^p-estimates.

Theorem 5.7.3. *For every $z \in \mathbb{C}$ the function*

$$\lambda \mapsto f^\lambda(z), \quad \lambda \in \mathbb{D},$$

is holomorphic.

Proof. We have

$$\partial_{\bar z} f^\lambda = (\mathbf{I} - \lambda \mu \mathcal{S})^{-1}(\lambda \mu) = \lambda \mu + \lambda^2 \mu \mathcal{S} \mu + \cdots \tag{5.55}$$

where the right-hand side of (5.55) defines a vector-valued holomorphic function of λ. The Cauchy transform $\mathcal{C}$ is a bounded linear operator from $L^2(\mathbb{C}) \to VMO(\mathbb{C})$ and hence preserves holomorphicity. (Note that any bounded linear operator can be seen to preserve holomorphicity by passing it through the limit quotient defining the complex derivative.) As

$$\mathcal{C}(\partial_{\bar z} f^\lambda)(z) = f^\lambda(z) - z,$$

we see that $\lambda \to f^\lambda$ is a holomorphic $VMO(\mathbb{C})$-valued function of λ. However, we cannot use this fact to show that $f^\lambda(z)$ depends holomorphically on λ since $VMO(\mathbb{C})$-functions need not be locally bounded.

Thus we need to turn to L^p-estimates. Recall that $(\mathbf{I} - \lambda \mu \mathcal{S})^{-1} \lambda \mu \in L^p(\mathbb{C})$ for some $p > 2$, in fact, for all p such that $k|\lambda|\|\mathcal{S}\|_p < 1$. Next, by Theorem 4.3.13, the Cauchy transform takes $L^p(\mathbb{C})$ into $C^\alpha(\mathbb{C})$, in particular, to continuous functions. Thus point evaluation on $C^\alpha(\mathbb{C})$ is a bounded linear functional which therefore preserves analyticity. In summary, we have shown that $\lambda \mapsto f^\lambda(z)$ is holomorphic for every z. $\square$

It is clear that the above argument works in much greater generality. We may for instance assume that

1. $\Omega \subset \mathbb{C}$ is a domain with $|\Omega| < \infty$.
2. $\Delta \subset \mathbb{C}$ is an open set.
3. $\lambda \mapsto \mu_\lambda$ is holomorphic $\Delta \to L^\infty(\mathbb{C})$ (in the sense of (5.46)).
4. $|\mu_\lambda| \leqslant k(\lambda) \chi_\Omega$, where for all $\lambda \in \Delta$ the bound $k(\lambda) < 1$.

Under these hypotheses we see that the principal solution $f^\lambda = f^{\mu(\cdot,\lambda)}$ to the corresponding Beltrami equation depends holomorphically on the parameter λ. Lemma 5.7.1 and the proof of Theorem 5.7.3 give us the following theorem.

Theorem 5.7.4. *There is a subset $\Omega' \subset \Omega$ of full measure such that for each fixed $z \in \Omega'$, for all $\lambda \in \Delta$, the complex derivatives $\partial_{\bar{z}} f^\lambda(z)$ and $\partial_z f^\lambda(z)$ of the principal solution exist, and*

$$\begin{aligned} \lambda &\mapsto \partial_{\bar{z}} f^\lambda(z) \\ \lambda &\mapsto \partial_z f^\lambda(z) \end{aligned} \qquad (5.56)$$

are holomorphic in the domain Δ.

Moreover, for every $z \in \mathbb{C}$ the function

$$\lambda \mapsto f^\lambda(z), \qquad \lambda \in \Delta,$$

is holomorphic.

Even when conformality at ∞ is lost, we still retain holomorphic dependence, but now for the normalized solutions.

Analytic Dependence on Parameters

Corollary 5.7.5. *Suppose $\lambda \mapsto \mu_\lambda$ defines a holomorphic map $\Delta \to L^\infty(\mathbb{C})$, where $\|\mu_\lambda\|_\infty < 1$ for $|\lambda| < 1$. Let f^λ be the normalized solution to the equation*

$$\frac{\partial f^\lambda}{\partial \bar{z}} = \mu_\lambda(z) \frac{\partial f^\lambda}{\partial z}, \qquad (5.57)$$

Then, for each $z \in \mathbb{C}$, the function

$$\lambda \mapsto f^\lambda(z), \qquad \lambda \in \Delta,$$

is holomorphic.

Proof. Suppose first that μ_λ are compactly supported in $\mathbb{D}(0, r)$. Then the normalized solution to the Beltrami equation (5.57) is

$$f^\lambda(z) = \frac{F^\lambda(z) - F^\lambda(0)}{F^\lambda(1) - F^\lambda(0)},$$

where F^λ is the principal solution to (5.57). As $\lambda \mapsto F^\lambda(z)$ is holomorphic for each z by Theorem 5.7.3, so too is $\lambda \mapsto f^\lambda(z)$. If μ_λ is not compactly

5.8. EXTENSION

supported, we approximate as in Theorem 5.3.4 and Lemma 5.3.5; since the normalized solutions are unique, $(f_r^\lambda)(z) \to f^\lambda(z)$ for every z and λ. For each z the normalized approximation varies holomorphically. This gives us, for each fixed z, a sequence of holomorphic maps of the disk $\phi_r(\lambda) = (f_r^\lambda)(z)$, whereupon the uniform convergence of $f_r^\lambda \to f^\lambda$ as $r \to \infty$ shows this family of analytic maps to be normal. It follows that $\lambda \mapsto f^\lambda(z)$ is holomorphic, as required. □

The reader should note, however, that holomorphic dependence does not carry over to the inverse mapping. As a simple example, let

$$f(z) = \begin{cases} z + \lambda \bar{z}, & |z| \leqslant 1 \\ z + \lambda/z, & |z| \geqslant 1 \end{cases}$$

with $\mu = \lambda \chi_\mathbb{D}$. For $|w|$ small we have

$$f^{-1}(w) = \frac{w - \lambda \bar{w}}{1 - |\lambda|^2},$$

and this is not holomorphic in λ.

5.8 Extension of Quasisymmetric Mappings of the Real Line

One of the basic properties of a quasisymmetric mapping of the real line $\mathbb{R}$ is that it is the restriction of a global quasiconformal mapping of the plane $\mathbb{C}$. This fact was first proven by Ahlfors and Beurling [59] and has important applications, in particular, to conformal welding which we discuss later.

In fact, there are equivariant versions of this theorem due to Tukia [357] and Douady-Earle [110]. These results state that if a quasisymmetric mapping commutes with a Fuchsian group, then the extension can be chosen to do so as well. This has important applications in Teichmüller theory, Kleinian groups and dynamics.

In this section we give an elementary proof of the extension theorem, following Kenig and Jerison [200]. This extension is not equivariant, but we shall subsequently discuss very briefly the more general Douady-Earle extension.

Theorem 5.8.1. Let $f : \mathbb{R} \to \mathbb{R}$ be an η-quasisymmetric mapping. Then there is a K-quasiconformal mapping $F : \mathbb{C} \to \mathbb{C}$ such that $F|\mathbb{R} = f$. Furthermore, $K \leqslant 8\eta^*(2)^3$.

Proof. We extend f to the upper half-plane $\mathbb{H}$. The result will then follow by a reflection argument we prove subsequently. We symmetrize η to simplify our formulas,

$$\eta^*(t) = \max\{\eta(t), 1/\eta(1/t)\}$$

We define our extension F on the dyadic points $D = \{k2^m + i2^m : k, m \in \mathbb{Z}\}$ by

$$F(k2^m + i2^m) = f(k2^m) + i[f((k+1)2^m) - f(k2^m)] \quad (5.58)$$

This is illustrated below.

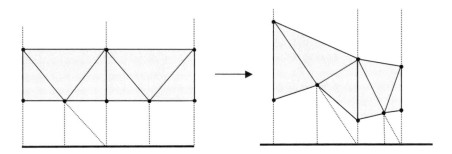

The piecewise linear extension

We then extend F linearly over each triangle T with vertices in D. By construction the image of each triangle is nondegenerate, and distinct triangles are mapped to distinct triangles. It follows that F is a homeomorphism. To see that F is quasiconformal it is enough to show that in each triangle F is quasisymmetric with an uniform bound. At each level there are three different types of triangles, those with vertices of the form

$$\begin{aligned} T_1 &= \{k2^m + i2^m, (k+1)2^m + i2^m, (2k+1)2^{m-1} + i2^{m-1}\} \\ T_2 &= \{k2^m + i2^m, k2^m + i2^{m-1}, (2k+1)2^{m-1} + i2^{m-1}\} \\ T_3 &= \{(k+1)2^m + i2^m, (2k+1)2^{m-1} + i2^{m-1}, (k+1)2^m + i2^{m-1}\} \end{aligned}$$

If we label the vertices $\{a, b, c\}$, we need to show that

$$\frac{|F(a) - F(b)|}{|F(a) - F(c)|}, \quad \frac{|F(a) - F(b)|}{|F(b) - F(c)|}, \quad \frac{|F(a) - F(c)|}{|F(b) - F(c)|} \quad (5.59)$$

are uniformly bounded away from 0 and ∞. For the horizontal side of T_1, the image side length has the estimates

$$|F((k+1)2^m + i2^m) - F(k2^m + i2^m)| \geq |f((k+1)2^m) - f(k2^m)|,$$
$$|F((k+1)2^m + i2^m) - F(k2^m + i2^m)| \leq (2 + \eta(1))|f((k+1)2^m) - f(k2^m)|$$

while for the other sidelengths of $F(T_1)$ note that

$$|F(k2^m + i2^m) - F((2k+1)2^{m-1} + i2^{m-1})| = \sqrt{2}\,|f((2k+1)2^{m-1}) - f(k2^m)|$$

and

$$\begin{aligned} &|f((k+1)2^m) - f((2k+1)2^{m-1})| \\ &\leq |F((k+1)2^m + i2^m) - F((2k+1)2^{m-1} + i2^{m-1})| \\ &\leq (2 + \eta(2))\,|f((k+1)2^m) - f((2k+1)2^{m-1})| \end{aligned}$$

5.8. EXTENSION

Hence, for T_1 the ratios in (5.59) are all bounded from above by $(2+\eta(1))\eta(2)$, and from below by the reciprocal of this number. For the vertices of T_3 lying on the same vertical line, the image side length is

$$|F((k+1)2^m + i2^m) - F((k+1)2^m + i2^{m-1})|$$
$$= |f((k+2)2^m) - f((2k+3)2^{m-1})|$$

The same analysis as above shows that for T_3 all the ratios in (5.59) are bounded from above by $(2+\eta(2))\eta(1/2)\eta(2)$. The analysis for T_2 is even simpler, and this latter estimate holds for all of its edge ratios too. It follows that in each triangle, hence in the entire upper half-plane, F is K-quasiconformal with K depending only on η, the bound we give being readily obtained from the estimates above, with a little manipulation, once we recall that $\eta^*(t) = \max\{\eta(t), 1/\eta(1/t)\}$. □

The extension we have just described has the virtue of being quite elementary, in fact proving quasiconformality amounted to little more than manipulating the definition of quasisymmetry. However, the estimate of the distortion K is not optimal and could be improved at the expense of greater complexity in the argument. Further, a highly desirable property of an extension, from the point of view of applications in Teichmüller theory and dynamics, is conformal naturality as discussed in the introduction to this section. Tukia [357] first constructed an extension with this property, however, the Douady-Earle extension [110] is easier to describe and also has these properties. We next briefly discuss this.

5.8.1 The Douady-Earle Extension

There is an operator E, that extends a quasisymmetric self-mapping f of the circle $\mathbb{S}$ to a quasiconformal self-mapping $E(f) = F$ of the disk $\mathbb{D}$, which has good distortion bounds (as $\eta_f(t) \to t$ we have $K(F) \to 1$) and has the following conformal naturality:

If $\phi, \psi : \mathbb{D} \to \mathbb{D}$ are linear fractional transformations, then

$$E(\phi \circ f \circ \psi) = \phi \circ E(f) \circ \psi \tag{5.60}$$

In fact, given a quasisymmetric map f, the extension $F = E(f) : \mathbb{D} \to \mathbb{D}$ is a real analytic diffeomorphism, bilipschitz in the hyperbolic metric (and therefore quasiconformal), and can be defined as follows. Let

$$G(z,w) = \frac{1}{2\pi} \int_{\mathbb{S}} \frac{f(\zeta) - w}{1 - \bar{w}f(\zeta)} \frac{1 - |z|^2}{|z - \zeta|^2} |d\zeta| \tag{5.61}$$

Given $z \in \mathbb{D}$, we find the unique point w such that $G(z,w) = 0$ and then set

$$E(f)(z) = w$$

Conformal naturality is a simple consequence of the identities at (2.13) for linear fractional transformations of the disk. Further, this conformal naturality means we need only make estimates at the origin. However, there are some subtleties. First, that w exists and is unique, second, that $E(f)|\mathbb{S} = f$, and, finally getting optimal distortion bounds. We will not discuss these further.

5.8.2 The Beurling-Ahlfors Extension

The Beurling-Ahlfors extension is also easy to write down if we skip the technical details involved in establishing the claimed properties; see [229] for proofs.

Given an increasing quasisymmetric mapping of $\mathbb{R}$, we have for every $t > 0$,

$$\frac{1}{\eta(1)} \leq \frac{f(x+t) - f(x)}{f(x) - f(x-t)} \leq \eta(1)$$

Set

$$u(z) = \int_0^1 f(x + ty)\, dt = \frac{1}{y}\int_x^{x+y} f(s)\, ds$$
$$v(z) = \int_0^1 f(x - ty)\, dt = \frac{1}{y}\int_{x-y}^{x} f(s)\, ds$$

Then

$$F(z) = \frac{1+i}{2}[u(z) - iv(z)]$$

defines a K-quasiconformal mapping of $\mathbb{C}$ with $F|\mathbb{R} = f$ and $K \leq \eta^2(1)$.

5.9 Reflection

We now discuss the reflection principle for quasiconformal mappings.

Theorem 5.9.1. *Let $f : \mathbb{H} \to \mathbb{H}$ be a K-quasiconformal mapping of the upper half-space. Then the mapping defined by*

$$F(z) = \begin{cases} f(z), & z \in \mathbb{H} \\ \overline{f(\bar{z})}, & \bar{z} \in \mathbb{H} \end{cases} \tag{5.62}$$

has a continuous extension to $\mathbb{R}$ after which it defines a K-quasiconformal homeomorphism of $\mathbb{C}$.

Proof We may assume $f(0) = 0$, $f(1) = 1$. Let $\mu(z)$ denote the Beltrami coefficient of f. We extend μ to the entire plane $\mathbb{C}$ by the rule

$$\tilde{\mu}(z) = \begin{cases} \mu(z), & z \in \mathbb{H} \\ \overline{\mu(\bar{z})}, & \bar{z} \in \mathbb{H} \end{cases} \tag{5.63}$$

Let $\tilde{F}(z)$ denote the unique normalized solution to the equation

$$\frac{\partial \tilde{F}}{\partial \bar{z}} = \tilde{\mu}(z) \frac{\partial \tilde{F}}{\partial z} \tag{5.64}$$

5.10. CONFORMAL WELDING

Uniqueness implies that $\overline{\tilde{F}(\bar{z})} = \tilde{F}(z)$ for all $z \in \mathbb{C}$, as both these maps are normalized solutions to (5.64). Thus $\tilde{F}(\mathbb{H}) = \mathbb{H}$ and $\tilde{F}(\mathbb{C} \setminus \mathbb{H}) = \mathbb{C} \setminus \mathbb{H}$. The Stoilow factorization Theorem 5.5.1 applied in $\mathbb{H}$ shows us that

$$F = \Phi \circ \tilde{F}$$

for a conformal-self mapping of $\mathbb{H}$. However, $\Phi(0) = 0, \Phi(1) = 1$ and Φ fixes ∞, so Φ is the identity and $\tilde{F}$ provides us with the desired K-quasiconformal extension. □

A *Jordan domain* in $\mathbb{C}$ is a domain bounded by a topological circle on the Riemann sphere. Such a domain might be unbounded in the plane if its boundary passes through ∞.

Corollary 5.9.2. *Let Ω and Ω' be Jordan domains. Let $f : \Omega \to \Omega'$ be quasiconformal. Then f has a continuous homeomorphic extension to the boundary.*

Proof. The classical theorem of Carathéodory [89] tells us that both Riemann maps $\phi : \mathbb{H} \to \Omega$ and $\psi : \mathbb{H} \to \Omega'$ have continuous homeomorphic extensions to the boundary. Then $\psi \circ f \circ \phi : \mathbb{H} \to \mathbb{H}$ is quasiconformal. This map therefore extends continuously and homeomorphically to the boundary by Theorem 5.9.1. The result follows. □

Another approach to Corollary 5.9.2 is to write $f = \varphi \circ g$, where $g : \mathbb{C} \to \mathbb{C}$ is quasiconformal, using Corollary 5.5.4. As $g(\Omega)$ is again a Jordan domain, the conformal map $\varphi : g(\Omega) \to \Omega'$ will extend homeomorphically to the boundary.

5.10 Conformal Welding

Here we discuss conformal welding or sewing. It is in effect a representation theorem for quasisymmetric mappings of $\mathbb{R}$ via Riemann mappings of $\mathbb{H}$ (or equivalently, the circle and disk). Apparently, this theorem was first proved by Pfluger [297], see also Lehto-Virtanen [229].

Theorem 5.10.1. *Let $f : \mathbb{R} \to \mathbb{R}$ be a quasisymmetric mapping. Then there is a Jordan domain Ω and conformal mappings $\phi : \mathbb{H} \to \Omega$ and $\psi : \mathbb{C} \setminus \overline{\mathbb{H}} \to \mathbb{C} \setminus \overline{\Omega}$, with homeomorphic extensions to the closures, such that*

$$f = \phi^{-1} \circ \psi | \mathbb{R} \tag{5.65}$$

The domain Ω is unique up to Möbius transformation and is a quasidisk.

Proof. Let F be a quasiconformal extension of f to $\mathbb{H}$ given by Theorem 5.8.1 and μ its Beltrami coefficient. Extend μ by 0 on $\mathbb{C} \setminus \mathbb{H}$ to get $\tilde{\mu}$ and assume $\tilde{F} : \mathbb{C} \to \mathbb{C}$ has $\tilde{\mu}$ as a Beltrami coefficient. Set $\Omega = \tilde{F}(\mathbb{H})$. The uniqueness up to conformal mapping of solutions to the Beltrami equation gives a conformal mapping $\phi : \mathbb{H} \to \Omega$ with

$$\tilde{F} = \phi \circ F : \mathbb{H} \to \Omega \tag{5.66}$$

By construction $\psi = \tilde{F}|(\mathbb{C} \setminus \overline{\mathbb{H}})$ is conformal.

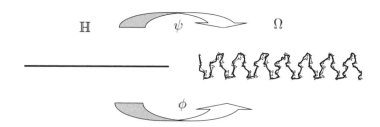

Welding of a quasisymmetric mapping via conformal maps of Jordan domains

All these maps have continuous homeomorphic extensions to the boundary of their domain of definition, and then (5.66) reads as

$$\psi|\mathbb{R} = (\phi|\mathbb{R}) \circ (F|\mathbb{R}),$$

as desired.

We now establish uniqueness. Suppose Ω_1 and Ω_2 are Jordan domains with conformal maps $\phi_i : \mathbb{H} \to \Omega_i$ and $\psi_i : \mathbb{C}\setminus\overline{\mathbb{H}} \to \mathbb{C}\setminus\overline{\Omega_i}$, with $\phi_i^{-1} \circ \psi = f : \mathbb{R} \to \mathbb{R}$. We may suppose that Ω_1 is the quasidisk constructed earlier and so ψ_1 is the restriction of a quasiconformal mapping $g : \mathbb{C} \to \mathbb{C}$. We may further suppose all the mappings in question extend homeomorphically to the boundaries of their domains of definition. We construct a homeomorphism of the plane taking Ω_1 to Ω_2 by

$$h(z) = \begin{cases} \phi_2 \circ \phi_1^{-1}, & z \in \Omega_1 \\ \psi_2 \circ \psi_1^{-1}, & z \in \mathbb{C}\setminus\Omega_1 \end{cases} \quad (5.67)$$

Now consider $F = h \circ g : \mathbb{C} \to \mathbb{C}$. This map is a homeomorphism that is quasiconformal in $\mathbb{H}$ and in $\mathbb{C}\setminus\overline{\mathbb{H}}$. We want to observe that F is quasiconformal. Absolute continuity of F on almost all horizontal lines is clear, for each such line lies entirely in $\mathbb{H}$ or $\mathbb{C}\setminus\overline{\mathbb{H}}$. For vertical lines the only issue is near $\mathbb{R}$, where we may use the uniform continuity of F on any compact neighborhood of a closed subinterval of $\mathbb{R}$. It follows that h is ACL and therefore quasiconformal. Then as g is quasiconformal, we see h is quasiconformal and $\mu_h = 0$ almost everywhere since $|\partial\Omega_1| = |g(\mathbb{R})| = 0$. Thus h is conformal on $\mathbb{C}$ and hence a Möbius transformation.

$\square$

It is not true in general that an arbitrary homeomorphism $f : \mathbb{R} \to \mathbb{R}$ can be identified as a conformal welding. The first counterexample is due to K. Oikawa [289], but see also J. Vainio [364]. The most recent developments are due to D. Hamilton; see [153, 154] and the references therein, and C. Bishop [61, 62]. There it is shown that every homeomorphism of $\mathbb{R}$ is "almost" a welding in a precise way.

Chapter 6

Parameterizing General Linear Elliptic Systems

In previous chapters we have established simple methods to identify all solutions to the $\mathbb{C}$-linear Beltrami equation $f_{\bar{z}} = \mu f_z$. We found that it suffices to find one global homeomorphic solution $f : \mathbb{C} \to \mathbb{C}$ to the equation and then every other solution $g : \Omega \to \mathbb{C}$ can be represented as the composition

$$g = \phi \circ f, \qquad \text{where } \phi : f(\Omega) \to \mathbb{C} \text{ is holomorphic}$$

This parameterization, the Stoilow factorization, is a powerful tool heavily used elsewhere in this book and in more general circumstances to develop a deeper understanding of the properties of solutions to the Beltrami equation. These include unique continuation, openess and discreteness, the argument principle, normal family properties, the isolation of branch points and many others.

However, in many concrete applications the framework we develop around the $\mathbb{C}$-linear Beltrami equation is not the correct one to operate within. For instance (as discussed in detail in Section 16.1.5 later), if we consider uniformly elliptic equations of divergence-type, with a measurable but not necessarily symmetric coefficient matrix,

$$\text{div } A(z)\nabla u = 0; \qquad u \in W^{1,2}_{loc}(\Omega),\ A : \Omega \to \mathbb{R}^{2\times 2}, \qquad (6.1)$$

then any solution u in a simply connected domain Ω admits a conjugate function $v \in W^{1,2}_{loc}(\Omega)$ with $\nabla v = *A(z)\nabla u$. Here $*(x,y) = (-y,x)$ is rotation by 90 degrees. Together the functions u and v define the complex valued function

$$f = u + iv,$$

which solves the $\mathbb{R}$-linear equation

$$\frac{\partial f}{\partial \bar{z}} = \mu(z)\frac{\partial f}{\partial z} + \nu(z)\overline{\frac{\partial f}{\partial z}}, \qquad (6.2)$$

where the coefficients $\mu(z)$, $\nu(z)$ are explicit functions of the elements of $A(z)$, independent of the particular solution u. For the precise relation here, see (16.20). Conversely, with these given coefficients μ and ν, the real part u of any solution $f = u + iv$ to (6.2) solves the second-order equation (6.1).

Another class of examples is given by the hodograph transformation of the complex gradient of a p-harmonic function. This takes one to a linear equation of the following type

$$\frac{\partial h}{\partial \overline{\xi}} = \left(\frac{1}{2} - \frac{1}{p}\right)\left(\frac{\xi}{\overline{\xi}}\frac{\partial h}{\partial \xi} + \frac{\overline{\xi}^n}{\xi^n}\frac{\overline{\partial h}}{\partial \xi}\right)$$

Similarly, the general linear system arises when performing the hodograph transform on the minimizers of energy integrals of the form

$$\mathcal{E}[u] = \int_\Omega E(\nabla u)$$

Together these examples indicate that there is a clear need to find and parameterize all possible solutions to the linear elliptic system (6.2). However, a simple observation shows that the classical Stoilow factorization theorem cannot work here. Indeed, for the solutions to the general linear Beltrami equations, the composition $\phi \circ f$ with a holomorphic function ϕ generally even fails to satisfy (6.2)!

It is in fact surprising that a complete parameterization of solutions to (6.2) is possible and it is the purpose of this short chapter to present and develop a method that generalizes the Stoilow factorization in a form applicable to the general linear elliptic system.

6.1 Stoilow Factorization for General Elliptic Systems

Consider the general elliptic operator

$$\mathcal{L} = \frac{\partial}{\partial \overline{z}} - \mu(z)\frac{\partial}{\partial z} - \nu(z)\overline{\frac{\partial}{\partial z}}, \qquad |\mu(z)| + |\nu(z)| \leqslant k < 1 \qquad (6.3)$$

We begin our considerations with the following factorization theorem associated with (6.3).

Theorem 6.1.1. *Let $f \in W^{1,2}_{loc}(\Omega)$ be a homeomorphic solution to the equation*

$$\mathcal{L}f = \frac{\partial f}{\partial \overline{z}} - \mu(z)\frac{\partial f}{\partial z} - \nu(z)\overline{\frac{\partial f}{\partial z}} = 0 \qquad (6.4)$$

with $\mathcal{L}$ as at (6.3), and so in particular f is K-quasiconformal with $K = \frac{1+k}{1-k}$.

6.1. FACTORIZATION FOR ELLIPTIC SYSTEMS

Then any other solution $g \in W^{1,2}_{loc}(\Omega)$ to this equation takes the form

$$g = F(f(z)), \tag{6.5}$$

where F is a K^2-quasiconformal mapping satisfying

$$\frac{\partial F}{\partial \bar{w}} = \lambda(w) \,\Im m \left(\frac{\partial F}{\partial w} \right), \qquad w \in f(\Omega), \tag{6.6}$$

where

$$\lambda(w) = \frac{-2i\,\nu(z)}{1 + |\nu(z)|^2 - |\mu(z)|^2} \qquad \text{with} \tag{6.7}$$

$$|\lambda(w)| \leqslant \frac{2k}{1+k^2} < 1, \qquad z = f^{-1}(w)$$

Conversely, if $F \in W^{1,2}_{loc}(f(\Omega))$ satisfies (6.6), then

$$g = F \circ f$$

solves (6.4).

We refer to (6.6) as the *reduced Beltrami equation*. We use this terminology as not every quasiconformal mapping satisfies such an equation. For instance, $f(z) = z + b\bar{z}$, $b \in \mathbb{D}$, is quasiconformal and has $f_{\bar{z}} = b$ and $\Im m\, f_z = 0$.

Note the explicit form of the coefficient $\lambda(w)$; in particular, it is remarkable that λ does not depend on the derivatives of f. Thus for instance the possible Hölder regularity of μ and ν is inherited by λ, and similarly for any other smoothness class invariant under quasiconformal mappings. Note, however, that in general λ does depend on f.

Proof. Since $\mathcal{L}f = 0$, we have $|f_{\bar{z}}| \leqslant k|f_z|$ almost everywhere and where $k < 1$. Therefore the homeomorphism f is a quasiconformal mapping. Similarly, g is quasiregular, and with Theorem 3.8.2,

$$F := g \circ f^{-1} \in W^{1,2}_{loc}(f(\Omega))$$

To prove that F satisfies the reduced Beltrami equation (6.6) we need to apply a lengthy, though elementary, use of chain rules. Substituting $g = F \circ f$ to the equation $\mathcal{L}f = 0$, we have

$$f_{\bar{z}}\, F_w + \overline{f_z}\, F_{\bar{w}} = \mu f_z\, F_w + \mu \overline{f_{\bar{z}}}\, F_{\bar{w}} + \nu \overline{f_z}\, F_w + \nu f_{\bar{z}}\, F_{\bar{w}}$$

With the help of (6.4) we can reformulate this as

$$\nu f_{\bar{z}}\, \overline{F_{\bar{w}}} = \nu \overline{f_z} \left(F_w - \overline{F_w} \right) + \left(\overline{f_z} - \mu \overline{f_{\bar{z}}} \right) F_{\bar{w}}$$

Taking the complex conjugate gives a second identity, and from this system of two equations we eliminate $\overline{F_{\bar{w}}}$. As a result,

$$\left(|f_z - \bar{\mu} f_{\bar{z}}|^2 - |\nu|^2 |f_{\bar{z}}|^2 \right) F_{\bar{w}} = \left(|\nu|^2 f_z f_{\bar{z}} + \nu \bar{\mu}\, \overline{f_z} f_{\bar{z}} - \nu |f_z|^2 \right) \left(F_w - \overline{F_w} \right)$$

Since f satisfies (6.4), a straightforward manipulation shows that both sides of the above identity have the common factor

$$\left(1 - |\mu|^2 - |\nu|^2\right) |f_z|^2 - 2\Re[\mu\bar{\nu}(f_z)^2]$$

This is nonzero whenever $f_z \neq 0$, that is, almost everywhere. Dividing out the factor then gives

$$F_{\bar{w}} = \frac{-2i\,\nu(z)}{1 + |\nu(z)|^2 - |\mu(z)|^2} \left(F_w - \overline{F_w}\right) = \lambda(w)\,\Im m(F_w), \qquad (6.8)$$

as claimed.

Clearly, we have the following ellipticity bounds:

$$
\begin{aligned}
|\lambda(w)| &= \frac{2|\nu(z)|}{1 + |\nu(z)|^2 - |\mu(z)|^2} \leqslant \frac{2|\nu(z)|}{1 + |\nu(z)|^2 - (k - |\nu(z)|)^2} \\
&= \frac{2|\nu(z)|}{1 + 2k|\nu(z)| - k^2} \leqslant \frac{2k}{1 + 2k^2 - k^2} = \frac{2k}{1 + k^2}
\end{aligned}
$$

Conversely, if F satisfies (6.6) and $\mathcal{L}f = 0$, then $g := F \circ f \in W^{1,2}_{loc}(\Omega)$ and, via the chain rule, $g_{\bar{z}} - \mu g_z - \nu \overline{g_z} = 0$. This completes the proof. $\square$

It is through the above factorization result that the study of general linear elliptic operators narrows down to the special case

$$\frac{\partial}{\partial \bar{z}} - \lambda(z)\,\Im m\left(\frac{\partial}{\partial z}\right) = \frac{\partial}{\partial \bar{z}} - \frac{\lambda(z)}{2}\left(\frac{\partial}{\partial z} - \overline{\frac{\partial}{\partial z}}\right)$$

We call this the reduced Beltrami operator. Note as another special case that if the Beltrami operator (6.3) happens to be $\mathbb{C}$-linear, equivalently if $\nu \equiv 0$, then the corresponding reduced operator is precisely the Cauchy-Riemann operator $\frac{\partial}{\partial \bar{z}}$.

On the other hand, the linear functions

$$f(z) = az + b, \quad \text{where } a \in \mathbb{R},$$

are always solutions to the reduced homogeneous Beltrami equation. These affine maps are precisely those holomorphic functions for which f_z is real, that is, those functions that satisfy $f_{\bar{z}} \equiv \Im m(f_z) \equiv 0$.

6.2 Linear Families of Quasiconformal Mappings

Given two quasiconformal or quasiregular mappings, their linear combinations are usually not quasiregular. However, the quasiregularity follows if the mappings are solutions to the same linear Beltrami equation

$$\frac{\partial f}{\partial \bar{z}} = \mu(z)\frac{\partial f}{\partial z} + \nu(z)\overline{\frac{\partial f}{\partial z}} \qquad (6.9)$$

6.2. LINEAR FAMILIES

That is, if f_1, f_2 satisfy (6.9), then all their $\mathbb{R}$-linear combinations

$$f_{\alpha,\beta} := \alpha f_1 + \beta f_2, \qquad \alpha, \beta \in \mathbb{R}$$

are solutions to (6.9). We may speak of a *linear family* of quasiregular mappings. Conversely, to be discussed in detail in Chapter 16, we will show that to a linear class $\mathcal{F}$ of (K-)quasiregular mappings there is associated a Beltrami-type equation (6.9) satisfied by the mappings $f \in \mathcal{F}$.

Suppose next that f_1, f_2 are $\mathbb{R}$-linearly independent solutions to (6.9) and in addition are homeomorphic in the entire complex plane $\mathbb{C}$. In this case we may ask even more: If $\alpha^2 + \beta^2 \neq 0$, are the mappings $f_{\alpha,\beta}$ homeomorphisms?

Somewhat unexpectedly, the answer to this question is yes. In fact, the question is closely related to the uniqueness of the normalized solutions to (6.9). For the $\mathbb{C}$-linear operator we proved the uniqueness of normalized solutions in Theorem 5.3.4. However, the general $\mathbb{R}$-linear case requires a more sophisticated approach, and in particular, we need a few simple topological facts which we now state.

Lemma 6.2.1. *Let γ be a Jordan curve and $f : \mathbb{C} \to \mathbb{C}$ a homeomorphism. Suppose that one of the curves γ or $f(\gamma)$ lies inside the other (that is, is separated from ∞). Then the increment of the argument*

$$\Delta_{0 \leq t \leq 2\pi} \operatorname{Arg}\left[f\big(\xi(t)\big) - \xi(t)\right] = \pm 2\pi \qquad (6.10)$$

with the $\pm$ sign depending on the orientation of f. Here $\xi = \xi(t)$ is any parameterization of γ.

The above increment of the argument equals zero if the curves γ and $f(\gamma)$ lie outside each other. Perhaps the simplest way to see (6.10) is by deforming the inside curve to a point via a homotopy deformation within the component bounded by the outer curve.

The next theorem is the core of our argument.

Theorem 6.2.2. *Let $f : \mathbb{C} \to \mathbb{C}$ be a homeomorphic solution to the reduced Beltrami equation*

$$\frac{\partial f}{\partial \bar{z}} = \lambda(z) \, \Im m \left(\frac{\partial f}{\partial z}\right), \qquad |\lambda(z)| \leq k < 1$$

If the function $f(z)$ has two fixed points, then

$$f(z) \equiv z, \qquad z \in \mathbb{C}$$

Proof. Assume that f has (at least) two fixed points $z_0 \neq z_1$. Conjugating with similarities, we may assume that $z_0 = 0$ and $z_1 = 1$. If

$$f(z) \not\equiv z, \qquad (6.11)$$

then consider the one-parameter family of mappings

$$f^t(z) := \frac{f(z) - tz}{1-t}; \qquad f^t(0) = 0, \ f^t(1) = 1, \ 0 \leq t < 1$$

We want to prove that all mappings $f^t : \mathbb{C} \to \mathbb{C}$ are K-quasiconformal homeomorphisms, where $K = \frac{1+k}{1-k}$. These mappings are certainly nonconstant and K-quasiregular, as

$$|f^t_{\bar{z}}| \leq k|\Im m\, f^t_z| \leq k|f^t_z| \tag{6.12}$$

Let $T \subset [0, 1)$ denote the set of parameters $t \in [0, 1)$ for which f^t is a homeomorphism. Clearly, $t = 0$ is one such parameter. By the quasiconformal version of the Hurwitz theorem, or more precisely by Theorem 3.9.4, we find that T is a relatively closed subset of $[0, 1)$. Thus we need to show that the set T is open. For this we take any $t \in T$, $t > 0$, and consider the perturbations $f^{t\pm\varepsilon}$ for small $\varepsilon > 0$.

In view of (6.12) the mapping $F^t(z) = f^t(z) - z$ is nonconstant and K-quasiregular. Moreover, it has at least two zeroes, $F^t(z_0) = F^t(z_1) = 0$. Thus by the argument principle, for all sufficiently large $R > 0$, the increment

$$\Delta_{|z|=R} \text{Arg}\left[f^t(z) - z\right] \geq 2 \cdot 2\pi$$

On the other hand, if the circle $\mathbb{S}_R = \{|z| = R\}$ does not intersect its image $f^t(\mathbb{S}_R)$, Lemma 6.2.1 allows the increment to be at most 2π. Therefore for every R large enough, there must be a point $z = z_R$ such that

$$|f^t(z_R)| = |z_R| = R \tag{6.13}$$

However, f^t is a quasiconformal homeomorphism of the plane. Hence (6.13) forces linear growth at ∞ on f^t. That is, with inequality (3.45), we have

$$\frac{1}{\lambda(K)}|z| \leq |f^t(z)| \leq \lambda(K)|z|, \quad \text{for } |z| \text{ large enough}$$

Further, as soon as $\varepsilon > 0$ is small enough, the perturbations $f^{t\pm\varepsilon}(z)$ exhibit the same linear growth. Precisely, we have

$$f^{t\pm\varepsilon}(z) = f^t(z) \pm \frac{\varepsilon}{(1-t)^2 \pm \varepsilon(1-t)}(f(z) - z) \tag{6.14}$$

Next, we fix $w \in \mathbb{C}$. Since by assumption f^t is a homeomorphism, the winding number of $f^t(\mathbb{S}_R)$ around w is 1, for $|z| = R$ large. On the other hand, from (6.14) we obtain

$$|f^{t\pm\varepsilon}(z)| \geq \frac{1}{2\lambda(K)}|z| \quad \text{whenever } \varepsilon \leq \frac{(1-t-\varepsilon)^2}{4\lambda(K)^2}$$

Therefore the winding number of $f^{t\pm\varepsilon}(\mathbb{S}_R)$ around w is 1, as soon as $|z| > 2\lambda(K)|w|$. It follows that the mappings $f^{t\pm\varepsilon}$ are homeomorphisms. In particular, $t \pm \varepsilon \in T$.

6.2. LINEAR FAMILIES

We have proven that for all $0 \leqslant t < 1$ the mappings $f(z) - tz$ are quasiconformal homeomorphisms of the plane. Hence by Theorem 3.9.4, their locally uniform limit $f(z) - z$ is either a homeomorphism or constant. But as f was assumed to have two distinct fixed points, we must have

$$f(z) \equiv z, \quad z \in \mathbb{C}$$

This completes the proof. □

Recall that a homeomorphism $f \in W^{1,2}_{loc}(\mathbb{C})$ satisfying (6.9) is called a normalized solution to the equation if $f(0) = 0$ and $f(1) = 1$.

Theorem 6.2.3. *Every linear Beltrami equation*

$$\frac{\partial f}{\partial \bar{z}} = \mu(z) \frac{\partial f}{\partial z} + \nu(z) \overline{\frac{\partial f}{\partial z}}, \qquad |\mu(z)| + |\nu(z)| \leqslant k < 1; \ z \in \mathbb{C}$$

admits exactly one normalized solution $f : \mathbb{C} \to \mathbb{C}$.

Proof. The existence of a normalized solution is established, in the more general setting of nonlinear equations, by Theorem 8.2.1. Hence it is the uniqueness that requires our attention here. For the complex linear case and Theorem 5.3.4, the uniqueness was an immediate consequence of the classical Stoilow factorization theorem. Similarly, here we use a factorization, now enabled by Theorem 6.1.1. Indeed, if f, g are normalized solutions to the above system, the theorem shows that

$$F = g \circ f^{-1}$$

is a homeomorphic solution to the reduced Beltrami equation (6.6). Since $F(z) - z$ vanishes at the points $z = 0$ and $z = 1$, Theorem 6.2.2 now gives $F(z) \equiv z$; thus $f \equiv g$. □

In fact, the same argument proves the following more general statement.

Corollary 6.2.4. *The global homeomorphic solutions $f : \mathbb{C} \to \mathbb{C}$ to the general linear system $f_{\bar{z}} = \mu(z) f_z + \nu(z) \overline{f_z}$ are determined by their values at two distinct points $a, b \in \mathbb{C}$.*

Next, let us return to linear families of quasiconformal mappings.

Corollary 6.2.5. *Suppose f and g are homeomorphic solutions to the linear uniformly elliptic system $h_{\bar{z}} = \mu(z) h_z + \nu(z) \overline{h_z}$ in the complex plane. Then every linear combination*

$$F_{\alpha,\beta} = \alpha f(z) + \beta g(z), \qquad \alpha, \beta \in \mathbb{R}$$

is either a homeomorphism or a constant. In fact, under the normalization $f(0) = g(0) = 0$, if $f(1)$ and $g(1)$ are $\mathbb{R}$-linearly independent, then $F_{\alpha,\beta} = \alpha f(z) + \beta g(z)$ is a homeomorphism whenever $\alpha^2 + \beta^2 \neq 0$.

Proof. Suppose $\alpha f(z) + \beta g(z)$ is not injective and $\alpha^2 + \beta^2 > 0$. Then clearly $\alpha, \beta \neq 0$, and we have a pair of points ζ, w with

$$\alpha f(\zeta) + \beta g(\zeta) = \alpha f(w) + \beta g(w), \qquad \zeta \neq w$$

Consider the homeomorphism $F = g \circ f^{-1}$. By Theorem 6.1.1, F satisfies the reduced Beltrami equation (6.6), hence also

$$\Phi(z) = \beta F(z) - \alpha f(w) - \beta g(w)$$

is a homeomorphic solution to the equation. But here $\Phi(z) = -\alpha z$ at least at the points $z = f(\zeta)$ and $z = f(w)$. Thus $\Phi(z) \equiv -\alpha z$ by Theorem 6.2.2. In other words, $\beta F(z) + \alpha z \equiv \alpha f(w) + \beta g(w)$, proving our claim. $\square$

In the case of Corollary 6.2.5 we speak of the *linear family* $\mathcal{F} = \{F_{\alpha,\beta}\}$ *generated* by the mappings f and g. We may always normalize $f(0) = g(0) = 0$. With a change of coordinates we may choose the values $f(1), g(1)$ at our wish. If, say, $g(1) = \alpha \neq 0$ is required, simply take

$$g = \alpha g_\alpha,$$

where g_α is the normalized solution to the equation

$$g_{\bar{z}} = \mu g_z + \frac{\bar{\alpha}}{\alpha} \nu \overline{g_z}$$

6.3 The Reduced Beltrami Equation

In the previous sections we have seen that solutions to the reduced Beltrami equation (6.6) play a role very similar to the one that the holomorphic functions have played in the study of $\mathbb{C}$-linear first-order elliptic systems. Therefore a deeper understanding of the reduced equations and their solutions is necessary for applications of the general linear elliptic systems to the theory of second-order PDEs such as that in (6.1). For this reason we consider in this section a few basic results and open conjectures concerning the reduced equation.

Recall that for $a \in \mathbb{R}$ and $b \in \mathbb{C}$ the affine functions $f(z) = az + b$ are always solutions to the reduced homogeneous Beltrami equation. These functions have the very special property that

$$\frac{\partial f}{\partial z} \equiv \overline{\frac{\partial f}{\partial z}}$$

We conjecture that, apart from this trivial case, for any solution to the reduced equation the imaginary part of the f_z-derivative is nonzero almost everywhere.

Conjecture 6.3.1. *Suppose* $f : \Omega \to \mathbb{C}$ *is a (quasiregular) solution to the reduced Beltrami equation*

$$\frac{\partial f}{\partial \bar{z}} = \lambda(z) \, \Im m \, \frac{\partial f}{\partial z}, \qquad |\lambda(z)| \leq k < 1 \tag{6.15}$$

Then either f_z *is a (real) constant or else* $\Im m \, f_z \neq 0$ *for almost every* $z \in \Omega$.

6.3. THE REDUCED BELTRAMI EQUATION

The conjecture asks for properties for $\Im m\, f_z$ very similar to those of the Jacobian $J(z, \phi)$ of a quasiregular mapping ϕ. Such properties are important, for instance, in determining the G-limits of Beltrami operators in Section 16.6.2.

In the next section we prove Conjecture 6.3.1 for solutions that are global homeomorphisms of the plane. We will see that for such solutions one has further information. We first prove the following closely related result.

Theorem 6.3.2. *If $f \in W^{1,2}_{loc}(\mathbb{C})$ is a homeomorphic solution to the reduced equation (6.15), then $\Im m\, f_z$ does not change sign:*

- *If $f(1) - f(0) \in \mathbb{R}$, then $\Im m\, f_z \equiv 0$.*
- *If $\Im m(f(1) - f(0)) > 0$, then $\Im m\, f_z \geqslant 0$ almost everywhere.*
- *If $\Im m(f(1) - f(0)) < 0$, then $\Im m\, f_z \leqslant 0$ almost everywhere.*

Proof. We may assume that $f(0) = 0$. If $f(z) \equiv tz$ for some $t \in \mathbb{R}$, then we have the first option. Otherwise, first note that

$$\frac{f(a) - f(b)}{a - b} \notin \mathbb{R} \quad \text{whenever } a \neq b \in \mathbb{C}$$

Namely, if $t \in \mathbb{R}$ is the value of the quotient, then $f(a) - ta = f(b) - tb$. But $f(z)$ and z are both solutions to the same equation (6.15) and hence by Corollary 6.2.5 generate a linear family of quasiconformal mappings. In particular, for any $t \in \mathbb{R}$ either $f(z) - tz \equiv 0$ or $f(z) - tz$ is injective.

For $\alpha, \beta \in \mathbb{R}$, $\alpha^2 + \beta^2 \neq 0$, we now view the quotient

$$\lambda = \lambda(a, b) = \Im m \left(\frac{f(a) - f(b)}{a - b} \right) \neq 0$$

as a real-valued function in two complex variables $(a, b) \in \mathbb{C} \times \mathbb{C}$ defined outside the diagonal set $\{(a, b) : a = b\}$. As the complement of the diagonal in $\mathbb{C} \times \mathbb{C}$ is connected and as λ is continuous, we see that it does not change sign. To determine the sign of λ, it is enough to compute $\lambda(1, 0) = \Im m(f(1) - f(0))$; for convenience we may assume that $\lambda(1, 0) > 0$. This number will determine the sign of $\Im m\, f_z$, too. In fact, let $z_0 \in \mathbb{C}$ be a point of differentiability of f. For sufficiently small complex numbers $h \in \mathbb{C} \setminus \{0\}$, we can use Taylor's first-order expansion to write

$$\Im m \left(\frac{f_z h + f_{\bar{z}} \bar{h} + o(|h|)}{h} \right) = \Im m \left(\frac{f(z+h) - f(z)}{h} \right) \geqslant 0$$

The claim follows choosing $h = |h|e^{i\theta}$ so that $f_{\bar{z}}(z_0)\bar{h}^2 \in \mathbb{R}$ and then letting $|h| \to 0$. $\square$

Theorem 6.3.2 does not hold for noninjective mappings. It also need not hold for homeomorphisms in proper subdomains.

Example 6.3.3. Let $f(z) = (2+k)z^2 + 2kz\bar{z} - k\bar{z}^2$, where $0 < k < 1$. As an even function f is noninjective. On the other hand, f is injective in the right halfplane $\{z : \Re(z) > 0\}$. Further,

$$\frac{\partial f}{\partial \bar{z}} = ik\,\Im m\, \frac{\partial f}{\partial z},$$

where $\Im m\, f_z = 4\,\Im m\, z$ attains both positive and negative values.

Finally, suppose we have, say, $\Im m\, f_z \geqslant 0$ in Theorem 6.3.2. Then f satisfies the reduced *inequality*

$$\left|\frac{\partial f}{\partial \bar{z}}\right| \leqslant k\,\Im m\, \frac{\partial f}{\partial z} \qquad (6.16)$$

This fact shows the injective solutions to the reduced Beltrami equation to have a number of fascinating properties. For instance, note that (6.16) is stable under all convex combinations. Or further, the inequality is stable under a convolution with any positive mollifier, thus allowing for smooth solutions.

The expression $\Im m\, f_z$ is actually a null Lagrangian; see Section 19.2. It therefore plays the same important role in the reduced distortion inequality (6.16) as the Jacobian determinant does for the full distortion inequality $|Df|^2 \leqslant KJ(z,f)$. However, in this case one should integrate along curves (instead of over domains) to obtain interesting estimates.

Finally, we draw the reader's attention to the interesting possible connections between the reduced equation (6.15) and the equation satisfied by a monotone mapping (3.66).

6.4 Homeomorphic Solutions to Reduced Equations

We now have one of the main results of this chapter. The proof we give here is motivated by earlier work of C. Sbordone [318] on G-compactness and extended by G. Alessandrini and V. Nesi who studied this problem in a slightly restricted case which basically requires that μ and ν are compactly supported, see [14]. These methods were refined and simplified by Astala and Jääskeläinen, [32], and we closely follow their approach. In addition, the proof draws upon a result of Fabes and Stroock [121] which is of independent interest and so we subsequently give a proof for this result as well.

Theorem 6.4.1. *Suppose* $f : \mathbb{C} \to \mathbb{C}$ *is a homeomorphic* $W^{1,2}_{loc}$-*solution to the reduced Beltrami equation*

$$\frac{\partial f}{\partial \bar{z}} = \lambda(z)\,\Im m\frac{\partial f}{\partial z}, \qquad |\lambda(z)| \leqslant k < 1, \qquad (6.17)$$

Then either f_z *is a constant* $c \in \mathbb{R}$, *or else* $\Im m(f_z) \neq 0$ *for almost every* $z \in \Omega$. *In particular, the inequalities established in Theorem 6.3.2 are almost everywhere strict.*

6.4. HOMEOMORPHIC SOLUTIONS

Proof. We write $f(z) = u(z) + iv(z)$, where u and v are real valued. We similarly write $\lambda(z) = a(z) + ib(z)$.

Now, taking the imaginary part of (6.17) shows us that $b(z)(v_x - u_y) = u_y + v_x$, that is

$$u_y = \frac{b(z) - 1}{b(z) + 1} v_x$$

Thus

$$2\,\Im m \frac{\partial f}{\partial z} = v_x - u_y = \frac{2}{b(z) + 1} v_x = \frac{2}{b(z) - 1} u_y \tag{6.18}$$

Since $|b(z)| \leq |\lambda(z)| \leq k < 1$, the coefficients $2/(b(z) \pm 1)$ in (6.18) are uniformly bounded below. Hence to prove Theorem 6.4.1 it suffices to show that either $u_y \equiv 0$ or else $u_y \neq 0$ almost everywhere.

The trick in proving Theorem 6.4.1 is that, for the reduced equation (6.17), the derivative u_y is a solution to the adjoint equation determined by a non-divergence-type operator. To state this more precisely, consider an operator

$$L = \sum_{i,j=1}^{2} \sigma_{ij}(z) \frac{\partial^2}{\partial x_i \partial x_j}, \tag{6.19}$$

where $\sigma_{ij} = \sigma_{ji}$ are measurable and the matrix

$$\sigma(z) := \begin{bmatrix} \sigma_{11}(z) & \sigma_{12}(z) \\ \sigma_{12}(z) & \sigma_{22}(z) \end{bmatrix} \tag{6.20}$$

is uniformly elliptic,

$$\frac{1}{K}|\xi|^2 \leq \langle \sigma(z)\xi, \xi \rangle = \sigma_{11}(z)\xi_1^2 + 2\sigma_{12}(z)\xi_1\xi_2 + \sigma_{22}(z)\xi_2^2 \leq K|\xi|^2 \tag{6.21}$$

for all $\xi, z \in \mathbb{C}$. Here K is the ellipticity constant. Then we say that the function $w \in L^2_{loc}(\mathbb{C})$ is a weak solution to the adjoint equation

$$L^* w = 0 \tag{6.22}$$

if

$$\int w L\phi = 0, \quad \text{for every } \phi \in C_0^\infty(\mathbb{C}) \tag{6.23}$$

It is useful to note, that if w is a solution to (6.22), then by approximating we see that (6.23) holds whenever $\phi \in W^{2,2}(\mathbb{C})$ and ϕ has compact support, that is, $\phi \in W_0^{2,2}(\Omega)$ for some bounded domain Ω.

To identify u_y as a solution to an equation of the type (6.22) we note that the components of solutions $f = u + iv$ to general Beltrami equations satisfy a divergence type second-order equation. We shall establish this in full generality later, in Section 16.1.5, and so do not discuss this here. In case of (6.17), however, the component u satisfies the equation

$$\text{div}\, A(z)\nabla u = 0, \quad A(z) := \begin{bmatrix} 1 & a_{12}(z) \\ 0 & a_{22}(z) \end{bmatrix}, \tag{6.24}$$

where the matrix elements are

$$a_{12} = \frac{\Re e(\lambda)}{1 - \Im m(\lambda)} = \frac{a(z)}{1 - b(z)}, \quad a_{22} = \frac{1 + \Im m(\lambda)}{1 - \Im m(\lambda)} = \frac{1 + b(z)}{1 - b(z)} > 0 \quad (6.25)$$

Precisely, (6.24) means that for every $\phi \in C_0^\infty(\mathbb{C})$ we have

$$0 = \int \nabla \phi \cdot A(z) \nabla u = \int \phi_x(u_x + a_{12} u_y) + \phi_y a_{22} u_y \quad (6.26)$$

But since derivatives of smooth test functions are again test functions, we can replace ϕ by $\phi_y \in C_0^\infty(\mathbb{C})$. In this case the identity (6.26) takes the form

$$\begin{aligned}
0 &= \int \phi_{yx} u_x + a_{12} \phi_{yx} u_y + a_{22} \phi_{yy} u_y \\
&= \int \phi_{xx} u_y + a_{12} \phi_{xy} u_y + a_{22} \phi_{yy} u_y \\
&= \int (\phi_{xx} + a_{12} \phi_{xy} + a_{22} \phi_{yy}) u_y
\end{aligned}$$

Thus u_y is indeed a distributional solution to the adjoint equation $L^* u_y = 0$, where

$$L = \frac{\partial^2}{\partial x^2} + a_{12} \frac{\partial^2}{\partial x \partial y} + a_{22} \frac{\partial^2}{\partial y^2} \quad (6.27)$$

and a_{12}, a_{22} are given by (6.25).

Note that the original matrix $A(z)$ is not symmetric. However, the operator L in (6.27) can be represented by the symmetric matrix

$$\sigma(z) := \begin{bmatrix} 1 & a_{12}(z)/2 \\ a_{12}(z)/2 & a_{22}(z) \end{bmatrix} \quad (6.28)$$

and from (6.25) we see that σ is uniformly elliptic.

Next, Theorem 6.3.2 informs us that for a homeomorphic solution f to the reduced equation, $\Im m(f_z)$ does not change sign, it is either everywhere non-negative or everywhere non-positive. In view the identity (6.18) the same holds for the derivative u_y also. It is precisely here that the assumption of homeomorphicity is used.

Thus we may assume that u_y is a non-negative solution to the adjoint equation, $L^*(u_y) = 0$, where L is defined in (6.27). Theorem 6.4.1 then follows from the theorem of Fabes and Stroock [121, Theorem 2.1] which we prove below at Theorem 6.4.2. It states that non-negative solutions to the adjoint equation satisfy a uniform reverse Hölder estimate. □

6.4.1 Fabes-Stroock Theorem

Here we prove the result we needed in our proof of Theorem 6.4.1

6.4. HOMEOMORPHIC SOLUTIONS

Theorem 6.4.2. *Consider the operator*

$$L = \sum_{i,j=1}^{2} \sigma_{ij}(z) \frac{\partial^2}{\partial x_i x_j}, \qquad (6.29)$$

where $\sigma_{ij} = \sigma_{ji}$ are measurable and L is uniformly elliptic with constant K,

$$\frac{1}{K}|\xi|^2 \leqslant \langle \sigma(z)\xi, \xi \rangle \leqslant K|\xi|^2, \qquad \xi, z \in \mathbb{C}$$

Then there exists a constant C, depending only on the ellipticity constant K, such that for all $w \geqslant 0$ satisfying

$$L^* w = 0$$

in $\mathbb{C}$, and for all disks $\mathbb{D}(a, r) \subset \mathbb{C}$ of radius r,

$$\left[\frac{1}{r^2} \int_{\mathbb{D}(a,r)} w(z)^2 \, dz \right]^{1/2} \leqslant \frac{C}{r^2} \int_{\mathbb{D}(a,r)} w(z) \, dz \qquad (6.30)$$

In particular, $w > 0$ almost everywhere, or else $w \equiv 0$.

We begin with a lemma of [121, Lemma 2.0].

Lemma 6.4.3. *Let $w \in L^2_{loc}(\mathbb{C})$, $w \geqslant 0$, satisfy*

$$\int w \, L\phi = 0 \qquad \text{for all } \phi \in C_0^\infty(\mathbb{C})$$

Then there exists a constant c, depending only on the ellipticity constant of L, such that for all disks $\mathbb{D}(a, r) \subset \mathbb{C}$ of radius r, we have

$$\int_{\mathbb{D}(a,r)} w(z) \, dz \leqslant c \int_{\mathbb{D}(a,r/2)} w(z) \, dz \qquad (6.31)$$

Proof. We may assume that the center of $\mathbb{D}(a, r)$ is the origin. For $\delta > 0$ we set

$$h(z) = (1 + \delta)^2 r^2 - |z|^2 \qquad (6.32)$$

and establish a few simple estimates of $L(h^2)$. First

$$(Lh^2)(z) = 8\langle \sigma(z)z, z \rangle - 4h(z) \operatorname{tr}(\sigma(z)),$$

so that $(Lh^2)(z) \geqslant \frac{8}{K}|z|^2 - 8K|h(z)|$. Next, if we choose δ sufficiently small, we can assume that

$$\begin{aligned} L(h^2) &\geqslant c_1 r^2 & \text{for } (1+\delta)r > |z| \geqslant (1-\delta)r \\ |L(h^2)| &\leqslant c_2 r^2 & \text{for } |z| < (1+\delta)r \end{aligned}$$

where, again, the constants $c_1, c_2 > 0$ depend only on the ellipticity constant of the operator L. Thus

$$\int_{\mathbb{D}_r \setminus \mathbb{D}_{(1-\delta)r}} w(z)\, dz \leqslant \frac{1}{c_1} \int_{\mathbb{D}_{(1+\delta)r} \setminus \mathbb{D}_{(1-\delta)r}} w(z) L\left(\frac{h^2}{r^2}\right) dz$$

On the other hand, $h^2 \in W_0^{2,2}(\mathbb{D}_{(1+\delta)r})$ as h is C^∞-smooth and both h^2 and $\nabla(h^2)$ vanish on $\partial \mathbb{D}_{(1+\delta)r}$. Therefore

$$\int_{\mathbb{D}_{(1+\delta)r}} w(z) L\left(\frac{h^2}{r^2}\right) dz = 0,$$

and hence

$$\int_{\mathbb{D}_r} w(z)\, dz \leqslant c \int_{\mathbb{D}_{(1-\delta)r}} w(z)\, dz$$

Now we iterate this estimate so as to obtain the claim (6.31). $\square$

Later, in Chapter 17, we will systematically study operators such as those defined at (6.29) using methods from quasiconformal analysis. In particular, will find the solution to following the Dirichlet problem in a bounded domain Ω,

$$L(u) = h, \qquad u \in W^{2,2}(\Omega) \text{ with } u = 0 \text{ on } \partial \Omega \qquad (6.33)$$

By one of the key estimates in this connection, the Alexandrov-Bakelman-Pucci maximum principle, see Theorem 17.3.1, we have

$$\|u\|_{L^\infty(\Omega)} \leqslant c \operatorname{diam}(\Omega) \|h\|_{L^2(\Omega)} \qquad (6.34)$$

where c depends only on the ellipticity constants of the coefficient matrix $\sigma = (\sigma_{ij})$. We will also make use of the following consequence of the Leibniz' rule,

$$L(\phi u) = 2\langle \sigma(z)\nabla \phi, \nabla u\rangle + \phi L(u) + u L(\phi), \qquad \phi, u \in W^{2,2}(\Omega) \qquad (6.35)$$

Lemma 6.4.4. *Suppose u solves the Dirichlet problem (6.33) where $h \in L^2(\Omega)$. Then every non-negative solution w to the adjoint equation $L^*w = 0$ satisfies*

$$\int_\Omega w|\nabla u|^2 \leqslant c \operatorname{diam}(\Omega) \|h\|_{L^2(\Omega)} \int_\Omega w|h|$$

Proof. Note first that since $u^2 = 0$ and $\nabla(u^2) = 0$ on the boundary $\partial \Omega$, we have $u^2 \in W^{2,2}(\Omega)$. Thus

$$\int_\Omega w L(u^2) = 0$$

Therefore (6.35) gives

$$\int_\Omega w|\nabla u|^2 \leqslant c \int_\Omega w \langle \sigma(z)\nabla u, \nabla u\rangle$$

$$= -c \int_\Omega w\, u L(u) \leqslant c \operatorname{diam}(\Omega) \|h\|_{L^2(\Omega)} \int_\Omega w|h|$$

6.4. HOMEOMORPHIC SOLUTIONS

where the last estimate is based on (6.33). □

Proof of Theorem 6.4.2. Let $\mathbb{D}_r = \mathbb{D}(a,r)$. We first show that for every non-negative function $g \in L^2(\mathbb{D}_r)$ we have

$$\int_{\mathbb{D}_r} wg \leqslant \frac{c}{r} \|g\|_{L^2} \int_{\mathbb{D}_{2r}} w \qquad (6.36)$$

Here and below the letter c will be a constant depending only on the ellipticity bound of L. It is likely to be different at each occurrence.

For the estimate, we solve the Dirichlet problem (6.33) in the domain $\Omega = \mathbb{D}(a, 2r)$, where we choose as h the zero extension of g, that is, $h = g\chi_{\mathbb{D}(a,r)}$. Next, let $\phi \in C_0^\infty(\mathbb{D}(a, 2r))$ satisfy $\phi \equiv 1$ on $\mathbb{D}(a,r)$ with $|\partial^\alpha \phi_r / \partial x_\alpha| \leqslant c_\alpha r^{-|\alpha|}$ for $|\alpha| \leqslant 2$. Now $\phi u \in W^{2,2}(\mathbb{D}(a, 2r))$ so that

$$\int_{\mathbb{D}_{2r}} wL(\phi u) = 0$$

Applying (6.35) we then find

$$\int_{\mathbb{D}_r} wg = \int_{\mathbb{D}_{2r}} w\, L(u)\phi \leqslant 2\int_{\mathbb{D}_{2r}} w|\langle \sigma(z)\nabla\phi, \nabla u\rangle| + \int_{\mathbb{D}_{2r}} w\, uL(\phi)$$

$$\leqslant c \left(\int_{\mathbb{D}_{2r}} w|\nabla\phi|^2\right)^{1/2} \left(\int_{\mathbb{D}_{2r}} w|\nabla u|^2\right)^{1/2} + \|u\|_{L^\infty(\mathbb{D}_{2r})} \int_{\mathbb{D}_{2r}} wL(\phi)$$

$$\leqslant \left(\frac{c}{r}\|g\|_{L^2} \int_{\mathbb{D}_{2r}} w\right)^{1/2} \left(\int_{\mathbb{D}_r} wg\right)^{1/2} + \frac{c}{r}\|g\|_{L^2} \int_{\mathbb{D}_{2r}} w$$

by rearranging terms and using Lemma 6.4.4. These estimates prove (6.36), which is, in fact, equivalent to

$$\left[\frac{1}{r^2}\int_{\mathbb{D}(a,r)} w(z)^2\, dz\right]^{1/2} \leqslant \frac{c}{r^2} \int_{\mathbb{D}(a,2r)} w(z)\, dz$$

Combining now with Lemma 6.4.3 we have the proof of the reverse Hölder inequalities (6.30).

It remains to show that if the set $E = \{z \in \mathbb{C} : w(z) = 0\}$ has positive measure, then $w \equiv 0$. If $a \in E$ is a point of density, we can find disks $\mathbb{D}_r = \mathbb{D}(a,r)$ with

$$|\mathbb{D}_r \setminus E| < \varepsilon r^2, \qquad \text{where } C\sqrt{\varepsilon} < 1,$$

and C is the constant of the reverse Hölder inequality (6.30). Now

$$\int_{\mathbb{D}_r} w = \int_{\mathbb{D}_r \setminus E} w \leqslant \left(\int_{\mathbb{D}_r} w^2\right)^{1/2} |\mathbb{D}_r \setminus E|^{1/2} \leqslant C\sqrt{\varepsilon} \int_{\mathbb{D}_r} w \qquad (6.37)$$

This is possible only if w vanishes identically in $\mathbb{D}(a,r)$, that is, $\mathbb{D}(a,r) \subset E$. But then we can replace the disk by a slightly larger one, $\mathbb{D}(a, \delta r)$ with $1 < \delta < (1-\varepsilon)^{-1/2}$, and argue as in (6.37). It follows that w vanishes in $\mathbb{D}(a, \delta r)$. Iterating the argument proves that w vanishes identically. □

Chapter 7

The Concept of Ellipticity

One of the most important and central themes of this monograph concerns the concept of ellipticity as it pertains to systems of partial differential equations. The types of equations we shall explore evolved from the first-order system of Cauchy-Riemann equations, the same equations that everyone meets in a first course in complex analysis [6] and that play such a central role,

$$\begin{aligned} u_x &= v_y \\ u_y &= -v_x, \end{aligned}$$

and of course there are the related second-order equations for u and v, namely, the Laplace equation

$$\Delta u = u_{xx} + u_{yy} = 0$$

A prototype of a nonlinear elliptic PDE is provided, for example, by

$$\text{div}(|\nabla u|^{p-2} \nabla u) = 0, \quad 1 < p < \infty$$

This is the well-known, much studied p-harmonic equation. In two dimensions the equation admits a clear-cut complex analytic reduction. Namely, if we consider the complex gradient of a solution,

$$f = u_x - iu_y = 2\frac{\partial u}{\partial z},$$

then, as we will see in Section 16.3.2, this function satisfies the quasilinear Beltrami equation

$$\frac{\partial f}{\partial \bar{z}} = \left(\frac{1}{p} - \frac{1}{2}\right)\left(\frac{\bar{f}}{f}\frac{\partial f}{\partial z} + \frac{f}{\bar{f}}\overline{\frac{\partial f}{\partial z}}\right) = \mu_1(f)\frac{\partial f}{\partial z} + \mu_2(f)\overline{\frac{\partial f}{\partial z}},$$

where $\mu_1(f) = \overline{\mu_2(f)}$ depend on f but not on the variable z. When $p = 2$, this reduces to the Cauchy-Riemann equations $f_{\bar{z}} = 0$. In general, the solutions will be $(p-1)^{\mp 1}$-quasiregular functions, where the sign is chosen according to whether $1 < p \leqslant 2$ or $p \geqslant 2$.

Another basic example of a nonlinear elliptic PDE is one that we have met already met in the introductory Section 2.1, namely, the equation

$$D^t f(z) H(f(z)) Df(z) = J(z,f) G(z), \quad \text{for almost every } z \in \Omega, \quad (7.1)$$

describing the mappings $f : \Omega \to \Omega'$ that are conformal with respect to the measurable structures (Ω, G) and (Ω', H). We will see later in Chapter 10 that (7.1) is equivalent to the equation

$$\frac{\partial f}{\partial \bar{z}} = \mu(z,f) \frac{\partial f}{\partial z} + \nu(z,f) \overline{\frac{\partial f}{\partial z}},$$

where the coefficients μ and ν are explicitly given in terms of the matrices $G(z) = (g_{ij}(z))$ and $H(w) = (h_{kl}(w))$,

$$\mu(z,w) = \frac{g_{11}(z) - g_{22}(z) + 2ig_{12}(z)}{g_{11}(z) + g_{22}(z) + h_{11}(w) + h_{22}(w)}$$

and

$$\nu(z,w) = \frac{h_{22}(w) - h_{11}(w) - 2ih_{12}(w)}{g_{11}(z) + g_{22}(z) + h_{11}(w) + h_{22}(w)}$$

These simple examples and many more similar ones suggest that in two dimensions the notions of ellipticity and the Beltrami-type systems are intimately related. In fact, we shall see that they are just two sides of the same coin. Subsequently, this connection suggests the development of a theory of nonlinear Beltrami equations, generalizing the linear theory from Chapter 5. It is in fact the combination of these two ideas that leads to a deep understanding of nonlinear elliptic PDE's and systems in two dimensions. In higher dimensions where the tools from geometric complex analysis are not available, much less is known.

Briefly, in many ways the modern concept of ellipticity is based on analysis of the geometric properties of both holomorphic mappings and harmonic functions, and this is a view that we shall adopt in here. To make these relations clear and explicit, we begin by first discussing in this chapter the different aspects of the notion of an elliptic system. Results concerning existence, uniqueness and regularity, as well as applications, will appear in later chapters.

In this book we shall be concerned only with equations in the plane and therefore do not strive for complete generality. However, it will be convenient in later applications to have in hand the basic concepts of ellipticity and notions about more general systems in more variables. This is what motivates the following discussion.

7.1 The Algebraic Concept of Ellipticity

Let $\mathcal{L}$ be a linear differential operator of order m with constant coefficients acting on functions of k real variables. Thus

$$(\mathcal{L} f)(x) = \sum_{|\alpha| \leqslant m} A_\alpha \left(\frac{\partial^{|\alpha|} f}{\partial x_1^{\alpha_1} \cdots \partial x_k^{\alpha_k}} \right) (x) \quad (7.2)$$

Here α_i, $i = 1, 2, \ldots, k$, are nonegative integers, $\alpha = (\alpha_1, \alpha_2, \ldots, \alpha_k)$ and
$$|\alpha| = \alpha_1 + \alpha_2 + \cdots + \alpha_k$$
The coefficients A_α are linear operators: We assume there are given finite-dimensional vector spaces $\mathbb{E}$ and $\mathbb{F}$ in the background, the range of f and $\mathcal{L}f$, respectively,
$$f : \Omega \to \mathbb{E}, \qquad \mathcal{L}f : \Omega \to \mathbb{F},$$
so that the coefficients $A_\alpha \in \hom(\mathbb{E}, \mathbb{F})$. Typically $\mathbb{E} = \mathbb{R}^p$ or $\mathbb{C}^p$ and the homomorphism A_α is either a real or complex matrix, or most often simply a real or complex scalar. The operator $\mathcal{L}$ works as follows. We take a suitably differentiable function f defined on a domain $\Omega \subset \mathbb{R}^k$ and valued in $\mathbb{E}$. Then for each $x \in \Omega$ we compute various partial derivatives of f at x, giving us a collection of elements of $\mathbb{E}$ on which each A_α acts giving, an element of $\mathbb{F}$, and we then sum all these terms. Thus $\mathcal{L}$ transforms $\mathbb{E}$-valued functions to $\mathbb{F}$-valued functions. In particular, identities such as $\mathcal{L}f = 0$ represent systems of linear differential equations, with $N_1 = \dim(\mathbb{F})$ equations and $N_2 = \dim(\mathbb{E})$ unknowns.

The classification of the linear operator $\mathcal{L}$ depends on the algebraic properties of the *principal symbol*
$$\mathcal{P}(\zeta) = \mathcal{P}_m(\zeta) = \sum_{|\alpha|=m} \zeta^\alpha A_\alpha, \qquad \zeta^\alpha = \zeta_1^{\alpha_1} \cdots \zeta_k^{\alpha_k} \in \mathbb{R}, \qquad (7.3)$$
for $\zeta \in \mathbb{R}^k$. This is a homogeneous polynomial of degree m in the variable ζ and has values in $\hom(\mathbb{E}, \mathbb{F})$.

Definition of Ellipticity

Definition 7.1.1. *The linear operator $\mathcal{L}$ defined in (7.2) is said to be elliptic if for each $\zeta \in \mathbb{R}^k \setminus \{0\}$ the linear transformation*
$$\mathcal{P}_m(\zeta) : \mathbb{E} \to \mathbb{F}$$
is an isomorphism.

Of course, by the homogeneity, the ellipticity condition only needs to be checked for those $\zeta \in \mathbb{S}^{k-1}$, that is, $|\zeta| = 1$.

In the cases of most interest we will have $\mathbb{E} = \mathbb{F}$. Then ellipticity can most easily be examined by considering the characteristic polynomial,
$$P(\zeta) = \det \mathcal{P}_m(\zeta) \neq 0, \qquad \zeta \in \mathbb{S}^{k-1} \qquad (7.4)$$

In the case where $\mathbb{E} = \mathbb{F}$ is a Hermitian inner-product space, there is another concept we will make use of. The linear partial differential operator $\mathcal{L}$ is said to be *strongly elliptic* if for every $\zeta \in \mathbb{S}^{k-1}$ and unit vector $\mathbf{e} \in \mathbb{E}$,
$$\Re\langle \mathcal{P}_m(\zeta)\mathbf{e}, \mathbf{e}\rangle_\mathbb{E} > 0 \qquad (7.5)$$

Notice that the left hand side of (7.5) is a polynomial of degree m, and as the polynomial evidently is non-negative, we must have m even.

7.2 Some Examples of First-Order Equations

Obviously, a number of examples are now in order. We have already observed that first-order equations can never be strongly elliptic.

Consider the single equation in $\mathbb{R}^k$, $k \geq 1$, for a function u,

$$\mathcal{L}u = \langle A, \nabla u \rangle = A_1 \frac{\partial u}{\partial x_1} + \cdots + A_k \frac{\partial u}{\partial x_k},$$

where $A = (A_1, \ldots, A_k) \in \mathbb{R}^k \setminus \{0\}$. The principal symbol is a linear map of $\mathbb{R}$ and can be identified with the scalar number $\mathcal{P}_1(\zeta) = \langle A, \zeta \rangle$. In dimensions $k > 1$ this vanishes for some $\zeta \in \mathbb{S}^{k-1}$. Thus $\mathcal{L}$ is not elliptic unless $k = 1$ and $\mathcal{L}$ is a multiple of $\frac{d}{dx}$.

Then consider the system of N equations in N unknown functions $U = (u_1, u_2, \ldots, u_N)$,

$$\mathcal{L}U = \sum_{\nu=1}^{k} A^\nu \frac{\partial U}{\partial x_\nu} = \sum_{j=1}^{N} \sum_{\nu=1}^{k} A_{i,j}^\nu \frac{\partial u_j}{\partial x_\nu},$$

where A^ν are given square $N \times N$ matrices for $\nu = 1, 2, \ldots, k$. The operator $\mathcal{L} = \sum_{\nu=1}^{n} A^\nu \frac{\partial}{\partial x_\nu}$ will be elliptic if and only if

$$P(\zeta) = \det\left(\sum_{\nu=1}^{k} \zeta_\nu A^\nu\right) \neq 0, \qquad \zeta \in \mathbb{S}^{k-1}$$

The determinant will be a homogeneous polynomial of degree N in the variables ζ, thus $P(-\zeta) = (-1)^N P(\zeta)$. As P does not vanish on $\mathbb{S}^{k-1}$ it must be of even order (unless $k = 1$). Therefore we have established the following corollary.

Corollary 7.2.1. *Suppose that the underlying space $\mathbb{R}^k$ has dimension $k \geq 2$. Then for a first-order partial differential linear operator to be elliptic, it is necessary that the number of equations and unknown functions be the same and that this number be even.*

Of particular importance are the Cauchy-Riemann operators

$$\partial^\pm = \begin{bmatrix} 1 & 0 \\ 0 & 1 \end{bmatrix} \frac{\partial}{\partial x} \pm \begin{bmatrix} 0 & -1 \\ 1 & 0 \end{bmatrix} \frac{\partial}{\partial y},$$

whose characteristic polynomials are

$$\begin{aligned} P^\pm(\zeta) &= \det\left(\begin{bmatrix} 1 & 0 \\ 0 & 1 \end{bmatrix} \zeta_1 \pm \begin{bmatrix} 0 & -1 \\ 1 & 0 \end{bmatrix} \zeta_2\right) \\ &= \det\left(\begin{bmatrix} \zeta_1 & \mp\zeta_2 \\ \pm\zeta_2 & \zeta_1 \end{bmatrix}\right) = \zeta_1^2 + \zeta_2^2 = |\zeta|^2 \end{aligned}$$

and have only the trivial root $\zeta = 0$. Thus these equations are elliptic. In fact, suitably differentiable solutions to the equations $\partial^+ f = 0$ are holomorphic functions, while the solutions to $\bar{\partial} f = 0$ are antiholomorphic functions (the complex conjugates of holomorphic functions).

We cannot resist giving another interesting example, furnished by the "quaternionic system." Here we have four unknown functions $U = (u^1, u^2, u^3, u^4)$ and four equations in $\mathbb{R}^4$,

$$\frac{\partial u^1}{\partial x_1} - \frac{\partial u^2}{\partial x_2} + \frac{\partial u^3}{\partial x_3} - \frac{\partial u^4}{\partial x_4} = 0$$

$$\frac{\partial u^2}{\partial x_1} + \frac{\partial u^1}{\partial x_2} + \frac{\partial u^4}{\partial x_3} + \frac{\partial u^3}{\partial x_4} = 0$$

$$\frac{\partial u^3}{\partial x_1} + \frac{\partial u^4}{\partial x_2} - \frac{\partial u^1}{\partial x_3} - \frac{\partial u^2}{\partial x_4} = 0$$

$$\frac{\partial u^4}{\partial x_1} - \frac{\partial u^3}{\partial x_2} - \frac{\partial u^2}{\partial x_3} + \frac{\partial u^1}{\partial x_4} = 0,$$

the principal symbol of which is the quaternion

$$\mathcal{P}(\zeta) = \begin{bmatrix} \zeta_1 & -\zeta_2 & \zeta_3 & -\zeta_4 \\ \zeta_2 & \zeta_1 & \zeta_4 & \zeta_3 \\ -\zeta_3 & -\zeta_4 & \zeta_1 & \zeta_2 \\ \zeta_4 & -\zeta_3 & -\zeta_2 & \zeta_1 \end{bmatrix}$$

for which the characteristic polynomial takes the form

$$P(\zeta) = \det \mathcal{P}(\zeta) = |\zeta|^4$$

There are four linear solutions of the form

$$Z_1 = \begin{bmatrix} x_1 \\ x_2 \\ x_3 \\ x_4 \end{bmatrix}, \quad Z_2 = \begin{bmatrix} -x_1 \\ -x_2 \\ x_3 \\ x_4 \end{bmatrix}, \quad Z_3 = \begin{bmatrix} x_1 \\ x_2 \\ -x_3 \\ -x_4 \end{bmatrix}, \quad Z_4 = \begin{bmatrix} -x_1 \\ -x_2 \\ -x_3 \\ -x_4 \end{bmatrix}$$

which give rise to the concept of principal solutions. Also note that a solution that depends only on the two variables $x = x_1$ and $y = x_2$ satisfies the coupled Cauchy-Riemann equations

$$\begin{cases} u_x^1 = u_y^2 \\ u_y^1 = -u_x^2 \end{cases} \quad \begin{cases} u_x^3 = -u_y^4 \\ u_y^3 = u_x^4 \end{cases}$$

7.3 General Elliptic First-Order Operators in Two Variables

The general elliptic first-order differential operator in two variables (acting on functions valued in $\mathbb{R}^{2n}$) can be written as

$$\mathcal{L} = A\frac{\partial}{\partial x} + B\frac{\partial}{\partial y},$$

7.3. OPERATORS IN TWO VARIABLES

where A and B are $2n \times 2n$ matrices. The principal symbol is $aA + bB$, $a, b \in \mathbb{R}$, and the ellipticity condition is

$$\det(aA + bB) \neq 0, \quad a^2 + b^2 \neq 0$$

Of course, as soon as we put $a = 0$ or $b = 0$, we find $A, B \in GL(2n, \mathbb{R})$. The ellipticity condition can be reinterpreted as saying that $\mathcal{L}$ does not annihilate affine vector fields of the form $f = (ax + by + c)\mathbf{e}$, $\mathbf{e} \in \mathbb{R}^{2n}$. That is, ellipticity is equivalent to requiring

$$\mathcal{L}f = (aA + bB)\mathbf{e} \neq 0$$

for each $\mathbf{e} \neq \mathbf{0}$.

7.3.1 Complexification

Of course, the fact that elliptic systems of first-order must have an even number of equations and unknowns strongly suggests a complexification of the problem. This is what we shall do here.

We begin with an identification $\mathbb{R}^{2n} \simeq \mathbb{C}^n$, and so our unknown functions become complex vector fields,

$$F = U + iV : \Omega \to \mathbb{C}^n,$$

and define the complex derivatives

$$\frac{\partial}{\partial z} = \frac{1}{2}\left(\frac{\partial}{\partial x} - i\frac{\partial}{\partial y}\right), \quad \frac{\partial}{\partial \bar{z}} = \frac{1}{2}\left(\frac{\partial}{\partial x} + i\frac{\partial}{\partial y}\right)$$

Our operator $\mathcal{L}$ then takes the form

$$\mathcal{L} = \alpha \frac{\partial}{\partial z} + \beta \overline{\frac{\partial}{\partial z}} + \gamma \frac{\partial}{\partial \bar{z}} + \delta \overline{\frac{\partial}{\partial \bar{z}}}, \tag{7.6}$$

where $\alpha, \beta, \gamma, \delta$ are complex $n \times n$ matrices. We can find the ellipticity condition in terms of these matrices by testing the operator $\mathcal{L}$ against linear complex vector fields of the type

$$F(z) = (\zeta \bar{z} + \bar{\zeta} z)H,$$

where $\zeta \in \mathbb{C} \setminus \{0\}$ and $H \in \mathbb{C}^n$. We compute that

$$\begin{aligned}\mathcal{L}F &= \alpha \bar{\zeta} H + \beta \zeta \overline{H} + \gamma \zeta H + \delta \bar{\zeta} \overline{H} \\ &= (\alpha \bar{\zeta} + \gamma \zeta) H + (\delta \bar{\zeta} + \beta \zeta) \overline{H}\end{aligned}$$

In particular, we have the following lemma.

Lemma 7.3.1. *The linear partial differential operator $\mathcal{L}$ as in (7.6) is elliptic if and only if the linear map*

$$H \mapsto (\alpha \bar{\zeta} + \gamma \zeta) H + (\delta \bar{\zeta} + \beta \zeta) \overline{H}$$

is nonsingular for all $\zeta \in \mathbb{C} \setminus \{0\}$.

We leave the following exercise in linear algebra to the reader.

Lemma 7.3.2. *Let P and Q be complex $n \times n$ matrices. The linear map*

$$H \mapsto PH + Q\overline{H}, \qquad H \in \mathbb{C}^n,$$

is nonsingular if and only if

$$\det \begin{bmatrix} P & Q \\ \overline{Q} & \overline{P} \end{bmatrix} \neq 0$$

Note that in fact this determinant of the $2n \times 2n$ matrix of Lemma 7.3.2 is a real number. Accordingly, the ellipticity of the linear differential operator $\mathcal{L}$ is now

$$\det \begin{bmatrix} \bar\zeta\,\alpha + \zeta\,\gamma & \bar\zeta\,\delta + \zeta\,\beta \\ \zeta\,\bar\delta + \bar\zeta\,\bar\beta & \zeta\,\bar\alpha + \bar\zeta\,\bar\gamma \end{bmatrix} \neq 0$$

whenever $\zeta \in \mathbb{C} \setminus \{0\}$. Equivalently,

$$\det \left(\zeta \begin{bmatrix} \gamma & \beta \\ \bar\delta & \bar\alpha \end{bmatrix} + \bar\zeta \begin{bmatrix} \alpha & \delta \\ \bar\beta & \bar\gamma \end{bmatrix} \right) \neq 0$$

A case of special interest is the operator

$$\mathcal{L} = \frac{\partial}{\partial \bar z} + \alpha \frac{\partial}{\partial z} + \beta \overline{\frac{\partial}{\partial z}},$$

where α and β are complex $n \times n$ matrices. We apply the above results with $\delta = 0$ and $\gamma = \mathbf{I}$. The ellipticity condition given by Lemma 7.3.2 is

$$\det \left(\zeta \begin{bmatrix} \mathbf{I} & \beta \\ 0 & \bar\alpha \end{bmatrix} + \bar\zeta \begin{bmatrix} \alpha & 0 \\ \bar\beta & \mathbf{I} \end{bmatrix} \right) \neq 0$$

or, once we reduce to the case $|\zeta| = 1$ and write $e^{i\theta} = \bar\zeta^2$,

$$\det \left(\mathbf{I} + \begin{bmatrix} e^{i\theta}\alpha & \beta \\ \bar\beta & e^{-i\theta}\bar\alpha \end{bmatrix} \right) \neq 0$$

Note that the operator norm of the matrix

$$\left\| \begin{bmatrix} e^{i\theta}\alpha & \beta \\ \bar\beta & e^{-i\theta}\bar\alpha \end{bmatrix} \right\| \leq |\alpha| + |\beta|$$

We have now established the result we have been aiming at.

Theorem 7.3.3. *Let $\alpha, \beta \in \mathbb{C}^{n \times n}$ be such that $|\alpha| + |\beta| < 1$. Then the linear differential operator*

$$\mathcal{L} = \frac{\partial}{\partial \bar z} + \alpha \frac{\partial}{\partial z} + \beta \overline{\frac{\partial}{\partial z}}$$

is elliptic.

We remark that this result is quite far from describing all the linear elliptic operators except in the case $n = 1$. We now examine this case in some detail.

7.3.2 Homotopy Classification

We are ultimately seeking a definition of ellipticity for nonlinear equations with measurable coefficients. Typically, one classifies operators up to homotopy, that is, up to continuous deformations. Our definition of ellipticity will depend on this classification. Thus in this section we introduce the homotopy classification and find all the components for first-order systems in the case $n = 1$.

We consider the general linear partial differential operator of first-order

$$\mathcal{L} = \alpha \frac{\partial}{\partial z} + \beta \overline{\frac{\partial}{\partial z}} + \gamma \frac{\partial}{\partial \bar{z}} + \delta \overline{\frac{\partial}{\partial \bar{z}}}, \tag{7.7}$$

Two such elliptic operators,

$$\begin{aligned}
\mathcal{L}_0 &= \alpha_0 \frac{\partial}{\partial z} + \beta_0 \overline{\frac{\partial}{\partial z}} + \gamma_0 \frac{\partial}{\partial \bar{z}} + \delta_0 \overline{\frac{\partial}{\partial \bar{z}}} \\
\mathcal{L}_1 &= \alpha_1 \frac{\partial}{\partial z} + \beta_1 \overline{\frac{\partial}{\partial z}} + \gamma_1 \frac{\partial}{\partial \bar{z}} + \delta_1 \overline{\frac{\partial}{\partial \bar{z}}},
\end{aligned}$$

are said to be *homotopic* if there are continuous functions

$$\alpha, \beta, \gamma, \delta : [0,1] \to \mathbb{C}^{n \times n}$$

with the boundary conditions

$$\begin{array}{llll}
\alpha(0) = \alpha_0, & \alpha(1) = \alpha_1, & \beta(0) = \beta_0, & \beta(1) = \beta_1 \\
\gamma(0) = \gamma_0, & \gamma(1) = \gamma_1, & \delta(0) = \delta_0, & \delta(1) = \delta_1
\end{array}$$

such that the linear partial differential operators

$$\mathcal{L}_t = \alpha(t) \frac{\partial}{\partial z} + \beta(t) \overline{\frac{\partial}{\partial z}} + \gamma(t) \frac{\partial}{\partial \bar{z}} + \delta(t) \overline{\frac{\partial}{\partial \bar{z}}}$$

are elliptic for all $t \in [0,1]$.

7.3.3 Classification; $n = 1$

Even the definition of the operator $\mathcal{L}$ as in (7.7) suggests that there are at least four components in the space of linear elliptic partial differential operators, namely, those belonging to the four operators

$$\frac{\partial}{\partial z}, \quad \overline{\frac{\partial}{\partial z}}, \quad \frac{\partial}{\partial \bar{z}}, \quad \overline{\frac{\partial}{\partial \bar{z}}}$$

Let us try to establish this for the case $n = 1$, so that $\alpha, \beta, \gamma, \delta \in \mathbb{C}$. Now Lemma 7.3.1 tells us that the principal symbol is represented by the linear map from $\mathbb{C} \to \mathbb{C}$ defined for $\zeta \in \mathbb{C} \setminus \{0\}$ by

$$\mathcal{P}(\zeta) w = (\alpha \bar{\zeta} + \gamma \zeta) w + (\beta \zeta + \delta \bar{\zeta}) \bar{w} \tag{7.8}$$

and that $\mathcal{L}$ is elliptic if and only if for each ζ this map vanishes only at the origin.

Next, motivated by Lemma 7.3.2, we compute that

$$\det\left(\zeta\begin{bmatrix}\gamma & \beta \\ \delta & \alpha\end{bmatrix} + \bar{\zeta}\begin{bmatrix}\alpha & \delta \\ \beta & \gamma\end{bmatrix}\right) = |\bar{\zeta}\alpha + \zeta\gamma|^2 - |\bar{\zeta}\delta + \zeta\beta|^2 \qquad (7.9)$$

Of course, we may use here the basic identity (2.24). Because of the assumed ellipticity of $\mathcal{L}$, this quantity is nonvanishing for all $\zeta \in \mathbb{C}\setminus\{0\}$. The determinant is a real-valued continuous function of ζ and so has constant sign, giving two cases. If positive, we define

$$\mathcal{L}_t = \alpha\frac{\partial}{\partial z} + \gamma\frac{\partial}{\partial \bar{z}} + (1-t)\beta\overline{\frac{\partial}{\partial z}} + (1-t)\delta\overline{\frac{\partial}{\partial \bar{z}}}$$

The formula in (7.9) gives

$$\det(\mathcal{L}_t(\zeta)) = |\alpha\bar{\zeta} + \gamma\zeta|^2 - (1-t)^2|\beta\zeta + \delta\bar{\zeta}|^2,$$

which is increasing in t for each fixed ζ. Thus each $\mathcal{L}_t$ is elliptic, $0 \leq t \leq 1$. As the sign of the determinant is constant in any homotopy component of elliptic operators, we may put $\zeta = 1$ to determine the component. We see that $\mathcal{L}$ is clearly homotopic to

$$\mathcal{L} \approx \alpha\frac{\partial}{\partial z} + \gamma\frac{\partial}{\partial \bar{z}} \quad \text{if } |\alpha + \gamma| > |\delta + \beta| \qquad (7.10)$$

Similarly, if the determinant is negative we have

$$\mathcal{L} \approx \beta\overline{\frac{\partial}{\partial z}} + \delta\overline{\frac{\partial}{\partial \bar{z}}}, \quad \text{if } |\alpha + \gamma| < |\delta + \beta| \qquad (7.11)$$

The two cases (7.10) and (7.11) are clearly distinct.

Each such case breaks into two further subcases as follows. As we have seen, the determinant of the principal symbol for the operator

$$\mathcal{L}_t = \alpha(t)\frac{\partial}{\partial z} + \gamma(t)\frac{\partial}{\partial \bar{z}} \qquad (7.12)$$

is $|\zeta\alpha(t) + \bar{\zeta}\gamma(t)|^2$. This is nonzero for all ζ if and only if $|\gamma(t)| \neq |\alpha(t)|$. Since any homotopy from $\mathcal{L}$ to either $\frac{\partial}{\partial z}$ or $\frac{\partial}{\partial \bar{z}}$ may be assumed to have the form (7.12) by our first observations, the intermediate value theorem then assures us that we can homotope $\mathcal{L}$ to $\frac{\partial}{\partial \bar{z}}$ through elliptic operators only if $|\gamma| > |\alpha|$. And of course, $|\gamma| < |\alpha|$ is necessary to homotope $\mathcal{L}$ to $\frac{\partial}{\partial z}$.

Theorem 7.3.4. *Suppose the linear partial differential operator*

$$\mathcal{L} = \alpha\frac{\partial}{\partial z} + \beta\overline{\frac{\partial}{\partial z}} + \gamma\frac{\partial}{\partial \bar{z}} + \delta\overline{\frac{\partial}{\partial \bar{z}}}, \qquad \alpha, \beta, \gamma, \delta \in \mathbb{C},$$

7.3. OPERATORS IN TWO VARIABLES

is elliptic. Then it is homotopic to exactly one of the following four operators:

$$\frac{\partial}{\partial \bar{z}} \quad \text{if and only if} \quad |\alpha + \gamma| > |\delta + \beta| \quad \text{and} \quad |\gamma| > |\alpha| \quad (7.13)$$

$$\overline{\frac{\partial}{\partial z}} \quad \text{if and only if} \quad |\alpha + \gamma| < |\delta + \beta| \quad \text{and} \quad |\delta| > |\beta| \quad (7.14)$$

$$\frac{\partial}{\partial z} \quad \text{if and only if} \quad |\alpha + \gamma| > |\delta + \beta| \quad \text{and} \quad |\gamma| < |\alpha| \quad (7.15)$$

$$\overline{\frac{\partial}{\partial z}} \quad \text{if and only if} \quad |\alpha + \gamma| < |\delta + \beta| \quad \text{and} \quad |\delta| < |\beta| \quad (7.16)$$

We remark that none of the conditions (7.13)–(7.16) are sufficient in and of themselves to guarantee ellipticity, as the reader may easily verify.

We will principally be interested in the Cauchy-Riemann component

$$\mathcal{E}_{CR} = \{\mathcal{L} \approx \frac{\partial}{\partial \bar{z}}\}$$

of the homotopy equivalence classes of elliptic first-order differential operators. We evidently have the following test for an operator to be in this class.

Lemma 7.3.5. *The linear partial differential operator*

$$\mathcal{L} = \alpha \frac{\partial}{\partial z} + \beta \overline{\frac{\partial}{\partial z}} + \gamma \frac{\partial}{\partial \bar{z}} + \delta \overline{\frac{\partial}{\partial \bar{z}}}, \qquad \alpha, \beta, \gamma, \delta \in \mathbb{C},$$

is elliptic and homotopic to $\frac{\partial}{\partial \bar{z}}$, that is, $\mathcal{L} \in \mathcal{E}_{CR}$ if and only if $|\gamma| > |\alpha|$ and

$$|\alpha + e^{i\theta}\gamma| > |\delta + e^{i\theta}\beta| \quad \text{for all } \theta \in \mathbb{R}$$

In particular, if

$$|\gamma| > |\alpha| + |\beta| + |\delta|, \tag{7.17}$$

then $\mathcal{L} \in \mathcal{E}_{CR}$. The condition (7.17) is necessary and sufficient if $\alpha\beta\delta = 0$.

The reformulations of the lemma necessary to identify the other components are clear.

Let us now look for a moment at the inhomogeneous equations $\mathcal{L}f = h$ for some function h. If $\mathcal{L} \in \mathcal{E}_{CR}$, then we can reformulate this equation so as to have a more desirable form, in particular, so as to have no terms involving $\overline{\partial/\partial \bar{z}}$.

Theorem 7.3.6. *Let $\mathcal{L} \in \mathcal{E}_{CR}$. Then the inhomogenous equation*

$$\mathcal{L}f = h$$

is equivalent to the equation

$$\frac{\partial f}{\partial \bar{z}} = \mu \frac{\partial f}{\partial z} + \nu \overline{\frac{\partial f}{\partial z}} + g, \tag{7.18}$$

where
$$\mu = \frac{\overline{\beta}\delta - \alpha\overline{\gamma}}{|\gamma|^2 - |\delta|^2}, \qquad \nu = \frac{\overline{\alpha}\delta - \beta\overline{\gamma}}{|\gamma|^2 - |\delta|^2}, \qquad g = \frac{\overline{\gamma}h - \delta\overline{h}}{|\gamma|^2 - |\delta|^2}$$
and
$$|\mu| + |\nu| \leqslant k < 1 \qquad (7.19)$$

Proof. We evidently have the two equations
$$\begin{aligned} \alpha\, f_z + \beta\, \overline{f_z} + \gamma\, f_{\bar{z}} + \delta\, \overline{f_{\bar{z}}} &= h \\ \overline{\alpha}\, \overline{f_z} + \overline{\beta}\, f_z + \overline{\gamma}\, \overline{f_{\bar{z}}} + \overline{\delta}\, f_{\bar{z}} &= \overline{h} \end{aligned}$$

If $\delta = 0$, we are done, for we write
$$f_{\bar{z}} = -\frac{\alpha}{\gamma}\, f_z - \frac{\beta}{\gamma}\, \overline{f_z} + \frac{h}{\gamma}$$

and $|\alpha/\gamma| + |\beta/\gamma| < 1$ by Lemma 7.3.5, the latter condition being necessary and sufficient in this case. Thus we suppose $\delta \neq 0$. Then we eliminate the $\overline{f_{\bar{z}}}$ term to find
$$(|\gamma|^2 - |\delta|^2) f_{\bar{z}} = (\overline{\beta}\delta - \alpha\overline{\gamma}) f_z + (\overline{\alpha}\delta - \beta\overline{\gamma})\overline{f_z} + \overline{\gamma}h - \delta\overline{h} \qquad (7.20)$$

As $\mathcal{L}$ is elliptic, Lemma 7.3.1 tells us that for each $\zeta \neq 0$ the map of $\mathbb{C}$,
$$\omega \mapsto (\alpha\zeta + \gamma\bar{\zeta})\,\omega + (\delta\zeta + \beta\bar{\zeta})\,\overline{\omega},$$

vanishes only at the origin. We put $|\zeta| = |\omega| = 1$, rearrange the terms and take absolute values, and find not only our previous observation,
$$|\alpha + \gamma\zeta| \neq |\delta + \beta\zeta|, \qquad |\zeta| = 1,$$

but also that
$$|\alpha - \delta\overline{\omega}| \neq |\beta - \gamma\omega|, \qquad |\omega| = 1$$

Again, as continuous functions of the variables, the difference between the left and right hand sides of each equation has constant sign in each component of ellipticity. As $\mathcal{L} \in \mathcal{E}_{CR}$, we may substitute $\gamma = 1$, $\alpha = \beta = \delta = 0$ to see what this sign is. We therefore obtain the two inequalities
$$|\alpha + \gamma\zeta| > |\delta + \beta\zeta|, \qquad |\zeta| = 1 \qquad (7.21)$$
$$|\beta - \gamma\omega| > |\alpha - \delta\overline{\omega}|, \qquad |\omega| = 1 \qquad (7.22)$$

We square (7.21) and rearrange terms to see for arbitrary ζ, with $|\zeta| = 1$,
$$2\,\mathfrak{Re}[(\overline{\alpha}\gamma - \overline{\delta}\beta)\zeta] > |\delta|^2 + |\beta|^2 - |\alpha|^2 - |\gamma|^2,$$

which yields the inequality
$$2|\overline{\alpha}\gamma - \overline{\delta}\beta| < -|\delta|^2 - |\beta|^2 + |\alpha|^2 + |\gamma|^2 \qquad (7.23)$$

7.4. MEASURABLE COEFFICIENTS

Similar manipulations of (7.22) yield

$$2|\overline{\delta}\alpha - \overline{\beta}\gamma| < -|\delta|^2 + |\beta|^2 - |\alpha|^2 + |\gamma|^2 \tag{7.24}$$

Addition of (7.23) and (7.24) results in

$$|\overline{\alpha}\gamma - \overline{\delta}\beta| + |\overline{\delta}\alpha - \overline{\beta}\gamma| < |\gamma|^2 - |\delta|^2 \tag{7.25}$$

Returning to (7.20), we find that not only have we established that $|\gamma|^2 > |\delta|^2$, so we can divide out the leading term, but also in fact (7.19) with

$$k = \frac{|\overline{\alpha}\gamma - \overline{\delta}\beta| + |\overline{\delta}\alpha - \overline{\beta}\gamma|}{|\gamma|^2 - |\delta|^2}, \tag{7.26}$$

which completes the proof of the theorem. □

7.4 Partial Differential Operators with Measurable Coefficients

We now have at hand enough technology to generalize the concept of ellipticity to operators and equations with measurable coefficients. We shall mainly study the case of first-order operators in the plane, but the generalizations to higher-orders and dimensions are clear. Thus

$$\mathcal{L} = \alpha(z)\frac{\partial}{\partial z} + \beta(z)\overline{\frac{\partial}{\partial z}} + \gamma(z)\frac{\partial}{\partial \bar{z}} + \delta(z)\overline{\frac{\partial}{\partial \bar{z}}},$$

defined on a domain $\Omega \subset \mathbb{C}$ and where $\alpha, \beta, \gamma, \delta : \Omega \to \mathbb{C}^{n \times n}$ are measurable functions, is said to be elliptic if for almost every $z_0 \in \Omega$ the linear constant coefficient partial differential operator

$$\mathcal{L}_{z_0} = \alpha(z_0)\frac{\partial}{\partial z} + \beta(z_0)\overline{\frac{\partial}{\partial z}} + \gamma(z_0)\frac{\partial}{\partial \bar{z}} + \delta(z_0)\overline{\frac{\partial}{\partial \bar{z}}}$$

is in the same component of homotopy equivalence classes of such operators. Alternatively, we could assume that for almost every $(z_0, z_1) \in \Omega \times \Omega$ the operators $\mathcal{L}_{z_0}$ and $\mathcal{L}_{z_1}$ are homotopic through elliptic operators.

As we are principally interested in the Cauchy-Riemann component, Theorem 7.3.6 asserts that, in the case $n = 1$, any first-order elliptic linear partial differential operator in $\mathcal{E}_{CR}$ with measurable coefficients is equivalent to the operator

$$\mathcal{L} = \frac{\partial}{\partial \bar{z}} - \mu(z)\frac{\partial}{\partial z} - \nu(z)\overline{\frac{\partial}{\partial z}}, \tag{7.27}$$

where ellipticity is equivalent to the condition

$$|\mu(z)| + |\nu(z)| < 1 \quad \text{for almost every } z \in \Omega \tag{7.28}$$

We shall see in practice that the condition (7.28) is very weak, even for operators of the form (7.27). Notice that for constant coefficient operators we have $|\mu| + |\nu| = k < 1$. Hence we make the following refinements of the concept of ellipticity. Operators of the form (7.27) with μ, ν measurable are said to be *uniformly elliptic* if there is a number $k < 1$ such that $|\mu(z)| + |\nu(z)| \leqslant k$ for almost every $z \in \Omega$. Otherwise, if only (7.28) holds, we say the operator is *degenerate elliptic*.

Note in particular that the Beltrami equations studied in Chapter 5 fall into the class of uniformly elliptic operators.

7.5 Quasilinear Operators

Operators of the form

$$\mathcal{L}f = \alpha(z,f)\frac{\partial f}{\partial z} + \beta(z,f)\overline{\frac{\partial f}{\partial z}} + \gamma(z,f)\frac{\partial f}{\partial \bar{z}} + \delta(z,f)\overline{\frac{\partial f}{\partial \bar{z}}}, \qquad (7.29)$$

in which the coefficients are functions of the *independent variable* z and the *dependent variable* $w = f(z)$, are called *quasilinear*. Our basic ellipticity assumption will be that should we freeze the coefficients at the two distinct points (z_1, w_1) and (z_2, w_2) in $\mathbb{C} \times \mathbb{C}$, then the resultant constant coefficient operators will be in the same homotopy component of elliptic operators. There is no loss of generality in assuming that this component is the Cauchy-Riemann component, and if we assume this is the case, we can rewrite $\mathcal{L}f = \psi(z)$, where $\mathcal{L}$ is defined at (7.29), in the form of an equation as

$$\frac{\partial f}{\partial \bar{z}} = \mu(z,f)\frac{\partial f}{\partial z} + \nu(z,f)\overline{\frac{\partial f}{\partial z}} + \psi(z) \qquad (7.30)$$

We shall be looking for solutions to (7.30) that define mappings between domains Ω and Ω' in $\mathbb{C}$, in which case the ellipticity constraint is equivalent to

$$|\mu(z,w)| + |\nu(z,w)| < 1 \qquad \text{for almost every } (z,w) \in \Omega \times \Omega'$$

To develop the theory, and before pressing on to the degenerate elliptic case, we shall usually make the *uniform ellipticity* assumption that there is a $k < 1$ such that

$$|\mu(z,w)| + |\nu(z,w)| \leqslant k \qquad \text{for almost every } (z,w) \in \Omega \times \Omega'$$

There is an important point we need to make here. The Lebesgue measurability of the functions μ, $\nu : \Omega \times \Omega' \to \mathbb{C}$ does not guarantee that the composed functions

$$z \mapsto \mu(z, f(z)), \qquad z \mapsto \nu(z, f(z))$$

are measurable functions of the independent variable z even if in fact f is quasiconformal. Measurability would of course be assured if μ and ν were continuous in the dependent variable $w = f(z)$, but important applications necessitate weaker assumptions, as we have already seen for instance in the introductory

7.6. LUSIN MEASURABILITY

examples at the beginning of this chapter. Hence we cannot restrict ourselves to equations with coefficients continuous in the dependent variable.

Another reason for these weaker assumptions concerns the duality between the z- and w-variables: We would like them to be interchangeable! The hodographic transformation, discussed later in Section 16.3, is a fundamental application of this interchangeability. The basic identity (2.51) is another aspect of the duality. We would also like to study the inverses of solutions to (7.29), compositions of solutions and so forth. There are many reasons for this, as the reader will come to see. Thus we need to build up a reasonably large class of elliptic equations that are closed under these operations.

As far as the aforementioned problem with measurability is concerned, Lusin's measurability criterion strikes us as the most natural path to seek optimal assumptions on the coefficients. Thus we state the following definition.

Definition of Uniformly Elliptic Equation; Quasilinear Case

Definition 7.5.1. *A quasilinear equation*

$$\frac{\partial f}{\partial \bar{z}} = \mu(z,f) \frac{\partial f}{\partial z} + \nu(z,f) \overline{\frac{\partial f}{\partial z}}$$

defined for Sobolev functions $f \in W^{1,1}(\Omega, \Omega')$ *is said to be* uniformly elliptic *if*

1. *both μ and ν are Lusin-measurable on $\Omega \times \Omega'$ and*

2. $|\mu(z,w)| + |\nu(z,w)| \leqslant k < 1$ *for almost every* $(z,w) \in \Omega \times \Omega'$.

We now take a few moments to discuss the concept of Lusin measurability.

7.6 Lusin Measurability

This section is basically on measure theory, though we do not gain any essential generalization by considering abstract measure spaces. For notational convenience we therefore work in a special setting. Let X, Y and Z be measurable subsets of $\mathbb{C}$. We will be concerned here with complex-valued functions

$$F : X \times Y \times Z \to \mathbb{C}$$

Let $L^0(X \times Y \times Z)$ denote the space of all such functions that are measurable with respect to the product measure. The well-known Lusin's criterion for measurability tells us that every measurable function $F = F(x,y,z)$ is continuous when restricted to suitable compact subsets, which can be chosen to be arbitrarily close in the sense of measure to the entire domain $X \times Y \times Z$. What is not always possible is that these compact approximating sets can be Cartesian products of compact subsets in the coordinate spaces, that is, of the form $X_1 \times Y_1 \times Z_1$. To see this the reader need only examine the function $F : \mathbb{R} \times \mathbb{R} \times \mathbb{R} \to \mathbb{R}$ defined by

$$F(x, y, z) = \begin{cases} 1, & y \geqslant x \\ 0, & y < x \end{cases}$$

Indeed, this example (and also a later example, see Section 8.1) leads us to state the following definition.

Definition of Lusin Measurability

Definition 7.6.1. *A function $F \in L^0(X \times Y \times Z)$ is said to be Lusin-measurable if there are increasing sequences of compact sets $X_1 \subset X_2 \subset \cdots \subset X_l \subset \cdots X$, $Y_1 \subset Y_2 \subset \cdots \subset Y_l \subset \cdots Y$ and $Z_1 \subset Z_2 \subset \cdots \subset Z_l \subset \cdots Z$ such that*

1. *For each $l = 1, 2, \ldots$, the restriction*

$$F : X_l \times Y_l \times Z_l \to \mathbb{C}$$

   is continuous.

2. *The unions $\cup_l X_l$, $\cup_l Y_l$ and $\cup_l Z_l$ have full measure in X, Y and Z, respectively.*

Roughly, the condition of Lusin measurability tells us that a function is measurable in each variable independently.

A useful class of Lusin-measurable functions is furnished by functions of the form

$$F(x, y, z) = \Phi(\alpha(x), \beta(y), \gamma(z)), \tag{7.31}$$

where $\alpha(x)$, $\beta(y)$ and $\gamma(z)$ are measurable functions defined in X, Y and Z, respectively, and the function $\Phi = \Phi(\alpha, \beta, \gamma)$ is continuous. In fact, these are typical cases.

For us, the most important aspect of Lusin-measurable functions is expressed in the following lemma.

Lemma 7.6.2. *Let $F : X \times Y \times Z \to \mathbb{C}$ be Lusin-measurable. Suppose we are given measurable functions*

$$\alpha : \Omega \to X, \qquad \beta : \Omega \to Y, \qquad \gamma : \Omega \to Z$$

defined on a measure space Ω, all satisfying the condition $\mathcal{N}^{-1}$.
   Then the function

$$f(\omega) = F(\alpha(\omega), \beta(\omega), \gamma(\omega)), \qquad \omega \in \Omega,$$

is measurable.

7.6. LUSIN MEASURABILITY

In the lemma the mere assumption of measurability of F is insufficient for f to be measurable.

Proof. The Tietze extension theorem allows us to extend $F : X_l \times Y_l \times Z_l \to \mathbb{C}$ to a continuous function $F_l : X \times Y \times Z \to \mathbb{C}$. Here of course X_l, Y_l and Z_l are the compact sets on which F is continuous, sets that are given to us by the definition of Lusin measurability.

Consider the functions of one variable,
$$f_l(\omega) = F_l\big(\alpha(\omega), \beta(\omega), \gamma(\omega)\big), \qquad f_l : \Omega \to \mathbb{C}$$
Since F_l is continuous, f_l is measurable. What remains is to show that for almost every $\omega \in \Omega$ we have $f_l(\omega) \to f(\omega)$. This is where we have to use the condition $\mathcal{N}^{-1}$ for α, β and γ together with the Lusin measurability for F. These properties imply that the set
$$E = \left(\bigcup_{j=1}^{\infty} \alpha^{-1}(X_j)\right) \cap \left(\bigcup_{k=1}^{\infty} \beta^{-1}(Y_k)\right) \cap \left(\bigcup_{l=1}^{\infty} \gamma^{-1}(Z_l)\right) \subset \Omega \qquad (7.32)$$
has full measure in Ω. If $\omega \in E$ then, since $\{X_j\}$, $\{Y_k\}$ and $\{Z_l\}$ are increasing sequences, for all sufficiently large m we have
$$\omega \in \alpha^{-1}(X_m) \cap \beta^{-1}(Y_m) \cap \gamma^{-1}(Z_m),$$
and hence
$$f_m(x) = F_l\big(\alpha(\omega), \beta(\omega), \gamma(\omega)\big) = F\big(\alpha(\omega), \beta(\omega), \gamma(\omega)\big) = f(\omega)$$
This ensures the convergence of f_m to f almost everywhere. $\square$

The reader should have no difficulty in generalizing the above result and the notion of Lusin measurability to functions defined on finite products of arbitrary measure spaces.

We note in passing that in light of our definition of uniform ellipticity for quasilinear systems and Lemma 7.6.2 above, the Lusin condition $\mathcal{N}^{-1}$ will be an important property of solutions to such equations. Indeed, any reasonable definition of a solution to the equation
$$\frac{\partial f}{\partial \bar{z}} = \mu(z, f) \frac{\partial f}{\partial z} + \nu(z, f) \overline{\frac{\partial f}{\partial z}}$$
implies that f is (weakly) quasiregular,
$$\left|\frac{\partial f}{\partial \bar{z}}\right| \leqslant (|\mu(z, f)| + |\nu(z, f)|) \left|\frac{\partial f}{\partial z}\right| \leqslant k \left|\frac{\partial f}{\partial z}\right|$$
Further, if f also has enough Sobolev regularity (for example, $f \in W^{1,2}_{loc}(\Omega)$), then by Corollary 5.5.2 the solution has the property $\mathcal{N}^{-1}$ guaranteeing the measurability of the composed coefficients $\mu(z, f(z))$ and $\nu(z, f(z))$.

It is clear from the above discussion that in the degenerate elliptic case, establishing the properties $\mathcal{N}$ and $\mathcal{N}^{-1}$ will be an important step in deducing improved regularity and other desirable properties of solutions.

7.7 Fully Nonlinear Equations

In this section we discuss what it means for a nonlinear equation of the form

$$\Phi(z, f, \frac{\partial f}{\partial z}, \frac{\partial f}{\partial \bar{z}}) = 0 \qquad (7.33)$$

to be elliptic. Here we are seeking a solution as a mapping defined on a domain $\Omega \subset \mathbb{C}$ and valued in $\Omega' \subset \mathbb{C}$, belonging to the Sobolev class $W^{1,1}_{loc}(\Omega)$. Actually, we will most often seek a solution in the better space $W^{1,2}_{loc}(\Omega)$, but for the definition this does not matter. Thus we consider the function

$$\Phi : \Omega \times \Omega' \times \mathbb{C} \times \mathbb{C} \to \mathbb{C}$$

Of course, there is great generality here, and some restrictions on Φ are going to be necessary to obtain a class of equations whose solutions enjoy good geometric properties and are of physical relevance. First, on any open set where both the z- and $\bar{z}$-derivatives vanish, we have f constant and the functional equation $\varphi(z) = \Phi(z, f, 0, 0) = 0$. Thus we desire that constants be solutions to the equations and therefore are led to the requirement

$$\Phi(z, w, 0, 0) \equiv 0, \qquad (z, w) \in \Omega \times \Omega' \qquad (7.34)$$

It will simplify our discussion, and it will subsequently be seen to cause no loss in generality, to temporarily assume that for fixed $(z, w) \in \Omega \times \Omega'$ the function

$$(\zeta, \xi) \mapsto \Phi(z, w, \zeta, \xi)$$

belongs to the space $C^1(\mathbb{C}^2)$ and is continuous on $\hat{\mathbb{C}} \times \hat{\mathbb{C}}$, with value ∞, whenever $|\zeta| + |\eta| = \infty$.

The approach to ellipticity we take follows Petrovsky. Accordingly, given a function $f \in W^{1,1}_{loc}(\Omega)$, the nonlinear equation (7.33) is said to be *elliptic on f* if the linearized operator

$$\mathcal{L}F = \frac{\partial \Phi}{\partial f_z} F_z + \frac{\partial \Phi}{\partial \overline{f_z}} \overline{F_z} + \frac{\partial \Phi}{\partial f_{\bar{z}}} F_{\bar{z}} + \frac{\partial \Phi}{\partial \overline{f_{\bar{z}}}} \overline{F_{\bar{z}}} \qquad (7.35)$$

is elliptic with the measurable coefficients given by the partials of $\Phi(z, w, \zeta, \xi)$:

$$\begin{aligned}
\alpha(z) &= \Phi_\zeta(z, f(z), f_z(z), f_{\bar{z}}(z)) \\
\beta(z) &= \Phi_{\bar{\zeta}}(z, f(z), f_z(z), f_{\bar{z}}(z)) \\
\gamma(z) &= \Phi_\xi(z, f(z), f_z(z), f_{\bar{z}}(z)) \\
\delta(z) &= \Phi_{\bar{\xi}}(z, f(z), f_z(z), f_{\bar{z}}(z))
\end{aligned}$$

In what follows we shall require that (7.33) be elliptic on every Sobolev function $f \in W^{1,1}_{loc}(\Omega)$. In particular, we can test the ellipticity of the operator on the linear functions

$$f(z) = w_0 + (z - z_0)\zeta_0 + (\bar{z} - \bar{z}_0)\xi_0,$$

7.7. FULLY NONLINEAR EQUATIONS

where $z_0 \in \Omega$ and $w_0 \in \Omega'$. Thus ellipticity implies that the operator

$$\mathcal{L}F = \alpha_0 \frac{\partial F}{\partial z} + \beta_0 \overline{\frac{\partial F}{\partial z}} + \gamma_0 \frac{\partial F}{\partial \bar{z}} + \delta_0 \overline{\frac{\partial F}{\partial \bar{z}}} \qquad (7.36)$$

with the constant coefficients

$$\begin{aligned} \alpha_0 &= \Phi_\zeta(z_0, w_0, \zeta_0, \xi_0) \\ \beta_0 &= \Phi_{\bar\zeta}(z_0, w_0, \zeta_0, \xi_0) \\ \gamma_0 &= \Phi_\xi(z_0, w_0, \zeta_0, \xi_0) \\ \delta_0 &= \Phi_{\bar\xi}(z_0, w_0, \zeta_0, \xi_0) \end{aligned}$$

is elliptic. As before, there is no loss in generality in assuming that this operator lies in the Cauchy-Riemann component and thus, in particular, $|\gamma_0|^2 - |\delta_0|^2 \neq 0$. This translates into the requirement on the Jacobian determinant of the mapping $\xi \mapsto \Phi(z_0, w_0, \zeta_0, \xi)$ that

$$\left|\frac{\partial \Phi}{\partial \xi}\right|^2 - \left|\frac{\partial \Phi}{\partial \bar\xi}\right|^2 \neq 0 \qquad (7.37)$$

Thus, recalling that Φ depends in a $C^1(\mathbb{C})$ manner on ξ, this map is a diffeomorphism of class $C^1(\mathbb{C})$. Moreover, as $\xi \to \infty$, we have this mapping converging to ∞ as well. Thus the extension to $\hat{\mathbb{C}}$ is a continuous local injection, which the monodromy theorem asserts is a global homeomorphism. In particular, this implies we can solve the equation

$$\Phi(z_0, w_0, \zeta_0, \xi) = 0$$

uniquely for ξ and express it as a function of the three variables z_0, w_0 and ζ_0, say

$$\xi_0 = H(z_0, w_0, \zeta_0) \qquad (7.38)$$

This means of course that our original equation (7.33) can be written in the much more convenient form

$$\frac{\partial f}{\partial \bar z} = H\left(z, f, \frac{\partial f}{\partial z}\right), \qquad H: \Omega \times \Omega' \times \mathbb{C} \to \mathbb{C} \qquad (7.39)$$

This alone should convince the reader of the strength of the ellipticity assumption. Now of course, condition (7.34) reads as

$$H(z, w, 0) \equiv 0, \qquad (z, w) \in \Omega \times \Omega' \qquad (7.40)$$

After writing the equation in the form (7.39), we can once again examine the ellipticity of the equation for

$$\Phi(z, w, \zeta, \xi) = \xi - H(z, w, \zeta)$$

The linearized operator in (7.35) now takes the form

$$\mathcal{L}F = F_{\bar z} - H_\zeta F_z - H_{\bar\zeta} \overline{F_z},$$

which will be elliptic if and only if

$$|H_\zeta| + |H_{\bar\zeta}| < 1$$

However, uniform ellipticity requires that

$$|H_\zeta| + |H_{\bar\zeta}| \leqslant k < 1, \tag{7.41}$$

which is easily seen to be equivalent to the Lipschitz condition

$$|H(z,w,\zeta_1) - H(z,w,\zeta_2)| \leqslant k|\zeta_1 - \zeta_2| \tag{7.42}$$

Notice that in fact if we were only to assume Lipschitz regularity of the map $\zeta \mapsto H(z,w,\zeta)$, then (7.42) would imply (7.41) for almost every $\zeta \in \mathbb{C}$. Thus we will stick with the condition (7.42) to retain greater generality.

Next, we must again confront the issue of measurablility of the function

$$z \mapsto H(z, f(z), f_z(z)),$$

and so we will need to assume the Lusin measurability condition. Thus we finally state the following definition.

Definition of Uniformly Elliptic Equation; Nonlinear Case

Definition 7.7.1. *The nonlinear system of equations*

$$\Phi\left(z, f, \frac{\partial f}{\partial z}, \frac{\partial f}{\partial \bar z}\right) = 0$$

defined for Sobolev functions $f \in W^{1,1}_{loc}(\Omega, \Omega')$ *is said to be* uniformly elliptic *if it can be written as*

$$\frac{\partial f}{\partial \bar z} = H\left(z, f, \frac{\partial f}{\partial z}\right), \tag{7.43}$$

where

1. *The function* $H : \Omega \times \Omega' \times \mathbb{C} \to \mathbb{C}$ *is Lusin-measurable.*

2. $H(z, w, 0) \equiv 0$ *for all* $(z, w) \in \Omega \times \Omega'$, *and*

3. *there is a number* $0 \leqslant k < 1$, *called the ellipticity constant, such that*

$$|H(z,w,\zeta_1) - H(z,w,\zeta_2)| \leqslant k|\zeta_1 - \zeta_2| \tag{7.44}$$

for all $(z, w) \in \Omega \times \Omega'$ *and* $\zeta_1, \zeta_2 \in \mathbb{C}$.

We also refer to the number

$$K = \frac{1+k}{1-k} \in [1, \infty)$$

as the distortion constant of H.

7.7. FULLY NONLINEAR EQUATIONS

The following fundamental composition result will be used throughout the book.

Theorem 7.7.2. *Suppose we are given a mapping $f : \Omega \to \Omega'$ satisfying Lusin's condition $\mathcal{N}^{-1}$ and a measurable function $\varphi : \Omega \to \mathbb{C}$. Then, under the conditions 1 – 3 in Definition 7.7.1, the function*

$$z \mapsto H\bigl(z, f(z), \varphi(z)\bigr), \qquad z \in \Omega,$$

is measurable.

Proof. The Lusin measurability criterion that we have imposed gives us compact sets Z_j, W_j and Ξ_j so that $H : Z_j \times W_j \times \Xi_j \to \mathbb{C}$ is continuous, $j = 1, 2, \ldots$. The sequences of sets $Z_1 \subset Z_2 \subset \cdots \mathbb{C}$, $W_1 \subset W_2 \subset \cdots \mathbb{C}$ and $\Xi_1 \subset \Xi_2 \subset \cdots \mathbb{C}$ can be chosen to each have union of full measure in $\mathbb{C}$. In fact, we shall next show that the factors Ξ_j can be replaced by the entire complex plane:

We claim that under the assumptions 1–3, for each $j = 1, 2, \ldots$, the restriction $H : Z_j \times W_j \times \mathbb{C} \to \mathbb{C}$ is continuous.

To prove the claim, it will be convenient to introduce the notation

$$\mathbb{F}_j = Z_j \times W_j \subset \mathbb{C}^2$$

Fix the set $\mathbb{F}_j$ and choose a point $a_0 \in \mathbb{F}_j$. For $\varepsilon > 0$ and $\xi \in \mathbb{C}$, we define

$$\Delta_j(\varepsilon, \xi) = \sup\{|H(a, \xi) - H(a_0, \xi)| : a \in \mathbb{F}_j \text{ and } |a - a_0| \leqslant \varepsilon\}$$

Since $\Delta_j(\varepsilon, \xi)$ is bounded by $2|\xi|$ and is nondecreasing in ε, we have a limit

$$\Delta_j(\xi) = \lim_{\varepsilon \to 0} \Delta_j(\varepsilon, \xi)$$

We also have the inequality $\Delta_j(\varepsilon, \xi) \leqslant \Delta_m(\varepsilon, \xi)$ for $m \geqslant j$, so that

$$\Delta_j(\xi) \leqslant \Delta_m(\xi) \quad \text{whenever } j \leqslant m$$

From the continuity of $H : \mathbb{F}_m \times \Xi_m \to \mathbb{C}$, we have $\Delta_m(\xi) = 0$ for every $\xi \in \Xi_m$. Thus

$$\Delta_j(\xi) = 0 \quad \text{whenever } \xi \in \Xi = \bigcup_{m=1}^{\infty} \Xi_m$$

Next, consider an arbitrary $\zeta \in \mathbb{C}$. The uniform ellipticity (7.44) gives the inequality $\Delta_j(\varepsilon, \zeta) \leqslant \Delta_j(\varepsilon, \xi) + 2k|\xi - \zeta|$ for arbitrary $\xi \in \mathbb{C}$. Taking $\xi \in \Xi$ and passing to the limit as $\varepsilon \to 0$, we find

$$\Delta_j(\zeta) \leqslant 2|\xi - \zeta| \quad \text{for every } \xi \in \Xi$$

As the set Ξ has full measure in $\mathbb{C}$, we can choose $\xi \in \Xi$ to make $|\xi - \zeta|$ as small as we wish. Hence

$$\Delta_j(\zeta) = 0 \quad \text{for every } \zeta \in \mathbb{C}$$

Now fix a point $(a_0, \xi_0) \in \mathbb{F}_j \times \mathbb{C}$ and let the points $(a, \xi) \in \mathbb{F}_j \times \mathbb{C}$ approach (a_0, ξ_0). Again, by the uniform ellipticity (7.44), we find that

$$|H(a,\xi) - H(a_0,\xi_0)| \leqslant |H(a,\xi_0) - H(a_0,\xi_0)| + k|\xi - \xi_0|$$

Passing to the limit as $(a, \xi) \to (a_0, \xi_0)$ with $a \in \mathbb{F}_j$, we see that $H(a, \xi) \to H(a_0, \xi_0)$. That is, $H : \mathbb{F}_j \times \mathbb{C} \to \mathbb{C}$ is continuous at (a_0, ξ_0), and so the above claim is established.

To complete the proof of the theorem we repeat, for the convenience of the reader, the argument applied in Lemma 7.6.2. With Tietze's theorem we extend $H : \mathbb{F}_j \times \mathbb{C} \to \mathbb{C}$ to $H_j : \mathbb{C}^2 \times \mathbb{C} \to \mathbb{C}$. Moreover, the set

$$E = \left(\bigcup_{j=1}^{\infty} Z_j\right) \cap \left(\bigcup_{k=1}^{\infty} f^{-1}(W_k)\right) \cap \left(\bigcup_{l=1}^{\infty} \varphi^{-1}(\mathbb{C})\right) \subset \Omega$$

has full measure in Ω since by assumption f satisfies the condition $\mathcal{N}^{-1}$. Note that no such condition needs to be required from φ since in the last factor we have the entire space $\Omega = \varphi^{-1}(\mathbb{C})$. It is precisely for this purpose that we established the above claim.

Now it suffices to note that, for any given $z \in E$, we have

$$H_m(z, f(z), \varphi(z)) = H(z, f(z), \varphi(z))$$

as soon as m is large enough. Hence $H_m(z, f(z), \varphi(z)) \to H(z, f(z), \varphi(z))$ pointwise. As each of the functions $H_m(z, f(z), \varphi(z))$ is measurable, we have proved the theorem. □

One of the most profound features of elliptic equations in the plane is that in spite of nonlinearity, the difference between two solutions to the same elliptic equation satisfies some other elliptic equation with the same ellipticity constants. This fact will prove useful in obtaining the a priori bounds necessary to establish existence and uniqueness, as well as to establish the regularity theory of solutions.

While it might seem obvious, once we solve the general nonlinear equation (7.33) we obtain the usual sorts of results for the linear case as corollaries. In fact, our approach takes a different route. We have already covered the linear case in the previous Chapter 5, and our strategy will be to apply these results as a priori bounds in the nonlinear setting: Every solution to a nonlinear elliptic equation as in (7.43) satisfies a linear Beltrami equation $f_{\bar{z}} = \mu f_z$, where $|\mu(z)| \leqslant k < 1$, since

$$\left|\frac{\partial f}{\partial \bar{z}}\right| = \left|H\left(z, f, \frac{\partial f}{\partial z}\right) - H(z, f, 0)\right| \leqslant k\left|\frac{\partial f}{\partial z}\right|$$

Therefore once we have existence, proved through general methods such as Schauder fixed-point theory, we automatically obtain regularity. These results, covered in the next few chapters, will culminate in a fairly general Riemann mapping theorem for the solutions to nonlinear elliptic equations.

7.8 Second-Order Elliptic Systems

We close this chapter with a brief discussion of the systems of two real equations of second-order. Such systems can always be written in the form of one complex equation for an unknown function $f = u + iv$. They take the form

$$\mathcal{L}f = \alpha \frac{\partial^2 f}{\partial^2 z} + \beta \frac{\partial^2 f}{\partial^2 \bar{z}} + \gamma \overline{\frac{\partial^2 f}{\partial^2 z}} + \delta \overline{\frac{\partial^2 f}{\partial^2 \bar{z}}} + \mu \frac{\partial^2 f}{\partial z \partial \bar{z}} + \nu \overline{\frac{\partial^2 f}{\partial z \partial \bar{z}}} \qquad (7.45)$$

Thus such systems are parameterized by points

$$(\alpha, \beta, \gamma, \delta, \mu, \nu) \in \mathbb{C}^6$$

The homotopy classification of operators determines an open set $\mathcal{E} \subset \mathbb{C}^6$, the set of parameters for which this operator is elliptic. This set, as was shown by Bojarski [68] (see also the references therein), consists of precisely six components, each represented by exactly one of the operators

$$\left.\begin{array}{cc} \dfrac{\partial^2}{\partial z \, \partial \bar{z}} & \overline{\dfrac{\partial^2}{\partial z \, \partial \bar{z}}} \\[2pt] \dfrac{\partial^2}{\partial z \, \partial z} & \overline{\dfrac{\partial^2}{\partial z \, \partial z}} \\[2pt] \dfrac{\partial^2}{\partial \bar{z} \, \partial \bar{z}} & \overline{\dfrac{\partial^2}{\partial \bar{z} \, \partial \bar{z}}} \end{array}\right\} \qquad (7.46)$$

The natural boundary value problem for second-order systems is that of prescribing the solution on the boundary of a domain, the Dirichlet problem. Following the theory of the Laplace equation

$$\Delta u = u_{xx} + u_{yy} = 0$$

for a scalar function $u : \Omega \to \mathbb{R}$, one might expect that the Dirichlet problem for elliptic systems is also well posed. Surprisingly, this is not always the case. For example, the system

$$f_{\bar{z}\bar{z}} = 0$$

on the unit disk $\mathbb{D}$ admits infinitely many solutions that vanish on the boundary circle, namely, all those of the form $f(z) = (1 - |z|^2)h(z)$, where h is any holomorphic function defined on $\mathbb{D}$ and continuous on $\overline{\mathbb{D}}$. Similar examples exist for the equation $f_{zz} = 0$. The last four operators in the homotopy classification in (7.46) are not strongly elliptic. Therefore from the point of view of ellipticity theory, the only reasonable elliptic second-order systems are those deformable to the complex Laplacian $\partial \bar{\partial}$ or its complex conjugate. Let us denote the component containing the Laplacian by $\mathcal{E}_L$ and call it the Laplacian component.

Despite many obvious analogies between the real and complex Laplacian operators, there are also many differences. The first obvious difference is that the real operators $-\Delta$ and Δ cannot be continuously deformed through elliptic operators from one to the other. They lie in different components. However, the homotopy

$$\mathcal{L}_t = e^{i\pi t} \frac{\partial^2}{\partial z \partial \bar{z}}$$

connects the two operators $\frac{\partial^2}{\partial z \partial \bar{z}}$ and $-\frac{\partial^2}{\partial z \partial \bar{z}}$ within the class of elliptic operators.

The theory of quasiconformal mappings again turns out to be particularly useful in the study of elliptic equations in the Laplacian component. Again, in [68] it is shown that operators in the component $\mathcal{E}_L$ can each be written in the form

$$\mathcal{L}f = \frac{\partial^2 f}{\partial z \partial \bar{z}} + \alpha \frac{\partial^2 f}{\partial^2 z} + \beta \frac{\partial^2 f}{\partial^2 \bar{z}} + \gamma \overline{\frac{\partial^2 f}{\partial^2 z}} + \delta \overline{\frac{\partial^2 f}{\partial^2 \bar{z}}}, \tag{7.47}$$

and ellipticity is characterized by the condition

$$|1 + \alpha \zeta + \beta \bar{\zeta}| > |\gamma \bar{\zeta} + \delta \zeta| \tag{7.48}$$

for all complex numbers ζ with $|\zeta| = 1$, together with the requirement that the point $(\alpha, \beta, \gamma, \delta) \in \mathbb{C}^4$ must belong to the component in $\mathbb{C}^4$ determined by the inequality (7.48) and containing $(0, 0, 0, 0)$.

Of course, the inequality

$$|\alpha| + |\beta| + |\gamma| + |\delta| \leqslant k < 1 \tag{7.49}$$

implies (7.48), but naturally the converse need not hold, and there are many operators in $\mathcal{E}_L$ of considerable interest both theoretical and practical that do not satisfy the inequality in (7.49). In particular, there are operators which are linear over the complex numbers, that is, $\mathcal{L}f = f_{z\bar{z}} + \alpha f_{zz} + \beta f_{\bar{z}\bar{z}}$.

Now, by (7.48) ellipticity means that the function $\zeta \mapsto 1 + \alpha \zeta + \beta \bar{\zeta}$ does not vanish on the unit circle. Equivalently, this says that the quadratic equation

$$\alpha \zeta^2 + \zeta + \beta = 0$$

has no solution on the circle. As this property must hold for all operators in the component $\mathcal{E}_L$, we may put $\alpha = \beta = 0$ to see that one root is always inside the circle. Then consider the operators

$$\frac{\partial^2}{\partial z \, \partial \bar{z}} + \varepsilon \frac{\partial^2}{\partial z \, \partial z} + \varepsilon \frac{\partial^2}{\partial \bar{z} \, \partial \bar{z}}$$

As $\varepsilon \to 0$, we see that there is always a root of modulus exceeding 1, that is, outside the circle. Hence we may assume that the roots are $-b$ and $-1/a$ with $a, b \in \mathbb{D}$. In this case

$$\alpha = \frac{a}{1 + ab}, \qquad \beta = \frac{b}{1 + ab}$$

This gives us the ellipticity criterion of the following theorem.

Theorem 7.8.1. *The elliptic operators in the Laplacian component that are linear over the field of complex numbers take the form*

$$\mathcal{L} = (1 + ab) \frac{\partial^2}{\partial z \, \partial \bar{z}} + a \frac{\partial^2}{\partial z \, \partial z} + b \frac{\partial^2}{\partial \bar{z} \, \partial \bar{z}}, \tag{7.50}$$

where $a, b \in \mathbb{D}$.

7.8. SECOND-ORDER SYSTEMS

It is interesting to observe that such $\mathcal{L}$ of the form (7.50) can be written as the composition of two first-order Beltrami operators,

$$\mathcal{L} = \left(\frac{\partial}{\partial \bar{z}} + a\frac{\partial}{\partial z}\right) \circ \left(\frac{\partial}{\partial z} + b\frac{\partial}{\partial \bar{z}}\right) \qquad (7.51)$$

Note that these operators belong to different components of the homotopy classification of elliptic first-order operators; the first is deformable to $\frac{\partial}{\partial \bar{z}}$, while the second can be deformed to $\frac{\partial}{\partial z}$.

7.8.1 Measurable Coefficients

Concerning equations with measurable coefficients, we shall discuss only those of the form given in (7.50). We state the following definition.

Definition 7.8.2. *The operator*

$$\mathcal{L} = [1 + a(z)b(z)]\frac{\partial^2}{\partial z \, \partial \bar{z}} + a(z)\frac{\partial^2}{\partial z \, \partial z} + b(z)\frac{\partial^2}{\partial \bar{z} \, \partial \bar{z}} \qquad (7.52)$$

with measurable coefficients a and b is said to be uniformly elliptic *if there are numbers, called the* ellipticity constants, $0 \leqslant k_1, k_2 < 1$ *such that*

$$|a(z)| \leqslant k_1, \quad |b(z)| \leqslant k_2 \qquad \text{for almost every } z \in \Omega \qquad (7.53)$$

Despite this simple and rather elegant characterization of the operators $\mathcal{L}$ in the Laplacian component, the complete analogy between such operators with measurable coefficients and the constant coefficient Laplace operator remains to be discovered. For instance, the decomposition in (7.51) cannot be made if both the coefficients a and b are only assumed measurable. Also, consider the C^∞-smooth function

$$f(z) = z\left(|z|^2 - 1\right), \qquad z \in \mathbb{D}$$

This satisfies an equation $\mathcal{L}f = 0$ with $\mathcal{L}$ as in (7.52) and (7.53) if we choose for instance

$$a(z) = -\gamma\frac{z}{\bar{z}}, \quad b(z) = \frac{1-\gamma}{\gamma}\frac{\bar{z}}{z} \quad \text{and} \quad \frac{1}{2} < \gamma < 1$$

However, f vanishes on the boundary $\partial \mathbb{D}$. Thus the Dirichlet problem on $\mathbb{D}$ with 0 boundary values admits a nonzero solution.

An interesting subclass of the general uniformly elliptic second-order equations in the Laplace component $\mathcal{E}_L$ are found in the form

$$\mathcal{L}f = f_{z\bar{z}} + \alpha(z)f_{zz} + \beta f_{\bar{z}\bar{z}} + \gamma(z)\overline{f_{zz}} + \delta(z)\overline{f_{\bar{z}\bar{z}}} \qquad (7.54)$$

with the uniform ellipticity assumption

$$\| |\alpha| + |\gamma| \|_\infty + \| |\beta| + |\delta| \|_\infty \leqslant k < 1 \qquad (7.55)$$

Solutions to (7.55) can be regarded as solutions to the equation

$$\frac{\partial^2 f}{\partial z\, \partial \bar{z}} + \mu(z)\frac{\partial^2 f}{\partial^2 z} + \nu(z)\frac{\partial^2 f}{\partial^2 \bar{z}} = 0,$$

where we have

$$\mu(z) = \alpha(z) + \gamma(z)\frac{\overline{f_{zz}}}{f_{zz}} \tag{7.56}$$

$$\nu(z) = \eta(z) + \delta(z)\frac{\overline{f_{\bar{z}\bar{z}}}}{f_{\bar{z}\bar{z}}} \tag{7.57}$$

Thus

$$\|\mu\|_\infty + \|\nu\|_\infty \leqslant k < 1 \tag{7.58}$$

Again we shall see that the Beurling transform plays a key role in investigating these equations.

Chapter 8

Solving General Nonlinear First-Order Elliptic Systems

The purpose of this chapter is to establish general results concerning the existence, uniqueness and regularity of the solutions to the nonlinear elliptic equations

$$\mathcal{F}(z, u, \nabla u) = 0 \tag{8.1}$$

in two dimensions. As we discussed in Section 7.7, under the assumption of ellipticity, (8.1) reduces to the system

$$\frac{\partial f}{\partial \bar{z}} = H\left(z, f, \frac{\partial f}{\partial z}\right)$$

The existence theory for such systems can be established under surprisingly general conditions even in the nonlinear setting. It will become clear that, together with ellipticity, it is the Lusin measurability that is the key notion here.

We are primarily interested in global solutions, that is, solutions defined in the entire plane $\mathbb{C}$. There are many ways to normalize a solution of the linear equation. In the nonlinear case matters are somewhat more problematic, as we shall see. We choose therefore normalizations that have proven to be most effective in the linear case. We will initially be concerned with nonlinear systems that reduce to the $\bar{\partial}$-equation near infinity. That is, the solutions are holomorphic near ∞.

A solution $f \in W^{1,2}_{loc}(\mathbb{C})$ to

$$\frac{\partial f}{\partial \bar{z}} = H\left(z, f, \frac{\partial f}{\partial z}\right) \tag{8.2}$$

normalized by the condition

$$f(z) = z + a_1 z^{-1} + a_2 z^{-2} + \cdots$$

outside a compact set will be called a *principal solution*. Of course, this concept is relevant only when $H(z, w, \zeta)$ vanishes for large values of $|z|$.

235

If this does not happen, the notion of the principal solution is no longer adequate and solutions need to be normalized by their values at three given points, just as in the case of conformal mappings. A homeomorphic $W^{1,1}_{loc}(\mathbb{C})$-solution to (8.2) is called *normalized* if $f(0) = 0$ and $f(1) = 1$. The normalization $f(\infty) = \infty$ follows automatically since f is a quasiconformal mapping of the entire plane, see Theorem 3.6.3. It should be noted that the normalized solutions actually belong to $W^{1,2}_{loc}(\mathbb{C})$.

Remark 8.0.3. *As will be apparent later, it is of crucial importance that every principal or normalized solution to (8.2), if it exists at all, solves its own linear Beltrami equation, defined by*

$$\frac{\partial f}{\partial \bar{z}} = \mu(z)\frac{\partial f}{\partial z}, \qquad \mu(z) = \frac{H(z, f(z), f_z(z))}{f_z(z)}$$

Therefore all properties of such solutions from the linear theory, established in Chapter 5, immediately apply. These facts will be frequently used in what follows. For instance, because of the linear growth at ∞, principal solutions are automatically quasiconformal homeomorphisms.

8.1 Equations Without Principal Solutions

Before moving on to solve the nonlinear Beltrami equation, wherever possible, we give an example showing that even for apparently nice equations, there need not be a principal solution. The example also serves to further motivate the Lusin measurability condition we shall put on Beltrami coefficients in order that principal solutions exist.

Theorem 8.1.1. *For every $0 < k < 1$ there exists a measurable function $\mu : \mathbb{C} \times \mathbb{C} \to \mathbb{C}$ such that*

- $|\mu(z, w)| \leqslant k\chi_{\mathbb{D}}(z)$ *and*

- *the quasilinear uniformly elliptic Beltrami equation*

$$\frac{\partial f}{\partial \bar{z}} = \mu(z, f)\frac{\partial f}{\partial z} \qquad (8.3)$$

has no principal solution.

Proof. We define $\mu(z, w)$ by the rule

$$\mu(z, w) = \begin{cases} 0, & |z| > 1 \\ 0, & |z| \leqslant 1 \text{ and } |w - z| > k/2 \\ k, & |z| \leqslant 1 \text{ and } |w - z| \leqslant k/2 \end{cases} \qquad (8.4)$$

Suppose that $f \in W^{1,2}_{loc}(\mathbb{C})$ is a principal solution. Then as f is certainly a homeomorphism by Theorem 5.3.2, we see that the set

$$F = \{z \in \mathbb{C} : |z - f(z)| \leqslant k/2\}$$

8.2. EXISTENCE OF SOLUTIONS

is closed. Outside the set F the mapping f is holomorphic as $\mu(z, f(z)) \equiv 0$, $z \notin F$. Note that the inequality

$$|f(z) - z| \leqslant k/2 \tag{8.5}$$

remains valid in $\Omega = \mathbb{C} \setminus F$ as (8.5) holds on the boundary of the open set Ω, while at ∞ we have $|f(z) - z| \to 0$ by virtue of the fact that f is a principal solution. Thus the maximum principle applies to give us (8.5) throughout $\mathbb{C}$. This then shows that $\mu(z, f) = k\chi_{\mathbb{D}}$. According to Theorem 5.1.2, the only principal solution to the linear elliptic equation $f_{\bar{z}} = k\chi_{\mathbb{D}} f_z$ is

$$f(z) = \begin{cases} z + k\bar{z}, & |z| \leqslant 1 \\ z + k/z, & |z| > 1 \end{cases}$$

However, this solution fails to satisfy the criterion in (8.5) as $|f(1) - 1| = k > k/2$. $\square$

We point out that the equation in (8.3) with $\mu(z, w)$ as in (8.4) has many solutions of the form $g(z) = z + a$ as soon as $a > k$.

The reader may wish to examine the function $H(z, w, \zeta) = \mu(z, w)\zeta$, with $\mu(z, w)$ defined as at (8.4), to see that even if H is fairly simple, the condition of Lusin measurability is not satisfied.

8.2 Existence of Solutions

When one is looking for solutions to the general nonlinear elliptic systems

$$\frac{\partial f}{\partial \bar{z}} = H\left(z, f, \frac{\partial f}{\partial z}\right), \qquad z \in \mathbb{C} \tag{8.6}$$

there are necessarily some constraints to be placed on the function H that we now discuss. We write

$$H : \mathbb{C} \times \mathbb{C} \times \mathbb{C} \to \mathbb{C},$$

and following the discussion in Section 7.7, we make these assumptions on H:

1. The homogeneity condition, that $f_{\bar{z}} = 0$ whenever $f_z = 0$, equivalently,

$$H(z, w, 0) \equiv 0, \quad \text{for almost every } (z, w) \in \mathbb{C} \times \mathbb{C} \tag{8.7}$$

2. The uniform ellipticity condition, that for almost every $z, w \in \mathbb{C}$ and all $\zeta, \xi \in \mathbb{C}$,

$$|H(z, w, \zeta) - H(z, w, \xi)| \leqslant k|\zeta - \xi|, \qquad 0 \leqslant k < 1 \tag{8.8}$$

3. H is Lusin-measurable: Thus

- There are compact $Z_1 \subset Z_2 \subset \cdots \subset \mathbb{C}$ whose union has full measure.
- There are compact $W_1 \subset W_2 \subset \cdots \subset \mathbb{C}$ whose union has full measure.
- There are compact $\Xi_1 \subset \Xi_2 \subset \cdots \subset \mathbb{C}$ whose union has full measure.
- For each $j = 1, 2, \ldots$, the map $H : Z_j \times W_j \times \Xi_j \to \mathbb{C}$ is continuous.

Let us make a few comments regarding the condition of Lusin measurablility in assumption 3 above. In what follows we shall often use Theorem 7.7.2 without explicitly mentioning it, as it ensures that the function $z \mapsto H(z, f(z), f_z(z))$ on the right hand side of our equation is measurable, at least when f is quasiregular.

Lusin measurability is a considerably weaker assumption than the classical Carathéodory condition. The latter requires $H(z, w, \xi)$ to be continuous in w for almost every z. Such a lack of symmetry in the hypotheses, concerning the regularity in z and w, would cause some difficulties in studying the equations for the inverse mapping. Lusin measurability seems to embrace all our needs. To illustrate, the elliptic system studied later in (16.69) for the complex gradient of a p-harmonic function fails to satisfy the Carathéodory condition though it enjoys the Lusin measurability property.

We have already discussed this earlier in (7.31), but it is worth repeating that by Egoroff's theorem, any measurable function of one variable is Lusin-measurable. Of course, continuity and similar conditions guarantee Lusin measurability. Further, we already saw in the previous section an example where no principal solution needs to exist when Lusin measurability fails.

Now we consider the main result and goal of this chapter.

Theorem 8.2.1. *Under the hypotheses 1–3 above, the equation*

$$\frac{\partial f}{\partial \bar{z}} = H\left(z, f, \frac{\partial f}{\partial z}\right) \tag{8.9}$$

admits a normalized solution. If, in addition, $H(z, w, \xi)$ is compactly supported in the z-variable, then the equation admits a principal solution.

The proof of Theorem 8.2.1 will be presented in four steps, and each of them will take a while to establish.

Under the condition of uniform ellipticity, we see that if f is any solution, then $|f_{\bar{z}}| = |H(z, f, f_z) - H(z, f, 0)| \leqslant k|f_z|$. This means that there is a function $\mu = \mu(z)$ such that

$$\frac{\partial f}{\partial \bar{z}} = \mu(z) \frac{\partial f}{\partial z}, \qquad |\mu(z)| \leqslant k < 1$$

In other words, f satisfies its own linear Beltrami equation and hence enjoys all the properties developed in the linear theory.

8.3 Proof of Theorem 8.2.1

As we have said, the proof of the theorem is divided into four subresults. In the first we prove the existence of solutions, assuming regularity in the equation but within a setup flexible and powerful enough to be applied in the later steps. These, on the other hand, will then develop the approximation methods necessary for finding the solution in more general situations.

8.3.1 Step 1: H Continuous, Supported on an Annulus

The normalized and the principal solution both have their own independent virtues. Therefore we have decided to give, in a uniform manner, the proof of existence for both cases. In fact, it is only in the present first step, where $H = H(z, w, \xi)$ is continuous, that the argument differs for these two cases.

Besides continuity we require that H be supported on the annulus

$$\mathbb{A} = \mathbb{A}_R = \{z \in \mathbb{C} : \frac{1}{R} < |z| < R\}$$

Theorem 8.3.1. *Suppose that $H : \mathbb{C} \times \mathbb{C} \times \mathbb{C} \to \mathbb{C}$ is continuous and that for some $R > 1$, $H(z, w, \xi) = 0$ whenever $z \notin \mathbb{A}_R$. Then, under the assumptions 1–3 of Theorem 8.2.1, the equation*

$$\frac{\partial f}{\partial \bar{z}} = H\left(z, f, \frac{\partial f}{\partial z}\right)$$

admits both a normalized and a principal solution.

Proof. We look for a solution f in the form

$$f(z) = ze^{\mathcal{A}\phi} + \mathcal{B}\phi, \qquad \phi \in L^p(\mathbb{C}), \tag{8.10}$$

where the exponent $p > 2$. Here the definition of the operators $\mathcal{A}, \mathcal{B}$ depends on the specific normalization chosen for f. To obtain a principal solution we set

$$\mathcal{A}\phi(z) = \mathcal{C}\phi(z) = \frac{1}{\pi}\int_\mathbb{C} \frac{\phi(\tau)\,d\tau}{z-\tau} \tag{8.11}$$

$$\mathcal{B}\phi(z) = -\frac{1}{\pi}\int_\mathbb{C} \phi(\tau)\,d\tau,$$

whereas for the normalized solution we define

$$\mathcal{A}\phi(z) = \mathcal{C}\phi(z) - \mathcal{C}\phi(1) = \frac{z}{\pi}\int_\mathbb{C} \frac{\phi(\tau)\,d\tau}{\tau(z-\tau)} \tag{8.12}$$

$$\mathcal{B}\phi(z) = 0 \tag{8.13}$$

Next, we choose and fix for the rest of this proof an exponent $p > 2$ such that

$$k\mathbf{S}_p < 1 \tag{8.14}$$

where k is identified as the ellipticity (or Lipschitz) constant in (8.8). In particular, p lies on the critical interval (q_k, p_k); see Definition 5.4.1.

Note that
$$f_{\bar{z}} = z\,\phi\,e^{\mathcal{A}\phi}, \qquad f_z = e^{\mathcal{A}\phi}\bigl(1 + z\,\mathcal{S}\phi\bigr)$$

The density function ϕ will be supported on the annulus $\mathbb{A}$. Concerning the normalization, we have for every ϕ,
$$f(z) = z + a_1 z^{-1} + a_2 z^{-2} + \cdots$$
for the principal solution and $f(0) = 0$, $f(1) = 1$ for the normalized solution. Thus we need to solve only the following integral equation:
$$\phi = \frac{1}{ze^{\mathcal{A}\phi}} H\bigl(z, ze^{\mathcal{A}\phi} + \mathcal{B}\phi, e^{\mathcal{A}\phi} + ze^{\mathcal{A}\phi}\mathcal{S}\phi\bigr) \tag{8.15}$$

To solve this equation we first associate with every given $\phi \in L^p(\mathbb{C})$ an operator $\mathbf{R} : L^p(\mathbb{C}) \to L^p(\mathbb{C})$ defined by
$$\mathbf{R}\Phi = \frac{1}{ze^{\mathcal{A}\phi}} H\bigl(z, ze^{\mathcal{A}\phi} + \mathcal{B}\phi, e^{\mathcal{A}\phi} + ze^{\mathcal{A}\phi}\mathcal{S}\Phi\bigr)$$

Through the ellipticity hypothesis we observe that $\mathbf{R}$ is a contractive operator on $L^p(\mathbb{C})$. Indeed, from (8.8) we have the pointwise inequality
$$|\mathbf{R}\Phi_1 - \mathbf{R}\Phi_2| \leqslant k\,|\mathcal{S}\Phi_1 - \mathcal{S}\Phi_2|$$
Hence
$$\|\mathbf{R}\Phi_1 - \mathbf{R}\Phi_2\|_p \leqslant k\,\mathbf{S}_p\|\Phi_1 - \Phi_2\|_p, \qquad k\mathbf{S}_p < 1$$
By the Banach contraction principle, $\mathbf{R}$ has a unique fixed point $\Phi \in L^p(\mathbb{C})$. In other words, with each $\phi \in L^p(\mathbb{C})$ we can associate a unique function $\Phi \in L^p(\mathbb{C})$ such that
$$\Phi = \frac{1}{ze^{\mathcal{A}\phi}} H\bigl(z, ze^{\mathcal{A}\phi} + \mathcal{B}\phi, e^{\mathcal{A}\phi} + ze^{\mathcal{A}\phi}\mathcal{S}\Phi\bigr) \tag{8.16}$$

In fact, the procedure (8.16), $\phi \mapsto \Phi$, gives a well-defined and nonlinear operator $\mathbf{T} : L^p(\mathbb{C}) \to L^p(\mathbb{C})$ by simply requiring that $\mathbf{T}\phi = \Phi$. Further, solving the original integral equation (8.15) means precisely that we have to find a fixed point for the operator $\mathbf{T}$. This, however, is more involved than in the case of the contraction $\mathbf{R}$. We will need to invoke the celebrated Schauder fixed-point theorem [320], which we next recall.

Theorem 8.3.2. *Let $\mathcal{D}$ be a closed convex subset of a Banach space $\mathbb{X}$ and $T : \mathcal{D} \to \mathcal{D}$ a continuous (possibly nonlinear) operator. If the image $T(\mathcal{D})$ has compact closure, then T has a fixed point.*

Of course, we now have to check how we can apply this theorem. We first need uniform L^p-bounds for $\mathbf{T}\phi$. For this observe from (8.16) that
$$\|\Phi\|_p \leqslant k\left\|\frac{1}{z} + \mathcal{S}\Phi\right\|_{L^p(\mathbb{A})} \leqslant k\left\|\frac{1}{z}\right\|_{L^p(\mathbb{A})} + k\,\mathbf{S}_p\|\Phi\|_p$$

8.3. EXISTENCE OF SOLUTIONS

Therefore
$$\|\mathbf{T}\phi\|_p = \|\Phi\|_p \leqslant \frac{k}{1-k\,\mathbf{S}_p}\left\|\frac{1}{z}\right\|_{L^p(\mathbb{A})} =: C_p(k,\mathbb{A})$$

Moreover, $\operatorname{supp}(\mathbf{T}\phi) \subset \mathbb{A}$.

These properties suggest that we should seek a fixed point for $\mathbf{T}$ from the closed convex set
$$\mathcal{D} = \{\phi \in L^p(\mathbb{C}) : \operatorname{supp}(\phi) \subset \mathbb{A} \text{ and } \|\phi\|_p \leqslant C_p(k,\mathbb{A})\}$$

Another important observation to make is that the operator $\mathbf{T} : \mathcal{D} \to \mathcal{D}$ is continuous and compact. Indeed, consider a sequence $\{\phi_n\}$ of functions $\phi_n \in \mathcal{D}$ converging weakly in $L^p(\mathbb{C})$ to some function $\phi_0 \in \mathcal{D}$. We need to show that $\mathbf{T}\phi_n \to \mathbf{T}\phi_0$ strongly in $L^p(\mathbb{C})$. The compactness of the Cauchy transform (see Theorem 4.3.14) shows that $\mathcal{A}\phi_n \to \mathcal{A}\phi_0$ uniformly on $\mathbb{C}$. Similarly, $\mathcal{B}\phi_n \to \mathcal{B}\phi_0$. Next, denote by $\Phi_0 = \mathbf{T}\phi_0$ the $L^p(\mathbb{C})$-solution to the equation
$$\Phi_0 = \frac{1}{ze^{\mathcal{A}\phi_0}} H\!\left(z, ze^{\mathcal{A}\phi_0} + \mathcal{B}\phi_0, e^{\mathcal{A}\phi_0} + ze^{\mathcal{A}\phi_0}\mathcal{S}\Phi_0\right),$$

whose existence and uniqueness was shown in (8.16). We claim that the functions $\Phi_n = \mathbf{T}\phi_n$ converge to Φ_0 in $L^p(\mathbb{C})$. Here we make use of the following pointwise estimates:
$$\begin{aligned}|\Phi_n - \Phi_0| &\leqslant \frac{1}{|ze^{\mathcal{A}\phi_n}|}\Big|H\!\left(z, ze^{\mathcal{A}\phi_n} + \mathcal{B}\phi_n, e^{\mathcal{A}\phi_n} + ze^{\mathcal{A}\phi_n}\mathcal{S}\Phi_n\right) \\ &\qquad - H\!\left(z, ze^{\mathcal{A}\phi_n} + \mathcal{B}\phi_n, e^{\mathcal{A}\phi_n} + ze^{\mathcal{A}\phi_n}\mathcal{S}\Phi_0\right)\Big| + \varepsilon_n(z) \\ &\leqslant k|\mathcal{S}\Phi_n - \mathcal{S}\Phi_0| + \varepsilon_n(z)\end{aligned}$$

Here
$$\begin{aligned}\varepsilon_n(z) &= \Big|\frac{1}{ze^{\mathcal{A}\phi_n}} H\!\left(z, ze^{\mathcal{A}\phi_n} + \mathcal{B}\phi_n, e^{\mathcal{A}\phi_n} + ze^{\mathcal{A}\phi_n}\mathcal{S}\Phi_0\right) \\ &\qquad - \frac{1}{ze^{\mathcal{A}\phi_0}} H\!\left(z, ze^{\mathcal{A}\phi_0} + \mathcal{B}\phi_0, e^{\mathcal{A}\phi_0} + ze^{\mathcal{A}\phi_0}\mathcal{S}\Phi_0\right)\Big|\end{aligned}$$

The above inequalities yield
$$\|\Phi_n - \Phi_0\|_p \leqslant \frac{1}{1 - k\,\mathbf{S}_p}\|\varepsilon_n\|_p$$

We need to show only that $\|\varepsilon_n\|_p \to 0$. First, notice that the ε_n have a L^p-majorant independent of n,
$$\varepsilon_n(z) \leqslant 2k\left|\frac{1}{z} + \mathcal{S}\Phi_0\right|\chi_\mathbb{A} \in L^p(\mathbb{C})$$

Moreover, from the convergence of $\mathcal{A}\phi_n$, $\mathcal{B}\phi_n$ we see that $\varepsilon_n(z) \to 0$ almost everywhere. It is precisely at this point that we have appealed to the continuity of the function H, which was one of the assumptions in step 1.

By Lebesgue's Dominated convergence theorem, we conclude that $\|\varepsilon_n\|_p \to 0$, establishing the L^p-convergence of $\{\Phi_n\}$ to Φ_0. This means that $\mathbf{T}\phi_n \to \mathbf{T}\phi_0$ in $L^p(\mathbb{C})$, proving that $\mathbf{T} : \mathcal{D} \to \mathcal{D}$ is both continuous and compact.

Now, summarizing, the Schauder fixed-point theorem tells us that the operator $\mathbf{T}$ has a fixed point in $\mathcal{D} \subset L^p(\mathbb{C})$ and this provides us with our solutions.

We need make a few observations to complete the proof of step 1. First, as noted in Remark 8.0.3, the principal solution we have found will be K-quasiconformal, $K = \frac{1+k}{1-k}$. Similarly, the normalized solution will be K-quasiconformal as it is K-quasiregular, has compactly supported Beltrami coefficient and has linear growth near ∞ by (8.12). This completes step 1. □

In order to free ourselves from the extra conditions on H that we utilized in step 1 we shall need appropriate approximation of H by continuous functions. This rather special approximation is the goal of our next step.

8.3.2 Step 2: Good Smoothing of H

Starting with the Lusin measurability of H, we have the increasing sequences of sets $\{Z_j\}, \{W_j\}$ and $\{\Xi_j\}$, each with union of full measure in $\mathbb{C}$, such that $H : Z_j \times W_j \times \Xi_j \to \mathbb{C}$ is continuous. Recall then from the proof of Theorem 7.7.2 that under assumptions 1–3 we can actually assume that $\Xi_j = \mathbb{C}$ for every index j. We also recall the notation

$$\mathbb{F}_j := Z_j \times W_j \subset \mathbb{C}^2$$

introduced in Theorem 7.7.2.

What we want to do now is to restrict H to the sets $\mathbb{F}_j \times \mathbb{C}$ and then extend this restriction continuously to the entire plane. This process will define continuous functions H_j,

$$H_j : \mathbb{C}^2 \times \mathbb{C} \to \mathbb{C}$$

Of course, the Tietze extension theorem would give us such an extension. However, the point here is that we would like H_j to inherit the global uniform ellipticity bounds of H in order to guarantee existence for the nonlinear equation associated with H_j. Clearly we will have $H_j(z, w, \zeta) \to H(z, w, \zeta)$ for each

$$(z, w, \zeta) \in \bigcup_\nu Z_\nu \times \bigcup_\nu W_\nu \times \mathbb{C},$$

in particular almost everywhere.

It is our next task to establish such a good extension of $H : \mathbb{F}_j \times \mathbb{C} \to \mathbb{C}$ to $\mathbb{C}^2 \times \mathbb{C}$. The approach we take is quite close to that of the Whitney extension theorem, as presented in [335, p. 171]. Thus we fix an index $j \in \mathbb{N}$ and consider the set $\mathbb{F} = \mathbb{F}_j = Z_j \times W_j \subset \mathbb{C}^2$.

We begin by recalling the celebrated Whitney decomposition of the open set $\mathbb{C}^2 \setminus \mathbb{F}$ into cubes,

$$\mathbb{C}^2 \setminus \mathbb{F} = \bigcup_{\nu=1}^\infty Q_\nu,$$

8.3. EXISTENCE OF SOLUTIONS

where Q_ν are nonoverlapping cubes each of size proportional to the distance of Q_ν to the set $\mathbb{F}$. Precisely, we have

$$\operatorname{diam}(Q_\nu) \leqslant \operatorname{dist}(Q_\nu, \mathbb{F}) \leqslant 4\operatorname{diam}(Q_\nu), \qquad \nu = 1, 2, \ldots \qquad (8.17)$$

The Whitney decomposition has another feature: If we enlarge Q_ν by the factor $6/5$ while keeping the center of Q_ν fixed, we obtain

$$\frac{6}{5} Q_\nu \subset \mathbb{C}^2 \setminus \mathbb{F}$$

There is an associated partition of unity given by functions $\eta_\nu \in C_0^\infty(\frac{6}{5} Q_\nu)$ so that

$$\sum_{\nu=1}^\infty \eta_\nu(a) \equiv 1, \qquad a \in \mathbb{C}^2 \setminus \mathbb{F} \qquad (8.18)$$

Finally, for each $\nu = 1, 2, \ldots$ we choose a point $a_\nu \in \mathbb{F}$ that is closest to Q_ν.

We now construct the extension of $H : \mathbb{F} \times \mathbb{C} \to \mathbb{C}$ to $\mathbb{C}^2 \times \mathbb{C}$ and denote it by $H_j : \mathbb{C}^2 \times \mathbb{C} \to \mathbb{C}$,

$$H_j(a, \zeta) = \begin{cases} \sum_{\nu=1}^\infty \eta_\nu(a) H(a_\nu, \zeta) & \text{for } a \in \mathbb{C}^2 \setminus \mathbb{F} \\ H(a, \zeta) & \text{for } a \in \mathbb{F} \end{cases}$$

That H_j is continuous is not too difficult to see; we refer the reader to [335, p. 173] for details. Clearly,

$$H_j(z, w, \zeta) \equiv 0, \quad \text{whenever } \zeta = 0$$

Also, the condition of uniform ellipticity remains valid, for if $a \in \mathbb{C}^2 \setminus \mathbb{F}$,

$$|H_j(a, \zeta_1) - H_j(a, \zeta_2)| \leqslant \sum_{\nu=1}^\infty \eta_\nu(a) |H(a_\nu, \zeta_1) - H(a_\nu, \zeta_2)|$$

$$\leqslant k \sum_{\nu=1}^\infty \eta_\nu(a) |\zeta_1 - \zeta_2| = k\, |\zeta_1 - \zeta_2|$$

The same holds for $a \in \mathbb{F}$, which gives us the estimate

$$|H_j(a, \zeta_1) - H_j(a, \zeta_2)| \leqslant k\, |\zeta_1 - \zeta_2|, \qquad a \in \mathbb{C}^2; \ \zeta_1, \zeta_2 \in \mathbb{C}$$

Remark 8.3.3. *If $H(z, w, \xi) = 0$ whenever $|z| \geqslant R$, we may assume that $H_j(z, w, \xi) = 0$ for $|z| \geqslant 2R$, and $j = 1, 2, \ldots$. To see this simply include the set $\{z : |z| \geqslant 2R\}$ to each Z_j before smoothing H.*

8.3.3 Step 3: Lusin-Egoroff Convergence

Let H_ν be the good smoothing of H as constructed in step 2. Thus $H_\nu \equiv H$ on the sets $Z_j \times W_j \times \mathbb{C}$ for all $\nu \geqslant j$. In particular, $H_\nu \to H$ uniformly on each set $Z_j \times W_j \times \mathbb{C}$, $j = 1, 2, \ldots$. In view of this observation we adopt for Lusin-measurable functions a more general and very natural concept of convergence. We recall Egoroff's theorem, which tells us that functions converging almost everywhere actually converge uniformly on suitable compact subsets. Now, if for a sequence of Lusin-measurable functions, these compact subsets can be chosen to be Cartesian products of closed sets in the coordinate spaces, then we say that the functions converge in the *Lusin-Egoroff sense*. It will be helpful for us to exploit such convergence in the following device.

Lemma 8.3.4. *Let $f_\nu : \mathbb{C} \to \mathbb{C}$ be K-quasiconformal mappings converging uniformly on compact subsets to a quasiconformal mapping $f : \mathbb{C} \to \mathbb{C}$. Then for every bounded domain $\Omega \subset \mathbb{C}$ and for every fixed number $T < \infty$, we have*

$$\sup_{\|\psi\|_p \leqslant T} \|H_\nu(z, f_\nu, \psi) - H(z, f, \psi)\|_{L^2(\Omega)} \to 0$$

as $\nu \to \infty$.

Proof. We first need a few preliminaries: By Hölder's inequality,

$$\int_X |\psi|^2 \leqslant \|\psi\|_p^2 |X|^{1-\frac{2}{p}} \leqslant T^2 |X|^{1-\frac{2}{p}} \qquad (8.19)$$

for every measurable set $X \subset \Omega$. Furthermore, consider the sets

$$\mathbb{U} = \bigcup_{\nu=1}^\infty f_\nu(\Omega)$$

This is certainly an open bounded set in $\mathbb{C}$. Since the inverse quasiconformal mappings f_ν^{-1} converge to f^{-1} uniformly on compact subsets, there is a constant $C > 0$ such that

$$f_\nu^{-1}(Y) \leqslant C|Y|^{1/K}$$

for every measurable subset $Y \subset \mathbb{U}$ and every $\nu = 1, 2, \ldots$. Here we may apply the area distortion inequality (13.22) for the explicit exponent should we wish to do so.

Given $\varepsilon > 0$, our goal is now to show that there is a natural number $N = N(\varepsilon)$ such that

$$\|H_\nu(z, f_\nu, \psi) - H(z, f, \psi)\|_{L^2(\Omega)} \leqslant 3\varepsilon \qquad (8.20)$$

whenever $\|\psi\|_p \leqslant T$ and $\nu \geqslant N$.

Recall that the function $H_j = H_j(z, w, \xi)$ coincides with $H = H(z, w, \xi)$ on the set $Z_j \times W_j \times \mathbb{C}$, where $Z_1 \subset Z_2 \subset \cdots \subset \mathbb{C}$ and $W_1 \subset W_2 \subset \cdots \subset \mathbb{C}$ have union of full measure in $\mathbb{C}$. Now, given $\nu = 1, 2, \ldots$, $j = 1, 2, \ldots$ and $\psi \in \mathcal{F}$, we consider the sets

$$E_{j,\nu}^\psi = \{z \in \Omega : z \in Z_j, \ f(z), \ f_\nu(z) \in W_j, \ |\psi(z)| \leqslant j\}$$

8.3. EXISTENCE OF SOLUTIONS

We aim to show that if j is sufficiently large, these sets have almost full measure in Ω, independently of $\nu = 1, 2, \ldots$ and the specific ψ with $\|\psi\|_p \leqslant T$. To see this we consider the complement of $E_{j,\nu}^{\psi}$ in Ω,

$$\Omega \setminus E_{j,\nu}^{\psi} = [\Omega \setminus Z_j] \cup [\Omega \setminus f^{-1}(W_j)] \cup [\Omega \setminus f_\nu^{-1}(W_j)] \cup \{z \in \Omega : |\psi(z)| > j\}$$

Thus

$$\left|\Omega \setminus E_{j,\nu}^{\psi}\right| \leqslant |\Omega \setminus Z_j| + |\Omega \setminus f^{-1}(W_j)| + |\Omega \setminus f_\nu^{-1}(W_j)| + |\{z \in \Omega : |\psi(z)| > j\}|$$

Then we make the following estimates,

$$\begin{aligned}
|\Omega \setminus f^{-1}(W_j)| &\leqslant |f^{-1}(\mathbb{U} \setminus W_j)| \leqslant C|\mathbb{U} \setminus W_j|^{1/K} \\
|\Omega \setminus f_\nu^{-1}(W_j)| &\leqslant C|\mathbb{U} \setminus W_j|^{1/K} \\
|\{z \in \Omega : |\psi(z)| > j\}| &\leqslant \frac{1}{j}\int_\Omega |\psi|^2 \leqslant \frac{T^2}{j}
\end{aligned}$$

Thus, choosing j sufficiently large, the sets $\Omega \setminus E_{j,\nu}^{\psi}$ have measure as small as we wish. In particular, with the aid of (8.19), we find a number j_0 such that

$$\|\psi\|_{L^2(\Omega \setminus E_{j_0,\nu}^{\psi})} \leqslant \varepsilon, \quad \text{whenever } \|\psi\|_p \leqslant T \tag{8.21}$$

For the rest of our consideration we fix this number j_0. Of course, j_0 depends on ε and T but not on $\nu = 1, 2, \ldots$ or the function ψ.

We are now ready to determine the integer N for which the inequality (8.20) holds for all $\nu \geqslant N$. To this aim we recall that the function $H : Z_{j_0} \times W_{j_0} \times \mathbb{D}_{j_0} \to \mathbb{C}$ is uniformly continuous, where $\mathbb{D}_{j_0} = \{z : |z| \leqslant j_0\}$. This means that there exists $\delta > 0$ such that

$$|H(z, w_1, \xi) - H(z, w_2, \xi)| \leqslant \varepsilon |\Omega|^{-1/2} \tag{8.22}$$

provided $z \in Z_{j_0}, w_1, w_2 \in W_{j_0}, |\xi| \leqslant j_0$ and $|w_1 - w_2| \leqslant \delta$.

We now choose $N \geqslant j_0$ so large that

$$\sup_\Omega |f_\nu - f| \leqslant \delta, \quad \nu \geqslant N$$

As a final step toward the proof of the inequality (8.20) we invoke the uniform ellipticity bounds to get the very crude pointwise estimate

$$|H_\nu(z, f_v, \psi) - H(z, f, \psi)| \leqslant 2|\psi|$$

We then split the L^2-norm at (8.20) as follows:

$$\begin{aligned}
\|H_\nu&(z, f_v, \psi) - H(z, f, \psi)\|_{L^2(\Omega)} \\
&\leqslant 2\|\psi\|_{L^2(\Omega \setminus E_{j_0,\nu}^{\psi})} + \|H_\nu(z, f_v, \psi) - H(z, f, \psi)\|_{L^2(E_{j_0,\nu}^{\psi})} \\
&\leqslant 2\varepsilon + \|H(z, f_v, \psi) - H(z, f, \psi)\|_{L^2(E_{j_0,\nu}^{\psi})} \\
&\leqslant 2\varepsilon + \varepsilon|\Omega|^{-1/2}\|\chi_\Omega\|_2 = 3\varepsilon
\end{aligned}$$

This is justified because, for $z \in E^{\psi}_{j_0,\nu} \subset Z_{j_0} \subset Z_\nu$, we have $f(z), f_\nu(z) \in W_{j_0} \subset W_\nu$, $|\psi(z)| \leqslant j_0$ and $|f_\nu(z) - f(z)| < \delta$, as required for (8.22) to hold. □

The stage is now set for the final step in the proof of Theorem 8.2.1.

8.3.4 Step 4: Passing to the Limit

For each integer $m \geqslant 2$ we consider the annulus
$$\mathbb{A}_m = \{z : \frac{1}{m} < |z| < m\}$$
and a function $\eta_m \in C_0^\infty(\mathbb{A}_{2m})$ such that $0 \leqslant \eta_m(z) \leqslant 1$ and $\eta_m(z) \equiv 1$ for $z \in \mathbb{A}_m$.

We then solve the equation
$$\frac{\partial f^m}{\partial \bar{z}} = \eta_m(z) H_m\left(z, f^m, \frac{\partial f^m}{\partial z}\right),$$
where $f^m : \mathbb{C} \to \mathbb{C}$ stands for either the principal solution or the normalized solution. Note that each f_m is $K = \frac{1+k}{1-k}$-quasiconformal. We shall continue to use the same notation in both cases since there is no difference in the proof. Note, though, that in the case of principal solutions the support of the function $z \mapsto \eta_m(z) H_m(z, f^m, f_z^m)$ is bounded independently of m; see Remark 8.3.3.

Normal family arguments (see Corollary 3.9.3) show that the sequence $\{f^m\}$ has a subsequence, converging locally uniformly to a mapping $f : \mathbb{C} \to \mathbb{C}$. Moreover, f is a principal or a normalized quasiconformal mapping, depending on the case being considered. Therefore we complete the proof of Theorem 8.2.1 once we show that f satisfies the original equation
$$\frac{\partial f}{\partial \bar{z}} = H\left(z, f, \frac{\partial f}{\partial z}\right)$$

For this purpose we need to establish the strong L^2-convergence of the derivatives of f^m on any annulus $\mathbb{A} = \mathbb{A}_R = \{z : \frac{1}{R} < |z| < R\}$. Thus we take $m, n > 2R$ and consider the equations
$$\frac{\partial f^m}{\partial \bar{z}} = H_m\left(z, f^m, \frac{\partial f^m}{\partial z}\right), \qquad z \in \mathbb{A}_{2R}$$
$$\frac{\partial f^n}{\partial \bar{z}} = H_n\left(z, f^n, \frac{\partial f^n}{\partial z}\right), \qquad z \in \mathbb{A}_{2R}$$

We subtract these equations to obtain a pointwise inequality
$$\begin{aligned}|f_{\bar{z}}^n - f_{\bar{z}}^m| &\leqslant |H_m(z, f^m, f_z^m) - H(z, f, f_z^m)| \\ &+ |H(z, f, f_z^m) - H(z, f, f_z^n)| \\ &+ |H(z, f, f_z^n) - H_n(z, f^n, f_z^n)| \end{aligned} \qquad (8.23)$$

8.3. EXISTENCE OF SOLUTIONS

Next, note the uniform bounds
$$\|f_{\bar{z}}^n\|_{L^p(\mathbb{A}_{2R})} + \|f_{\bar{z}}^m\|_{L^p(\mathbb{A}_{2R})} \leqslant C = C_R,$$

independent of m and n. These follow from either (5.23) or the explicit bounds in (13.27). Making use of the ellipticity condition for the middle term we can now write inequality (8.23) in the form

$$|f_{\bar{z}}^n - f_{\bar{z}}^m| \leqslant k|f_z^n - f_z^m| + \varepsilon^m(z) + \varepsilon^n(z), \tag{8.24}$$

where
$$\varepsilon^j(z) = \left|H_j(z, f^j, f_z^j) - H(z, f, f_z^j)\right|, \qquad j = n, m$$

Here Lemma 8.3.4 implies that
$$\|\varepsilon^m\|_{L^2(\mathbb{A}_{2R})} \leqslant \sup_{\|\psi\|_p \leqslant C_R} \|H_m(z, f^m, \psi) - H(z, f, \psi)\|_{L^2(\mathbb{A}_{2R})} \to 0$$

as $m \to \infty$. Similarly, $\|\varepsilon^n\|_{L^2(\mathbb{A}_{2R})} \to 0$ when $n \to \infty$.

At this point the Caccioppoli estimates are particularly useful. It is convenient to reformulate inequality (8.24) as

$$\left|(f^m - f^n)_{\bar{z}} - \mu(f^m - f^n)_z\right| \leqslant \varepsilon^m(z) + \varepsilon^n(z), \qquad |\mu(z)| \leqslant k < 1,$$

and then apply Theorem 5.4.3. Choosing a test function $\eta \in C_0^\infty(\mathbb{A}_{2R})$ with $\eta(z) \equiv 1$ on $\mathbb{A}$, we obtain

$$\|Df^n - Df^m\|_{L^2(\mathbb{A})} \leqslant C(k)\|\nabla \eta\|_2 \sup_{\mathbb{A}_{2R}} |f^n - f^m|$$
$$+ C(k)\|\eta\|_\infty \left(\|\varepsilon^m\|_{L^2(\mathbb{A}_{2R})} + \|\varepsilon^n\|_{L^2(\mathbb{A}_{2R})}\right)$$

As $f^m \to f$ uniformly on $\mathbb{A}_{2R}$, we therefore have

$$Df^m \to Df \quad \text{in } L^2(\mathbb{A})$$

We are now ready to complete the proof. For our last estimate we make use of Lemma 8.3.4 once more. Combining this with the ellipticity bounds gives us

$$\|f_{\bar{z}}^m - H(z, f, f_z)\|_{L^2(\mathbb{A})}$$
$$\leqslant \|H_m(z, f^m, f_z^m) - H(z, f, f_z^m)\|_{L^2(\mathbb{A})} + \|H(z, f, f_z^m) - H(z, f, f_z)\|_{L^2(\mathbb{A})}$$
$$\leqslant \sup_{\|\psi\|_p \leqslant C_R} \|H_m(z, f^m, \psi) - H(z, f, \psi)\|_{L^2(\mathbb{A}_{2R})} + k\|f_z^m - f_z\|_{L^2(\mathbb{A})} \to 0$$

as $m \to \infty$.

Thus
$$\frac{\partial f}{\partial \bar{z}} = H\!\left(z, f, \frac{\partial f}{\partial z}\right),$$

as desired. $\square$

8.4 Equations with Infinitely Many Principal Solutions

We now turn to themes related to the uniqueness of the Beltrami systems

$$\frac{\partial f}{\partial \bar{z}} = H\left(z, f, \frac{\partial f}{\partial z}\right)$$

Of course, uniqueness is not to be expected in complete generality, however it is a little surprising how spectacularly things can go wrong.

Theorem 8.4.1. *For each $0 < k < 1$, there is a measurable function $\mu : \mathbb{C} \times \mathbb{C} \to \mathbb{D}$ such that*

- $|\mu(z, w)| \leqslant k \chi_\mathbb{D}(z)$

- *The quasilinear uniformly elliptic Beltrami equation*

$$f_{\bar{z}} = \mu(z, f) f_z \qquad (8.25)$$

has a continuous family of, and in particular, infinitely many, principal solutions.

Proof. We define $\mu(z, w)$ by the rule

$$\mu(z, w) = \begin{cases} 0, & |z| > 1 \\ k, & |z| \leqslant 1 \text{ and } |w - z| > k|z| \\ |w - z|/|z|, & |z| \leqslant 1 \text{ and } |w - z| \leqslant k|z| \end{cases} \qquad (8.26)$$

Next, let λ be any real number with $0 < \lambda < k$. Define

$$f^\lambda(z) = \begin{cases} z + \lambda \bar{z}, & |z| \leqslant 1 \\ z + \lambda/z, & |z| > 1 \end{cases}$$

Then certainly $f \in W^{1,2}_{loc}(\mathbb{C})$, and near ∞ we have the correct normalization for f to be a principal solution. We need only check that f is indeed a solution. But this is trivial. □

Notice too that $f_a(z) = a(z + k\bar{z})$ is a solution as soon as $|a - 1| \geqslant k$. It is possibly the case that the family f^λ defined above give all the principal solutions, while for $|a-1| \geqslant k$ is possibly the case that these solutions are unique with the normalization $f(z) = az + a_1 z^{-1} + \cdots$ near ∞. If so, this would be a quite surprising dichotomy.

Note that $\mu(z, w)$ is a C^∞-smooth function on $\Omega \times \mathbb{C}$, where

$$\Omega = \{z \in \mathbb{C} : z \neq 0, |z| \neq 1\},$$

and as such is not a very bad function, certainly Lusin-measurable. Notice that μ is not Lipschitz as a function of the second variable. We shall see in Section

8.5. LIOUVILLE THEOREMS

8.6 that the smoothings of $\mu(z,w)$, defined for $m = 1, 2, \ldots$ by

$$\mu_m(z,w) = \begin{cases} 0, & |z| > 1 \\ k, & 1/m \leqslant |z| \leqslant 1 \text{ and } |w-z| > k|z| \\ |w-z|/|z|, & 1/m \leqslant |z| \leqslant 1 \text{ and } |w-z| \leqslant k|z| \\ k \sin m & |z| < 1/m, \end{cases}$$

admit unique principal solutions. This is because we have smoothed away the bad singularity at the origin. Indeed, one can check that the unique principal solution is

$$f_m(z) = \begin{cases} z + (k \sin m)\bar{z}, & |z| \leqslant 1 \\ z + (k \sin m)/z, & |z| > 1 \end{cases} \quad (8.27)$$

8.5 Liouville Theorems

We have already seen generalizations of the classical Liouville theorem in Chapter 5. These were based on the Caccioppoli inequalities. Here we wish to find further growth and regularity criteria that are sufficient to imply a Liouville-type theorem. We will need these sorts of results for achieving uniqueness for solutions of nonlinear systems.

A little consideration of the example in the previous sections suggests that the local behavior at a single point, despite uniform ellipticity, can destroy uniqueness. In order to set up and state fairly precise results, we need to introduce the space $L^{2\pm}(\mathbb{C})$,

$$L^{2\pm}(\mathbb{C}) = \bigcup_{0 < \varepsilon < 1} L^{2-\varepsilon}(\mathbb{C}) \cap L^{2+\varepsilon}(\mathbb{C}) \quad (8.28)$$

The following theorem and its later L^p-generalizations will be a key component of many of our uniqueness results.

Theorem 8.5.1. *Suppose* $F \in W^{1,2}_{loc}(\mathbb{C})$ *satisfies the homogeneous distortion inequality*

$$|F_{\bar{z}}| \leqslant k|F_z| + \sigma(z)|F|, \quad (8.29)$$

where $\sigma \in L^{2\pm}(\mathbb{C})$ *and* $0 \leqslant k < 1$. *Then F is continuous.*

If, moreover, F is bounded, then $F(z) = C_1 e^{\theta(z)}$, *where* $\theta \in C_0(\hat{\mathbb{C}})$. *In particular, if*

$$\lim_{z \to \infty} F(z) = 0,$$

then $F \equiv 0$.

Proof. We express (8.29) in the form of a linear equation

$$F_{\bar{z}} = \mu(z)F_z + A(z)F, \quad (8.30)$$

where $|\mu(z)| \leqslant k < 1$ and $A \in L^{2\pm}(\mathbb{C})$. It is a fairly simple matter to see that we may choose a pair (p,q) of Hölder conjugate exponents within the critical interval, so $q_k < q$ and $p < p_k$, such that

$$A \in L^p(\mathbb{C}) \cap L^q(\mathbb{C})$$

We now solve the auxiliary equation
$$\theta_{\bar z} = \mu(z)\theta_z + A(z)$$
for θ as the Cauchy transform of a density function w,
$$\theta(z) = -\frac{1}{2\pi i}\int_{\mathbb{C}}\frac{w(\zeta)}{z-\zeta}d\zeta = (\mathcal{C}w)(z).$$

Therefore w satisfies the integral equation $w = \mu\mathcal{S}w + A$, and hence $w = (\mathbf{I} - \mu\mathcal{S})^{-1}A$. And so, in particular, $w \in L^p(\mathbb{C}) \cap L^q(\mathbb{C})$. By Theorem 4.3.11 we see that in fact θ is a continuous function vanishing at ∞, $\theta \in C_0(\hat{\mathbb{C}})$. If we write
$$F(z) = g(z)e^{\theta(z)},$$
we immediately see $g(z) = F(z)e^{-\theta(z)} \in W^{1,2}_{\mathrm{loc}}(\mathbb{C})$. We compute $g_{\bar z}$ and g_z to find that g satisfies the Beltrami equation
$$\frac{\partial g}{\partial \bar z} = \mu(z)\frac{\partial g}{\partial z}.$$

By Theorem 5.4.2 we find that g is a continuous function. Thus F is continuous.

Finally, if $F \in L^\infty(\mathbb{C})$, we see that g is bounded and therefore constant by the Liouville theorem. If $\lim_{z\to\infty} F(z) = 0$, this constant must of course be equal to 0. Hence $F \equiv 0$, as claimed. □

To examine the necessity of the various hypotheses of Theorem 8.5.1, we consider the following example.

Theorem 8.5.2. *Let $\varepsilon > 0$ and set*
$$\theta(z) = [\log(1+|z|^2)]^{\frac{1}{2}-\varepsilon}, \qquad F(z) = e^{-\theta(z)}.$$
Then $F_{\bar z} = A(z)F$, $A \in L^2(\mathbb{C})$ and $F(z) \to 0$ as $|z| \to \infty$.

Proof. We first compute that
$$\theta_{\bar z} = \frac{(1-2\varepsilon)z}{2(1+|z|^2)\log^{\varepsilon+1/2}(1+|z|^2)} \equiv A(z)$$

We next want to show that $A \in L^2(\mathbb{C})$. We therefore estimate
$$\int_{\mathbb{C}}|\theta_{\bar z}|^2 \leq C\int_0^\infty \left(\frac{r}{1+r^2}\right)^2 \frac{r\,dr}{\log^{2\varepsilon+1}(1+r^2)} < \infty.$$

Thus $\theta_{\bar z} \in L^2(\mathbb{C})$. As $F_{\bar z} = \theta_{\bar z}F$, the result is clear. □

Theorem 8.5.2 shows that the hypothesis $\sigma \in L^{2\pm}(\mathbb{C})$ cannot be removed from Theorem 8.5.1 without further modification. In particular, it is not enough that $\sigma \in L^2(\mathbb{C})$. However, when we later study the nonlinear $\bar\partial$-problem for an

8.5. LIOUVILLE THEOREMS

important application, the theory of holomorphic motions, we shall need to have at hand a result similar to Theorem 8.5.1 where σ is merely assumed to be in $L^2(\mathbb{C})$, as uniqueness in this setting will be a crucial ingredient. We next show that $\sigma \in L^2(\mathbb{C})$ will suffice to guarantee uniqueness provided we make stronger decay assumptions on either σ or F. In fact, we can also relax the assumptions on Sobolev regularity as long as we stay on the critical interval.

Theorem 8.5.3. *Suppose that $F \in W^{1,q}_{loc}(\mathbb{C})$ satisfies the distortion inequality*

$$|F_{\bar{z}}| \leq k|F_z| + o(z)|F|, \qquad 0 \leq k < 1, \tag{8.31}$$

where $\sigma \in L^2(\mathbb{C})$ and the Sobolev regularity exponent q lies in the critical interval $q_k < q < p_k$.

Then $F = e^\theta g$, where g is quasiregular and $\theta \in VMO$. If, moreover, one of the following additional hypotheses holds,

1. *σ has compact support and $\lim_{z \to \infty} F(z) = 0$, or*

2. *$F \in L^p(\mathbb{C})$ for some $p > 0$ and $\limsup_{z \to \infty} |F(z)| < \infty$,*

then $F \equiv 0$.

Proof. It is enough to consider the cases $q_k < q < 2$. Again, we write (8.31) in the form (8.30), however, this time $A \in L^2(\mathbb{C})$. We solve, as before, the auxilliary equation $\theta_{\bar{z}} = \mu(z)\theta_z + A$ by setting $\theta = \mathcal{C}\omega$, $\omega = (\mathbf{I} - \mu\mathcal{S})^{-1}A \in L^2(\mathbb{C})$, where $\mathcal{C}$ is the Cauchy transform defined in (4.42). The reader may wish to note at this point that if σ has compact support, then $\theta(z) \to 0$ as $z \to \infty$.

According to Theorem 4.3.9, we have $\theta \in VMO$. We may write θ as the sum

$$\theta = \theta^0 + \theta^1, \tag{8.32}$$

with the BMO-norm $\|\theta^0\|_*$ as small as we wish and $\theta^1 \in L^\infty(\mathbb{C})$, simply by splitting $\omega = \omega_0 + \omega_1$ with $\|\omega_0\|_2$ small and $\omega_1 \in C_0^\infty(\mathbb{C})$. The John-Nirenberg inequality (4.93) implies that for every $r > 0$ and disk $B \subset \mathbb{C}$,

$$\frac{1}{|B|}\int_B e^{r|\theta - \theta_B|} \leq 2e^{2r\|\theta^1\|_\infty} < \infty \tag{8.33}$$

when $\|\theta^0\|_*$ is chosen small enough.

We now seek to find for every $\varepsilon > 0$ a constant C_ε such that

$$|\theta_B| \leq C_\varepsilon + \varepsilon \log|B| \tag{8.34}$$

for every disk B of radius greater than 1. For this we recall the inequality for a general function $u \in BMO$,

$$|u_B - u_{2B}| \leq \frac{1}{|B|}\int_{2B}|u - u_{2B}| \leq 4\|u\|_*$$

Iterating this inequality a number of times, starting with the unit disk, gives

$$|u_B| \leq (C_0 + C_1 \log|B|)\|u\|_*,$$

and combining with the decomposition (8.32), where $\|\theta^0\|_* \leqslant \varepsilon/C_1$, we have the bound (8.34).

Let us then consider the function

$$g = e^{-\theta} F$$

and note that (8.33) gives us $e^{-\theta} \in \bigcap_{p<\infty} L^p_{loc}(\mathbb{C})$. Similarly, from Sobolev's inequality (Theorem A.6.1) we obtain

$$F \in L^{\frac{2q}{2-q}}_{loc}(\mathbb{C})$$

As $D\theta \in L^2(\mathbb{C})$, for all bounded domains Ω,

$$\int_\Omega |FD\theta|^q \leqslant \left(\int_\Omega |D\theta|^2\right)^{q/2} \left(\int_\Omega |F|^{\frac{2q}{2-q}}\right)^{1-q/2} < \infty$$

In particular, the derivatives $g_z = e^{-\theta}(F\theta_z + F_z)$ and $g_{\bar{z}} = e^{-\theta}(F\theta_{\bar{z}} + F_{\bar{z}})$ are locally integrable to any power less than q. Thus

$$g \in W^{1,q-\varepsilon}_{loc}(\mathbb{C}), \quad \text{for all } \varepsilon > 0 \tag{8.35}$$

Next, choose ε so small that $q - \varepsilon$ lies in the critical interval (q_k, p_k). As

$$g_{\bar{z}} - \mu g_z = (-AF + F_{\bar{z}} - \mu F_z)e^{-\theta} = 0 \tag{8.36}$$

and $\|\mu\|_\infty \leqslant k < 1$, we see that g is weakly quasiregular, and hence by Theorem 5.4.2 g is actually quasiregular. In particular, g is continuous.

Suppose now that the first additional hypothesis is satisfied; that is, σ has compact support and $F(z) \to 0$ as $|z| \to \infty$. In this case we already also know that $\theta(z) \to 0$ when $|z| \to \infty$. It follows that g is a bounded quasiregular mapping, thus a constant, and as g vanishes at ∞, we have $g \equiv 0$. Thus $F \equiv 0$.

For the case of the second additional hypothesis, we choose an exponent $2 < s < p_k$ such that $s < \frac{2q}{2-q}$, and define

$$t = \frac{2q}{s(2-q)} > 1 \quad \text{and} \quad r = \frac{2q}{(2-q)(t-1)}$$

By Hölder's inequality, for each disk $B \subset \mathbb{C}$,

$$\frac{1}{|B|}\int_B |g|^s \leqslant e^{s\theta_B}\left(\frac{1}{|B|}\int_B e^{r|\theta - \theta_B|}\right)^{1-1/t}\left(\frac{1}{|B|}\int_B |F|^{\frac{2q}{2-q}}\right)^{1/t} \tag{8.37}$$

As $F \in L^{\frac{2q}{2-q}}_{loc}(\mathbb{C})$ and $\limsup_{|z|\to\infty} |F(z)| < \infty$, it is not difficult to see that

$$\limsup_{R\to\infty} \frac{1}{|B|}\int_B |F|^{\frac{2q}{2-q}} < \infty, \quad B = B_R$$

8.6. UNIQUENESS

The middle term $\frac{1}{|B|}\int_B e^{r|\theta-\theta_B|}$ remains bounded by (8.33), while for the first term we apply (8.34) and estimate

$$e^{s\theta_B} \leqslant e^{sC_\varepsilon}|B|^{\varepsilon s} \leqslant C(s,\varepsilon)R^{s-2},$$

where we choose $\varepsilon = 1 - 2/s$ in (8.34). Thus we have bounded the growth of g. Theorem 5.4.4 tells us that this bound is so strong that our function g must be a constant. In other words,

$$F = C_0 e^\theta$$

If $F \in L^p(\mathbb{C})$ for some $p > 0$ and if $C_0 \neq 0$, we have of course $e^{p\theta} \in L^1(\mathbb{C})$. But then for every disk $B \subset \mathbb{C}$,

$$e^{p\theta_B} \leqslant \frac{1}{|B|}\int_B e^{p\theta} \leqslant \frac{C_p}{|B|},$$

by Jensen's inequality. The last estimate actually contradicts (8.34). Indeed,

$$e^{p\theta_B} \geqslant e^{-pC_\varepsilon - p\varepsilon \log|B|} = \frac{e^{-pC_\varepsilon}}{|B|^{p\varepsilon}} > \frac{C_p}{|B|}$$

whenever we fix $\varepsilon < 1/p$ and let $|B| \to \infty$. Hence $C_0 = 0$. □

It is worthwhile pointing out here that even with decay conditions at ∞ the requirement $\sigma \in L^2(\mathbb{C})$ cannot be seriously relaxed. For instance, if

$$F(z) = \bar{z}\chi_\mathbb{D} + \frac{1}{z}\chi_{\mathbb{C}\setminus\mathbb{D}}, \qquad \sigma(z) = \frac{1}{z}\chi_\mathbb{D}, \tag{8.38}$$

we have $F_{\bar{z}} = \sigma F$ and not only is σ compactly supported, but also $\sigma \in weak - L^2(\mathbb{C})$. In particular, σ lies in $L^p(\mathbb{C})$ for all $p < 2$. It is a moment's work to verify that the function $F \in W^{1,\infty}(\mathbb{C})$.

Finally, we point out an interesting related question concerning the higher-dimensional analogs of the Liouville theorem, which at the time of this writing remain unsolved. Suppose $f \in W^{1,n}_{loc}(\mathbb{R}^n)$ satisfies the distortion inequality

$$|Df|^n \leqslant KJ(x,f) + \sigma(x)|f|^n$$

with $\sigma \in L^{n\pm}(\mathbb{R}^n)$. Is f continuous? Additionally, if $f \to 0$ as $x \to \infty$, is it true that $f \equiv 0$?

8.6 Uniqueness

We have seen examples showing that neither uniqueness nor existence will follow in general without some regularity assumptions on the behavior of H with respect to the z and w variables. For existence we need only assume that the function H is Lusin-measurable with respect to these variables. For uniqueness we will have to be content with the following Lipschitz regularity assumption:

$$|H(z, w_1, \zeta) - H(z, w_2, \zeta)| \leq C|\zeta|\,|w_1 - w_2| \tag{8.39}$$

for some absolute constant C independent of z and ζ.

Theorem 8.6.1. *Suppose $H : \mathbb{C} \times \mathbb{C} \times \mathbb{C} \to \mathbb{C}$ satisfies (8.39) and the conditions 1–3 of Theorem 8.2.1 and is compactly supported in the z-variable. Then the uniformly elliptic nonlinear differential equation*

$$\frac{\partial f}{\partial \bar{z}} = H\left(z, f, \frac{\partial f}{\partial z}\right) \tag{8.40}$$

admits exactly one principal solution.

Proof. Suppose that both f and g are principal solutions to (8.40), so

$$\begin{aligned}\frac{\partial f}{\partial \bar{z}} &= H\left(z, f, \frac{\partial f}{\partial z}\right) \\ \frac{\partial g}{\partial \bar{z}} &= H\left(z, g, \frac{\partial f}{\partial z}\right)\end{aligned}$$

We set

$$F = f - g$$

and estimate

$$\begin{aligned}|F_{\bar{z}}| &= |H(z, f, f_z) - H(z, g, g_z)| \\ &\leq |H(z, f, f_z) - H(z, f, g_z)| + |H(z, f, g_z) - H(z, g, g_z)| \\ &\leq k|f_z - g_z| + C\chi_R|g_z|\,|f - g|,\end{aligned}$$

where we remind the reader here of our notation that χ_R denotes the characteristic function of the disk $\mathbb{D}(0, R)$. Put briefly, F satisfies the differential inequality

$$|F_{\bar{z}}| \leq k|F_z| + C\chi_R|g_z|\,|F|$$

By assumption, the principal solutions $f, g \in W^{1,2}_{loc}(\mathbb{C})$ with

$$\lim_{z \to \infty} f(z) - g(z) = 0$$

Once we observe that

$$\sigma = C\chi_R(z)|g_z| \in L^2(\mathbb{C})$$

and has compact support, Theorem 8.5.3 shows us that $F \equiv 0$, as desired. $\square$

8.6. UNIQUENESS

8.6.1 Uniqueness for Normalized Solutions

The strong control at ∞ quickly gives the principal solutions their uniqueness, even in the case of general nonlinear systems, as soon as the Lipschitz condition (8.39) holds. However, for the normalized solutions the situation is considerably more subtle. Here we can discuss only a special situation where there is an interesting connection between the reduced distortion inequality and nonlinear PDEs.

Consider the nonlinear Beltrami operator

$$\frac{\partial}{\partial \bar{z}} - H\left(z, \frac{\partial}{\partial z}\right) \quad \text{in } \Omega \subset \mathbb{C},$$

where $H(z, \xi)$ satisfies the usual ellipticity conditions and, in addition, is real homogeneous; that is

$$H(z, t\xi) = t\, H(z, \xi) \quad \text{for } t \in \mathbb{R} \tag{8.41}$$

Theorem 8.6.2. *Suppose f is a homeomorphic solution to the equation*

$$\frac{\partial f}{\partial \bar{z}} = H\left(z, \frac{\partial f}{\partial z}\right), \qquad f \in W^{1,2}_{loc}(\Omega).$$

Then any other $W^{1,2}_{loc}$-solution g to this equation in Ω takes the form

$$g = F\big(f(z)\big)$$

where $F : f(\Omega) \to \mathbb{C}$ satisfies the reduced distortion inequality

$$\left|\frac{\partial F}{\partial \bar{w}}\right| \leqslant k' \left|\Im m\, \frac{\partial F}{\partial w}\right|, \qquad w \in f(\Omega) \tag{8.42}$$

Remark 8.6.3. *In the distortion inequality (8.42) the constant k' depends only on the ellipticity constant k at (8.8).*

Proof. As before, we represent $g = F\big(f(z)\big)$, and substitute this to the equation to get the nonlinear relation between F_w and $F_{\bar{w}}$. Thus

$$f_{\bar{z}} F_w + \overline{f_z}\, F_{\bar{w}} = H\left(z,\, f_z F_w + \overline{f_{\bar{z}}}\, F_{\bar{w}}\right), \tag{8.43}$$

where $z = f^{-1}(w)$ and $f_{\bar{z}} = H(z, f_z)$. We solve this for $F_{\bar{w}}$, in terms of F_w, using the contraction mapping theorem. For the precise computation we refer the reader to Section 9.1. As a result, we obtain $\mathcal{H}$ so that

$$\frac{\partial F}{\partial \bar{w}} = \mathcal{H}\left(w, \frac{\partial F}{\partial w}\right), \tag{8.44}$$

where $\mathcal{H} : f(\Omega) \times \mathbb{C} \to \mathbb{C}$ satisfies the usual ellipticity bounds. In particular,

$$|\mathcal{H}(w, \xi_1) - \mathcal{H}(w, \xi_2)| \leqslant k'|\xi_1 - \xi_2| \tag{8.45}$$

In addition, we note that
$$\mathcal{H}(w, t\xi) = t\mathcal{H}(w, \xi)$$

This follows from the real homogeneity of (8.43) with respect to complex variables F_w and $F_{\bar{w}}$ and from the uniqueness of the solution for $F_{\bar{w}}$ in (8.44).

Another important observation is that $\mathcal{H}(w, 1) \equiv 0$. Namely, consider a particular case of (8.44) in which we let $\xi = F_w = 1$ and solve it for $F_{\bar{w}} = \mathcal{H}(w, 1)$. This is equivalent to the equation
$$f_{\bar{z}} + \overline{f_z}\, F_{\bar{w}} = H\left(z, f_z + \overline{f_{\bar{z}}}\, F_{\bar{w}}\right)$$

We then find that
$$|f_z|\,|F_{\bar{w}}| = \left|H\left(z, f_z + \overline{f_{\bar{z}}}\, F_{\bar{w}}\right) - H(z, f_z)\right| \leqslant k\,|f_{\bar{z}}|\,|F_{\bar{w}}| \leqslant k^2\,|f_z|\,|F_{\bar{w}}|$$

Hence $F_{\bar{w}} = 0$, as desired.

Now, let $\xi = a + ib$ be any complex number with $a, b \in \mathbb{R}$. Then $\mathcal{H}(w, a) = a\mathcal{H}(w, 1) = 0$, so
$$\mathcal{H}(w, a + ib) = \mathcal{H}(w, a + ib) - \mathcal{H}(w, a)$$

Hence $|\mathcal{H}(w, a + ib)| \leqslant k'|b|$; that is, $|\mathcal{H}(w, \zeta)| \leqslant k\big|\Im m\,\zeta\big|$.

The proof is complete. $\square$

With Theorem 6.2.2, any global homeomorphic solution to the reduced distortion inequality, that fixes at least two points must be the identity. Hence we obtain the following corollary.

Corollary 8.6.4. *Under the homogeneity (8.41) and the usual ellipticity conditions, the nonlinear equation*
$$\frac{\partial f}{\partial \bar{z}} = H\left(z, \frac{\partial f}{\partial z}\right) \tag{8.46}$$
admits a unique normalized solution.

It is not clear whether the $\mathbb{R}$-homogeneity, condition (8.41), is a necessary restriction on H in general. However, Corollary 8.6.4 holds for some inhomogeneous cases as well, for instance, if $|\mathcal{H}(z, \zeta)| \leqslant k(z)|\zeta|$ with $|k(z)| \leqslant \varepsilon$ small enough near infinity. Indeed, it is an interesting question as to how small ε must be here: Are the normalized solutions to (8.46) always unique?

8.7 Lipschitz $H(z, w, \zeta)$

If we were to assume, in addition to the hypotheses of Theorem 8.6.1, that the function $H(z, w, \zeta)$ is Lipschitz with respect to z and w, then it will turn out, and indeed it is the purpose of this section to prove, that the principal

8.7. LIPSCHITZ $H(z, w, \zeta)$

solution will lie in the Sobolev space $W^{2,s}_{loc}(\mathbb{C})$ for every s in the critical interval $q_k < s < p_k$. In particular, if we take $s > 2$, then the Sobolev embedding theorem implies that $f \in C^{1,\alpha}_{loc}(\mathbb{C})$ with $\alpha = 1 - \frac{2}{s}$. This additional regularity is in many ways extremely useful, but of course it is not always possible to achieve this for solutions to general uniformly elliptic equations.

So motivated, we suppose that

$$|H(z_1, w_1, \zeta) - H(z_2, w_2, \zeta)| \leqslant C|\zeta|(|z_1 - z_2| + |w_1 - w_2|), \qquad (8.47)$$

where C is an absolute constant and $z_1, z_2, w_1, w_2, \zeta \in \mathbb{C}$ are arbitrary.

Theorem 8.7.1. *The uniformly elliptic nonlinear differential equation*

$$\frac{\partial f}{\partial \bar{z}} = H\left(z, f, \frac{\partial f}{\partial z}\right), \qquad (8.48)$$

with H satisfying (8.47), admits exactly one principal solution. Moreover, this solution f belongs to the class

$$f \in W^{2,s}_{loc}(\mathbb{C}) \subset C^{1,\alpha}_{loc}(\mathbb{C}), \qquad \alpha = 1 - \frac{2}{s},$$

for every $2 < s < p_k$.

Proof. Theorem 8.6.1 leaves us in no doubt that (8.48) admits a unique principal solution f. We shall make use of the difference quotient characterization of Sobolev functions. Thus we consider

$$F(z) = \frac{f(z+h) - f(z)}{h} - 1$$

for sufficiently small complex numbers $h \in \mathbb{D}$, $h \neq 0$. Clearly, the complex derivatives

$$\frac{\partial F}{\partial \bar{z}} = \frac{f_{\bar{z}}(z+h) - f_{\bar{z}}(z)}{h}$$
$$\frac{\partial F}{\partial z} = \frac{f_z(z+h) - f_z(z)}{h}$$

belong to the space $L^s_{loc}(\mathbb{C})$, as both $f(z)$ and $f(z+h)$ are K-quasiregular functions of $\mathbb{C}$ with distortion $K = \frac{1+k}{1-k}$. Note also that as f is holomorphic near ∞ the normalization of the principal solution f implies that $F(z) = \mathcal{O}(|z|^{-2})$ as $z \to \infty$. In particular $F \in C_0(\hat{\mathbb{C}})$ and, moreover,

$$\frac{\partial F}{\partial \bar{z}}, \frac{\partial F}{\partial z} \in L^s(\mathbb{C})$$

We shall have established the theorem once we provide uniform bounds for the s-norms of F_z and $F_{\bar{z}}$, that is, a bound independent of h. See Lemma A.5.4.

We begin with the pointwise inequality

$$|f_{\bar{z}}(z+h) - f_{\bar{z}}(z)|$$
$$\leq |H(z+h, f(z+h), f_z(z+h)) - H(z+h, f(z+h), f_z(z))|$$
$$+ |H(z+h, f(z+h), f_z(z)) - H(z, f(z), f_z(z))|$$
$$\leq k|f_z(z+h) - f_z(z)| + C(|f(z+h) - f(z)| + |h|)|f_z(z)| \qquad (8.49)$$

Since the function $z \mapsto H(z, w, \zeta)$ is supported in the disk $\mathbb{D}(0, R)$ and since $|h| \leq 1$, we find that $f_{\bar{z}}(z+h) - f_{\bar{z}}(z)$ vanishes for all $z \in \mathbb{C}$ with $|z| \geq R + 1$. Thus the inequality in (8.49) is unaffected if we multiply the right hand side by the characteristic function χ_{R+1}. Upon division by h we have the inhomogeneous distortion inequality

$$|F_{\bar{z}}(z)| \leq k\chi_{R+1}(z)|F_z(z)| + C\chi_{R+1}(z)|f_z(z)||F(z)| + C\chi_{R+1}(z)|f_z(z)|$$

As we have done many times since Theorem 8.5.1, we express this inequality in the form of an equation,

$$F_{\bar{z}} = \mu(z)F_z + A(z)F + B(z), \qquad (8.50)$$

where the coefficients μ, A and B enjoy uniform bounds that are independent of h, namely,

$$|A| \leq C\chi_{R+1}|f_z| \in L^s(\mathbb{C}) \cap L^{s/(s-1)}(\mathbb{C})$$
$$|B| \leq C\chi_{R+1}|f_z| \in L^s(\mathbb{C})$$

We again solve the auxiliary equation

$$\theta_{\bar{z}} = \mu(z)\theta_z + A(z)$$

for $\theta \in C_0(\hat{\mathbb{C}}) \cap W^{1,s}_{loc}(\mathbb{C})$ and consider

$$g(z) = F(z)e^{-\theta(z)} \in W^{1,s}(\mathbb{C})$$

Then (8.50) reduces to the inhomogeneous Beltrami equation for g,

$$g_{\bar{z}} = \mu(z)g_z + B(z)e^{-\theta(z)} \qquad (8.51)$$

As g vanishes at ∞, we may represent it by means of its Cauchy transform,

$$g(z) = \mathcal{C}w(z) = \frac{1}{2\pi} \int_{\mathbb{C}} \frac{w(\zeta)}{z - \zeta} d\zeta$$

with the density function $w = g_{\bar{z}} \in L^s(\mathbb{C})$ and $s > 2$. This gives us the integral equation for w,

$$w = \mu \mathcal{S}w + Be^{-\theta},$$

which we can solve with the inverse $(\mathbf{I} - \mu \mathcal{S})^{-1}$ of the Beltrami operator. Thus

$$w = (\mathbf{I} - \mu \mathcal{S})^{-1}(Be^{-\theta}),$$

which shows, among other things, that the s-norm of w does not depend on h. This now gives the uniform bounds for $\|F_{\bar{z}}\|_s$ and $\|F_z\|_s$ and completes the proof of the theorem. $\square$

Chapter 9

Nonlinear Riemann Mapping Theorems

Let $\Omega \subset \mathbb{C}$ be a domain. We say that Ω is a *Jordan domain* if $\partial\Omega$ is a topological circle in $\hat{\mathbb{C}}$. More generally, we will use the following terminology: We say that the domain Ω is a Jordan n-connected domain if $\partial\Omega$ has n disjoint connected components, each of which is a Jordan curve.

The most important theorem about Jordan domains is the celebrated Riemann mapping theorem, which asserts that all Jordan domains are conformally equivalent to the disk.

Theorem 9.0.2. *A domain $\Omega \subset \mathbb{C}$ is a Jordan domain if and only if there is a homeomorphism $\varphi : \overline{\Omega} \to \overline{\mathbb{D}}$ that is conformal on Ω. Moreover, given arbitrary points*

$$z_0 \in \Omega, \qquad a \in \partial\Omega,$$

there is exactly one such Riemann map with $\varphi(z_0) = 0$ and $\varphi(a) = 1$.

Of course, a conformal map is a homeomorphic solution to the Cauchy-Riemann equations, our prototype of a uniformly elliptic first-order PDE, and so the Riemann mapping theorem can be viewed as asserting the existence of a homeomorphic transformation of a domain to a disk effected by a solution to a uniformly elliptic PDE. It is in this setting that we seek our generalization.

Often in the literature one chooses another normalization of the form $\varphi(z_0) = 0$ and $\varphi'(z_0) > 0$. A normalization of this sort will be impossible to achieve for more general nonlinear equations whose solutions are only differentiable almost everywhere. This is why we have chosen the route via a boundary normalization. Such a normalization will cause difficulty should the boundary of a domain be highly irregular. Therefore, for existence and uniqueness in the normalized situation, we will formulate the results in terms of Jordan domains, avoiding the apparently necessary technical machinery of prime ends and their impressions necessary to formulate and establish more general results. However, existence in the general setting without a normalization will be ensured. There is yet another reason to study the problem for Jordan domains, and that is to ensure that the mapping extends homeomorphically to the boundary. This is another property of solutions to uniformly elliptic PDEs that we shall establish. Here precisely is the formulation of the problem we shall study.

The Mapping Problem for Elliptic Equations

Suppose $H(z, w, \zeta)$ satisfies the conditions of Definition 7.7.1. Let Ω and Ω' be simply connected planar domains. A homeomorphism $f : \Omega \to \Omega'$ is said to be *H-conformal* if f is a solution to the uniformly elliptic equation

$$\frac{\partial f}{\partial \bar{z}} = H\left(z, f, \frac{\partial f}{\partial z}\right), \qquad H : \Omega \times \Omega' \times \mathbb{C} \to \mathbb{C} \qquad (9.1)$$

Given

$$z_0 \in \Omega, \quad a \in \partial \Omega \quad \text{and} \quad w_0 \in \Omega', \quad b \in \partial \Omega',$$

the mapping problem is to find a homeomorphic H-conformal map $f : \Omega \to \Omega'$ such that

$$f(z_0) = w_0, \qquad f(a) = b.$$

Further, if Ω and Ω' are Jordan domains, we are to show that f extends homeomorphically to the boundary.

We now solve the above mapping problem.

Theorem 9.0.3. *Let Ω and Ω' be simply connected domains and $z_0 \in \Omega$, $a \in \partial \Omega$, $w_0 \in \Omega'$ and $b \in \partial \Omega'$. Then for every uniformly elliptic system as in (9.1), there is a H-conformal homeomorphism $f : \Omega \to \Omega'$ with $f(z_0) = w_0$. Moreover, if Ω and Ω' are Jordan domains, then f extends homeomorphically to the boundary and we may choose f so that $f(a) = b$.*

If, in addition, H is Lipschitz with respect to the w-variable,

$$|H(z, w_1, \zeta) - H(z, w_2, \zeta)| \leqslant C|\zeta||w_1 - w_2|, \qquad (9.2)$$

for every $z \in \Omega$, $w_1, w_2 \in \Omega'$ and $\zeta \in \mathbb{C}$, then the solution to the mapping problem is unique.

After the simply connected case we establish the n-connected counterparts. We show that any finitely connected domain can be mapped H-conformally onto a canonical domain. For n-connected domains there are several natural choices for canonical domains; here we have chosen the Schottky annuli, see Section 9.3.

In fact, with a change of variables provided in the next section, it is only necessary to solve the mapping problem in canonical domains.

9.1 Ellipticity and Change of Variables

One further remarkable property of nonlinear elliptic equations in the plane is that they remain elliptic after a quasiconformal change of the variables z and w. To see this let us consider a uniformly elliptic nonlinear system, with ellipticity constant k,

$$\frac{\partial f}{\partial \bar{z}} = H\left(z, f, \frac{\partial f}{\partial z}\right) \tag{9.3}$$

defined for functions $f : \Omega \to \Omega'$. Let us fix two quasiconformal mappings

$$\phi : U \to \Omega, \qquad \psi : \Omega' \to U'$$

Given f, we may define a new mapping

$$F = \phi \circ f \circ \psi : U \to U'$$

Diagramatically, we have

$$\begin{array}{ccc} \Omega & \xrightarrow{f} & \Omega' \\ \phi \uparrow & & \downarrow \psi \\ U & \xrightarrow{F} & U' \end{array}$$

Then (9.3) can be written as an equation for $F = F(u)$ having the form

$$\frac{\partial F}{\partial \bar{u}} = H^\sharp\left(u, F, \frac{\partial F}{\partial u}\right) \tag{9.4}$$

We want to see that the equation (9.4) is also elliptic. We must expect that the constant of ellipticity, should this equation turn out to be uniformly elliptic, will depend not only on the ellipticity constant k but also on the maximal distortions of the two functions ϕ and ψ.

We do the calculation in two steps, the first step being when $\psi(z) = z$, the identity mapping. Thus we write

$$F(u) = f(\phi(u)), \quad \text{where } \phi : U \to \Omega \text{ is } K'\text{-quasiconformal}$$

In particular, ϕ satisfies a Beltrami equation $\phi_{\bar{u}} = \nu(u)\phi_u$. Equivalently, we have $f = F(h(z))$, where $h(z) = \phi^{-1}(z)$ and h is also K'-quasiconformal. With the chain rule we relate the derivatives and compute

$$\begin{aligned} f_{\bar{z}} &= F_u h_{\bar{z}} + F_{\bar{u}} \overline{h_z} \\ f_z &= F_u h_z + F_{\bar{u}} \overline{h_{\bar{z}}} \end{aligned}$$

We substitute these values into (9.3) to obtain the equation

$$F_u h_{\bar{z}} + F_{\bar{u}} \overline{h_z} = H(z, F, F_u h_z + F_{\bar{u}} \overline{h_{\bar{z}}}) \tag{9.5}$$

This equation can of course be written as

$$\Phi\left(u, F, \frac{\partial F}{\partial u}, \frac{\partial F}{\partial \bar{u}}\right) = 0 \tag{9.6}$$

as soon as we express z, $h_z(z)$ and $h_{\bar{z}}(z)$ in terms of u by the formula $z = \phi(u)$. Now $h = \phi^{-1}$ is K'-quasiconformal, it is the solution to the equation

$$\frac{\partial h}{\partial \bar{z}} = -\nu(h) \overline{\frac{\partial h}{\partial z}};$$

see (2.51), the point being that the function $|\nu| \leq \frac{K'-1}{K'+1} < 1$. With this notation, (9.5) becomes

$$F_{\bar{u}} - \nu F_u = \frac{1}{\overline{h_z}} H(z, F, (F_u - \bar{\nu} F_{\bar{u}}) h_z) \tag{9.7}$$

At this point it is important to realize that, since $h_z \neq 0$ almost everywhere, the presence of the factor $1/\overline{h_z}$ causes no change in the Lipschitz constant with respect to the variable $(F_u - \bar{\nu} F_{\bar{u}})$. While for $h_z = 0$, we have $H = 0$.

This shows us that (9.7) is in fact uniquely solvable for $F_{\bar{u}}$ by the contraction mapping principle. Therefore we may write it in the form (9.4). The only point remaining to be discussed concerns the Lipschitz constant of the map

$$\zeta \mapsto H^\sharp(u, F, \zeta)$$

If we put $\xi_1 = H^\sharp(u, F, \zeta_1)$ and $\xi_2 = H^\sharp(u, F, \zeta_2)$ and then return to (9.7), we see that

$$\xi_1 - \nu \zeta_1 = \frac{1}{\overline{h_z}} H(z, F, (\zeta_1 - \bar{\nu} \xi_1) h_z)$$

$$\xi_2 - \nu \zeta_2 = \frac{1}{\overline{h_z}} H(z, F, (\zeta_2 - \bar{\nu} \xi_2) h_z)$$

If we subtract these two equations and use the Lipschitz condition on H, we have

$$|(\xi_1 - \xi_2) - \nu(\zeta_1 - \zeta_2)| \leq k |(\zeta_1 - \zeta_2) - \bar{\nu}(\xi_1 - \xi_2)|,$$

or equivalently,

$$\left| \frac{\frac{\xi_1 - \xi_2}{\zeta_1 - \zeta_2} - \nu}{1 - \bar{\nu} \frac{\xi_1 - \xi_2}{\zeta_1 - \zeta_2}} \right| \leq k \tag{9.8}$$

Then to see the estimate for the ellipticity of $H^\sharp$, we notice that the map

$$z \mapsto \frac{z - \nu}{1 - \bar{\nu} z}$$

is an isometry of the hyperbolic disk, and use of the triangle inequality there results in the estimate

$$\left| \frac{\xi_1 - \xi_2}{\zeta_1 - \zeta_2} \right| \leq \frac{k + k'}{1 + kk'} = k^\sharp < 1, \tag{9.9}$$

where $|\nu| \leq k' = \frac{K'-1}{K'+1}$. Hence

$$|H^\sharp(u, F, \zeta_1) - H^\sharp(u, F, \zeta_2)| \leq k^\sharp |\zeta_1 - \zeta_2|, \tag{9.10}$$

9.2. SIMPLY CONNECTED DOMAINS

which gives us the estimate we wanted.

In the special case where ϕ is conformal, we have that $k' = 0$ and so $k^\sharp = k$, and we have the same uniform ellipticity constant as the original equation. In fact, it is interesting to write out the distortion constants for the two equations. We simply find that

$$K^\sharp = K_1 K_2, \tag{9.11}$$

where K_1 is the distortion constant for the original equation and ϕ is K_2-quasiconformal.

The second computation, where $\psi(z)$ is a K''-quasiconformal mapping, follows in much the same manner. There is only one obvious outcome to this calculation, and we summarize these results in the following theorem.

Theorem 9.1.1. *Let $\phi : U \to \Omega$ be K_ϕ-quasiconformal and $\psi : \Omega' \to U'$ be K_ψ-quasiconformal. Consider the uniformly elliptic nonlinear equation*

$$\frac{\partial f}{\partial \bar{z}} = H\left(z, f, \frac{\partial f}{\partial z}\right)$$

with distortion constant K. Then, the change of variables $z = \phi(u)$ and $w = \psi(f)$ results in the uniformly elliptic equation for $F = \psi \circ f \circ \phi : U \to U'$ given by

$$\frac{\partial F}{\partial \bar{u}} = H^\sharp\left(u, F, \frac{\partial F}{\partial u}\right)$$

whose distortion constant $K^\sharp \leqslant K_\phi K K_\psi$ and ellipticity constant

$$k^\sharp \leqslant \frac{K^\sharp - 1}{K^\sharp + 1}$$

In particular, a conformal change of variables does not change the ellipticity constants of the resulting equations.

9.2 The Nonlinear Mapping Theorem in Simply Connected Domains

We begin by showing that the H-conformal Riemann mapping problem has a solution in the disk $\mathbb{D}$.

Theorem 9.2.1. *For every $H(z, w, \zeta)$ satisfying the conditions of Definition 7.7.1, there is a homeomorphic $W^{1,2}(\mathbb{D})$-solution $f : \mathbb{D} \to \mathbb{D}$ to the differential equation*

$$\frac{\partial f}{\partial \bar{z}} = H\left(z, f, \frac{\partial f}{\partial z}\right) \quad \text{for almost every } z \in \mathbb{D} \tag{9.12}$$

The mapping f extends homeomorphically to the boundary.

If in addition H is Lipschitz with respect to the w-variable,

$$|H(z, w_1, \zeta) - H(z, w_2, \zeta)| \leqslant C|\zeta||w_1 - w_2|, \tag{9.13}$$

for every $z \in \Omega$, $w_1, w_2 \in \Omega'$, $\zeta \in \mathbb{C}$, then the solution to the mapping problem is unique.

This result will immediately imply the general Theorem 9.0.3.

Proof of Theorem 9.0.3. Let $\varphi : \Omega \to \mathbb{D}$, $\varphi(z_0) = 0$ be conformal, and if Ω is a Jordan domain, we choose $\varphi(a) = 1$. Similarly we choose conformal $\varphi' : \Omega' \to \mathbb{D}$, $\varphi'(w_0) = 0$, and if Ω' is a Jordan domain, we choose $\varphi'(b) = 1$. The change of variables φ for z and φ' for w induces a new uniformly elliptic PDE on the disk. We even have the same ellipticity constants; see Theorem 9.1.1. Thus we need only solve this new equation on the disk. Further, the maps φ and φ' extend homeomorphically to the boundaries if Ω and Ω' are Jordan domains. Once solved in the disk, we map back to our original domains to find a solution to the posed problem. □

9.2.1 Existence

In proving Theorem 9.2.1 we may argue as in the proof of Theorem 9.0.3 and thus may assume that $z_0 = w_0 = 0$ and that $a = b = 1$ and establish existence. As a preliminary step, we impose the additional requirement of H, namely, that

$$H : \overline{\mathbb{D}} \times \overline{\mathbb{D}} \times \mathbb{C} \to \mathbb{C} \tag{9.14}$$

is continuous and vanishes in a neighborhood of the origin in the z-variable, say for $|z| < r$. We may further use the Whitney extension given in Section 8.3.2 to assume that H is defined and continuous on $\overline{\mathbb{D}} \times \mathbb{C} \times \mathbb{C}$, retaining (for later use) the Lipschitz condition in (9.13). These assumptions have not affected the uniform ellipticity bounds on H.

We shall seek our H-conformal mapping $f : \overline{\mathbb{D}} \to \overline{\mathbb{D}}$, $f(0) = 0$ and $f(1) = 1$ in the form

$$f(z) = z\, e^{\Upsilon \omega}, \tag{9.15}$$

where $\omega \in L^s(\mathbb{D})$, $2 < s < s_k$, is a density function for the Cauchy transform on $\mathbb{D}$,

$$\begin{aligned}(\Upsilon \omega)(z) &= (\mathcal{C}_\mathbb{D} \omega)(z) - (\mathcal{C}_\mathbb{D} \omega)(1) \\ &= \frac{1}{\pi} \int_\mathbb{D} \left(\frac{\omega(\tau)}{z - \tau} - \frac{z\overline{\omega(\tau)}}{1 - z\overline{\tau}} \right) d\tau - \frac{1}{\pi} \int_\mathbb{D} \left(\frac{\omega(\tau)}{1 - \tau} - \frac{\overline{\omega(\tau)}}{1 - \overline{\tau}} \right) d\tau\end{aligned}$$

The upper bound for the exponent s is s_k determined by the equation

$$k \|\mathcal{S}_\mathbb{D}\|_{s_k} = 1, \tag{9.16}$$

and in view of the facts that $\|\mathcal{S}_\mathbb{D}\|_2 = 1$, the norms are continuous and $k < 1$, we have $s_k > 2$.

Recall that $\Upsilon : L^s(\mathbb{D}) \to W^{1,s}(\mathbb{D}) \subset C(\overline{\mathbb{D}})$ and that the latter inclusion is compact. Therefore the operator Υ has the following properties:

$$\begin{aligned}\Re(\Upsilon \omega) &= 0 \quad \text{on the boundary of } \mathbb{D} \\ (\Upsilon \omega)(1) &= 0 \\ \frac{\partial}{\partial \bar{z}} \Upsilon \omega &= \omega, \quad \frac{\partial}{\partial z} \Upsilon \omega = \mathcal{S}_\mathbb{D} \omega\end{aligned}$$

9.2. SIMPLY CONNECTED DOMAINS

In particular, with f in (9.15) we certainly have
$$f(0) = 0, \quad f(1) = 1, \quad |f(z)| = 1 \text{ for } |z| = 1$$

We now compute the derivatives of f,
$$\begin{aligned} f_{\bar{z}} &= ze^{\Upsilon\omega}\omega \\ f_z &= e^{\Upsilon\omega} + ze^{\Upsilon\omega}\mathcal{S}_\mathbb{D}\omega, \end{aligned}$$

and upon substituting into (9.1) we have
$$\omega = \frac{1}{z} e^{-\Upsilon\omega} H(z, ze^{\Upsilon\omega}, e^{\Upsilon\omega} + ze^{\Upsilon\omega}\mathcal{S}_\mathbb{D}\omega) \tag{9.17}$$

Thus our mapping f will be the $W^{1,s}(\mathbb{D})$-solution to (9.12) as soon as we solve the equation (9.17) for $\omega \in L^s(\mathbb{D})$. Then, as such it, will be quasiregular and we may write $f(z) = \Phi(g(z))$ for some quasiconformal homeomorphism $g : \mathbb{D} \to \mathbb{D}$ with $g(0) = 0$ and conformal map $\Phi : \mathbb{D} \to \mathbb{C}$. All this follows from the Stoïlow factorization theorem; see Corollary 5.5.3. As $|f(z)| = |g(z)| = 1$ when $|z| = 1$, we have $\Phi : \mathbb{D} \to \mathbb{D}$. We also have $\Phi(0) = 0$. We now use an elementary degree argument to show that Φ is the identity, using the fact that the degrees of f and Φ are the same because g is a homeomorphism. We shall discuss this application of the variation of the argument in somewhat greater generality when we come to the proof of uniqueness, so we do not give too many details here. We compute the increment of the argument
$$\begin{aligned} \Delta \arg \Phi(z) &= \Delta \arg f(z) = \Delta(\arg ze^{\Upsilon\omega}) \\ &= \Delta \arg z + \Delta(\arg e^{\Upsilon\omega}) = 2\pi + 0 = 2\pi \end{aligned}$$

Thus Φ is conformal, and hence $\Phi(z) = z$ by our normalization. Thus f is quasiconformal and therefore admits an extension to the boundary circle. In particular, f will be the H-conformal mapping we are looking for.

We are left with the problem of finding ω, that is, of solving the integral equation in (9.17). As in our earlier existence results, we shall make use of the fixed-point principle. First, the contraction mapping principle shows that for every $\omega \in L^s(\mathbb{D})$ there exists a unique $\Psi \in L^s(\mathbb{D})$ such that
$$\Psi = \frac{1}{z} e^{-\Upsilon\omega} H(z, ze^{\Upsilon\omega}, e^{\Upsilon\omega} + ze^{\Upsilon\omega}\mathcal{S}_\mathbb{D}\Psi) \tag{9.18}$$

We also have a uniform bound for the map Ψ independent of ω. To get this bound, first write χ_r for the characteristic function of $\mathbb{D} \setminus D(r)$ and then begin with the pointwise inequality
$$\begin{aligned} |\Psi| &\leq k\chi_r(z) \frac{1}{|z|} \left|e^{-\Upsilon\omega}\right| \left|e^{\Upsilon\omega} + ze^{\Upsilon\omega}\mathcal{S}_\mathbb{D}\Psi\right| \\ &\leq k|\mathcal{S}_\mathbb{D}\Psi| + \chi_r(z) \frac{1}{|z|}, \end{aligned}$$

where we use the ellipticity of H. We have chosen $\|\mathcal{S}_\mathbb{D}\|_s < 1/k$, and hence (9.16) gives

$$\|\Psi\|_s \leqslant \frac{k}{1-k\|\mathcal{S}_\mathbb{D}\|_s} \left\| \chi_r(z) \frac{1}{|z|} \right\|_{L^s(\mathbb{D})} = M \quad (9.19)$$

for some constant $M = M(r)$. As usual, we set up a fixed-point problem. We consider the nonlinear map

$$\mathbf{T}: L^s(\mathbb{D}) \to L^s(\mathbb{D})$$

defined by the rule

$$\mathbf{T}\omega = \Psi$$

The estimate in (9.19) indicates that we should look for a solution to (9.17) as a fixed point of $\mathbf{T}$ in the closed convex set

$$B = \{\omega \in L^s(\mathbb{D}) : \|\omega\|_s \leqslant M\}$$

since $\mathbf{T}$ maps the entire space $L^s(\mathbb{C})$ into this ball. We argue as follows to show that $\mathbf{T}$ is both continuous and compact, so as to use the Schauder fixed-point theorem.

Consider a sequence $\{\omega_n\}_{n=1}^\infty$ converging weakly in $L^s(\mathbb{D})$. Thus $\Upsilon\omega_n \to \Upsilon\omega$ uniformly on $\mathbb{D}$. We need to show that the sequence

$$\Psi_n = \mathbf{T}\omega_n \to \mathbf{T}\omega = \Psi$$

strongly in $L^s(\mathbb{D})$. Recall that Ψ_n is defined as the unique solution to the equation

$$\Psi_n = \frac{1}{z} e^{-\Upsilon\omega_n} H(z, ze^{\Upsilon\omega_n}, e^{\Upsilon\omega_n} + ze^{\Upsilon\omega_n}\mathcal{S}_\mathbb{D}\Psi_n) \quad (9.20)$$

We subtract (9.18) from this equation, use the triangle inequality and telesope the sum to obtain

$$|\Psi_n - \Psi| \leqslant k|\mathcal{S}_\mathbb{D}(\Psi_n - \Psi)| + \alpha_n(z) + \beta_n(z) + \gamma_n(z), \quad (9.21)$$

where

$$\alpha_n \leqslant \frac{k\chi_r|e^{-\Upsilon\omega_n}|}{|z|} |e^{\Upsilon\omega_n} - e^{\Upsilon\omega}| |1 + z\mathcal{S}_\mathbb{D}\Psi|$$
$$\leqslant C(1 + |\mathcal{S}_\mathbb{D}\Psi|) \in L^s(\mathbb{D})$$
$$\beta_n \leqslant \frac{\chi_r|e^{-\Upsilon\omega_n}|}{|z|} |H(z, ze^{\Upsilon\omega_n}, e^{\Upsilon\omega}(1+z\mathcal{S}_\mathbb{D}\Psi)) - H(z, ze^{\Upsilon\omega}, e^{\Upsilon\omega}(1+z\mathcal{S}_\mathbb{D}\Psi))|$$
$$\leqslant C(1 + |\mathcal{S}_\mathbb{D}\Psi|) \in L^s(\mathbb{D})$$
$$\gamma_n \leqslant \frac{\chi_r|e^{-\Upsilon\omega_n} - e^{-\Upsilon\omega}|}{|z|} |H(z, ze^{\Upsilon\omega}, e^{\Upsilon\omega} + ze^{\Upsilon\omega}\mathcal{S}_\mathbb{D}\Psi)|$$
$$\leqslant C(1 + |\mathcal{S}_\mathbb{D}\Psi|) \in L^s(\mathbb{D}),$$

9.2. SIMPLY CONNECTED DOMAINS

where in each case the constant C is independent of n. Next, note that as $\Upsilon w_n \to \Upsilon w$ at every point of $\mathbb{D}$ (and in fact uniformly), we have

$$\alpha_n(z) \to 0, \quad \beta_n(z) \to 0, \quad \gamma_n(z) \to 0, \quad \text{almost everywhere in } \mathbb{D}$$

Similarly, as in the proof of Theorem 8.2.1, it is at this point in establishing that $\beta_n \to 0$ almost everywhere that we have used our preliminary assumption on the continuity of the function $w \mapsto H(z, w, \zeta)$ for every $z \in \mathbb{D}$ and $\zeta \in \mathbb{C}$.

Now we may use the Lebesgue dominated convergence theorem to conclude that

$$\alpha_n(z) \to 0, \quad \beta_n(z) \to 0, \quad \gamma_n(z) \to 0 \text{ in } L^s(\mathbb{D})$$

The inequality in (9.21) yields the estimate

$$\|\Psi_n - \Psi\|_s \leqslant \frac{k}{1 - k\|S_\mathbb{D}\|_s} \left(\|\alpha_n(z)\|_s + \|\beta_n(z)\|_s + \|\gamma_n(z)\|_s\right) \to 0,$$

finally showing that **T** is continuous and compact. We may now apply the Schauder fixed-point theorem to complete the existence proof modulo the assumption of the continuity of H and the assumption in (9.14) that $H(z, w, \zeta)$ vanishes for $|z| < r$.

Finally, we make use of the approximation argument precisely as in the proof of Theorem 8.2.1 to remove these assumptions, as the reader can easily verify. With this argument we complete the proof of the existence of the H-conformal Riemann mapping. □

9.2.2 Uniqueness

The arguments we are about to present to establish uniqueness are quite topological and geometric in nature. Therefore it is possible to exploit them in many different situations, once we have established various topological properties of solutions, such as openness.

We recall from our example in Section 8.4 that uniqueness is not always guaranteed and that some regularity of H in the dependent variable w is necessary. We have made the assumption that

$$|H(z, w_1, \zeta) - H(z, w_2, \zeta)| \leqslant C|\zeta||w_1 - w_2| \tag{9.22}$$

for $z \in \Omega$, $w_1, w_2 \in \Omega'$ and all $\zeta \in \mathbb{C}$.

Thus our situation is that we have two homeomorphic mappings $f^1, f^2 : \overline{\mathbb{D}} \to \overline{\mathbb{D}}$ with $f^1(0) = f^2(0) = 0$ and $f^1(1) = f^2(1) = 1$, each satisfying the uniformly elliptic equation (9.1), and we are aiming to establish that $f^1 \equiv f^2$. We of course consider the function

$$f(z) = f^1(z) - f^2(z) \tag{9.23}$$

defined and continuous on $\overline{\mathbb{D}}$. We have $f(0) = 0$ and $f(1) = 0$. If H were in fact independent of the w-variable, then this difference would be a quasiregular mapping and therefore either open (by the Stoilow factorization theorem) or constant. We shall give an argument to show that this difference cannot be an open mapping. In general, instead of an equation for f, we use the distortion inequality

$$|f_{\bar{z}}| \leqslant k|f_z| + C|f|(|f_z^1| + |f_z^2|), \qquad (9.24)$$

which can therefore be written as an inhomogeneous Beltrami equation

$$\frac{\partial f}{\partial \bar{z}} = \mu(z)\frac{\partial f}{\partial z} + \sigma(z)f, \qquad (9.25)$$

where $|\mu(z)| < k$ for $z \in \mathbb{D}$. Moreover, we have the estimate,

$$|\sigma(z)| \leqslant C(|f_z^1| + |f_z^2|) \in L^s(\mathbb{D}), \qquad s > 2 \qquad (9.26)$$

Again we first solve the auxiliary equation

$$\theta_{\bar{z}} = \mu(z)\theta_z + \sigma(z), \qquad (9.27)$$

and then for $\theta \in W_{\Re}^{1,s}(\mathbb{D}) \subset C(\overline{\mathbb{D}})$. Thus the real part of θ vanishes on the boundary. We then find that

$$f(z) = g(z)e^{\theta(z)}, \qquad (9.28)$$

where g is a quasiregular mapping, a solution to the Beltrami equation $g_{\bar{z}} = \mu(z)g_z$ normalized so that $g(0) = 0$ and $g(1) = 0$. The remainder of the proof is purely topological. We wish to show that (9.28) implies that $g \equiv 0$.

In order to proceed further we need to establish a couple of notions. Thus let $\varphi : \mathbb{S}^1 \to \mathbb{C}$ be a continuous function defined on the unit circle. Let X denote the compact set where φ vanishes and $U = \mathbb{S}^1 \setminus X$ its complement. The set U is a countable union of open arcs

$$U = \bigcup_{k=1}^{\infty} U_k,$$

and on each open arc we can define a continuous branch of the argument of φ,

$$\arg \varphi : U_k \to \mathbb{R},$$

unless $U = \mathbb{S}^1$. In any case we may speak of the total variation of the argument of φ on each U_k defined by

$$V_k = \sup \sum_{\nu=1}^{\infty} |\arg \varphi(e^{it_{\nu+1}}) - \arg \varphi(e^{it_\nu})|,$$

where the supremum is taken over all partitions $0 \leqslant t_1 < t_2 < \cdots < 2\pi$ with $e^{it_\nu} \in U_k$, $\nu = 1, 2, \ldots$. It is clear that the total variation does not depend on

9.3. MULTIPLY CONNECTED DOMAINS

the particular branch of argument we choose. We now define the *total variation of the argument* of φ as

$$\text{Varg}(\varphi) = \sum_{k=1}^{\infty} V_k,$$

and we have the following Lemma.

Lemma 9.2.2. *Let $f^1, f^2 : \mathbb{S}^1 \to \mathbb{S}^1$ be homeomorphisms of the same orientation. Then*

$$\text{Varg}(f^1 - f^2) \leqslant 2\pi$$

In fact, somewhat more is true, that under the hypotheses of the lemma one can show that

$$\text{Varg}(f^1 - f^2) = 2\pi - |X|,$$

where $|X|$ is the measure of the set $X = \{f^1(\zeta) = f^2(\zeta)\}$.

A hint for the proof of the lemma is to set $\varphi = f^1 - f^2$, decompose $\mathbb{S}^1$ into U_k as above and note that on each U_k the function $t \mapsto \arg \varphi(e^{it})$ is monotone.

With this lemma in hand we can return to (9.28). The factor $e^{\theta(z)}$ simply does not affect the total variation of f as $|e^{\theta(z)}| = 1$, θ being imaginary on the circle. In particular, we conclude that

$$\text{Varg}(g) \leqslant 2\pi$$

Now we state another lemma.

Lemma 9.2.3. *Let $g : \mathbb{D} \to \mathbb{C}$ be open and suppose g is continuous on $\overline{\mathbb{D}}$ and vanishes at $a \in \mathbb{D}$. Then*

$$\text{Varg}(g) \geqslant 2\pi$$

If in addition g vanishes on $\partial \mathbb{D}$, then $\text{Varg}(g) > 2\pi$.

In fact, for the lemma we may assume that the map g is holomorphic in $\mathbb{D}$ by the general topological version of the Stoilow factorization theorem.

Finally, as far as uniqueness is concerned, we find a contradiction as follows: $\text{Varg}(g) > 2\pi$ since g vanishes on the boundary. On the other hand, $\text{Varg}(g) \leqslant 2\pi$, as the difference of homeomorphisms.

9.3 Mappings onto Multiply Connected Schottky Domains

This section is dedicated to a detailed discussion of the nonlinear Riemann mapping theorem for multiply connected domains. Recall that the term H-conformal transformation $f : \Omega \to \Omega'$ pertains to a homeomorphic solution to the elliptic equation

$$\frac{\partial f}{\partial \bar{z}} = H\left(z, f, \frac{\partial f}{\partial z}\right) \tag{9.29}$$

taking Ω onto Ω'. We emphasize that in the multiply connected domains the range $\Omega' = f(\Omega)$ depends not only upon the topology of Ω but also on the equation being solved. Annuli with different conformal moduli are typical examples of this—there is no solution to the Cauchy-Riemann equations mapping one to the other. Therefore, not knowing Ω' explicitly demands that we must work with H defined on $\Omega \times \mathbb{C} \times \mathbb{C}$. We shall demonstrate the existence of an H-conformal transformation of a given Jordan n-connected domain Ω onto a canonical domain Ω'.

Let us reflect for a moment on the concept of a canonical domain. First, consider the doubly connected or Jordan 2-domains.

Theorem 9.3.1. *Let $\Omega \subset \hat{\mathbb{C}}$ be a doubly connected domain with nondegenerate boundary components. Then Ω can be conformally transformed onto the annulus*

$$\Omega' = \{z : r < |z| < R\},$$

where the ratio R/r is determined uniquely by the domain Ω. We call

$$\mathcal{M}_\Omega = \frac{1}{2\pi} \log\left(\frac{R}{r}\right)$$

the conformal module of Ω.

Concerning multiconnected domains, Jordan n-domains with $n \geqslant 3$, we only study conformal transformations of Ω onto a Schottky domain.

Definition 9.3.2. *An annulus with concentric slits, or briefly a Schottky domain, is an n-connected domain obtained from a round annulus $\{z : r < |z| < R\}$ by removing $(n-2)$ disjoint closed circular arcs.*

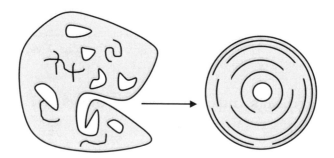

A Riemann mapping to a canonical domain

An n-connected generalization of Theorem 9.3.1 can be given as follows.

9.3. MULTIPLY CONNECTED DOMAINS

Theorem 9.3.3. *Every Jordan n-domain, $n \geqslant 2$, is conformally equivalent to a Schottky domain.*

Our goal is to extend this result to solutions of general nonlinear elliptic equations. We again assume that H satisfies the natural conditions of Definition 7.7.1.

Theorem 9.3.4. *Let $\Omega \subset \hat{\mathbb{C}}$ be a Jordan n-domain, $n \geqslant 2$. Then the elliptic equation*

$$\frac{\partial f}{\partial \bar{z}} = H\left(z, f, \frac{\partial f}{\partial z}\right) \tag{9.30}$$

admits a quasiconformal solution $f : \Omega \to \Omega'$, which maps Ω onto a Schottky domain.

9.3.1 Some Preliminaries

We make a few remarks to clarify the proof of Theorem 9.3.4. First, notice that any solution $f \in W^{1,2}_{loc}(\mathbb{C})$ to (9.30) is a quasiregular mapping and as such enjoys the argument principle. The following lemma, essentially of topological nature, will be crucial in establishing the required mapping properties of the solution. It is a classical theorem that such a mapping exists, see for instance R. Courant [98, 99].

Lemma 9.3.5. *Let $\Omega \subset \mathbb{C}$ be a Jordan n-domain and $f : \Omega \to \mathbb{C}$ a quasiregular mapping continuous up to the boundary $\partial \Omega = \Gamma_1 \cup \cdots \cup \Gamma_n$. Assume, moreover, that $f \neq 0$ on $\overline{\Omega}$ and that*

- *There are positive constants $0 < \delta_1 \leqslant \cdots \leqslant \delta_n$ such that*

$$|f(z)| = \delta_j \text{ for } z \in \Gamma_j, \quad j = 1, 2, \ldots, n$$

- *There are two components Γ_{j_1} and Γ_{j_2} of Γ for which we have*

$$\Delta_{\Gamma_j} \arg f = \begin{cases} 2\pi & \text{if } j = j_1 \\ -2\pi & \text{if } j = j_2 \\ 0 & \text{if } j_1 \neq j \neq j_2 \end{cases}$$

Under these assumptions f is a homeomorphic mapping of Ω onto a Schottky annulus $\{\delta_1 < |f| < \delta_n\}$ with $(n-2)$ slits on the circles $\{|w| = \delta_j\}$, $j = 2, \ldots, n-1$.

An example of a mapping satisfying the above conditions is the conformal mapping

$$\chi : \Omega \to \Omega' \tag{9.31}$$

of Ω onto the annulus $\{z : r < |z| < 1\}$ with $(n-2)$ circular slits.

We now see that for Theorem 9.3.4 it is enough to solve the following nonlinear boundary value problem.

Problem. Find a function $\varphi = \varphi(z)$ continuous in $\overline{\Omega}$ and satisfying the following two conditions.

- There are constants λ_j, $j = 1, \ldots, n$ such that
$$\Re \varphi(z) = \lambda_j \quad \text{for } z \in \Gamma_j, \quad \lambda_n = 0$$

- The function $f(z) = \chi(z) e^{\varphi(z)}$ solves the equation
$$\frac{\partial f}{\partial \bar{z}} = H\left(z, f, \frac{\partial f}{\partial z}\right)$$

Above, $\chi(z)$ denotes the conformal mapping of (9.31).

It will be convenient to use a quasiconformal change of coordinates in the z-variable. In this way we may assume that Ω is a circular or Schottky domain, that is, the unit disk $\mathbb{D}$ with $n - 1$ mutually disjoint closed disks removed. The configuration and size of these disks can be chosen more or less up to our discretion, while the associated ellipticity constants vary depending on such choices, they will remain finite with just a little care.

We shall look for the function $\varphi(z)$ in the following integral form,
$$\varphi(z) = -\frac{1}{\pi} \int_\Omega \left(A(z, \tau) \omega(\tau) + B(z, \tau) \overline{\omega(\tau)} \right) d\tau = (\mathcal{C}_\Omega \omega)(z)$$

where A and B are given kernels defined on $\overline{\Omega} \times \Omega$ and $\omega = \omega(\tau)$ is a density function in $L^p(\Omega)$ for a certain $p > 2$. We call $\mathcal{C}_\Omega$ the Cauchy transform of Ω.

Here are five basic properties of the Cauchy transform that we need to apply.
First,
$$|\mathcal{C}_\Omega \omega(z_1) - \mathcal{C}_\Omega \omega(z_2)| \leqslant C \|\omega\|_p |z_1 - z_2|^{1-2/p} \tag{9.32}$$

for $z_1, z_2 \in \overline{\Omega}$ and $p > 2$.

Second, to each boundary circle Γ_j, $j = 1, \ldots, n$, there corresponds a bounded linear functional $\Lambda_j : L^p(\Omega) \to \mathbb{R}$, $1 < p < \infty$, such that
$$(\Re \mathcal{C}_\Omega \omega)(z) \equiv \Lambda_j \omega \quad \text{for } z \in \Gamma_j, \quad \Lambda_n \equiv 0$$

Third,
$$\partial_{\bar{z}} \mathcal{C}_\Omega \omega = \omega \quad \text{for all } \omega \in L^p(\Omega)$$

The Beurling transform in Ω, defined by
$$\mathcal{S}_\Omega \omega = \partial_z \mathcal{C}_\Omega \omega,$$

is to be a bounded operator $\mathcal{S}_\Omega : L^p(\Omega) \to L^p(\Omega)$ for all $1 < p < \infty$, with the $L^p(\Omega)$-norms satisfying
$$\lim_{p \to 2} \|\mathcal{S}_\Omega\|_p = \|\mathcal{S}_\Omega\|_2 = 1 \tag{9.33}$$

9.3. MULTIPLY CONNECTED DOMAINS

We have already encountered two examples of such Cauchy transforms, one for the unit disk $\mathbb{D}$ in (4.134) and Theorem 9.2.1,

$$(\mathcal{C}_{\mathbb{D}}\omega)(z) = -\frac{1}{\pi}\int_{\mathbb{D}}\left(\frac{\omega(\tau)}{\tau - z} + \frac{z\,\overline{\omega(\tau)}}{1 - z\overline{\tau}}\right)d\tau,$$

and one for the annulus $\Omega = \{z : r < |z| < 1\}$ in Section 4.9,

$$\begin{aligned}(\mathcal{C}_{\Omega}\omega)(z) &= -\frac{1}{\pi}\int_{\Omega}\left(\frac{\omega(\tau)}{\tau - z} - \frac{z\,\overline{\omega(\tau)}}{1 - z\overline{\tau}}\right)d\tau \\ &+ \frac{1}{2\pi}\int_{\Omega}\frac{\omega(\tau)}{\tau}\sum_{k=1}^{\infty}\left(\frac{z + r^{2k}\tau}{z - r^{2k}\tau} + \frac{zr^{2k} + \tau}{zr^{2k} - \tau}\right)d\tau \\ &- \frac{1}{2\pi}\int_{A}\frac{\overline{\omega(\tau)}}{\tau}\sum_{k=1}^{\infty}\left(\frac{r^{2k} + z\overline{\tau}}{r^{2k} - z\overline{\tau}} + \frac{1 + r^{2k}\,z\overline{\tau}}{1 - r^{2k}\,z\overline{\tau}}\right)d\tau\end{aligned}$$

For the annulus the boundary functional $\Lambda_2 = 0$ while Λ_1 is given by

$$\Lambda_1\omega = -\frac{1}{\pi}\,\mathfrak{Re}\int_{\Omega}\frac{\omega(\tau)}{\tau}d\tau$$

In [170], by an analogy with the Cauchy operator of the annulus, the operators $\mathcal{C}_{\Omega}$ are constructed for all n-connected circular domains. Furthermore, the five requirements between (9.32) and (9.33) are also shown to hold.

9.3.2 Proof of the Mapping Theorem 9.3.4

With the aid of a good smoothing of H, in much the same way as we established the principal solutions or the nonlinear Riemann mapping theorem for simply connected domains, we can assume that the function

$$H : \Omega \times \mathbb{C} \times \mathbb{C} \to \mathbb{C}$$

is continuous.

We have therefore reduced the problem to solving the integral equation

$$\chi(z)e^{\mathcal{C}_{\Omega}\omega}\,\omega(z) = H\big(z,\,\chi e^{\mathcal{C}_{\Omega}\omega},\,(\chi' + \chi\,\mathcal{S}_{\Omega}\omega)\,e^{\mathcal{C}_{\Omega}\omega}\big) \qquad (9.34)$$

for $\omega \in L^p(\Omega)$ and for a certain $p > 2$. We choose this exponent $p > 2$ so that

$$k\,\|\mathcal{S}_{\Omega}\|_p < 1,$$

where $k < 1$ is the ellipticity constant of H.

In (9.34) the mapping $\chi(z)$ is conformal from Ω onto an annulus $\{z : r < |z| < 1\}$ with slits along concentric circles.

By the Schwarz reflection principle, χ has a holomorphic extension beyond the boundary of Ω. Thus we have the following estimates

$$r \leqslant |\chi(z)| \leqslant 1 \quad \text{and} \quad \|\chi'\|_\infty < \infty$$

Hence by the uniform ellipticity of H, every solution $\omega \in L^p(\Omega)$ to the equation (9.34), if it exists, satisfies

$$|\chi| e^{\Re \mathcal{C}_\Omega \omega} |\omega| \leqslant k|\chi' + \chi \mathcal{S}_\Omega \omega| e^{\Re \mathcal{C}_\Omega \omega},$$

so that

$$|\omega| \leqslant k|\mathcal{S}_\Omega \omega| + k\frac{|\chi'|}{|\chi|}$$

Taking the p-norms and rearranging gives

$$\|\omega\|_p \leqslant \frac{k\,\|\chi'/\chi\|_p}{1-k\|\mathcal{S}_\Omega\|_p} = M_{k,p}(\Omega) < \infty \tag{9.35}$$

This suggests that we should look for a solution in the closed L^p-ball

$$\mathbf{B} = \{\omega \in L^p(\Omega) : \|\omega\|_p \leqslant M_{k,p}(\Omega)\}$$

We define a mapping

$$\mathbf{T} : \mathbf{B} \to \mathbf{B}, \qquad \mathbf{T}\varphi = \Phi$$

by solving the equation

$$\chi\, e^{\mathcal{C}_\Omega \varphi}\, \Phi(z) = H\!\left(z,\, \chi e^{\mathcal{C}_\Omega \varphi},\, (\chi' + \chi \mathcal{S}_\Omega \Phi)\, e^{\mathcal{C}_\Omega \varphi}\right)$$

for $\Phi \in \mathbf{B}$. The existence and uniqueness of Φ are settled via the Banach contraction principle, as in the proof of Theorem 8.3.1.

We then need only to find a fixed point of $\mathbf{T}$ by using the Schauder fixed-point theorem. For this observe that the nonlinear operator $\mathbf{T} : \mathbf{B} \to \mathbf{B}$ is both continuous and compact. Indeed, suppose $\varphi_n \in \mathbf{B}$ converges weakly in $L^p(\Omega)$ to the function $\varphi \in \mathbf{B}$. By the property (9.32) the Cauchy transforms converge uniformly on $\overline{\Omega}$. Since $H : \Omega \times \mathbb{C} \times \mathbb{C} \to \mathbb{C}$ is continuous, we easily see, as in the case of simply connected domains, that $\Phi_n = \mathbf{T}\varphi_n$ converge strongly in $L^p(\mathbb{C})$ to a function $\Phi \in \mathbf{B}$ that satisfies the equation

$$\chi\, e^{\mathcal{C}_\Omega \varphi}\, \Phi = H\!\left(z,\, \chi e^{\mathcal{C}_\Omega \varphi},\, (\chi' + \chi \mathcal{S}_\Omega \Phi)\, e^{\mathcal{C}_\Omega \varphi}\right)$$

This, in view of the uniqueness, shows that $\Phi = \mathbf{T}\varphi$. The proof of Theorem 9.3.4 is complete. □

Chapter 10

Conformal Deformations and Beltrami Systems

It is now time to return to the study of Riemannian and conformal structures in planar domains, one of the starting themes of our discussion in Section 2.1.

Let Ω and Ω' be planar domains with measurable conformal structures $G : \Omega \to \mathbf{S}(2)$ and $H : \Omega' \to \mathbf{S}(2)$. We have already seen that a conformal mapping f between the surfaces (Ω, G) and (Ω', H) can be viewed as a solution to the first-order system of differential equations

$$D^t f(z) H(f(z)) Df(z) = J(z,f) G(z), \quad \text{for almost every } z \in \Omega \qquad (10.1)$$

Therefore a complex interpretation of the Beltrami system (10.1) is required. It will be useful to study this theme in some depth.

One of the main purposes of this chapter is to prove Theorem 10.1.1, which shows that (10.1) is equivalent to a quasilinear Beltrami equation

$$\frac{\partial f}{\partial \bar{z}} = \mu(z,f) \frac{\partial f}{\partial z} + \nu(z,f) \overline{\frac{\partial f}{\partial z}}$$

Moreover, we describe the explicit relations between the coefficients μ and ν with G and H. In fact, we will allow dependence of H on the z-variable and G on the function f since the calculations are effectively algebraic manipulations.

10.1 Quasilinearity of the Beltrami System

A priori the Beltrami system (10.1) is nonlinear, involving the partials in quadratic form. However, if we write $z = x + iy$ and $f(z) = u(z) + iv(z)$, we have

$$Df(z) = \begin{bmatrix} u_x & u_y \\ v_x & v_y \end{bmatrix},$$

so that $J(z,f) = u_x v_y - u_y v_x$, and

$$J(z,f)[Df(z)]^{-1} = \begin{bmatrix} v_y & -u_y \\ -v_x & u_x \end{bmatrix},$$

so that (10.1) reads as

$$\begin{bmatrix} u_x & u_y \\ v_x & v_y \end{bmatrix}^t H(f) = G(z) \begin{bmatrix} v_y & -u_y \\ -v_x & u_x \end{bmatrix},$$

from which linearity in the first-order derivatives of f is obvious. Indeed, if H is constant, the particular case of interest here is $H \equiv \mathbf{I}$; then (10.1) is linear. We write

$$G = \begin{bmatrix} g_{11} & g_{12} \\ g_{21} & g_{22} \end{bmatrix}, \quad g_{12} = g_{21}, \quad g_{11} g_{22} - g_{12}^2 = 1$$

$$H = \begin{bmatrix} h_{11} & h_{12} \\ h_{21} & h_{22} \end{bmatrix}, \quad h_{12} = h_{21}, \quad h_{11} h_{22} - h_{12}^2 = 1$$

Then we find the following four equations

$$\begin{aligned} h_{11} u_x + h_{12} v_x - g_{11} v_y + g_{12} v_x &= 0 \\ h_{12} u_x + h_{22} v_x + g_{11} u_y - g_{12} u_x &= 0 \\ h_{11} u_y + h_{12} v_y - g_{12} v_y + g_{22} v_x &= 0 \\ h_{12} u_y + h_{22} v_y + g_{12} u_y - g_{22} u_x &= 0, \end{aligned}$$

which we write as the two systems of two equations,

$$\begin{bmatrix} u_y \\ v_y \end{bmatrix} = \frac{1}{g_{11}} \begin{bmatrix} g_{12} - h_{12} & -h_{22} \\ h_{11} & g_{12} + h_{12} \end{bmatrix} \begin{bmatrix} u_x \\ v_x \end{bmatrix} \tag{10.2}$$

$$\begin{bmatrix} u_x \\ v_x \end{bmatrix} = \frac{1}{g_{22}} \begin{bmatrix} g_{12} + h_{12} & h_{22} \\ -h_{11} & g_{12} - h_{12} \end{bmatrix} \begin{bmatrix} u_y \\ v_y \end{bmatrix}, \tag{10.3}$$

and now it is easy to see that (10.2) and (10.3) are equivalent systems as the associated matrices are inverses of each other. Thus the Beltrami system (10.1) is equivalent to either of the linear systems (10.2) or (10.3).

10.1.1 The Complex Equation

We now shall write the Beltrami system (10.1) in complex notation. Of course,

$$u_x + i v_x = \frac{\partial f}{\partial z} + \frac{\partial f}{\partial \bar{z}}$$

$$v_y - i u_y = \frac{\partial f}{\partial z} - \frac{\partial f}{\partial \bar{z}}$$

10.1. QUASILINEARITY

for the derivatives of u and v, the real and imaginary parts of f. Substituting these in to either (10.2) or (10.3) we find two (equivalent) equations for $\frac{\partial f}{\partial z}$ and $\frac{\partial f}{\partial \bar{z}}$. For instance, (10.3) takes the form

$$(2g_{22} - 2ig_{12} - h_{22} - h_{11})\frac{\partial f}{\partial z} + (h_{11} - h_{22} + 2ih_{12})\overline{\frac{\partial f}{\partial z}}$$
$$+ (2g_{22} + 2ig_{12} + h_{11} + h_{22})\frac{\partial f}{\partial \bar{z}} + (h_{22} - h_{11} - 2ih_{12})\overline{\frac{\partial f}{\partial \bar{z}}} = 0$$

If we now take the complex conjugate of this equation we get another equation for the derivatives of f. From these two equations we choose to eliminate $\overline{\frac{\partial f}{\partial \bar{z}}}$, and obtain

$$\frac{\partial f}{\partial \bar{z}}(z) = \mu(z,f)\frac{\partial f}{\partial z} + \nu(z,f)\overline{\frac{\partial f}{\partial z}} \qquad (10.4)$$

where

$$\mu = \frac{g_{11} - g_{22} + 2ig_{12}}{g_{11} + g_{22} + h_{11} + h_{22}} \qquad (10.5)$$

and

$$\nu = \frac{h_{22} - h_{11} - 2ih_{12}}{g_{11} + g_{22} + h_{11} + h_{22}} \qquad (10.6)$$

Notice that g_{ij} is evaluated at z while h_{ji} is evaluated at $f(z)$ as in (10.1). Both μ and ν are functions of the dependent and independent variables.

We have established the equivalence of the Beltrami system (10.1) with the complex Beltrami equation (10.4), with μ and ν as defined in (10.5) and (10.6). We can now invert these last two equations so as to determine the coefficient matrices G and H in terms of the complex coefficients μ and ν from (10.4).

We put $\mu = a + ib$ and $\nu = \alpha + i\beta$ and set

$$T = g_{11} + g_{22} + h_{11} + h_{22} \qquad (10.7)$$

so that

$$aT = g_{11} - g_{22}, \qquad bT = 2g_{12},$$
$$\alpha T = h_{22} - h_{11}, \qquad \beta T = -2h_{12},$$

and then define the auxiliary variable

$$t = \frac{g_{11} + g_{22}}{g_{11} + g_{22} + h_{11} + h_{22}}, \qquad 1 - t = \frac{h_{11} + h_{22}}{g_{11} + g_{22} + h_{11} + h_{22}} \qquad (10.8)$$

Then

$$T(a+t) = 2g_{11}, \qquad T(a-t) = -2g_{22}, \qquad Tb = 2g_{12},$$
$$T(\alpha + 1 - t) = 2h_{22}, \qquad T(\alpha - 1 + t) = -2h_{11}, \qquad T\beta = -2h_{12}$$

Thus $a^2 + b^2 - t^2 = \alpha^2 + \beta^2 - (1-t)^2$, and so

$$2t = 1 + |\mu|^2 - |\nu|^2 \qquad (10.9)$$

Then $t^2 - a^2 - b^2 = 4T^{-2}$ yields

$$T = \frac{4}{\sqrt{1-(|\mu|+|\nu|)^2}\sqrt{1-(|\mu|-|\nu|)^2}}, \quad (10.10)$$

and then it is straightforward matter to see that

$$g_{11} = \tfrac{T}{4}\left(|1+\mu|^2 - |\nu|^2\right), \qquad g_{22} = \tfrac{T}{4}\left(|1-\mu|^2 - |\nu|^2\right), \qquad g_{12} = \tfrac{T}{2}\Im m(\mu)$$

$$h_{11} = \tfrac{T}{4}\left(|1-\nu|^2 - |\mu|^2\right), \qquad h_{22} = \tfrac{T}{4}\left(|1+\nu|^2 - |\mu|^2\right), \qquad h_{12} = \tfrac{-T}{2}\Im m(\nu),$$

and this gives the following formulas for G and H.

$$G = \frac{1}{\Delta}\begin{bmatrix} |1+\mu|^2 - |\nu|^2 & 2\Im m(\mu) \\ 2\Im m(\mu) & |1-\mu|^2 - |\nu|^2 \end{bmatrix} \quad (10.11)$$

$$H = \frac{1}{\Delta}\begin{bmatrix} |1-\nu|^2 - |\mu|^2 & -2\Im m(\nu) \\ -2\Im m(\nu) & |1+\nu|^2 - |\mu|^2 \end{bmatrix} \quad (10.12)$$

where

$$\Delta = \sqrt{1-(|\mu|+|\nu|)^2}\sqrt{1-(|\mu|-|\nu|)^2}$$

It is somewhat interesting to note that Δ involves all the sums and differences of the modulus of μ and ν, namely,

$$\Delta^2 = (1-|\mu|-|\nu|)(1+|\mu|-|\nu|)(1-|\mu|+|\nu|)(1+|\mu|+|\nu|)$$

Further, the dependence of G on ν and of H on μ can be written succinctly as

$$G = \frac{1}{\Delta}\left(\begin{bmatrix} |1+\mu|^2 & 2\Im m(\mu) \\ 2\Im m(\mu) & |1-\mu|^2 \end{bmatrix} - |\nu|^2 \mathbf{I}\right)$$

$$H = \frac{1}{\Delta}\left(\begin{bmatrix} |1-\nu|^2 & -2\Im m(\nu) \\ -2\Im m(\nu) & |1+\nu|^2 \end{bmatrix} - |\mu|^2 \mathbf{I}\right)$$

Actually, the reader will easily see that the foregoing algebraic calculations work in a little more generality. This we record in the following theorem.

Theorem 10.1.1. *The nonlinear Beltrami system*

$$D^t f(z) H(z,f) Df(z) = J(z,f) G(z,f),$$

for (Lusin) measurable conformal structures $G : \Omega \times \Omega' \to \mathbf{S}(2)$ *and* $H : \Omega \times \Omega' \to \mathbf{S}(2)$, *is equivalent to the quasilinear Beltrami equation*

$$f_{\bar{z}} = \mu(z,f) f_z + \nu(z,f) \overline{f_z},$$

where G, H *are related to* μ, ν *by (10.11) and (10.12), and conversely,* μ, ν *are given by (10.5) and (10.6).*

Of course if G is to depend only on z and H on f as per (10.1), then μ and ν must have a rather more special form. We will come to this in a moment.

10.2 Conformal Equivalence of Riemannian Structures

The powerful tools from Chapter 9 on the nonlinear measurable Riemann mapping theorem are now at our disposal. As a consequence, we will find that with great generality, any two Riemannian structures of $\mathbb{C}$ are conformally equivalent. Let $(G, \mathbb{C})$ and $(H, \mathbb{C})$ be conformal structures on the plane. Section 2.1 has already pointed out the connection with conformal isomorphisms,

$$f : \mathbb{C}_G \to \mathbb{C}_H \qquad (10.13)$$

Hence in view of (2.10), to find a conformal isomorphism we should establish a homeomorphic solution to the nonlinear differential equation

$$D^t f(z) H(f(z)) Df(z) = J(z, f) G(z) \qquad (10.14)$$

Here we find the (very weak) restrictions: The equation is solvable as soon as the structures H and G are measurable and bounded.

We recall that a measurable conformal structure is bounded if $\|\log |G|\|_\infty < \infty$, see (2.20) for the metric and related facts used here. We can reinterpret our solution to the measurable Riemann mapping Theorem 9.0.3 in the following way.

Theorem 10.2.1. *Let Ω and Ω' be simply connected domains in $\mathbb{C}$. Suppose that $G : \Omega \times \Omega' \to \mathbf{S}(2)$ and $H : \Omega \times \Omega' \to \mathbf{S}(2)$ are measurable and bounded. Then the nonlinear Beltrami system*

$$D^t f(z) H(z, f) Df(z) = J(z, f) G(z, f) \qquad (10.15)$$

admits a $W^{1,2}_{loc}$-homeomorphic solution $f : \Omega \to \Omega'$.

Within the generality of Theorem 10.2.1 we cannot hope for uniqueness or classification of the homeomorphic solutions. However, in the original setting of (10.13) or (10.14) a complete classification of the conformal isomorphisms is possible.

Theorem 10.2.2. *Let $G = G(z)$ and $H = H(w)$ be measurable conformal structures on $\mathbb{C}$ satisfying the ellipticity conditions*

$$1/K \leqslant \langle \xi, G(z)\xi \rangle \leqslant K, \qquad |\xi| = 1 \qquad (10.16)$$
$$1/K \leqslant \langle \zeta, H(w)\zeta \rangle \leqslant K, \qquad |\zeta| = 1 \qquad (10.17)$$

Then the equation

$$D^t f(z) H(f(z)) Df(z) = J(z, f) G(z) \quad \text{for almost every } z \in \mathbb{C}, \qquad (10.18)$$

admits exactly one normalized solution f. Its inverse h satisfies the equation
$$D^t h(w) G(h(w)) Dh(w) = J(w,h) H(w), \quad \text{for almost every } w \in \mathbb{C}$$
Moreover, every solution $g \in W^{1,2}_{loc}(\Omega)$ to the equation
$$D^t g(z) H(g(z)) Dg(z) = J(z,g) G(z) \quad \text{for almost every } z \in \Omega,$$
has the form
$$g(z) = (\phi \circ f \circ \psi)(z),$$
for holomorphic maps $\psi : \Omega \to \psi(\Omega)$ and $\phi : f(\psi(\Omega)) \to g(\Omega)$.

Proof. We simply need to recall that (10.18) is obtained as a system of equations for the composition $f = f^1 \circ f^2$, where f^2 solves the equation
$$D^t f^2(z) Df^2(z) = J(z,f^2) G(z),$$
and f^1 solves
$$D^t f^1(w) H(f^1(w)) Df^1(w) = J(w,f^1) \mathbf{I}$$
The first equation here is nothing other than the complex linear Beltrami equation, and the second equation yields the Beltrami equation for the inverse map. We can now apply Stoilow's factorization theorem to complete the proof. □

Later, in Section 20, we will consider generalizations of the measurable Riemann mapping theorem for degenerate equations. With these results even the boundedness of H and G can be somewhat relaxed.

10.3 Group Properties of Solutions

In this section we shall study the group properties of solutions and equations, namely, we find the equations satisfied by the inverse of a solution and the composition of two solutions.

Suppose that $f : \Omega \to \Omega'$ is a suitably differentiable homeomorphism and that $h : \Omega' \to \Omega$ is the inverse of f. If we write $f = f(z)$ and $h = h(w)$, from the equation
$$D^t f(z) H(z,f) Df(z) = J(z,f) G(z,f),$$
we put $z = h(w)$ and multiply on the left by $D^t h(w)$ and on the right by $Dh(w)$ and find that (away from where the Jacobian might vanish)
$$\begin{aligned} J(z,f) D^t h(w) G(h,w) Dh(w) &= D^t h(w) D^t f(z) H(z,f) Df(z) Dh(w) \\ &= H(w,h) \end{aligned}$$
Since $J(z,f) = 1/J(w,h)$, we now find that h satisfies the equation
$$D^t h(w) G(h,w) Dh(w) = J(w,h) H(w,h) \tag{10.19}$$

10.3. GROUP PROPERTIES

This of course simply expresses the fact that if $f : (\Omega, G) \to (\Omega', H)$ is conformal, then $h = f^{-1} : (\Omega', H) \to (\Omega, G)$ is also conformal.

Alternatively, using (2.49) and (2.50) we see (similarly to (2.51)) that if f satisfies the equation $f_{\bar{z}} = \mu(z, f) f_z + \nu(z, f) \overline{f_z}$, then its inverse h satisfies the equation

$$-h_{\bar{w}}(w) = \mu(z, f) \, \overline{h_w(w)} + \nu(z, f) \, h_w(w), \quad w = f(z)$$

Finally observe that $\mu(h, w) = \mu(z, f)$, which establishes the following lemma. For the proof we take for granted the nontrivial facts, established at Corollary 5.5.2, that the chain rule applies to quasiregular mappings and that, when nonconstant, quasiregular mappings have nonvanishing Jacobian almost everywhere.

Lemma 10.3.1. *Let $f \in W_{loc}^{1,2}(\Omega)$ be a homeomorphic solution to*

$$\frac{\partial f}{\partial \bar{z}} = \mu(z, f) \frac{\partial f}{\partial z} + \nu(z, f) \overline{\frac{\partial f}{\partial z}}$$

with $|\mu(z, f)| + |\nu(z, f)| \leqslant k < 1$. Then $h = f^{-1}$ is differentiable almost everywhere in $\Omega' = f(\Omega)$ and satisfies the equation

$$\frac{\partial h}{\partial \bar{w}} = -\nu(h, w) \frac{\partial h}{\partial w} - \mu(h, w) \overline{\frac{\partial h}{\partial w}} \tag{10.20}$$

Next is the formula for the Beltrami coefficient of the composition of quasiregular maps which the reader might compare with our earlier composition formula in (5.5.6).

Lemma 10.3.2. *Suppose that f and h are quasiregular and*

$$f_{\bar{z}} = \mu(z, f) f_z, \qquad h_{\bar{w}} = \nu(w, h) h_w$$

and $g = h \circ f$. Then

$$g_{\bar{z}} = \sigma(z, g) \, g_z,$$

where

$$\sigma(z, g) = \frac{\mu(z, f) + \nu(f, g) \frac{\overline{f_z}}{f_z}}{1 + \overline{\mu}(z, f) \nu(f, g) \frac{\overline{f_z}}{f_z}}, \tag{10.21}$$

and we have the ellipticity estimate

$$|\sigma| \leqslant \frac{|\mu| + |\nu|}{1 + |\mu||\nu|}$$

Proof. The formulas for the chain rule, (2.49) and (2.50), give

$$\begin{aligned} g_{\bar{z}} &= h_w(f) f_{\bar{z}} + h_{\bar{w}}(f) \overline{f_z} = \mu(z, f) h_w(f) f_z + \nu(f, g) \overline{f_z} \, \overline{h_w(f)} \\ g_z &= h_w(f) f_z + h_{\bar{w}}(f) \overline{f_{\bar{z}}} = f_z h_w(f) + \nu(f, g) \overline{\mu}(z, f) \, \overline{f_z} \, \overline{h_w(f)}, \end{aligned}$$

from which the first part is clear. To see the remainder we recall that the map $w \mapsto (\mu-w)/(1-\bar{\mu}w)$ is a hyperbolic isometry of the disk. The triangle inequality together with (2.15) gives

$$\rho_{\mathbb{D}}(0, \sigma(z,g)) = \rho_{\mathbb{D}}\left(\mu(z,f), \nu(f,g)\frac{\overline{f_z}}{f_z}\right)$$
$$\leqslant \rho_{\mathbb{D}}(0,\mu) + \rho_{\mathbb{D}}(0,\nu),$$
$$|\sigma(z,f)| \leqslant \tanh\frac{1}{2}\left(\rho_{\mathbb{D}}(0,\mu) + \rho_{\mathbb{D}}(0,\nu)\right)$$
$$= \frac{|\mu|+|\nu|}{1+|\mu||\nu|}$$

Here we have used the sum formula for the hyperbolic tangent. $\square$

It is a little surprising at first that the formula (10.21) does not involve the derivatives of h, even in the nonlinear setting. A little more surprising are some of its ramifications.

Lemma 10.3.3. *Suppose that f and h are quasiregular mappings satisfying*

$$\frac{\partial f}{\partial \bar{z}} = \mu(z,f)\frac{\partial f}{\partial z}, \tag{10.22}$$

$$\frac{\partial h}{\partial \bar{w}} = -\nu(w,h)\overline{\frac{\partial h}{\partial w}}, \tag{10.23}$$

and $g = h \circ f$. Then

$$\frac{\partial g}{\partial \bar{z}} = \alpha(z,g)\frac{\partial g}{\partial z} + \beta(z,g)\overline{\frac{\partial g}{\partial z}},$$

where

$$\alpha = \alpha(z,g) = \frac{\mu(1-|\nu|^2)}{1-|\mu|^2|\nu|^2} = \frac{\mu(z,f)(1-|\nu(f,g)|^2)}{1-|\mu(z,f)|^2|\nu(f,g)|^2} \tag{10.24}$$

and

$$\beta = \beta(z,g) = -\frac{\nu(1-|\mu|^2)}{1-|\mu|^2|\nu|^2} = -\frac{\nu(f,g)(1-|\mu(z,f)|^2)}{1-|\mu(z,f)|^2|\nu(f,g)|^2}, \tag{10.25}$$

where $f = f(z)$, and we have the ellipticity estimate

$$|\alpha| + |\beta| \leqslant \frac{|\mu|+|\nu|}{1+|\mu||\nu|} < 1$$

Proof. Suppose f satisfies (10.22) and h satisfies (10.23). We put $g(z) = (h \circ f)(z)$ and seek an equation for g that does not involve the derivatives of f or h. Again, we begin with the formulas for the chain rule (2.49) and (2.50), which give

$$g_{\bar{z}} = h_w f_{\bar{z}} + h_{\bar{w}}\overline{f_z} = \mu(z,f)f_z h_w - \nu(h,w)\overline{f_z}\,\overline{h_w}$$
$$g_z = h_w f_z + h_{\bar{w}}\overline{f_{\bar{z}}} = f_z h_w - \nu(h,w)\overline{\mu(z,f)}\,\overline{f_z} h_w$$
$$\overline{g_z} = \overline{h_w}\,\overline{f_z} - \nu(h,w)\mu(z,f)f_z h_w$$

10.3. GROUP PROPERTIES

Hence we are led to compute the following combination of g_z and $\overline{g_{\bar z}}$:

$$\begin{aligned}\mu(1-|\nu|^2)g_z - \nu(1-|\mu|^2)\overline{g_{\bar z}} &= \mu(1-|\nu|^2|\mu|^2)f_z h_w - \nu(1-|\nu|^2|\mu|^2)\overline{f_{\bar z}}\,\overline{h_w}\\ &= (1-|\nu|^2|\mu|^2)g_{\bar z}\end{aligned}$$

Thus the composition g satisfies the equation

$$\frac{\partial g}{\partial \bar z} = \alpha(z,g)\frac{\partial g}{\partial z} + \beta(z,g)\overline{\frac{\partial g}{\partial z}} \tag{10.26}$$

with α and β as given in (10.24) and (10.25). Notice that

$$|\alpha|+|\beta| = \frac{|\mu|+|\nu|}{1+|\mu||\nu|} \tag{10.27}$$

Recall from (2.15) that, in terms of the hyperbolic metric of the unit disk, we have the pointwise equality $|\nu| = \tanh \frac{1}{2}\rho_{\mathbb{D}}(0,\nu)$. Then (10.27) is the sum formula for the hyperbolic tangent, namely,

$$\begin{aligned}|\alpha|+|\beta| &= \frac{|\tanh \frac{1}{2}\rho_{\mathbb{D}}(0,\mu)| + |\tanh \frac{1}{2}\rho_{\mathbb{D}}(0,\nu)|}{1+|\tanh \frac{1}{2}\rho_{\mathbb{D}}(0,\mu)||\tanh \frac{1}{2}\rho_{\mathbb{D}}(0,\nu)|}\\ &= \tanh \frac{1}{2}\left(\rho_{\mathbb{D}}(0,\mu)+\rho_{\mathbb{D}}(0,\nu)\right),\end{aligned}$$

providing an analogy with Lemma 10.3.2. □

In the case where $\mu = \nu$, we observe that

$$\alpha(z,z) + \beta(z,z) \equiv 0$$

In particular, this shows that the identity function $g(z) = z$ is a solution to (10.26) in this case.

10.3.1 Semigroups

One of the fundamental properties of the set of holomorphic functions is that it is closed under composition. In other words, the space of $W^{1,2}_{loc}(\Omega,\Omega)$ solutions to the Cauchy-Riemann equations is closed under composition, and moreover, the space of homeomorphic solutions forms a Lie group. In this section we identify those equations with these two properties.

The equation

$$D^t f(z) H(z,f) Df(z) = J(z,f) G(z,f), \tag{10.28}$$

for $G, H : \Omega \times \Omega \to \mathbb{C}$, admits one formulation from which it is clear that the space of solutions forms a semigroup, namely, the case $G(z,f) = G(z)$ and $H(z,f) = G(f)$, giving the equation

$$D^t f(z) G(f) Df(z) = J(z,f) G(z) \tag{10.29}$$

In terms of the associated Beltrami equation, we recall that (10.28) is equivalent to the system
$$f_{\bar{z}} = \mu(z, f)f_z + \nu(z, f)\overline{f_z}, \tag{10.30}$$
and we find that (10.29) implies (using (10.5) and (10.6))
$$\mu(z, f) = -\nu(f, z), \tag{10.31}$$
while (10.30) becomes
$$f_{\bar{z}} = \mu(z, f)f_z - \mu(f, z)\overline{f_z} \tag{10.32}$$
The associated equation for $h = f^{-1}$ is
$$h_{\bar{z}} = \mu(z, h)h_z - \mu(h, z)\overline{h_z}, \tag{10.33}$$
which is evidently the same as (10.32). Thus (10.32) has the property that, if f is a solution, then so is f^{-1}. Moreover, the identity is always a solution.

The formulas in (10.11) and (10.12) imply further constraints on μ if the system in (10.28) is to be equivalent to a system such as (10.32). Indeed, we find from (10.5) that
$$\mu(z, w) = \frac{g_{11}(z) - g_{22}(z) + 2ig_{12}(z)}{g_{11}(z) + g_{22}(z) + g_{11}(w) + g_{22}(w)} \tag{10.34}$$
If we put $a = (g_{11} - g_{22})/2$ and $b = g_{12}$, then a, b are real-valued functions of z and
$$g_{11} + g_{22} = 2\sqrt{a^2(z) + b^2(z) + 1}$$
Upon setting
$$\eta(z) = a(z) + ib(z) = \frac{1}{2}(g_{11}(z) - g_{22}(z)) + ig_{12}(z),$$
we find
$$\mu(z, w) = \frac{\eta(z)}{\sqrt{|\eta(z)|^2 + 1} + \sqrt{|\eta(w)|^2 + 1}} = -\nu(w, z) \tag{10.35}$$
Of course, we have the ellipticity condition as
$$|\mu(z, w)| + |\nu(z, w)| = \frac{|\eta(z)| + |\eta(w)|}{\sqrt{|\eta(z)|^2 + 1} + \sqrt{|\eta(w)|^2 + 1}} < 1,$$
and we have uniform ellipticity if and only if η is bounded.

Conversely, every bounded function $\eta(z) : \Omega \to \mathbb{C}$ defines a conformal structure $G : \Omega \to \mathbf{S}(2)$ as follows,
$$g_{11} = \left(\Re e\{\eta\} + \sqrt{|\eta|^2 + 1}\right)$$
$$g_{22} = \left(-\Re e\{\eta\} + \sqrt{|\eta|^2 + 1}\right)$$
$$g_{12} = \Im m\{\eta\}$$

It is clear that now defining $\mu(z, w)$ by (10.35) gives us back (10.34). We summarize this discussion in the following theorem.

10.3. GROUP PROPERTIES

Theorem 10.3.4. *Let $\eta(z)$ be a bounded complex-valued function defined on a domain Ω. Then the solutions $f \in W^{1,2}_{loc}(\Omega, \Omega)$ to the quasilinear uniformly elliptic equation*

$$\frac{\partial f}{\partial \bar{z}} = \frac{1}{\sqrt{|\eta(z)|^2 + 4} + \sqrt{|\eta(f)|^2 + 4}} \left(\eta(z) \frac{\partial f}{\partial z} - \eta(f) \overline{\frac{\partial f}{\partial z}} \right) \qquad (10.36)$$

form a semigroup closed under composition. The family of homeomorphic solutions forms a uniformly quasiconformal group.

To compare the set of solutions to (10.36) with the set of holomorphic functions, we consider finding a conformal map $\varphi : (\Omega, G) \to (\Omega, \mathbf{I})$, that is, solving the equation

$$\varphi_{\bar{z}} = \sigma(z) \varphi_z, \qquad (10.37)$$

with $\varphi \in W^{1,2}_{loc}(\Omega, \Omega)$ a homeomorphism, and from (10.34)

$$\sigma(z) = \frac{\eta}{1 + \sqrt{|\eta|^2 + 1}}$$

Notice then that for any solution to (10.30) and hence (10.36), we see that $\varphi \circ f \circ \varphi^{-1}$ is conformal as a map of the usual structures and therefore holomorphic. We have the diagram of conformal maps

$$\begin{array}{ccc} (\Omega, G) & \xrightarrow{f} & (\Omega, G) \\ \varphi \downarrow & & \downarrow \varphi \\ (\Omega, \mathbf{I}) & \xrightarrow{\varphi \circ f \circ \varphi^{-1}} & (\Omega, \mathbf{I}) \end{array}$$

In particular, after a change of coordinates by φ, each solution to (10.36) is holomorphic. Conversely, every function of the form $f = \varphi^{-1} \circ h \circ \varphi$, with h holomorphic, satisfies (10.36).

10.3.2 Sullivan-Tukia Theorem

The affirmative solution to the two-dimensional Hilbert-Smith conjecture [266] asserts that in the plane any connected, locally compact topological transformation group is isomorphic as a topological group to a Lie group. Thus local compactness and connectedness of the space of homeomorphic solutions become important topological properties to establish. However, in the uniformly elliptic case, these properties follow from a more general observation of Sullivan [340] and Tukia [356] concerning groups of quasiconformal homeomorphisms.

We say a group G of quasiconformal self-homeomorphisms of a domain Ω is *uniformly* quasiconformal if there is a $K < \infty$ so that each element $g \in G$ is K-quasiconformal. As an example, the homeomorphic solutions to (10.36) form a uniformly quasiconformal group. Sullivan and Tukia show that, in two dimensions, every uniformly quasiconformal group is quasiconformally conjugate to a Möbius group. That is, $G = f \circ \Gamma \circ f^{-1}$ where Γ is a group of Möbius transformations. This result is false in higher dimensions. Conversely, and in any dimension, if $G = f \circ \Gamma \circ f^{-1}$ for some Möbius group, it is clearly a K^2-quasiconformal group.

The situation is not so clear-cut for semigroups of transformations. However, this is an important question to study for applications in the theory of iteration of rational functions. A semigroup G is said to be *uniformly*-quasiregular if there is a $K < \infty$ so that each element $g \in G$ is K-quasiregular. One way to construct a uniformly quasiregular semigroup is to conjugate a family of analytic mappings that is closed under composition, by a quasiconformal mapping. For instance, if R is a rational map of the Riemann sphere, then $G = \{f \circ R^n \circ f^{-1} : n \in \mathbb{N}\}$ is a uniformly quasiregular semigroup; here R^n denotes the nth iterate under composition. Only minor modification of Sullivan and Tukia's arguments are necessary to show that a uniformly quasiregular semigroup of the Riemann sphere is the conjugate of rational semigroup when the semigroup is cyclic or abelian. However, without these strong algebraic assumptions, Hinkkanen has shown the result to be false in general, see [167].

We now turn to the proof of the Sullivan-Tukia theorem.

Theorem 10.3.5. *Let G be a K-quasiconformal group acting on a domain Ω. Then there is a K-quasiconformal map $f : \Omega \to \mathbb{C}$ such that $\Gamma = f \circ G \circ f^{-1}$ is a group of conformal transformations of $f(\Omega)$.*

Proof. Theorem 3.1.3 establishes the local compactness of any family of K-quasiconformal mappings. Therefore G is separable, and further, if we establish the theorem for a countable dense subgroup of G, then the general result will follow since, of course, if a dense subgroup of G is conformal, then G is conformal. Hence, by considering the group generated by a countable dense subset, we may assume, without loss of generality, that G is countable.

Suppose we find a quasiconformal mapping f such that $\mu_{f \circ g}(z) = \mu_f(z)$ for each $g \in G$, where μ_f is the Beltrami coefficient of f. Then, by the uniqueness established in the Stoilow factorization Theorem 5.5.1 we see that $f \circ g \circ f^{-1}$ is conformal for every $g \in G$. Clearly $\Gamma = f \circ G \circ f^{-1}$ is a group.

From Lemma 10.3.2 and (10.21) we have

$$\mu_{f \circ g}(z) = \frac{\mu_g(z) + \mu_f(g(z))\frac{\overline{g_z}}{g_z}}{1 + \overline{\mu_g(z)}\mu_f(g(z))\frac{\overline{g_z}}{g_z}} = T_g\big(\mu_f(g(z))\big) \tag{10.38}$$

where for each fixed z, $T_g(\zeta) = \frac{\mu_g + \zeta \frac{\overline{g_z}}{g_z}}{1 + \zeta \overline{\mu_g} \frac{\overline{g_z}}{g_z}}$ is a linear fractional transformation of the disk—a hyperbolic isometry. We are then asked to find a Beltrami coefficient μ such that

$$T_g\big(\mu(g(z))\big) = \mu(z) \quad \text{almost everywhere}$$

There is a set function solution to this equation. Namely

$$U(z) = \{\mu_h(z) : h \in G\} \tag{10.39}$$

10.3. GROUP PROPERTIES

We see

$$T_g(U(g(z))) = \left\{ \frac{\mu_g + \mu_h(g(z))\frac{\overline{g_z}}{g_z}}{1 + \mu_h(g(z))\overline{\mu_g}\frac{g_z}{g_z}} : h \in G \right\}$$
$$= \{\mu_{g \circ h}(z) : h \in G\} = \{\mu_h(z) : h \in G\} = U(z),$$

where this penultimate fact follows since G is a group, so that $G \circ h = G$. Finally, as T_g acts isometrically on the hyperbolic disk and as $U(z) \subset \mathbb{D}(k)$, $k = \frac{K-1}{K+1}$, is always a bounded subset of the disk we need to pick up an isometric invariant of the set $U(z)$ to define μ. Thus we set

$\mu(z) =$ center of the smallest closed hyperbolic disk containing the set $U(z)$

Using elementary facts from hyperbolic geometry one can show that a smallest disk exists and is unique, see eg. [356]. Therefore its center is a well defined function of a bounded set.

The function μ is measurable, since G is countable, and bounded, $|\mu| \leqslant k$—clearly much better estimates are possible using hyperbolic trigonometry and noting that $0 \in U(z)$ for all z, but this would lead us astray. By construction $T_g(\mu \circ g) = \mu$ for each $g \in G$. We obtain the function f we seek by solving the appropriate Beltrami equation $f_{\bar{z}} = \mu f_z$. This completes the proof. □

Remark 10.3.6. *Maskit [249] has shown that any group of conformal transformations of a domain is conformally conjugate to a Möbius group. Using this result we can therefore further assert that each uniformly quasiconformal group acting on a domain in $\hat{\mathbb{C}}$ is quasiconformally conjugate to a Möbius group.*

Next, if G is assumed to be a uniformly quasiregular semigroup, a similar argument as given above will work. Away from the branching loci we consider all the branches of inverses of elements of G and define $U(z) = \{\mu_h : h \in G \cup G^{-1}\}$. There is a slight problem in that we will ultimately require $\{\mu_{g \circ h}(z) : h \in G \cup G^{-1}\} = \{\mu_h(z) : h \in G \cup G^{-1}\}$. This will not be true in general unless we impose some additional algebraic structure on G such as being cyclic (generated by one element) or abelian. As we have noted, such an assumption will be necessary for the result to be true.

10.3.3 Ellipticity Constants

Here, for completeness, we just want to explicitly relate the ellipticity constants between the two forms of the Beltrami system we have looked at. Recall that the matrices G and H given in the Beltrami system (10.1) are both positive definite and have determinant 1. The ellipticity constant for the Beltrami system is $K = |G| \, |H|$. We compute from (10.11) and (10.12) the traces of the matrices G and H to be

$$\text{tr}(G) = 2 \frac{1 + |\mu|^2 - |\nu|^2}{\sqrt{1 - (|\mu| + |\nu|)^2} \sqrt{1 - (|\mu| - |\nu|)^2}}$$

$$\text{tr}(H) = 2\frac{1 - |\mu|^2 + |\nu|^2}{\sqrt{1 - (|\mu| + |\nu|)^2}\sqrt{1 - (|\mu| - |\nu|)^2}}$$

We also have $|G^{-1}| = |G|^{-1}$ and

$$\text{tr}(G) = |G| + |G|^{-1}, \qquad \text{tr}(H) = |H| + |H|^{-1},$$

so that

$$|G| = \frac{1}{2}\left(\text{tr}(G) + \sqrt{\text{tr}^2(G) - 4}\right)$$
$$|H| = \frac{1}{2}\left(\text{tr}(H) + \sqrt{\text{tr}^2(H) - 4}\right)$$

Next, we recall that from (10.7) and (10.5),

$$T^2|\mu|^2 = (g_{11} - g_{22})^2 + 4g_{12}^2 = \text{tr}^2(G) - 4$$
$$T^2|\nu|^2 = (h_{22} - h_{11})^2 + 4h_{12}^2 = \text{tr}^2(H) - 4$$

and that from (10.8), (10.9) and (10.7)

$$\frac{T}{2}(1 + |\mu|^2 - |\nu|^2) = \text{tr}(G)$$
$$\frac{T}{2}(1 - |\mu|^2 + |\nu|^2) = \text{tr}(H)$$

Then

$$|G| = \frac{(1+|\mu|)^2 - |\nu|^2}{\sqrt{1 - (|\mu|+|\nu|)^2}\sqrt{1 - (|\mu|-|\nu|)^2}}$$
$$= \frac{\sqrt{1+|\mu|+|\nu|}}{\sqrt{1-|\mu|-|\nu|}}\frac{\sqrt{1+|\mu|-|\nu|}}{\sqrt{1-|\mu|+|\nu|}}$$

$$|H| = \frac{(1+|\nu|)^2 - |\mu|^2}{\sqrt{1 - (|\mu|+|\nu|)^2}\sqrt{1 - (|\mu|-|\nu|)^2}}$$
$$= \frac{\sqrt{1+|\mu|+|\nu|}}{\sqrt{1-|\mu|-|\nu|}}\frac{\sqrt{1-|\mu|+|\nu|}}{\sqrt{1+|\mu|-|\nu|}}$$

We define

$$k = |\mu| + |\nu| \tag{10.40}$$

and find

$$K = |G|\,|H| = \frac{1+k}{1-k} \tag{10.41}$$

$$k = |\mu| + |\nu| = \frac{K-1}{K+1} \tag{10.42}$$

Chapter 11

A Quasilinear Cauchy Problem

In this chapter we shall study existence and uniqueness problems for a quasilinear Cauchy problem. We will not aim for maximal generality even though a large class of equations is covered with the results obtained here. Our goal is rather to develop tools for our subsequent discussion of the theory of holomorphic motions, although for that application it is, in fact, sufficient to work with the smooth case.

11.1 The Nonlinear $\bar{\partial}$-Equation

The differential equation we are interested in here is

$$\frac{\partial g}{\partial \bar{z}} = \Psi(z, g) \tag{11.1}$$

$$g(z) \to z_0 \quad \text{as } z \to \infty \tag{11.2}$$

This equation lies slightly outside our theme of ellipticity, yet the reader will see that it plays a crucial role in a central result in the modern theory of quasiconformal mappings. Still (11.1) is of the form $\mathcal{A}(z, g, g_z, g_{\bar{z}}) = 0$, where $\mathcal{A}$ is Petrovsky-elliptic; see (7.35). However, the homogeneity condition we have required, $\mathcal{A}(z, g, 0, 0) = 0$, is no longer satisfied and so the general theory developed in the previous chapters will not apply without modification.

The classical linear Cauchy problem has the right hand side of (11.1) in the form

$$\Psi(z, g) = \sigma_1(z) g(z) + \sigma_2(z) \overline{g(z)}$$

with $\sigma_i \in L^{2\pm}(\mathbb{C})$, the spaces defined in (8.28). The solutions to the linear equation are the pseudo-analytic or generalized analytic functions studied by Bers, Bers-Nirenberg, Vekua and others. We refer the reader to Vekua's book [366] for a comprehensive treatment of the linear theory. It is clear that (11.1) will not have a solution, unique or otherwise, without some hypothesis on the function Ψ. Therefore we shall make the following assumptions on the function Ψ.

- $\Psi(z,\cdot): \mathbb{C} \to \mathbb{C}$ is continuous for almost all $z \in \mathbb{C}$. (11.3)
- There is a compactly supported $\sigma \in L^2(\mathbb{C})$ with $\Psi(z,\zeta) \leqslant \sigma(z)|\zeta|$. (11.4)

The last condition suggests we are near the case of homogeneous equations. It also tells us that Ψ has compact support in the z-variable. In view of the normalization (11.2), this means that our solutions will be holomorphic and bounded outside a disk $\mathbb{D}(0,R)$ for some sufficiently large R.

We begin our investigation with the following fixed-point theorem.

11.2 A Fixed-Point Theorem

The following lemma will be quite useful to us.

Lemma 11.2.1. *Let*

$$T\varphi = \mathcal{C}\left(\Psi(z, e^{\varphi(z)})e^{-\varphi(z)}\right),$$

where $\mathcal{C}$ is the Cauchy transform. Then there is a $\phi \in VMO(\mathbb{C})$ such that

$$T\phi = \phi \qquad (11.5)$$

Proof. If φ is a measurable function finite almost everywhere, then (11.4) gives

$$|\Psi(z, e^{\varphi(z)})e^{-\varphi(z)}| \leqslant \sigma(z) \in L^2(\mathbb{C})$$

In particular, as the Cauchy transform of an $L^2(\mathbb{C})$-function, we have from Theorem 4.3.9 that $T\varphi \in VMO(\mathbb{C})$. Thus any fixed-point φ of the operator T is a $VMO(\mathbb{C})$ function for which, by (11.4), we have $\varphi(\infty) = 0$. We may assume without loss of generality that σ is supported in the disk. Consider the operator

$$T: L^2(\mathbb{D}) \to L^2(\mathbb{D})$$

We first show that T is continuous. Suppose that $\varphi_n \to \varphi$ in $L^2(\mathbb{D})$. Then a subsequence converges, $\varphi_n(z) \to \varphi(z)$, almost everywhere in $\mathbb{D}$ and the continuity with respect to the second variable implies that for almost every z,

$$\Psi(z, e^{\varphi_n(z)})e^{-\varphi_n(z)} \to \Psi(z, e^{\varphi(z)})e^{-\varphi(z)}$$

as $n \to \infty$. Next, as $|\Psi(z, e^{\varphi_n(z)})e^{-\varphi_n(z)}| \leqslant \sigma(z)$ and $|\Psi(z, e^{\varphi(z)})e^{-\varphi(z)}| \leqslant \sigma(z)$, the Lebesgue dominated convergence theorem shows us that

$$\|\Psi(\cdot, e^{\varphi_n})e^{-\varphi_n} - \Psi(\cdot, e^{\varphi})e^{-\varphi}\|_{L^2(\mathbb{D})} \to 0$$

as $n \to \infty$. The continuity of the Cauchy transform, Theorem 4.3.9, now implies that

$$\|T\varphi_n - T\varphi\|_{L^2(\mathbb{D})} \to 0$$

Moreover, the operator $\mathcal{C} : L^2(\mathbb{D}) \to L^2(\mathbb{D})$ is compact, and so if

$$B = \{\eta \in L^2(\mathbb{D}) : \|\eta\|_2 \leqslant \|\mathcal{C}\|_{L^2(\mathbb{D}) \to L^2(\mathbb{D})} \|\sigma\|_2\},$$

then $T : B \to B$ is both continuous and compact. Now, as usual, we appeal to the Schauder fixed-point theorem to find a fixed-point $\phi \in L^2(\mathbb{D})$. However, we have already observed that any fixed-point lies in $VMO(\mathbb{C})$. □

11.3 Existence and Uniqueness

The growth condition (11.4) in the above fixed-point theorem has an almost perfect match with the Liouville theorems established in Section 8.5. As a consequence, one obtains existence and uniqueness in the quasilinear Cauchy problem.

Theorem 11.3.1. *Suppose that Ψ satisfies conditions (11.3) and (11.4) above and $z_0 \in \mathbb{C}$.*

Then there is a solution g to the Cauchy problem (11.1) and (11.2) such that $g \in W^{1,q}_{loc}(\mathbb{C})$ for all $q < 2$. If, moreover, Ψ satisfies the Lipschitz bound

$$|\Psi(z, \zeta) - \Psi(z, \xi)| \leqslant \sigma(z)|\zeta - \xi| \tag{11.6}$$

for some compactly supported $\sigma(z) \in L^2(\mathbb{C})$, then the solution g is unique.

Recall that even in the linear case, $g_{\bar{z}} = \sigma g$, we do not necessarily have uniqueness even if it is assumed that $\sigma \in \text{weak-}L^2(\mathbb{C})$, see (8.38).

Proof. We begin with existence. If $z_0 = 0$, then $g(z) \equiv 0$ is such a solution. If $z_0 \neq 0$, suppose we have a solution g_0 for the problem with $z_0 = 1$. Then

$$g(z) = z_0 g_0(z)$$

certainly has the property that $g(z) \to z_0$ as $z \to \infty$. Moreover, g satisfies the nonlinear Cauchy problem $g_{\bar{z}} = \tilde{\Psi}(z, g)$ for the function

$$\tilde{\Psi}(z, f) = z_0 \Psi(z, \frac{1}{z_0} f),$$

and it is easily seen that $\tilde{\Psi}$ again satisfies the conditions (11.3), (11.4) required of it. Thus we may assume without loss of generality that $z_0 = 1$.

We apply Lemma 11.2.1 to find a fixed-point of the integral equation $T\phi = \phi$ and set

$$g = e^{\phi}$$

We now show that g has all the requisite properties. First, as $\phi \in VMO(\mathbb{C})$ we have $g \in L^p_{loc}(\mathbb{C})$ for all $p < \infty$ as a simple consequence of the John-Nirenberg lemma. Next, as a fixed-point, we have identified ϕ as the Cauchy transform of the compactly supported function $\Psi(z, g(z))/g(z)$ whose modulus is dominated by $\sigma(z) \in L^2(\mathbb{C})$. Thus both ϕ_z and $\phi_{\bar{z}}$ are $L^2(\mathbb{C})$ functions, which certainly implies that $Dg \in L^q_{loc}(\mathbb{C})$ for all $q < 2$. We may also make the calculation

$$\begin{aligned} \frac{\partial g}{\partial \bar{z}} &= e^\phi \frac{\partial \phi}{\partial \bar{z}} \\ &= e^\phi \Psi(z, e^\phi) e^{-\phi} \\ &= \Psi(z, e^\phi), \end{aligned}$$

so g is indeed a solution to our equation. Finally, that $\Psi(z, g(z))/g(z)$ is compactly supported gives $\phi(z) = \mathcal{O}(1/z)$ near ∞ and hence that $g(z) \to 1$ as $z \to \infty$.

Last, we discuss uniqueness. Of course as with our earlier uniqueness results we want to apply a suitable version of the Liouville theorem. To do this we apparently need the additional Lipschitz hypothesis in (11.6), which will then assert that the difference between two solutions g and h satisfies

$$|g_{\bar{z}} - h_{\bar{z}}| = |\Psi(z, g) - \Psi(z, h)| \leqslant \sigma(z)|g - h|,$$

so we may apply Theorem 8.5.3 with $k = 0$. $\square$

Chapter 12

Holomorphic Motions

The notion of holomorphic motions, introduced by Mañé, Sad and Sullivan [237], explains in a striking manner the many connections quasiconformal mappings have to holomorphic dynamics, Teichmüller theory and many other areas of complex analysis. It shows that a holomorphic deformation of the identity mapping, in the space of injections of a given set A into $\mathbb{C}$, yields only quasisymmetric mappings. An important fact concerning holomorphic motions is Słodkowski's generalized λ-lemma [329], which shows that any holomorphic motion of any set $A \subset \hat{\mathbb{C}}$ admits an extension to a holomorphic motion of the entire Riemann sphere $\hat{\mathbb{C}}$. Thus the quasisymmetric mappings obtained from a holomorphic deformation of the identity are the restrictions of global quasisymmetric maps of $\mathbb{C}$ no matter how complicated the initial set A is or how bad the deformation might be.

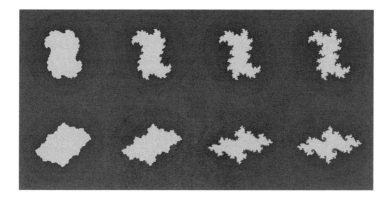

Holomorphic flow of Julia sets in the primary component of the Mandelbrot set
Each boundary is the quasisymmetric image of a circle.

We examine fairly carefully this example of Julia sets varying with a holomorphic parameter a bit later in Section 12.2.

As another simple instance, this shows that one cannot holomorphically deform a domain with an inward directed cusp on the boundary to a domain with an outward cusp on the boundary, even though the boundaries can be mapped one to another by an isometry (say a reflection) because these domains are not equivalent by an entire quasiconformal mapping of $\mathbb{C}$. Of course, the domains are conformally equivalent—it is the global obstruction that prevents the deformation.

Here we shall establish an important connection with the nonlinear Cauchy problem discussed in the previous chapter and using this connection we shall give a complete proof of the extended λ-lemma.

12.1 The λ-Lemma

Basically a holomorphic motion is an isotopy of a subset A of the extended complex plane $\hat{\mathbb{C}}$ analytically parameterized by a complex variable λ in the unit disk $\mathbb{D} = \{\lambda \in \mathbb{C} : |z| < 1\}$. A useful feature of holomorphic motions is that the continuity assumptions can be dismissed from the definition, and in fact, analyticity alone forces strong regularity and extendability properties on the motion.

Definition of Holomorphic Motion

Definition 12.1.1. *Let A be a subset of $\hat{\mathbb{C}}$. A holomorphic motion of A is a map $\Phi : \mathbb{D} \times A \to \hat{\mathbb{C}}$ such that*

- *For any fixed $a \in A$, the map $\lambda \to \Phi(\lambda, a)$ is holomorphic in $\mathbb{D}$.*

- *For any fixed $\lambda \in \mathbb{D}$, the map*

$$a \to \Phi(\lambda, a) = \Phi_\lambda(a)$$

  is an injection.

- *The mapping Φ_0 is the identity on A,*

$$\Phi(0, a) = a, \quad \text{for every } a \in A$$

Note especially that there is no assumption regarding the measurability of A or the continuity of Φ as a function of a or of the pair (λ, a). That such continuity occurs is a consequence of the following remarkable λ-lemma of Mañé-Sad-Sullivan [237].

Theorem 12.1.2. *If $\Phi : \mathbb{D} \times A \to \hat{\mathbb{C}}$ is a holomorphic motion, then Φ has an extension to $\tilde{\Phi} : \mathbb{D} \times \overline{A} \to \hat{\mathbb{C}}$ such that*

12.1. THE λ-LEMMA

- $\tilde{\Phi}$ is a holomorphic motion of $\overline{A}$.
- Each $\tilde{\Phi}_\lambda(\cdot) : \overline{A} \to \hat{\mathbb{C}}$ is quasisymmetric
- $\tilde{\Phi}$ is jointly continuous in (λ, a).

Proof. We begin with a normalization. We may assume that A has at least three points, and after changing things by an initial linear fraction transformation, we can assume 0, 1 and $\infty \in A$. We want to normalize the motion in such a way that these points are fixed. To do this we consider the motion

$$(\lambda, a) \mapsto \frac{\Phi(\lambda, a) - \Phi(\lambda, 0)}{\Phi(\lambda, a) - \Phi(\lambda, \infty)} \frac{\Phi(\lambda, 1) - \Phi(\lambda, \infty)}{\Phi(\lambda, 1) - \Phi(\lambda, 0)},$$

which we continue to denote by Φ.

Let ρ denote the hyperbolic metric of the triply punctured sphere $\mathbb{C} \setminus \{0, 1\}$, as described in Section 2.9.2. Since this metric is complete, given $z, w \in \mathbb{C} \setminus \{0, 1\}$ a bounded hyperbolic distance apart, we see that $|z| \to 0$ implies that $|w| \to 0$. Thus there is a continuous function $\eta : \mathbb{R}_+ \times \mathbb{R}_+ \to \mathbb{R}_+$ such that for each fixed $M < \infty$, $\eta(M, \varepsilon) \to 0$ as $\varepsilon \to 0$ and

$$|w| \leq \eta(M, |z|)$$

whenever $z, w \in \mathbb{C} \setminus \{0, 1\}$ and $\rho(z, w) < M$.

Thus if a_1, a_2 and a_3 are distinct points of A, the holomorphic motion Φ gives rise to the holomorphic function

$$g(\lambda) = \frac{\Phi_\lambda(a_1) - \Phi_\lambda(a_2)}{\Phi_\lambda(a_1) - \Phi_\lambda(a_3)},$$

which is holomorphic in $\mathbb{D}$ with values in $\mathbb{C} \setminus \{0, 1\}$. The generalised Schwarz-Pick lemma of Ahlfors (see Theorem 2.9.6) shows that the mapping g is a contraction of the respective hyperbolic metrics. That is,

$$\rho(g(\lambda), g(0)) \leq \rho_\mathbb{D}(\lambda, 0) = \log \frac{1 + |\lambda|}{1 - |\lambda|}, \tag{12.1}$$

where $\rho_\mathbb{D}$ is the Poincaré (hyperbolic) metric of the disk $\mathbb{D}$. Since

$$g(0) = \frac{a_1 - a_2}{a_3 - a_4},$$

we find

$$\left| \frac{\Phi_\lambda(a_1) - \Phi_\lambda(a_2)}{\Phi_\lambda(a_1) - \Phi_\lambda(a_3))} \right| \leq \eta\left(M, \left| \frac{a_1 - a_2}{a_3 - a_4} \right|\right) \tag{12.2}$$

with $M = \log \frac{1 + |\lambda|}{1 - |\lambda|}$. This in fact shows that $\Phi_\lambda = \Phi(\lambda, \cdot) : A \to \mathbb{C}$ is quasisymmetric and hence is uniformly continuous in A. It therefore extends continuously to $\tilde{\Phi}_\lambda : \overline{A} \to \hat{\mathbb{C}}$. Permuting the points a_1, a_2 and a_3 in (12.2) shows this extension to be injective and that $\tilde{\Phi}_\lambda(\cdot)$ is a homeomorphism onto its image. For each

$a \in \overline{A} \setminus A$ the function $\tilde{\Phi}(\cdot, a)$ is holomorphic since it is the local uniform limit of holomorphic functions; the joint continuity in (λ, a) follows since for every $r < 1$ the family $\{\tilde{\Phi}_\lambda(\cdot) : \lambda \in r\mathbb{D}\}$ is equicontinuous. The proof is complete. □

The reader should be aware that complex analyticity is a crucial ingredient of Theorem 12.1.2. The result does not hold even for real analytic motions. Consider

$$\Phi(t,z) = \begin{cases} x + iy, & y \leq 0 \\ x + i(1+t^2)y, & y > 0, \end{cases}$$

which is injective in $z = x + iy$, real analytic in t but not even continuous in z.

12.2 Two Compelling Examples

In this section we present two compelling examples as applications of the theory of holomorphic motions. The details, concerned as they are with the theories of discrete groups and holomorphic dynamics, would lead us too far astray from our central theme in this book. Therefore we merely try to give the reader a glimpse of what is going on.

12.2.1 Limit Sets of Kleinian Groups

A Kleinian group is a discrete group Γ of linear fractional transformations

$$\gamma(z) = \frac{az+b}{cz+d}$$

of the Riemann sphere $\hat{\mathbb{C}}$. Consider a parameterized family of Kleinian groups

$$\Gamma_\lambda = \langle \gamma_{i,\lambda} : i = 1, \ldots, n \rangle, \qquad \lambda \in \mathbb{D}$$

where

$$\gamma_{i,\lambda}(z) = \frac{a_i(\lambda)z + b_i(\lambda)}{c_i(\lambda)z + d_i(\lambda)}$$

and the coefficients $a_i(\lambda)$, etc, are holomorphic functions of λ. Such a family may arise by varying the holomorphic structure of a Riemann surface (Teichmüller theory) as Kleinian groups uniformize such surfaces.

According to a well-known theorem of Jørgensen [202], members of a continuously parameterized family of discrete groups are all canonically isomorphic to Γ_0. It is also well-known that the fixed-point sets of elements of a Kleinian group are dense in the limit set $\Lambda(\Gamma)$. As solutions of the equation $\gamma_\lambda(z) = z$, $\gamma_\lambda \in \Gamma_\lambda$, these fixed-points move holomorphically with the parameters involved. That two different elements share a fixed-point in a Kleinian group has algebraic implications [48], which in turn imply that fixed-points of a varying family cannot collide unless both points belong to the same element. Such deformations produce parabolic elements and change the geometry of the orbit space $(\hat{\mathbb{C}} - \Lambda_\lambda)/\Gamma_\lambda$, (for instance it may become noncompact).

12.2. EXAMPLES

The density of fixed-points in the limit set therefore implies via the λ-lemma that the limit set moves holomorphically and the isotopy is through quasisymmetric mappings, as long as no new parabolic elements are produced (and this restriction is easily seen to be necessary). From this we deduce that holomorphic deformations of Kleinian groups are canonically quasisymmetrically conjugate on their respective limit sets. Thus for instance a holomorphic deformation of a Fuchsian group produces a quasi-Fuchsian group. The limit set moves holomorphically from a round circle to a quasicircle—the quasiconformal image of a round circle under a quasiconformal mapping of $\mathbb{C}$. Using this fact we are able, for instance, to bound the change of Hausdorff dimension of this family of limit sets.

12.2.2 Julia Sets of Rational Maps

In the case of the Julia set $J(R)$ of a rational map $R : \hat{\mathbb{C}} \to \hat{\mathbb{C}}$, we illustrate holomorphic deformations with a special case that has most of the interesting features. Any quadratic polynomial is conjugate by a Möbius transformation to a mapping of the form $p_c(z) = z^2 + c$. We recall that the Fatou set of a rational mapping is the open (possibly empty) region where the dynamics are stable, that is, the maximal domain where the family $\{R^n = R \circ \cdots \circ R : n \in \mathbb{N}\}$ is normal. Then $J(R) = \mathbb{C} \setminus F(R)$. The well-known Mandelbrot set consists of those complex parameters $c \in \mathbb{C}$ such that the Julia set $J(z^2 + c)$ is connected.

Inside the Mandelbrot set we have the hyperbolic regions consisting of those c's for which p_c has an attracting periodic cycle; that is, for some $n \in \mathbb{N}$ the mapping $p_c^n(z)$ has a fixed-point z_0 such that $|(p_c^n)'(z_0)| < 1$. It is conjectured that the Mandelbrot set is the closure of these hyperbolic regions. Each hyperbolic region is simply connected and so is holomorphically parameterized by the disk. Let us fix a hyperbolic region U, vary $c \in U$ and then study the associated dynamical system. Every attracting cycle attracts a critical point of our mapping. Thus for a quadratic polynomial there is exacly one finite attracting cycle. As with Kleinian groups, the repelling periodic points, those points z_0 such that for some $n \in \mathbb{N}$, $p_c^n(z_0) = z_0$ and $|(p_c^n)'(z_0)| > 1$, are dense in the Julia set. As we move c in U these points remain repelling: They cannot become attracting as there is already one such cycle, nor can they become indifferent, $|(p_c^n)'(z_0)| = 1$, as such cycles also "attract" critical points as well. It is clear that the repelling points move holomorphically with the parameter c. We need to see that they do not collide. Observe that if repelling points of period m and n collide ($m < n$), then by continuity two points on the period-n cycle collide. At such a point w, $p_c^n(z) - w$ has a double root and derivative zero. Thus for nearby time the derivative was less than 1 in modulus giving an extra attracting cycle and a contradiction. This shows that the repelling points cannot collide. It follows that they move holomorphically (while the parameters lie within the given component), and therefore, by density, so does the Julia set. This is the reason why Julia sets are quasisymmetrically similar. It follows that we may apply all the geometric and analytic information we have obtained concerning quasiconformal mappings to the dynamical theory of complex iteration.

12.3 The Extended λ-Lemma

The most surprising fact about holomorphic motions is that they always extend to ambient holomorphic motions (that is, holomorphically parameterized isotopies of $\hat{\mathbb{C}}$) and that at each "time" λ the associated homeomorphism of the plane is quasiconformal. This is the extended λ-lemma. The extended λ-lemma was first proven by Z. Słodkowski [329] using techniques from several complex variables, in particular, the structure of polynomial hulls of sets that fiber over the circle. The result had earlier been conjectured by Mañé-Sad-Sullivan and Sullivan-Thurston [342]. They, along with Bers-Royden [58], had proved partial results.

In this chapter we shall give a complete and self-contained proof of the extended λ-lemma from the point of view of one complex variable. The proof we present is based on a sketch given to us by L. Lempert of a new proof of this result due to E. Chirka [92]. It is based around the solution of the nonlinear Cauchy problem and we feel is the simplest and most compelling proof available. The earlier proofs in [329] and [33] were based around the solution of the nonlinear Riemann-Hilbert problem by Snirelman in 1972 and extended by a result of Forstnerič [128]. Another proof given by Douady [109] was based on an ordinary differential equation and continuity methods.

Theorem 12.3.1. *Every holomorphic motion of a set $A \subset \mathbb{C}$ is the restriction of a holomorphic motion of $\mathbb{C}$.*

As a consequence, one directly obtains the complete version of the extended λ-lemma.

> The Extended λ-Lemma

Theorem 12.3.2. *If $\Phi : \mathbb{D} \times A \to \hat{\mathbb{C}}$ is a holomorphic motion of $A \subset \mathbb{C}$, then Φ has an extension to $\tilde{\Phi} : \mathbb{D} \times \hat{\mathbb{C}} \to \hat{\mathbb{C}}$ such that*

- *$\tilde{\Phi}$ is a holomorphic motion of $\hat{\mathbb{C}}$.*

- *Each $\tilde{\Phi}_\lambda(\cdot) : \hat{\mathbb{C}} \to \hat{\mathbb{C}}$ is a K-quasiconformal self-homeomorphism with*

$$K \leqslant \frac{1+|\lambda|}{1-|\lambda|}$$

- *$\tilde{\Phi}$ is jointly continuous in (λ, a).*

- *For $\lambda_1, \lambda_2 \in \mathbb{D}$, $\tilde{\Phi}_{\lambda_1} \circ \tilde{\Phi}_{\lambda_2}^{-1}$ is K-quasiconformal with*

$$\log K \leqslant \rho_{\mathbb{D}}(\lambda_1, \lambda_2)$$

Remark 12.3.3. *The last statement implies continuity in the best possible sense, for as $\lambda_1 \to \lambda_2$ not only does $\tilde{\Phi}_{\lambda_1} \circ \tilde{\Phi}_{\lambda_2}^{-1}$ converge to the identity, but its distortion also tends to 1.*

12.3.1 Holomorphic Motions and the Cauchy Problem

The result we want to establish here is the fact that parameterized families of solutions to the nonlinear Cauchy problem form holomorphic motions. This is all that is needed to complete our proof of the extended λ-lemma.

Theorem 12.3.4. *Let $\Psi : \mathbb{C} \times \mathbb{C} \to \mathbb{C}$ satisfy*

1. $|\Psi(\lambda, \zeta) - \Psi(\lambda, \xi)| \leqslant \chi_{\mathbb{D}}(\lambda)\sigma(\lambda)|\zeta - \xi|$ *for some function* $\sigma \in L^2(\mathbb{C})$.
2. $\Psi(\lambda, 0) \equiv 0$.

Let $g = g^z \in W^{1,q}_{loc}(\mathbb{C})$, $q > 1$, be the unique solution to the nonlinear Cauchy problem

$$\frac{\partial}{\partial \bar{\lambda}} g(\lambda) = \Psi(\lambda, g(\lambda)), \qquad g(\infty) = z \in \mathbb{C},$$

provided by Theorem 11.3.1. Define $\Phi : \mathbb{C} \times \mathbb{C} \to \mathbb{C}$ by the rule

$$\Phi(\lambda, z) = g^z(1/\lambda), \qquad \lambda \in \mathbb{D}$$

Then Φ gives a holomorphic motion of $\mathbb{C}$.

Proof. The first thing to observe is that, by 1,

$$\frac{\partial}{\partial \bar{\lambda}} (\Phi(\lambda, z_0)) = \frac{\partial}{\partial \bar{\lambda}} g^{z_0}(1/\lambda) = \Psi(1/\lambda, g^{z_0}(\lambda)) = 0, \qquad \lambda \in \mathbb{D},$$

for $\Psi(w, \zeta) = 0$ if $|w| \geqslant 1$. Thus $\lambda \mapsto \Phi(\lambda, z_0)$ is holomorphic in the unit disk $\mathbb{D}$. We also have $g^z(\lambda) = z + \mathcal{O}(1/\lambda)$ as $\lambda \to \infty$, so that

$$\Phi(0, z) \equiv z$$

We now need only check injectivity. Suppose that for some $\lambda \in \mathbb{D}$ we have

$$\Phi(\lambda, z_1) = \Phi(\lambda, z_2)$$

This means that $g^{z_0}(1/\lambda) = g^{z_1}(1/\lambda)$. Consider the Möbius transformation

$$\gamma_\lambda(z) = \frac{z + \bar{\lambda}}{1 + \lambda z}, \qquad \gamma_\lambda(\infty) = \frac{1}{\lambda}$$

We see that

$$h(w) = (g^z \circ \gamma_\lambda)(w)$$

satisfies the equation

$$\frac{\partial}{\partial \bar{w}} h = \overline{\gamma'_\lambda(w)} \Psi(\gamma_\lambda(w), h(w)) \qquad (12.3)$$

$$h(\infty) = g^z(1/\lambda) \qquad (12.4)$$

Now we have

$$\gamma'_\lambda(w) = \frac{1 - |\lambda|^2}{(1 + \lambda w)^2},$$

and if we put
$$\Psi_\lambda(w,\zeta) = \overline{\gamma'_\lambda(w)}\Psi(\gamma_\lambda(w),\zeta),$$
we have the bounds
$$|\Psi_\lambda(w,\zeta) - \Psi_\lambda(w,\xi)| \leqslant \chi_{\mathbb{D}}(w)|\gamma'_\lambda(w)|\sigma(\gamma_\lambda(w))\,||\zeta - \xi|$$
Since
$$\int_{\mathbb{D}} |\gamma'_\lambda(w)|^2 \sigma(\gamma_\lambda(w))^2\, dz = \int_{\mathbb{D}} \sigma(w)^2\, dz,$$
we in fact have the same $L^2(\mathbb{C})$-bounds. With Theorem 11.3.1 this of course means that the problem
$$\frac{\partial}{\partial \overline{w}} h = \Psi_\lambda(w, h(w)) \qquad (12.5)$$
$$h(\infty) = g^z(1/\lambda) \qquad (12.6)$$
has a unique solution. But now $h_1(w) = (g^{z_1} \circ \gamma_\lambda)(w)$ and $h_2(w) = (g^{z_2} \circ \gamma_\lambda)(w)$ both satisfy (12.5) and (12.6). Thus $h_1 \equiv h_2$, hence $g^{z_1} \equiv g^{z_2}$, which certainly implies $z_1 = z_2$ by looking at these functions near ∞. □

12.3.2 Holomorphic Axiom of Choice

A key idea in the proof of the extended λ-lemma is the reduction to finite-point sets using the compactness properties of families of analytic maps. Indeed, we shall see in a moment that families of holomorphic motions themselves are precompact in the topology of local uniform convergence.

Lemma 12.3.5. *Let $\{0, 1, \infty\} \subset A \subset \hat{\mathbb{C}}$ and $\Phi_n : \mathbb{D} \times A \to \mathbb{C}$, $n = 1, 2, \ldots$, be holomorphic motions normalized so that, for each n and $\lambda \in \mathbb{D}$,*
$$\Phi_n(\lambda, 0) = 0, \qquad \Phi_n(\lambda, 1) = 1, \qquad \Phi_n(\lambda, \infty) = \infty$$
Then there is a subsequence $\Phi_{n_j} : \mathbb{D} \times A \to \mathbb{C}$ and a holomorphic motion $\Phi : \mathbb{D} \times A \to \mathbb{C}$ such that
$$\Phi_{n_j}(\lambda, a) \to \Phi(\lambda, a)$$
for all $(\lambda, a) \in \mathbb{D} \times A$. Moreover, this convergence is uniform on compact subsets.

Proof. First, we see that the family of maps $\Phi_n : \mathbb{D} \times A \to \mathbb{C}$ is equicontinuous. If $k < 1$, then from the λ-lemma for each $|\lambda| < k$ the map $\Phi_n(\lambda, \cdot) : A \to \mathbb{C}$ is η-quasisymmetric with quasisymmetry function depending only on $\frac{1+k}{1-k}$. Since each map fixes 0, 1 and ∞, this family is equicontinuous by Theorem 3.9.1. Next, $\Phi_n(\cdot, a_0)$ is a family of analytic functions from the disk valued in $\mathbb{C} \setminus \{0, 1\}$ and as such is again equicontinuous, as we have noted earlier. The triangle inequality
$$|\Phi(\lambda, a) - \Phi(\lambda_0, a_0)| \leqslant |\Phi(\lambda, a_0) - \Phi(\lambda_0, a_0)| + |\Phi(\lambda, a) - \Phi(\lambda, a_0)|$$
then easily imples that $\{\Phi_n\}$ is equicontinuous and thus normal. In particular, there is a subsequence Φ_{n_j} converging to a limit mapping Φ.

12.3. THE EXTENDED λ-LEMMA

We have to show that Φ is a holomorphic motion. For each fixed a, the mapping $\Phi(\cdot, a)$ is analytic as it is the limit of the analytic maps $\Phi_{n_j}(\cdot, a)$. Certainly, $\Phi(0, a) = a$ as $\Phi_{n_j}(0, a) = a$. Finally, each $\Phi_{n_j}(\lambda, \cdot)$ is quasisymmetric with quasisymmetry function depending only on $|\lambda|$. The limit is therefore quasisymmetric (and therefore injective) since the normalization assures us the limit is nonconstant, see Corollary 3.9.3. □

The compactness property described in Lemma 12.3.5 allows us to reduce the proof of the extended λ-lemma to the case where A is a finite-point set. We take a countable dense subset of A, say $\{a_0, a_1, \ldots\}$, and for each n we extend the motion $\Phi|A_n$ of the set $A_n = \{a_0, a_1, \ldots, a_n\}$ to a motion Φ_n of $\mathbb{C}$. Note that $\Phi_{n+1}|A_n = \Phi_n|A_n = \Phi|A_n$. Thus any limit motion will agree with Φ on a dense subset of A and therefore on A by uniform continuity.

The problem of extending a holomorphic motion of a finite set of points has been dubbed the "holomorphic axiom of choice" by Bers. Actually, he formulated the problem by asking if a motion of a finite number of points can be extended to a single further point. We shall prove an apparently stronger result, which is of course easily seen to be equivalent.

Theorem 12.3.6. *Let $\varphi_i : \mathbb{D} \to \hat{\mathbb{C}}$, $i = 1, 2, \ldots, n$, be holomorphic functions such that for each $\lambda \in \mathbb{D}$ we have $\varphi_i(\lambda) \neq \varphi_j(\lambda), i \neq j$. Then there is a holomorphic motion $\tilde{\Phi} : \mathbb{D} \times \mathbb{C} \to \mathbb{C}$ such that*

$$\tilde{\Phi}(\lambda, z_i) = \varphi_i(\lambda),$$

where $z_i = \varphi_i(0)$, $i = 1, 2, \ldots, n$.

Proof. Let $E = \{z_1, z_2, \ldots, z_n\}$. The hypotheses of course amount to saying that we are given a holomorphic motion of E. We want to extend this motion to all of $\mathbb{C}$.

Set $\Phi : \mathbb{D} \times E \to \mathbb{C}$,

$$\Phi(\lambda, z_i) = \varphi_i(\lambda), \qquad i = 1, 2, \ldots, n$$

Then Φ is a holomorphic motion of the finite-point set E. We want to extend Φ to a motion of the entire complex plane. Choose $R > 1$ and define

$$g^{z_k}(\lambda) = \Phi(1/\lambda, z_k), \qquad k = 1, 2, \ldots, n \quad \text{and} \quad |\lambda| \geq R \qquad (12.7)$$

Notice that g^{z_k} is holomorphic in the disk $\{|\lambda| > R\}$ and that $g^{z_k}(\lambda) \to z_k$ as $\lambda \to \infty$.

The reader will quickly see that as $R > 1$ is arbitrary, an easy application of Lemma 12.3.5 shows it will be enough to extend the motion $\Phi|\mathbb{D}(0, 1/R) \times E$ to $\Phi|\mathbb{D}(0, 1/R) \times \mathbb{C}$. We shall do this using the solution to the nonlinear Cauchy problem. In effect we find a problem that the already given g^{z_k} solve and then extend it.

First, extend g^{z_k} to the plane via the rule

$$g^{z_k}(\lambda) = \overline{g^{z_k}(R^2/\bar{\lambda})}, \qquad |\lambda| \leq R \qquad (12.8)$$

On the set $\{(\lambda, g^{z_k}(\lambda)) \in \mathbb{C} \times \mathbb{C} : k = 1, 2, \ldots, n, \, \lambda \in \mathbb{C}\}$ define

$$\Psi(\lambda, g^{z_k}(\lambda)) = \frac{\partial g^{z_k}(\lambda)}{\partial \bar\lambda} \qquad (12.9)$$

Each g^{z_k} is then a solution to the Cauchy problem (12.9), however, the function Ψ is not globally defined. We want to extend Ψ in such a way that existence and uniqueness in the Cauchy problem are guaranteed. Given the formulation of our uniqueness statements, we want to extend Ψ so as to have a global Lipschitz estimate. To do this we must first check that Ψ as defined by (12.9) is Lipschitz. First, note that for $j \neq k$ the functions $|g^{z_k}(\lambda) - g^{z_j}(\lambda)| = |\varphi_k(1/\lambda) - \varphi_j(1/\lambda)|$ are uniformly bounded below on the relatively compact set $\{|\lambda| < 1/R\}$ of the unit disk $\mathbb{D}$ as the functions φ are disjoint by hypothesis. Therefore if we write

$$\sigma(\lambda) = \sup_{j \neq k} \frac{|\Psi(\lambda, g^{z_k}(\lambda)) - \Psi(\lambda, g^{z_j}(\lambda))|}{|g^{z_k}(\lambda) - g^{z_j}(\lambda)|},$$

then

$$|\Psi(\lambda, g^{z_k}(\lambda)) - \Psi(\lambda, g^{z_j}(\lambda))| \leqslant \sigma(\lambda)|g^{z_k}(\lambda) - g^{z_j}(\lambda)| \qquad (12.10)$$

Furthermore, the function σ is actually continuous and compactly supported, and hence certainly in $L^2(\mathbb{C})$.

Next, for each $\lambda \in \mathbb{C}$ we consider the function $F_\lambda : E_\lambda \to \mathbb{C}$, where $E_\lambda = \{g^{z_k}(\lambda) : k = 1, \ldots, n\}$, a finite set, and

$$F_\lambda(g^{z_k}(\lambda)) = \Psi(\lambda, g^{z_k}(\lambda)) \qquad (12.11)$$

The functions F_λ are uniformly Lipschitz by the estimate in (12.10). We now use the Whitney extension theorem [335, chap. VI] to extend the function $F_\lambda : E_\lambda \to \mathbb{C}$ to $\tilde{F}_\lambda : \mathbb{C} \to \mathbb{C}$. Not only is this extension Lipschitz, but the Lipschitz constant is increased only by a bounded amount, independent of the set E_λ [335, Theorem 4, p. 177]. For each λ this gives us a function $\Psi(\lambda, z) = \tilde{F}_\lambda(z)$ that satisfies the Lipschitz estimate

$$|\Psi(\lambda, z) - \Psi(\lambda, w)| \leqslant \tilde\sigma(\lambda)|z - w|$$

with $\tilde\sigma$ bounded and compactly supported. The only remaining issue is the measurability of $\Psi(\lambda, z)$ as a function of λ. However, this is a relatively straightforward consequence of the explicit Whitney extension construction.

Suppose now that we solve the nonlinear Cauchy problem

$$\Psi(\lambda, g^z(\lambda)) = \frac{\partial}{\partial \bar\lambda} g^z(\lambda), \qquad (12.12)$$

where g^z is the unique solution with $g^z(\infty) = z$. By construction and uniqueness, g^{z_k} is such a solution. Therefore application of Theorem 12.3.4 completes the proof. $\square$

As we have noted above, this theorem together with the holomorphic axiom of choice gives a proof of the extended λ-lemma. A few more words are needed in order to complete the proof of Theorem 12.3.2.

12.3. THE EXTENDED λ-LEMMA

Proof of Theorem 12.3.2

In view of what we have already proved, the only issue is the distortion bounds on the quasiconformal functions $\Phi_\lambda = \Phi(\lambda, \cdot) : \mathbb{C} \to \mathbb{C}$ (recall that the λ-lemma already implies that Φ_λ is quasisymmetric and therefore quasiconformal). That is, we wish to show that

$$K(\Phi_\lambda) \leq \frac{1+|\lambda|}{1-|\lambda|}$$

For this we need to prove, using the analytic dependence of $\Phi(\lambda, z)$ on λ, that the quotient of derivatives $\mu_\lambda = \partial_{\bar{z}} \Phi_\lambda / \partial_z \Phi_\lambda$ is an analytic function of λ, valued in the unit ball of $L^\infty(\mathbb{C})$. Since $L^\infty(\mathbb{C})$ is the dual space to $L^1(\mathbb{C})$ to obtain the analyticity that we require, it suffices to show that for every $\omega \in L^1(\mathbb{C})$ the integral

$$\Psi(\lambda) = \int_\mathbb{C} \omega(z) \mu_\lambda(z) = \int_\mathbb{C} \omega(z) \frac{\Phi_{\bar{z}}(\lambda, z)}{\Phi_z(\lambda, z)}$$

is holomorphic for $\lambda \in \mathbb{D}$. As in many of our earlier arguments, it suffices to consider the case where ω has compact support.

We have

$$\frac{\Phi_{\bar{z}}(\lambda, z)}{\Phi_z(\lambda, z)} = \lim_{h \in \mathbb{R}, h \to 0} \frac{1 + i\sigma_\lambda(z, h)}{1 - i\sigma_\lambda(z, h)},$$

where

$$\sigma_\lambda(z, h) = \frac{\Phi(\lambda, z+ih) - \Phi(\lambda, z)}{\Phi(\lambda, z+h) - \Phi(\lambda, z)}$$

For fixed $z \in \mathbb{C} \setminus \{0, 1\}$ and h, the function σ_λ is clearly holomorphic in λ and never assumes the values $0, 1$ of ∞. Moreover, $\sigma_0(z, h) \equiv i$. Therefore the generalized Schwarz-Pick lemma, Theorem 2.9.6, gives the uniform bounds for the hyperbolic distance,

$$\rho(\sigma_\lambda(z, h), i) \leq \log \frac{1+|\lambda|}{1-|\lambda|}$$

We see that, for $|\lambda| < r$ small enough, $|\sigma_\lambda(z, h) + i|$ is bounded away from zero, hence $\frac{1+i\sigma_\lambda(z,h)}{1-i\sigma_\lambda(z,h)}$ is uniformly bounded on a disk $|\lambda| < r$. It follows now from the dominated convergence theorem that

$$\Psi_n(\lambda) = \int_\mathbb{C} \omega(z) \frac{1 + i\sigma_\lambda(z, 1/n)}{1 - i\sigma_\lambda(z, 1/n)} \, dz$$

converges boundedly to $\Psi(\lambda)$. Thus $\Psi(\lambda)$ is holomorphic in a neighborhood $\mathbb{D}(0, r)$ of the origin.

Other points of the λ-disk can be reached by a change of variables. Namely, let $\lambda_0 \in \mathbb{D}$ and

$$\tilde{\Phi}(\lambda, z) = \Phi\left(\frac{\lambda + \lambda_0}{1 + \bar{\lambda}_0 \lambda}, \Phi_{\lambda_0}^{-1}(z)\right), \qquad \lambda \in \mathbb{D}, z \in \mathbb{C}$$

This is a new holomorphic motion with

$$\lambda \mapsto \int_{\mathbb{C}} \alpha(z) \frac{\tilde{\Phi}_{\bar{z}}(\lambda, z)}{\tilde{\Phi}_z(\lambda, z)}$$

analytic in the disk $|\lambda| < r$ for any compactly supported $\alpha \in L^1(\mathbb{C})$. Since

$$\Phi\left(\frac{\lambda + \lambda_0}{1 + \overline{\lambda}_0 \lambda}, z\right) = \tilde{\Phi}(\lambda, \Phi_{\lambda_0}(z)),$$

we may use change of variables and the identity (5.45) to conclude that the complex dilatation μ_λ of Φ_λ is a holomorphic function of λ in a neighborhood of λ_0, hence in the whole disk.

Now μ_λ is a $L^\infty(\mathbb{C})$-valued analytic function with $\mu_0 \equiv 0$ and $\|\mu_\lambda\|_\infty < 1$ for all $\lambda \in \mathbb{D}$. If $\omega \in L^1(\mathbb{C})$ with $\|\omega\|_1 \leqslant 1$, we may apply the usual Schwarz lemma to

$$\lambda \mapsto \langle \omega, \mu_\lambda \rangle$$

As a result, $\|\mu_\lambda\|_\infty = \sup\{|\langle \omega, \mu_\lambda \rangle| : \omega \in L^1(\mathbb{C}), \|\omega\|_1 \leqslant 1\}$ is at most $|\lambda|$. In particular, the dilatation at time λ is at most

$$K_\lambda = \frac{1 + \|\mu_\lambda\|_\infty}{1 - \|\mu_\lambda\|_\infty} \leqslant \frac{1 + |\lambda|}{1 - |\lambda|}$$

The only thing left to observe is the last claim of continuity. For this note that the Beltrami coefficients $\mu_\lambda(z) = \mu_{\tilde{\Phi}_\lambda}(z)$ are an analytic function of λ for almost every $z \in \mathbb{C}$ (Theorem 5.7.1). Given $\lambda_1, \lambda_2 \in \mathbb{D}$ and setting

$$\Psi = \tilde{\Phi}_{\lambda_1} \circ \tilde{\Phi}_{\lambda_2}^{-1},$$

the composition formula, Theorem 13.37, shows us that

$$\log K_\Psi = \|\rho_{\mathbb{D}}(\mu_{\lambda_1}, \mu_{\lambda_2})\|_\infty \leqslant \|\rho_{\mathbb{D}}(\lambda_1, \lambda_2)\|_\infty,$$

since, for almost every z, $\mu : \mathbb{D} \to \mathbb{D}$ is analytic and therefore a contraction in the hyperbolic metric. We have now completed the proof of the generalized λ-lemma, Theorem 12.3.2. □

Often holomorphic motions have special symmetries that we would like to retain in extending them. We present here two such examples that will be useful later.

Suppose that the set A is invariant under complex conjugation and that we have a holomorphic motion of A with the following mirror symmetry,

$$\overline{\Phi(\overline{\lambda}, \overline{z})} = \Phi(\lambda, z), \qquad \lambda \in \mathbb{D}, \ z \in A \tag{12.13}$$

Such motions arise naturally when a quasiconfomal mapping preserving the upper half-plane is embedded in a holomorphic motion.

12.3. THE EXTENDED λ-LEMMA

Theorem 12.3.7. *Suppose we have a set A, closed under complex conjugation, and a holomorphic motion $\Phi : \mathbb{D} \times A \to \mathbb{C}$ that is symmetric in the sense of (12.13). Then we can extend Φ to a symmetric motion of the whole complex plane $\mathbb{C}$.*

Proof. The only point we need to clarify here is that the above proof of the generalized λ-lemma gives the symmetric extension if the motion $\Phi : \mathbb{D} \times A \to \mathbb{C}$ is symmetric to start with. Indeed, let $E = \{z_k : k = 1, \ldots, n\}$ be a finite set closed under complex conjugation and suppose $\Phi : \mathbb{D} \times E \to \mathbb{C}$ is a holomorphic motion satisfying (12.13). The functions $g^z(\lambda)$ defined in (12.7) and (12.8) are symmetric as well,

$$g^{\bar{z}}(\bar{\lambda}) = \overline{g^z(\lambda)}, \qquad z \in E, \ \lambda \in \mathbb{C}$$

It follows that sets $E_\lambda = \{g^{z_k}(\lambda) : k = 1, \ldots, n\}$ satisfy $\overline{E_\lambda} = E_{\bar{\lambda}}$ and that (see (12.9)) the Lipschitz functions of (12.11) have the symmetry

$$F_{\bar{\lambda}}(\bar{\zeta}) = \overline{F_\lambda(\zeta)}, \qquad \zeta \in E_\lambda, \ \lambda \in \mathbb{C}$$

Now the Whitney extension operator is easily seen to preserve this symmetry; this should be compared with [335, Theorem 4, p. 172]. Hence the symmetry is inherited by the vector fields $\Psi(\lambda, z)$ and hence by the solutions $g^z(\lambda)$ for all $z \in \mathbb{C}$ and $\lambda \in \mathbb{C}$. It follows that the extended motion $\Phi : \mathbb{D} \times \mathbb{C} \to \mathbb{C}$ satisfies (12.13) globally. □

Typical further examples of holomorphic motions with symmetries are equivariant motions arising in the theory of Kleinian groups. Here we are given a set A invariant under a discrete group Γ of Möbius transformations and a holomorphic motion $\Phi : \mathbb{D} \times A \to \mathbb{C}$ of A that conjugates elements of Γ to Möbius transformations $g_\lambda \in \Gamma_\lambda$,

$$\Phi_\lambda \circ g(a) = g_\lambda \circ \Phi_\lambda(a), \qquad g \in \Gamma, \ \lambda \in \mathbb{D}, \ a \in A \tag{12.14}$$

Theorem 12.3.8. *If Γ is a discrete Möbius group and A is a Γ-invariant set, then any Γ-equivariant holomorphic motion $\Phi : \mathbb{D} \times A \to \mathbb{C}$ can be extended to an equivariant motion of $\mathbb{C}$.*

Proof. We just use the holomorphic axiom of choice and extend the motion orbit by orbit, applying the action of Γ. For simplicity let us assume Γ contains no elliptic elements, that is, elements of finite order. We may assume that A contains at least three points. Then the Möbius transformations g_λ are uniquely determined in the entire plane $\mathbb{C}$ by the relation (12.14), and the coefficients of g_λ depend holomorphically on the parameter λ.

Further, the Mañé, Sad and Sullivan λ-lemma, Theorem 12.1.2, extends the motion (equivariantly) to the closure of A. Thus we may take A closed. Since A is a Γ-invariant closed set containing at least three points, the limit set $L(\Gamma) \subset A$. In particular, since there are no elliptic elements, A contains the fixed-points of all $\mathbf{I} \neq g \in \Gamma$. Similarly, $A_\lambda = \Phi_\lambda(A)$ contains the fixed-points of the g_λ's.

Choose a point $z_0 \in \mathbb{C} \setminus A$. Via the holomorphic axiom of choice, the motion extends to $\{z_0\} \cup A$. But now the group action determines the motion of the orbit $\Gamma(z_0)$,

$$\Phi(\lambda, g(z_0)) = g_\lambda \circ \Phi_\lambda(z_0), \qquad g \in \Gamma, \ \lambda \in \mathbb{D} \qquad (12.15)$$

Since $g_\lambda \circ \Phi_\lambda(z_0) \in \mathbb{C} \setminus A_\lambda$ and A_λ contains the fixed-points of the group elements, $\zeta \mapsto \Phi_\lambda(\zeta)$ as defined in (12.15) is injective on $\Gamma(z_0) \cup A$. Since the dependence on λ is clearly holomorphic, Φ is thus extended to an equivariant motion of $\Gamma(z_0) \cup A$.

Continuing in this manner with a dense sequence $\{z_n\} \subset \mathbb{C} \setminus A$ we extend Φ equivariantly to a holomorphic motion of the plane $\mathbb{C}$. □

This and other equivariant versions of the extended λ-lemma for holomorphic motions were established in [116, 330], where important applications, particularly with regard to Teichmüller theory and Riemann surfaces, are discussed.

12.4 Distortion of Dimension in Holomorphic Motions

A holomorphic motion Φ_λ of a set E will in general distort the metric characteristics of the set. The examples of Julia sets show that holomorphic motions of a circle can produce nowhere-differentiable curves for all time $\lambda \neq 0$. However, one quickly realizes that the deformations cannot be arbitrarily large, as the distortion of the ambient quasiconformal mapping is well controlled. A natural question is then for instance how much can the Hausdorff dimension change under a holomorphic motion and is it any more or less than that of an arbitrary quasiconformal mapping with the same distortion bounds?

Here we give one example from [21]. In a later section we shall see that the example is actually optimal in a number of ways.

We first construct a holomorphic motion of a Cantor set as follows. Choose a large integer $n \in \mathbb{N}$ and find disks $\mathbb{D}(z_j, r) \subset \mathbb{D}$ with disjoint closures, all of the same radius $r = r_n$, such that

$$\frac{1}{2} \leqslant nr^2 \leqslant 1, \qquad (12.16)$$

Consider the holomorphic family of similarities

$$\gamma_{j,\lambda}(z) = a(\lambda)z + z_j,$$

where $a(\lambda)$ is a holomorphic function with values $a(\mathbb{D}) \subset (r\mathbb{D}) \setminus \{0\}$. Each of the $\gamma_{j,\lambda}$'s is a strict contraction for each fixed $\lambda \in \mathbb{D}$. Therefore (see [168]) there is a unique compact set $E_\lambda \subset \mathbb{C}$ such that

$$\bigcup_{j=1}^{n} \gamma_{j,\lambda}(E_\lambda) = E_\lambda \qquad (12.17)$$

12.4. DISTORTION OF DIMENSION

Moreover, since $\gamma_{j,\lambda}(\mathbb{D}) \subset \mathbb{D}(z_j, r) \subset \mathbb{D}$ with the $\mathbb{D}(z_j, r)$ disjoint, it follows that E_λ is a (self-similar) Cantor set. It is also clear that $E_\lambda \subset \mathbb{D}$.

For simplicity, write $B_{j,\lambda} := \gamma_{j,\lambda}(\mathbb{D})$. If one reverses the above picture, one can define an expanding mapping $g_\lambda : \bigcup_j B_{j,\lambda} \to \mathbb{D}$ by $g_\lambda|_{B_{j,\lambda}} = \gamma_{j,\lambda}^{-1}$. Clearly,

$$E_\lambda = \bigcap_{k=0}^{\infty} g_\lambda^{-k}(\mathbb{D})$$

Furthermore, in a natural manner we can identify the points $x \in E_\lambda$ with sequences $(j_k(\lambda))_{k=0}^{\infty} \in \{1, \ldots, n\}^{\mathbb{N}}$. We simply define $j_k = j_k(\lambda) = j$ if $g_\lambda^k(x) \in B_{j,\lambda}$. This gives a one-to-one correspondence with E_λ and $\{1, \ldots, n\}^{\mathbb{N}}$. In this identification $g_\lambda : (j_k)_{k=0}^{\infty} \mapsto (j_{k+1})_{k=0}^{\infty}$. In other words, we see that g_λ acts as a shift on E_λ.

The important observation here is that conjugating the corresponding shifts on E_0 and E_λ gives a holomorphic motion

$$\Phi : \mathbb{D} \times E_0 \to \mathbb{C}$$

such that $\Phi_\lambda \circ g_0 = g_\lambda \circ \Phi_\lambda$ on the set E_0. In fact, the fixed-points of the iterates $\gamma_{j_1,\lambda} \circ \gamma_{j_2,\lambda} \circ \cdots \circ \gamma_{j_m,\lambda}$, $m \in \mathbb{N}$, correspond to periodic sequences of integers in the above identification. For such points z, it is easy to see that $\lambda \mapsto \Phi_\lambda(z)$ is holomorphic. Since the fixed-points are dense in E_λ, the λ-lemma, Theorem 12.1.2, implies that we have a holomorphic motion for which $\Phi_\lambda(E_0) = E_\lambda$.

The choice of the coefficients $a(\lambda)$ will then determine the properties of the motion Φ_λ. To obtain a motion with dimension distortion as large as possible, we notice that as a function of λ, $\log a_t(\lambda)$ is holomorphic with values in the half-plane $\{z : \Re e\, z < \log r\}$. To have maximal growth under these restrictions we choose

$$a_t(\lambda) = \exp\left(\log r - \frac{1-\lambda}{1+\lambda} \frac{\log(nr^t)}{t}\right), \quad \lambda \in \mathbb{D}, \qquad (12.18)$$

where

$$0 < t < 2$$

is a parameter to be chosen later. Note that by (12.16) for each fixed t, and for n large enough, $\log(nr^t) > 0$. For such $n \in \mathbb{N}$, a_t is indeed holomorphic in $\mathbb{D}$ with $a_t(\mathbb{D}) = (r\mathbb{D})\setminus\{0\}$.

Next, we want to estimate the dimension of E_λ. This is determined by the scaling ratios $|\gamma_{j,\lambda}'|$ in (12.17). Since the scaling is the same for each index j, the dimension $d(\lambda) = \dim(E_\lambda)$ is obtained [168] from the equation

$$n(|a_t(\lambda)|)^{d(\lambda)} = 1$$

By (12.18), $n(|a_t(0)|)^t = 1$. This gives $d(0) = t$ and hence also determines the value of the parameter t. Further, from (12.16) we see that

$$\left|\frac{1}{2} + \log r / \log n\right| < \delta_n$$

Hence if $0 < \lambda < 1$, the choice (12.18) leads to

$$\frac{d(0)}{d(\lambda)} = \frac{t \log(|a_t(\lambda)|)}{\log(1/n)} = \frac{d(0)}{2} + \left(1 - \frac{d(0)}{2}\right)\frac{1-\lambda}{1+\lambda} + \varepsilon, \qquad (12.19)$$

where $|\varepsilon| \leqslant 4\delta_n \to 0$ as $n \to \infty$, as long as (12.16) is satisfied.

Collecting the information above, we have the following theorem.

Theorem 12.4.1. *Suppose $0 < t < 2$. Then there is a set $E \subset \mathbb{C}$ with $\dim(E) = t$ and a holomorphic motion $\Phi : \mathbb{D} \times E \to \mathbb{C}$ such that*

$$\frac{1}{\dim(E_\lambda)} - \frac{1}{2} = \frac{1-\lambda}{1+\lambda}\left(\frac{1}{\dim(E)} - \frac{1}{2}\right)$$

for all $0 < \lambda < 1$. Here $E_\lambda = \Phi_\lambda(E)$.

Proof. Choose a countable collection $\{B_k\}_1^\infty$ of pairwise disjoint subdisks of $\mathbb{D}$ and define, using the argument above, in each disk B_k a holomorphic motion Φ of a Cantor set E_k with $\Phi_\lambda(E_k) \subset B_k$. If $d(\lambda) = \dim(\Phi_\lambda(E_k))$, we may assume that $d(0) = t$ and that for each k the estimate (12.19) holds with $\varepsilon = \frac{1}{k}$.

Clearly, this construction determines a holomorphic motion Ψ of the union $E = \bigcup_k E_k$. It is equally clear that for this set E we have the equality in (12.19) with $\varepsilon = 0$. $\square$

Naturally, one may combine the above result with (the strongest form of) the extended λ-lemma, Theorem 12.3.2. This yields the following result.

Corollary 12.4.2. *Let $0 < t < 2$ and $K \geqslant 1$. Then there is a set $E \subset \mathbb{C}$ with $\dim(E) = t$ and K-quasiconformal mappings f_1, f_2 of $\hat{\mathbb{C}}$ such that*

$$\frac{1}{\dim(f_1 E)} - \frac{1}{2} = K\left(\frac{1}{\dim(E)} - \frac{1}{2}\right) \qquad (12.20)$$

and

$$\frac{1}{\dim(f_2 E)} - \frac{1}{2} = \frac{1}{K}\left(\frac{1}{\dim(E)} - \frac{1}{2}\right) \qquad (12.21)$$

Proof. When one extends the motion Φ_λ of Theorem 12.4.1 to $\mathbb{C}$ using the extended λ-lemma, the extension has dilatation $K \leqslant \frac{1+|\lambda|}{1-|\lambda|}$. Hence the Theorem gives a set E and a K-quasiconformal map for which, say,

$$\frac{1}{\dim(f_2 E)} - \frac{1}{2} \leqslant \frac{1}{K}\left(\frac{1}{\dim(E)} - \frac{1}{2}\right) \qquad (12.22)$$

In the next chapter we prove the opposite inequality for any set E and any K-quasiconformal map of $\mathbb{C}$. Hence we must have the equality in (12.22). The equality (12.20) follows similarly. $\square$

12.5 Embedding Quasiconformal Mappings in Holomorphic Flows

We have seen in the previous sections that a holomorphic flow of a set A gives rise to a holomorphically parameterized family of quasiconformal mappings of $\mathbb{C}$, extending the given flow. In this section we prove the converse, namely, that any quasiconformal mapping of the plane embeds in a holomorphic flow.

The proof follows quickly from the holomorphic dependence on parameters described in Theorems 5.7.3 and 5.7.5. We shall then use this fact, together with the estimates derived above, to give elementary proofs of such things as distortion estimates for quasiconformal mappings (e.g., Teichmüller's theorem) and other results concerning the deformation of quasiconformal mappings that we shall use elsewhere.

These results, together with those from our subsequent chapters on higher integrability, clearly demonstrate the power of the circle of ideas connecting quasiconformal mappings and holomorphic motions.

Let us look first at an elementary but nevertheless interesting example. Recall that a K-quasicircle is the image of the unit circle under a quasiconformal mapping of $\mathbb{C}$. We will have a lot more to say about quasicircles in Section 13.3, where we will give conjecturally sharp estimates of their dimension.

Theorem 12.5.1. *Let* $\varphi : \mathbb{D} \to \mathbb{C}$ *be a conformal mapping. Then for all* $r < 1$, $\varphi(\mathbb{S}(0,r))$ *is a* $\frac{1+r}{1-r}$*-quasicircle.*

Proof. We may assume $\varphi(0) = 0$. For $\lambda \in \mathbb{D}$ set

$$\varphi_\lambda(z) = \frac{1}{\phi'(0)} \frac{\varphi(\lambda z)}{\lambda}$$

Note that since φ is differentiable at 0, $\varphi_\lambda(z) \to \varphi_0(z) \equiv z$ uniformly on compact subsets of $\mathbb{D}$ as $\lambda \to 0$. The removable singularity theorem for bounded analytic functions shows that for each $z \in \mathbb{D}$ the map $\lambda \mapsto \varphi_\lambda(z) : \mathbb{D} \to \mathbb{C}$ is holomorphic. Clearly, $\varphi_\lambda : \mathbb{D} \to \mathbb{C}$ is an injection (indeed conformal). Thus φ_λ is the restriction of a holomorphic motion of $\mathbb{C}$, say $\Phi_\lambda : \mathbb{C} \to \mathbb{C}$, which is $\frac{1+|\lambda|}{1-|\lambda|}$-quasiconformal at time $\lambda \in \mathbb{D}$. The only thing left to observe is that

$$\varphi(\mathbb{S}(0,r)) = \rho\,\varphi_r(\mathbb{S}) = \rho\,\Phi_r(\mathbb{S}),$$

where $\rho = r\varphi'(0)$. Hence the curve is a $\frac{1+r}{1-r}$-quasicircle. $\square$

Notice the further implication of the proof:

Theorem 12.5.2. *Let* $\varphi : \mathbb{D} \to \mathbb{C}$ *be a conformal mapping. Then for all* $r < 1$, $\varphi|\mathbb{D}(0,r)$ *extends to a* $\frac{1+r}{1-r}$*-quasiconformal mapping of* $\mathbb{C}$.

The theorem points to how we might embed a quasiconformal map of a general planar domain in a holomorphic flow. It is clear that there must be some restrictions on the map since even conformal mappings of the disk need not admit homeomorphic extensions to the plane; consider for instance the Riemann map $\mathbb{D} \to \mathbb{C} \setminus [0, \infty)$.

On the other hand, every global mapping embeds into a flow.

Theorem 12.5.3. *Let* $f : \mathbb{C} \to \mathbb{C}$ *be* $\frac{1+k}{1-k}$-*quasiconformal. Then there is a holomorphic motion* $F : \mathbb{D} \times \mathbb{C} \to \mathbb{C}$ *such that* $f(z) = F(k, z)$, $z \in \mathbb{C}$.

Proof. Let $\mu(z)$ be the Beltrami coefficient of f and $k = \|\mu\|_\infty$. For $\lambda \in \mathbb{D}$ set

$$\mu_\lambda(z) = \frac{\lambda}{k} \mu(z)$$

The equation

$$\frac{\partial g^\lambda}{\partial \bar{z}} = \mu_\lambda(z) \frac{\partial g^\lambda}{\partial z}$$

for $g^\lambda : \mathbb{C} \to \mathbb{C}$ has exactly one solution normalized by $g^\lambda(0) = 0$ and $g^\lambda(1) = 1$. By Theorem 5.7.5, this solution depends holomorphically on the parameter λ. Also, $g^0(z) \equiv z$. We therefore set

$$F(\lambda, z) = \left(1 + \frac{\lambda}{k} (f(1) - f(0) - 1)\right) g^\lambda(z) + \frac{\lambda}{k} f(0)$$

to define a holomorphic motion of $\mathbb{C}$. Note that $F_k(0) = f(0)$ and $F_k(1) = f(1)$, so that again by uniqueness $F_k \equiv f$. □

A combination of the above also applies for quasiconformal $f : \mathbb{D} \to \Omega$ of the disk. However, we first need to look at the inverse $f^{-1} : \Omega \to \mathbb{D}$. Let $\mu = \chi_\Omega \mu_{f^{-1}}$ so that μ is defined on $\mathbb{C}$ and $k = \|\mu\|_\infty$. Let $F : \mathbb{C} \to \mathbb{C}$ be the unique normalized solution to the Beltrami equation $F_{\bar{z}} = \mu \, F_z$. As $\mu_F(z) = \mu_{f^{-1}}(z)$ for $z \in \Omega$ by uniqueness Theorem 5.5.1, there is a conformal mapping

$$\psi : F(\Omega) \to \mathbb{D}$$

so that

$$f^{-1} = \psi \circ F : \Omega \to \mathbb{C}, \quad \text{that is, } f = F^{-1} \circ \psi^{-1}$$

We arrive at the following easy corollary of our discussion.

Theorem 12.5.4. *Let* $f : \mathbb{D} \to \mathbb{C}$ *be* $\frac{1+k}{1-k}$-*quasiconformal. Then* $f|\mathbb{D}(0,r)$ *has an extension to* $\mathbb{C}$ *that is* $\frac{1+r}{1-r} \cdot \frac{1+k}{1-k}$-*quasiconformal.*

12.6 Distortion Theorems

We have already seen and used some distortion theorems for quasiconformal mappings. Here we give elementary proofs of these results based on the theory of holomorphic motions. While there are also other approaches not requiring this deep theory of holomorphic motions, once we have it in hand its consequences

12.6. DISTORTION THEOREMS

must be exploited. Further, these proofs offer surprising connections. Let us first draw a connection between distortion theorems for normalized quasiconformal mappings and Schotty's classical theorem on the rate of growth of analytic functions omitting two points.

We begin by recalling the family of normalized quasiconformal mappings. For $k \in [0, 1)$,

$$\mathcal{QC}(k) = \{f : \mathbb{C} \to \mathbb{C} : f(0) = 0, f(1) = 1 \text{ and } f \text{ is } \tfrac{1+k}{1-k}\text{-quasiconformal}\}$$

Next consider the family of analytic functions omitting 0 and 1, where for $r > 0$ we normalize

$$\mathcal{A}(r) = \{\varphi : \mathbb{D} \to \mathbb{C} \setminus \{0, 1\} : \varphi \text{ is analytic and } |\varphi(0)| = r\},$$

and then the two extremal functions

$$\Psi_{\mathcal{QC}}(k, r) = \sup\{|f(z)| : f \in \mathcal{QC}(k), |z| = r\}$$

and

$$\Psi_{\mathcal{A}}(k, r) = \sup\{|\varphi(k)| : \varphi \in \mathcal{A}(r)\}$$

We have the following theorem.

Theorem 12.6.1. *For each $k \in [0, 1)$ and $r > 0$,*

$$\Psi_{\mathcal{QC}}(k, r) = \Psi_{\mathcal{A}}(k, r)$$

Proof. Let us first show $\Psi_{\mathcal{QC}}(k, r) \geqslant \Psi_{\mathcal{A}}(k, r)$. Choose $\varphi \in \mathcal{A}(r)$. Evidently the function

$$h(\lambda, a) = \begin{cases} \phi(\lambda), & a = \phi(0) \\ 0, & a = 0 \\ 1, & a = 1 \\ \infty, & a = \infty \end{cases}$$

is a holomorphic motion of the four-point set $\{0, 1, \varphi(0), \infty\}$. This motion extends to a holomorphic motion $H : \mathbb{D} \times \mathbb{C} \to \mathbb{C}$. Fix λ with $|\lambda| = k$. Then $H(\lambda, \cdot) \in \mathcal{QC}(k)$, and as $|\phi(0)| = r$ we have

$$|\phi(\lambda)| = |H(\lambda, \phi(0))| \leqslant \Psi_{\mathcal{QC}}(k, r)$$

To see $\Psi_{\mathcal{QC}}(k, r) \leqslant \Psi_{\mathcal{A}}(k, r)$ we argue as follows. Let $f \in \mathcal{QC}(k)$. Theorem 12.5.3 embeds f in a holomorphic motion $F : \mathbb{D} \times \mathbb{C} \to \mathbb{C}$ with $F(k, z) = f(z)$. This flow is normalized so that $F(\lambda, 0) = 0$ and $F(\lambda, 1) = 1$. In particular, since the flow is injective at each time $\lambda \in \mathbb{D}$, we have $\varphi_z(\lambda) = F(\lambda, z) \in \mathbb{C} \setminus \{0, 1\}$. Moreover, $\varphi_z(0) = F(0, z) = z$. Then if $|z| = r$,

$$|f(z)| = |F(k, z)| = |\varphi_z(k)| \leqslant \Psi_{\mathcal{A}}(k, |\varphi(0)|) = \Psi_{\mathcal{A}}(k, r),$$

which completes the proof of the theorem. $\square$

While of theoretical interest, this theorem is not useful unless one can estimate one of the functions involved. Fortunately, in [164] Hempel interprets $\Psi = \Psi(k,r)$ in terms of hyperbolic geometry and gives the following implicit formula:
$$\int_r^\Psi \rho(-t)\,dt = \log\frac{1+k}{1-k},$$
where $\rho(z)$ is the hyperbolic density of the thrice-punctured sphere $\mathbb{C}\setminus\{0,1\}$. The left hand side of this equation is simply the hyperbolic distance between $-r$ and $-\Psi$ in $\mathbb{C}\setminus\{0,1\}$. The density is defined from the hyperbolic metric of the disk via the Riemann covering map, in this case the elliptic modular function $\mathbf{w}(\zeta)$, so
$$\rho(\mathbf{w}(\zeta))|\mathbf{w}'(\zeta)| = \frac{1}{\Im m(\zeta)}.$$
From this and knowledge of the elliptic modular function, one is able to obtain bounds. Hempel gave explict bounds for instance (with $K = \frac{1+k}{1-k}$)
$$\Psi(k,1) \leqslant \frac{1}{16}e^{\pi K}$$
but there are slightly more refined estimates. For instance, [18] gives
$$\Psi(k,1)\min\{r^K, r^{1/K}\} \leqslant \Psi(k,r) \leqslant \Psi(k,1)\max\{r^K, r^{1/K}\}$$
and
$$\Psi(k,1) = \frac{1}{16}e^{\pi K} - \frac{1}{2} + o(1),$$
where $0 < o(1) < 2e^{-\pi\frac{1+k}{1-k}}$. Furthermore, Ahlfors and Beurling give the estimate
$$\Psi(k,1) \leqslant e^{ak/(1-k)} = e^{a(K-1)/2},$$
with $a = 2\frac{d}{dk}\Psi(k,1) \approx 8.76\ldots$ [11]. This estimate is of course better when k is small. Many other estimates of this type can be found in the book by G. Anderson, M. Vamanamurthy and M. Vuorinen [18]. Putting the most elementary of these estimates together with what we have proved gives the following distortion theorem, a generalization of Mori's theorem on circular distortion (the case $r=1$).

Theorem 12.6.2. *Let $f : \mathbb{C} \to \mathbb{C}$ be K-quasiconformal and $s > 0$. Then*

$$\frac{\max_{\theta\in[0,2\pi)}\{|f(z_0+rse^{i\theta})-f(z_0)|\}}{\min_{\theta\in[0,2\pi)}\{|f(z_0+se^{i\theta})-f(z_0)|\}} \leqslant e^{\pi K}\max\{r^K, r^{1/K}\} \quad (12.23)$$

Proof. Choose $\theta_0 \in [0,2\pi)$ so that
$$|f(z_0+se^{i\theta_0})-f(z_0)| = \min_{\theta\in[0,2\pi)}\{|f(z_0+se^{i\theta})-f(z_0)|\}$$

12.7. DEFORMATIONS OF MAPS

and put $w = z_0 + se^{i\theta_0}$. We set

$$g(z) = \frac{f(z_0 + zse^{i\theta_0}) - f(z_0)}{f(w) - f(z_0)}$$

so that $g : \mathbb{C} \to \mathbb{C}$ has $g(0) = 0$ and $g(1) = 1$. If $z \in \mathbb{C}$ with $|z| = r$, we have

$$|g(z)| \leqslant \Psi(k,r) \leqslant e^{\pi K} \max\{r^K, r^{1/K}\}$$

Substituting the formula for g and taking the maximum proves (12.23). □

Notice that the argument, when using the Ahlfors-Beurling bound, also gives the following, which is a much better estimate when K is close to 1.

Theorem 12.6.3. *Let $f : \mathbb{C} \to \mathbb{C}$ be K-quasiconformal and $s > 0$. Then*

$$\frac{\max_{\theta \in [0,2\pi)}\{|f(z_0 + rse^{i\theta}) - f(z_0)|\}}{\min_{\theta \in [0,2\pi)}\{|f(z_0 + se^{i\theta}) - f(z_0)|\}} \leqslant e^{5(K-1)} \max\{r^K, r^{1/K}\} \qquad (12.24)$$

Briefly, this describes another of the close relationships between quasiconformality and quasisymmetry.

Corollary 12.6.4. *Suppose $f : \mathbb{C} \to \mathbb{C}$ is K-quasiconformal. Then f is η-quasisymmetric with*

$$\eta(t) = \min\{e^{\pi K}, e^{5(K-1)}\} \max\{t^K, t^{1/K}\}$$

Notice that this estimate has the property that $\eta(t) \to t$ as $K \to 1$ uniformly on compact subsets.

In a similar vein, if $f : \mathbb{C} \to \mathbb{C}$ is K-quasiconformal with $f(0) = 0$ and letting $z, w \in \mathbb{C}$ with $0 < |w| \leqslant |z|$, then Theorem 12.6.2 has the form

$$\left|\frac{f(z)}{f(w)}\right| \leqslant e^{\pi K} \left(\frac{|z|}{|w|}\right)^K$$

Again notice that we are free to use the better estimate

$$\left|\frac{f(z)}{f(w)}\right| \leqslant e^{5(K-1)} \left(\frac{|z|}{|w|}\right)^K \qquad (12.25)$$

should K be close to 1. The reader will be aware that all of the above results give Hölder continuity estimates on f, however this is only for global mappings. Local estimates reduce to these as described in Section 3.10.

12.7 Deformations of Quasiconformal Mappings

In this last section we prove a result that will later have an important application in estimating the dimension of quasicircles. Given a K-quasiconformal homeomorphism f of $\mathbb{C}$, it allows us to deform f to the identity on a set of a definite size while leaving f unchanged sufficiently far away, so that the distortion of f is not increased very much. Of course all this needs to be quantified.

Theorem 12.7.1. Let $f : \mathbb{C} \to \mathbb{C}$ be $\frac{1+k}{1-k}$-quasiconformal and normalized so that $f(0) = 0$ and $f(1) = 1$. Let $k < \alpha < 1$. Then there is $g : \mathbb{C} \to \mathbb{C}$ such that

1. g is $\frac{1+\alpha}{1-\alpha}$-quasiconformal.

2. $g_{|\mathbb{D}} = f$.

3. $g(z) \equiv z$ for $z \in \mathbb{C} \setminus \mathbb{D}(0, R)$, where $R = e^{\pi \frac{1+\alpha/k}{1-\alpha/k}}$.

Proof. Let μ be the Beltrami coefficient of f, $k = \|\mu\|_\infty$. We obtain a holomorphic motion via the normalized solutions to

$$g_{\bar{z}}^\lambda = \frac{\lambda}{\alpha} \mu g_z^\lambda,$$

where $g^\lambda(0) = 0$, $g^\lambda(1) = 1$ and $\lambda \in \mathbb{D}$. Note that $\|\mu g^\lambda\|_\infty \leqslant k/\alpha < 1$, so g^λ is $\frac{1+k/\alpha}{1-k/\alpha}$-quasiconformal independent of λ. Further,

$$g^\alpha = f$$

Theorem 12.6.2 gives us the uniform estimate

$$|g^\lambda(z)| \leqslant e^{\pi \frac{1+\alpha/k}{1-\alpha/k}}, \qquad z \in \mathbb{D}, \lambda \in \mathbb{D}$$

Therefore

$$H(\lambda, z) = \begin{cases} g^\lambda(z), & z \in \mathbb{D} \\ z, & z \in \mathbb{C} \setminus \mathbb{D}(0, R) \end{cases}$$

is a holomorphic motion of the set $\mathbb{D} \cup (\mathbb{C} \setminus \mathbb{D}(0, R))$. This motion extends to $\tilde{H} : \mathbb{D} \times \mathbb{C} \to \mathbb{C}$. Set

$$g(z) = \tilde{H}(\alpha, z)$$

Then g is $\frac{1+\alpha}{1-\alpha}$-quasiconformal in $\mathbb{C}$, and for $z \in \mathbb{D}$,

$$g(z) = \tilde{H}(\alpha, z) = H(\alpha, z) = g^\alpha(z) = f(z),$$

while for $z \in \mathbb{C} \setminus \mathbb{D}(0, R)$,

$$g(z) = \tilde{H}(\alpha, z) = z$$

Thus g is the quasiconformal mapping required to prove the theorem. □

Translation, scaling and reflection in the unit disk, gives the following corollary in which form the deformation theorem is mostly used.

Corollary 12.7.2. Let $f : \mathbb{C} \to \mathbb{C}$ be $\frac{1+k}{1-k}$-quasiconformal. Let $k < \alpha < 1$, $z_0 \in \mathbb{C}$ and $r > 0$. If $s = e^{-\pi \frac{1+\alpha/k}{1-\alpha/k}} < 1$, then there is $g : \mathbb{C} \to \mathbb{C}$ such that

1. g is $\frac{1+\alpha}{1-\alpha}$-quasiconformal.

2. $g_{|\mathbb{C} \setminus \mathbb{D}(z_0, r)} = f$.

12.7. DEFORMATIONS OF MAPS

3. For $|z - z_0| \leqslant sr$ we have $g(z) = \alpha z + \beta = A(z)$, a similarity with
$$A(z_0) = f(z_0), \qquad A(z_0 + r) = f(z_0 + r)$$

Proof. Apply Theorem 12.7.1 to the mapping
$$F(z) = \frac{f(z_0 + r) - f(z_0)}{f(z_0 + \frac{r}{z}) - f(z_0)}, \qquad z \in \overline{\mathbb{C}}$$

This is a normalized $\frac{1+k}{1-k}$-quasiconformal mapping of $\mathbb{C}$, fixing $0, 1$ and ∞.

Changing variables $w = z_0 + r/z$, we have
$$f(w) = f(z_0) + \frac{f(z_0 + r) - f(z_0)}{F\left(\frac{r}{w - z_0}\right)},$$

where
$$F\left(\frac{r}{w - z_0}\right) = \frac{f(z_0 + r) - f(z_0)}{f(w) - f(z_0)}$$

As $|w - z_0| < sr$ if and only if $|z| > 1/s$, a deformation of F using Theorem 12.7.1 gives a deformation of f that has the required properties 1–3. $\square$

Finally, we want to give a result that is similar in nature though it is presented more as an existence theorem. The point here is that we perturb a principal solution to the Beltrami equation to the identity on a definite-sized set without increasing the distortion.

Theorem 12.7.3. *Let $\mu : \mathbb{D} \to \mathbb{D}$, $\|\mu\|_\infty = k < 1$, be a Beltrami coefficient. Then there is a quasiconformal map $f : \mathbb{C} \to \mathbb{C}$ such that*

- *f is $\frac{1+k}{1-k}$-quasiconformal.*
- *$\mu_f(z) = \mu(z)$ for $z \in \mathbb{D}$.*
- *$f(z) \equiv z$ for all $|z| > 2$.*

Proof. As a consequence of Theorem 5.7.3, we now know that that the principal solutions to the Beltrami equation
$$(f^\lambda)_{\bar{z}}(z) = \frac{\lambda}{k} \mu(z) \chi_\mathbb{D}(z) (f^\lambda)_z(z)$$

move holomorphically. Each such solution is conformal in $\mathbb{C} \setminus \mathbb{D}$ and has expansion $(f^\lambda)(z) = z + o(1)$. Applying Theorem 2.10.4 we see that for all $\lambda \in \mathbb{D}$,
$$f^\lambda(\mathbb{D}) \subset \mathbb{D}(0, 2)$$

We now argue exactly as before. The motion
$$H(\lambda, z) = \begin{cases} f^\lambda(z), & z \in \mathbb{D} \\ z, & z \in \mathbb{C} \setminus \mathbb{D}(0, 2) \end{cases}$$

is a holomorphic motion of $\mathbb{D} \cup (\mathbb{C} \setminus \mathbb{D}(0, 2))$ that extends to $\tilde{H}$. If we set $f(z) = \tilde{H}(k, z)$, then all the claims of the theorem are satisfied. $\square$

Chapter 13

Higher Integrability

In this chapter we will meet some of the most important recent advances in the theory of planar quasiconformal mappings. The results will be used in wide and diverse settings and will provide a bridge between complex dynamics and the regularity theory and singularity structure of second-order elliptic equations.

Finding the exact degree of smoothness of a K-quasiconformal mapping is the primary question we shall address in this chapter. It should not be surprising that knowing such estimates provides bounds on how a mapping might distort the measure of a given set. Bojarski's Theorem 5.4.2 shows that the differential of a quasiconformal mapping is integrable to some power $p = p(K)$ larger than the *natural* exponent $p = 2$. Using Calderón-Zygmund theory [87, 86] and following Bojarski, Gehring and Reich [138] gave bounds for this distortion and made the "area distortion conjecture", which we shall prove here in Theorem 13.1.5 following [21]. This theorem describes the precise degree of integrability one can expect for the differential—we will even pay close attention to the borderline cases for later applications. We shall later see in Chapter 16 that complex gradients of solutions to second-order PDEs of divergence-type are quasiconformal mappings, and from this, using the area distortion theorem, we will obtain sharp regularity results for such equations.

It is not at all obvious but is nevertheless the case that knowledge of the degree of integrability of the differential of a mapping also controls the distortion of Hausdorff dimension—even for Cantor sets of zero measure. This is of particular interest in areas such as holomorphic dynamics where the dimension of such concepts as Julia sets or the limit sets of Kleinian groups convey considerable geometric information concerning the dynamical system or discrete group being studied.

13.1. DISTORTION OF AREA

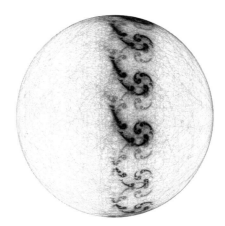

Limit set of a Kleinian group on the Riemann sphere.
This Cantor set is quasiconformally equivalent to the middle thirds set.

The theory of holomorphic motions discussed in Chapter 12 has already shown us the surprising fact that an analytically parameterized family of injections of an arbitrary set is the restriction of a global holomorphic flow of quasiconformal mappings. From knowledge of the precise regularity, we shall be able to optimally bound the distortion of dimension under such a flow. These bounds are far better than one might achieve from Hölder estimates. Following this basic circle of ideas, we will present a result of Smirnov that gives (conjecturally optimal) bounds for the distortion of lines and circles (where there is a topological obstruction to lowering the dimensions).

We will also pay close attention to what happens *below* the natural exponent $p = 2$. There are two reasons for this. First, we will meet the interesting phenomenon of unexpected regularity gain. A weak solution to a Beltrami (or related) equation will in fact be a strong solution—provided it is not too weak. Here we will formulate and prove sharp results. Second, it is an unexpected feature of the analysis of singularities of mappings (either quasiconformal or solutions to second-order equations and so forth) that uniform bounds below the ambient dimension (and in all dimensions, see [189]) seem to be crucial for proving removable singularity theorems—dubbed Painlevé's problem by analogy with the classical result for analytic functions. Again, we will formulate and prove precise theorems relating the dimension of a set and the class of mappings for which it is removable.

13.1 Distortion of Area

As mentioned above, Bojarski's Theorem 5.4.2 concerning the higher integrability of quasiconformal mappings shows that $f \in W_{loc}^{1,p}(\Omega)$ for all p such that

$\|\mu_f\|_\infty \mathbf{S}_p < 1$, where $\mathbf{S}_p$ are the p-norms of the Beurling transform. As we have pointed out, it is conjectured that $\mathbf{S}_p = p - 1$ for $p > 2$. If this conjecture were true, for K-quasiconformal mappings the inequality would imply $\frac{K-1}{K+1}(p-1) < 1$. Rearranging this, we find the value

$$p < \frac{2K}{K-1}$$

Thus we might expect a K-quasiconformal mapping to lie in the Sobolev space $W^{1,p}_{loc}(\Omega)$ for all $p < \frac{2K}{K-1}$. The formulas in (2.44) show the radial stretching $f : z \mapsto z|z|^{1/K-1}$ to have differential

$$|Df| = \frac{1}{K}|z|^{1/K-1}$$

Then

$$|Df|^{2K/(K-1)} = K^{2K/(1-K)}|z|^{-2},$$

which is not locally integrable near 0. However, $Df \in \text{weak-}L^{\frac{2K}{K-1}}(\mathbb{C})$. Thus we expect that the optimal degree of regularity is precisely the number $p_K = 2K/(K-1)$, and it is the main aim of this chapter to establish this fact.

In this first section we study how a quasiconformal mapping can distort the area of a subset $E \subset \mathbb{C}$. We will see that holomorphic deformations give optimal bounds on the distortion. These bounds will lead to a number of fundamental consequences, including the optimal higher integrability.

We already know from previous chapters that the solutions to the Beltrami equation depend holomorphically on the parameters. Here we will also need the holomorphic dependence for the derivatives. The following key observation is our starting point; we extract from Theorem 5.7.2 and (5.52) the following.

Theorem 13.1.1. *Suppose μ is a compactly supported measurable function with $\|\mu\|_\infty = k < 1$. For $\lambda \in \mathbb{D}$, let $f(\lambda, z) = f^\lambda(z)$ be the principal solution to the Beltrami equation*

$$\frac{\partial f}{\partial \bar{z}} = \frac{\lambda}{k}\mu(z)\frac{\partial f}{\partial z} \qquad (13.1)$$

If μ vanishes in a neighborhood of a point $z \in U \subset \mathbb{C}$, then the derivative $f_z^\lambda(z)$ of the analytic function $z \mapsto f^\lambda(z)$, $z \in U$, depends holomorphically on $\lambda \in \mathbb{D}$.

For convenience, we shall use the term *principal mapping* to mean a quasiconformal mapping with compactly supported complex dilatation and normalized by

$$f(z) = z + \mathcal{O}\left(\frac{1}{z}\right) \qquad \text{as } z \to \infty$$

13.1.1 Initial Bounds for Distortion of Area

We begin by building techniques for estimating the area $|f(E)|$ for general K-quasiconformal mappings f. It turns out to be fruitful to study this question in two special complementary situations.

13.1. DISTORTION OF AREA

In the first and more straightforward case, let us assume that $E \subset \mathbb{C}$ is compact and that f is conformal on $\mathbb{C} \setminus E$.

Theorem 13.1.2. *Suppose f is a K-quasiconformal principal mapping of $\mathbb{C}$ that is conformal outside a compact subset E. Then we have*

$$|f(E)| \leq K|E| \tag{13.2}$$

Proof. We shall estimate the area $|f(E)|$ with the help of the identity in (5.9), namely,

$$f_z = 1 + \mathcal{S}(f_{\bar{z}})$$

This gives

$$\begin{aligned} |f(E)| &= \int_E J(z,f) = \int_E \left(|f_z|^2 - |f_{\bar{z}}|^2\right) \\ &= |E| + \int_E 2\mathfrak{Re}\left(\mathcal{S}\, f_{\bar{z}}\right) + \int_E \left(|\mathcal{S}(f_{\bar{z}})|^2 - |f_{\bar{z}}|^2\right) \end{aligned}$$

Since $f_{\bar{z}}$ vanishes outside E by assumption, the latter integral

$$\int_E |\mathcal{S}(f_{\bar{z}})|^2 - |f_{\bar{z}}|^2 \leqslant \int_{\mathbb{C}} |\mathcal{S}(f_{\bar{z}})|^2 - |f_{\bar{z}}|^2 = 0$$

since the Beurling transform is an isometry in $L^2(\mathbb{C})$. In the first integral we represent $f_{\bar{z}}$ as a power series in μ. As in (5.8), we see

$$f_{\bar{z}} = \mu\, f_z = \mu + \mu\mathcal{S}\mu + \mu\mathcal{S}\mu\mathcal{S}\mu + \cdots$$

with convergence in for instance the L^2-norm. Next, as $|\mu| \leqslant \frac{K-1}{K+1}\chi_E$ by assumption, the integral over E of this series may be estimated by iterating the inequalities

$$\int_E |\mathcal{S}(\mu g)| \leqslant \sqrt{|E|} \left(\int_{\mathbb{C}} |\mathcal{S}(\mu g)|^2\right)^{1/2} \leqslant \sqrt{|E|}\|\mu\|_{\infty} \left(\int_E |g|^2\right)^{1/2}$$

with $g = 1$, $g = \mathcal{S}\mu$, $g = \mathcal{S}\mu\mathcal{S}\mu$ and so on. This now yields

$$|f(E)| \leq |E| + 2\|\mu\|_{\infty}|E| + 2\|\mu\|_{\infty}^2|E| + \cdots = |E|\left(1 + \frac{2\|\mu\|_{\infty}}{1 - \|\mu\|_{\infty}}\right)$$

As $K = \frac{1+\|\mu\|_{\infty}}{1-\|\mu\|_{\infty}}$, we have shown that $|f(E)| \leqslant K|E|$. $\square$

13.1.2 Weighted Area Distortion

Let us next turn to the case that is complementary to Theorem 13.1.2. That is, we aim at obtaining estimates for $|f(E)|$ when the complex dilatation of f vanishes on the set E. For later purposes it will be useful to consider the situation in a slightly more general setup, and therefore we introduce weights in the estimates. For us a *weight* is simply a non-negative measurable function. We will need to study the distortion of a *weighted* area (that is, $\int_E w(z) dz$) under a K-quasiconformal mapping. For most of the results in this chapter we will take the unit weight $w(z) \equiv 1$, but in establishing optimal removability results for quasiregular mappings, the more general weighted case, first presented in [34], will be needed.

To obtain sharp bounds we need to consider a principal mapping f. We prove the following theorem.

Theorem 13.1.3. *Suppose f is a principal K-quasiconformal mapping of $\mathbb{C}$ that is conformal outside the unit disk $\mathbb{D}$. Assume also that we are given a measurable set $E \subset \mathbb{D}$ and a weight $w(z) \geqslant 0$, $z \in E$. If $f|E$ is conformal, meaning that $f_{\bar{z}}(z) = 0$ for almost every $z \in E$, then*

$$\left(\frac{1}{\pi} \int_E w(z)^{1/K} \right)^K \leqslant \frac{1}{\pi} \int_E J(z, f) \, w(z) \leqslant \left(\frac{1}{\pi} \int_E w(z)^K \right)^{1/K} \qquad (13.3)$$

As we shall later see, Theorem 13.1.3 is sharp; Example 13.1.3 will present a large class of weights and K-quasiconformal mappings f for which one of the bounds in (13.3) holds as an equality.

Proof of Theorem 13.1.3. We first establish the distortion estimate for open sets E and prove the general case subsequently by approximation. Moreover, we need to consider only weight functions $w \geqslant 0$ that are bounded away from 0 and ∞ on the set E. The argument for a general w follows by an obvious limiting argument.

Suppose that the weight $w(z)$ and the mapping f are given, with $f_{\bar{z}}(z) = 0$ for almost every point z of the open set E.

Our goal is to show that

$$\frac{1}{\pi} \int_E w(z) J(z, f) \leqslant \left(\frac{1}{\pi} \int_E w(z)^K \right)^{1/K} \qquad (13.4)$$

For this let μ be the complex dilatation of f, so that

$$\frac{\partial f}{\partial \bar{z}}(z) = \mu(z) \frac{\partial f}{\partial z}(z), \qquad |\mu| \leqslant \frac{K-1}{K+1} \chi_{\mathbb{D} \setminus E} \qquad (13.5)$$

Then for each complex number $|\lambda| < 1$, we write

$$\mu_\lambda(z) = \lambda \frac{K+1}{K-1} \mu(z),$$

$z \in \mathbb{C}$, and consider the principal solution $f^\lambda \in W^{1,2}_{loc}(\mathbb{C})$ to

$$\frac{\partial f^\lambda}{\partial \bar{z}}(z) = \mu_\lambda(z) \frac{\partial f^\lambda}{\partial z}(z)$$

13.1. DISTORTION OF AREA

In particular, for $\lambda = k = \frac{K-1}{K+1}$, we have $f^\lambda = f$. Furthermore, by Theorem 13.1.1, for each fixed $z \in E$ the function $\lambda \mapsto (f^\lambda)'(z) = f_z^\lambda(z)$ is holomorphic in $\mathbb{D}$. Note also that by conformality on (the open set) E,

$$(f^\lambda)'(z) \neq 0 \quad \text{for all } z \in E \text{ and } \lambda \in \mathbb{D} \tag{13.6}$$

In fact, as it is enough to prove the claim for all compact subsets $E' \subset E$, we may assume that on E the derivatives $|(f^\lambda)'(z)|$ are bounded away from 0 and ∞, with constants depending on λ.

We will approach (13.4) by expressing it in a logarithmic form, which allows a decomposition of the integral. In fact, given a function $a(z) > 0$, $z \in E$, we use the concavity of the logarithm in conjunction with Jensen's inequality to find

$$\log \int_E a(z) = \sup_p \left[\int_E p(z) \log \left(\frac{a(z)}{p(z)} \right) \right] \tag{13.7}$$

Here the supremum is taken over the functions p such that

- $p(z) > 0$ for almost every $z \in E$ and
- $\int_E p = 1$.

Note that the supremum is attained when $p(z) \equiv a(z)/\int_E a$.

In our case we take

$$a(z) = \frac{1}{\pi} w(z) J(z, f^\lambda) = \frac{1}{\pi} w(z) |(f^\lambda)'(z)|^2$$

for $z \in E$. Then (13.7) gives

$$\log \left(\frac{1}{\pi} \int_E w(z) |(f^\lambda)'(z)|^2 \right) \tag{13.8}$$

$$= \sup_p \left[\int_E p(z) \log w(z) + \int_E p(z) \log \left(\frac{1}{\pi} \frac{|(f^\lambda)'(z)|^2}{p(z)} \right) \right]$$

If we take a close look at this identity, we see that the latter integral

$$h_p(\lambda) = \int_E p(z) \log \left(\frac{1}{\pi} \frac{|(f^\lambda)'(z)|^2}{p(z)} \right) \tag{13.9}$$

is harmonic in λ, by (13.6). Moreover, we can use the area formula Theorem 2.10.1 to deduce

$$h_p(\lambda) \leq \log \left(\frac{1}{\pi} \int_E |(f^\lambda)'(z)|^2 \right) \leq \log \left(\frac{1}{\pi} \int_\mathbb{D} J(z, f^\lambda) \right) \leq 0$$

Consequently, $\lambda \mapsto h_p(\lambda)$ is harmonic and nonpositive in $\mathbb{D}$.

We are now in a position to use Harnack's inequality; it turns out that it is precisely here that the sharpness of the bounds is forged. We have

$$h_p(\lambda) \leq \frac{1-|\lambda|}{1+|\lambda|} h_p(0) = \frac{1-|\lambda|}{1+|\lambda|} \int_E p(z) \log\left(\frac{1}{\pi}\frac{1}{p(z)}\right) \tag{13.10}$$

Combining (13.8) and (13.10), we obtain

$$\log\left(\frac{1}{\pi}\int_E w(z)|(f^\lambda)'(z)|^2\right)$$
$$\leq \sup_p \left[\int_E p(z)\log w(z) + \frac{1-|\lambda|}{1+|\lambda|}\int_E p(z)\log\left(\frac{1}{\pi}\frac{1}{p(z)}\right)\right]$$
$$= \frac{1}{K}\sup_p\left[\int_E p(z)\log\left(\frac{1}{\pi}\frac{w(z)^K}{p(z)}\right)\right]$$

when $\lambda = \frac{K-1}{K+1}$. It remains to choose the function p, and naturally we take $p(z) = \frac{w(z)^K}{\int_E w}$, the function maximizing the last expression. As a result,

$$\log\left(\frac{1}{\pi}\int_E w(z)|(f^\lambda)'(z)|^2\right) \leq \frac{1}{K}\log\left(\frac{1}{\pi}\int_E w(z)^K\right) \tag{13.11}$$

Exponentation gives the estimate (13.4) for open sets E.

To deduce the lower bound in (13.3) one needs to use Harnack's inequality in the form

$$h_p(\lambda) \geq \frac{1+|\lambda|}{1-|\lambda|} h_p(0), \tag{13.12}$$

rather than (13.10), but otherwise argue in a similar fashion.

We now consider the case of arbitrary measurable sets E for which $f|E$ is conformal. Choose a decreasing sequence $\{E_n\}_{n=1}^\infty$ of open sets such that $E \subset E_n$ and

$$|E_n \setminus E| \to 0 \quad \text{as } n \to \infty$$

Set $\mu_n = \mu\chi_{\mathbb{D}\setminus E_n}$, with μ as in (13.5), and let f_n be the principal solution to the Beltrami equation $(f_n)_{\bar z} = \mu_n(z)(f_n)_z$.

It follows from Lemma 5.3.1 that

$$\|(f_n)_z - f_z\|_{L^2(\mathbb{C})} = \|\mathcal{S}(f_n)_{\bar z} - \mathcal{S}f_{\bar z}\|_{L^2(\mathbb{C})} \to 0$$

as $n \to \infty$. Also recall that on the sets E_n and E, we have $J(z, f_n) = |(f_n)_z|^2$ and $J(z,f) = |f_z|^2$, respectively. Since w can be assumed to be bounded from above on the E_n's, we deduce

$$\int_E w(z)J(z,f) = \lim_{n\to\infty}\int_{E_n} w(z)J(z,f_n) \tag{13.13}$$

13.1. DISTORTION OF AREA

Clearly, for any $q > 0$,

$$\int_E w(z)^q = \lim_{n\to\infty} \int_{E_n} w(z)^q \qquad (13.14)$$

As the inequalities (13.3) hold for each of the open sets E_n, using (13.13) and (13.14) gives the proof of Theorem 13.1.3. □

As claimed, the inequalities of Theorem 13.1.3 are optimal, and this can be seen from a variety of different examples such as that from [34] we present below. Moreover, the distribution of w can be chosen in an essentially arbitrary manner, as we can see in (13.15).

13.1.3 An Example

Suppose $0 < w_1 < \cdots < w_n$ and $0 < p_j < 1$, $j = 1, \ldots, n$, satisfy

$$\sum_{j=1}^n p_j w_j^K \leq w_1^K$$

Then we claim that there is an open set $E \subset \mathbb{D}$ with $|E| = \pi \sum_{j=1}^n p_j$, a K-quasiconformal mapping f with $f|E$ conformal and a weight w, positive on E, such that

$$|\{z \in E : w(z) = w_j\}| = \pi p_j, \qquad j = 1, \ldots, n, \qquad (13.15)$$

$$\frac{1}{\pi} \int_E w(z) J(z, f) = \left(\frac{1}{\pi} \int_E w(z)^K\right)^{1/K} \qquad (13.16)$$

We may also arrange that f is a principal mapping conformal outside $\mathbb{D}$. To construct such mappings and weights satisfying (13.15) and (13.16), we need some notation. First, we may normalize the numbers w_j so that

$$\sum_{j=1}^n p_j w_j^K = 1 \qquad (13.17)$$

Then, as our assumption is that $w_1 \geq 1$ and $w_{j+1} > w_j$, we can find numbers $0 < R_j \leq 1$ so that

$$w_j = \left(\prod_{l=1}^j R_l\right)^{-2/K}, \qquad j = 1, \ldots, n \qquad (13.18)$$

Consider the the radial stretching defined by

$$f_R(z) = \begin{cases} R^{\frac{1}{K}-1} z, & 0 \leq |z| \leq R \\ z|z|^{\frac{1}{K}-1}, & R < |z| \leq 1 \\ z, & 1 < |z| \end{cases} \qquad (13.19)$$

Also, set $f_R^\rho(z) = \rho f_R(\frac{z}{\rho})$, $z \in \mathbb{C}$. To complete the notation let ρ_j, $j = 1, \ldots, n$, be numbers defined by the following rule,

$$R_n^2 \rho_n^2 = p_n, \qquad R_j^2 \rho_j^2 - \rho_{j+1}^2 = p_j, \qquad 1 \leq j \leq n-1 \qquad (13.20)$$

Then by (13.18) and (13.17) we have $\rho_1 = 1$.

After all these preparations we simply set

$$f = f_{R_1}^{\rho_1} \circ \cdots \circ f_{R_n}^{\rho_n}, \qquad w = \sum_{j=1}^{n} w_j \chi_{E_j},$$

where for $1 \leq j \leq n-1$,

$$E_n = \{z : |z| < \rho_n R_n\}, \qquad E_j = \{z : \rho_{j+1} < |z| < \rho_j R_j\}$$

Then by (13.20), $|E_j| = \pi p_j$ for each $j = 1, \ldots, n$, while (13.15) is clearly satisfied and (13.16) follows since by the chain rule $J_{f|E} \equiv w^{K-1}$ on the set $E = \cup_{j=1}^n E_j$. □

Similarly, when we compare the example with the estimates of the previous section, we note that the mapping f_R^{-1} is conformal off the set $E = \{z : R^{1/K} < |z| < 1\}$ with

$$|f_R^{-1}(E)| = \pi - \pi R^2 = \pi - \pi(1 + R^{2/K} - 1)^K = K|E| + \mathcal{O}(|E|^2)$$

Hence in (13.2) the multiplicative constant in front of $|E|$ cannot be improved.

13.1.4 General Area Estimates

In Section 3.7 we discussed Lusin's condition $\mathcal{N}$ and saw that K-quasiconformal mappings preserve sets of Lebesgue measure zero. Later, with Bojarski's theorem we saw that the maps in fact distort the measure in a Hölder-continous manner. Here we shall establish that the correct Hölder exponent is $1/K$ for all K-quasiconformal mappings $f : \mathbb{C} \to \mathbb{C}$.

It will be useful to express the result under various normalizations. In the case when f is conformal outside the unit disk, we obtain the following very sharp estimate.

Theorem 13.1.4. *Suppose f is a K-quasiconformal principal mapping of $\mathbb{C}$ that is conformal outside the unit disk $\mathbb{D}$. Let $E \subset \mathbb{D}$ be measurable. Then*

$$\frac{|f(E)|}{\pi} \leq K \left(\frac{|E|}{\pi}\right)^{1/K} \qquad (13.21)$$

Conversely, the K-quasiconformal principal mapping f_0, defined by

$$f_0(z) = z|z|^{1/K-1} \quad \text{for } |z| \leq 1; \quad f(z) = z \quad \text{for } |z| \geq 1,$$

satisfies

$$\frac{|f_0(E)|}{\pi} = \left(\frac{|E|}{\pi}\right)^{1/K}, \qquad E = \mathbb{D}(0, \rho), \quad 0 < \rho < 1$$

13.1. DISTORTION OF AREA

Proof. First, any measurable set can be approximated from above by open sets G and from below by closed sets F so that $|G \setminus F|$ is arbitrarily small. We have already used such approximations earlier. From Theorem 3.7.6 we know that the measure $|f(E)|$ is absolutely continuous with respect to the two-dimensional Lebesgue measure. Therefore we may assume that E is in fact a finite union of open disks. In particular, $|\partial E| = 0$.

Next, we shall use the measurable Riemann mapping Theorem 5.3.2 to reduce the claim to Theorems 13.1.2 and 13.1.3. Let μ be the complex dilatation of f. We construct a new dilatation μ_0 by requiring that $\mu_0(z) \equiv 0$ in E and $\mu_0(z) = \mu(z)$ elsewhere, thus $\mu_0 = \mu \chi_{\mathbb{C} \setminus E}$. If g is the principal quasiconformal mapping of $\mathbb{C}$ with complex dilatation μ_0, then by Theorem 5.5.6 we have

$$f = h \circ g,$$

where h is K-quasiconformal in $\mathbb{C}$, conformal outside $g(\overline{E})$ and normalized by $h(z) = z + \mathcal{O}(\frac{1}{z})$.

At this point we may apply Theorem 13.1.2, which gives

$$|f(E)| = |h \circ g(E)| = |h \circ g(\overline{E})| \leqslant K|g(\overline{E})| = K|g(E)|$$

since $|\partial E| = 0$ and both h and g preserve sets of measure zero. On the other hand, the auxiliary mapping g satisfies all the assumptions of Theorem 13.1.3. Namely, g is conformal in $(\mathbb{C} \setminus \mathbb{D}) \cup E$ since f is conformal outside the unit disk. Moreover, it has the correct normalization at infinity, and so we can apply Theorem 13.1.3 with the weight $w(z) \equiv 1$ to obtain

$$\frac{|f(E)|}{\pi} \leqslant K \frac{|g(E)|}{\pi} \leqslant K \left(\frac{|E|}{\pi} \right)^{1/K}$$

This completes the proof. $\square$

In the general case we formulate the following invariant formulation.

Area Distortion Theorem

Theorem 13.1.5. *For every $K \geqslant 1$ there is a constant C_K, depending only on K, such that for any K-quasiconformal mapping $f : \mathbb{C} \to \mathbb{C}$, for any disk $B \subset \mathbb{C}$ and for any subset $E \subset B$, we have*

$$\frac{1}{C_K} |f(B)| \left(\frac{|E|}{|B|} \right)^K \leqslant |f(E)| \leqslant C_K |f(B)| \left(\frac{|E|}{|B|} \right)^{1/K} \tag{13.22}$$

Proof. Composing with similarities, we can assume that B is the unit disk $\mathbb{D}$. Further, as global quasiconformal mappings are η-quasisymmetric, with $\eta = \eta_K$, Lemma 3.7.4 in particular applies. Therefore $|f(\mathbb{D})| \leqslant C_K |f(\mathbb{D}(0, \frac{1}{2}))|$. We can thus assume that $E \subset \mathbb{D}(0, \frac{1}{2})$.

Let us first consider the inequality on the right hand side. As in proving Theorem 13.1.4, we factorize $f = \gamma \circ \phi$, where ϕ is chosen to be a principal quasiconformal mapping with complex dilatation $\mu_\phi = \chi_\mathbb{D} \mu_f$. It follows that

$$\gamma : \phi(\mathbb{D}) \to f(\mathbb{D})$$

is conformal.

We now need Lipschitz bounds for this function in the subdomain $\phi(\mathbb{D}(0, \frac{1}{2}))$. Indeed, the Koebe distortion theorem, as formulated in (2.70), provides us with the estimate

$$|\gamma'(w)| \leqslant 4 \frac{\operatorname{dist}(\gamma \circ \phi(z), \partial f(\mathbb{D}))}{\operatorname{dist}(\phi(z), \partial \phi(\mathbb{D}))} \quad \text{for } w = \phi(z),\ |z| \leqslant 1/2$$

Here $\operatorname{dist}(\gamma \circ \phi(z), \partial f(\mathbb{D})) = \operatorname{dist}(f(z), \partial f(\mathbb{D})) \leqslant \operatorname{diam} f(\mathbb{D})$. For the denominator, with the quasisymmetry of ϕ,

$$|\phi(\zeta) - \phi(z)| = \frac{|\phi(\zeta) - \phi(z)|}{|\phi(\zeta) - \phi(0)|} |\phi(\zeta) - \phi(0)| \geqslant \frac{1}{\eta(2)} |\phi(\zeta) - \phi(0)|$$

whenever $|\zeta| = 1$ and $|z| = 1/2$. To estimate the last expression consider the function $\Psi(z) = (\phi(1/z) - \phi(0))^{-1}$ holomorphic in $\mathbb{D}$ with $\Psi(0) = 0$ and $\Psi'(0) = 1$. With the Schwarz lemma we see that $\min_{|\zeta|=1} |\Psi(\zeta)| \leqslant 1$, and this gives

$$\min_{|\zeta|=1} |\phi(\zeta) - \phi(0)| \geqslant \frac{1}{\eta(1)} \max_{|\zeta|=1} |\phi(\zeta) - \phi(0)| \geqslant \frac{1}{\eta(1)}$$

Thus, finally,

$$\operatorname{dist}(\phi(z), \partial \phi(\mathbb{D})) \geqslant \frac{1}{\eta(1)\eta(2)}$$

We therefore have

$$|\gamma'(z)|^2 \leqslant C_K \operatorname{diam}(f(\mathbb{D}))^2 \leqslant C_K |f(\mathbb{D})|$$

for all $z \in \phi(\mathbb{D}(0, 1/2))$. Hence

$$\frac{|f(E)|}{|f(\mathbb{D})|} = \frac{|\gamma(\phi E)|}{|f(\mathbb{D})|} \leqslant C_K |\phi(E)| \leqslant C_K |E|^{1/K},$$

by Theorem 13.1.4 applied to ϕ. Then the right-hand-side inequality in (13.22) is established.

For the other inequality let us first look at the domain $\Omega = f(\mathbb{D})$. Recall from Section 3.4 that as a quasiconformal image of the unit disk, Ω is roughly of the size of a disk of radius $R \simeq \operatorname{diam}(\Omega)$. More precisely, if $z_0 = f(0)$, let

$$B = \mathbb{D}(z_0, R), \qquad R = \max_{|\zeta|=1} |f(\zeta) - f(0)|$$

Then, as in Lemma 3.4.5,

$$\delta B \subset \Omega = f(\mathbb{D}) \subset B, \qquad \delta = \frac{1}{\eta(1)}$$

13.2. HIGHER INTEGRABILITY

The inverse mapping $g = f^{-1}$ is quasisymmetric with distortion

$$\eta_g(t) = 1/\eta^{-1}(1/t)$$

Thus

$$|g(B)| \leqslant C_0 |g(\delta B)| \leqslant C_0 |g(\Omega)|, \qquad C_0 = \frac{1}{\eta^{-1}(\delta)^2},$$

where $g(\Omega) = \mathbb{D}$. Since we already have the right hand side of (13.22), we can apply this to the set $f(E) \subset B$ and obtain

$$|E|^K = |g(f(E))|^K \leqslant C_0 \frac{|g(f(E))|}{|g(B)|} \leqslant C_1 \left(\frac{|f(E)|}{|B|}\right)^{1/K} \leqslant C_1 \left(\frac{|f(E)|}{|f(\mathbb{D})|}\right)^{1/K}$$

Exponentiating the estimate gives the left hand side of (13.22). □

13.2 Higher Integrability

The distortion bounds of Theorem 13.1.5 are equivalent to the fact that for all K-quasiconformal mappings of $\mathbb{C}$, the Jacobian $J(\cdot, f) \in \text{weak-}L_{loc}^p$ with $p = \frac{K}{K-1}$. Here, however, we prove only the following direction, which suffices for our purposes.

Theorem 13.2.1. *Let $f : \mathbb{C} \to \mathbb{C}$ be a K-quasiconformal mapping and let $B \subset \mathbb{C}$ be a disk. Then there is a finite constant C_K, depending only on K, such that*

$$|\{z \in B : J(z,f) > t\}| \leqslant C_K |B|^{1/(1-K)} \left(\frac{|f(B)|}{t}\right)^{K/(K-1)} \tag{13.23}$$

for all $t > 0$.

Proof. Let $E_t = \{z \in B : J(z,f) > t\}$. From Theorem 13.1.5 we have

$$t|E_t| \leqslant \int_{E_t} J(z,f) = |f(E_t)| \leqslant C_K |f(B)| \left(\frac{|E_t|}{|B|}\right)^{1/K}$$

Solving for $|E_t|$ proves the required estimate. □

In fact, the invariant form of the area distortion Theorem 13.1.5 has a natural interpretation in terms of the A_p-weights from Section 4.5.3. Namely,

$$\frac{\int_E J(z,f)}{\int_B J(z,f)} \leqslant C_K \left(\frac{|E|}{|B|}\right)^{1/K},$$

which shows us that the weight $w(z) = J(z,f)$ lies in the class A_∞, hence in A_p for some $p < \infty$. A little later, in Section 13.4.1, we will discuss in finer detail the relations between quasiconformal Jacobians and A_p-weights.

Corollary 13.2.2. *The Jacobian derivative $J(z, f)$ of a K-quasiconformal mapping $f : \mathbb{C} \to \mathbb{C}$ belongs to the space of weights $A_\infty = \bigcup_{p<\infty} A_p$.*

We now come to one of the main results of this chapter. Following [21], we establish the optimal L^p-regularity of the derivatives of quasiconformal mappings.

> **Optimal L^p-Regularity for Derivatives of Quasiconformal Mappings**

Theorem 13.2.3. *Suppose $f : \mathbb{C} \to \mathbb{C}$ is K-quasiconformal. Then*
$$f \in W^{1,p}_{loc}(\mathbb{C}) \quad \text{for all } p < \frac{2K}{K-1}$$
Moreover, for each $K > 1$, there are K-quasiconformal mappings f such that $f \notin W^{1,2K/(K-1)}_{loc}(\mathbb{C})$.

Proof. Let $B \subset \mathbb{C}$ be a disk. Since $|Df(z)|^2 \leqslant KJ(z, f)$ almost everywhere and f is continuous, it suffices to show that $\int_B |J(z,f)|^q < \infty$ for all $0 < q < \frac{K}{K-1}$. The previous Theorem 13.2.1 gave bounds for the distribution function of the Jacobian derivative. Integrating these, we have for any $0 < T < \infty$,

$$\begin{aligned}
\int_B |J(z,f)|^q &= q \int_0^\infty t^{q-1} |\{z \in B : |J(z,f)| > t\}| dt \\
&\leqslant q \int_0^T t^{q-1} |B| + qC_K |B|^{1/(1-K)} \int_T^\infty t^{q-1} \left(\frac{|f(B)|}{t}\right)^{K/(K-1)} dt \\
&\leqslant T^q |B| + C'_K |B|^{1/(1-K)} |f(B)|^{K/(K-1)} \int_T^\infty t^{q-1-K/(K-1)} dt \\
&= T^q |B| + \frac{C'_K}{\frac{K}{K-1} - q} |B|^{1/(1-K)} |f(B)|^{K/(K-1)} T^{q-K/(K-1)} < \infty
\end{aligned}$$

Choosing as the parameter $T = |f(B)|/|B|$ gives the quantitative bounds

$$\int_B |J(z,f)|^q \leqslant \frac{C(K)}{\frac{K}{K-1} - q} |B| \left(\frac{|f(B)|}{|B|}\right)^q, \qquad 0 \leqslant q < \frac{K}{K-1} \qquad (13.24)$$

For the last statement of the theorem, concerning sharpness, take the K-quasiconformal mapping

$$f(z) = z|z|^{\frac{1}{K}-1}, \qquad z \in \mathbb{C} \qquad (13.25)$$

With this mapping,

$$|\{z \in \mathbb{D} : J(z,f) > t\}| = \pi(Kt)^{\frac{-K}{K-1}}, \qquad t > 1/K$$

In particular, for this f the estimate (13.23) holds as an equality, with the constant C_K replaced by $C = K^{-K/(K-1)}$. Similarly

$$\int_\mathbb{D} |Df|^{2q} \geqslant \int_\mathbb{D} J(z,f)^q = \frac{C}{\frac{K}{K-1} - q} \qquad (13.26)$$

13.2. HIGHER INTEGRABILITY

for all $q \in [1, \frac{K}{K-1})$. Here $C = 2\pi K^{1-q}/(K-1)$. □

It is often useful to express the higher integrability in terms of reverse Hölder inequalities. Also, in this setup the above approach describes the sharp results valid for the derivatives of planar quasiconformal mappings. From Theorem 13.2.3 and (13.24), we extract the following useful corollary.

Corollary 13.2.4. *There is a constant C_K, depending only on K, such that if $B \subset \mathbb{C}$ is a disk and f is a K quasiconformal homeomorphism of $\mathbb{C}$ then*

$$\frac{1}{|B|} \int_B \left(\left|\frac{\partial f}{\partial z}\right| + \left|\frac{\partial f}{\partial \bar{z}}\right| \right)^p \leq \frac{C_K}{\frac{2K}{K-1} - p} \left(\frac{|f(B)|}{|B|} \right)^{p/2} \quad (13.27)$$

for all $p \in [0, \frac{2K}{K-1})$.

This estimate too is optimal, up to the determination of the multiplicative constant C_K, as the example (13.26) shows.

Next, via the Stoilow factorization theorem, higher integrability for quasiconformal mappings leads immediately to improved reqularity for quasiregular mappings.

Corollary 13.2.5. *Let f be a K-quasiregular mapping in the domain Ω. Then*

$$f \in W^{1,p}_{loc}(\Omega) \quad \text{for all } p < \frac{2K}{K-1}$$

Furthermore,

$$Df \in \text{weak-}L^{2K/(K-1)}_{loc}(\Omega)$$

More precisely, for any compact $\Omega' \subset \Omega$ there is a constant C, depending only on K, Ω, Ω' and $\int_\Omega |Df|^2$, such that

$$\int_E |Df|^2 \leq C|E|^{1/K}, \quad E \subset \Omega'$$

Proof. By the Stoilow factorization Theorem 5.5.1, $f = h \circ g$, where g is K-quasiconformal in $\mathbb{C}$ and h is holomorphic in the open set $g(\Omega)$. Since the derivatives of h are locally bounded, the claim follows. □

In an analogous manner we derive precise lower-integrability results, that is, integrability of negative powers of the Jacobian $J(z, f)^{-s}$. Let us start with the following estimate.

Theorem 13.2.6. *Let $f : \mathbb{C} \to \mathbb{C}$ be a K-quasiconformal mapping. Then there is a finite constant C_K, depending only on K, such that for any disk $B \subset \mathbb{C}$,*

$$|\{z \in B : J(z,f) < t\}| \leq C_K \frac{|B|^{K/(K-1)}}{|f(B)|^{1/(K-1)}} t^{1/(K-1)} \quad (13.28)$$

for all $t > 0$.

Proof. We argue as in Theorem 13.2.1. Let $E = \{z \in B : J(z, f) < t\}$. Then

$$t|E| \geqslant \int_E J(z, f) = |f(E)| \geqslant \frac{1}{C_K} |f(B)| \left(\frac{|E|}{|B|}\right)^K,$$

where we used Theorem 13.1.5. Unwinding the estimate for $|E|$, we obtain (13.28). $\square$

One may interpret the theorem also as a weak-type estimate for the negative powers of the Jacobian derivative. If $B \subset \mathbb{C}$ is a disk, then by Theorem 13.2.6,

$$\left|\left\{z \in B : \frac{1}{J(z,f)} > t\right\}\right| \leqslant C_K |f(B)|^{1/(1-K)} \left(\frac{|B|^K}{t}\right)^{1/(K-1)}$$

Integrating such bounds and reasoning as in (13.24) or Corollary 13.2.4 gives the following theorem.

Theorem 13.2.7. *If $f : \mathbb{C} \to \mathbb{C}$ is K-quasiconformal and $B \subset \mathbb{C}$ is a disk, then for every*

$$\frac{-2}{K-1} < q \leqslant 0$$

we have

$$\frac{1}{|B|} \int_B (|f_z| + |f_{\bar{z}}|)^q \leqslant \frac{C_K}{2 + q(K-1)} \left(\frac{|f(B)|}{|B|}\right)^{q/2} \quad (13.29)$$

The constant C_K depends only on K.

In this last result a radial stretching shows that (13.29) fails at $q = \frac{-2}{K-1}$. This time, however, we must take $f(z) = z|z|^{K-1}$, $z \in \mathbb{C}$.

13.2.1 Integrability at the Borderline

We have seen that each K-quasiconformal mapping f is contained in the Sobolev class $f \in W_{loc}^{1,p}$ as long as $p < \frac{2K}{K-1}$, while the simple example of radial stretching showed that in general

$$f \notin W_{loc}^{1, 2K/(K-1)}$$

Somewhat surprisingly, when $f|E$ is conformal as in Theorem 13.1.3, then on E the derivatives of f do integrate at the borderline exponent $p = \frac{2K}{K-1}$ even if the set E is of quite irregular character. In later sections this fact from [34] will be the key to establishing precise removability results. It also appears essential in the understanding of quasiconformal deformations of Hausdorff measures as in Section 13.5.1.

Theorem 13.2.8. *Let $K > 1$ and assume that $f : \mathbb{C} \to \mathbb{C}$ is a K-quasiconformal principal mapping and that f is conformal outside $\mathbb{D}$. Let $E \subset \mathbb{D}$ be a measurable set. If $f_{\bar{z}}(z) = 0$ for almost every $z \in E$, then*

$$\frac{1}{\pi} \int_E J(z, f)^{K/(K-1)} \leqslant 1 \quad (13.30)$$

13.2. HIGHER INTEGRABILITY

There are sets E, of positive measure, and principal mappings f, conformal on E, such that (13.30) holds as an equality.

Proof. We choose a sequence of weights such that $w_0 = 1$ and
$$w_n = J(z,f)^{1/K + \cdots + 1/K^n},$$
$n \geq 1$. By Theorem 13.1.3,
$$\frac{1}{\pi} \int_E w_n(x) J(z,f) \leq \left(\frac{1}{\pi} \int_E w_n(x)^K \right)^{1/K}$$
$$= \left(\frac{1}{\pi} \int_E w_{n-1}(x) J(x,f) \right)^{1/K}$$
for each $n \geq 1$. Using this argument inductively, we arrive at
$$\frac{1}{\pi} \int_E J(x,f)^{1 + 1/K + \cdots + 1/K^n}$$
$$= \frac{1}{\pi} \int_E w_n(x) J(x,f) \leq \left(\frac{1}{\pi} \int_E J(x,f) \right)^{1/K^n} \leq \left(\frac{|E|}{\pi} \right)^{1/K^{n+1}} \quad (13.31)$$

With Fatou's lemma we can pass to the limit $n \to \infty$ in (13.31). This proves (13.30).

To prove the second part take for instance $E = \mathbb{D}(0,R)$ with $R < 1$ and let f_R be the K-quasiconformal mapping defined in (13.19). □

As always, Stoilow factorization provides similar results for quasiregular mappings.

Corollary 13.2.9. *Suppose $f : \mathbb{C} \to \mathbb{C}$ is a K-quasiregular mapping, $K > 1$. Let $E \subset \mathbb{C}$ be measurable and bounded. If $f_{\bar{z}}(z) = 0$ for almost every $z \in E$, then*
$$\int_E |Df(z)|^{2K/(K-1)} < \infty \quad (13.32)$$

Proof. Using a linear change of variables, we first assume that $E \subset \mathbb{D}(0, \frac{1}{2})$. Again we have the decomposition $f = h \circ g$, where g is a K-quasiconformal principal mapping, conformal outside $\mathbb{D}$, and h is K-quasiregular in $\mathbb{C}$ with
$$h_{\bar{z}}(z) = 0 \quad \text{for all } z \in g(\mathbb{D})$$

Since h is holomorphic in a neighborhood of the set $g(E)$, $\sup_{g(E)} |h'(x)| \leq C_h < \infty$ for a constant C_h depending on the function h. Hence by Theorem 13.2.8 we have
$$\int_E |Df(z)|^p \leq C_h^p \int_E |Dg(z)|^p \leq C(h,K,p) < \infty$$
with the exponent $p = \frac{2K}{K-1}$. □

On the other hand, it is not too difficult to find examples where

$$\int_E |Df|^p = \infty \quad \text{for all} \quad p > \frac{2K}{K-1}$$

even if $f_{\bar{z}} = 0$ on E. For details, see for instance [34].

13.2.2 Distortion of Hausdorff Dimension

In this section we show how determining the precise regularity for quasiregular and quasiconformal mappings, in Theorems 13.2.3 and 13.2.5, will lead in turn to a further understanding of their basic geometric properties.

Let us start here with the distortion of Hausdorff dimension $\dim_{\mathcal{H}}(E)$ under a K-quasiconformal mapping f. We have seen previously that quasiconformal and quasiregular mappings preserve null sets and sets of positive area. In addition, we shall see here that sets E of Hausdorff dimension $\dim_{\mathcal{H}}(E) = 0$ and $\dim_{\mathcal{H}}(E) = 2$ are also preserved.

For the sets of the intermediate dimensions, however, the Hausdorff dimension is not necessarily preserved. Instead, we have the following elegant estimates.

Theorem 13.2.10. *Let f be a nonconstant K-quasiregular mapping in a domain Ω and suppose $E \subset \Omega$ is compact. Then*

$$\frac{1}{K}\left(\frac{1}{\dim_{\mathcal{H}}(E)} - \frac{1}{2}\right) \leqslant \frac{1}{\dim_{\mathcal{H}}(f(E))} - \frac{1}{2} \leqslant K\left(\frac{1}{\dim_{\mathcal{H}}(E)} - \frac{1}{2}\right) \quad (13.33)$$

As shown in Corollary 12.4.2, these inequalities are the best possible. For both we have sets E and maps f where equality holds.

Proof. Suppose $f \in W^{1,2}_{loc}(\Omega)$ is K-quasiregular and nonconstant and let $E \subset \Omega$ be a compact subset with $\dim_{\mathcal{H}}(E) < 2$. There is no restriction to assume $E \subset \mathbb{D}(0, 1/2)$. Let us first use Stoilow factorization, $f = h \circ g$, where $g : \mathbb{C} \to \mathbb{C}$ is a principal K-quasiconformal mapping, conformal outside $\mathbb{D}$, and h holomorphic in $g(\Omega)$. Since $\dim_{\mathcal{H}}(hF) = \dim_{\mathcal{H}}(F)$ for any compact subset of $F \subset g(\Omega)$, we can actually assume that $f = g$. With this setup let us cover E by a finite number of disks B_i, each of diameter $\text{diam}(B_i) < \delta$. With Vitali-type covering arguments we may assume that the disks have pairwise disjoint interiors. On the image side quasisymmetry restricts the distortion and tells us that

$$\text{diam}(f(B_i))^2 \leqslant C_0 |f(B_i)|,$$

where the constant C_0 depends only on K. We also know from Theorem 13.2.3 that $J(z, f) \in L^p_{loc}(\mathbb{C})$ for each $1 < p < K/(K-1)$. Hence for each $t < 1$,

$$\sum_i \text{diam}(f(B_i))^{2t} \leqslant C_0^t \sum_i |f(B_i)|^t$$

$$\leqslant C_0^t \sum_i \left(\int_{B_i} J(z,f)^p\right)^{t/p} |B_i|^{t(1-1/p)}$$

13.3. DIMENSION OF QUASICIRCLES

$$\leqslant C_0^t \left(\sum_i \int_{B_i} J(z,f)^p \right)^{t/p} \left(\sum_i |B_i|^{\frac{p}{p-t}t(1-\frac{1}{p})} \right)^{1-t/p} \quad (13.34)$$

The disks B_i are all contained in $\mathbb{D}$ with disjoint interiors. This means that for every $p < K/(K-1)$, we have uniform bounds

$$\sum_i \int_{B_i} J(z,f)^p \leqslant \int_{E_\delta} J(z,f)^p \leqslant C_1(p) < \infty.$$

We have therefore shown that

$$\sum_i \operatorname{diam}(fB_i)^{2t} \leqslant C_2 \left(\sum_i \operatorname{diam}(B_i)^{\frac{2t(p-1)}{p-t}} \right)^{1-t/p}, \quad p < \frac{K}{K-1}.$$

Because of this uniform bound, this estimate holds for any (finite or infinite) covering of E by disks B_i with disjoint interiors and with $\operatorname{diam}(B_i) < \delta$.

In this estimate, if $\dim_\mathcal{H}(E) < 2t\frac{p-1}{p-t}$, with a proper choice of the covering $\{B_i\}$ the sum on the right hand side can be made arbitrarily small and we deduce $\dim_\mathcal{H}(f(E)) < 2t$. Solving in terms of $\dim_\mathcal{H}(E)$, we find

$$\dim_\mathcal{H}(f(E)) \leqslant \frac{2p\,\dim_\mathcal{H}(E)}{2(p-1) + \dim_\mathcal{H}(E)} \to \frac{2K\,\dim_\mathcal{H}(E)}{2 + (K-1)\,\dim_\mathcal{H}(E)}$$

as $p \to K/(K-1)$.

Here in fact is precisely the left hand inequality in the claim of Theorem 13.2.10. The right hand inequality follows immediately since the inverse of a K-quasiconformal mapping is also K-quasiconformal. $\square$

We wish to point out the following immediate consequence of the extended λ-lemma for holomorphic motions.

Theorem 13.2.11. *Given a holomorphic motion* $\Phi : \mathbb{D} \times E \to \mathbb{C}$ *of a set* $E \subset \mathbb{C}$, *write* $E_\lambda = \Phi_\lambda(E)$. *Then*

$$\frac{1-|\lambda|}{1+|\lambda|} \left(\frac{1}{\dim_\mathcal{H}(E)} - \frac{1}{2} \right) \leqslant \frac{1}{\dim_\mathcal{H}(E_\lambda)} - \frac{1}{2} \leqslant \frac{1+|\lambda|}{1-|\lambda|} \left(\frac{1}{\dim_\mathcal{H}(E)} - \frac{1}{2} \right)$$

Moreover, for both inequalities there are holomorphic motions Φ *where the estimate holds as an equality.*

13.3 The Dimension of Quasicircles

A quasicircle is the image of the unit circle under a quasiconformal homeomorphism of $\mathbb{C}$. Quasicircles have been studied intensely for many years because of the good function theoretic properties of the domains they bound, relationships with Teichmüller theory and Kleinian groups and recently interesting applications in dynamics, see [135] for a general survey. Perhaps the best known geometric criteria for a quasicircle is Ahlfors' three-point property [4].

Theorem 13.3.1. *Let $\mathcal{C}$ be a Jordan curve in the plane. Then $\mathcal{C}$ is a quasicircle if and only if there is a constant d such that for each pair of points $z_1, z_2 \in \mathcal{C}$ we have*
$$\min_{j=1,2} \operatorname{diam}(\gamma_j) \leqslant d\,|z_1 - z_2|,$$
where γ_1, γ_2 are the two components of $\mathcal{C} \setminus \{z_1, z_2\}$.

This theorem derives its name from the fact that it implies
$$|z_1 - z_3| \leqslant d|z_1 - z_2|$$
for each point z_3 in the component of $\mathcal{C} \setminus \{z_1, z_2\}$ with smallest diameter. In fact this theorem is quantitative in the sense that one can bound the distortion K of the corresponding quasiconformal mapping in terms of d and conversely.

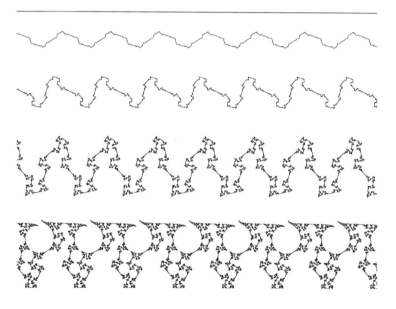

A holomorphic variation of the limit set of a Fuchsian group giving a family of quasilines of increasing dimension and distortion

One of the nice consequences of the area distortion theorem is the best possible estimate of the general distortion of Hausdorff dimension under quasiconformal mappings (Theorem 13.2.10). However, the topological dimension of a set cannot be changed by a homeomorphism, and so it is likely that this brings extra constraints for the distortion of the Hausdorff dimension of circles. These

13.3. DIMENSION OF QUASICIRCLES

estimates are important in the theory of Kleinian groups and elsewhere; the deformations of a Fuchsian group of the first kind give parameterized families of quasi-Fuchsian groups whose limit sets are quasiconformal images of round circles. Similarly, the Julia sets of quadratic polynomials $z^2 + c$, with c small, are quasicircles (actually this holds in the whole primary component of the Mandelbrot set). These two examples are actually simple consequences of the theory of holomorphic motions as discussed earlier; moreover, in each of these cases the Hausdorff dimension of the limit set or Julia set conveys geometric information about the group or mapping.

The earlier Theorem 13.2.10 gives the following estimate of dimension distortion: If $\mathcal{C}$ is a K-quasicircle, it is a K-quasiconformal image of a set of dimension 1, hence

$$\dim_{\mathcal{H}}(\mathcal{C}) \leqslant 1 + \frac{K-1}{K+1} = 1 + k$$

Our interest here lies in deciding the correct estimate for the right hand side. Since topology shows that the trivial lower bound 1 is a minimum for the dimension of quasicircles, one might expect an upper bound of the form $1 + \mathcal{O}(k^2)$ for the dimension of a K-quasicircle with K close to 1 (k close to 0). Indeed, in the above-mentioned example of quasicircle Julia sets, there is the well-known result of Ruelle, building on work of Bowen, which states that if $\mathcal{J}_c$ is the Julia set of $z^2 + c$, then for $c \neq 0$ sufficiently small we have

$$1 < \dim_{\mathcal{H}}(\mathcal{J}_c) = 1 + \frac{|c|^2}{4 \log 2} + o(|c|^2)$$

This is all explained in the very interesting book by Zinsmeister on thermodynamic formalism and holomorphic dynamical systems [377]. The general theory of holomorphic motions shows that the Julia set of the quadratic $z \mapsto P_\lambda(z) = \lambda z + z^2$ is a $\frac{1+|\lambda|}{1-|\lambda|}$-quasicircle for all $|\lambda| < 1$, while with theorem 13.3.5 we can improve the distortion bound to $\sqrt{\frac{1+|\lambda|}{1-|\lambda|}}$. The map P_λ is conjugate by a linear fractional transformation to $z^2 + c$, where

$$\lambda = 1 - \sqrt{1 - 4c}$$

Thus $\mathcal{J}_c$ is a $\sqrt{\frac{1+|1-\sqrt{1-4c}|}{1-|1-\sqrt{1-4c}|}}$-quasicircle. For $|c|$ small we see that $\mathcal{J}_c$ is approximately a $\frac{1+|c|}{1-|c|}$-quasicircle. This family of quasicircles (with $k = |c|$) shows we should expect that

$$\dim_{\mathcal{H}}(\mathcal{C}_k) \geqslant 1 + k^2/(4 \log 2) \approx 1 + 0.36 \, k^2$$

for certain families of $\frac{1+k}{1-k}$-quasicircles. The first general substantive result in this direction was obtained by Becker and Pommerenke [49], who showed that for k close to 0 the bounds

$$1 + 0.36 k^2 \leqslant \sup\{\dim_{\mathcal{H}}(\mathcal{C}_k)\} \leqslant 1 + 37 k^2$$

hold. Astala, Rohde and Schramm have improved the lower bound to $1+0.69k^2$ by considering holomorphic variations of the von Koch snowflake, and the following question (which we formulate here as a conjecture) was posed in [21].

Conjecture 13.3.2. *Let $\mathcal{C}_k$ be a $K = \frac{1+k}{1-k}$-quasicircle. Then*

$$\dim_{\mathcal{H}}(\mathcal{C}_k) \leqslant 1 + k^2,$$

and this bound is sharp.

Dimension of Quasicircles

The rest of this section is devoted to proving an unpublished result of S. Smirnov showing that the upper bound in Conjecture 13.3.2 holds. Thus now the conjecture asks only about sharpness. We follow some notes written up on Smirnov's work by Astala and Prause in proving the following theorem.

Theorem 13.3.3. *Let $\mathcal{C}_k$ be a $K = \frac{1+k}{1-k}$-quasicircle. Then*

$$\dim_{\mathcal{H}}(\mathcal{C}_k) \leqslant 1 + k^2$$

In what follows we use an elementary normalization so as to consider a quasicircle to be the image of the real line under a quasiconformal mapping of $\hat{\mathbb{C}}$—thus a quasiline. Let $\mathcal{C} = f(\mathbb{R})$, where $f : \mathbb{C} \to \mathbb{C}$ is K-quasiconformal. Let μ_f denote the Beltrami coefficient of f, $\|\mu_f\|_\infty = k < 1$.

13.3.1 Symmetrization of Beltrami Coefficients

Given a quasiline $\mathcal{C} = f(\mathbb{R})$, the first thing we want to do is construct another quasiconformal mapping of the plane that takes $\mathbb{R}$ to $\mathcal{C}$ but has the smallest possible distortion among such mappings. In this section we give such a construction, following the ideas of Smirnov, via symmetrization of the Beltrami coefficient.

Suppose first that $h : \mathbb{C} \to \mathbb{C}$ and

$$h(\bar{z}) = \overline{h(z)} \tag{13.35}$$

Then h preserves the real line and

$$\mu_h(\bar{z}) = \overline{\mu_h(z)} \tag{13.36}$$

Conversely, if we choose a $\mu = \mu_h$ that satisfies (13.36), we may integrate it, that is, solve the associated Beltrami equation. A little computation shows that if $g(z) = \overline{h(\bar{z})}$, then $\mu_g = \mu_h$ and hence by Stoilow factorization the mappings differ only by a normalizing similarity. If for instance both h and g are principal mappings or $g(0) = 0 = h(0)$ with $g(1) = 1 = h(1)$, then $g \equiv h$, whereupon (13.35) holds. Assuming now (13.35), let us set

$$\tilde{f} = f \circ h^{-1} : \mathbb{C} \to \mathbb{C}$$

13.3. DIMENSION OF QUASICIRCLES

Then $\tilde{f}(\mathbb{R}) = f(h(\mathbb{R})) = f(\mathbb{R}) = \mathcal{C}$. Next, the composition formula for Beltrami coefficients gives

$$\mu_{\tilde{f}}(h(z)) = \frac{\mu_f(z) - \mu_h(z)}{1 - \mu_f(z)\overline{\mu_h(z)}} \left(\frac{h_z(z)}{|h_z(z)|}\right)^2 \tag{13.37}$$

We would like to choose h, or more precisely μ_h, so as to minimize the supremum of this quantity under the restriction $\mu_h(z) = \overline{\mu_h(\bar{z})}$.

In view of (13.37) it is natural to apply the transformation

$$\Phi: w \mapsto \frac{w - \mu_h(z)}{1 - w\overline{\mu_h(z)}} \left(\frac{h_z(z)}{|h_z(z)|}\right)^2, \tag{13.38}$$

which is a hyperbolic isometry of the disk. The constraints (13.35) and (13.36) give

$$\mu_{\tilde{f}}(w) = \Phi(\mu_f(z)) \quad \text{and} \quad \overline{\mu_{\tilde{f}}(\bar{w})} = \Phi\left(\overline{\mu_f(\bar{z})}\right), \quad w = h(z)$$

Hence the best choice for the Beltrami coefficient $\mu_h(z)$ appears to be the bisector of the hyperbolic geodesic segment joining the points $\mu_f(z)$ and $\overline{\mu_f(\bar{z})}$ in the disk. Explicit formulas for this bisector can be found in, for instance, [48].

If we choose μ_h in this way, (13.36) holds and we may solve the associated Beltrami equation with $h(\bar{z}) = \overline{h(z)}$. Letting $w = h(z)$ gives

$$\mu_{\tilde{f}}(w) = -\overline{\mu_{\tilde{f}}(\bar{w})}, \quad w \in \mathbb{C}, \tag{13.39}$$

and we compute in the hyperbolic metric $\rho_\mathbb{D}$ of the disk,

$$2\rho_\mathbb{D}(0, \mu_{\tilde{f}}(w)) = \rho_\mathbb{D}(\mu_{\tilde{f}}(w), \overline{\mu_{\tilde{f}}(\bar{w})}) = \rho_\mathbb{D}(\mu_f(z), \overline{\mu_f(\bar{z})}) \leqslant 2\rho_\mathbb{D}(0, \|\mu_f\|_\infty)$$

Here the first equality holds from (13.39), and the middle equality holds by virtue of the fact that Φ is an isometry. As $\rho_\mathbb{D}(0, a) = 2\log\frac{1+|a|}{1-|a|}$, we see that $\|\mu_{\tilde{f}}(h(z))\|_\infty = \|\mu_{\tilde{f}}(z)\|_\infty \leqslant \|\mu_f\|_\infty$, so the dilatation has not increased.

Definition 13.3.4. *We say that the dilatation μ is antisymmetric if it satisfies (13.39).*

Well-known results about quasiconformal reflections going back to Ahlfors [4] show that if $\mathcal{C}$ is a K-quasicircle, then there is a K^2-quasiconformal map $f: \mathbb{C} \to \mathbb{C}$ with $f(\mathbb{S}) = \mathcal{C}$ that is conformal on $\mathbb{D}$. If $\mathcal{C}$ is unbounded, we can replace $\mathbb{D}$ by the upper half-plane $\mathbb{H}$ with $\mu_f|\mathbb{H} = 0$. Now in the above construction

$$\rho_\mathbb{D}(0, \mu_{\tilde{f}}(w)) \leqslant \frac{1}{2}\rho_\mathbb{D}(0, \|\mu_f\|_\infty), \quad w \in \mathbb{C},$$

so that the converse is also true. With Möbius invariance we formulate the following theorem.

Theorem 13.3.5. *Let $f: \mathbb{C} \to \mathbb{C}$ be K^2-quasiconformal and $f|\mathbb{D}$ conformal. Then $f(\mathbb{S})$ is a K-quasicircle.*

13.3.2 Distortion of Dimension

We will approach the dimension distortion of quasicircles with much the same machinery as that used for the area distortion and Theorem 13.1.3 in particular. Also here the idea is to embed the given quasiconformal mapping into a holomorphic motion. Now, however, following Smirnov, one makes a very effective use of the antisymmetry described in (13.39). It is the following, slightly technical, distortion theorem that will give the major part of the result we want. It is of course clear that sums such as those appearing in (13.41) below are going to be used to control the Hausdorff dimension.

Theorem 13.3.6. *Let $f : \mathbb{C} \to \mathbb{C}$ be a $\frac{1+k}{1-k}$-quasiconformal mapping for which*

- *f has antisymmetric dilatation.*
- *f is a principal mapping, conformal outside the unit disk.*
- *There is a finite union of disjoint disks $E = \mathbb{D}(z_1, r_1) \cup \cdots \cup \mathbb{D}(z_n, r_n) \subset \mathbb{D}$ and $z_1, \ldots, z_n \in \mathbb{R}$, on which f is also conformal.*

Given $0 < t \leq 2$, set

$$t(k) = \frac{(1+k^2)t}{1-k^2+k^2 t} \in (0, 2] \tag{13.40}$$

Then

$$\sum_{j=1}^{n} (|f'(z_j)| \, r_j)^{t(k)} \leq 8^{t(k)} \left(\sum_{j=1}^{n} r_j^t \right)^{\frac{1-k^2}{1+k^2} \frac{t(k)}{t}} \tag{13.41}$$

Proof. We embed f in a holomorphic motion. For $\lambda \in \mathbb{D}$, set

$$\mu_\lambda(z) = \frac{\lambda}{k} \mu_f(z)$$

Let f_λ be the principal solution to the Beltrami equation $f_{\bar{z}} = \mu_\lambda f_z$. Uniqueness of the solution implies that $f_k = f$. Since μ vanishes on the set E, it follows that $f_\lambda | E$ is conformal, and we argue as in the proof of Theorem 13.1.3 that for each $z \in E$ the map $\lambda \mapsto (f_\lambda)'(z)$ is well defined, nonzero and holomorphic. To use the antisymmetry of the dilatation μ, note that

$$\mu_{\overline{\lambda}}(\bar{z}) = -\overline{\mu_\lambda(z)}, \qquad z \in \mathbb{C}, \, \lambda \in \mathbb{D} \tag{13.42}$$

By uniqueness of the principal solution,

$$f_{\overline{\lambda}}(\bar{z}) = \overline{f_{-\lambda}(z)} \tag{13.43}$$

Analyticity then implies the same relation for the derivatives $(f_\lambda)'(z)$ on the set E. To connect the two values of f_λ we take z real, whereupon

$$|(f_{-\overline{\lambda}})'(z)| = |(f_\lambda)'(z)|, \qquad z \in E \cap \mathbb{R}, \, \lambda \in \mathbb{D} \tag{13.44}$$

Next we want to use the logarithmic decomposition and harmonic functions as in (13.7) and (13.9). However, here it is easiest to use a discrete version of the argument and simply formulate the following lemma.

13.3. DIMENSION OF QUASICIRCLES

Lemma 13.3.7. Let $a_1, a_2, \ldots, a_n$ be non-negative. Then

$$\log\left(\sum_{j=1}^n a_j\right) = \sup\left\{\sum p_j \log \frac{a_j}{p_j} : \sum_{j=1}^n p_j = 1, \, p_j > 0\right\}$$

We apply Lemma 13.3.7 with

$$a_j(\lambda) = \frac{1}{8} |(f_\lambda)'(z_i)| r_i, \qquad \lambda \in \mathbb{D} \qquad (13.45)$$

Given the numbers $p = \{p_j\}$, define as in (13.9),

$$h_p(\lambda) = \sum_{j=1}^n p_j \log\left(\frac{a_j(\lambda)^2}{p_j}\right) = -\sum_{j=1}^n p_j \log p_j + \sum_{j=1}^n p_j \log\left(\frac{1}{64} |(f_\lambda)'(z_j)|^2 r_j^2\right)$$

Since the derivatives $(f_\lambda)'(z_j)$ are analytic and nonvanishing in λ, for each fixed $p = \{p_j\}$ the function $h_p(\lambda)$ is harmonic.

Next note that each mapping f_λ is conformal in the disks B_j and outside $\mathbb{D}$. Applying Koebe's $\frac{1}{4}$-theorem to each of these disks in turn shows that

$$B_{\lambda, i} = \mathbb{D}(f_\lambda(z_i), \frac{1}{4}|(f_\lambda)'(z_i)|r_i) \subset f_\lambda(\mathbb{D}(z_i, r_i))$$

while Theorem 2.10.4 gives

$$f_\lambda(\mathbb{D}) \subset \mathbb{D}(0, 2) \qquad (13.46)$$

Hence the disks $B_{\lambda, j}$, $j = 1, 2, \ldots, n$, are disjoint and lie in $\mathbb{D}(0, 2)$. Thus

$$\sum_{j=1}^n |(f_\lambda)'(z_j)|^2 r_j^2 \leq 64, \qquad \lambda \in \mathbb{D} \qquad (13.47)$$

With (13.44) and Lemma 13.3.7 we have now shown that the harmonic function h_p satisfies

$$h_p(\lambda) = h_p(-\bar{\lambda}), \quad h_p(\lambda) \leq 0, \qquad \lambda \in \mathbb{D} \qquad (13.48)$$

As a result, we may use a refinement of Harnack's inequality.

Lemma 13.3.8. Suppose that u is harmonic in $\mathbb{D}$, negative and even with respect to the x-variable; that is, $u(z) = u(-\bar{z})$. Then

$$\frac{1 + |x|^2}{1 - |x|^2} u(0) \leq u(x) \leq \frac{1 - |x|^2}{1 + |x|^2} u(0), \qquad x \in (-1, 1)$$

Proof. We may assume $u(0) = -1$; otherwise, replace u by $u(z)/u(0)$. There is a real-valued harmonic conjugate function v such that $\varphi = u + iv$ is holomorphic and valued in the left half-plane $\mathbb{H}_L$ (since u is negative) and $\varphi(0) = -1$.

Since $u(z) = u(-\bar{z})$, the auxiliary function

$$g(z) = \frac{1}{2}\bigl(\varphi(z) + \overline{\varphi(\bar{z})}\bigr)$$

is even with respect to the complex variable; $g(z) = g(-z)$ for $z \in \mathbb{D}$. The Riemann map $\Phi(\zeta) = \frac{\zeta+1}{\zeta-1} : \mathbb{H}_L \to \mathbb{D}$, so that $\Psi = \Phi \circ g : \mathbb{D} \to \mathbb{D}$ is holomorphic and $\Psi(0) = 0$. Further, $\Psi'(0) = 0$ since it is even. Thus $|\Psi(z)| \leq |z|^2$ by the usual Schwarz inequality applied to the holomorphic $\Psi(z)/z$.

Since the function g is real on the real axis,

$$u(x) = g(x) = \frac{\Psi(x) + 1}{\Psi(x) - 1} \geq -\frac{1 + |x|^2}{1 - |x|^2} = u(0)\frac{1 + |x|^2}{1 - |x|^2}, \qquad x \in (-1, 1)$$

The other inequality is similar (using the bound $\Psi(x) \geq -|x|^2$, $x \in (-1, 1)$). $\square$

Returning to the claim (13.41), it will be enough show that

$$\frac{1}{t(k)} \log \left(\sum_{j=1}^{n} a_j(k)^{t(k)} \right) \leq \frac{1 - k^2}{1 + k^2} \frac{1}{t} \log \left(\sum_{j=1}^{n} a_j(0)^t \right), \qquad (13.49)$$

where

$$\frac{1}{t(k)} - \frac{1}{2} = \frac{1 - k^2}{1 + k^2} \left(\frac{1}{t} - \frac{1}{2} \right) > 0$$

In fact, $t(k)$ is precisely the exponent defined in (13.40).

Toward this inequality, let us temporarily fix the positive numbers $p = \{p_j\}$ with $\sum_j p_j = 1$ and consider the harmonic functions $h_p(\lambda)$. The specific parameter value $\lambda = k$ of interest to us is real, and thus we may use the symmetries of h_p with Lemma 13.3.8 to conclude that

$$\frac{1}{t(k)} \sum_{j=1}^{n} p_j \log \left(\frac{a_j(k)^{t(k)}}{p_j} \right) = \left(\frac{1}{t(k)} - \frac{1}{2} \right) \sum_{j=1}^{n} p_j \log \frac{1}{p_j} + \frac{1}{2} h_p(k)$$

$$\leq \frac{1 - k^2}{1 + k^2} \left[\left(\frac{1}{t} - \frac{1}{2} \right) \sum_{j=1}^{n} p_j \log \frac{1}{p_j} + \frac{1}{2} h_p(0) \right]$$

$$= \frac{1 - k^2}{1 + k^2} \frac{1}{t} \sum_{j=1}^{n} p_j \log \left(\frac{a_j(0)^t}{p_j} \right)$$

$$\leq \frac{1 - k^2}{1 + k^2} \frac{1}{t} \log \left(8^{-t} \sum_{i=1}^{n} r_i^t \right),$$

where the last inequality uses Lemma 13.3.7. Now, taking the supremum over the sequences $p = \{p_j\}$, with yet another application of Lemma 13.3.7, finally proves (13.49) and hence also the theorem. $\square$

13.3. DIMENSION OF QUASICIRCLES

The only unfortunate part of Theorem 13.3.6 that we shall need to remove is the assumption of conformality on the family of disks E. Such a family will ultimately be used to cover a segment of $\mathbb{R}$ and estimate the Hausdorff dimension of the image via sums of the diameters of their images. Removal of this assumption will be achieved in part via a deformation of f using Corollary 12.7.2.

So now let us complete the proof of Theorem 13.3.3. We have a $K = \frac{1+k}{1-k}$-quasiline $\mathcal{C} = f(\mathbb{R}) \subset \mathbb{C}$. The translation invariance of the problem shows us that it is enough to find the claimed distortion estimate on some subinterval of $\mathbb{R}$. We will get the estimate on $I = [-1/2, 1/2]$. Let g be the principal solution of the Beltrami equation $g_{\bar{z}} = \mu_f(z)\chi_\mathbb{D}(z)g_z$. Since g and f have the same Beltrami coefficient on $\mathbb{D}$ they differ there by a conformal map and so, in particular,

$$\dim_\mathcal{H}(f([-1/2, 1/2])) = \dim_\mathcal{H}(g([-1/2, 1/2]))$$

Consequently, there is no loss of generality if we replace f by g. We may further alter the distortion while still retaining the fact that f is a principal solution with $f(\mathbb{R}) = \mathcal{C}$, so as to assume that $\mu = \mu_f$ is antisymmetric,

$$\mu(\bar{z}) = -\overline{\mu(z)}$$

We may also embed f in a holomorphic motion, but we wish at the same time to make a deformation as in Corollary 12.7.2. In particular, we choose arbitrary α with

$$k < \alpha < 1$$

For $\lambda \in \mathbb{D}$, set $\mu_\lambda = \frac{\lambda}{\alpha}\mu_f$. Then for f_λ, the principal solution to the Beltrami equation $f_{\bar{z}} = \mu_\lambda f_z$ we have $f_\alpha = f$. Note that $\|\mu_\lambda\|_\infty \leq k/\alpha < 1$, so f_λ is $\frac{1+k/\alpha}{1-k/\alpha}$-quasiconformal independent of λ.

Next, take a family of disjoint disks $E = B_1(z_1, r_1) \cup \cdots \cup B_n(z_n, r_n)$ whose centers lie in our interval I. As in Corollary 12.7.2, we deform f_λ on these disk $B_j(z_j, r_j)$, while retaining its values outside E. We get $\tilde{f}_\lambda : \mathbb{C} \to \mathbb{C}$ a $\frac{1+|\lambda|}{1-|\lambda|}$-quasiconformal mapping such that

- $\tilde{f}_\lambda|\mathbb{C} \setminus E = f_\lambda$.
- $\tilde{f}_\lambda(z_j + r_j) = f_\lambda(z_j + r_j)$, $\tilde{f}_\lambda(z_j) = f_\lambda(z_j)$.
- $\tilde{f}_\lambda|\mathbb{D}(z_j, \ell r_j)$ is a similarity.
- $(\tilde{f}_\lambda)'(z_j) = \frac{f_\lambda(z+r_j) - f_\lambda(z)}{r_j}$.

Recall here that

$$\ell = e^{-\pi \frac{1+k/\alpha}{1-k/\alpha}}$$

In the above deformation we may have lost the holomorphicity in the annuli $\mathbb{D}(z_j, r_j) \setminus \mathbb{D}(z_j, \ell r_j)$, but what is important, $\lambda \mapsto (\tilde{f}_\lambda)'(z_j)$ is still analytic and

$$|(\tilde{f}_{-\bar{\lambda}})'(z_j)| = |(\tilde{f}_\lambda)'(z_j)|, \qquad \lambda \in \mathbb{D},$$

as seen from (13.43). Further, the disks $\tilde{f}_\lambda \mathbb{D}(z_j, \ell r_j)$ are disjoint and all contained in $\mathbb{D}(0, 2)$. Thus

$$\sum_{j=1}^{n} |(\tilde{f}_\lambda)'(z_j)|^2 \, \ell^2 \, r_j^2 \leqslant 64, \qquad \lambda \in \mathbb{D}$$

We are now set to follow the argument of Theorem 13.3.6. As a result,

$$\sum_{j=1}^{n} \left(|\tilde{f}_\alpha'(z_j)| \, r_j\right)^{t(\alpha)} \leqslant \left(\frac{8}{\ell}\right)^{t(\alpha)} \left(\sum_{j=1}^{n} r_j^t\right)^{\frac{1-\alpha^2}{1+\alpha^2} \frac{t(\alpha)}{t}}, \tag{13.50}$$

where $f = f_\alpha$. Because of quasiconformality, we also have the easy distortion estimate of Theorem 12.6.2:

$$\begin{aligned}
\operatorname{diam}(f(B_j(z_j, r_j))) &\leqslant e^{\pi \frac{1+\alpha}{1-\alpha}} |f(z_j) - f(z_j + r_j)| \\
&= e^{\pi \frac{1+\alpha}{1-\alpha}} |(\tilde{f}_\alpha)'(z_j)| r_j
\end{aligned} \tag{13.51}$$

Putting this together gives us the inequality

$$\sum_{j=1}^{n} \operatorname{diam}^{t(\alpha)}(f(\mathbb{D}(z_j, r_j))) \leqslant 2^6 \ell^{-t(\alpha)} \, e^{\pi t(\alpha) \frac{1+\alpha}{1-\alpha}} \left(\sum_{j=1}^{n} r_j^t\right)^{\frac{1-\alpha^2}{1+\alpha^2} \frac{t(\alpha)}{t}} \tag{13.52}$$

Inequality (13.52) holds whenever $\{\mathbb{D}(z_j, r)\}_{j=1}^n$ is a family of disjoint disks with $z_i \in \mathbb{R}$ and contained in $\mathbb{D}$. Obviously, one can cover the segment $[-1/2, 1/2]$ with two such families of arbitrarily small radii. We recall that

$$t(\alpha) = \frac{(1+\alpha^2)t}{1 - \alpha^2 + \alpha^2 t}$$

Then for every $t > 1$ we can make the sum on the right hand side of (13.52) tend to 0, which implies that for every $\alpha > k$ we have $\dim_\mathcal{H}(f([-1/2, 1/2])) \leqslant t(\alpha)$. We then let $t \to 1$ and observe that $\dim_\mathcal{H}(f([-1/2, 1/2])) \leqslant 1 + \alpha^2$. Finally, $\alpha > k$ was arbitrary, and so for the dimension of the $\frac{1+k}{1-k}$-quasicircle $\mathcal{C} = f(\mathbb{R})$,

$$\dim_\mathcal{H} \mathcal{C} \leqslant 1 + k^2$$

The proof of Theorem 13.3.3 is complete. $\square$

It is clear from (13.52) that the above argument gives improved dimension bounds, in comparison with (13.33), for all sets E lying on a circle or even on any rectifiable curve. More refined distortion bounds for such sets require a deeper understanding of the multifractal spectrum of the harmonic measure. For the first steps in this direction see [302].

13.4 Quasiconformal Mappings and BMO

Quasiconformal mappings also arise naturally in the context of function spaces. As an example, we have already seen in Section 3.8 that the composition operators
$$T_g : \phi \mapsto \phi \circ g, \qquad g : \Omega \to \Omega' \text{ quasiconformal}$$
are precisely the algebra isomorphisms between the Royden algebras $R(\Omega')$ and $R(\Omega)$. We shall next show that a similar relation holds for the function space $BMO(\mathbb{C})$. For convenience, we recall from (4.3.1), where we established global John-Nirenberg estimates, that $BMO(\mathbb{C})$ is the space of functions u with
$$\|u\|_* = \sup_B \frac{1}{|B|} \int_B |u - u_B| < \infty$$
Here the supremum is taken over disks $B \subset \mathbb{C}$, and the notation u_B stands for the average of u over the disk B.

Let us give some motivation for this relation. Note first that the class of BMO functions is invariant under rescaling; If $t > 0$, then $u_t(x) = u(tx)$ has the same BMO-norm as u does. We see that composing with any similarity gives an exact isometry of $BMO(\mathbb{C})$. A quasisymmetry may be viewed as a perturbed similarity. It should thus come as no surprise that composing with a quasisymmetric mapping gives a bounded operator on BMO. This in fact is the content of our first theorem, due to M. Reimann [309], relating the theory of quasiconformal mappings and BMO.

Reimann's Theorem

Theorem 13.4.1. *A quasiconformal homeomorphism $f : \mathbb{C} \to \mathbb{C}$ induces a linear isomorphism*
$$T_f : BMO(\mathbb{C}) \to BMO(\mathbb{C})$$
by the rule
$$T_f[u] = u \circ f$$
The norm of T_f depends only on the distortion of f.

Proof. The proof is based on the John-Nirenberg Lemma A.8.3. That lemma states that for any disk $B \subset \mathbb{C}$,
$$\left|\{z \in B : |u(z) - u_B| > t\}\right| \leqslant C_0 |B| e^{-\delta t}, \qquad 0 < t < \infty, \tag{13.53}$$
whenever $u \in BMO(\mathbb{C})$. Here $\delta = C_1/\|u\|_*$ and C_0, C_1 are universal constants.

Given a quasiconformal map as in our hypothesis, and given a disk $B_0 = \mathbb{D}(z_0, r)$, let
$$R = \max\{|f(\zeta) - f(z_0)| : |\zeta - z_0| = r\}$$
Lemma 3.4.5 tells us that for $B = B(f(z_0), R)$ we have
$$\delta B \subset f(B_0) \subset B, \qquad \delta = \delta_K = \frac{1}{\eta(1)}$$

In particular, using the quasisymmetry of the mapping f,

$$|B_0| \leqslant |f^{-1}(B)| \leqslant C_\delta |f^{-1}(\delta B)| \leqslant C_\delta |B_0|, \qquad C_\delta = \frac{1}{\eta^{-1}(\delta)^2}$$

We may now compare the distribution functions in the disks,

$$|\{z \in B_0 : |u \circ f(z) - u_B| > \sigma\}| \leqslant |\{z \in f^{-1}B : |u \circ f(z) - u_B| > \sigma\}|$$

$$\leqslant C_\delta \frac{|B_0|}{|f^{-1}B|} |f^{-1}\{w \in B : |u(w) - u_B| > \sigma\}|$$

The last expression is controlled by the invariant form of the area distortion, Theorem 13.1.5, which applies equally well to the inverse function f^{-1},

$$\frac{|f^{-1}\{w \in B : |u(w) - u_B| > \sigma\}|}{|f^{-1}B|} \leqslant C_K \left(\frac{|\{w \in B : |u(w) - u_B| > \sigma\}|}{|B|} \right)^{1/K}$$

At this point it is time to invoke the John-Nirenberg Lemma (13.53) and apply it to the function $u \in BMO$. Combining the lemma with the distribution estimates proves that

$$|\{z \in B : |u \circ f(z) - u_B| > \sigma\}| \leqslant C |B_0| e^{-\delta \sigma/K}, \qquad 0 < t < \infty$$

Last, an integration with respect to σ gives

$$\|u \circ f\|_* \leqslant C \|u \circ f\|_*,$$

where the constant C depends only on K. □

There is also a converse to Theorem 13.4.1. Namely, Reimann [309] proved that if a homeomorphism $f : \mathbb{C} \to \mathbb{C}$ preserves BMO and if, in addition, $f \in W^{1,1}_{loc}$, then f is a quasiconformal mapping, with $K(f)$ depending only on the operator norm $\|T_f\|_{BMO}$. It remains an open question whether the converse direction holds for general homeomorphisms preserving BMO, that is, without an a priori assumption on the Sobolev class $W^{1,1}_{loc}$.

If f is a homeomorphism that actually preserves BMO in all subdomains $\tilde{\Omega} \subset \Omega$, then f is quasiconformal [20].

Reimann's theorem also has a local formulation: Any quasiconformal composition operator

$$T_g : \phi \mapsto \phi \circ g, \qquad g : \Omega \to \Omega' \text{ quasiconformal},$$

gives an isomorphism $BMO(\Omega') \to BMO(\Omega)$. In fact, as pointed out quantitatively in Theorem 3.6.2, locally any quasiconformal map is quasisymmetric. We leave it to the reader to modify the argument for Reimann's theorem to obtain results in the local setting.

13.4. QUASICONFORMAL MAPPINGS AND BMO

13.4.1 Quasiconformal Jacobians and A_p-Weights

We now refer back to our earlier discussion of A_p-weights in (4.90). Our aim in this section is to express quasiconformal Jacobians as A_p-weights. Besides their intrinsic interest, these results will be essential in the proof of Theorem 14.0.4 and in determining the critical interval for integrability of the differential.

Earlier in this chapter we interpreted the higher integrability of derivatives of quasiconformal mappings as a reverse Hölder estimate. This explains the close relationship to the theory of A_p-weights. These connections may be viewed in a variety of ways, and in particular it will be convenient to determine simultaneously the A_p-range for all powers $\omega(z) = J(z.f)^s$. Recall that the classes A_p increase with p, and therefore we shall look for the smallest indices p or classes A_p possible.

Theorem 13.4.2. *Suppose $f : \mathbb{C} \to \mathbb{C}$ is a K-quasiconformal homeomorphism. Let*

$$\omega(z) = J(z,f)^s, \qquad \frac{-1}{K-1} < s < \frac{K}{K-1} \tag{13.54}$$

Then ω is locally integrable. Moreover, $\omega \in A_p$ whenever

$$p > \begin{cases} 1 + s(K-1), & 0 \leqslant s < \frac{K}{K-1} \\ 1 - \frac{s}{K}(K-1), & \frac{-1}{K-1} < s \leqslant 0 \end{cases} \tag{13.55}$$

The result does not hold for any p outside this range.

Proof. Let B be a disk. We apply Corollary 13.2.4 and Theorem 13.2.7; for later purposes we shall also need to keep track of the constants involved. Assuming that s lies in the range (13.54),

$$\frac{1}{|B|}\int_B \omega \leqslant \frac{1}{|B|}\int_B (|f_z| + |f_{\bar{z}}|)^{2s} \leqslant C(K,s) \left(\frac{|f(B)|}{|B|}\right)^s$$

Moreover, the constant here has the bound

$$C(K,s) \leqslant \frac{C_K}{\min\{1 - s\frac{K-1}{K}, 1 + s(K-1)\}}, \tag{13.56}$$

where C_K depends only on K. Next, a second application of Corollary 13.2.4 and Theorem 13.2.7 yields

$$\left(\frac{1}{|B|}\int_B \omega^{1/(1-p)}\right)^{p-1} = \left(\frac{1}{|B|}\int_B J(z,f)^{s/(1-p)}\right)^{p-1}$$
$$\leqslant C\left(K, \frac{s}{1-p}\right)^{p-1} \left(\frac{|f(B)|}{|B|}\right)^{-s}$$

That in turn holds as long as

$$-1/(K-1) < \frac{s}{1-p} < K/(K-1) \tag{13.57}$$

But this is just another equivalent way to state condition (13.55).

Assuming the condition holds, we may multiply the inequalities for ω and obtain

$$|\omega|_{A_p} = \sup_B \left(\frac{1}{|B|}\int_B \omega\right)\left(\frac{1}{|B|}\int_B \omega^{1/(1-p)}\right)^{p-1} \leqslant C(\omega) < \infty,$$

where it is useful to record the upper bound

$$C(\omega) \leqslant C(K,s)\, C\left(K, \frac{s}{1-p}\right)^{p-1} \tag{13.58}$$

To see that for no other exponents do we have $\omega(z) = J(z,f)^s \in A_p$, simply consider the radial stretchings $f_1(z) = z|z|^{K-1}$ and $f_2(z) = z|z|^{\frac{1}{K}-1}$. If p and s do not satisfy (13.55), then one of the functions

$$J(z,f_j)^{s/(1-p)}, \qquad j=1,2,$$

fails to be integrable in any neighborhood of the origin. In this case the corresponding weight $\omega(z) = J(z,f_j)^s$ cannot belong to the class A_p. $\square$

In particular, we see that $w = w_s \in A_2$ whenever $|s| < 1/(K-1)$. This will imply another theorem of Reimann [309], that the logarithm of the Jacobian of a quasiconformal mapping lies in BMO. We shall consider this from the point of view of general inverse Hölder inequalities.

Lemma 13.4.3. *Let u be a positive and measurable function in $\mathbb{C}$. Suppose there is $s > 0$ such that for each disk $B \subset \mathbb{C}$,*

$$\left(\frac{1}{|B|}\int_B u^s\right)^{1/s} \leqslant A \left(\frac{1}{|B|}\int_B u^{-s}\right)^{-1/s}, \tag{13.59}$$

where the constant $A \geqslant 1$ is independent of B. Then the function

$$\log u \in BMO(\mathbb{C}) \quad \text{and} \quad \|\log u\|_* \leqslant \frac{1}{s}\sqrt{A^s - 1}$$

Proof. With the substitution $u = e^v$ the bound (13.59) takes the form of a double integral

$$\frac{1}{|B|^2}\int_B\int_B e^{s(v(z)-v(\zeta))} \leqslant A^s$$

Now interchange z and ζ and add up the inequalities. This gives

$$\frac{1}{|B|^2}\int_B\int_B \{e^{s(v(z)-v(\zeta))} + e^{s(v(\zeta)-v(z))}\} \leqslant 2A^s$$

Using the inequality $e^a + e^{-a} \geqslant 2 + a^2$ and a Hölder estimate, we obtain

$$\left(\frac{1}{|B|}\int_B |v - v_B|\right)^2 \leqslant \frac{1}{|B|}\int_B |v - v_B|^2$$
$$= \frac{1}{2s^2|B|^2}\int_B\int_B |sv(z) - sv(\zeta)|^2 \leqslant \frac{A^s - 1}{s^2},$$

13.5. REMOVABLE SINGULARITIES

which proves the claim. □

If $\omega \in A_p$, then the definition (4.90) shows that for $s = \min\{1, 1/(p-1)\}$,

$$\left(\frac{1}{|B|}\int_B \omega^s\right)^{1/s} \leqslant \left(\frac{1}{|B|}\int_B \omega\right) \leqslant A\left(\frac{1}{|B|}\int_B \omega^{-s}\right)^{-1/s}$$

Thus each weight $\omega \in A_p$ has logarithm $\log \omega \in BMO$. In the special case of quasiconformal Jacobians we obtain the following corollary.

Corollary 13.4.4. *If $f : \mathbb{C} \to \mathbb{C}$ is a K-quasiconformal homeomorphism, then $\log J(z, f) \in BMO$ with*

$$\|\log J(\cdot, f)\|_* \leqslant C(K),$$

where $C = C(K)$ depends only on K.

For simplicity of presentation we have stated Theorem 13.4.2 and Corollary 13.4.4 only for global mappings $f : \mathbb{C} \to \mathbb{C}$. However, again the corresponding statements hold for all quasiconformal mappings $f : \Omega \to \Omega'$ between domains $\Omega, \Omega' \subset \mathbb{C}$. The key point here is that for the these local theorems it is enough to find uniform bounds in hyperbolic disks, roughly for disks $\mathbb{D}(z, r)$ with $\mathbb{D}(z, 2r) \subset \Omega$. Therefore Theorem 3.6.2 with the above argument gives the local versions, too. We leave the details to the reader.

13.5 Painlevé's Theorem: Removable Singularities

Painlevé's theorem, a classical result in complex function theory, states that sets of zero length are removable for bounded holomorphic functions. More precisely, suppose $E \subset \mathbb{C}$ is a compact subset with $\mathcal{H}^1(E) = 0$. Then according to Painlevé, whenever Ω is a domain containing E and f is a bounded function holomorphic in $\Omega \setminus E$, then f extends to a bounded holomorphic function of Ω.

On the other hand, given a set of dimension $\dim_\mathcal{H}(E) > 1$, it carries a Hausdorff measure whose Cauchy transform is bounded and holomorphic outside E. Hence E cannot be removable.

In this section we shall establish the counterpart, conjectured in [189], of Painlevé's theorem for quasiregular mappings. We show that sets of $\mathcal{H}^s$-measure 0 for $s = 2/(K+1)$ are removable for bounded K-quasiregular mappings. This result is a refinement of the solution presented in [21] which missed only the borderline case. For a further refinement for $K > 1$, see Theorem 13.5.5 below.

It is not difficult to see that for each $K \geqslant 1$ and each $t > \frac{2}{K+1}$ there are nonremovable sets E with $\dim_\mathcal{H}(E) = t$. Indeed, using Corollary 12.4.2 we can find sets E and F together with a K-quasiconformal $g : \mathbb{C} \to \mathbb{C}$ such that

$g(E) = F$ and $\dim_{\mathcal{H}}(E) = t$, $\dim_{\mathcal{H}}(F) = 1 + \delta > 1$. Let ν denote the $(1 + \delta)$-Hausdorff measure defined on compact subsets of $\mathbb{C}$. We set

$$\varphi(w) = \frac{1}{2\pi i} \int_F \frac{d\nu(\zeta)}{w - \zeta}$$

as the Cauchy transform of ν. Now φ is bounded and analytic in $\mathbb{C} \setminus F$; in fact, φ is δ-Hölder-continuous in $\mathbb{C}$ by [132, Theorem III.4.4]. Hence

$$f = \varphi \circ g \qquad (13.60)$$

is a bounded and K-quasiregular mapping in $\mathbb{C} \setminus E$. If it could extend quasiregularly to $\mathbb{C}$, Liouville's theorem would force it to be a constant.

To have nonremovable sets of dimension equal to $2/(K + 1)$ takes a little more effort even if the principles of the construction are similar. The details of such a counterexample are described in Section 13.6.

For the positive direction and for a proof of the quasiregular version of Painlevé's theorem, we begin by recalling some basic facts concerning analytic capacity. If φ is holomorphic in a domain $\Omega \subset \hat{\mathbb{C}}$ with $\infty \in \Omega$ and $\varphi(\infty) = 0$, then we define

$$\varphi'(\infty) = \lim_{z \to \infty} z\varphi(z)$$

Analytic Capacity

Definition 13.5.1. *If $E \subset \mathbb{C}$ is compact, then the* analytic capacity *of E is*

$$\gamma(E) = \sup\{|\varphi'(\infty)| : \varphi \text{ holomorphic in } \mathbb{C} \setminus E, \; \varphi(\infty) = 0 \text{ and } \|\varphi\|_\infty \leq 1\}$$

The connection between the problem we are examining and analytic capacity is the following simple consequence of Definition 13.5.1.

Lemma 13.5.2. *Let $E \subset \mathbb{C}$ be compact. The analytic capacity $\gamma(E) = 0$ if and only if every bounded holomorphic function φ on $\mathbb{C} \setminus E$ is constant.*

More generally [132], $\gamma(E) = 0$ if and only if for every open $\Omega \supset E$, each bounded and holomorphic function in $\Omega \setminus E$ extends to a holomorphic function of Ω. The following facts are clear from the definition of analytic capacity:

- If $E \subset E'$ are compact, then

$$\gamma(E) \leq \gamma(E') \qquad (13.61)$$

- If h is a homeomorphism of $\mathbb{C}$, conformal in $\mathbb{C} \setminus E$ and normalized by $h(z) = z + \mathcal{O}(1/z)$ at $z = \infty$, then

$$\gamma(hE) = \gamma(E) \qquad (13.62)$$

- By the Schwarz lemma, $\gamma\big(\mathbb{D}(w, r)\big) = r$.

13.5. REMOVABLE SINGULARITIES

We shall also need the following result, a version of Painlevé's theorem.

Lemma 13.5.3. *Suppose* $E \subset \bigcup_{j=1}^{\infty} \overline{\mathbb{D}}(z_j, r_j)$ *is compact. Then,* $\gamma(E) \leq \sum_{j=1}^{\infty} r_j$.

Note in particular that it follows from Lemma 13.5.3 that $\gamma(E) \leq \mathcal{H}^1(E)$. So analytic capacity is dominated by linear measure.

Proof. Suppose φ is holomorphic in $\mathbb{C} \setminus E$, $\varphi(\infty) = 0$ and $\|\varphi\|_\infty \leq 1$. Then, near ∞ we have the Laurent series expansion $\varphi(z) = \sum_{j=1}^{\infty} c_j z^{-j}$ for z large enough. Let Ω be the interior of the unbounded component of $\mathbb{C} \setminus \bigcup_{j=1}^{\infty} \mathbb{D}(z_j, (1+\varepsilon)r_j)$. By our assumption $E \cap \Omega = \emptyset$. Hence if $z \in \Omega$, the Cauchy integral formula gives

$$\varphi(z) = \frac{1}{2\pi i} \int_{\partial \Omega} \frac{\varphi(\zeta)}{\zeta - z} \, d\zeta, \quad \text{hence} \quad \varphi'(\infty) = \frac{1}{2\pi i} \int_{\partial \Omega} \varphi(\zeta) \, d\zeta$$

Therefore

$$|\varphi'(\infty)| \leq \frac{1}{2\pi} \mathcal{H}^1(\partial \Omega) \|\varphi\|_\infty \leq \sum_{j=1}^{\infty} (1+\varepsilon) r_j$$

Taking the supremum over φ and the infimum over ε gives the proof. □

We can now state and prove the following theorem.

Painlevé's Theorem for Quasiregular Mappings

Theorem 13.5.4. *Compact sets of Hausdorff $\mathcal{H}^s$-measure zero, $s = 2/(K+1)$, are removable for bounded K-quasiregular mappings. Thus if E is compact with $\mathcal{H}^s(E) = 0$, then any bounded K-quasiregular mapping $f : \mathbb{C} \setminus E \to \mathbb{C}$ is constant.*

As a simple corollary, we find that sets of Hausdorff dimension 0 are removable irrespective of the distortion.

Proof. Let E be a compact set with $\mathcal{H}^s(E) = 0$ and let $\Omega \subset \mathbb{C}$ be a domain containing E. Assume, furthermore, that f is a bounded K-quasiregular mapping defined in $\Omega \setminus E$. Our goal is then to show that f extends to a K-quasiregular mapping of Ω. We may clearly assume that Ω is simply connected. The Riemann mapping of Ω takes E to a set of Hausdorff $\mathcal{H}^s$-measure zero. Hence we can assume that $\Omega = \mathbb{D}$. Since E has zero area, $\mu = f_{\bar{z}}/f_z$ is defined almost everywhere in $\mathbb{D}$ and satisfies $\|\mu\|_\infty \leq \frac{K-1}{K+1}$. Moreover, let $g : \mathbb{C} \to \mathbb{C}$ be the principal solution to

$$g_{\bar{z}} = \mu \chi_{\mathbb{D}} g_z \tag{13.63}$$

By Stoilow's factorization theorem, $f = \phi \circ g$, where ϕ is bounded and holomorphic in $g(\mathbb{D}) \setminus g(E)$. If we can now show that the analytic capacity

$$\gamma(g(E)) = 0, \tag{13.64}$$

then ϕ extends holomorphically to $g(\mathbb{D})$, hence f extends K-quasiregularly to Ω and we are done.

It therefore remains to show (13.64) for the principal solution to (13.63). To this purpose, fix a positive $\varepsilon > 0$ and cover E with a finite number of open disks $\mathbb{D}(z_j, r_j)$, $j = 1, \ldots, n$, so that

$$\sum_{j=1}^{n} r_j^s < \varepsilon, \qquad s = \frac{2}{K+1} \tag{13.65}$$

This is evidently possible since $\mathcal{H}^s(E) = 0$.

Next, let $\Omega_1 = \bigcup_{j=1}^{n} \mathbb{D}(z_j, r_j)$. We shall argue as in Theorem 13.1.4. Namely, we have the decomposition

$$g = h \circ g_1,$$

where both g_1 and h are principal K-quasiconformal mappings in $\mathbb{C}$. Moreover, we may require that g_1 be conformal in $\Omega_1 \cup (\mathbb{C} \setminus \mathbb{D})$ and that h be conformal outside $g_1(\overline{\Omega_1})$. By (13.61) and (13.62),

$$\gamma(g(E)) \leqslant \gamma(g(\overline{\Omega_1})) = \gamma(h(g_1(\overline{\Omega_1}))) = \gamma(g_1(\overline{\Omega_1}))$$

Thus we have been able to reduce the problem to estimating $\gamma(g_1(\overline{\Omega_1}))$. For this, set $\mathbb{D}_j = \mathbb{D}(z_j, r_j)$. Arguing as in the estimate (13.34) with $t = \frac{1}{2}$, we get

$$\sum_j \operatorname{diam}(g_1(\mathbb{D}_j)) \leqslant C_0(K) \left(\int_{\Omega_1} J(z, g_1)^p \right)^{1/2p} \left(\sum_j |\mathbb{D}_j|^{\frac{p-1}{2p-1}} \right)^{1-1/2p}, \tag{13.66}$$

where the constant $C_0(K)$ depends only on K. Now we are in the special situation of Theorem 13.2.8 with $g_1|\Omega_1$ conformal. Therefore we may take $p = K/(K-1)$ in (13.66) to obtain

$$\frac{1}{\pi} \int_{\Omega_1} J(z, g_1)^p \leqslant 1 \tag{13.67}$$

We wish to emphasize that it is precisely inequality (13.67) that enables the generalization from sets with $\dim_{\mathcal{H}}(E) < 2/(K+1)$ as in [21, Theorem 1.5] to the estimates for sets with $\mathcal{H}^s(E) = 0$, $s = 2/(K+1)$.

With the above choice of p we have $(p-1)/(2p-1) = \frac{1}{K+1}$. Thus (13.65) and (13.66) give

$$\sum_j \operatorname{diam}(g_1(\mathbb{D}_j)) < \pi C_0(K) \sum_{j=1}^{n} r_j^{2/(K+1)} < \pi C_0(K) \varepsilon \tag{13.68}$$

Finally, since Ω_1 was a finite union of disks $\mathbb{D}_j$, we have $g_1(\overline{\Omega_1}) = \bigcup_{j=1}^{n} \overline{g_1(\mathbb{D}_j)}$. According to Lemma 13.5.3, $\gamma(g(E)) \leqslant \pi C_0(K) \varepsilon$ for each $\varepsilon > 0$, and this proves Theorem 13.5.4. □

The above argument can in fact be further refined. Via the Stoilow factorization, it asks for conditions on sets E such that for any K-quasiconformal

13.5. REMOVABLE SINGULARITIES

mapping the image set $g(E)$ has vanishing analytic capacity. Here we can improve this in two ways. First, we can employ the recent remarkable work of Z. Tolsa [353] giving a geometric description of sets of zero analytic capacity, and second, we can analyze the quasiconformal images of such sets.

For instance, if a set $F \subset \mathbb{C}$ has finite or σ-finite length, then it has positive analytic capacity if and only if $F = F_1 \cup F_2$, where F_1 is rectifiable, that is, contained in a countable union of C^1-curves, and where $\mathcal{H}^1(F_2) = 0$. Here the part F_2 falls under the control of Painlevé's Theorem, while from Theorem 13.3.3 we know that a quasiconformal image of a line segment or a smooth curve cannot attain the extremal dimension distortion. It follows from [23] that this fact also holds true for those subsets of rectifiable curves that have positive length. In particular, we see that if a K-quasiconformal image of a set of dimension $\frac{2}{K+1}$ is rectifiable, then it must have length zero. Combining these observations with arguments similar to those in Section 13.5.1 and Theorem 13.5.13, one obtains, as in [23], the following improved version of Painlevé's theorem for quasiregular mappings.

Theorem 13.5.5. *Let $K > 1$. If $E \subset \mathbb{C}$ is compact with*
$$\mathcal{H}^{2/(K+1)}(E) < \infty,$$
or, more generally, if E has σ-finite $\frac{2}{K+1}$-measure, then E is removable for all bounded K-quasiregular mappings.

For the complete details of the proof we refer the reader directly to [23]. Note, however, the remarkable fact that Theorem 13.5.5 holds only for $K > 1$: For instance, a line segment has finite $\mathcal{H}^1$-measure but is not removable for bounded analytic functions.

13.5.1 Distortion of Hausdorff Measure

Having established in Subsection 13.2.2 the range of distortion of Hausdorff dimension under K-quasiconformal deformations, the next problem is to see if at the critical dimensions the result can be improved to the level of Hausdorff measures. That is, choosing a dimension $s \in (0,2)$, let us set

$$s(K) = \frac{2Ks}{2+(K-1)s} \tag{13.69}$$

Hence $s(K) \geqslant s$ gives precisely the extremal relation on the left hand side of (13.33),

$$\frac{1}{s(K)} - \frac{1}{2} = \frac{1}{K}\left(\frac{1}{s} - \frac{1}{2}\right)$$

Then we ask if, for every K-quasiconformal mapping $f : \mathbb{C} \to \mathbb{C}$ and every compact $E \subset \mathbb{C}$ it holds,

$$\mathcal{H}^s(E) = 0 \quad \Rightarrow \quad \mathcal{H}^{s(K)}(fE) = 0$$

We expect that this is true and state the following conjecture.

Conjecture 13.5.6. *Given $K > 1$, let $0 < s < 2$ and $s(K)$ be related by rule (13.69). Then the pullback $f^\# \mathcal{H}^{s(K)}$ of the $s(K)$-dimensional Hausdorff measure under a K-quasiconformal mapping $f : \Omega \to \Omega'$ is absolutely continuous with respect to the s-dimensional Hausdorff measure.* [1]

We will give a proof of the conjecture when the dimension $s = 1$. Since quasiconformal mappings have the Lusin property $\mathcal{N}$, the required absolute continuity holds in dimension $s = 2$. Naturally, the counting measure $\mathcal{H}^0$ is preserved by any quasiconformal homeomorphism.

It will be handy to introduce the s-dimensional *content* of a compact set $E \subset \mathbb{C}$,

$$\mathcal{M}^s(E) = \inf \left\{ \sum_{j=1}^\infty r_j^s : E \subset \bigcup_{j=1}^\infty \mathbb{D}(z_j, r_j) \right\},$$

where the infimum is taken over all countable coverings of E by families of disks $\mathbb{D}(z_j, r_j)$. The s-content $\mathcal{M}^s$ is not a measure but has the convenient property that $\mathcal{M}^s(E) < \infty$ for every compact set E. Moreover, in comparison with the s-dimensional Hausdorff measure, for any compact set E,

$$\mathcal{M}^s(E) = 0 \quad \Leftrightarrow \quad \mathcal{H}^s(E) = 0$$

Hence the content provides a useful tool in estimating the Hausdorff measures.

Returning to the quasiconformal distortion of Hausdorff measures, a glance at the proof of Theorem 13.2.10 shows that the critical bounds on Hausdorff dimension result from the extremal higher integrability of the Jacobian derivative. Hence it is from here that one must proceed as far as possible. This takes us immediately to questions of integrability at the borderline exponent, that is, to Theorem 13.2.8. In particular, we see that understanding the distortion of Hausdorff measures under the quasiconformal deformations requires a decomposition such as that used in proving the area distortion estimates (13.21). Precisely, we want a representation of the form

$$f = h \circ g, \tag{13.70}$$

where, roughly, g is conformal on E and h is conformal outside $g(E)$. By analogy with the (second) area bound Theorem 13.1.3, we have the general estimate for the inner factor.

Theorem 13.5.7. *Suppose $g : \mathbb{C} \to \mathbb{C}$ is a principal K-quasiconformal mapping, conformal outside the unit disk $\mathbb{D}$. Suppose further that $\{B_j\}_{j=1}^n$ is a finite family of disjoint subdisks of $\mathbb{D}$ and that g is conformal on the union*

$$E = \bigcup_{j=1}^n B_j$$

---

[1] During the editing of this book this conjecture was established by M. Lacey, E. Sawyer and I. Uriarte-Tuero [223]

13.5. REMOVABLE SINGULARITIES

Then for every $0 < s \leqslant 2$, we have

$$\mathcal{M}^{s(K)}(g(E)) \leqslant \sum_{j=1}^{n} \operatorname{diam}\bigl(g(B_j)\bigr)^{s(K)} \leqslant C_K \left(\sum_{j=1}^{n} \operatorname{diam}(B_j)^s \right)^{\frac{1}{s}\frac{s(K)}{K}},$$

where $s(K)$ is determined from (13.69). The constant C_K depends only on K.

Proof. From Theorem 13.2.8 we have

$$\int_E J(z,g)^{K/(K-1)} \leqslant 1$$

Arguing now with Hölder estimates precisely as in (13.34), we have for $p = \frac{K}{K-1}$ and for any $t < 1$,

$$\sum_i \operatorname{diam}^{2t} g(B_i) \leqslant C_0^t \sum_i |B_i|^{t(1-\frac{1}{p})} \left(\int_{B_i} J(z,g)^p \right)^{\frac{t}{p}} \leqslant C_0^t \left(\sum_i |B_i|^{\frac{(p-1)t}{p-t}} \right)^{1-\frac{t}{p}}$$

Choosing

$$t = \frac{s(K)}{2} = \frac{sK}{2+(K-1)s}$$

leads to

$$t\frac{p-1}{p-t} = \frac{s}{2}, \qquad 1 - \frac{t}{p} = \frac{s(K)}{sK},$$

and these prove the lemma. $\square$

On the other hand, it is in estimating the distortion under the outer factor in (13.70) that difficulties now arise. In the case of area, the two-dimensional Hausdorff measure, the required estimate was a consequence of (13.2), that for any principal K-quasiconformal mapping h,

$$|h(E)| \leqslant K|E| \quad \text{whenever } h|\mathbb{C} \setminus E \text{ is conformal} \tag{13.71}$$

Proving the counterpart of (13.71) for general dimensions s presently appears to be a difficult task. However, in the special case that the dimension $s = 1$, the following result from [23] provides a required analogy.

Theorem 13.5.8. *Suppose $E \subset \mathbb{C}$ is a compact set and let $h : \mathbb{C} \to \mathbb{C}$ be a principal K-quasiconformal mapping that is conformal on $\mathbb{C} \setminus E$. Then*

$$\mathcal{M}^1(h(E)) \leqslant C_K \mathcal{M}^1(E)$$

with C_K depending only on K.

We will reduce the theorem to the invariance of the following BMO-capacity. For a compact subset $E \subset \mathbb{C}$, let

$$\gamma_{BMO}(E) = \sup\{|\varphi'(\infty)| : \varphi \text{ holomorphic in } \mathbb{C} \setminus E, \ \varphi(\infty) = 0 \text{ and } \|\varphi\|_* \leq 1\},$$

where $\|\varphi\|_*$ stands for the BMO-norm of φ in $\mathbb{C}$. In contrast to the analytic capacity, by the following result of Král [220] and Kaufman [208], sets of vanishing BMO-capacity admit a simple description.

Theorem 13.5.9. *There are constants $C_1, C_2 < \infty$ such that for any compact subset $E \subset \mathbb{C}$,*

$$C_1 \mathcal{M}^1(E) \leq \gamma_{BMO}(E) \leq C_2 \mathcal{M}^1(E)$$

Proof. First, let E be a compact set and $\varphi \in BMO(\mathbb{C})$ a function that is analytic outside E. Fix a number $\varepsilon > 0$ and take a covering of E by disks B_j, with radius r_j, such that $\sum_j \text{diam}(B_j) \leq \mathcal{M}^1(E) + \varepsilon$. Consider a partition of unity associated with this covering; that is, for each j take an infinitely differentiable function η_j supported on $2B_j$ with

$$\|\nabla \eta_j\|_\infty \leq \frac{C_0}{r_j}$$

and such that $\sum_{j=1}^n \eta_j = 1$ in a neighborhood of E.

Using Cauchy's integral formula, we see that for all circles large enough to contain E in their interiors,

$$\varphi'(\infty) = \frac{1}{2\pi i} \int_{|\zeta|=R} \varphi(\zeta) d\zeta = \frac{1}{2\pi i} \int_{|\zeta|=R} \varphi(\zeta)\Big(1 - \sum_{j=1}^n \eta_j(\zeta)\Big) d\zeta$$

Here the latter integrand is smooth in the entire plane (setting it as zero on E) and hence we may apply Green's formula (2.52). As a result,

$$\varphi'(\infty) = -\frac{1}{\pi} \sum_{j=1}^n \int_\mathbb{C} \varphi(z) \frac{\partial \eta_j(z)}{\partial \bar{z}} = -\frac{1}{\pi} \sum_{j=1}^n \int_\mathbb{C} [\varphi(z) - \varphi_{2B_j}] \frac{\partial \eta_j(z)}{\partial \bar{z}}$$

$$\leq 4C_0 \sum_{j=1}^n \frac{1}{|2B_j|} \int_{2B_j} |\varphi(z) - \varphi_{2B_j}| \, r_j \leq 2C_0 \|\varphi\|_* [\mathcal{M}^1(E) + \varepsilon]$$

Thus we have shown that

$$\gamma_{BMO}(E) \leq 2C_0 \mathcal{M}^1(E)$$

For the converse, if $\mathcal{M}^1(E) > 0$, then by Frostman's lemma (see for instance [251, p. 112]), there exists a positive measure μ, supported on E, such that $\mu(\mathbb{D}(x,r)) \leq r$ and $\mu(E) \geq C \mathcal{M}^1(E)$. The Cauchy transform

$$\varphi(w) = \frac{1}{2\pi i} \int_E \frac{d\mu(\zeta)}{w - \zeta}$$

13.5. REMOVABLE SINGULARITIES

is analytic outside E, $\varphi(\infty) = 0$ and $\varphi'(\infty) = \mu(E)$. It remains only to estimate the BMO-norm of φ. Fix a ball $B = \mathbb{D}(z_0, r)$ and set $c_B = \int_{\mathbb{C}\setminus 2B} \frac{d\mu(w)}{w-z_0}$. We have

$$\frac{1}{|B|} \int_B |\varphi(z) - c_B|\, dm(z)$$

$$\leq \frac{1}{|B|} \int_B \left(\int_{2B} \frac{1}{|w-z|}\, d\mu(w) + \int_{\mathbb{C}\setminus 2B} \left| \frac{1}{w-z} - \frac{1}{w-z_0} \right| d\mu(w) \right) dm(z)$$

For the first integral, using Fubini's theorem, we see that

$$\frac{1}{|B|} \int_{2B} \int_B \frac{1}{|w-z|}\, dm(z)\, d\mu(w) \leq 4$$

Since $2|w-z| \geq |w-z_0|$ for $z \in B$ and $w \in \mathbb{C} \setminus 2B$, the second integral is bounded by

$$\frac{2}{|B|} \int_B \int_{\mathbb{C}\setminus 2B} \frac{|z-z_0|}{|w-z_0|^2}\, d\mu(w)\, dm(z) = \frac{2}{|B|} \int_B |z-z_0|\, dm(z) \int_{\mathbb{C}\setminus 2B} \frac{d\mu(w)}{|w-z_0|^2}$$

$$= \frac{2}{3} r \sum_{j=1}^{\infty} \int_{2^{j+1}B \setminus 2^j B} \frac{d\mu(w)}{|w-z_0|^2} \leq \frac{2}{3} r \sum_{j=1}^{\infty} \frac{2^{j+1} r}{(2^j r)^2} = \frac{4}{3}$$

Thus $\|\varphi\|_* \leq 16/3$. In particular,

$$\mu(E) \leq \frac{16}{3} \gamma_{BMO}(E),$$

and this estimate completes the proof. $\square$

Proof of Theorem 13.5.8. Let $h : \mathbb{C} \to \mathbb{C}$ be a principal K-quasiconformal mapping that is conformal on $\mathbb{C} \setminus E$. Choose a function $\varphi \in BMO(\mathbb{C})$ analytic outside $h(E)$ and vanishing at ∞ with derivative $\varphi'(\infty) = 1$. Then $\tilde{\varphi} = \varphi \circ h$ is analytic outside E and $\tilde{\varphi}(\infty) = 0$ with $\tilde{\varphi}'(\infty) = 1$. Furthermore, by Reimann's Theorem 13.4.1,

$$\|\tilde{\varphi}\|_* \leq C(K) \|\varphi\|_*$$

Since φ was arbitrary, we have $\gamma_{BMO}(E) \leq C(K) \gamma_{BMO}(h(E))$. A converse inequality follows by applying the argument to the inverse h^{-1}. Hence h preserves the BMO-capacity up to the constant $C(K)$ in Reimann's theorem, and this depends only on K. Hence the claim follows from Theorem 13.5.9. $\square$

We conjecture that actually in any dimension $0 < s \leq 2$ one has the corresponding estimates and propose the following.

Conjecture 13.5.10. *Suppose we are given a real number $s \in (0, 2]$. Then for any compact set $E \subset \mathbb{C}$ and for any principal K-quasiconformal mapping h that is conformal on $\mathbb{C} \setminus E$, we have*

$$\mathcal{M}^s(h(E)) \leq C \mathcal{M}^s(E) \tag{13.72}$$

with a constant C that depends only on K and s.

From the above results, the answer to this conjecture is positive for $s = 1$ and for $s = 2$, but at the moment the conjecture remains open for general dimensions s.

Last, collecting the above arguments, we have absolute continuity of the quasiconformal pullback of the Hausdorff measure $\mathcal{H}^s$ when $s = \frac{2}{K+1}$. That is, we have a proof of Conjecture 13.5.6 when the target dimension $= 1$.

The argument will also show that proving Conjecture 13.5.10 for some dimension s_0 implies Conjecture 13.5.6 for the case of the target dimension s_0.

Theorem 13.5.11. *Let $f : \Omega \to \Omega'$ be a K-quasiconformal mapping and suppose $E \subset \Omega$ is compact. If $\mathcal{H}^{2/(K+1)}(E) = 0$, then the length $\mathcal{H}^1(f(E)) = 0$.*

Proof. It suffices to show that if $E \subset \mathbb{D}$ is compact and $\phi : \mathbb{C} \to \mathbb{C}$ is a principal K-quasiconformal mapping conformal outside the unit disk $\mathbb{D}$, then we have the (quantitative) bound

$$\mathcal{M}^1(\phi(E)) \leqslant C \left(\mathcal{M}^{\frac{2}{K+1}}(E)\right)^{(K+1)/2K}, \qquad (13.73)$$

where the constant $C = C(K)$ depends only on K. In particular, the estimate will show that

$$\mathcal{H}^{2/(K+1)}(E) = 0 \quad \Rightarrow \quad \mathcal{H}^1(\phi(E)) = 0,$$

and the case of a general K-quasiconformal mapping f follows by applying the Stoilow factorization.

Suppose now that $E \subset \cup_{j=1}^n B_j$, where $B_j = \mathbb{D}(z_j, r_j)$, $j = 1, 2, \ldots, n$. As before, factor $\phi = h \circ g$, where g is principal K-quasiconformal mapping conformal in $(\mathbb{C} \setminus \mathbb{D}) \cup \left(\cup_{j=1}^n B_j\right)$. From Theorems 13.5.7 and 13.5.5 we have

$$\mathcal{M}^1(\phi(E)) = \mathcal{M}^1(h \circ g(E)) \leqslant C(K) \mathcal{M}^1(g(E))$$

$$\leqslant C_1(K) \left(\sum_{j=1}^n \operatorname{diam}(B_j)^{2/(K+1)}\right)^{(K+1)/2K}$$

Taking the infimum over the coverings of E gives (13.73) and thus proves the theorem. $\square$

One may interpret the Král-Kaufmann Theorem 13.5.9 as characterizing removable sets for analytic functions in BMO. Indeed, we see from the result that every function contained in $BMO(\mathbb{C})$ and analytic outside the compact set E extends to an analytic function of $\mathbb{C}$ if and only if E has zero length. Combining this view with Theorem 13.5.11, we have the following improvement of Theorem 13.5.4.

Corollary 13.5.12. *Sets of Hausdorff measure $\mathcal{H}^{2/(K+1)}(E) = 0$ are removable for all K-quasiregular mappings of $\mathbb{C} \setminus E$ that belong to $BMO(\mathbb{C})$.*

13.6. NONREMOVABLE SETS

The result is the best possible in the sense that I. Uriarte-Tuero [363] has constructed sets E, of positive and finite $\frac{2}{K+1}$-measure, that are not removable for all K-quasiregular mappings in BMO.

Remark. In addition to the BMO-capacity discussed above, one may also consider the corresponding quantity defined in terms of the space VMO. The capacity is again invariant under quasiconformal mappings, and this time the null sets of the capacity turn out to be the sets of σ-finite length. One obtains the following theorem.

Theorem 13.5.13. *If* $f\Omega \to \Omega'$ *is K-quasiconformal and $E \subset \Omega$ is compact, then*

$$E \text{ has } \sigma\text{-finite } \mathcal{H}^{2/(K+1)}\text{-measure} \;\Rightarrow\; f(E) \text{ has } \sigma\text{-finite length}$$

Again, for the details we refer the reader to [23].

13.6 Examples of Nonremovable Sets

The previous sections provided a delicate analysis of the distortion of one-dimensional sets under quasiconformal mappings but still leave open the cases where $\dim(E) = \frac{2}{K+1}$ precisely but E does not have σ-finite $\frac{2}{K+1}$-measure. Hence we are faced with the natural question: Are there compact sets E, with $\dim(E) = \frac{2}{K+1}$, that are not removable for all bounded K-quasiregular mappings ?

In this last section, following [23, 26], we give a positive answer and show that our results are sharp in a quite strong sense. Indeed, to compare with analytic removability first recall that, by Mattila's theorem [252], if a compact set E supports a probability measure with $\mu(\mathbb{D}(z,r)) \leqslant r\varepsilon(r)$ and

$$\int_0 \frac{\varepsilon(t)^2}{t} \, dt < \infty, \tag{13.74}$$

then the analytic capacity $\gamma(E) > 0$. On the other hand, by a result of Tolsa, if the integral in (13.74) diverges, then there are compact sets E of vanishing analytic capacity supporting a probability measure with $\mu(\mathbb{D}(z,r)) \leqslant r\varepsilon(r)$ [353]. In a complete analogy we prove the following theorem.

Theorem 13.6.1. *Let $K \geqslant 1$. Suppose $\beta(t) = t^{2/(K+1)} \varepsilon(t)$ is a gauge function such that*

$$\int_0 \frac{\varepsilon(t)^{1+1/K}}{t} \, dt < \infty \tag{13.75}$$

Then there is a compact set E that is not removable for bounded K-quasiregular mappings and yet supports a probability measure μ, with $\mu(\mathbb{D}(z,r)) \leqslant \beta(r)$ for every z and $r > 0$.

In particular, if we choose $\varepsilon(t)$ so that, in addition, for every $\alpha > 0$ we have $t^\alpha/\varepsilon(t) \to 0$ as $t \to 0$, then we obtain a set E that is not removable for bounded K-quasiregular mappings and such that $\dim(E) = \frac{2}{K+1}$.

Proof. We will construct a compact set E and a K-quasiconformal mapping ϕ such that $\mathcal{H}^\beta(E) \simeq 1$, and at the same time $\phi(E)$ has a positive and finite $\mathcal{H}^{\beta'}$-measure for some measure function $\beta'(t)$, where

$$\beta'(t) = t\,\varepsilon'(t) \quad \text{with} \quad \int_0 \frac{\varepsilon'(t)^2}{t}\,dt < \infty$$

Mattila's theorem then shows that $\gamma(\phi(E)) > 0$, so there exists nonconstant bounded functions h holomorphic on $\mathbb{C} \setminus \phi(E)$. Thus setting $f = h \circ \phi$, we see that E is not removable for bounded K-quasiregular mappings.

We shall construct the K-quasiconformal mapping ϕ as a limit of a sequence ϕ_N of K-quasiconformal mappings. E will be a Cantor-type set. To reach the optimal estimates we will need to change, at every step in the construction of E, both the size and the number m_j of the generating disks.

Without loss of generality we may assume that for every $\alpha > 0$, $t^\alpha/\varepsilon(t) \to 0$ as $t \to 0$. First, choose m_1 disjoint disks $D(z_i, R_1) \subset \mathbb{D}$, $i = 1, \ldots, m_1$, so that

$$\frac{1}{2} < c_1 = m_1\, R_1^2 < 1$$

The function $\kappa(t) = m_1\,\beta(tR_1)$ is continuous with $\kappa(0) = 0$ and, moreover, for each fixed t,

$$\kappa(t) = m_1(tR_1)^{\frac{2}{K+1}}\,\varepsilon(tR_1) = \varepsilon\left(t\sqrt{c_1/m_1}\right)\left(t\sqrt{c_1/m_1}\right)^{\frac{-2K}{K+1}} t^2 c_1 \to \infty$$

as $m_1 \to \infty$. Hence for any $t < 1$ we may choose m_1 so large that there exists $\sigma_1 \in (0, t)$ satisfying $m_1\,\beta(\sigma_1^K R_1) = 1$. In fact,

$$m_1\,\sigma_1 R_1\,\varepsilon(\sigma_1^K R_1)^{(K+1)/2K}\,c_1^{(1-K)/2K} = 1$$

Next, let $r_1 = R_1$. For each $i = 1, \ldots, m_1$, let $\varphi_i^1(z) = z_i + \sigma_1^K R_1\, z$ and, using the notation $\alpha D(z, \rho) = D(z, \alpha\rho)$, set

$$D_i = \frac{1}{\sigma_1^K}\,\varphi_i^1(\mathbb{D}) = D(z_i, r_1), \qquad D_i' = \varphi_i^1(\mathbb{D}) = D(z_i, \sigma_1^K r_1) \subset D_i$$

As the first approximation of the mapping, define

$$g_1(z) = \begin{cases} \sigma_1^{1-K}(z - z_i) + z_i, & z \in D_i' \\ \left|\frac{z-z_i}{r_1}\right|^{\frac{1}{K}-1}(z - z_i) + z_i, & z \in D_i \setminus D_i' \\ z, & z \notin \bigcup D_i \end{cases} \tag{13.76}$$

This is a K-quasiconformal mapping conformal outside of $\bigcup_{i=1}^{m_1}(D_i \setminus D_i')$. It maps each D_i onto itself and D_i' onto $D_i'' = D(z_i, \sigma_1 r_1)$, while the rest of the plane remains fixed. Set $\phi_1 = g_1$.

We now inductively construct the mapping as follows. After $N - 1$ steps we

13.6. NONREMOVABLE SETS

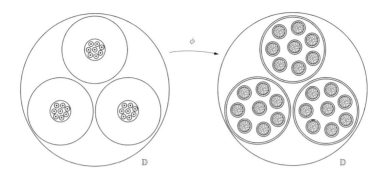

Construction of the Cantor set E.

take m_N disjoint disks of radius R_N, with union of $D(z_l^N, R_N)$ covering at least half of the area of $\mathbb{D}$,

$$c_N = m_N R_N^2 > \frac{1}{2} \tag{13.77}$$

As before, we may choose m_N so large that $m_1 \ldots m_N \, \beta(\sigma_1^K \ldots \sigma_N^K R_1 \ldots R_N) = 1$ holds for a unique σ_N as small as we wish. Note that $\lim_{N \to \infty} \sigma_N = 0$ and

$$m_1 \ldots m_N \, \sigma_1 R_1 \ldots \sigma_N R_N \, \varepsilon (\sigma_1^K R_1 \ldots \sigma_N^K R_N)^{(K+1)/2K} (c_1 \ldots c_N)^{(1-K)/2K} = 1$$

Set $\varphi_j^N(z) = z_j^N + \sigma_N^K R_N z$ and $r_N = R_N \sigma_{N-1} r_{N-1}$. For any multi-index $J = (j_1, ..., j_N)$, where $1 \leqslant j_k \leqslant m_k$, $k = 1, ..., N$, let

$$D_J = D(z_J, r_N) \phi_{N-1}\left(\frac{1}{\sigma_N^K} \varphi_{j_1}^1 \circ \cdots \circ \varphi_{j_N}^N(\mathbb{D})\right)$$

$$D'_J = D'(z_J, \sigma_N^K r_N) = \phi_{N-1}\left(\varphi_{j_1}^1 \circ \cdots \circ \varphi_{j_N}^N(\mathbb{D})\right)$$

and let

$$g_N(z) = \begin{cases} \sigma_N^{1-K}(z - z_J) + z_J, & z \in D'_J \\ \left|\frac{z-z_J}{r_N}\right|^{\frac{1}{K}-1}(z - z_J) + z_J, & z \in D_J \setminus D'_J \\ z, & \text{otherwise} \end{cases} \tag{13.78}$$

Clearly, g_N is K-quasiconformal, is conformal outside of the set

$$\bigcup_{J=(j_1,\ldots,j_N)} (D_J \setminus D'_J),$$

maps D_J onto itself and D'_J onto $D''_J = D(z_J, \sigma_N r_N)$ and elsewhere is the identity. We therefore obtain the map necessary for our inductive step by setting

$$\phi_N = g_N \circ \phi_{N-1} \tag{13.79}$$

Since $\{\phi_N\}_{N=1}^\infty$ is a sequence of K-quasiconformal mappings equal to the identity outside the unit disk $\mathbb{D}$, there is a nonconstant K-quasiconformal limit

$$\phi = \lim_{N \to \infty} \phi_N$$

with convergence in $W^{1,p}_{loc}(\mathbb{C})$ for any $p < \frac{2K}{K-1}$. However, ϕ maps the compact set

$$E = \bigcap_{N=1}^\infty \left(\bigcup_{j_1,\ldots,j_N} \varphi^1_{j_1} \circ \cdots \circ \varphi^N_{j_N}(\mathbb{D}) \right)$$

to the set

$$\phi(E) = \bigcap_{N=1}^\infty \left(\bigcup_{j_1,\ldots,j_N} \psi^1_{j_1} \circ \cdots \circ \psi^N_{j_N}(\mathbb{D}) \right),$$

where we have written $\psi^i_j(z) = z^i_j + \sigma_i R_i z$, $j = 1, \ldots, m_i$, $i \in \mathbb{N}$. Next, write

$$s_N = (\sigma^K_1 R_1) \ldots (\sigma^K_N R_N) \quad \text{and} \quad t_N = (\sigma_1 R_1) \ldots (\sigma_N R_N) \quad (13.80)$$

Observe that we have chosen the parameters R_N, m_N, σ_N so that

$$m_1 \ldots m_N \, \beta(s_N) = 1 \quad (13.81)$$
$$m_1 \ldots m_N \, t_N \, \varepsilon(s_N)^{(K+1)/2K} (c_1 \ldots c_N)^{(1-K)/2K} = 1 \quad (13.82)$$

We claim that $\mathcal{H}^\beta(E) \simeq 1$. Since $\operatorname{diam}(\varphi^1_{j_1} \circ \cdots \circ \varphi^N_{j_N}(\mathbb{D})) \leqslant \delta_N \to 0$ when $N \to \infty$, we have by (13.81),

$$\mathcal{H}^\beta(E) = \lim_{\delta \to 0} \mathcal{H}^\beta_\delta(E) \leqslant \lim_{\delta \to 0} \sum_{j_1,\ldots,j_N} \beta(\operatorname{diam}(\varphi^1_{j_1} \circ \cdots \circ \varphi^N_{j_N}(\mathbb{D})))$$
$$= m_1 \ldots m_N \, \beta(s_N) = 1$$

For the converse inequality, take a finite covering (U_j) of E by open disks of diameter $\operatorname{diam}(U_j) \leqslant \delta$ and let $\delta_0 = \inf_j(\operatorname{diam}(U_j)) > 0$. Denote by N_0 the minimal integer such that $s_{N_0} \leqslant \delta_0$. By construction, the family $(\varphi^{N_0}_{j_{N_0}} \circ \cdots \circ \varphi^1_{j_1}(\mathbb{D}))_{j_1,\ldots,j_{N_0}}$ is a covering of E with the $\mathcal{M}^\beta$-packing condition of Mattila [252]. Thus

$$\sum_j \beta(\operatorname{diam}(U_j)) \geqslant C \sum_{j_1,\ldots,j_{N_0}} \beta(\operatorname{diam}(\varphi^{N_0}_{j_{N_0}} \circ \cdots \circ \varphi^1_{j_1}(\mathbb{D}))) = C$$

Hence $\mathcal{H}^\beta_\delta(E) \geqslant C$, and letting $\delta \to 0$, we obtain $C \leqslant \mathcal{H}^\beta(E) \leqslant 1$, as desired.

A quite similar argument based on (13.82) shows that $\mathcal{H}^{\beta'}(\phi(E)) \simeq 1$ for a measure function $\beta'(t) = t\varepsilon'(t)$ as soon as, for all indexes N,

$$\varepsilon'(t_N) = \varepsilon(s_N)^{(K+1)/2K} (c_1 \ldots c_N)^{(1-K)/2K} \quad (13.83)$$

13.6. NONREMOVABLE SETS

We can find a continuous and nondecreasing function $\varepsilon'(t)$ satisfying (13.83) and

$$\int_0^{\cdot} \frac{\varepsilon'(t)^2}{t} \, dt < \infty \tag{13.84}$$

Indeed, let us first choose a continuous nondecreasing function $v(t)$ so that $v(t) \to 0$ as $t \to 0$ and so that (13.75) still holds in the form

$$\int_0^{\cdot} \frac{\varepsilon(t)^{1+1/K}}{t\,v(t)} \, dt < \infty$$

In the above inductive construction we can then choose the parameters σ_j so that $v(\sigma_1^K \cdots \sigma_N^K) \leqslant 2^{-N(1-1/K)}$ for every index N. Now (13.77) and (13.83) imply

$$\varepsilon'(t_N)^2 \leqslant \varepsilon(s_N)^{1+1/K}\, 2^{N(1-1/K)} \leqslant \frac{\varepsilon(s_N)^{1+1/K}}{v(s_N)}$$

On the other hand, by (13.80) we also have $t_{N-1}/t_N \leqslant s_{N-1}/s_N$, and so we may extend $\varepsilon'(t)$, determined by (13.83) only at the t_N's, so that it is continuous, is nondecreasing and satisfies

$$\int_0^{\cdot} \varepsilon'(t)^2 \, \frac{dt}{t} \leqslant \int_0^{\cdot} \frac{\varepsilon(s)^{1+1/K}}{v(s)} \, \frac{ds}{s} < \infty$$

Again, combining this estimate with Mattila's theorem [252] completes the proof of the theorem. □

Note that if we do not care about the analytic capacity of the target set, a straightforward modification of the construction given above, normalizing the disks of the construction so that $m_N t_N \eta(t_N) = 1$, yields the following corollary.

Corollary 13.6.2. *Let $K \geqslant 1$ and let $\beta(t) = t\eta(t)$ be a gauge function such that*

- *η is continuous and nondecreasing, $\eta(0) = 0$ and $\eta(t) = 1$ whenever $t \geqslant 1$.*
- *$\lim_{t \to 0} \dfrac{t^\alpha}{\eta(t)} = 0$ for all $\alpha > 0$*

Then there is a compact set $E \subset \mathbb{D}$ and a K-quasiconformal mapping $\phi : \mathbb{C} \to \mathbb{C}$ such that

$$\dim(E) = \frac{2}{K+1} \quad \text{and} \quad \mathcal{H}^\beta(\phi(E)) = 1$$

Concerning the deformation of sets with positive and finite $\frac{2}{K+1}$-measure Uriarte-Tuero [363] was recently able to reach the optimal example by considerably refining the above construction.

Theorem 13.6.3. *For every $K > 1$ there is a compact set E with*

$$0 < \mathcal{H}^{2/(K+1)}(E) < \infty$$

and a K-quasiconformal mapping $g : \mathbb{C} \to \mathbb{C}$ such that the length $\mathcal{H}^1(g(E)) > 0$. We refer the reader to [363] for the details.

Chapter 14

L^p-Theory of Beltrami Operators

Classically the homeomorphic solutions to the Beltrami equation

$$\frac{\partial f}{\partial \bar{z}} = \mu(z) \frac{\partial f}{\partial z}$$

are constructed via the Neumann iteration procedure such as that by Bojarski [64], although there do exist other, more function theoretic approaches, for example, in Lehto-Virtanen [229].

The L^p-boundedness of the Beurling transform was a crucial ingredient in the success of the Neumann iteration method and provided the first results toward the self-improving regularity of quasiconformal mappings.

Now that we have been able to establish, through the theory of holomorphic motions, the precise higher-integrability properties of quasiconformal mappings, we wish to return to the Neumann procedure and see what improvements the knowledge of optimal Sobolev regularity can bring.

More precisely, in this chapter we will study quite thoroughly the invertibility properties of the operators $\mathcal{B} = \mathbf{I} - \mu \mathcal{S}$ and its transpose $\mathcal{B}' = \mathbf{I} - \mathcal{S}\mu$ for $\|\mu\|_\infty = k < 1$. In particular, we shall now determine the exact value of the critical interval (q_k, p_k) that was introduced earlier in Definition 5.4.1. Hence, a number of themes initiated in Sections 5.4 and 5.5, such as the Caccioppoli inequalities, Stoilow factorization and regularity weakly quasiregular mappings will attain their optimal form. The results of this chapter are mostly based on the joint work [31] with E. Saksman.

In fact, nonlinearity will bring no restrictions on the study of the invertibility of Beltrami-type operators. We will consider inhomogeneous elliptic systems of the type

$$\frac{\partial f}{\partial \bar{z}} = H\left(z, \frac{\partial f}{\partial z}\right) + \phi(z), \tag{14.1}$$

where $\phi \in L^p(\mathbb{C})$. The problem is to determine when one can find a solution among the functions with $Df \in L^p(\mathbb{C})$.

We may rewrite (14.1) as

$$\begin{aligned}\phi(z) &= f_{\bar{z}} - H(z, f_z) \\ &= f_{\bar{z}} - H(z, \mathcal{S}f_{\bar{z}})\end{aligned}$$

with the aid of the Beurling transform. We therefore are led to consider the nonlinear singular integral operator

$$(\mathcal{B}g)(z) = g(z) - H(z, \mathcal{S}g), \tag{14.2}$$

which we call a *Beltrami operator*. We will give the name "Beltrami operators" to their adjoints as well.

Our first question can be reformulated as determining in which $L^p(\mathbb{C})$-spaces the operators $\mathcal{B}$ are invertible. This is completely answered by the following theorem, which is the main goal of this chapter.

Invertibility of Beltrami Operators

Theorem 14.0.4. *Assume that the measurable function $H : \mathbb{C} \times \mathbb{C} \to \mathbb{C}$ satisfies the ellipticity condition*

$$|H(z, \zeta) - H(z, \xi)| \leqslant k|\zeta - \xi| \quad \text{with } k < 1, \tag{14.3}$$

together with the normalization $H(z, 0) = 0$ for all $z \in \mathbb{C}$. Then the operator $\mathcal{B} : L^p(\mathbb{C}) \to L^p(\mathbb{C})$ defined in (14.2) is invertible whenever

$$1 + k < p < 1 + \frac{1}{k} \tag{14.4}$$

Moreover, in this range of exponents, $\mathcal{B}$ is a bilipschitz homeomorphism of $L^p(\mathbb{C})$. In particular, there are constants $c_p(k)$ and $C_p(k)$ such that for all $g_1, g_2 \in L^p(\mathbb{C})$,

$$c_p(k)\|g_1 - g_2\|_{L^p(\mathbb{C})} \leqslant \|\mathcal{B}g_1 - \mathcal{B}g_2\|_{L^p(\mathbb{C})} \leqslant C_p(k)\|g_1 - g_2\|_{L^p(\mathbb{C})} \tag{14.5}$$

In the case where $\mathcal{B}$ is linear, Theorem 14.0.4 gives the important consequence that the operators

$$\mathbf{I} - \mu \mathcal{S}$$

are invertible on $L^p(\mathbb{C})$ whenever $\|\mu\|_\infty \leqslant k < 1$ and $1 + k < p < 1 + 1/k$.

Conversely, we will show that Theorem 14.0.4 fails whenever $p \leqslant 1 + k$ or $p \geqslant 1 + 1/k$, even in the simplest case when $H(z, w)$ is C^∞-smooth and linear with respect to w. Note that then the corresponding Beltrami operator has the form

$$\mathcal{B}(g) = g - \mu_1(z)\mathcal{S}g - \mu_2(z)\overline{\mathcal{S}g},$$

where $|\mu_1(z)| + |\mu_2(z)| \leq k < 1$. For instance, we will show that for each $p \geq 1 + 1/k$ there are $h \in L^p(\mathbb{C}) \cap C^\infty(\mathbb{C})$ and $\mu \in C^\infty(\mathbb{C})$ with $\|\mu\|_\infty = k$ that oscillate so badly at ∞ that the inhomogeneous Beltrami equation $w_{\bar{z}} - \mu w_z = h$ admits no solution w with $\nabla w \in L^p(\mathbb{C})$. For details, see the example in (14.26) and (14.27).

In fact, we have the following theorems.

Theorem 14.0.5. *For any $p \in [2, \infty)$, there are $\mu \in L^\infty$ with $p = 1 + 1/\|\mu\|_\infty$ such that neither of the operators*

$$\mathbf{I} - \mathcal{S}\mu, \qquad \mathbf{I} - \mu\mathcal{S}$$

is invertible on $L^p(\mathbb{C})$ or $L^q(\mathbb{C})$. Here, $\frac{1}{p} + \frac{1}{q} = 1$.

Section 14.2 will be devoted to the proof of Theorem 14.0.4, while the examples leading to Theorem 14.0.5 are described in Section 14.3. As a immediate corollary to Theorems 14.0.4 and 14.0.5, we have the following key theorem: For each $0 \leq k < 1$ we determine explicitly the critical interval, defined in Section 5.4.

The Critical Interval

Theorem 14.0.6. *For each $0 \leq k < 1$ the critical interval is*

$$(q_k, p_k) = (1 + k, 1 + 1/k)$$

The result has a variety of consequences on quasiconformal mappings and elliptic PDEs. We will consider a few of them in Section 14.3. The reader will find many more throughout this monograph.

If μ is compactly supported and f is a quasiconformal $W_{loc}^{1,2}(\mathbb{C})$-solution to the Beltrami equation normalized by $f(z) = z + \mathcal{O}(|z|^{-1})$ near ∞, then from (5.8) we have the useful formulas

$$f_{\bar{z}} = (\mathbf{I} - \mu\mathcal{S})^{-1}\mu \qquad (14.6)$$
$$f_z = 1 + \mathcal{S}(f_{\bar{z}}) \qquad (14.7)$$
$$= 1 + (\mathbf{I} - \mathcal{S}\mu)^{-1}(\mathcal{S}\mu) \qquad (14.8)$$

An immediate consequence of Theorem 14.0.4 is that every quasiconformal map with maximal complex dilatation $\|\mu\|_\infty = k$ belongs to $W_{loc}^{1,p}(\Omega)$ for each $p < 1 + 1/k$.

It is worthwhile noting that our proof works in precisely the converse direction: Results on area distortion from the previous chapter imply the correct L^p-integrability for the partial derivatives of quasiconformal maps, and we will use this fact as a starting point in our proof. It would be very interesting indeed to find more direct methods for establishing invertibility and in particular for identifying p-norms of Beltrami operators. Theorem 14.0.4 would in fact be an immediate consequence of the conjectural values $\|\mathcal{S}\|_p = \max\{p - 1, 1/(p - 1)\}$.

14.1 Spectral Bounds and Linear Beltrami Operators

As a first simple exercise, let us identify the spectrum of the Beurling transform in $L^p(\mathbb{C})$ for $p \in (1, \infty)$. That is, we seek to determine those constants μ for which $\mathcal{S} - \mu$ is invertible.

Theorem 14.1.1. *Let $p \in (1, \infty)$. Then the spectrum*
$$\sigma(\mathcal{S} : L^p(\mathbb{C}) \to L^p(\mathbb{C})) = \mathbb{S} = \{\lambda \in \mathbb{C} : |\lambda| = 1\}$$

Proof. Let $|\zeta| \neq 1$ and denote by T_ζ the operator given by a change of variables
$$(T_\zeta h)(z) = h(\bar{z} - \bar{\zeta} z)$$

The theorem will follow as soon as we can verify the indentity
$$(\mathcal{S} - \zeta \mathbf{I})^{-1} = \frac{1}{1 - |\zeta|^2} \left(\bar{\zeta} \mathbf{I} + T_\zeta^{-1} \mathcal{S} T_\zeta \right) \tag{14.9}$$

To see this we solve the inhomogeneous Beltrami equation
$$f_z - \zeta f_{\bar{z}} = h$$

for f and set $\phi(z) = T_\zeta(f)(z) = f(\bar{z} - \bar{\zeta}z)$. We compute
$$\phi_{\bar{z}}(z) = h(\bar{z} - \bar{\zeta}z)$$

and then
$$\begin{aligned}
(1 - |\zeta|^2) T_\zeta f_{\bar{z}} &= (1 - |\zeta|^2) f_{\bar{z}}(\bar{z} - \bar{\zeta}z) \\
&= \phi_z(z) + \bar{\zeta} \phi_{\bar{z}}(z) \\
&= \mathcal{S} T_\zeta h + \bar{\zeta} T_\zeta h = T_\zeta(\bar{\zeta} h + T_\zeta^{-1} \mathcal{S} T_\zeta h),
\end{aligned}$$

which in light of our choice $(\mathcal{S} - \zeta \mathbf{I})^{-1} h = f_{\bar{z}}$ is enough to establish the identity in (14.9). Moreover, we see from the above argument (after interchanging ζ for $\bar{\zeta}$) for $\zeta \in \mathbb{S}$ that $T_{\bar{\zeta}} h$ lies in the kernel of $\mathcal{S} + \bar{\zeta} \mathbf{I}$, which is therefore nontrivial for all $\zeta \in \mathbb{S}$. This completes the proof. $\square$

Thus the spectral behavior of the Beurling transform resembles that of its one-dimensional analog, the Hilbert transform H. Namely, the relation $H^2 = -\mathbf{I}$ shows that $\sigma(H : L^p(\mathbb{R}) \to L^p(\mathbb{R})) = \{i, -i\}$ for all $p \in (1, \infty)$.

Theorem 14.1.1 has the unfortunate consequence that the spectrum $\sigma(\mathcal{S})$ does not yield any direct information on the growth of the L^p-norm

$$\|\mathcal{S}\|_{L^p(\mathbb{C}) \to L^p(\mathbb{C})}$$

14.2 Invertibility of the Beltrami Operators

In this section we prove the main result of this chapter (Theorem 14.0.4) and discuss in more detail the system of nonlinear first-order partial differential equations in the plane covered by it. The proof of Theorem 14.0.4 is based on a fundamental auxiliary result yielding a priori bounds for solutions of the *linear* Beltrami operator $\mathbf{I} - \mu \mathcal{S}$, which we first establish. Later we will return to the general result and show that it can be reduced to the linear case.

The strategy for the proof is to apply a quasiconformal change of variables for solving the inhomogeneous equation

$$f - \mu \mathcal{S} f = g$$

This leads us in a natural way to consider the Beurling transform $\mathcal{S}$ on weighted L^p-spaces with weights constructed from the Jacobians of quasiconformal mappings.

Lemma 14.2.1. *Assume that $\mu \in L^\infty(\mathbb{C})$ with $\|\mu\|_\infty \leqslant k < 1$. Suppose that $1 + k < p < 1 + 1/k$. Then the operator $\mathbf{I} - \mu \mathcal{S}$ is bounded below on $L^p(\mathbb{C})$; that is,*

$$\|(\mathbf{I} - \mu \mathcal{S})g\|_p \geqslant C_0(k,p) \|g\|_p, \qquad g \in L^p(\mathbb{C}) \tag{14.10}$$

Proof. Choose an arbitrary function g on $\mathbb{C}$ such that

$$g \in C_0^\infty(\mathbb{C}) \tag{14.11}$$

and write

$$h = g - \mu \mathcal{S} g \tag{14.12}$$

Since such functions like g are dense in $L^p(\mathbb{C})$, it is enough to prove the claim under the restriction of (14.11). Furthermore, if

$$w = \mathcal{C} g \tag{14.13}$$

is the Cauchy transform of g, then $w \in C^\infty(\mathbb{C})$ with derivatives $w_z, w_{\bar{z}} \in L^p(\mathbb{C})$ for $p > 1$. Moreover, as $w_{\bar{z}} = g$ and $w_z = \mathcal{S} g$, we see that w must satisfy the inhomogeneous Beltrami equation

$$\frac{\partial w}{\partial \bar{z}} = \mu \frac{\partial w}{\partial z} + h \tag{14.14}$$

We have thus reduced the question of establishing the inequality in (14.10) to showing that in fact

$$\left\| \frac{\partial w}{\partial \bar{z}} \right\|_p \leqslant C_0^{-1} \|h\|_p$$

This amounts to proving an a priori bound for the differential equation (14.14). In order to achieve this set $K = (1+k)/(1-k)$ so that the assumed restriction on p takes the form

$$\frac{2K}{K+1} < p < \frac{2K}{K-1} \tag{14.15}$$

14.2. INVERTIBILITY

Next, choose a K-quasiconformal homeomorphism $f : \mathbb{C} \to \mathbb{C}$ satisfying the Beltrami equation
$$f_{\bar{z}} = \mu f_z \qquad (14.16)$$
Thus, upon setting
$$u = w \circ f^{-1},$$
we may calculate
$$\begin{aligned} w_{\bar{z}} &= (u_z \circ f) f_{\bar{z}} + (u_{\bar{z}} \circ f) \overline{f_z} \\ \mu w_z + h &= \mu \big((u_z \circ f) f_z + (u_{\bar{z}} \circ f) \overline{f_{\bar{z}}} \big) + h \end{aligned} \qquad (14.17)$$

Next an application of the Beltrami equation (14.16) enables us to eliminate the terms involving u_z. This leads to the equation
$$(u_{\bar{z}} \circ f) \overline{f_z} = \frac{h}{1 - |\mu|^2} \qquad (14.18)$$

Consequently,
$$\int_{\mathbb{C}} |(u_{\bar{z}} \circ f) \overline{f_z}|^p \leqslant (1 - k^2)^{-p} \int_{\mathbb{C}} h^p \qquad (14.19)$$

We shall then use the A_p-properties of the derivatives $|f_z|$. Note that $|f_{\bar{z}}|^2 \leqslant |f_z|^2 \leqslant (1-k^2)^{-1} J(z, f)$. Therefore a change of variables gives
$$\int_{\mathbb{C}} |(u_z \circ f) f_{\bar{z}}|^p \leqslant \int_{\mathbb{C}} |(u_z \circ f) f_z|^p \leqslant (1-k^2)^{-1} \int_{\mathbb{C}} |u_z|^p \omega, \qquad (14.20)$$

where $\omega = |f_z \circ f^{-1}|^{p-2}$. The weight ω is comparable to $J(z, f^{-1})^{1-p/2}$. From (14.15) we see that the exponent
$$s = 1 - \frac{p}{2} \in \left(-\frac{1}{K-1}, \frac{K}{K-1} \right)$$

In particular, the criteria of Theorem 13.4.2 are satisfied, and we obtain
$$J(z, f^{-1})^s = J(z, f^{-1})^{1-p/2} \in A_p$$
whenever $1 + k < p < 1 + 1/k$.

We are now in a position to use the theorem of Coifman and Fefferman [94] on the weighted spaces $L^p(\omega)$ explained in Theorem 4.5.6. As the Beurling transform is a regular Calderón-Zygmund operator, we may apply the result in the form
$$\int_{\mathbb{C}} |u_z|^p \omega \leqslant \lambda_\omega^p \int_{\mathbb{C}} |u_{\bar{z}}|^p \omega, \qquad (14.21)$$
where λ_ω is the norm of $\mathcal{S}$ on the Banach space L_ω^p. Furthermore, λ_ω has a bound $\lambda_\omega \leqslant \alpha(|\omega|_{A_p}) < \infty$ depending only on the A_p-norm of ω. The Coifman-Fefferman theorem hence gives
$$\begin{aligned} \int_{\mathbb{C}} |u_z|^p \omega &\leqslant \lambda_\omega^p \int_{\mathbb{C}} |u_{\bar{z}}|^p \omega \leqslant \lambda_\omega^p \int_{\mathbb{C}} |(u_{\bar{z}} \circ f) f_z|^p \\ &\leqslant \frac{\lambda_\omega^p}{(1-k^2)^p} \int_{\mathbb{C}} h^p \end{aligned}$$

Combining these estimates with (14.20) and inserting them into (14.17) leads to the inequality

$$\|g\|_p \equiv \|w_{\bar{z}}\|_p \leqslant \|(u_{\bar{z}} \circ f)\overline{f_{\bar{z}}}\|_p + \|(u_z \circ f)f_{\bar{z}}\|_p \leqslant 2K^2\lambda_w\|h\|_p, \qquad (14.22)$$

which shows that the operator $\mathbf{I} - \mu \mathcal{S}$ is bounded from below on $L^p(\mathbb{C})$ for all $1+k < p < 1+\frac{1}{k}$. Furthermore, we find that the constant $C_0(k,p) = (2K^2\lambda_w)^{-1}$ depends only on k and p. $\square$

14.2.1 Proof of Invertibility; Theorem 14.0.4

We now apply the above to prove the main result of this chapter, Theorem 14.0.4. We assume that the numbers k, p and the function H satisfy the assumptions (14.3) and (14.4) of the theorem. We shall first establish the bilipschitz property for the operator $\mathcal{B}$.

By hypothesis, $\mathcal{B}$ maps the zero function to the zero function and, moreover, for $g_1, g_2 \in L^p(\mathbb{C})$ we have the estimate

$$\begin{aligned}\|\mathcal{B}g_1 - \mathcal{B}g_2\|_{L^p(\mathbb{C})} &\leqslant \|g_1 - g_2\|_{L^p(\mathbb{C})} + k\|\mathcal{S}g_1 - \mathcal{S}g_2\|_{L^p(\mathbb{C})} \\ &\leqslant (k\mathbf{S}_p + 1)\|g_1 - g_2\|_{L^p(\mathbb{C})}, \end{aligned} \qquad (14.23)$$

where, as before, $\mathbf{S}_p$ denotes the norm of the Beurling transform $\mathcal{S} : L^p(\mathbb{C}) \to L^p(\mathbb{C})$. The inequality in (14.23) shows $\mathcal{B}$ to be a well-defined Lipschitz map.

Let us assume then that $g_1, g_2 \in L^p(\mathbb{C})$. We set

$$h = \mathcal{B}g_1 - \mathcal{B}g_2$$

According to (14.3), we may write

$$H(z, \mathcal{S}g_1) - H(z, \mathcal{S}g_2) = \mu(z)(\mathcal{S}g_1 - \mathcal{S}g_2),$$

where the function $\mu = \mu(g_1, g_2) : \mathbb{C} \to \mathbb{C}$ satisfies $\|\mu\|_\infty \leqslant k < 1$. Hence the difference $g = g_1 - g_2$ solves the inhomogeneous Beltrami equation

$$h = g - \mu \mathcal{S}g$$

We now invoke the a priori estimate in (14.10) to find that

$$\|g\|_{L^p(\mathbb{C})} \leqslant C_0(k,p)^{-1}\|h\|_{L^p(\mathbb{C})} \qquad (14.24)$$

When put together with the above, we have proven the required bilipschitz estimate (14.5).

Finally, in order to obtain the invertibility of the nonlinear operator $\mathcal{B}$, we must observe that since $\mathcal{B}$ is a bilipschitz map, it will suffice to show that the image $\mathcal{B}(L^p(\mathbb{C}))$ is dense in $L^p(\mathbb{C})$. To that end let $h_1 \in L^2(\mathbb{C})$ be arbitrary and denote $\widetilde{\mathcal{B}}g = h_1 + H(z, \mathcal{S}g)$ for $g \in L^2(\mathbb{C})$. As $\mathcal{S}$ is an isometry on $L^2(\mathbb{C})$, we see from condition (14.3) that $\widetilde{\mathcal{B}}$ is a strict contraction on $L^2(\mathbb{C})$. Hence the Banach fixed-point theorem yields an element $g \in L^2(\mathbb{C})$ with $\widetilde{\mathcal{B}}g = g$, that is, $\mathcal{B}g = h_1$.

Now assume that $h_1 \in L^2(\mathbb{C}) \cap L^p(\mathbb{C})$. The solution $g \in L^2(\mathbb{C})$ for $\mathcal{B}g = h_1$ also satisfies $g \in L^p(\mathbb{C})$ according to the bilipschitz inequality (14.5). We thus infer that $\mathcal{B}(L^p(\mathbb{C}))$ contains the set $L^2(\mathbb{C}) \cap L^p(\mathbb{C})$, which is in fact dense in $L^p(\mathbb{C})$. The proof of Theorem 14.0.4 is complete. □

For later purposes we note that in the linear case $\mathcal{B} = \mathbf{I} - \mu \mathcal{S}$ we obtain from (14.21) and (14.22) the norm estimate

$$\|(\mathbf{I} - \mu \mathcal{S})^{-1}\|_{L^p} \leqslant 2K^2 \|\mathcal{S}\|_{L^p_\omega(\mathbb{C})}, \qquad (14.25)$$

where $\|\mathcal{S}\|_{L^p_\omega(\mathbb{C})}$ is the operator norm of the Beurling operator $\mathcal{S}$ on the weighted space L^p_ω.

Linear Inhomogeneous Equations: Existence of Solutions

We will make frequent use of Theorem 14.0.4. Often it has the most natural interpretation in terms of elliptic PDEs. Here is a first immediate consequence.

Theorem 14.2.2. *Let $\mu, \nu \in L^\infty(\mathbb{C})$ with $|\mu| + |\nu| \leqslant k < 1$ almost everywhere. Then the equation*

$$\frac{\partial f}{\partial \bar{z}} - \mu(z) \frac{\partial f}{\partial z} - \nu(z) \overline{\frac{\partial f}{\partial z}} = h(z)$$

has a solution f, locally integrable with gradient in $L^p(\mathbb{C})$, whenever $1 + k < p < 1 + 1/k$ and $h \in L^p(\mathbb{C})$. Further, f is unique up to an additive constant.

In the next section we have a look at what happens outside the L^p-interval $1 + k < p < 1 + 1/k$. For a more systematic discussion on the related PDEs, see Chapter 16.

14.3 Determining the Critical Interval

The critical interval has played an important role in many (if not most) of the previous sections. As promised, we are finally in a position to identify them precisely for each $0 \leqslant k < 1$.

Indeed, by definition the critical interval was the largest open interval of the exponents p where all the operators $\mathbf{I} - \mu \mathcal{S}$ with $\|\mu\|_\infty \leqslant k$ were invertible on $L^p(\mathbb{C})$, with norms of the inverses depending only on p and k. Thus from Theorem 14.0.4 we see that the critical interval always contains the open interval $(1 + k, 1 + 1/k)$. It therefore remains to show that the invertibility fails at the end points $1 + k$ and $1 + 1/k$.

Here again we shall reflect on the fact that for elliptic differential operators the regularity of the coefficients always reflects the local smoothness and growth properties of the solutions to the corresponding equations.

For example, if the coefficient μ of the Beltrami equation $f_{\bar{z}} = \mu f_z$ is continuously differentiable, the same holds for f, and therefore in particular the derivatives of f are locally in L^p for all exponents p.

However, the oscillations of the coefficients, even at one point, can destroy this L^p-regularity. The following example will give us a $\mu, h \in C^\infty(\mathbb{C})$ such that the corresponding inhomogeneous Beltrami equation does not have appropriate L^p-solutions; this appears already in the borderline case $\|\mu\|_\infty = 1/(p-1)$ in Theorem 14.0.4. Consequently, the behavior of the coefficients near infinity may also play an important role in the global L^p-estimates.

To be more specific let $p > 2$ and define

$$\mu(z) = \frac{z^2}{p + (p-1)|z|^2} \tag{14.26}$$

Then $\mu \in C^\infty(\mathbb{C})$ and $\|\mu\|_\infty = 1/(p-1)$. However, we show that the operator $\mathbf{I} - \mu \mathcal{S} : L^p(\mathbb{C}) \to L^p(\mathbb{C})$ is not surjective, hence certainly not invertible.

Lemma 14.3.1. *Suppose $p > 2$ and the coefficient μ are as in (14.26). Then there is $\phi \in L^p(\mathbb{C})$ for which the system $w_{\bar{z}} - \mu w_z = \phi$ does not admit any solutions w with $\nabla w \in L^p(\mathbb{C})$.*

Proof. Notice first that the quasiconformal diffeomorphism $f : \mathbb{C} \to \mathbb{C}$, $f(z) = z(1+|z|^2)^{-1/p}$ satisfies the Beltrami equation $f_{\bar{z}} = \mu f_z$. The derivatives f_z and $f_{\bar{z}}$ are in *weak*-$L^p(\mathbb{C})$ but not in $L^p(\mathbb{C})$. Consider the function h defined by

$$h(z) = z(1+|z|^2)^{-1/p} \log^{-1/p}(1+|z|^2)$$

Define ϕ by the equation

$$h_{\bar{z}} - \mu h_z = \phi$$

and note that $\phi \in L^p(\mathbb{C})$. However, we have $h_{\bar{z}}, h_z \notin L^p(\mathbb{C})$.

We claim that the equation

$$F_{\bar{z}} - \mu F_z = \phi, \tag{14.27}$$

has no solution such that $F_z, F_{\bar{z}} \in L^p(\mathbb{C})$. Upon assuming the contrary we observe that $F - h \in W^{1,2}_{loc}(\mathbb{C})$ solves the homogeneous Beltrami equation with coefficient μ. By Stoilow factorization this implies that

$$F(z) = h(z) + \Phi \circ f(z),$$

where Φ is an entire analytic function. For large $|z|$ our construction shows that

$$|h(z)| \leq |z|^{1-2/p} \leq 2|f(z)|$$

On the other hand, from Theorem 4.3.13 we deduce $|F(z)| \leq C(1 + |z|^{1-2/p})$. Once we combine these bounds, we soon see that

$$|\Phi(w)| \leq (2C+2)|w|$$

for large $|w|$.

14.3. DETERMINING THE CRITICAL INTERVAL

Therefore, according to the classical Liouville theorem, $\Phi(w) = c_1 w + c_2$ for some constants c_1 and c_2. Consequently, $F = h + c_1 f + c_2$. However, if we calculate the derivatives, we see that

$$F_{\bar{z}} = h_{\bar{z}} + c_1 f_{\bar{z}} = \frac{-z^2}{p(1+|z|^2)^{1+1/p}} \left(c_1 + \frac{1 + \log(1+|z|^2)}{\log^{1+1/p}(1+|z|^2)} \right)$$

If $c_1 \neq 0$ here, the constant term is dominant in the latter factor and then $F_{\bar{z}} \notin L^p(\mathbb{C})$. But if $c_1 = 0$, then $F_{\bar{z}} = h_{\bar{z}} \notin L^p(\mathbb{C})$. Therefore no choice of the constant c_1 will yield the gradient of the solution in $L^p(\mathbb{C})$. □

With the above lemma we now have for $p = 1 + 1/k$ examples of Beltrami operators $\mathbf{I} - \mu \mathcal{S}$ that fail to be surjective on $L^p(\mathbb{C})$. For the lower critical exponent $p = 1 + k$ it is sufficient to note that the adjoint of $\mathbf{I} - \mu \mathcal{S} : L^p(\mathbb{C}) \to L^p(\mathbb{C})$ is the operator

$$\mathbf{I} - \mathcal{S}^{-1} \bar{\mu} : L^q(\mathbb{C}) \to L^q(\mathbb{C}),$$

where $p^{-1} + q^{-1} = 1$. The identities

$$\overline{(\mathbf{I} - \mathcal{S}\mu)\omega} = (\mathbf{I} - \mathcal{S}^{-1}\bar{\mu})\bar{\omega} \quad \text{and} \quad \mathbf{I} - \mu \mathcal{S} = \mathcal{S}^{-1}(\mathbf{I} - \mathcal{S}\mu)\mathcal{S} \qquad (14.28)$$

show that for a fixed μ the three operators

$$\mathbf{I} - \mu \mathcal{S} : L^p(\mathbb{C}) \to L^p(\mathbb{C}), \quad \mathbf{I} - \mathcal{S}\mu : L^q(\mathbb{C}) \to L^q(\mathbb{C}), \quad \mathbf{I} - \mu \mathcal{S} : L^q(\mathbb{C}) \to L^q(\mathbb{C})$$

are all simultaneously invertible. As a consequence, the above Beltrami operator $\mathbf{I} - \mu \mathcal{S}$ with $\|\mu\|_\infty = k$ is noninvertible also on $L^q(\mathbb{C})$, $q = 1 + k$.

In fact, we see that $\mathbf{I} - \mu \mathcal{S}$ cannot be surjective on $L^q(\mathbb{C})$, $q = 1 + k$, either. Otherwise, its adjoint would be bounded below. By the above identities this would hold also for $\mathbf{I} - \mu \mathcal{S}$ on $L^p(\mathbb{C})$, $p = 1 + 1/k$. But then the proof of Theorem 14.0.4 would show that the operator is invertible on $L^p(\mathbb{C})$ for $p = 1 + 1/k$.

As a further remark, the above considerations also show that the operators

$$\mathbf{I} - \bar{\mu} \mathcal{S} \quad \text{and} \quad \mathcal{S} - \mu \mathbf{I} = \mathcal{S}(\mathbf{I} - \mathcal{S}^{-1}\mu)$$

are simultaneously L^p-invertible.

With this example we have completed the proof of Theorem 14.0.5. This was the last step in determining the precise form of the critical interval,

$$I = (q_k, p_k), \quad \text{where} \quad q_k = 1 + k \quad \text{and} \quad p_k = 1 + \frac{1}{k}$$

Now that the critical intervals have been identified, we have a variety of consequences. For instance, the result provides us with the optimal form of the Caccioppoli inequalities.

Theorem 14.3.2. *Let $1 + k < s < 1 + 1/k$ and let $f \in W^{1,s}_{loc}(\Omega)$ satisfy the distortion inequality*

$$\left|\frac{\partial f}{\partial \bar{z}}(z)\right| \leqslant k \left|\frac{\partial f}{\partial z}(z)\right| \quad \text{for almost every } z \in \Omega \tag{14.29}$$

Then we have the Caccioppoli estimate

$$\|\eta\, Df\|_s \leqslant C_s(k)\|fD\eta\|_s$$

whenever η is a compactly supported Lipschitz function in Ω.

Similarly, as in Theorem 5.4.2, the proof is reduced to the invertibility of the Beltrami operators. The mappings

$$f_0(z) = \frac{1}{z}|z|^{1-1/K} \quad \text{and} \quad f_1(z) = z|z|^{1/K-1}, \tag{14.30}$$

where $K = \frac{1+k}{1-k}$, show that the Caccioppoli inequalities fail at both borderline cases $s = 1 + k$ and $s = 1 + 1/k$.

The Caccioppoli inequalities in turn provide regularity for weakly quasiregular mappings.

Corollary 14.3.3. *Suppose $K \geqslant 1$ and*

$$\frac{2K}{K+1} < q, p < \frac{2K}{K-1}$$

Then every weakly K-quasiregular mapping $F \in W^{1,q}_{loc}(\Omega)$ belongs to $W^{1,p}_{loc}(\Omega)$. In particular, the Jacobian determinant will be locally integrable and the mapping F will be K-quasiregular. Moreover, we have the uniform estimate

$$\|F_{\bar{z}}\|_{L^p(\Omega')} + \|F_z\|_{L^p(\Omega')} \leqslant C\|F_z\|_{L^q(\Omega')} \tag{14.31}$$

for every relatively compact subdomain $\Omega' \subset \Omega$, where $C = C(p,q,\Omega,\Omega')$.

The examples in (14.30) show that Corollary 14.3.3 fails for $q < 2K/(K+1)$ and for $p \geqslant 2K/(K-1)$.

It is a curious fact that settling the lower borderline case, or determining the minimal regularity that implies every weakly K-quasiregular mapping f to be continuous and quasiregular, depends on properties slightly weaker than the invertibility of the operators $\mathbf{I} - \mu \mathcal{S}$. As we shall see in the next section, the regularity scales are determined by the L^p-injectivity of the appropriate Beltrami operators. That, on the other hand, may hold even in the borderline cases.

14.4 Injectivity in the Borderline Cases

In the previous section we established the basic properties of the Beltrami operators $\mathbf{I} - \mu \mathcal{S}$, proving that they are invertible on $L^p(\mathbb{C})$ whenever p lies in the critical interval $(1+k, 1+1/k)$. We also showed that at end points $p = 1+k$ and $p = 1+1/k$ the operators fail to be surjective, hence certainly are not invertible. There is, however, some L^p-stability at the end points: The Beltrami operators remain injective.

The L^p-injectivity of the Beltrami operators has interesting concrete interpretations for p both at the lower and at the upper end of the critical interval. The lower end is connected to the optimal Stoilow theorem and optimal regularity results on weakly quasiregular mappings. The upper end, on the other hand, relates to sharp Liouville theorems.

We consider first the lower borderline $p = 1+k$. Here we reduce the injectivity on $L^p(\mathbb{C})$ to weighted estimates on the Beurling transform. Indeed, we will have to use the fact that the operator norm of $\mathcal{S}$ on the weighted space $L^p_\omega(\mathbb{C})$ depends linearly on the A_p-norm of the weight ω, in other words, that

$$\|\mathcal{S}\phi\|_{L^p_\omega} \leqslant C(p) \|\omega\|_{A_p} \|\phi\|_{L^p_\omega}, \qquad 1 < p < \infty, \tag{14.32}$$

holds for all $\omega \in A_p$ and $\phi \in L^p_\omega$, with a constant $C(p)$ depending only on p. This linear dependence was shown by Petermichl and Volberg [294]; see Theorem 4.5.7 earlier. Applying this result to the weights $w(z) = J(z,f)^s$ we have the following theorem.

Theorem 14.4.1. *Suppose that $\mu \in L^\infty(\mathbb{C})$ with $\|\mu\|_\infty = k < 1$. Let $q = 1 + \|\mu\|_\infty$. Then the operator*

$$\mathbf{I} - \mu \mathcal{S} : L^q(\mathbb{C}) \to L^q(\mathbb{C}) \tag{14.33}$$

is injective.

Proof. Let $q < 2$. We begin with a duality argument. Since $I - \mathcal{S}\mu = \mathcal{S}(I - \mu\mathcal{S})\mathcal{S}^{-1}$, the claim is equivalent to showing that the adjoint operator $(I - \mathcal{S}\mu)' = I - \mu\mathcal{S} : L^p(\mathbb{C}) \to L^p(\mathbb{C})$ has a dense range in the weak topology of $L^p(\mathbb{C})$. Here $\frac{1}{q} + \frac{1}{p} = 1$ and $q = 1 + \|\mu\|_\infty = 1 + k$.

To prove the density, fix $\phi \in C_0^\infty(\mathbb{C})$ and let $0 < \varepsilon < 1$. Since $p < 1 + \frac{1}{(1-\varepsilon)k}$, from Theorem 14.0.4 we have

$$\phi_\varepsilon = \big(\mathbf{I} - (1-\varepsilon)\mu\mathcal{S}\big)^{-1} \phi \in L^p(\mathbb{C})$$

Furthermore,

$$\phi_\varepsilon - \mu\mathcal{S}\phi_\varepsilon = \phi - \mu\mathcal{S}(\varepsilon\phi_\varepsilon)$$

By the density of C_0^∞ in $L^p(\mathbb{C})$ it is enough to show that our assumptions imply that $\varepsilon\phi_\varepsilon$ tends weakly to 0 as $\varepsilon \to 0$.

For this we claim that the linear dependence (14.32) on the A_p-norm yields

$$\|(I - (1-\varepsilon)\mathcal{S}\mu)^{-1}\|_{L^p(\mathbb{C}) \to L^p(\mathbb{C})} \leqslant \frac{C(q)}{\varepsilon} \qquad \text{for all small } \varepsilon > 0 \tag{14.34}$$

Once the estimate (14.34) has been proven, it gives the uniform bounds
$$\|\varepsilon\phi_\varepsilon\|_p \leqslant C(p)\|\phi\|_p \tag{14.35}$$
On the other hand,
$$\|\phi_\varepsilon\|_2 \leqslant \frac{1}{1+\varepsilon k - k} \|\phi_\varepsilon - (1-\varepsilon)\mu\mathcal{S}\phi_\varepsilon\|_2 = \frac{1}{1+\varepsilon k - k} \|\phi\|_2$$
This shows that $\varepsilon\phi_\varepsilon \to 0$ in L^2, and combined with the uniform bound (14.35) it also implies that $\varepsilon\phi_\varepsilon \to 0$ weakly in $L^p(\mathbb{C})$.

Thus to conclude the proof we have to show that the linear dependence (14.32) on the A_p-norm implies (14.34). This takes us back to our proof of Theorem 14.0.4 and the quasiconformal estimates it is based on. Here, in fact, we must squeeze the last drop of information available from the area distortion Theorem 13.2.1: The derivatives of a $\frac{1+k}{1-k}$-quasiconformal mapping are in weak-$L^{1+1/k}$, and hence the corresponding L^p-bounds blow up like $\left(1 + 1/k - p\right)^{-1}$ when $p \to 1 + 1/k$.

We have already recorded in Theorem 13.4.2 the estimates the distortion of area gives for the A_p-norms of the weights $\omega(z) = J(z, f)^s$. If f_ε is now the quasiconformal homeomorphism with dilatation $(1-\varepsilon)\mu$, $k = \|\mu\|_\infty$, consider the weight
$$\omega_\varepsilon(z) = J(z, f_\varepsilon^{-1})^{1-2/p}, \qquad p = 1 + 1/k \tag{14.36}$$
Then from (13.56) and (13.58),
$$\|\omega_\varepsilon\|_{A_p} \leqslant \frac{C(K)}{\varepsilon}, \qquad K = \frac{1+\|\mu\|_\infty}{1-\|\mu\|_\infty}$$
On the other hand, according to the inequality (14.25), the norm of the operator $(I - (1-\varepsilon)\mu\mathcal{S})^{-1} : L^p(\mathbb{C}) \to L^p(\mathbb{C})$ is bounded by $2K^2\|\mathcal{S}\|_{L^p_{\omega_\varepsilon}}$, with the weight defined in (14.36). Combining with the Petermichl-Volberg Theorem 4.5.7 we end up with
$$\|(I - (1-\varepsilon)\mu\mathcal{S})^{-1}\|_{L^p} \leqslant C_0 \|\mathcal{S}\|_{L^p_{\omega_\varepsilon}} \leqslant C_1 \|\omega_\varepsilon\|_{A_p} \leqslant \frac{C(K)}{\varepsilon}$$
Therefore (14.34) is established, and the proof of the theorem is complete. □

We now consider applications of the injectivity of the Beltrami operators, in particular to the regularity theory. We view weakly quasiregular mappings as solutions $f \in W^{1,q}_{loc}(\Omega)$ to the Beltrami equation
$$\frac{\partial f}{\partial \bar{z}}(z) = \mu(z) \frac{\partial f}{\partial z}(z) \quad \text{for almost every } z \in \Omega, \tag{14.37}$$
where $\|\mu\|_\infty \leqslant k = \frac{K-1}{K+1}$ and $1 \leqslant q \leqslant 2$. We have seen that in general, for q small enough, a solution of (14.37) need even not be continuous. However, if the solution f is contained in the space $W^{1,2}_{loc}(\Omega)$, then f will be quasiregular and hence continuous, open and discrete.

14.4. BORDERLINE INJECTIVITY

Lemma 14.4.2. *Suppose μ is a measurable function with compact support and $\|\mu\|_\infty < 1$. Let $f \in W^{1,q}_{loc}(\Omega)$ be a weakly quasiregular mapping in a domain $\Omega \subset \mathbb{C}$, where $1 < q < 2$, such that the Beltrami coefficient $\mu_f(z) \equiv \mu(z)$, $z \in \Omega$.*

Then, if the Beltrami operator $\mathbf{I} - \mu \mathcal{S} : L^q(\mathbb{C}) \to L^q(\mathbb{C})$ is injective, it follows that f is quasiregular.

Proof. It is enough to prove that for each $\phi \in C_0^\infty(\Omega)$ the function $g = \phi f$ has partial derivatives in $L^2(\mathbb{C})$.

According to our assumptions, $g_z, g_{\bar{z}} \in L^q(\mathbb{C})$, and we have the inhomogeneous Beltrami equation

$$g_{\bar{z}} - \mu g_z = f(\phi_{\bar{z}} - \mu \phi_z) \in L^2(\mathbb{C}) \tag{14.38}$$

On the other hand, since $\mathbf{I} - \mu \mathcal{S}$ is invertible in $L^2(\mathbb{C})$, we can solve this equation also for a function $\widetilde{g}$ whose derivatives are in $L^2(\mathbb{C})$, that is

$$\widetilde{g}_{\bar{z}} - \mu \widetilde{g}_z = f(\phi_{\bar{z}} - \mu \phi_z), \qquad \widetilde{g}_z, \widetilde{g}_{\bar{z}} \in L^2(\mathbb{C})$$

Furthermore, since ϕ and μ have compact support, it follows that $\widetilde{g}_{\bar{z}} \in L^q(\mathbb{C})$ and $\widetilde{g}_z = \mathcal{S} \widetilde{g}_{\bar{z}} \in L^q(\mathbb{C})$. Observe that $(g - \widetilde{g})_{\bar{z}} - \mu(g - \widetilde{g})_z = 0$ in $L^q(\mathbb{C})$.

The assumed uniqueness or the injectivity of the Beltrami operator $\mathbf{I} - \mu \mathcal{S}$ now implies that $\widetilde{g} = g$ modulo a constant, establishing the lemma. $\square$

Put together, Theorem 14.4.1 and Lemma 14.4.2 give two very important consequences. First, we have the following result.

Corollary 14.4.3. *Suppose $K \geqslant 1$ and $f \in W^{1,q}_{loc}(\Omega)$ is weakly K-quasiregular, with $q = \frac{2K}{K+1}$. Then $f \in W^{1,2}_{loc}(\Omega)$ and, consequently, f is a quasiregular mapping. In particular, f is continuous, open and discrete.*

The reader should note that Corollary 14.4.3 holds for no exponent $q < \frac{2K}{K+1}$. Indeed, examples of weakly quasiregular mappings show that they can be discontinuous with bad singularities. The following construction originates from [189]. Let $\mathbb{D} = \{z : |z| < 1\}$ and choose a countable number of disjoint disks $B_j = \mathbb{D}(x_j, r_j) \subset \mathbb{D}$ such that the measure $|\mathbb{D} \setminus \bigcup_j B_j| = 0$. Define

$$f(z) = r_j^{1+1/K} \frac{|z - x_j|^{1-1/K}}{z - x_j} + \bar{x}_j, \qquad z \in B_j, \tag{14.39}$$

and on $\mathbb{D} \setminus \bigcup_j B_j$ set $f(z) = \bar{z}$, while for $|z| \geqslant 1$ let $f(z) = 1/z$. Even if f has a "pole" at each center x_i, f is weakly K-quasiregular and it is not difficult to show that f is contained in all Sobolev classes $W^{1,q}_{loc}(\mathbb{C})$ with $q < \frac{2K}{K+1}$. Indeed, we see that $f(z) - \bar{z} \in W^{1,q}_0(B_j)$ for every index j. We then propose that the reader verify the following general fact.

Lemma 14.4.4. *Suppose $\Omega_j \subset \Omega$ are pairwise disjoint open subsets of a domain Ω. Assume we are given functions $g_j \in W^{1,q}_0(\Omega_j)$ for each $j = 1, 2, \cdots$ and define*

$$g(z) = g_j(z), \quad z \in \Omega_j \qquad \text{and} \quad g(z) = 0 \text{ otherwise}$$

If $\sum_{j=1}^{\infty} \int_{\Omega_j} |Dg_j|^q < \infty$, then $g \in W^{1,q}(\Omega)$ and

$$\int_\Omega |Dg|^q = \sum_{j=1}^{\infty} \int_{\Omega_j} |Dg_j|^q$$

As a further corollary of (14.39), we see that if μ is the Beltrami coefficient of f, then according to Lemma 14.4.2 the operator $\mathbf{I} - \mu \mathcal{S}$ is not injective on $L^q(\mathbb{C})$, $1 < q < 1 + \|\mu\|_\infty$.

A remarkable feature of the above Corollary 14.4.3 is that one has self-improving regularity even in the borderline case $q = \frac{2K}{K+1}$, where the usual tool, the Caccioppoli estimates, do not hold!

Similarly, the Stoilow factorization theorem can now be presented in its optimal form. In fact, the following theorem is just a reformulation of Corollary 14.4.3.

Corollary 14.4.5. *Suppose $K \geqslant 1$ and $f \in W^{1,q}_{loc}(\Omega)$ is a weakly K-quasiregular mapping, where $q = \frac{2K}{K+1}$. Then there is a K-quasiconformal mapping $g : \mathbb{C} \to \mathbb{C}$ and a function ϕ holomorphic in $g(\Omega)$ such that*

$$f(z) = \phi \circ g(z), \qquad z \in \Omega$$

Proof. It follows from Corollary 14.4.3 that f is quasiregular. Therefore the standard Stoilow factorization Theorem 5.5.1 applies. □

The above example (14.39) of a weakly K-quasiregular mapping contained in $W^{1,q}_{loc}(\mathbb{C})$ for all $q < \frac{2K}{K+1}$ exhibits the failure of Stoilow factorization with this regularity. However, the mappings of that example are discontinuous, they are not even locally bounded. Hence one might wonder if, to get Stoilow factorization, one could lower the regularity requirements under the assumption of continuity. However, in the next section we show the factorization fails even for continuous weakly K-quasiregular mappings in $W^{1,q}_{loc}(\mathbb{C})$ for any $q < \frac{2K}{K+1}$. In other words, for mappings of bounded distortion, in Corollary 14.4.3 we have reached the minimal regularity possible.

14.4.1 Failure of Factorization in $W^{1,q}$, $q < \frac{2K}{K+1}$

In this section we present an example showing that even fairly nice weakly quasiregular mappings in $W^{1,q}$ may fail to have a Stoilow factorization when the exponent $q < \frac{2K}{K+1}$.

Theorem 14.4.6. *Let $K > 1$ and $q_0 < \frac{2K}{K+1}$. Then there is a Beltrami coefficient μ supported in the unit disk with the following properties.*

- $\|\mu\|_\infty = \frac{K-1}{K+1}$.

14.4. BORDERLINE INJECTIVITY

- The Beltrami equation $h_{\bar{z}} = \mu h_z$ admits a Hölder-continuous solution $h \in W^{1,q_0}_{loc}(\mathbb{C})$ that vanishes at infinity,

$$h(z) = \mathcal{O}\left(\frac{1}{z}\right) \quad \text{as } z \to \infty$$

- The solution h is not quasiregular, it fails to be in $W^{1,2}_{loc}(\mathbb{C})$ and it cannot be obtained from a holomorphic function by factoring with a quasiconformal mapping.

Proof. Choose Cantor sets E and F, both contained in the unit disk $\mathbb{D}$, and a K-quasiconformal mapping $g : \mathbb{C} \to \mathbb{C}$ with $g(z) = z$ outside the unit disk, such that $g(F) = E$, the Hausdorff dimension $\dim_{\mathcal{H}}(E) = 1 + \varepsilon > 1$, as well as

$$\frac{1}{\dim_{\mathcal{H}}(E)} - \frac{1}{2} = \frac{1}{K}\left(\frac{1}{\dim_{\mathcal{H}}(F)} - \frac{1}{2}\right) \tag{14.40}$$

According to Corollary 12.4.2, such choices are possible by using holomorphic motions and the generalized λ-lemma. The $0 < \varepsilon < 1$ is to be determined later. If we write $\dim_{\mathcal{H}}(F) = 1 + \alpha$, where $-1 < \alpha < \varepsilon$, then (14.40) can be rearranged as

$$K = \frac{1-\alpha}{1+\alpha} \frac{1+\varepsilon}{1-\varepsilon} > 1$$

or

$$\alpha = \frac{1 - \frac{1-\varepsilon}{1+\varepsilon} K}{1 + \frac{1-\varepsilon}{1+\varepsilon} K}$$

Let ν denote the $(1+\varepsilon)$-Hausdorff measure defined on compact subsets of $\mathbb{C}$. We set

$$\varphi(w) = \frac{1}{2\pi i} \int_E \frac{d\nu(\zeta)}{w - \zeta}$$

as the Cauchy transform of ν.

Then φ is a Hölder-continuous function (with exponent ε; see [131, Theorem III.4.4]) defined in $\mathbb{C}$ and holomorphic in $\mathbb{C} \setminus E$ and $\varphi(w) = \frac{C}{w} + \mathcal{O}(1/|w|^2)$ as $w \to \infty$. Hölder continuity gives the estimate

$$|\varphi'(w)| \leq \frac{C}{\text{dist}(w, E)^{1-\varepsilon}}$$

for some positive constant C. We are now going to improve the regularity of φ at the cost of adding some distortion.

With an explicit construction, such as in (13.78)-(13.79), the mapping g also comes with the estimates

$$\text{dist}(g(z), E) \approx \text{dist}(z, F)^{1/K}, \quad |Dg(z)| \approx \text{dist}(z, F)^{-1+1/K}$$

We now set

$$h(z) = (\varphi \circ g)(z)$$

Then $h(z) = \mathcal{O}(1/z)$ near ∞. Now g is quasiconformal and so is in $W_{loc}^{1,2}(\mathbb{C})$. The composition of the two mappings φ and g is readily seen to have the property ACL. Therefore we need only check the degree of integrability of the differential of h. Pointwise, we have

$$|Dh(z)| = |\varphi'(g(z))|\,|Dg(z)|,$$

so near the set F we have

$$|Dh(z)| \leq \left(\frac{c_1}{\text{dist}(z,F)}\right)^{(1-\varepsilon)/K + 1 - 1/K} = \left(\frac{c_1}{\text{dist}(z,F)}\right)^{1-\varepsilon/K}$$

Now as F is a regular Cantor set, the function $\text{dist}^{-1}(z,F)$ is integrable to any power smaller than $2 - \dim(F) = 1 - \alpha$. We solve

$$q_\varepsilon(1 - \varepsilon/K) = (1-\alpha)$$

to obtain

$$q_\varepsilon = \frac{2K}{(K + \frac{1+\varepsilon}{1-\varepsilon})(1 - \frac{\varepsilon}{K})}$$

Then we have $h \in W_{loc}^{1,q}$ for all $q < q_\varepsilon$.

We now have to adjust ε so that $q_\varepsilon > q_0$. This is possible since $q_\varepsilon \to \frac{2K}{K+1}$ as $\varepsilon \to 0$. Then the derivatives of h belong to $L^{q_0}(\mathbb{C})$. It is clear that h is Hölder continuous and decays at infinity as required. Outside E, thus almost everywhere, it satisfies the same Beltrami equation as g. However, h is bounded in the entire plane and hence it cannot be quasiregular. In particular it does not belong to $W_{loc}^{1,2}(\mathbb{C})$, nor can it be written as a composition of a holomorphic function and a quasiconformal mapping. $\square$

14.4.2 Injectivity and Liouville-Type Theorems

We have already seen in a number of different circumstances the utility of Liouville-type theorems in establishing the uniqueness properties for solutions of linear and nonlinear partial differential equations. A bit later in this book we shall study certain degenerate elliptic equations, that is, equations that are not uniformly elliptic. Solutions to such equations, should they exist, will not in general be quasiconformal or quasiregular mappings. They will fall into the more general class of mappings of finite distortion, which we study in more depth in Chapter 20. Actually, the theory of such mappings is already well established in all dimensions $n \geq 2$; see [26, 180, 191, 279]. In the present section we will consider only a specific Liouville-type result that will be used to prove the $L^p(\mathbb{C})$-injectivity of the Beltrami operators for the upper borderline $p = 1 + 1/k$.

Let us consider (orientation-preserving) homeomorphisms $f \in W_{loc}^{1,1}(\mathbb{C})$ such that $|Df(z)|^2 \leq K(z) J(z,f)$ for some function $K(z)$, finite almost everywhere. Hence we simply require that at almost every point z, if the Jacobian $J(z,f) = 0$, all partial derivatives of f vanish.

14.4. BORDERLINE INJECTIVITY

For such mappings we may define the distortion function by

$$K(z,f) := |Df(z)|^2/J(z,f) \quad \text{if } J(z,f) \neq 0 \tag{14.41}$$

and set $K(z,f) = 1$ elsewhere. We will prove a Liouville-type result for such general mappings assuming the distortion satisfies the integral bound

$$\frac{1}{\pi R^2} \int_{\mathbb{D}(0,R)} K(z,f) \leq K_\infty, \tag{14.42}$$

for all *sufficiently large* disks $\mathbb{D}(0,R)$ centered at the origin. The constant K_∞ is, of course, supposed to be independent of the disk.

By combining the isoperimetric inequality with appropriate differential inequalities, we shall obtain the following.

Theorem 14.4.7. *Suppose $f \in W^{1,1}_{loc}(\mathbb{C})$ is a homeomorphism such that the distortion function $K(z,f) < \infty$ almost everywhere. Suppose that $K(z,f)$ satisfies the bound (14.42) for all $R > R_0 > 1$ and that $J(z,f) \in L^q(\mathbb{C})$ for some $1 < q \leq \frac{K_\infty}{K_\infty - 1}$. Then f is constant.*

Proof. Let $t > 0$ and set $B = \mathbb{D}(0,t)$. Abbreviate $J = J(z,f)$ and $K = K(z,f)$. By the isoperimetric inequality from Theorem 3.43 and Hölder's inequality we see that for almost all $t \geq 0$

$$\begin{aligned}
\frac{1}{|B|} \int_B J &\leq \left(\frac{1}{|\partial B|} \int_{\partial B} |Df| \right)^2 \\
&\leq \left(\frac{1}{|\partial B|} \int_{\partial B} \frac{|Df|^2}{K} \right) \left(\frac{1}{|\partial B|} \int_{\partial B} K \right) \\
&\leq \left(\frac{1}{|\partial B|} \int_{\partial B} J \right) h(t),
\end{aligned}$$

where we have denoted

$$h(t) = \frac{1}{|\partial B|} \int_{\partial B} K$$

Next, if we define the increasing function

$$\phi(t) = \int_{\mathbb{D}(0,t)} J(z,f),$$

then the previous estimate may be written in the form

$$\phi(t) \leq \frac{t}{2} \phi'(t) h(t) \quad \text{for almost every } t > 0 \tag{14.43}$$

We first claim that

$$\phi(t) = o(t^{2/K_\infty}) \quad \text{as } t \to \infty \tag{14.44}$$

To see this, fix an arbitrary bounded and measurable set $E \subset \mathbb{C}$ and consider the disks $B = \mathbb{D}(0,t) \supset E$. Since $J \in L^q(\mathbb{C})$ by assumption, we have

$$\phi(t) = \int_B J = \int_{B \setminus E} J + \int_E J$$

$$\leqslant |B \setminus E|^{1-1/q} \Big(\int_{B \setminus E} J^q \Big)^{1/q} + |E|^{1-1/q} \Big(\int_E J^q \Big)^{1/q}$$

Multiplying by $|B|^{-1+1/q}$ and letting t grow to ∞ we conclude that

$$\limsup_{t \to \infty} (\pi t^2)^{(-1+1/q)} \phi(t) \leqslant \Big(\int_{\mathbb{C} \setminus E} J^q \Big)^{1/q},$$

which clearly yields (14.44) since the set E was arbitrary and $1 - 1/q \leqslant 1/K_\infty$.

In order to simplify our forthcoming computations, we introduce new variables. Set

$$s = t^2, \qquad h(t) = K_\infty H(t^2), \qquad \phi(t) = (\psi(t^2))^{1/K_\infty}$$

Now (14.43) and (14.44) take the somewhat more pleasant form

$$\psi(s) \leqslant s\psi'(s) H(s) \quad \text{for almost every } s > 0 \tag{14.45}$$

together with

$$\psi(s) = o(s) \quad \text{as } s \to \infty \tag{14.46}$$

In addition, our assumption (14.42) on the distortion function K shows that

$$K_\infty \geqslant \frac{1}{\pi r^2} \int_{\mathbb{D}(0,r)} K = \frac{1}{\pi r^2} \int_0^r \Big(\int_{S_t} K \, dt \Big)$$

$$\geqslant \frac{2}{\pi r^2} \int_0^r \pi t \, h(t) \, dt = r^{-2} \int_0^{r^2} K_\infty H(s) \, ds$$

We have obtained the simple bound

$$\frac{1}{s} \int_0^s H(u) \, du \leqslant 1, \tag{14.47}$$

which is valid, say, for all $s \geqslant s_0$.

Let $a \geqslant s_0$ be arbitrary. It is enough to show that $\psi(a) = 0$. Since ϕ is absolutely continuous, we may integrate (14.45) to deduce

$$\psi(a) \leqslant \psi(s) \exp\Big(- \int_a^s \frac{du}{uH(u)} \Big)$$

The desired conclusion follows from (14.46) as soon as we prove that

$$\int_a^s \frac{du}{uH(u)} \geqslant \log(s) - C(a)$$

14.4. BORDERLINE INJECTIVITY

Set $G(s) = \int_a^s H(u)du$. According to (14.47), $0 \leqslant G(u) \leqslant u$, and hence an application of the elementary inequality $H + 1/H \geqslant 2$ together with an integration by parts leads to the estimate

$$\int_a^s \frac{du}{uH(u)} \geqslant 2\int_a^s \frac{du}{u} - \int_a^s \frac{H(u)du}{u} = 2\log(s/a) - \frac{G(s)}{s} - \int_a^s \frac{G(u)du}{u^2}$$

$$\geqslant 2\log(s/a) - 1 - \int_a^s \frac{du}{u} = \log(s) - (1 + \log(a))$$

This then completes the proof. $\square$

The above Liouville-type theorem is sharp in a quite strong sense: If $K_\infty > 1$, then there is a K_∞-quasiconformal (and hence nonconstant) mapping $f : \mathbb{C} \to \mathbb{C}$ such that

$$J(z, f) \in L^p(\mathbb{C}) \quad \text{for all } p > \frac{K_\infty}{K_\infty - 1}$$

For example, the mapping

$$f(z) = z\,\chi_\mathbb{D} + z|z|^{\frac{1}{K_\infty} - 1}\chi_{\mathbb{C}\setminus\mathbb{D}}$$

satisfies all these requirements. Note further that the K_∞-quasiconformal $g(z) = z|z|^{\frac{1}{K_\infty}-1}$ satisfies

$$J(z,g) \in \text{weak-}L^q(\mathbb{C}) \quad \text{for } q = \frac{K_\infty}{K_\infty - 1}$$

The general Liouville-type theorem can also be formulated as an injectivity property of an appropriate Beltrami operator. Indeed, the injectivity at the higher borderline exponent $p = 1 + 1/\|\mu\|_\infty$ reduces to Theorem 14.4.7.

Theorem 14.4.8. *Suppose that $p > 2$ and $f : \mathbb{C} \to \mathbb{C}$ solves the equation $f_{\bar{z}} - \mu_1 f_z - \mu_2 \overline{f_z} = 0$, where $\||\mu_1| + |\mu_2|\|_\infty \leqslant 1/(p-1)$. If both $f_z, f_{\bar{z}}$ are in $L^p(\mathbb{C})$, then f is a constant.*

Proof. Since $\||\mu_1| + |\mu_2|\|_\infty \leqslant 1/(p-1)$, we have $|f_{\bar{z}}| \leqslant |f_z|/(p-1)$ and the distortion $K(z,f) \leqslant K := \frac{p}{p-2}$ at almost every $z \in \mathbb{C}$. Accordingly, $J(z,f) \in L^{K/(K-1)}(\mathbb{C})$ and thus Theorem 14.4.7 implies the claim. $\square$

Next, for noninjectivity, for $p < 1 + \|\mu\|_\infty$ and $\mathbf{I} - \mu S$ examples are provided by the functions f in (14.39). Similarly, if we let

$$g(z) = z\chi_\mathbb{D} + z|z|^{-2k/(k+1)}(1 - \chi_\mathbb{D}), \tag{14.48}$$

$\mathbb{D} = \{z : |z| < 1\}$, then $g_z, g_{\bar{z}} \in L^p(\mathbb{C})$ for every $p > 1 + \frac{1}{k}$ and $(\mathbf{I} - \mu_1 S)g_{\bar{z}} = 0$ for $\mu_1 = -k\frac{z}{\bar{z}}\chi_{\mathbb{D}^c}$.

Combining these facts, we observe the following consequence.

Theorem 14.4.9. *The Beltrami operator* $\mathbf{I} - \mu_1 \mathcal{S} - \mu_2 \overline{\mathcal{S}}$ *is injective in* $L^p(\mathbb{C})$ *whenever*

$$\||\mu_1| + |\mu_2|\|_\infty \leqslant \frac{1}{p-1} < 1$$

There are Beltrami operators $\mathbf{I} - \mu_1 \mathcal{S} - \mu_2 \overline{\mathcal{S}}$ *for which the injectivity fails whenever* $\||\mu_1| + |\mu_2|\|_\infty > \frac{1}{p-1}$. *Similarly, if*

$$\||\mu_1| + |\mu_2|\|_\infty \leqslant p - 1 < 1,$$

then $\mathbf{I} - \mu_1 \mathcal{S} - \mu_2 \overline{\mathcal{S}}$ *is injective on* $L^p(\mathbb{C})$ *and there are* μ_1, μ_2 *for which the injectivity fails for every* $p < 1 + \||\mu_1| + |\mu_2|\|_\infty$.

Proof. Theorems 14.4.7 and 14.4.8 show that $\mathbf{I} - \mu_1 \mathcal{S} - \mu_2 \overline{\mathcal{S}}$ is injective on $L^p(\mathbb{C})$ if $1 + k \leqslant p \leqslant 1 + 1/k$, $k = \||\mu_1| + |\mu_2|\|_\infty$. The noninjectivity part was established in (14.39) and (14.48). □

14.5 Beltrami Operators; Coefficients in VMO

The example in (14.26) in the previous section shows that even for C^∞-smooth coefficients the elliptic system

$$\frac{\partial f}{\partial \bar{z}} - \mu(z) \frac{\partial f}{\partial z} = h, \qquad h \in L^p(\mathbb{C}),$$

may not admit any solutions in the natural smoothness class $Df \in L^p(\mathbb{C})$. On the other hand, slight global control of the regularity of the coefficient μ, even in the very weak sense of vanishing mean oscillations (that is, $\mu \in VMO(\mathbb{C})$) changes the situation dramatically.

We show that this VMO condition will force the operators $\mathbf{I} - \mu \mathcal{S}$ to be invertible in all L^p-spaces. Here we must also assume that μ is defined at ∞ in the following average sense:

$$\mu(\infty) = \lim_{R \to \infty} \frac{1}{\pi R^2} \int_{\mathbb{D}(0,R)} \mu(z) \qquad (14.49)$$

We simply abbreviate these assumptions by saying that $\mu \in VMO(\hat{\mathbb{C}})$.

The definition of the class VMO seems to vary slightly in the literature. As is Chapter 4, we define $VMO(\mathbb{C})$ to be the closure of $C_0^\infty(\mathbb{C})$ in $BMO(\mathbb{C})$. This is equivalent to taking the closure of $C(\hat{\mathbb{C}})$ in BMO.

First, the assumption $\mu \in VMO(\mathbb{C})$ with $\|\mu\|_\infty < 1$ makes $\mathbf{I} - \mu \mathcal{S} : L^p(\mathbb{C}) \to L^p(\mathbb{C})$ a Fredholm operator with index zero regardless of the value of $p \in (1, \infty)$. Namely, Theorem 14.1.1 together with the spectral radius formula shows that for every $\varepsilon > 0$, $\|\mathcal{S}^j\|_{p \to p} \leqslant (1 + \varepsilon)^j$ as soon as $j \geqslant N = N(\varepsilon)$. We deduce the existence of an integer $m = m_p$ such that

$$\|\mu^m \mathcal{S}^m\|_p < 1,$$

14.5. COEFFICIENTS IN VMO

which certainly implies that the operator $\mathbf{I} - \mu^m \mathcal{S}^m$ is invertible on $L^p(\mathbb{C})$.

At this point one invokes the result of A. Uchiyama, Theorem 4.6.14, that the commutator of $\mathcal{S}$ with $\mu \in VMO(\mathbb{C})$ is a compact operator on each $L^p(\mathbb{R}^n)$, $p \in (1, \infty)$. As $\mu \mathcal{S} - \mathcal{S}\mu$ is a compact operator, we may apply this repeatedly and see that the difference $K = \mu^{j_0} \mathcal{S}^{j_0} - (\mu \mathcal{S})^{j_0}$ is a compact operator on $L^p(\mathbb{C})$. Writing $P = 1 + \mu\mathcal{S} + (\mu\mathcal{S})^2 + \cdots + (\mu\mathcal{S})^{j_0-1}$, it follows that

$$(\mathbf{I} - \mu\mathcal{S})P = P(\mathbf{I} - \mu\mathcal{S}) = \mathbf{I} - (\mu\mathcal{S})^{j_0} = (\mathbf{I} - \mu^{j_0}\mathcal{S}^{j_0}) + K$$

This shows that $\mathbf{I} - \mu\mathcal{S}$ is a Fredholm operator. Finally, the continuous homotopy $t \mapsto \mathbf{I} - t\mu\mathcal{S}$, $t \in [0,1]$, shows that in fact its index is zero.

A Fredholm operator of index zero is invertible as soon as it is injective. Hence the following theorem is our next step.

Theorem 14.5.1. *Let* $2 \leqslant p < \infty$ *and suppose that* $\mu \in VMO(\hat{\mathbb{C}})$ *satisfies* $\|\mu\|_\infty < 1$. *Then the operator* $\mathbf{I} - \mu\mathcal{S}$ *is injective on* $L^p(\mathbb{C})$.

Proof. There is no loss of generality in assuming that $p > 2$. Showing that the kernel of the operator

$$\mathbf{I} - \mu\mathcal{S} : L^p(\mathbb{C}) \to L^p(\mathbb{C})$$

is trivial amounts to proving that every solution of the Beltrami equation

$$\frac{\partial f}{\partial \bar{z}} = \mu(z) \frac{\partial f}{\partial z} \tag{14.50}$$

with $f_z, f_{\bar{z}} \in L^p(\mathbb{C})$ is a constant.

Assume therefore that f has partial derivatives in L^p and solves (14.50), $1 < p < \infty$. From the definition of $VMO(\hat{\mathbb{C}})$ there is a constant $\mu(\infty)$, defined in (14.49), such that

$$\lim_{R \to \infty} \frac{1}{\pi R^2} \int_{\mathbb{D}(0,R)} |\mu(z) - \mu_\infty| \, dz = 0 \tag{14.51}$$

Indeed, for each $\varphi \in C_0^\infty(\mathbb{C})$, we can write

$$\frac{1}{\pi R^2} \int_{\mathbb{D}(0,R)} |\mu - \mu(\infty)| \leqslant \frac{1}{\pi R^2} \int_{\mathbb{D}(0,R)} |\mu - \mu_{\mathbb{D}(0,R)}| + |\mu_{\mathbb{D}(0,R)} - \mu(\infty)|$$

$$\leqslant \|\mu - \varphi\|_{BMO} + \frac{1}{\pi R^2} \int_{\mathbb{D}(0,R)} |\varphi - \varphi_{\mathbb{D}(0,R)}|$$

$$+ |\mu_{\mathbb{D}(0,R)} - \mu(\infty)|$$

As $\mu \in VMO(\hat{\mathbb{C}})$, we can make this first term, $\|\mu - \varphi\|_{BMO}$, as small as we like with a suitable choice of φ. Clearly, the remaining two terms tend to 0 as well when $R \to \infty$.

Next note that $|\mu(\infty)| < 1$, and consider the linear map $\phi(z) = z - \mu(\infty)\bar{z}$. Write $g = f \circ \phi$ so that $\nabla g \in L^p(\mathbb{C})$ and so that g satisfies the Beltrami equation $g_{\bar{z}} = \tilde{\mu} g_z$, where

$$\tilde{\mu} = \frac{\mu \circ \phi - \mu_\infty}{1 - \overline{\mu_\infty} \mu \circ \phi}$$

Clearly, $\widetilde{\mu} \in VMO(\mathbb{C})$ with $\|\widetilde{\mu}\|_\infty < 1$, and $\widetilde{\mu}$ satisfies the relation analogous to (14.51),

$$\lim_{R \to \infty} \frac{1}{\pi R^2} \int_{\mathbb{D}(0,R)} |\widetilde{\mu}| \, dz = 0$$

Instead of introducing new notation, we now simply assume $\mu(\infty) = 0$.

The distortion function K, defined by $|Dg(z)|^2 = K(z)J(z,g)$, satisfies the pointwise estimates $1 \leqslant K(z) = 1 + 2|\mu(z)|(1 - |\mu(z)|)^{-1} \leqslant 1 + \frac{2}{1-k}|\mu(z)|$. Therefore

$$\lim_{R \to \infty} \frac{1}{\pi R^2} \int_{\mathbb{D}(0,R)} K(z) \, dz = 1,$$

and since $J(z,g) \in L^q(\mathbb{C})$, $1 < q = \frac{p}{2} < \infty$, by assumption Theorem 14.4.7 implies that g is constant. Therefore the kernel of $\mathbf{I} - \mu \mathcal{S}$ must be trivial. $\square$

We now present the main result of this section.

Theorem 14.5.2. *Suppose that* $\mu \in VMO(\hat{\mathbb{C}})$ *and that* $\|\mu\|_\infty < 1$. *Then the operators* $\mathbf{I} - \mu \mathcal{S}$ *and* $\mathbf{I} - \mathcal{S}\mu$ *are invertible in* $L^p(\mathbb{C})$ *for all* $p \in (1, \infty)$.

Proof. The above argument shows that for all $p \in (1, \infty)$ the operator $\mathbf{I} - \mu \mathcal{S}$ is Fredholm on L^p with index zero, and by Theorem 14.5.1 it is also injective when $p \geqslant 2$. This proves that $\mathbf{I} - \mu \mathcal{S}$ is invertible for $2 \leqslant p < \infty$.

For the case $1 < p \leqslant 2$ we note that the adjoint to $\mathbf{I} - \mu \mathcal{S} : L^p(\mathbb{C}) \to L^p(\mathbb{C})$ is the invertible operator

$$\mathbf{I} - \mathcal{S}^{-1}\bar{\mu} : L^{p'}(\mathbb{C}) \to L^{p'}(\mathbb{C})$$

with $p + p' = pp'$. The identity

$$\overline{(\mathbf{I} - \mathcal{S}\mu)\omega} = (\mathbf{I} - \mathcal{S}^{-1}\bar{\mu})\bar{\omega}$$

shows that $\mathbf{I} - \mathcal{S}\mu : L^{p'}(\mathbb{C}) \to L^{p'}(\mathbb{C})$ is also invertible. So too therefore are the operators

$$\mathbf{I} - \mu\mathcal{S} = \mathcal{S}^{-1}(\mathbf{I} - \mathcal{S}\mu)\mathcal{S} : L^{p'}(\mathbb{C}) \to L^{p'}(\mathbb{C})$$
$$\mathbf{I} - \mathcal{S}\mu = \mathcal{S}(\mathbf{I} - \mu\mathcal{S})\mathcal{S}^{-1} : L^p(\mathbb{C}) \to L^p(\mathbb{C})$$

The theorem is proved. $\square$

As a last related aspect of weak quasiregularity, we see that in the VMO setting we again obtain strong regularity consequences; Theorem 14.5.2 and Lemma 14.4.2 have the following immediate corollary.

Corollary 14.5.3. *Assume that* $q > 1$ *and* $f \in W^{1,q}_{loc}(\Omega)$ *has dilatation* μ *and that* f *is weakly quasiregular. If* μ *is in* $VMO(\Omega)$, *then* f *is quasiregular.*

14.6 Bounds for the Beurling Transform

The L^p-estimates of the previous sections fail in the limiting case $p \to 1$. However, the results from Chapter 13 still have consequences for the Beurling operator around the space $L^1(\mathbb{C})$. We briefly discuss these in this last section.

As we have mentioned on several occasions, the exact value of the L^p-norm of the Beurling transform remains a challenging open problem. However, correct bounds in the extremal cases $p = \infty$ and $p = 1$ are possible to achieve. As a Calderón-Zygmund operator, $\mathcal{S}$ cannot be bounded on L^1 or L^∞. Instead, $\mathcal{S}$ takes the smaller space $L^1 \log L^1$ to L^1 while images of L^∞-functions are exponentially integrable BMO functions.

In these borderline cases we wish to establish precise estimates, following [21]. We begin with the $L^1 \log L^1$-bounds.

Theorem 14.6.1. *Suppose w is a nonnegative function with support in a measurable set $E \subset B$. Then*

$$\int_{B \setminus E} |\mathcal{S}w| \leqslant \int_E w(z) \log\left(\frac{w(z)}{|w|_B}\right) \tag{14.52}$$

for every disk B containing the set E. Here $|w|_B = \frac{1}{|B|}\int_B |w|$. Moreover, we have the equality, e.g., when $w = \chi_E$, $E = \mathbb{D}(r)$ and $B = \mathbb{D}$.

Proof. With a change of variables one can assume that $B = \mathbb{D}$. We first choose a function μ such that $|\mu| = \chi_{\mathbb{D} \setminus E}$ and

$$\mu(z)\mathcal{S}w(z) = |\mathcal{S}w(z)|, \quad \text{for almost every } z \in \mathbb{D} \setminus E \tag{14.53}$$

For each λ we then solve the Beltrami equation

$$f_{\bar{z}}(z) = \lambda \mu(z) f_z(z) \tag{14.54}$$

Let $f = f^\lambda$ be the principal solution to (14.54). We know from Theorem 13.1.3 that

$$\frac{1}{\pi}\int_E w(z) J(z,f) \leqslant \left(\frac{1}{\pi}\int_E w(z)^K\right)^{1/K}, \quad K = \frac{1+|\lambda|}{1-|\lambda|} \tag{14.55}$$

Next we develop the estimate (14.55) in powers of λ. Recall that $J(z,f) = |f_z|^2 - |f_{\bar{z}}|^2 = |1 + \mathcal{S}(f_{\bar{z}})|^2 - |f_{\bar{z}}|^2$. The identity can be combined with the representation $f_{\bar{z}} = (\mathbf{I} - \lambda \mu \mathcal{S})^{-1} \lambda \mu = \lambda \mu(z) + h_\lambda(z)$, where $\|h_\lambda\|_{L^2} \leqslant C|\lambda|^2$. This means that

$$\frac{1}{\pi}\int_E w(z) J(z,f) = \frac{1}{\pi}\int_E w(1 + 2\Re(\lambda \mathcal{S}\mu)) + \mathcal{O}(|\lambda|^2) \tag{14.56}$$

On the other hand,

$$\left(\frac{1}{\pi}\int_E w(z)^K\right)^{1/K} = \frac{1}{\pi}\int_E w + \frac{2|\lambda|}{\pi}\int_E w(z) \log\left(\frac{w(z)}{|w|_\mathbb{D}}\right) + \mathcal{O}(|\lambda|^2)$$

Comparing the coefficients of the terms linear in λ and choosing $0 < \lambda < 1$ gives

$$\frac{1}{\pi}\int_{\mathbb{D}\setminus E} \mu \mathcal{S}(w\chi_E) = \frac{1}{\pi}\int_E w\mathcal{S}\mu \leq \frac{1}{\pi}\int_E w(z)\log\left(\frac{w(z)}{|w|_\mathbb{D}}\right)$$

The choice (14.53) then gives the required estimate (14.52).

On the other hand, for $E = \mathbb{D}(0,r)$ we have $\mathcal{S}\chi_E(z) = \frac{r^2}{z^2}\chi_{\mathbb{C}\setminus E}$, and integrating we have

$$\int_{\mathbb{D}\setminus E} |\mathcal{S}(\chi_E)| = |E|\log\left(\frac{\pi}{|E|}\right)$$

We thus have functions for which the equality holds in (14.52). □

For the characteristic functions $w = \chi_E$ one can combine (14.52) with the simple estimate $\int_E |\mathcal{S}(\chi_E)| \leq |E|^{1/2}(\int_\mathbb{C} |\mathcal{S}(\chi_E)|^2)^{1/2} = |E|$ since $\mathcal{S}$ is an isometry on $L^2(\mathbb{C})$. This yields

$$\int_\mathbb{D} |\mathcal{S}(\chi_E)| \leq |E|\log\left(\frac{\alpha}{|E|}\right), \qquad E \subset \mathbb{D}, \tag{14.57}$$

where $\alpha = e\pi$. The general case of $w \in L^1\log L^1$ is studied in Theorem 14.6.4 below. But before that we describe the following counterpart in the space L^∞.

Theorem 14.6.2. *Let $B \subset \mathbb{C}$ be a disk. If w is a measurable function such that $|w(z)| \leq \chi_B(z)$ almost everywhere, then*

$$|\{z \in B : |\mathcal{S}\omega(z)| > t\}| \leq C_0(1+2\pi t)|B|e^{-t}, \qquad t > 0 \tag{14.58}$$

Proof. Write $E_t = \{z \in B : |\mathcal{S}\omega(z)| > t\}$. We may take $t \geq 1$ since for $t < 1$ the claim (14.58) is obvious. Let $n \in \mathbb{N}$ be determined by the condition

$$2\pi t < n \leq 1 + 2\pi t,$$

and for $k = 0, 1, \ldots, n-1$, set

$$A_k = \left\{z \in E_t : \frac{2\pi k}{n} \leq \arg \mathcal{S}\omega(z) < \frac{2\pi(k+1)}{n}\right\}$$

Then

$$\int_{A_k} |\mathcal{S}\omega| = e^{-2\pi ki/n}\int_{A_k} \mathcal{S}\omega + \int_{A_k}\left(e^{-i\arg\mathcal{S}\omega} - e^{-2\pi ki/n}\right)\mathcal{S}\omega$$

$$\leq \left|\int_{A_k} \mathcal{S}\omega\right| + \frac{2\pi}{n}\int_{A_k} |\mathcal{S}\omega|$$

Hence by (14.57) and a change of variables,

$$t\left(1-\frac{2\pi}{n}\right)|A_k| \leq \left(1-\frac{2\pi}{n}\right)\int_{A_k} |\mathcal{S}\omega| \leq \left|\int_B \omega \mathcal{S}\chi_{A_k}\right| \leq |A_k|\log\left(\frac{e|B|}{|A_k|}\right)$$

14.6. BEURLING TRANSFORM BOUNDS

We see that $|A_k| \leqslant e|B|e^{t(\frac{2\pi}{n}-1)} \leqslant e^2|B|e^{-t}$. Summing up,

$$|E_t| \leqslant \sum_{k=0}^{n-1} |A_k| \leqslant C_0(1 + 2\pi t)|B|e^{-t}, \qquad t > 0,$$

where $C_0 = e^2$. □

The estimate (14.58) has the correct growth. For example, the function $w = (z/\bar{z})\chi_{\mathbb{D}}(z)$ satisfies
$$\mathcal{S}w = (1 + 2\log|z|)\chi_{\mathbb{D}}(z)$$
It remains an open problem whether the linear term $2\pi t$ in Theorem 14.6.2 can be replaced by a constant. This is true for the slightly weaker estimate

$$|\{z \in B : |\Re\,\mathcal{S}w(z)| > t\}| \leqslant 2\alpha|B|e^{-t}, \qquad (14.59)$$

which follows directly as above from (14.57).

Integrating the distribution function and applying Theorem 14.6.2 now gives the precise exponential integrability.

Theorem 14.6.3. *Suppose B is a disk and w a function supported in B. If $\|w\|_\infty < 1$, then*

$$\int_B e^{|\mathcal{S}w|} < \infty$$

For the function $w_0 = \frac{\bar{z}}{z}\chi_{\mathbb{D}}(z)$ with $\|w_0\|_\infty = 1$, the integral diverges on $B = \mathbb{D}$.

Finally, we return to the $L^1 \log L^1$-bounds.

Theorem 14.6.4. *For each $\delta > 1$ there is a constant $M(\delta) < \infty$ such that*

$$\int_B |\mathcal{S}v| \leqslant \delta \int_B |v(z)| \log\left(1 + M(\delta)\frac{|v(z)|}{|v|_B}\right)$$

for all disks B and for all functions $v \in L^1 \log L^1(\mathbb{C})$.

The reader should note that is not possible to have $M(\delta) \leqslant C < \infty$ independently of $\delta > 1$. We will prove this in Example 14.6.5 below.

Proof. Let ω be a function, unimodular in B and vanishing in $\mathbb{C}\setminus B$, such that

$$\int_B |\mathcal{S}v| = \int_B \omega \mathcal{S}v = |v|_B \int_B \mathcal{S}(\omega)\frac{v}{|v|_B}$$

We then apply the elementary inequality $ab \leqslant a\log(1+a) + \exp(b) - 1$. Since we have from Theorem 14.6.2

$$\int_B e^{|\mathcal{S}\omega|/\delta} - 1 \leqslant |B|\frac{C_0}{\delta - 1}\left(1 + \frac{2\pi\delta}{\delta - 1}\right) = M_1(\delta)|B|,$$

it follows that
$$\int_B |\mathcal{S}v| \leq M_1(\delta) \int_B |v| + \delta \int_B |v| \log\left(1 + \delta \frac{|v|}{|v|_B}\right)$$

Now define $E_0 = \{z \in B : |v(z)| < \frac{1}{e}|v|_B\}$. As $t \mapsto t \log \frac{1}{t}$ is increasing on $(0, \frac{1}{e})$, $-e \int_{E_0} |v| \log(\frac{|v|}{|v|_B}) \leq |v|_B |E_0| \leq \int_B |v|$. Thus
$$\int_B |v| \log\left(1 + \delta \frac{|v|}{|v|_B}\right) \leq \int_{B \setminus E_0} |v| \log\left(\frac{|v|}{|v|_B}(e+\delta)\right) + \int_{E_0} |v| \log\left(1 + \frac{\delta}{e}\right)$$
$$\leq \int_B |v| \log\left(M_2 \frac{|v|}{|v|_B}\right), \tag{14.60}$$
where $M_2 = e^2 + e\delta$. In conclusion, if $M = M_2 \exp(M_1(\delta))$,
$$\int_B |\mathcal{S}v| \leq \delta \int_B |v| \log\left(M \frac{|v|}{|v|_B}\right) \leq \delta \int_B |v| \log\left(1 + M \frac{|v|}{|v|_B}\right),$$
completing the estimates. □

We complete the argument with the following example.

Example 14.6.5. *For each $M < \infty$, there is an $\eta = \eta(M) > 0$ and a nonnegative function $v \in L \log L(\mathbb{D})$ such that*
$$\int_\mathbb{D} |\mathcal{S}v| > (1+\eta) \int_\mathbb{D} |v(z)| \log\left(1 + M \frac{|v(z)|}{|v|_\mathbb{D}}\right)$$

Proof. It suffices to show (see (14.60)) that for no $M < \infty$ does the estimate
$$\int_\mathbb{D} |\mathcal{S}v| \leq \int_\mathbb{D} |v(z)| \log\left(M \frac{|v(z)|}{|v|_\mathbb{D}}\right)$$
hold for all non-negative functions $v \in L \log L(\mathbb{D})$. Let $A_\varepsilon = \{z : \varepsilon < |z| < 1\}$ and
$$v(z) = \frac{1}{|z|^2} \chi_{A_\varepsilon}(z)$$
Then $v = f_{\bar{z}}$, where $f(z) = \frac{1}{z} \log(\frac{|z|^2}{\varepsilon^2}) \chi_{A_\varepsilon}(z) + \frac{1}{z} \log(\frac{1}{\varepsilon^2}) \chi_{\mathbb{C} \setminus \mathbb{D}}(z)$. In particular, for $z \in A_\varepsilon$ we have $\mathcal{S}v = f_z = \frac{1}{z^2}(1 + \log(\frac{\varepsilon^2}{|z|^2}))$. A calculation gives
$$\frac{1}{\pi} \int_\mathbb{D} |\mathcal{S}v| = 2(\log \frac{1}{\varepsilon})^2 - 2 \log \frac{1}{\varepsilon} + \frac{1}{2},$$
while
$$\frac{1}{\pi} \int_\mathbb{D} |v(z)| \log\left(M \frac{|v(z)|}{|v|_\mathbb{D}}\right) = 2(\log \frac{1}{\varepsilon})^2 + 2M \log \frac{1}{\varepsilon} - 2(\log \frac{1}{\varepsilon}) \log \log(\frac{1}{\varepsilon^2})$$

Therefore (14.6.5) would lead to $\log \log(\frac{1}{\varepsilon^2}) \leq M + 1$, which is not possible for arbitrary $\varepsilon > 0$. □

Chapter 15

Schauder Estimates for Beltrami Operators

The fundamental ideas concerning the Hölder regularity of solutions to PDEs started with J. Schauder's work [321, 322] in the early 1930s. He was largely concerned with linear, quasilinear and nonlinear elliptic equations of second order. Suppose we have a linear elliptic equation

$$\sum_{j,l=1}^{n} a_{jl}(x) u_{x_j x_l} = h(x), \qquad x \in \Omega \subset \mathbb{R}^n,$$

where the coefficients and the right hand side are Hölder-continuous of exponent $0 < \alpha < 1$; that is, $a_{jl}, h \in C^\alpha_{loc}(\Omega)$. Schauder proved that the solutions u lie in the space $C^{2,\alpha}(\Omega)$ using singular integrals acting in the space $C^\alpha(\mathbb{R}^n)$; the L^p-bounds for singular integrals came later in 1952 in the pioneering work of Calderón and Zygmund [86]. Schauder's basic philosophy was that equations with C^α-coefficients can be viewed locally as perturbations of equations with constant coefficients. We shall adhere to this philosophy when studying first-order Beltrami-type equations. The key tool will be the Beurling operator acting on $C^\alpha(\mathbb{C})$, see Theorem 4.7.1.

Our fundamental Hölder estimate, from which we shall derive a variety of results, concerns the global solution to the inhomogeneous Beltrami equation

$$\frac{\partial f}{\partial \bar{z}} - \mu(z) \frac{\partial f}{\partial z} - \nu(z) \overline{\frac{\partial f}{\partial z}} = h(z), \qquad (15.1)$$

where $h \in L^s(\mathbb{C}) \cap C^\alpha(\mathbb{C})$ for some $1 \leqslant s \leqslant 2$ and $0 < \alpha < 1$, while the coefficients $\mu, \nu \in C^\alpha(\mathbb{C})$. We also assume the ellipticity $|\mu(z)| + |\nu(z)| \leqslant k < 1$ for $z \in \mathbb{C}$.

According to Theorem 14.2.2, (15.1) always admits a solution f with square-integrable derivatives. The point is that now the derivatives also belong to the Hölder space $C^\alpha(\mathbb{C})$. The precise result, which exhibits the correct dependence on $\|h\|_s, \|h\|_{C^\alpha}, \|\mu\|_{C^\alpha}$ and $\|\nu\|_{C^\alpha}$, is contained in the following global estimate.

Theorem 15.0.6. *Suppose $h \in L^s(\mathbb{C}) \cap C^\alpha(\mathbb{C})$, where $1 \leqslant s \leqslant 2$ and $0 < \alpha < 1$, and f is a solution to (15.1) with $f_{\bar{z}}, f_z \in L^2(\mathbb{C})$. Then*

$$\|f_z\|_{C^\alpha} + \|f_{\bar{z}}\|_{C^\alpha} \leqslant C_k(\alpha, s) \left[\|h\|_{C^\alpha} + (\|\mu\|_{C^\alpha} + \|\nu\|_{C^\alpha})^{1+2/\alpha s} \|h\|_s\right] \quad (15.2)$$

Note that the estimate in (15.2) is invariant under rescaling of the z-variable. In fact, such invariance requires this specific form for the inequality.

The main purpose of this chapter is to establish this fundamental estimate. It has far-reaching consequences. One of them is the following theorem.

$C^{\ell,\alpha}$-Regularity of Solutions to Quasilinear Equations

Theorem 15.0.7. *Suppose that $f \in W^{1,2}_{loc}(\Omega, \Omega')$ is the solution to*

$$\frac{\partial f}{\partial \bar{z}} - \mu(z,f)\frac{\partial f}{\partial z} - \nu(z,f)\overline{\frac{\partial f}{\partial z}} = h, \qquad z \in \Omega, \quad (15.3)$$

where the coefficients $\mu, \nu \in C^{\ell,\alpha}_{loc}(\Omega \times \Omega')$ and satisfy the ellipticity condition

$$|\mu(z,\zeta)| + |\nu(z,\zeta)| \leqslant k < 1, \qquad (z,\zeta) \in \Omega \times \Omega', \quad (15.4)$$

while the data $h \in C^{\ell,\alpha}_{loc}(\Omega)$. Then f belongs to $C^{\ell+1,\alpha}_{loc}(\Omega)$.

15.1 Examples

The $C^{\ell,\alpha}$-regularity fails at the borderline cases $\alpha = 0$ and $\alpha = 1$, and we start by describing a family of examples that illustrate this phenomenon. More precisely, let $C^{\ell,0}$ or $C^{\ell,1}$ stand for the space of functions whose ℓ^{th}-order derivatives are continuous or Lipschitz-continuous, respectively. For $\alpha = 0$ or $\alpha = 1$, the $C^{\ell+1,\alpha}$-regularity fails even if we assume slightly more regularity of the coefficients and the data h. Namely, for each non-negative integer ℓ, there is a solution to the elliptic Beltrami equation $f_{\bar{z}} = \mu(z)f_z$ with $\mu \in C^\ell(\mathbb{C})$, whose $(\ell+1)^{\text{th}}$-order derivatives are unbounded.

Of course, it suffices to construct an example on any small ball $\Omega = \mathbb{D}(0,r)$. For $\ell = 0, 1, 2, \ldots$ set

$$f(z) = z + z^{\ell+1}\sqrt{\log|z|}$$

For r sufficiently small, f is a homeomorphism contained in $W^{1,2}(\Omega)$. We compute

$$\frac{\partial f}{\partial \bar{z}}(z) = \frac{z^{\ell+1}}{4\bar{z}\sqrt{\log|z|}}, \qquad \text{hence } f_{\bar{z}} \in C^\ell(\Omega)$$

$$\frac{\partial f}{\partial z}(z) = 1 + (\ell+1)z^\ell\sqrt{\log|z|} + \frac{z^\ell}{4\sqrt{\log|z|}}, \qquad \text{hence } f_z \notin C^\ell(\Omega)$$

As a matter of fact, the ℓth-order derivatives of f_z are not bounded. However,

$$\mu(z) = \frac{f_{\bar{z}}}{f_z} \in C^\ell(\Omega)$$

Obviously, $f \notin C^{\ell,1}_{loc}(\Omega)$.

In the case when μ is only continuous, a simpler example is provided by $f(z) = -z \log|z|^2$, where $|z| \leqslant r = e^{-2}$. Then $f : \mathbb{D}(0,r) \to \mathbb{D}(0,4r)$ is a homeomorphism and

$$\frac{\partial f}{\partial \bar{z}} = -\frac{z}{\bar{z}}, \qquad \frac{\partial f}{\partial z} = -1 - \log|z|^2, \qquad \mu_f(z) = \frac{z}{\bar{z}(1 + \log|z|^2)}$$

Thus f is quasiconformal with a continuous Beltrami coefficient, and yet f is not C^1.

15.2 The Beltrami Equation with Constant Coefficients

Here we establish the existence solutions for the inhomogeneous Beltrami equation with constant coefficients and data in the Hölder class.

Theorem 15.2.1. *Let $\mu_0, \nu_0 \in \mathbb{C}$ such that*

$$|\mu_0| + |\nu_0| \leqslant k < 1$$

Then for each $h \in C^\alpha(\mathbb{C})$, $0 < \alpha < 1$, the equation

$$\frac{\partial f}{\partial \bar{z}} - \mu_0 \frac{\partial f}{\partial z} - \nu_0 \overline{\frac{\partial f}{\partial z}} = h \tag{15.5}$$

has a solution $f \in C^{1,\alpha}(\mathbb{C})$ that is unique up to a linear function and, moreover, satisfies the estimate

$$\|Df\|_{C^\alpha} = \|f_z\|_{C^\alpha} + \|f_{\bar{z}}\|_{C^\alpha} \leqslant \frac{20}{\alpha(1-\alpha)(1-k)^3} \|h\|_{C^\alpha} \tag{15.6}$$

Proof. By a linear change of variables, we shall reduce the result to the inhomogeneous Cauchy-Riemann equation. Namely, for any given constants $a, b \in \mathbb{D}$ we may use the transformation

$$f(\zeta) = \frac{g(z) - b\overline{g(z)}}{1 - |b|^2}, \qquad \zeta = z + a\bar{z}, \tag{15.7}$$

or in other words,

$$g(z) = f(\zeta) + b\overline{f(\zeta)}, \qquad z = \frac{\zeta - a\bar{\zeta}}{1 - |a|^2} \tag{15.8}$$

Accordingly, if $h \in C^\alpha(\mathbb{C})$ is given, let $u(z) = h(\zeta) = h(z + a\bar{z})$.
One can now use Theorem 4.7.2 to find a solution $g \in C^{1,\alpha}(\mathbb{C})$ to

$$\frac{\partial g}{\partial \bar{z}}(z) = u(z) + ab\,\overline{u(z)}$$

with

$$\|Dg\|_{C^\alpha} \leqslant \frac{5\|u\|_{C^\alpha}}{\alpha(1-\alpha)} \leqslant \frac{5(1+|a|)^\alpha \|h\|_{C^\alpha}}{\alpha(1-\alpha)} \qquad (15.9)$$

We are to choose a and b so that the function f, defined through (15.7) and (15.8), satisfies (15.5). Hence we let

$$a = \frac{-2\mu_0}{1 + |\mu_0|^2 - |\nu_0|^2 + \sqrt{(1 + |\mu_0| - |\nu_0|^2)^2 - 4|\mu_0|^2}}$$

$$b = \frac{-2\nu_0}{1 + |\nu_0|^2 - |\mu_0|^2 + \sqrt{(1 + |\nu_0| - |\mu_0|^2)^2 - 4|\mu_0|^2}}$$

Using the ellipticity bound $|\mu_0| + |\nu_0| \leqslant k$, we easily verify that indeed

$$|a| \leqslant \frac{|\mu_0|}{1 - |\nu_0|^2} \leqslant k, \qquad |b| \leqslant \frac{|\nu_0|}{1 - |\mu_0|^2} \leqslant k$$

We are left with the task of estimating the Hölder norms of the derivatives. An elementary manipulation shows that

$$f_{\bar{\zeta}}(\zeta) = \frac{g_{\bar{z}} + a(b-1)g_z - b\overline{g_z}}{(1 - |a|^2)(1 - |b|^2)},$$

which gives

$$\|f_{\bar{\zeta}}\|_{C^\alpha} \leqslant \frac{(1 + |a||b|)\|g_{\bar{z}}\|_{C^\alpha} + (|a| + |b|)\|g_z\|_{C^\alpha}}{(1 - |a|^2)(1 - |b|^2)(1 - |a|)^\alpha}$$

Similarly,

$$\|f_\zeta\|_{C^\alpha} \leqslant \frac{(1 + |a||b|)\|g_z\|_{C^\alpha} + (|a| + |b|)\|g_{\bar{z}}\|_{C^\alpha}}{(1 - |a|^2)(1 - |b|^2)(1 - |a|)^\alpha},$$

so that

$$\|Df\|_{C^\alpha} \leqslant \frac{\|g_z\|_{C^\alpha} + \|g_{\bar{z}}\|_{C^\alpha}}{(1 - |a|)^{1+\alpha}(1 - |b|)},$$

which when combined with (15.9) completes the proof of the theorem. $\square$

15.3 A $C^{1,1}$-Partition of Unity

Here we give a construction of a partition of unity that is adapted to suit our subsequent proof of a version of Theorem 15.2.1 for more general coefficients.

Let $Q = [-1, 1) \times [-1, 1)$ denote a square that is left closed and right open whose center lies at the origin and has side length 2.

15.3. PARTITION OF UNITY

Consider a function $\eta \in C_0^{1,1}(\mathbb{C})$ defined at $z = x + iy$ by the rule

$$\eta(z) = \begin{cases} \cos^2(\pi x/2)\cos^2(\pi y/2), & z \in Q \\ 0, & z \notin Q \end{cases} \tag{15.10}$$

We note that the first-order derivatives of η are Lipschitz-continuous, and in fact we have the following easy estimates:

$$|\eta_z| = |\eta_{\bar{z}}| \leq \pi, \qquad |\eta_{zz}| = |\eta_{\bar{z}\bar{z}}| \leq \frac{3\pi^2}{4}, \qquad |\eta_{z\bar{z}}| \leq \frac{\pi^2}{4} \tag{15.11}$$

Now we fix a parameter t whose specific value will be chosen in applications. Next to each integer lattice point $\mathbf{j} = m + in \in \mathbb{Z} + i\mathbb{Z}$ we denote by

$$Q_\mathbf{j} = t\mathbf{j} + tQ = \{z = x + iy : -t \leq x - tm < t, \ -t \leq y - tn < t\} \tag{15.12}$$

Each point $z \in \mathbb{C}$ is covered by exactly four squares, and this observation can be recorded as

$$\sum_{\mathbf{j} \in \mathbb{Z} + i\mathbb{Z}} \chi_{Q_\mathbf{j}}(z) \equiv 4 \tag{15.13}$$

We now set out to define a specific partition of unity subordinate to this family of squares. For an index $\mathbf{j} = m + in$, we set

$$\phi^\mathbf{j}(z) = \eta\left(\mathbf{j} + \frac{z}{t}\right) = \cos^2\left(\frac{\pi x}{2t} + \frac{\pi m}{2}\right) \cos^2\left(\frac{\pi y}{2t} + \frac{\pi n}{2}\right), \tag{15.14}$$

while $\phi^\mathbf{j}$ vanishes outside $Q_\mathbf{j}$. The infinite series $\sum_\mathbf{j} \phi^\mathbf{j}$ has at most four nonzero terms. In fact, we can compute

$$\begin{aligned}
\sum_\mathbf{j} \phi^\mathbf{j}(z) &= \sum_{m \& n \text{ odd}} + \sum_{m \& n \text{ even}} + \sum_{m \text{ odd \& } n \text{ even}} + \sum_{m \text{ even \& } n \text{ odd}} \\
&= \sin^2\left(\frac{\pi x}{2t}\right)\sin^2\left(\frac{\pi y}{2t}\right) + \sin^2\left(\frac{\pi x}{2t}\right)\cos^2\left(\frac{\pi y}{2t}\right) \\
&\quad + \cos^2\left(\frac{\pi x}{2t}\right)\sin^2\left(\frac{\pi y}{2t}\right) + \cos^2\left(\frac{\pi x}{2t}\right)\cos^2\left(\frac{\pi y}{2t}\right) \\
&= 1
\end{aligned}$$

We also record the following estimates of the first derivatives of $\phi^\mathbf{j}$:

$$|\phi^\mathbf{j}_z| = |\phi^\mathbf{j}_{\bar{z}}| \leq \frac{\pi}{t}, \qquad |\phi^\mathbf{j}_{zz}| = |\phi^\mathbf{j}_{\bar{z}\bar{z}}| \leq \frac{3\pi^2}{4t^2}, \qquad |\phi^\mathbf{j}_{z\bar{z}}| \leq \frac{\pi^2}{4t^2} \tag{15.15}$$

The C^α-estimates of the derivatives of the functions $\phi^\mathbf{j}$ follow from the general geometric fact that if a function $\eta \in \text{Lip}(\mathbb{C})$ is supported in a square of side length s, then

$$\|\eta\|_{C^\alpha} \leq 5(\|\eta_z\|_\infty + \|\eta_{\bar{z}}\|_\infty) s^{1-\alpha}$$

Therefore, in view of (15.15), we have

$$\|\phi^\mathbf{j}\|_{C^\alpha} \leq \frac{10\pi}{t^\alpha} \tag{15.16}$$

$$\|\phi^\mathbf{j}_z\|_{C^\alpha} = \|\phi^\mathbf{j}_{\bar{z}}\|_{C^\alpha} \leq \frac{5\pi^2}{t^{1+\alpha}} \tag{15.17}$$

Given a function $f: \mathbb{C} \to \mathbb{C}$, we can decompose f as

$$f = \sum_{\mathbf{j} \in \mathbb{Z}+i\mathbb{Z}} f^{\mathbf{j}}, \qquad f^{\mathbf{j}} = \phi^{\mathbf{j}} f \tag{15.18}$$

The components $f^{\mathbf{j}}$ of the function f are supported in the square $Q_{\mathbf{j}}$. We also have a similar decomposition of the derivative,

$$f_z = \sum_{\mathbf{j} \in \mathbb{Z}+i\mathbb{Z}} f_z^{\mathbf{j}}, \qquad f_{\bar{z}} = \sum_{\mathbf{j} \in \mathbb{Z}+i\mathbb{Z}} f_{\bar{z}}^{\mathbf{j}}, \tag{15.19}$$

where $f_z^{\mathbf{j}} = \frac{\partial f^{\mathbf{j}}}{\partial z}$ and $f_{\bar{z}}^{\mathbf{j}} = \frac{\partial f^{\mathbf{j}}}{\partial \bar{z}}$, respectively. We now prove the following lemma.

Lemma 15.3.1. *Suppose that $f \in C^{1,\alpha}(\mathbb{C})$, $0 < \alpha \leqslant 1$. Then*

$$\|f_z\|_{C^\alpha} \leqslant 8 \max_{\mathbf{j} \in \mathbb{Z}+i\mathbb{Z}} \|f_z^{\mathbf{j}}\|_{C^\alpha}, \qquad \|f_{\bar{z}}\|_{C^\alpha} \leqslant 8 \max_{\mathbf{j} \in \mathbb{Z}+i\mathbb{Z}} \|f_{\bar{z}}^{\mathbf{j}}\|_{C^\alpha} \tag{15.20}$$

Proof. Given two distinct points $z_1, z_2 \in \mathbb{C}$, there are four squares that contain z_1 and four squares that contain z_2 (these two families of squares may not be disjoint). Therefore we can find 8 indices $\mathbf{j}_1, \mathbf{j}_2, \ldots, \mathbf{j}_8 \in \mathbb{Z} + i\mathbb{Z}$ such that

$$f_z(z_1) = \sum_{i=1}^{8} f_z^{\mathbf{j}_i}(z_1), \qquad f_z(z_2) = \sum_{i=1}^{8} f_z^{\mathbf{j}_i}(z_2)$$

Then

$$\frac{|f_z(z_1) - f_z(z_2)|}{|z_1 - z_2|^\alpha} \leqslant \sum_{i=1}^{8} \frac{|f_z^{\mathbf{j}_i}(z_1) - f_z^{\mathbf{j}_i}(z_2)|}{|z_1 - z_2|^\alpha}$$

$$\leqslant \sum_{i=1}^{8} \|f_z^{\mathbf{j}_i}\|_{C^\alpha} \leqslant 8 \max_{\mathbf{j} \in \mathbb{Z}+i\mathbb{Z}} \|f_z^{\mathbf{j}}\|_{C^\alpha}$$

Thus $\|f_z\|_{C^\alpha} \leqslant 8 \max_{\mathbf{j}} \|f_z^{\mathbf{j}}\|_{C^\alpha}$. In the same way we achieve the desired estimates for the other derivative $f_{\bar{z}}$. □

15.4 An Interpolation

We will need the following interpolation lemma in our subsequent discussion.

Lemma 15.4.1. *Let $1 \leqslant s \leqslant q \leqslant \infty$ and $0 < \alpha < 1$. Then, assuming that all the norms involved in the estimate below are finite, we have*

$$\|F\|_\infty \leqslant 2\|F\|_s^{\frac{\alpha s}{2+\alpha s}} \|F\|_{C^\alpha}^{\frac{2}{2+\alpha s}} \leqslant 2\left(\|F\|_s + \|F\|_{C^\alpha}\right) \tag{15.21}$$

$$\|F\|_q \leqslant 2\|F\|_s^{\frac{2s+\alpha sq}{2q+\alpha sq}} \|F\|_{C^\alpha}^{\frac{2q-2s}{2q+\alpha sq}} \leqslant 2\left(\|F\|_s + \|F\|_{C^\alpha}\right) \tag{15.22}$$

Proof. Given a disk $\mathbb{D}(z, R)$, we have

$$|F(z)| \leqslant \inf_{\mathbb{D}(z,R)} |F| + R^\alpha \|F\|_{C^\alpha}$$
$$\leqslant (\pi R^2)^{-1/s} \|F\|_s + R^\alpha \|F\|_{C^\alpha},$$

and so we choose R such that

$$R^{\alpha+2/s} = \|F\|_s / \|F\|_{C^\alpha}$$

to obtain (15.21). Now the easy estimate

$$\|F\|_q \leqslant \|F\|_\infty^{1-s/q} \|F\|_s^{s/q}$$

gives (15.22). □

15.5 Hölder Regularity for Variable Coefficients

We fix $0 < \alpha < 1$ and are now going to consider the inhomogeneous Beltrami equation with C^α-coefficients,

$$\frac{\partial f}{\partial \bar{z}}(z) - \mu(z) \frac{\partial f}{\partial z}(z) - \nu(z) \overline{\frac{\partial f}{\partial z}}(z) = h(z), \qquad z \in \mathbb{C}, \qquad (15.23)$$

with data $h \in C^\alpha(\mathbb{C})$ and

$$|\mu(z)| + |\nu(z)| \leqslant k < 1$$

We denote
$$\mathbf{M}_\alpha = \mathbf{M}_\alpha(\mu, \nu) = \|\mu\|_{C^\alpha} + \|\nu\|_{C^\alpha} \qquad (15.24)$$

In order to ensure existence and uniqueness we shall have to assume that our data h in fact lies in some Lebesgue space $L^s(\mathbb{C})$ for some $1 \leqslant s \leqslant 2$. It follows from the interpolation in Lemma 15.4.1 that then $h \in L^q(\mathbb{C})$ for every $q \in [s, \infty]$. In particular, $h \in L^2(\mathbb{C})$, and we have a unique solution to (15.23) satisfying $\|f_z\|_2 + \|f_{\bar{z}}\|_2 < \infty$.

For $\mu, \nu, h \in C_0^\infty(\mathbb{C})$ in (15.23), it is easy to see that in fact $f \in C^\infty(\mathbb{C})$; one can for instance repeat the argument of Lemma 5.2.1 applied to the equation $\sigma_{\bar{z}} = \mu \sigma + \nu \overline{\sigma_z} + h$. However, the quantitative C^α-bounds need more care.

We begin the proof of Theorem 15.0.6 with a few observations. First, note that if $\mathbf{M}_\alpha = 0$, then the coefficients μ, ν are constant, a case we have already dealt with in Theorem 15.2.1. Hence we may assume $\mathbf{M}_\alpha > 0$. Furthermore, it suffices to prove the theorem under the extra assumption that $f \in C^{1,\alpha}(\mathbb{C})$, for we can approximate the equation by

$$f^n_{\bar{z}} - \mu_n(z) f^n_z - \nu_n(z) \overline{f^n_z} = h_n, \qquad n = 1, 2, \ldots$$

This is because we may construct $h_n \in C_0^\infty(\mathbb{C})$ and $h_n \to h$ in $L^2(\mathbb{C})$ with $\|h_n\|_{C^\alpha} \leq \|h\|_{C^\alpha}$ as in (4.49). Similarly, we could find $\mu_n, \nu_n \in C_0^\infty(\mathbb{C})$ with $\mu_n \to \mu$ and $\nu_n \to \nu$ uniformly on compact subsets of $\mathbb{C}$, while retaining the bounds $|\mu_n| + |\nu_n| \leq k$ and $\mathbf{M}_\alpha(\mu_n, \nu_n) \leq \mathbf{M}_\alpha(\mu, \nu)$. The solutions f^n have $Df^n \in L^2(\mathbb{C})$ and are C^∞-smooth. Moreover, $f_{\bar{z}}^n$ has compact support with $f_z^n = Sf_{\bar{z}}^n = \mathcal{O}(1/z^2)$ at infinity. In particular, $f^n \in C^{1,\alpha}(\mathbb{C})$. On the other hand, we can argue as in (5.16) to show that the derivatives $f_z^n \to f_z$, $f_{\bar{z}}^n \to f_{\bar{z}}$ weakly in $L^2(\mathbb{C})$, where f is a solution to (15.1). This is enough to see that if we can establish the uniform bound in (15.2) for f^n, we may pass to the limit to obtain the result.

Therefore we assume $f \in C^{1,\alpha}$. Having Lemma 15.3.1 in hand, we see that we need only estimate $\|Df^{\mathbf{j}}\|_{C^\alpha} = \|f_z^{\mathbf{j}}\|_{C^\alpha} + \|f_{\bar{z}}^{\mathbf{j}}\|_{C^\alpha}$ appropriately for each $\mathbf{j} \in \mathbb{Z} + i\mathbb{Z}$. We declare here the choice of parameter in this decomposition as

$$t = \left(\frac{\alpha(1-\alpha)(1-k)^3}{1 + 100\,\mathbf{M}_\alpha}\right)^{1/\alpha} < \infty \qquad (15.25)$$

Each "component" function $f^{\mathbf{j}} = f\phi^{\mathbf{j}}$ satisfies its own Beltrami equation,

$$f_{\bar{z}}^{\mathbf{j}} - \mu(z)f_z^{\mathbf{j}} - \nu(z)\overline{f_z^{\mathbf{j}}} = h_{\mathbf{j}}, \qquad z \in \mathbb{C}, \qquad (15.26)$$

where

$$h_{\mathbf{j}} = \phi^{\mathbf{j}} h + f\phi_{\bar{z}}^{\mathbf{j}} - \mu f \phi_z^{\mathbf{j}} - \nu \overline{f \phi_z^{\mathbf{j}}}$$

We shall view (15.26) as a perturbation of the constant coefficient case on the scale give by t. Accordingly, we set

$$\mu_0 = \mu(t\mathbf{j}), \qquad \nu_0 = \nu(t\mathbf{j}),$$

which are precisely the values of the C^α-functions μ and ν at the center of the square $Q_{\mathbf{j}}$. Then we observe that

$$f_{\bar{z}}^{\mathbf{j}} - \mu_0 f_z^{\mathbf{j}} - \nu_0 \overline{f_z^{\mathbf{j}}} = h_{\mathbf{j}} + (\mu - \mu_0)f_z^{\mathbf{j}} + (\nu - \nu_0)\overline{f_z^{\mathbf{j}}}$$

We have two very useful facts in hand. Namely, we expect each of the terms $(\mu - \mu_0)$ and $(\nu - \nu_0)$ to be small on $Q_{\mathbf{j}}$ and note that $f^{\mathbf{j}}$ is supported only on $Q_{\mathbf{j}}$. Using the $C^{1,\alpha}$-estimate in (15.6) for the Beltrami operator with constant coefficients, we find

$$\alpha(1-\alpha)(1-k)^3 \|Df^{\mathbf{j}}\|_{C^\alpha}$$
$$\leq 20\|h_{\mathbf{j}}\|_{C^\alpha} + 20\|(\mu - \mu_0)f_z^{\mathbf{j}} + (\nu - \nu_0)\overline{f_z^{\mathbf{j}}}\|_{C^\alpha}$$
$$\leq 20\|h_{\mathbf{j}}\|_{C^\alpha} + 20\left(\|\mu - \mu_0\|_{L^\infty(\mathbf{Q}_{\mathbf{j}})} + \|\nu - \nu_0\|_{L^\infty(\mathbf{Q}_{\mathbf{j}})}\right)\|f_z^{\mathbf{j}}\|_{C^\alpha} + 20\mathbf{M}_\alpha \|f_z^{\mathbf{j}}\|_\infty$$
$$\leq 20\|h_{\mathbf{j}}\|_{C^\alpha} + 20(1 + \sqrt{2})\, t^\alpha \mathbf{M}_\alpha \|f_z^{\mathbf{j}}\|_{C^\alpha}$$

since of course $|\mu(z) - \mu_0| \leq (\sqrt{2}t)^\alpha \|\mu\|_{C^\alpha}$, $|\nu(z) - \nu_0| \leq (\sqrt{2}t)^\alpha \|\nu\|_{C^\alpha}$ and $\|f^{\mathbf{j}}\|_\infty \leq t^\alpha \|f^{\mathbf{j}}\|_{C^\alpha}$. Here is a first moment where we make use of the particular

15.5. VARIABLE COEFFICIENTS

scale t chosen in (15.25). We want t so small that we can absorb the last term into the left hand side. Having done so, we arrive at the estimate

$$\|Df^{\mathbf{j}}\|_{C^\alpha} \leqslant \frac{40}{\alpha(1-\alpha)(1-k)^3} \|h_{\mathbf{j}}\|_{C^\alpha} \tag{15.27}$$

We need to estimate the C^α-norm of $h_{\mathbf{j}}$ by that of h. As before,

$$\begin{aligned}\|h_{\mathbf{j}}\|_{C^\alpha} &\leqslant \|h\,\phi^{\mathbf{j}}\|_{C^\alpha} + \|\phi^{\mathbf{j}}_{\bar{z}} f\|_{C^\alpha} + \|\mu\phi^{\mathbf{j}}_{\bar{z}} f\|_{C^\alpha} + \|\nu\overline{\phi^{\mathbf{j}}_{\bar{z}} f}\|_{C^\alpha} \\ &\leqslant \|h\|_{C^\alpha} + \|\phi^{\mathbf{j}}\|_{C^\alpha}\|h\|_\infty + (\|\phi^{\mathbf{j}}_{\bar{z}}\|_\infty + \|\phi^{\mathbf{j}}_{\bar{z}}\|_\infty)\|f\|_{C^\alpha} \\ &\quad + (\|\phi^{\mathbf{j}}_{\bar{z}}\|_\infty \|\mu\|_{C^\alpha} + \|\phi^{\mathbf{j}}_{\bar{z}}\|_{C^\alpha} + \|\phi^{\mathbf{j}}_{\bar{z}}\|_\infty\|\nu\|_{C^\alpha} + \|\phi^{\mathbf{j}}_{\bar{z}}\|_{C^\alpha})\|f\|_\infty\end{aligned}$$

Next, we use the estimates of (15.15)–(15.17) to get

$$\|h_{\mathbf{j}}\|_{C^\alpha} \leqslant \|h\|_{C^\alpha} + \frac{10\pi}{t^\alpha}\|h\|_\infty + \frac{25\pi}{t}\|f\|_{C^\alpha} + \left(\frac{\pi}{t}\mathbf{M}_\alpha + \frac{10\pi^2}{t^{1+\alpha}}\right)\|f\|_\infty$$

Substituting this back into (15.27) and using the specific choice of t at (15.25), we arrive at the estimate

$$\|Df\|_{C^\alpha} \leqslant C_{\alpha,k}\left[\|h\|_{C^\alpha} + \mathbf{M}_\alpha\|h\|_\infty + (\|f\|_{C^\alpha} + \mathbf{M}_\alpha\|f\|_\infty)\mathbf{M}_\alpha^{1/\alpha}\right] \tag{15.28}$$

What we have left to do is estimate $\|f\|_{C^\alpha}$, $\|f\|_\infty$ and $\|h\|_\infty$ in terms of $\|h\|_{C^\alpha}$ and $\|h\|_s$. Thus we fix an exponent q in the critical interval

$$s \leqslant q < 2, \qquad 1+k < q < 1 + \frac{1}{k}$$

The L^p-invertibility of the Beltrami operators in Theorem 14.2.2 and the interpolation in Lemma 15.4.1 give us the estimate

$$\|Df\|_q \leqslant C_{k,q}\|h\|_q \leqslant C_{k,q}\|h\|_s^{\frac{2s+\alpha sq}{2q+\alpha sq}} \|h\|_{C^\alpha}^{\frac{2q-2s}{2q+\alpha sq}}, \tag{15.29}$$

and this same estimate holds should we care to replace q by its Hölder conjugate exponent p, $1/p + 1/q = 1$. Thus

$$\|Df\|_p \leqslant C_{k,p}\|h\|_p \leqslant C_{k,p}\|h\|_s^{\frac{2s+\alpha sp}{2p+\alpha sp}} \|h\|_{C^\alpha}^{\frac{2p-2s}{2p+\alpha sp}} \tag{15.30}$$

We now multiply (15.29) and (15.30) together and take a square root to obtain

$$(\|Df\|_p\|Df\|_q)^{1/2} \leqslant C_{k,p,\alpha}\|h\|_s^{\frac{s+\alpha s}{2+\alpha s}} \|h\|_{C^\alpha}^{\frac{2-s}{2+\alpha s}}$$

As f is the Cauchy transform of $f_{\bar z}$, we use Theorem 4.3.11, that is, $\|f\|_\infty^2 \leqslant C_p\|f_{\bar z}\|_p\|f_{\bar z}\|_q$, to give us the following L^∞-bound for f.

Lemma 15.5.1. *For $1 \leqslant s < 2$ and $0 < \alpha < 1$ we have*

$$\|f\|_\infty \leqslant C_{k,s,\alpha}\|h\|_s^{\frac{s+\alpha s}{2+\alpha s}} \|h\|_{C^\alpha}^{\frac{2-s}{2+\alpha s}} \tag{15.31}$$

We next turn to estimating $\|f\|_{C^\alpha}$.

Lemma 15.5.2. *For $1 \leqslant s < 2$ and $0 < \alpha < 1$ we have*

$$\|f\|_{C^\alpha}^{1+\alpha} \leqslant C_\alpha \|Df\|_2 \|Df\|_{C^\alpha}^\alpha \leqslant C_{k,s,\alpha} \|h\|_2 \|Df\|_{C^\alpha}^\alpha$$

Proof. Sobolev embedding, or Theorem 4.3.13, gives $\|f\|_{C^\alpha} \leqslant C_r \|f_{\bar z}\|_r$ for $\alpha = 1 - \frac{2}{r}$. Using our interpolation again,

$$\begin{aligned}\|f\|_{C^\alpha}^{1+\alpha} &\leqslant C_\alpha \|Df\|_r^{1+\alpha} \leqslant C_\alpha \|Df\|_2^{\alpha+2/r} \|Df\|_{C^\alpha}^{1-2/r} \\ &= C_\alpha \|Df\|_2 \|Df\|_{C^\alpha}^\alpha\end{aligned}$$

Here we have used (15.22) with q replaced by r and with $s = 2$. $\square$

We now return to (15.28):

$$\|Df\|_{C^\alpha} \leqslant C_{k,s,\alpha} \left[\|h\|_{C^\alpha} + \mathbf{M}_\alpha \|h\|_\infty + (\|f\|_{C^\alpha} + \mathbf{M}_\alpha \|f\|_\infty) \mathbf{M}_\alpha^{1/\alpha} \right]$$

The term $\|f\|_{C^\alpha} \leqslant C \|h\|_2^{\frac{1}{1+\alpha}} \|Df\|_{C^\alpha}^{\frac{\alpha}{1+\alpha}}$ on the right hand side can be absorbed back onto the left hand side once we apply the inequality

$$xy^{\alpha/(1+\alpha)} \leqslant \varepsilon^{-\alpha} x^{1+\alpha} + \varepsilon y$$

with a sufficiently small choice of ε. This gives

$$\|Df\|_{C^\alpha} \leqslant C_{k,s,\alpha} \left[\|h\|_{C^\alpha} + \mathbf{M}_\alpha \|h\|_\infty + \mathbf{M}_\alpha^{1+1/\alpha} \left(\|h\|_2 + \|h\|_s^{\frac{s+\alpha s}{2+\alpha s}} \|h\|_{C^\alpha}^{\frac{2-s}{2+\alpha s}} \right) \right]$$

We can finally complete the proof of Theorem 15.0.6 by using the two interpolation inequalities in Lemma 15.4.1 again to estimate both $\|h\|_\infty$ and $\|h\|_2$ in terms of the $\|h\|_s$ and $\|h\|_{C^\alpha}$ norms in the above inequality. $\square$

15.6 Hölder-Caccioppoli Estimates

Theorem 15.0.6 can be used to derive a number of different $C^{1,\alpha}$-estimates. For instance, let us suppose that $f \in W^{1,2}_{loc}(\Omega)$ is a local solution to the Beltrami equation

$$\frac{\partial f}{\partial \bar z} - \mu(z) \frac{\partial f}{\partial z} - \nu(z) \overline{\frac{\partial f}{\partial z}} = h, \qquad z \in \Omega \subset \mathbb{C}, \tag{15.32}$$

where the coefficients and data $\mu, \nu, h \in C^\alpha(\Omega)$ and $|\mu(z)| + |\nu(z)| \leqslant k < 1$. We set

$$\mathbf{M}_\alpha = \|\mu\|_{C^\alpha} + \|\nu\|_{C^\alpha}$$

Consider an arbitrary test function $\varphi \in C_0^\infty(\Omega)$. If we multiply (15.32) by φ, we can obtain an equation for $F = \varphi f$ that has the form

$$F_{\bar z} - \mu F_z - \nu \overline{F_z} = H, \tag{15.33}$$

15.6. HÖLDER-CACCIOPPOLI ESTIMATES

where
$$H = \varphi h + \varphi_{\bar{z}} f - \mu \varphi_z f - \nu \overline{\varphi_z f} \tag{15.34}$$

Here we are faced with the problem that the coefficients μ, ν are defined and Hölder-continuous only on the domain Ω. For this purpose we choose another test function $\eta \in C_0^\infty(\Omega)$ with $|\eta| \leqslant 1$ everywhere such that $\eta(z) \equiv 1$ in the support of φ and define new coefficients

$$\mu_1(z) = \eta(z)\mu(z), \qquad \nu_1(z) = \eta(z)\nu(z),$$

which are Hölder-continuous in the entire plane with

$$\mathbf{M}_\alpha(\mu_1, \nu_1) \leqslant \|\eta\|_{C^\alpha} + \mathbf{M}_\alpha(\mu, \nu)$$

It is clear that F also satisfies the global Beltrami equation $F_{\bar{z}} - \mu_1 F_z - \nu_1 \overline{F_z} = H$. Accordingly, we have the following estimates:

$$\|DF\|_{C^\alpha} \leqslant C_{\alpha,s} \left[\|H\|_{C^\alpha} + (\|\eta\|_{C^\alpha} + \mathbf{M}_\alpha)^{1+2/\alpha s} \|H\|_s \right]$$
$$\|\varphi Df\|_{C^\alpha} \leqslant \|f\nabla\varphi\|_{C^\alpha} + \|DF\|_{C^\alpha}$$
$$\|H\|_{C^\alpha} \leqslant C_\alpha (\|\varphi h\|_{C^\alpha} + \|f\nabla\varphi\|_{C^\alpha} + \mathbf{M}_\alpha \|f\nabla\varphi\|_\infty)$$
$$\|H\|_s \leqslant C_s (\|\varphi h\|_s + \|f\nabla\varphi\|_s)$$

Using the interpolation Lemma 15.4.1, we also have

$$\mathbf{M}_\alpha \|f\nabla\varphi\|_\infty \leqslant C_{\alpha,s} \mathbf{M}_\alpha \|f\nabla\varphi\|_s^{\frac{\alpha s}{2+\alpha s}} \|f\nabla\varphi\|_{C^\alpha}^{\frac{2}{2+\alpha s}}$$
$$\leqslant C_{\alpha,s} \left[\mathbf{M}_\alpha^{1+2/\alpha s} \|f\nabla\varphi\|_s + \|f\nabla\varphi\|_{C^\alpha} \right]$$

These estimates yield the following Caccioppoli inequalities

Theorem 15.6.1.
$$\|\varphi Df\|_{C^\alpha} \leqslant C_{\alpha,s,k} \left[\|f\nabla\varphi\|_{C^\alpha} + (\|\eta\|_{C^\alpha} + \mathbf{M}_\alpha)^{1+2/\alpha s} \|f\nabla\varphi\|_s \right]$$
$$+ C_{\alpha,s,k} \left[\|\varphi h\|_{C^\alpha} + (\|\eta\|_{C^\alpha} + \mathbf{M}_\alpha)^{1+2/\alpha s} \|\varphi h\|_s \right]$$

With Caccioppoli estimates at our disposal we reach the local $C^{\ell,\alpha}$-regularity of solutions to the linear Betrami equations.

Theorem 15.6.2. *Let $\Omega \subset \mathbb{C}$ be a domain, ℓ a non-negative integer and $0 < \alpha < 1$. Let $|\mu(z)| + |\nu(z)| \leqslant k < 1$ and $\mu, \nu, h \in C_{loc}^{\ell,\alpha}(\Omega)$. That is, the ℓth derivatives of these functions are locally Hölder-continuous of exponent α. Then every solution $f \in W_{loc}^{1,2}(\Omega)$ of the inhomogeneous Beltrami equation*

$$\frac{\partial f}{\partial \bar{z}} - \mu(z)\frac{\partial f}{\partial z} - \nu(z)\overline{\frac{\partial f}{\partial z}} = h(z), \qquad z \in \Omega, \tag{15.35}$$

actually lies in $C_{loc}^{\ell+1,\alpha}(\Omega)$.

Proof. Let us first establish the case $\ell = 0$ and prove that $f \in C^{1,\alpha}_{loc}(\Omega)$ whenever μ, ν and $h \in C^{\alpha}_{loc}(\Omega)$. To start with, as $f \in W^{1,2}_{loc}(\Omega)$ by assumption, Theorem A.6.6 takes f to $L^p_{loc}(\mathbb{C})$ for any $p > 0$. If we multiply f by a test function $\varphi \in C^\infty_0(\mathbb{C})$ as after (15.32), then $F = \varphi f$ satisfies the Beltrami equation (15.33), where the inhomogeneity H lies in every $L^p(\mathbb{C})$-space, $p > 0$. In particular, we may choose $2 < p < 1 + \frac{1}{k}$ from the critical interval, and then Theorem 14.2.2 shows that $(\varphi f)_{\bar{z}} \in L^p(\mathbb{C})$. Thus, via the embedding Theorem 4.3.13, we have f locally Hölder-continuous with any exponent $0 < \beta < \frac{1-k}{1+k} < 1$.

We now apply the Caccioppoli estimates of Theorem 15.6.1. First, these prove that f belongs to $C^{1,\beta}_{loc}(\Omega)$, and in particular to $C^{\alpha}_{loc}(\Omega)$. With another use of the Caccioppoli estimates, we obtain $f \in C^{1,\alpha}_{loc}(\Omega)$.

The stage is now set for an induction argument, with the case $\ell = 0$ established. Thus consider the Beltrami equation

$$f_{\bar{z}} - \mu(z) f_z - \nu(z) \overline{f_z} = h(z), \qquad z \in \Omega, \tag{15.36}$$

with $\mu, \nu, h \in C^{\ell,\alpha}(\Omega)$. We differentiate (15.36) with respect to x and use continuity of the derivatives to interchange the order of differentiation to see that

$$(f_x)_{\bar{z}} - \mu(z)(f_x)_z - \nu(z)\overline{(f_x)_z} = h_x(z) - \mu_x f_z - \nu_x \overline{f_z}, \qquad z \in \Omega$$

The induction hypothesis puts the right hand side here in $C^{\ell-1,\alpha}$, and so f_x in $C^{\ell,\alpha}$. Similarly for the y-derivative. Thus $f \in C^{\ell+1,\alpha}$ and so forth. $\square$

15.7 Quasilinear Equations

We next consider the quasilinear equation in the entire plane.

Theorem 15.7.1. *Suppose that $f \in W^{1,2}_{loc}(\mathbb{C})$ is a solution to*

$$\frac{\partial f}{\partial \bar{z}} - \mu(z,f) \frac{\partial f}{\partial z} - \nu(z,f) \overline{\frac{\partial f}{\partial z}} = h \in C^\alpha(\mathbb{C}) \cap L^s(\mathbb{C}) \tag{15.37}$$

with $0 < \alpha < 1$ and $1 \leq s \leq 2$. Suppose too that following assumptions on ellipticity and continuity,

$$|\mu(z,\zeta)| + |\nu(z,\zeta)| \leq k < 1, \qquad (z,\zeta) \in \mathbb{C} \times \mathbb{C}, \tag{15.38}$$

$$|\mu(z_1,\zeta_1) - \mu(z_2,\zeta_2)| + |\nu(z_1,\zeta_1) - \nu(z_2,\zeta_2)| \leq A \left(|z_1 - z_2|^\alpha + |\zeta_1 - \zeta_2|^\alpha\right) \tag{15.39}$$

hold. Then $f \in C^{1,\alpha}$, and for some constant $C = C(\alpha, s, k, A)$, we have the uniform estimate

$$\|f_z\|_{C^\alpha} + \|f_{\bar{z}}\|_{C^\alpha} \leq C \max\{\mathbf{H}, \mathbf{H}^{13}\}, \tag{15.40}$$

where $\mathbf{H} = \|h\|_{C^\alpha} + \|h\|_s$.

15.7. QUASILINEAR EQUATIONS

Proof. For $p \geqslant 2$ it follows from the interpolation Lemma 15.4.1 that

$$\|h\|_\infty + \|h\|_p \leqslant 4\mathbf{H}$$

We choose $2 < p < 1 + \frac{1}{k}$ on the critical interval and thereby obtain

$$\|Df\|_p \leqslant C_p \|h\|_p \leqslant C_p \mathbf{H}$$

Next, if $\beta = 1 - \frac{2}{p}$, we have the Sobolev embedding $C^\beta(\mathbb{C}) \subset W^{1,p}(\mathbb{C})$, and so

$$\|f\|_{C^\beta} \leqslant C_{\beta,p} \|Df\|_p \leqslant 4C_{\beta,p} \mathbf{H} \tag{15.41}$$

Since both μ and ν are bounded functions, it follows from (15.39) that

$$\|\mu(z,f(z))\|_{C^{\alpha\beta}} + \|\nu(z,f(z))\|_{C^{\alpha\beta}} \leqslant C_A(1 + \mathbf{H}^\alpha),$$

and also, since h is bounded,

$$\|h\|_{C^{\alpha\beta}} \leqslant 2\|h\|_{C^\alpha}^\beta \|h\|_\infty^{1-\beta} \leqslant 8\mathbf{H} \tag{15.42}$$

We may now use Theorem 15.0.6 applied with exponent $\alpha\beta$ in place of α to obtain

$$\begin{aligned}\|Df\|_{C^{\alpha\beta}} &\leqslant C\left(\|h\|_{C^{\alpha\beta}} + (1 + \mathbf{H}^\alpha)^{1 + \frac{2}{\alpha s}}\|h\|_s\right)\\&\leqslant C\left(\mathbf{H} + \mathbf{H}(1 + \mathbf{H}^\alpha)^{1 + \frac{2}{\alpha s}}\right),\end{aligned}$$

and this together with the interpolation in (15.21) gives

$$\begin{aligned}\|Df\|_\infty &\leqslant \|Df\|_2^{\frac{\alpha\beta}{1+\alpha\beta}} \|Df\|_{C^{\alpha\beta}}^{\frac{1}{1+\alpha\beta}}\\&\leqslant C\|h\|_2^{\frac{\alpha\beta}{1+\alpha\beta}} \left(\mathbf{H} + \mathbf{H}(1 + \mathbf{H}^\alpha)^{1+\frac{2}{\alpha s}}\right)^{\frac{1}{1+\alpha\beta}}\end{aligned}$$

This actually shows that (15.42) can be improved to

$$\begin{aligned}\|\mu(z,f(z))\|_{C^\alpha} + \|\nu(z,f(z))\|_{C^\alpha} &\leqslant C(1 + \|Df\|_\infty^\alpha)\\&\leqslant C\left(1 + \mathbf{H}^\alpha(1+\mathbf{H}^{\alpha+\frac{2}{s}})^{\frac{\alpha}{1+\alpha\beta}}\right)\end{aligned}$$

Using Theorem 15.0.6, we conclude that

$$\|Df\|_{C^\alpha} \leqslant C \max\{\mathbf{H}, \mathbf{H}^{13}\},$$

and here the constant depends on α, k, s, A. $\square$

It remains to prove the local $C^{\ell,\alpha}$-regularity of solutions to quasilinear equations. With the above tools this is quickly achieved.

Proof of Theorem 15.0.7. Suppose that

$$f_{\bar{z}} - \mu(z,f)f_z - \nu(z,f)\overline{f_z} = h, \qquad z \in \Omega,$$

where the data $h \in C^{\ell,\alpha}_{loc}(\Omega)$, the coefficients $\mu, \nu \in C^{\ell,\alpha}_{loc}(\Omega \times \Omega')$ and we have the uniform ellipticity $|\mu(z,\zeta)| + |\nu(z,\zeta)| \leqslant k < 1$ for all $(z,\zeta) \in \Omega \times \Omega'$. Precisely as in the proof of Theorem 15.6.2, alone the uniform ellipticity and local boundedness of h imply that $f \in C^{\beta}_{loc}(\Omega)$ for any exponent $0 < \beta < \frac{1-k}{1+k} < 1$. But then

$$z \mapsto \mu(z, f(z)), \ \nu(z, f(z)) \in C^{\alpha\beta}_{loc}(\Omega)$$

Now, applying Theorem 15.6.2, we obtain $f \in C^{1,\alpha\beta}_{loc}(\Omega)$, hence f has locally bounded derivatives. Therefore the functions $\mu(z, f(z)), \nu(z, f(z))$ are in fact locally α-Hölder-continuous. Another use of Theorem 15.6.2 thus gives $f \in C^{1,\alpha}_{loc}(\Omega)$. This brings the coefficent functions $\mu(z, f(z)), \nu(z, f(z))$ to $C^{1,\alpha}_{loc}(\Omega)$, and thus $f \in C^{2,\alpha}_{loc}(\Omega)$ by Theorem 15.6.2. We may in fact continue this procedure up to the ℓth derivatives and thus prove the theorem. □

Chapter 16

Applications to Partial Differential Equations

As we pointed out in the introduction to this book, in physics and mathematics conservation laws and equations of motion or state are typically described by divergence-type second-order differential equations. In that introduction we carefully worked through an example from the calculus of variations—essentially a Lagrangian approach to deformations. There are many other classical examples with some interesting modern overtones. For instance, Noether's theorem [288] is a basic result from theoretical physics, and it implies that a conservation law, and therefore a second-order equation of divergence-type, can be derived from any continuous symmetry of a physical system. For example, the laws of physics do not change from one moment to the next and are therefore symmetric with respect to time, hence there is conservation of energy. This is true for all physical laws based upon the action principle, and it expresses a classical mechanics relationship between pairs of conjugate variables such as position-momentum and time-energy.

These varied and compelling applications provide considerable motivation to study the existence and regularity properties of these equations in general, and in particular the topological and analytic properties of the minimizers of functionals and the solutions of second-order equations in divergence form. We will see the methods of complex analysis in two dimensions: Conformal geometry and quasiconformal mappings provide powerful techniques, not available in other dimensions, to solve these highly nonlinear partial differential equations, especially those in divergence form. Our aim in this chapter is to describe in depth two principal techniques—the Hodge $*$ (or duality) method and the method of hodographic transformation. Although both these techniques are of quite classical origin, they are surprisingly powerful. The Hodge $*$ method works in great generality for equations with measurable coefficients, but they must be in divergence form. The hodographic method, however, requires regularity assumptions on the structural field (the coefficients) but works for more general types of equations than those of divergence-type. One can see both of these approaches at work in the simplest of examples—constructing analytic functions from solutions to Laplace's equation (harmonic functions). We could either construct a harmonic conjugate (and so form an analytic function) or we could take the complex gradient of this harmonic function, which again is analytic. One can upon close inspection even see how much regularity is necessary for each approach.

In this chapter we will give a modern interpretation of these methods and extend the ideas associated with them to a considerably general setting.

16.1 The Hodge * Method

As outlined above, the ideas here are vaguely akin to the classical methods of constructing analytic functions as pairs of conjugate harmonic functions. Basically, given an equation, we associate with it another "conjugate" equation. These two systems are linked by a sort of duality.

We will see a truly remarkable connection with quasiconformal mappings here. The solutions to the equation and its conjugate together form the components of a quasiregular mapping that is far more regular than one has any right to expect of the solution to the original equation (just as a harmonic function and its conjugate form an analytic function). From this fact, optimal regularity and other geometric properties of solutions readily follow.

In fact, the duality mentioned above is quite concrete, for instance, linking divergence and curl. We may well ask whether the divergence and curl of a vector field determine it (up to constants), divergence measuring the local expansion and curl measuring the local rotation of the generated flow. For example, Maxwell's equations for electrostatics (in two dimensions) determine an electric field $\mathbf{E}(z)$ and a magnetic field $\mathbf{B}(z)$ by specifying their divergence and curl. In a steady state the curl of the electric field is zero, and the divergence of the electric field is a multiple of the electric charge density $\rho(z)$. Thus

$$\operatorname{div} \mathbf{E}(z) = \rho(z) \quad \text{and} \quad \operatorname{curl} \mathbf{E}(z) = 0$$

The electrostatic equations for the magnetic field are

$$\operatorname{div} \mathbf{B}(z) = 0 \quad \text{and} \quad \operatorname{curl} \mathbf{B}(z) = j(z),$$

where $j(z)$ is a current density. To eliminate constants one relies on boundary conditions to pick out the physical solution. If the total charge is finite, then the electric field $\mathbf{E}(z)$ must tend to zero near infinity.

16.1. THE HODGE * METHOD

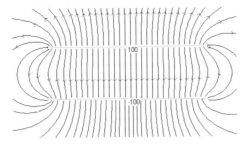

Electrostatic potential lines

Solving these equations gives the electric field in terms of the charge density or the magnetic field in terms of the current density. The well-known fact that any bounded vector field, defined on the entire plane whose divergence and curl both vanish identically, is constant is a type of Liouville theorem. The philosophy here is that it is possible to find the "correct" field if we are given both its divergence and its curl. Specifying the divergence alone leaves us with an underdetermined system and too many solutions to choose from to make meaningful statements.

Thus in this section we will show how to solve similar Hodge conjugate systems both in the linear case and in quite far-reaching generality including nonlinear equations. The techniques from quasiconformal mappings enable us to establish optimal regularity and other quite fine properties of solutions in this generality.

16.1.1 Equations of Divergence Type: The A-Harmonic Operator

As the prototype of a divergence-type equation, we start with the linear equation

$$\operatorname{div} A(z)\nabla u = 0, \qquad z \in \Omega \subset \mathbb{R}^2 \tag{16.1}$$

The theme of the present section is a thorough discussion of the different features of (16.1), from basic definitions to recent results. The following sections will then concentrate on general aspects of the complex analytic methods in nonlinear divergence-type equations. A good illustration of these results appears in Section 16.3.2, where we present a detailed study of the nonlinear prototype, the p-Laplace equation.

Throughout this section, apart from 16.1.5, we assume that the coefficient matrix $A = A(z) \in M^{2\times 2}(\mathbb{C})$ is symmetric. For (strong) ellipticity of the differential operator we require that

$$\frac{1}{K}|\xi|^2 \leqslant \langle A(z)\xi, \xi \rangle \leqslant K|\xi|^2, \qquad \xi \in \mathbb{R}^2, \tag{16.2}$$

almost everywhere in Ω. The factor K here can be either a constant or a measurable function on Ω with $1 \leqslant K(z) < \infty$ almost everywhere in Ω, but until further notice we will assume that K is a constant. For many purposes it is convenient to express the above ellipticity condition, equivalently, in terms of the following single inequality:

$$|\xi|^2 + |A(z)\xi|^2 \leqslant \left(K + \frac{1}{K}\right) \langle A(z)\xi, \xi \rangle \tag{16.3}$$

for almost every $z \in \Omega$ and all $\xi \in \mathbb{R}^2$. For the symmetric matrix $A(z)$ this is seen via construction of the eigenbasis.

16.1.2 The Natural Domain of Definition

As mentioned in the introduction to this section, in the case of divergence and curl, we will be able to identify solutions only up to constants. For this reason we introduce the following space defined on a domain Ω,

$$\mathbb{W}^{1,p}(\Omega) = \{f : |\nabla f| \in L^p(\Omega)\}/\mathbb{R}$$

This is the Banach space of locally integrable functions whose distributional gradient lies in $L^p(\Omega)$ and where we identify two functions should they differ by an additive constant. Notice that on bounded domains Ω we find that $\mathbb{W}^{1,p}(\Omega) + \mathbb{R}$ and $W^{1,p}(\Omega)$ coincide (the Poincaré inequality).

Equation (16.1) is interpreted in the distributional sense. That is, a function u is a (weak) solution to the equation if it has locally integrable derivatives with

$$\int_\Omega \langle A(z)\nabla u, \nabla \varphi \rangle = 0 \quad \text{for all } \varphi \in C_0^\infty(\Omega) \tag{16.4}$$

To make this meaningful we ask at least that $u \in W^{1,1}_{loc}(\Omega)$. However, for operators with nonsmooth coefficients, this condition is too weak for the basis of a viable theory. Hence we start with a little more regularity, with solutions of finite energy. Suppose that A satisfies the uniform ellipticity condition in (16.2). Then, associated with the operator div $A\nabla$ is the A-harmonic energy

$$\mathcal{E}_A[u] = \frac{1}{2} \int_\Omega \langle A(z)\nabla u, \nabla u \rangle$$

The smallest linear vector space, complete with respect to this energy $\mathcal{E}_A[u]$ and containing all smooth functions of finite energy, is actually $\mathbb{W}^{1,2}(\Omega)$. The closure of $C_0^\infty(\Omega)$, denoted by $\mathbb{W}_0^{1,2}(\Omega)$, is a Banach space equipped with the norm

$$\|u\|_{\mathbb{W}_0^{1,2}(\Omega)}^2 = \int_\Omega |\nabla u|^2$$

16.1. THE HODGE * METHOD

equivalent to the A-harmonic energy of u by (16.2). Via (16.4) the spaces $\mathbb{W}^{1,2}(\Omega)$ and $\mathbb{W}^{1,2}_0(\Omega)$ become the natural domains of definition of the A-harmonic operator

$$\operatorname{div} A\nabla : \mathbb{W}^{1,2}(\Omega) \to \left[\mathbb{W}^{1,2}_0(\Omega)\right]^*, \tag{16.5}$$

which assigns to every $u \in \mathbb{W}^{1,2}(\Omega)$ a bounded linear functional on the space $\mathbb{W}^{1,2}_0(\Omega)$ by the rule

$$\langle \operatorname{div} A\nabla u, \eta \rangle = -\int_\Omega \langle A\nabla u, \nabla \eta \rangle, \qquad \eta \in \mathbb{W}^{1,2}_0(\Omega)$$

The map (16.5) is surjective. Even more, if we restrict $\operatorname{div} A\nabla$ to $\mathbb{W}^{1,2}_0(\Omega)$, then we obtain a bijection. This fact is usually stated in the following weak form of the Dirichlet problem.

Theorem 16.1.1. *Given* $h \in \left[\mathbb{W}^{1,2}_0(\Omega)\right]^*$ *and* $u_0 \in \mathbb{W}^{1,2}(\Omega)$, *the Dirichlet problem*

$$\left.\begin{array}{r}\operatorname{div} A\nabla u = h \\ u \in u_0 + \mathbb{W}^{1,2}_0(\Omega)\end{array}\right\} \tag{16.6}$$

admits a unique solution.

The classical proof, which we omit here, is based on variational arguments. As a sketch, one considers the energy integral

$$\mathcal{E}[v] = \int_\Omega \langle A\nabla v, \nabla v \rangle - 2\langle h, v - u_0 \rangle, \quad \text{for } v \in u_0 + \mathbb{W}^{1,2}_0(\Omega)$$

The solution to (16.6) is the one with the least energy.

Solving (16.6) for $u \in \mathbb{W}^{1,2}_0(\Omega)$ provides us with a vector field $F = A(z)\nabla u \in L^2(\Omega)$. In this way we obtain the following characterization of the dual space to $\mathbb{W}^{1,2}_0(\Omega)$,

$$\left[\mathbb{W}^{1,2}_0(\Omega)\right]^* = \{\operatorname{div} F : F \in L^2(\Omega, \mathbb{R}^2)\}$$

Here the term "$\operatorname{div} F$", the divergence of F, is a notational formalism used to denote the linear functional on $\mathbb{W}^{1,2}_0(\Omega)$ defined by the rule

$$\langle \operatorname{div} F, \eta \rangle = -\int_\Omega \langle F, \nabla \eta \rangle, \qquad \eta \in \mathbb{W}^{1,2}_0(\Omega)$$

Remark 16.1.2. *Taking* $A(z) \equiv \mathbf{I}$ *arguments just like those above work to define the dual to* $\mathbb{W}^{1,q}_0(\Omega)$, *with* $1 < q < \infty$. *That is, for the dual exponent* p,

$$\left[\mathbb{W}^{1,q}_0(\Omega)\right]^* = \{\operatorname{div} F : F \in L^p(\Omega, \mathbb{R}^2)\}$$

16.1.3 The A-Harmonic Conjugate Function

A special feature of divergence equations in two dimensions is the ability to use complex analytic methods in obtaining solutions and regularity. This is especially pronounced in, for instance, harmonic functions–the solutions to the Laplace equation. Harmonic functions are components of analytic functions, and through this route complex analysis enters naturally.

For solutions to the divergence equation (16.1) a similar structure is provided by the *Hodge star* operator $*$, which here really is nothing more than the (counterclockwise) rotation by 90 degrees,

$$* = \begin{bmatrix} 0 & -1 \\ 1 & 0 \end{bmatrix} : \mathbb{R}^2 \to \mathbb{R}^2, \qquad ** = -\mathbf{I} \qquad (16.7)$$

There are two vector fields associated with each solution to the homogeneous equation

$$\operatorname{div} A(z)\nabla u = 0, \qquad u \in W^{1,2}_{loc}(\Omega)$$

The first, $E = \nabla u$, has zero curl (in the sense of distributions, the curl of any gradient field is zero), while the second, $B = A(z)\nabla u$, is divergence-free as a solution to the equation.

It is the Hodge star $*$ operator that transforms curl-free fields into divergence-free fields, and vice versa. In particular, if

$$E = \nabla w = (w_x, w_y), \qquad w \in W^{1,1}_{loc}(\Omega),$$

then $*E = (-w_y, w_x)$ and hence

$$\operatorname{div}(*E) = \operatorname{div}(*\nabla w) = 0,$$

at least in the distributional sense. We recall here the following well-known fact from calculus (the Poincaré lemma).

Lemma 16.1.3. *Let $E \in L^p(\Omega, \mathbb{R}^2)$, $p \geqslant 1$, be a vector field defined on a simply connected domain Ω. If $\operatorname{curl} E = 0$, then E is a gradient field; that is, there exists a real-valued function $u \in \mathbb{W}^{1,p}(\Omega)$ such that $\nabla u = E$.*

Thus in simply connected domains the A-harmonic equation $\operatorname{div} A(z)\nabla u = 0$ implies that the field $*A\nabla u$ is curl-free and may be rewritten as

$$\nabla v = *A(z)\nabla u, \qquad (16.8)$$

where $v \in W^{1,2}_{loc}(\Omega)$ is some Sobolev function unique up to an additive constant. This function v we call the *A-harmonic conjugate* of u. Sometimes in the literature one also finds the term *stream function* used for v. The stream function is defined for two-dimensional flows of various kinds. Level curves of the stream function are streamlines, which for an isotropic A are perpendicular to equipotential lines. For the particular case of fluid dynamics the difference between the stream function values at any two points gives the volumetric flow rate (or

16.1. THE HODGE ∗ METHOD

flux) through a line connecting the two points. Since streamlines are tangent to the flow, the value of the stream function must be the same along a streamline.

On the other hand, from the point of view of the conjugate function, (16.8) implies that if we define
$$A^* = *^t A^{-1} * = - * A^{-1} *,$$
then A^* satisfies the ellipticity bounds (16.2) whenever A does and
$$A^*(z)\nabla v = *\nabla u \qquad (16.9)$$
Hence
$$\operatorname{div} A^*(z)\nabla v = 0, \qquad v \in W^{1,2}_{loc}(\Omega) \qquad (16.10)$$
Equation (16.10), that is associated with (16.8) by $*$, is called the *conjugate A^*-harmonic equation*. Note that when the determinant $\det A \equiv 1$, we have
$$A^* = (\det A)\, A^* = A^t = A,$$
and so in this case the conjugate function v actually satisfies the same equation
$$\operatorname{div} A(z)\nabla v = 0, \qquad \det A \equiv 1$$

16.1.4 Regularity of Solutions

Having discussed the existence of the linear inhomogeneous problem in Theorem 16.1.1, we now turn to the regularity properties of solutions to these equations and the remarkable connection with quasiconformal mappings.

Given a divergence equation $\operatorname{div} A(z)\nabla u = 0$ with $A = A(z)$ measurable, symmetric and positive definite, we assume we are given a solution $u \in W^{1,2}_{loc}(\Omega)$. Supposing that v is an A-harmonic conjugate, we consider the complex function
$$f = u + iv \in W^{1,2}_{loc}(\Omega)$$
If we further assume the proposed ellipticity bounds of A in the formulation in (16.3), we see that the Hilbert-Schmidt norm of the derivative
$$\|Df\|^2 = |\nabla u|^2 + |\nabla v|^2 \leqslant \left(K + \frac{1}{K}\right) \langle A(z)\nabla u, \nabla u \rangle = \left(K + \frac{1}{K}\right) \langle -*\nabla v, \nabla u \rangle$$
We calculate
$$-\langle *\nabla v, \nabla u \rangle = u_x v_y - u_y v_x = J(z, f), \qquad z = x + iy$$
Thus the function f satisfies the differential inequality
$$\|Df\|^2 \leqslant \left(K + \frac{1}{K}\right) J(z, f)$$

almost everywhere. This inequality may also be written in terms of the partial derivatives, and with this we come to the equation

$$\max_\alpha |\partial_\alpha f(z)| \leq K \min_\alpha |\partial_\alpha f(z)|$$

for almost every $z \in \Omega$. Therefore the complex function $f = u + iv \in W^{1,2}_{loc}(\Omega)$ is a K-quasiregular mapping, where K is precisely the ellipticity constant from (16.2).

At this point we may use the Stoilow factorization Theorem 5.5.1 to write f as the composition of a quasiconformal homeomorphism and an analytic function. As an immediate corollary, the quasiconformal regularity Theorems 3.10.3, 13.2.3 and 14.4.3 give the following result.

Theorem 16.1.4. *Suppose $q \geq \frac{2K}{K+1}$ and $u \in W^{1,q}_{loc}(\Omega)$ is a solution to*

$$\operatorname{div} A(z) \nabla u = 0 \quad \text{in } \Omega \tag{16.11}$$

Then $u = w \circ \phi$, where $\phi : \mathbb{C} \to \mathbb{C}$ is K-quasiconformal and w is harmonic in the domain $\Omega' = \phi(\Omega)$. In particular,

- $u \in W^{1,p}_{loc}(\Omega)$ *for every* $2 \leq p < \frac{2K}{K-1}$.
- $u \in C^\alpha_{loc}(\Omega)$, *where* $\alpha = \frac{1}{K}$.

Note here that $\frac{2K}{K+1} < 2$, and so the hypotheses of the theorem require less regularity than one might routinely suppose.

The Stoilow factorization described above also has a number of further natural consequences, for instance, on unique continuation or the structure of level sets, and so on, for the $W^{1,2}_{loc}$-solutions to the divergence equation (16.11). We leave these for the reader to explore, but there is one we want to point out. It is a triviality that there are no entire quasiconformal mappings with a bounded real part. Thus we have the Liouville-theorem.

Corollary 16.1.5. *Suppose $q \geq \frac{2K}{K+1}$ and $u \in W^{1,q}_{loc}(\Omega)$ is a bounded entire solution to*

$$\operatorname{div} A(z) \nabla u = 0 \quad \text{in } \mathbb{C}$$

Then u is constant.

An important feature of Theorem 16.1.4 is its sharpness. The radial stretchings

$$f(z) = |z|^{\alpha-1} z, \qquad \alpha = \frac{1}{K},$$

give the complex Beltrami equation

$$\frac{\partial f}{\partial \bar{z}} = \mu(z) \frac{\partial f}{\partial z}, \qquad \mu(z) = \frac{1-K}{1+K} \frac{z}{\bar{z}} \tag{16.12}$$

16.1. THE HODGE * METHOD

We shall show later (see (16.38)) that a function f satisfies this equation if and only if its real and imaginary parts solve the divergence equation

$$\text{div } A(z) \nabla u = 0, \qquad (16.13)$$

where

$$A(z) = \frac{1}{K} \mathbf{I} + \left(K - \frac{1}{K} \right) \frac{z}{|z|} \otimes \frac{z}{|z|} \qquad (16.14)$$

In our case the real and imaginary parts of f,

$$u = |z|^{\alpha-1} \Re e\, z \quad \text{and} \quad v = |z|^{\alpha-1} \Im m\, z,$$

are precisely those found by Serrin [325] for the divergence equation (16.13) with coefficient matrix (16.14) and so exhibit this sharpness. The ellipticity constant of $A(z)$ is precisely K, and the solutions u, v belong to $W^{1,p}_{loc}(\mathbb{R}^2)$ for all $p < \frac{2K}{K-1}$ but not for $p \geqslant \frac{2K}{K-1}$.

Further, the Beltrami equation (16.12) also holds for the function $g(z) = \frac{1}{f(z)}$, which provides us with another set of solutions to (16.13), namely,

$$u_1 = |z|^{-\alpha-1} \Re e\, z \quad \text{and} \quad v_1 = -|z|^{-\alpha-1} \Im m\, z,$$

as we simply replace α by $-\alpha$. However, these solutions u_1, v_1 belong to $W^{1,q}_{loc}(\mathbb{R}^2)$ for all $1 \leqslant q < \frac{2K}{K+1}$, but not for $q \geqslant \frac{2K}{K+1}$, and most of our previous regularity results (such as Stoilow factorization) fail for these mappings.

16.1.5 General Linear Divergence Equations

One might equally well consider general linear elliptic equations with a bounded measurable but not necessarily symmetric coefficient matrix

$$A = A(z) = \begin{bmatrix} \alpha_{11} & \alpha_{12} \\ \alpha_{21} & \alpha_{22} \end{bmatrix}, \qquad \det(\mathbf{I} + A) > 0 \qquad (16.15)$$

The ellipticity conditions for A can be formulated so that they imply bounded distortion for the induced complex function $f = u + iv$. For this let us consider the inhomogeneous equation

$$\text{div } A \nabla u = \text{div } H, \qquad H = (h_1, h_2) \in L^1_{loc}(\Omega, \mathbb{R}^2), \qquad (16.16)$$

for u in the Sobolev class $W^{1,1}_{loc}(\Omega)$. We assume for simplicity that Ω is simply connected. Using the Hodge $*$ operator, as earlier, we can then find the corresponding stream function $v \in W^{1,1}_{loc}(\Omega)$ determined by the rule

$$\nabla v = *(A \nabla u - H) \qquad (16.17)$$

The identity in (16.17) concerns vectors in $\mathbb{R}^2$ and therefore can be expressed in complex notation. This will simplify things a bit. By setting $h = h_1 + ih_2$, we arrive, after a lengthy but quite routine purely algebraic manipulation, at the equivalent complex first-order equation for $f = u + iv$, which we record in the following theorem.

Theorem 16.1.6. *Let Ω be a simply connected domain, let $H = (h_1, h_2) \in L^1_{loc}(\Omega, \mathbb{R}^2)$ and let $u \in W^{1,1}_{loc}(\Omega)$ be a solution to*

$$\operatorname{div} A \nabla u = \operatorname{div} H \qquad (16.18)$$

If $v \in W^{1,1}(\Omega)$ is a solution to the conjugate A-harmonic equation (16.17), then the function $f = u + iv$ satisfies the inhomogeneous Beltrami equation

$$\frac{\partial f}{\partial \bar{z}} - \mu(z) \frac{\partial f}{\partial z} - \nu(z) \overline{\frac{\partial f}{\partial z}} = a(z)h(z) + b(z)\overline{h(z)} \qquad (16.19)$$

The coefficients are given by

$$\left. \begin{array}{rcl} \mu & = & \frac{1}{\det(\mathbf{I}+A)}(\alpha_{22} - \alpha_{11} - i(\alpha_{12} + \alpha_{21})) \;=\; 2b \\[4pt] \nu & = & \frac{1}{\det(\mathbf{I}+A)}(1 - \det A + i(\alpha_{12} - \alpha_{21})) \;=\; 2a - 1 \end{array} \right\}, \qquad (16.20)$$

and $h = h_1 + ih_2$.

Conversely, if $f \in W^{1,1}_{loc}(\Omega, \mathbb{C})$ is a mapping satisfying (16.19), then $u = \Re(f)$ and $v = \Im(f)$ satisfy (16.17) with A given by solving the complex equations in (16.20).

From our assumption of ellipticity A defined in (16.15), the denominator here is positive, $\det(\mathbf{I} + A) > 0$, which in terms of the coefficients says

$$(1 + \alpha_{11})(1 + \alpha_{22}) > \alpha_{12}\alpha_{21}$$

Having reduced (16.16) to a first-order complex system, it is natural to define the ellipticity parameters through this representation. That is, we require

$$|\mu(z)| + |\nu(z)| \leqslant k = \frac{K-1}{K+1} \qquad \text{almost everywhere} \qquad (16.21)$$

Conversely, every elliptic complex system (16.19) comes from a second-order equation (16.16), where the entries of the matrix $A = A(z)$ are explicitly given by the formulas

$$\alpha_{11}(z) \;=\; \frac{|1-\mu|^2 - |\nu|^2}{|1+\nu|^2 - |\mu|^2} \;\geqslant\; \frac{1}{K} \qquad (16.22)$$

$$\alpha_{22}(z) \;=\; \frac{|1+\mu|^2 - |\nu|^2}{|1+\nu|^2 - |\mu|^2} \;\geqslant\; \frac{1}{K} \qquad (16.23)$$

$$\alpha_{12}(z) \;=\; \frac{2\,\Im m(\nu - \mu)}{|1+\nu|^2 - |\mu|^2}, \qquad 2|\alpha_{12}(z)| \leqslant K^2 - 1 \qquad (16.24)$$

$$\alpha_{21}(z) \;=\; \frac{-2\,\Im m(\nu + \mu)}{|1+\nu|^2 - |\mu|^2}, \qquad 2|\alpha_{21}(z)| \leqslant K^2 - 1 \qquad (16.25)$$

We now have a very clear-cut and explicit one-to-one correspondence between the general linear first-order complex system (16.19) and the general linear divergence-type equation (16.16). From this picture we immediately see that

16.1. THE HODGE * METHOD

1. A is symmetric if and only if ν is real-valued. In this case
$$\nu = \frac{1 - \det A}{\det(\mathbf{I} + A)}$$

2. A has determinant 1 if and only if ν is purely imaginary.

3. A is symmetric with determinant 1 if and only if $\nu = 0$ (that is, we have a complex Beltrami equation).

4. A is diagonal,
$$A = \begin{bmatrix} \sigma_1 & 0 \\ 0 & \sigma_2 \end{bmatrix},$$
if and only if both μ and ν are real-valued. In this case
$$\mu = \frac{\sigma_2 - \sigma_1}{(1 + \sigma_1)(1 + \sigma_2)} \quad \text{and} \quad \nu = \frac{1 - \sigma_2 \sigma_1}{(1 + \sigma_1)(1 + \sigma_2)}$$

5. A is isotropic, meaning $A = \sigma(z)\mathbf{I}$ with $\sigma(z) \in \mathbb{R}$, if and only if $\mu = 0$ and ν is real-valued. The complex equation now takes the form
$$\frac{\partial f}{\partial \bar{z}} - \frac{1 - \sigma}{1 + \sigma} \overline{\frac{\partial f}{\partial z}} = a h + b \bar{h} \tag{16.26}$$

The complex analytic interpretation of planar second-order elliptic PDEs of divergence-type leads to some very powerful tools. For instance, as in Theorem 16.1.4, one obtains optimal regularity for solutions to (16.16). Let us state the following result, which is based on the invertibility of Beltrami operators (see in particular Theorem 14.2.2) applied to the equation for f given in (16.19) and the iterative technique using the Sobolev embedding theorem used earlier in the proof of the Caccioppoli estimate of Theorem 5.4.2. In fact, we will later prove a more general result in the nonlinear setting and therefore only sketch the proof here.

Theorem 16.1.7. *Let A be a bounded measurable coefficient matrix with ellipticity constant determined from the coefficients μ and ν defined in (16.20): Suppose that*
$$|\mu| + |\nu| \leqslant k < 1 \quad \text{and} \quad 1 + k < q < p < 1 + 1/k$$
If $H \in L^p_{loc}(\Omega, \mathbb{R}^2)$, then any solution $u \in W^{1,q}_{loc}(\Omega)$ to the general linear divergence equation
$$\operatorname{div} A \nabla u = \operatorname{div} H$$
lies in $W^{1,p}_{loc}(\Omega)$.

Proof. Note that by (16.17) the stream function v also belongs to $W^{1,q}_{loc}(\Omega)$. Let $K \subset \Omega$ be compact and choose a real-valued cutoff function $\eta \in C_0^\infty(\Omega)$, $\eta|K \equiv 1$. Now

$$\frac{\partial(\eta f)}{\partial \bar{z}} - \mu \frac{\partial(\eta f)}{\partial z} - \nu \overline{\frac{\partial(\eta f)}{\partial z}}$$

$$= \eta\left(\frac{\partial f}{\partial \bar{z}} - \mu \frac{\partial f}{\partial z} - \nu \overline{\frac{\partial f}{\partial z}}\right) + f\frac{\partial \eta}{\partial \bar{z}} - \mu f \frac{\partial \eta}{\partial z} - \nu \overline{\left(f\frac{\partial \eta}{\partial z}\right)}$$

$$= \eta(ah + b\bar{h}) + f\frac{\partial \eta}{\partial \bar{z}} - \mu f \frac{\partial \eta}{\partial z} - \nu \overline{\left(f\frac{\partial \eta}{\partial z}\right)}$$

Here denote the right hand side by g. Initially, $f \in W^{1,q}_{loc}(\Omega)$, and so by the Sobolev embedding theorem $g \in L^s(\mathbb{C})$ for some $q < s < p$. Invertibility of the Beltrami operators puts $\eta f \in W^{1,s}(\Omega)$ and $g \in L^{q'}(\Omega)$, again by Sobolev embedding theorem, and so we have the iteration set up as in the proof of Theorem 5.4.2. This leads ultimately to $\eta f \in W^{1,p}(\mathbb{C})$, so $f \in W^{1,p}_{loc}(\Omega)$. □

Notice that existence in Theorem 16.1.7 is straightforward. We extend the coefficients μ, ν, a, b outside of Ω by 0, assume $\Omega = \mathbb{C}$ and apply Theorem 14.2.2 providing a unique locally integrable solution whose gradient is in $L^p(\mathbb{C})$ whenever $2 \leq p < 1 + 1/k$ and $h \in L^p(\mathbb{C})$. We again take the liberty of pointing out that this bound is sharp.

For another example, of a quite different character, the complex method and the equivalence of (16.18) and (16.19) is the key to Chapter 18, solving Calderón's problem in impedance tomography. Both from the very first steps in finding exponentially increasing solutions of the isotropic conductivity equation $\operatorname{div} \sigma \nabla u = 0$, to the final detailed analysis of their various properties, the approach via complex analysis and quasiconformal mappings seems essential.

In fact, there is a wealth of similar applications in nonlinear PDEs and these applications will be the theme of the next part of this chapter.

16.1.6 A-Harmonic Fields

Here we present a slightly alternative view of the interplay of divergence-type operators with the Beltrami equations and quasiconformal mappings. This approach has the considerable virtue of also working in higher dimensions. For the higher-dimensional case see [196].

Recall that to each solution $u \in W^{1,2}_{loc}(\Omega)$ of the homogeneous equation $\operatorname{div} A(z)\nabla u = 0$ in Ω we managed to associate two vector fields, the curl-free field $E = \nabla u$ and the divergence-free field

$$B = A(z)\nabla u, \qquad \operatorname{div} B = 0 \qquad (16.27)$$

16.1. THE HODGE * METHOD

More generally, one can consider nonlinear elliptic equations of the form

$$\operatorname{div} \mathcal{A}(z, \nabla u) = 0, \tag{16.28}$$

where we will formulate the ellipticity by

$$|\xi|^2 + |\mathcal{A}(z, \xi)|^2 \leqslant \left(K + \frac{1}{K}\right) \langle \mathcal{A}(z, \xi), \xi \rangle, \qquad 1 \leqslant K < \infty \tag{16.29}$$

For the nonlinear setting this turns out to be the correct formulation. Also in this case we obtain a pair

$$E = \nabla u, \qquad B = \mathcal{A}(z, \nabla u), \quad \text{where } \operatorname{curl} E = 0, \operatorname{div} B = 0 \tag{16.30}$$

Conversely, by analogy with electrostatics, we shall refer to a pair $\mathcal{F} = [E, B]$ of arbitrary vector fields $E, B \in L^2_{loc}(\Omega, \mathbb{R}^2)$ with $\operatorname{div} B = 0$ and $\operatorname{curl} E = 0$, as a *div-curl couple*.

In order to use the ideas from quasiconformal theory that we have developed, we introduce the following quantities:

$$|\mathcal{F}(z)|^2 = |B(z)|^2 + |E(z)|^2 \quad \text{and} \quad J(z, \mathcal{F}) = \langle B(z), E(z) \rangle$$

A div-curl couple $\mathcal{F} = [E, B]$ that satisfies the distortion inequality

$$|\mathcal{F}(z)|^2 \leqslant \left(K + \frac{1}{K}\right) J(z, \mathcal{F}) \tag{16.31}$$

will be called a *K-quasiconformal field*. In what follows we shall closely mimic the familiar language of the complex Beltrami equation in order to stress these connections. Thus we define the $\pm$ components of $\mathcal{F}$ by

$$\mathcal{F}^+ = \frac{1}{2}(E + B) \quad \text{and} \quad \mathcal{F}^- = \frac{1}{2}(E - B) \tag{16.32}$$

The reader should view these components as corresponding to the Cauchy-Riemann partials $\overline{f}_z$ and $f_{\bar{z}}$ of a complex function. In a while we shall look at the analogy a little closer, but we recall here that in the case of a complex function f, the Hilbert-Schmidt norm of its derivative

$$\|Df(z)\|^2 = 2|f_z|^2 + 2|f_{\bar{z}}|^2,$$

while the operator norm $|Df(z)| = |f_z| + |f_{\bar{z}}|$. Hence the inequalities

$$\|Df(z)\|^2 \leqslant \left(K + \frac{1}{K}\right) J(z, f) \quad \text{and} \quad |Df(z)| \leqslant K J(z, f) \tag{16.33}$$

are pointwise equivalent.

The distortion inequality (16.31) takes the equivalent form

$$|\mathcal{F}^-(z)| = k(z)|\mathcal{F}^+(z)|, \quad \text{where } 0 \leqslant k(z) \leqslant \frac{K-1}{K+1} < 1 \tag{16.34}$$

We want to write this even more explicitly in a form close to that of the Beltrami equation. As the two vectors $\mathcal{F}^-$ and $k(z)\mathcal{F}^+$ in (16.34) evidently have the same length, they are isometric images of each other. Let $\mathbf{n} = \mathbf{n}(z)$ be the unit normal to the line bisecting the angle between these vectors,

$$\mathbf{n}(z) = \frac{\mathcal{F}^- - k\mathcal{F}^+}{|\mathcal{F}^- - k\mathcal{F}^+|}, \qquad (16.35)$$

provided $\mathcal{F}^-(z) \neq k(z)\mathcal{F}^+(z)$. In the latter case we choose $\mathbf{n}(z)$ to be any unit vector orthogonal to both $\mathcal{F}^-(z)$ and $\mathcal{F}^+(z)$. In this way we arrive at the Beltrami equation

$$\mathcal{F}^-(z) = \mathcal{M}(z)\mathcal{F}^+(z), \quad \text{where } \mathcal{M}(z) = k(z)[\mathbf{I} - 2\mathbf{n} \otimes \mathbf{n}] \qquad (16.36)$$

Note that $\mathbf{I} - 2\mathbf{n} \otimes \mathbf{n}$ is a symmetric orthogonal matrix of determinant -1.

In simply connected domains we can always write the curl-free field E as $E = \nabla u$ for some function $u \in W^{1,2}_{loc}(\Omega)$. The Beltrami equation (16.36) takes the form

$$\mathbf{A}(z)\nabla u = B(z), \qquad \operatorname{div} B = 0, \qquad (16.37)$$

where from (16.32) and (16.36) we calculate

$$\mathbf{A}(z) = \frac{\mathbf{I} - \mathcal{M}(z)}{\mathbf{I} + \mathcal{M}(z)} = \frac{1 - k(z)}{1 + k(z)} I + \frac{4k(z)}{1 - k(z)^2} \mathbf{n}(z) \otimes \mathbf{n}(z) \qquad (16.38)$$

We thus obtain the $\mathbf{A}$-harmonic equation for the potential function u of the vector field E,

$$\operatorname{div} \mathbf{A}(z)\nabla u = 0 \qquad (16.39)$$

The reader should compare this equation with (16.27). In fact, if E and B happen to arise from such a divergence-type equation, with (16.39) we have obtained a new equation for the solution u.

The point to this computation is that we gain not only the linearity of the equation but also a rather special algebraic structure for the coefficient matrix $\mathbf{A}$, a form that resembles many extremal situations concerning divergence equations and quasiconformal mappings. Another useful advantage is that

$$\det \mathbf{A}(z) \equiv 1,$$

and further it is rather surprising that we have not lost any of the ellipticity bounds

$$\frac{1}{K}|\xi|^2 \leqslant \langle \mathbf{A}(z)\xi, \xi \rangle \leqslant K|\xi|^2 \qquad (16.40)$$

Namely, we have

$$\langle \mathbf{A}(z)\xi, \xi \rangle = \frac{1-k}{1+k}|\xi|^2 + \frac{4k}{1-k^2}\langle \mathbf{n}(z), \xi \rangle^2,$$

and hence

$$\frac{1-k}{1+k}|\xi|^2 \leqslant \langle \mathbf{A}(z)\xi, \xi \rangle \leqslant \left(\frac{1-k}{1+k} + \frac{4k}{1-k^2}\right)|\xi|^2 = \frac{1+k}{1-k}|\xi|^2,$$

16.1. THE HODGE $*$ METHOD

as claimed.

Given now a matrix field $\mathbf{A}(z)$ as in (16.38) and a solution $u \in W^{1,2}_{loc}(\Omega)$ to (16.39), we want to relate them to quasiconformal mappings. For Ω simply connected we find the conjugate function $v \in W^{1,2}_{loc}(\Omega)$ with

$$\mathbf{A}(z)\nabla u = B = - * \nabla v, \qquad E = \nabla u$$

Set
$$f = u + iv$$

Of particular relevance to quasiconformal mappings (and to us here) is to relate the coefficient matrix of the divergence equation to the distortion tensor of f. We have already seen the strong connections in this case in our introduction while looking at minima of variational problems. To complete the proposed analogy we consider the distortion tensor $G : \Omega \to \mathbb{R}^{2\times 2}$ defined by

$$G(z) = J(z, f)^{-1} D^t f(z) Df(z), \qquad (16.41)$$

in particular, $\det G \equiv 1$. We have $\nabla v = *\mathbf{A}\nabla u$. Using the conditions $\det \mathbf{A} = 1$ and $\mathbf{A}^t = \mathbf{A}$ gives $\mathbf{A}* = *\mathbf{A}^{-1}$, and thus $\mathbf{A}\nabla v = *\nabla u$. In other words, $\mathbf{A} D^t f(z) = J(z,f)\bigl[Df(z)\bigr]^{-1}$, so that the coefficient matrix

$$G = \mathbf{A}^{-1} = \frac{1+k}{1-k}\mathbf{I} - \frac{4k}{1-k^2}\mathbf{n}\otimes\mathbf{n}$$

pointwise. We see that the real and imaginary parts of $f = u + iv$ satisfy the same second-order PDE, namely,

$$\operatorname{div} G^{-1}(z)\nabla u = 0, \qquad \operatorname{div} G^{-1}(z)\nabla v = 0$$

The relation $f = u + iv$ also enables us to identify $\overline{f_z}$ and $f_{\bar z}$ as the "Cauchy-Riemann" derivatives ($\pm$ components) of the quasiconformal field

$$\mathcal{F} = [\nabla u, - * \nabla v]$$

Following our calculations above, we obtain the Beltrami matrix $\mathcal{M} = k(z)(\mathbf{I} - 2\mathbf{n}(z)\otimes\mathbf{n}(z))$. Such matrices have the special form

$$\mathcal{M} = \begin{pmatrix} \alpha & \beta \\ \beta & -\alpha \end{pmatrix}, \qquad \alpha^2 + \beta^2 = k(z)^2 < 1$$

From the usual identification of $\mathbb{R}^2$ with the complex plane, $z = x + iy$, we see that the linear transformation $\mathcal{M}(z) : \mathbb{C} \to \mathbb{C}$ is anticonformal,

$$\mathcal{M}(z)\zeta = (\alpha(z) + i\beta(z))\overline{\zeta}$$

Therefore the identity $\mathcal{F}^-(z) = \mathcal{M}(z)\mathcal{F}^+(z)$ gives us back our usual Beltrami equation $f_{\bar z} = \mu(z)f_z$ with $\mu = \alpha + i\beta$ and $|\mu| \leqslant \frac{K-1}{K+1} = k < 1$. Thus the circle is complete.

We summarize these considerations in the following theorem.

Theorem 16.1.8. Let $\Omega \subset \mathbb{C}$ be a domain and suppose that
$$\mathcal{A}: \Omega \times \mathbb{R}^2 \to \mathbb{R}^2$$
is measurable in $z \in \Omega$ and continuous in $\zeta \in \mathbb{R}^2$ and satisfies the ellipticity condition
$$|\zeta|^2 + |\mathcal{A}(z,\zeta)|^2 \leqslant \left(K + \frac{1}{K}\right) \langle \zeta, \mathcal{A}(z,\zeta) \rangle, \qquad (z,\zeta) \in \Omega \times \mathbb{R}^2, \tag{16.42}$$
for some constant $1 \leqslant K < \infty$. Then every solution $u \in W^{1,2}_{loc}(\Omega)$ of the nonlinear equation
$$\mathrm{div}\,\mathcal{A}(z, \nabla u) = 0 \tag{16.43}$$
actually solves a linear elliptic equation
$$\mathrm{div}\,\mathbf{A}(z)\nabla u = 0, \tag{16.44}$$
where $\mathbf{A}(z)$ is a positive definite symmetric measurable matrix field of determinant 1 satisfying the ellipticity bounds
$$\frac{1}{K}|\zeta|^2 \leqslant \langle \mathbf{A}(z)\zeta, \zeta \rangle \leqslant K|\zeta|^2 \tag{16.45}$$

We shall see in the next section that this fact has quite remarkable consequences.

16.2 Topological Properties of Solutions

In this section we describe the topological properties of solutions to elliptic equations in divergence form. Most notably, we establish that any solution becomes harmonic after a quasiconformal change of coordinates.

We first observe that, as a consequence of Theorem 16.1.8, any solution $u \in \mathbb{W}^{1,2}(\Omega)$ to the nonlinear divergence equation (16.43) is an equilibrium solution for the variational energy integral
$$\mathcal{E}_{\mathbf{A}}[u] = \int_\Omega \langle \mathbf{A}(z)\nabla u, \nabla u \rangle \tag{16.46}$$
with $\mathbf{A}(z)$ as in (16.38). We now change variables by a K-quasiconformal mapping $f: \Omega \to \Omega'$, which solves the Beltrami equation
$$D^t f(z) Df(z) = G(z) J(z,f), \qquad G(z) = \mathbf{A}^{-1}(z) \tag{16.47}$$
The solution f is guaranteed by the measurable Riemann mapping theorem, see for instance Theorem 10.2.1. Note here that we also have a choice of Ω'. We now simply write $u(z) = h(f(z))$, where $h = h(w)$ is a function defined in Ω'. An elementary calculation reveals
$$\nabla u(z) = D^t f(z)(\nabla h)(w), \qquad w = f(z) \tag{16.48}$$

16.2. THE HODGE ∗ METHOD   419

In view of (16.46) we see that then the Dirichlet integral of h is precisely that of the energy $\mathcal{E}[u]$ of u,

$$\int_{\Omega'} |\nabla h(w)|^2 \, dw = \int_{\Omega} |(Df^t)^{-1}(z)\nabla u|^2 J(z,f) \, dz$$
$$= \int_{\Omega} \langle \mathbf{A}(z)\nabla u, \nabla u \rangle \, dz$$

This implies in particular that h is an equilibrium solution to the variation of the Dirichlet integral and as such is a harmonic function. We have therefore proven the following theorem.

Harmonic Factorization for Divergence Equations

Theorem 16.2.1. *Every solution $u \in W^{1,2}_{loc}(\Omega)$ of the nonlinear equation*

$$\mathrm{div}\mathcal{A}(z, \nabla u) = 0$$

with ellipticity bounds in (16.42) can be expressed as

$$u(z) = h(f(z)),$$

where $f : \Omega \to \Omega'$ is K-quasiconformal and h is harmonic on Ω'.

Here Ω' can be chosen to be any domain conformally equivalent to Ω; in particular, we could chose $\Omega = \Omega'$. There are further interesting consequences of this factorization theorem that follow from the properties of harmonic functions. For instance, we have the following Harnack inequality. Note that it is sharp in the sense that as the ellipticity constant $K \to 1$, we obtain the classical Harnack inequality for harmonic functions.

Harnack Inequality for Divergence Equations

Theorem 16.2.2. *Let $u \in W^{1,2}_{loc}(\mathbb{D})$ be a non-negative solution of the nonlinear equation*

$$\mathrm{div}\mathcal{A}(z, \nabla u) = 0,$$

with ellipticity bounds in (16.42). Then

$$\frac{1-\varphi_K(|z|)}{1+\varphi_K(|z|)} u(0) \leqslant u(z) \leqslant \frac{1+\varphi_K(|z|)}{1-\varphi_K(|z|)} u(0), \tag{16.49}$$

where $\varphi_K(t)$, $t \in [0,1)$, are distortion functions depending only on $K \geqslant 1$. The functions φ_K are strictly increasing, $\varphi_1(t) = t$ and $\varphi_K(t) < 1$. We have the explicit estimate

$$\varphi_K(t) \leqslant 4^{1-1/K} \, t^{1/K}, \qquad t \in [0,1) \tag{16.50}$$

Proof. We set

$$\varphi_K(t) = \sup\{|f(t)| : f \text{ is } K\text{-quasiconformal on } \mathbb{D}, f(0) = 0 \text{ and } f(\mathbb{D}) \subset \mathbb{D}\}$$

The functions $\varphi_K(t)$ are clearly increasing. From Theorem 12.6.3 obtain the estimate $\varphi_K(t) \leqslant e^{5(K-1)}$, while the more precise bound (16.50) follows from the quasiconformal version of the Schwarz lemma given in [229, Theorem II.3.1]. The bound is based on the modulus method, and it is sharp in the sense that $\lim_{t\to 0} \varphi_K(t)\, t^{-1/K} = 4^{1-1/K}$.

We now apply Theorem 16.2.1 to write

$$u(z) = h(f(z)), \qquad f : \mathbb{D} \to \mathbb{D}, \ f(0) = 0,$$

where h is positive and harmonic. The Poisson integral formula gives

$$\frac{1-|w|}{1+|w|} h(0) \leqslant h(w) \leqslant \frac{1+|w|}{1-|w|} h(0),$$

and so

$$\frac{1-|f(z)|}{1+|f(z)|} u(0) \leqslant u(z) \leqslant \frac{1+|f(z)|}{1-|f(z)|} u(0),$$

The result follows. $\square$

16.3 The Hodographic Method

16.3.1 The Continuity Equation

Since we will be studying nonlinear elliptic systems of first-order PDEs it is appropriate to have a motivational example in the back of one's mind. Further, this example is a good illustration of the sorts of connections between second-order elliptic equations in the plane and elliptic systems of first-order PDEs that we have been espousing.

The continuity equation comes from fluid flow and gas dynamics, a much-studied area in the 1950s when aircraft were first attempting to break the sound barrier and the previous elliptic equations for subsonic flow had become less useful in modeling the flow as the equations degenerated, hence Bers' very nice book [55]. The mathematics surrounding these equations is quite beautiful in and of itself, and so we will spend a little time going into some of the details.

The continuity equation expresses conservation of matter—clearly, if matter flows away from a point, there must be a decrease in the quantity remaining. If the fluid velocity is $\mathbf{u}$ and the density is ρ, then the mass that flows in a unit time across a unit length of boundary of a region is the component $\rho \mathbf{u} \cdot \nu$ (here ν denotes the outer normal). The change of mass must be balanced by the flow out of the region, hence there is conservation of mass. If Ω denotes the region, we have

$$\frac{d}{dt} \int_\Omega \rho = -\int_{\partial\Omega} \rho \mathbf{u} \cdot \nu$$

16.3. THE HODOGRAPHIC METHOD

(the minus sign is necessary because when u and ν point in the same direction, e.g., if $\mathbf{u}$ is outward and $\mathbf{u} \cdot \nu > 0$, the mass within decreases.) We differentiate under the integral sign and use Gauss's divergence theorem to get to

$$\int_\Omega \frac{d\rho}{dt} = -\int_\Omega \nabla \cdot (\rho \mathbf{u})$$

We let Ω shrink and apply the Lebesgue density theorem to obtain the continuity equation

$$\frac{d}{dt}\rho + \nabla \cdot (\rho \mathbf{u}) = 0$$

Thus the velocity field of an incompressible fluid has zero divergence. Electrodynamic equations are closely analogous to hydrodynamic equations, but electrodynamical properties are often easier to describe than hydrodynamics. By Bernoulli's theorem for potential flows, density is a (decreasing) function of the speed of the flow. Thus steady-state flows with density are concerned with the *continuity equation*

$$\operatorname{div} A(z, |\nabla U|)\nabla U = 0 \qquad (16.51)$$

Here we have slightly generalized to allow A to be a density function that may also describe various physical properties of the media, and U is the *potential function*, so $\mathbf{u} = \nabla U$. In the case of subsonic flows we will see that the complex gradient $f(z) = U_x - iU_y$ is a quasiregular mapping. However, the uniform ellipticity is lost when one approaches a critical value of velocity ($|f| = |\nabla U|$); this motivates our later study of such equations in the degenerate elliptic case.

A complete description also requires two other thermodynamic variables, such as pressure p and temperature T. Typically, one deals with the adiabatic and isentropic flows in which pressure is a definite function of density. The derivative, taken with respect to adiabatic change, of pressure with respect to density $\frac{dp}{d\rho}$ is positive and denoted c^2. It is interpreted as the speed of propagation of small disturbances in a flow and is referred to as the *local speed of sound*. As noted, Bernoulli's theorem for potential flows shows density to be a decreasing function of the speed $q = |\mathbf{u}|$ of the flow, and this leads (see [55]) to the following relation,

$$c^2 := -\frac{q\rho(q)}{\dot{\rho}(q)}$$

The type of quasilinear second-order equation one studies depends on the Mach number M, the ratio of an object's speed to the speed of sound in the medium and which can be interpreted as an ellipticity constant. The equation is elliptic at subsonic speeds ($M < 1$) and hyperbolic at supersonic speeds. Typically, one considers "nice" subsonic flows with possible regions where the flow becomes supersonic at a "shock" boundary—thus avoiding the delicate issues of degeneracy of the equations, which we shall later approach, albeit to a limited extent.

For an ideal gas $p = \alpha\rho^\gamma$, where $\gamma > 1$ is the ratio of specific heats and α is constant, but also more general pressure-density relations are important. Let us derive the equation for the potential function from the continuity equation (16.51).

The most common notation has the potential function ϕ, so $\phi = U$ from (16.51). Then, with $u = \phi_x$, $v = \phi_y$ and $q = \sqrt{u^2 + v^2}$, we find

$$q\, q_x = u\, u_x + v\, v_x, \qquad q\, q_y = u\, u_y + v\, v_y,$$

and we calculate

$$(\rho u)_x + (\rho v)_y = 0$$
$$\rho(q)\phi_{xx} + \rho(q)\phi_{yy} + \dot\rho(q) q_x\, u + \dot\rho(q) q_y\, v = 0$$
$$\phi_{xx} + \phi_{yy} + \frac{\dot\rho(q)}{q\,\rho(q)}\left(u^2\phi_{xx} + u v\, \phi_{xy}\right) + \frac{\dot\rho(q)}{q\,\rho(q)}\left(u v\, \phi_{xy} + v^2\phi_{yy}\right) = 0$$
$$(c^2 - u^2)\phi_{xx} + (c^2 - v^2)\phi_{yy} - 2 u v\, \phi_{xy} = 0$$

Therefore the equation for the potential function takes the form

$$(c^2 - u^2)\phi_{xx} - 2 u v\, \phi_{xy} + (c^2 - v^2)\phi_{yy} = 0 \qquad (16.52)$$

The continuity equation implies that there is a *stream function* ψ such that $\rho u = \psi_y$ and $\rho v = -\psi_x$. Thus the velocity potential ϕ and the stream function ψ are coupled by the equations

$$\rho\, \phi_x = \psi_y, \qquad \rho\, \phi_y = -\psi_x$$

Notice that for constant density (say $\rho = 1$) corresponding to the case of an incompressible flow, we have the Cauchy-Riemann equations. With a little manipulation these equations give

$$\phi_x \psi_y - \phi_y \psi_x = \frac{1}{\rho}(\psi_x^2 + \psi_y^2)$$
$$\phi_x \psi_y - \phi_y \psi_x = \rho(\phi_x^2 + \phi_y^2),$$

from which we deduce that the mapping $f = (\phi, \psi)$ has

$$\phi_x^2 + \phi_y^2 + \psi_x^2 + \psi_y^2 = \left(\rho + \frac{1}{\rho}\right)(\phi_x \psi_y - \phi_y \psi_x),$$

giving the distortion relation

$$\|Df\|^2 = \left(\rho + \frac{1}{\rho}\right) J(z, f),$$

and so f is quasiregular.

The Stoilow factorization Theorem 5.5.1 now asserts that an $W^{1,2}$-solution is of the form $f = h(g(z))$, where g is a homeomorphic solution and h is analytic. Thus, after a change of coordinates, the flow is the same as that for an incompressible flow (an analytic function), and therefore we obtain the topological nature of streamlines and so forth of these various nonstandard flows about objects.

16.3. THE HODOGRAPHIC METHOD

One of the most remarkable and useful techniques developed for the study of nonlinear equations is the hodographic transformation. While the idea seems to go back at least to Gauss, its utility in this area is amply demonstrated in the work of Lavrentiev and Bers. Roughly speaking, given a system of first-order PDEs and solutions $w = f(z)$ to this system, the hodographic transformation asks us to simply regard w as an independent variable in the hodographic plane, and z as a function of w. If the original system is quasilinear, this simple trick converts the equation into a linear system with variable coefficients.

If we study our potential equation in the hodograhic plane, we see similar features. We consider an irrotational flow, so

$$u_y = v_x, \tag{16.53}$$

as well as (16.52). If $f = (u, v)$, it can be shown that the Hilbert-Schmidt norm of the differential satisfies

$$(c^2 - u^2 - v^2)\|Df\|^2 = (2c^2 - u^2 - v^2)J(z, f) - (uv_y - vu_x)^2 - (uu_x - vu_y)^2$$

so that

$$\|Df\|^2 \leq \left(1 + \frac{c^2}{c^2 - u^2 - v^2}\right) J(z, f), \tag{16.54}$$

and once again f is quasiregular, with the same consequences as above. Here the distortion function is

$$K(z, f) = \left(1 + \frac{c^2}{c^2 - |f|^2}\right)$$

This time the critical case occurs when we approach the speed of sound. Indeed, more appears here. We must also have

$$(uv_y - vu_x)^2 + (vu_x - vu_y)^2 \to 0$$

on the critical set $(c^2 = |f|^2)$. From this we can identify the geometric structure of the critical set.

16.3.2 The p-Harmonic Operator $\text{div}|\nabla|^{p-2}\nabla$

The prototypical nonlinear equation in divergence form is the nonlinear analog of the Laplacian known as the the p-harmonic operator or p-Laplacian. In fact, the p-Laplacian exhibits many of the most interesting features of nonlinear equations, and its theory contains many of the most interesting nuances. The aim of this part of the chapter is to get to the core of this theory.

We shall assume without saying it everytime that $1 < p, q < \infty$ is a pair of Hölder conjugate exponents, that is, $p + q = pq$. To start we consider the inhomogeneous p-harmonic equation

$$\text{div}\left(|\nabla u|^{p-2}\nabla u\right) = \text{div} F, \quad \text{where } u \in \mathbb{W}^{1,p}(\Omega), \; F \in L^q(\Omega, \mathbb{R}^2) \tag{16.55}$$

The equation, together with its natural domain of definition, comes from consideration of the problem of minimizing the variational integral

$$\mathcal{E}_p[u] = \int_\Omega |\nabla u|^p - p\langle F, \nabla u\rangle \tag{16.56}$$

for $u \in u_0 + \mathbb{W}_0^{1,p}(\Omega)$, subject to given Dirichlet data $u_0 \in \mathbb{W}^{1,p}(\Omega)$.

As in the case of the Laplacian, the p-harmonic operator

$$\Delta_p = \mathrm{div}\left(|\nabla u|^{p-2}\nabla u\right) : \mathbb{W}^{1,p}(\Omega) \to [W_0^{1,p}(\Omega)]^*,$$

acting on $\varphi \in W_0^{1,p}(\Omega)$ by the rule

$$(\Delta_p u)(\varphi) = \int_\Omega \langle |\nabla u|^{p-2}\nabla u, \nabla \varphi\rangle,$$

is a surjective (nonlinear) map. Here, note that $|\nabla u|^{p-1} \in L^q(\Omega)$. Furthermore, restricted to functions vanishing on $\partial\Omega$, the operator is actually a bijection

$$\Delta_p : \mathbb{W}_0^{1,p}(\Omega) \to [W_0^{1,p}(\Omega)]^*,$$

where $\mathbb{W}_0^{1,p}(\Omega) = W_0^{1,p}(\Omega)$. Let us take a moment to explain these facts. Unwinding the definitions, we find with Remark 16.1.2 that we are required to solve $\mathrm{div}(|\nabla u|^{p-2}\nabla u) = \mathrm{div}\, F$. Young's inequality gives us an a priori lower bound on energy,

$$\mathcal{E}_p[u] \geqslant c\|\nabla u\|_p^p - C\|F\|_q^q$$

Take a minimizing sequence $u_j \in u_0 + W_0^{1,p}$ for (16.56). Evidently, we have the bound

$$\|\nabla u_j\|_p^p \leqslant c^{-1}(\mathcal{E}_p[u_1] + C\|F\|_q^q),$$

whereupon u_j has a weakly converging subsequence in $u_0 + W_0^{1,p}(\Omega)$ with $\nabla u_j \rightharpoonup \nabla u$ in $L^p(\Omega)$. Then,

$$\mathcal{E}_p[u] \leqslant \liminf \int_\Omega |\nabla u_j|^p - p\langle F, \nabla u_j\rangle$$

and since u_j is minimizing, the result follows. A similar argument based on the strict convexity of the functional establishes the injectivity, or the uniqueness in the Dirichlet problem, for (16.55).

16.3.3 Second-Order Derivatives

A function $u \in W_{loc}^{1,p}(\Omega)$ satisfying the equation $\mathrm{div}\left(|\nabla u|^{p-2}\nabla u\right) = 0$ in Ω is called p-*harmonic*. We next show that the divergence operator, which we understand in the distributional sense, can actually act pointwise on the function

$$F = |\nabla u|^{(p-2)/2}\nabla u \in L_{loc}^2(\Omega, \mathbb{R}^2)$$

16.3. THE HODOGRAPHIC METHOD

Proving that the expression F is a Sobolev function will involve the second-order derivatives of the solution $u \in W^{1,p}_{loc}(\Omega)$ only in an implicit manner. The following theorem is a first step toward a deeper understanding of the regularity properties of p-harmonic functions.

Theorem 16.3.1. *Let* $u \in W^{1,p}_{loc}(\Omega)$ *be p-harmonic in Ω, $1 < p < \infty$. Then the vector field*
$$F = |\nabla u|^{(p-2)/2} \nabla u$$
lies in the Sobolev space $W^{1,2}_{loc}(\Omega)$ and satisfies the Caccioppoli inequality
$$\int_\Omega \eta^2 |DF|^2 \leqslant C_p \int_\Omega |\nabla \eta|^2 |F|^2 \tag{16.57}$$
for every $\eta \in C_0^\infty(\Omega)$.

Proof. Let us start with the case $p \geqslant 2$. We fix a function $\eta \in C_0^\infty(\Omega)$ supported on a compact subset $G \subset \Omega$. For each vector $h \in \mathbb{R}^2$ such that $|h| < \text{dist}(G, \partial\Omega)$, we consider the test functions
$$\phi(z) = \eta^2(z)\Big(u(z+h) - u(z)\Big) \quad \text{and} \quad \phi(z-h)$$
As both lie in $W^{1,p}_0(\Omega)$, it is legitimate to write the identities
$$\int_\Omega \langle |\nabla u(z)|^{p-2} \nabla u(z), \nabla \phi(z) \rangle \, dz = 0$$
$$\int_\Omega \langle |\nabla u(z)|^{p-2} \nabla u(z), \nabla \phi(z-h) \rangle \, dz = 0$$
since the products are integrable.

After a change of variables (translation) in the second integral, we obtain the following formla
$$\int_\Omega \langle |\nabla u(z+h)|^{p-2} \nabla u(z+h) - |\nabla u(z)|^{p-2} \nabla u(z), \nabla \phi(z) \rangle \, dz = 0 \tag{16.58}$$

Next, using the product rule, we express the gradient of ϕ in the form
$$\nabla \phi(z) = 2\eta(z)\Big(u(z+h) - u(z)\Big)\nabla \eta(z) + \eta^2(z)\Big(\nabla u(z+h) - \nabla u(z)\Big),$$
and substituting the expression in (16.58) gives
$$\int_\Omega \eta^2(z) \langle |\nabla u(z+h)|^{p-2} \nabla u(z+h) - |\nabla u(z)|^{p-2} \nabla u(z), \nabla u(z+h) - \nabla u(z) \rangle \, dz$$
$$\leqslant 2 \int_\Omega |\eta| |\nabla \eta| |u(z+h) - u(z)| \Big| |\nabla u(z+h)|^{p-2} \nabla u(z+h) - |\nabla u(z)|^{p-2} \nabla u(z) \Big| \, dz$$

At this point we shall make use of the following two inequalities that are valid for $p \geqslant 2$ and for vectors ξ and ζ in any inner-product space. Both estimates are easy to obtain via a homogeneity argument, that is, restricted to the compact set $|\zeta|^2 + |\xi|^2 = 1$:

$$\langle |\xi|^{p-2}\xi - |\zeta|^{p-2}\zeta, \xi - \zeta \rangle \geqslant \alpha_p \Big| |\xi|^{(p-2)/2}\xi - |\zeta|^{(p-2)/2}\zeta \Big|$$

and

$$\Big| |\xi|^{p-2}\xi - |\zeta|^{p-2}\zeta \Big| \leqslant \beta_p \left(|\xi|^p + |\zeta|^p \right)^{(p-2)/2} \Big| |\xi|^{(p-2)/2}\xi - |\zeta|^{(p-2)/2}\zeta \Big|$$

We use these estimates with the choices $\xi = \nabla u(z+h)$ and $\zeta = \nabla u(z)$. Applying Hölder's inequality and rearranging the terms finally provides us with the bounds

$$\int_\Omega \eta^2(z) |F(z+h) - F(z)|^2 \tag{16.59}$$

$$\leqslant \frac{\alpha_p}{\beta_p} \left(\int_\Omega |\nabla \eta|^2 |u(z+h) - u(z)|^p \right)^{\frac{2}{p}} \left(\int_\Omega |\nabla \eta|^2 (|F(z+h)|^2 + |F(z)|^2) \right)^{1 - \frac{2}{p}}$$

It is precisely at this point that we have used the assumption $p \geqslant 2$ in order to have $1 - \frac{2}{p} \geqslant 0$.

Recall now that Sobolev functions may be characterized by a number of alternate ways. In particular, for dealing with nonlinear expressions of weak derivatives, the integrable difference quotients are particularly useful. Here, applying Lemmas A.5.3 and A.5.4, dividing (16.59) by $|h|^2$ and taking the limit will show that F lies in the Sobolev class $W^{1,2}_{loc}(\Omega)$. Moreover, the same reasoning gives the Caccioppoli inequalities

$$\int_\Omega \eta^2 |DF|^2 \leqslant C_p \left(\int_\Omega |\nabla \eta|^2 |\nabla u(z)|^p \right)^{2/p} \left(\int_\Omega |\nabla \eta|^2 |F|^2 \right)^{1-2/p}$$

$$= C_p \int_\Omega |\nabla \eta|^2 |F|^2$$

This proves Theorem 16.3.1 for all exponents $p \geqslant 2$. □

The case $1 < p \leqslant 2$ can be treated by a very elegant argument involving the dual q-harmonic equation (following the methods introduced in the previous section on Hodge $*$ coupling). We write the p-harmonic equation in the form of a first-order system,

$$\left. \begin{array}{rcl} |\nabla u(z)|^{p-2} u_x & = & v_y \\ |\nabla u(z)|^{p-2} u_y & = & -v_x \end{array} \right\}, \tag{16.60}$$

where the function $v \in W^{1,q}_{loc}(\Omega)$ is the *p-harmonic conjugate* of u. Here of course we need Ω to be simply connected. This is no restriction since we are studying local regularity.

16.3. THE HODOGRAPHIC METHOD

We solve (16.60) for the gradient of u,

$$\left. \begin{array}{rcl} |\nabla v(z)|^{q-2} v_x & = & -u_y \\ |\nabla v(z)|^{q-2} v_y & = & u_x \end{array} \right\}$$

Hence the conjugate v is q-harmonic,

$$\text{div}\left(|\nabla v|^{q-2}\nabla v\right) = 0$$

Since $1 < p \leqslant 2$, we have $q \geqslant 2$. In other words, what we have proven above shows that the vector $|\nabla v|^{(q-2)/2}\nabla v$ lies in $W^{1,2}_{loc}(\Omega)$ and satisfies the Caccioppoli estimate (16.57). In fact, $|\nabla v|^{(q-2)/2}\nabla v$ is obtained by rotating $|\nabla u|^{(p-2)/2}\nabla u$ by 90 degrees. Therefore we even have $C_q = C_p$ for all $1 < p < \infty$, where C_p is the constant in (16.57). □

16.3.4 The Complex Gradient

Following the example of the continuity equation, the next step in the study of the p-harmonic operator will be for us to examine the gradient map or, precisely, the complex gradient of the p-harmonic function u,

$$f = \frac{\partial u}{\partial z} = \frac{1}{2}\left(\frac{\partial u}{\partial x} - i\frac{\partial u}{\partial y}\right) \tag{16.61}$$

Recall that for a harmonic function u the complex gradient u_z is analytic. Our goal is to show for a general p-harmonic function that $f = u_z$ is a quasiregular mapping. A subtle point here is the smoothness, since quasiregularity requires that the mapping $f \in W^{1,2}_{loc}$. Recall from Chapter 5 that, in particular, many of the most useful properties of quasiregular mappings such as Stoilow factorization, higher integrability and so on, need not hold for solutions to the Beltrami equation of low regularity.

On the other hand, there is no immediately apparent reason why p-harmonic functions should have square-integrable second-order distributional derivatives. Therefore we turn to our earlier Theorem 16.3.1.

The key is to study not only the complex gradient in (16.61) but all its powers,

$$F^\delta = |f|^{\delta-1} f, \qquad \delta > 0 \tag{16.62}$$

We already know from Theorem 16.3.1 that $F^{p/2} \in W^{1,2}_{loc}(\Omega)$.

We aim to find a first-order elliptic systems satisfied by the functions F^δ. The following calculation we shall perform on this family is, at the moment, legitimate only when $\delta = p/2$. However, we will later justify this calculation for all $\delta > 0$, and so to avoid repetition we will make the presentation for a general exponent δ.

We begin with the formulas

$$\frac{\partial u}{\partial x} = |F^\delta|^{\frac{1}{\delta}-1}\left(F^\delta + \overline{F^\delta}\right), \qquad \frac{\partial u}{\partial y} = i|F^\delta|^{\frac{1}{\delta}-1}\left(F^\delta - \overline{F^\delta}\right) \tag{16.63}$$

Hence in the sense of distributions

$$\frac{\partial}{\partial y}|F^\delta|^{\frac{1}{\delta}-1}\left(F^\delta+\overline{F^\delta}\right)=i\frac{\partial}{\partial x}|F^\delta|^{\frac{1}{\delta}-1}\left(F^\delta-\overline{F^\delta}\right)$$

It is convenient to formulate this identity in terms of the Cauchy-Riemann operators. This gives

$$\frac{\partial}{\partial\bar{z}}|F^\delta|^{\frac{1}{\delta}-1}F^\delta=\frac{\partial}{\partial z}|F^\delta|^{\frac{1}{\delta}-1}\overline{F^\delta}$$

Now, in the case $F^\delta\in W^{1,2}_{loc}$, we can apply the product rule and reduce the equation to

$$\Im m\left(\frac{\partial}{\partial\bar{z}}F^\delta\right)=\frac{1-\delta}{1+\delta}\Im m\left(\frac{\overline{F^\delta}}{F^\delta}\frac{\partial}{\partial z}F^\delta\right) \qquad (16.64)$$

Note that this identity is well defined on the set where F^δ vanishes since almost everywhere on this set the derivatives of F^δ also vanish.

Equation (16.64) gives us some control of the imaginary part of F^δ. To control the real parts we return to the p-harmonic function u. Since the vector field $|\nabla u|^{p-2}\nabla u$ is divergence-free, we have (in the distributional sense)

$$\frac{\partial}{\partial x}|F^\delta|^{\frac{p-1-\delta}{\delta}}\left(F^\delta+\overline{F^\delta}\right)+i\frac{\partial}{\partial y}|F^\delta|^{\frac{p-1-\delta}{\delta}}\left(F^\delta-\overline{F^\delta}\right)=0$$

Once again we write the expression in complex notation, in which case it takes the simple form

$$\Re e\left(\frac{\partial}{\partial\bar{z}}F^\delta|F^\delta|^{\frac{p-1-\delta}{\delta}}\right)=0$$

As above, if $F^\delta\in W^{1,2}_{loc}$, in particular when $p=2\delta$, then we can differentiate the terms. After rearrangement we arrive at

$$\Re e\left(\frac{\partial}{\partial\bar{z}}F^\delta\right)=-\frac{p-1-\delta}{p-1+\delta}\Re e\left(\frac{\overline{F^\delta}}{F^\delta}\frac{\partial}{\partial z}F^\delta\right) \qquad (16.65)$$

Adding together (16.64) and (16.65), we have the following first-order elliptic equation for F^δ,

$$\frac{\partial F^\delta}{\partial\bar{z}}=\mu\frac{\partial F^\delta}{\partial z}+\nu\overline{\frac{\partial F^\delta}{\partial z}}, \qquad (16.66)$$

where

$$\mu=\frac{1}{2}\left(\frac{1-\delta}{1+\delta}-\frac{p-1-\delta}{p-1+\delta}\right)\frac{\overline{F^\delta}}{F^\delta} \qquad (16.67)$$

$$\nu=\frac{1}{2}\left(\frac{\delta-1}{\delta+1}-\frac{p-1-\delta}{p-1+\delta}\right)\frac{F^\delta}{\overline{F^\delta}} \qquad (16.68)$$

16.3. THE HODOGRAPHIC METHOD

The system we have thereby obtained is uniformly elliptic for all $\delta > 0$. In fact, we have
$$|\mu| + |\nu| = \max\left\{\left|\frac{p-1-\delta}{p-1+\delta}\right|, \left|\frac{1-\delta}{1+\delta}\right|\right\} < 1$$

This bound and the argument we have presented show that F^δ is a quasiregular mapping as soon as we can establish $F^\delta \in W^{1,2}_{loc}(\Omega)$. However, we do know at least one exponent, namely, $\delta = p/2$, where the required smoothness holds.

The power and flexibility of the complex analytic approach now become apparent. To prove the regularity for $\delta \neq p/2$, observe that each F^δ, $\delta > 0$, is the composition of $F^{p/2}$ with the quasiconformal radial stretching
$$\zeta \mapsto \zeta|\zeta|^{a-1}, \qquad a = \frac{2\delta}{p} > 0$$

Thus, as a composition of two quasiregular mappings, F^δ is also quasiregular, and as such it does lie in $W^{1,2}_{loc}(\Omega)$. Of course, it was this simple idea that motivated the definition in (16.62) and the subsequent arguments.

As a particular consequence, the computations (16.64) and (16.65) are now justified for all $\delta > 0$. The case $\delta = 1$ is naturally of greater interest.

Theorem 16.3.2. *The complex gradient $f = u_z = \frac{1}{2}(u_x - iu_y)$ of a p-harmonic function u is a quasiregular mapping and satisfies the equation*
$$\frac{\partial f}{\partial \bar{z}} = \left(\frac{1}{p} - \frac{1}{2}\right)\left(\frac{\bar{f}}{f}\frac{\partial f}{\partial z} + \frac{f}{\bar{f}}\overline{\frac{\partial f}{\partial z}}\right) \tag{16.69}$$

Now, the regularity is controlled by the dilatation bounds. For the complex gradient, Theorem 16.3.2 tells us that
$$\left|\frac{\partial f}{\partial \bar{z}}\right| \leq \left|1 - \frac{2}{p}\right|\left|\frac{\partial f}{\partial z}\right| \quad \text{almost everywhere} \tag{16.70}$$

Note the sharpness at $p = 2$.

Further, if we set $\delta = \sqrt{p-1}$, we obtain an equation that is $\mathbb{C}$-linear with respect to the complex derivatives, that is, the Beltrami equation for $f = F^{\sqrt{p-1}}$,
$$\frac{\partial f}{\partial \bar{z}} = \mu \frac{\partial f}{\partial z}, \qquad \mu = \mu(f) = \frac{1-\delta}{1+\delta}\frac{\bar{f}}{f}$$

Perhaps it is also interesting that for no $0 < \delta < 1$ does the coefficient μ in (16.67) vanish. In other words, we cannot obtain an equation of the form $F^\delta_{\bar{z}} = \nu(F^\delta)\,\overline{F^\delta_z}$. Consequently, we cannot obtain the usual $\mathbb{C}$-linear Beltrami equation for the inverse mapping.

As a result of Hölder continuity, Stoilow factorization and so forth for quasiregular mappings, we state the following theorem.

Theorem 16.3.3. *Every p-harmonic function in Ω belongs to $C^{1,\alpha}_{loc}(\Omega)$ with some $\alpha = \alpha(p) > 0$. The critical points of u (where $\nabla u = 0$) are isolated unless the function is constant.*

Next we shall continue to exploit the theory of quasiregular mappings to obtain explicit and sharp regularity results for all p-harmonic functions.

16.3.5 Hodograph Transform for the p-Laplacian

Suppose that 0 is a critical point of the p-harmonic function u, equivalently, we suppose that $f(0) = 0$ for the complex gradient $f = \frac{1}{2}(u_x - iu_y)$.

Near 0, the Stoilow factorization Theorem 5.5.1 shows we may express $f(z)$ as
$$f(z) = (\chi(z))^n,$$
where $n \in \mathbb{N}$ and where χ is a quasiconformal homeomorphism defined in a neighborhood of $z = 0$ with $\chi(0) = 0$. On substituting such a representation into (16.69), we get the following quasilinear equation for χ:
$$\frac{\partial \chi}{\partial \bar{z}} = \left(\frac{1}{p} - \frac{1}{2}\right)\left(\frac{\chi}{\bar{\chi}} \overline{\frac{\partial \chi}{\partial z}} + \frac{\overline{\chi^n}}{\chi^n} \frac{\partial \chi}{\partial z}\right)$$

Via the hodograph transform, that is, looking to the inverse map, this becomes the *linear* equation
$$\frac{\partial h}{\partial \bar{\xi}} = \left(\frac{1}{2} - \frac{1}{p}\right)\left(\frac{\xi}{\bar{\xi}} \frac{\partial h}{\partial \xi} + \frac{\bar{\xi}^n}{\xi^n} \overline{\frac{\partial h}{\partial \xi}}\right) \tag{16.71}$$

for the inverse mapping $h(\xi) = \chi^{-1}(\xi)$.

If $p = 2$, we obtain the Cauchy-Riemann equation $h_{\bar{\xi}} = 0$, and here the powers $\{\xi^l\}_{l\in\mathbb{N}}$ form a basis in the space of all solutions h to the equation. Our immediate goal is to find an analogous basis for solutions to the elliptic equation (16.71).

Rescalings $\xi \mapsto h(t\xi)$ of a solution are again solutions; hence we may assume that the solutions are defined in the unit disk $\mathbb{D}$; that is, $h \in \mathbb{W}^{1,2}(\mathbb{D})$ with $h(\xi) \neq 0$ except for $\xi = 0$.

Lemma 16.3.4. *Every solution $h \in \mathbb{W}^{1,2}(\mathbb{D})$ to (16.71) has an infinite series expansion of the form*
$$h(\xi) = \sum_{l \geq n} \left(A_l \xi^l + \varepsilon_l \overline{A_l}\, \bar{\xi}^l\right) |\xi|^{\lambda_l + n - l} \xi^{-n}, \tag{16.72}$$

where $\lambda_l = \lambda_l(n,p)$ and $\varepsilon_l = \varepsilon_l(n,p)$ are determined by
$$2\lambda_l = -np + \sqrt{4l^2(p-1) + n^2(p-2)^2}, \qquad \varepsilon_l = \frac{\lambda_l + n - l}{\lambda_l + n + l}$$

16.3. THE HODOGRAPHIC METHOD

The complex coefficients A_l, $l \geq n$, satisfy

$$\sum_{l \geq n+1} l |A_l|^2 \leq C(n,p) \int_{\mathbb{D}} |Dh|^2 < \infty$$

Moreover, the series (16.72) converges in $\mathbb{W}^{1,2}(\mathbb{D})$.

Conversely, given any complex numbers A_l, $l \geq n$, satisfying $\sum_{l \geq n+1} l|A_l|^2 < \infty$, then the series (16.72) converges in $\mathbb{W}^{1,2}(\mathbb{D})$ to a solution h for which we have

$$\int_{\mathbb{D}} |Dh|^2 \leq C(n,p) \sum_{l \geq n+1} l|A_l|^2$$

Remark. If $p = 2$, we see that $\lambda_l = l - n$ and $\varepsilon_l = 0$, thus $h(\xi) = \sum_{l \geq n} A_l \xi^{l-n}$.

We shall not give a proof of Lemma 16.3.4 or subsequent remarks but only point out that the proof follows by analysis of the Fourier series expansion

$$h(re^{i\theta}) = \sum_{l=-\infty}^{\infty} a_l(r) e^{i(l-n)\theta}$$

The lengthy though elementary computational details can be found in [185]. Obviously, away from the origin the series represents a C^∞-function. Thus from the point of view of regularity, we need only examine the behavior of h and its inverse map χ near the origin.

16.3.6 Sharp Hölder Regularity for p-Harmonic Functions

As might be expected, the least degree of regularity of solutions to the hodograph equation comes from a single term in the series development (16.72) at $0 = f(0)$. Actually, the analysis carried out in [185] reveals that the strongest singularity occurs when $n = 1$, $A_1 = 0$, $A_2 = 1$ and $A_3 = A_4 = \cdots = 0$. In that case we have

$$\lambda_2 = -p + \sqrt{16(p-1) + (p-1)^2}, \qquad \varepsilon_2 = \frac{\lambda_2 - 1}{\lambda_2 + 3},$$

and for the corresponding solution,

$$h(\xi) = |\xi|^{\lambda_2 - 1} \xi + \varepsilon_2 |\xi|^{\lambda_2 - 3} \bar{\xi}^3$$

Let $f(z)$ be the inverse of this mapping. As such, it is the complex gradient of a p-harmonic function. In fact, that p-harmonic function, say u, is the worst possible from the point of view of regularity. We refer the reader to [185] for the details and report only the final conclusion.

Theorem 16.3.5. *Let $u \in W^{1,p}_{loc}(\Omega)$ be p-harmonic, $1 < p < \infty$. Then*

$$u \in C^{k,\alpha}_{loc}(\Omega), \tag{16.73}$$

where the integer $k \geq 1$ and the Hölder exponent $\alpha \in (0,1]$ are determined by the equation

$$k + \alpha = \frac{1}{6}\left(7 + \frac{1}{p-1} + \sqrt{1 + \frac{14}{p-1} + \frac{1}{(p-1)^2}}\right) \tag{16.74}$$

For $p \neq 2$ this regularity class in (16.73) is optimal. That is, for each $1 < p < \infty$, $p \neq 2$, there exists a p-harmonic function that cannot even be approximated by smooth functions in the $C^{k,\alpha}$-norm, with k and α as in (16.73) and (16.74).

16.3.7 Removing the Rough Regularity in the Gradient

One interesting feature of the regularity results we have seen so far is that $k + \alpha$ decreases from ∞ to $\frac{4}{3}$ as p runs from 1 to ∞. In particular, this implies the slightly surprising fact that every p-harmonic function defined in a planar domain Ω belongs to the class $C^{1,\beta}_{loc}(\Omega)$ with some $\beta = \beta(p) > \frac{1}{3}$.

On the other hand, the δ-powers of the complex gradient f, defined earlier in (16.62) by $F^\delta = |f|^{\delta-1}f$, provide us with more and more regular vector fields when δ approaches ∞. For example, take a p-harmonic function u and apply the sharp regularity Theorem 16.3.5 to the conjugate function v. We find that the divergence-free vector field

$$F^{p-1} = |\nabla u|^{p-2}\nabla u = *\nabla v \tag{16.75}$$

lies in $C^{k-1,\alpha}$ with

$$\begin{aligned} k + \alpha &= \frac{1}{6}\left(7 + \frac{1}{q-1} + \sqrt{1 + \frac{14}{q-1} + \frac{1}{(q-1)^2}}\right) \\ &= \frac{1}{6}\left(7 + p - 1 + \sqrt{1 + 14(p-1) + (p-1)^2}\right) \end{aligned}$$

Thus $k - 1 + \alpha$, the regularity of the nonlinear expression (16.75), increases to infinity as $p \to \infty$.

This observation raises some interesting general questions regarding nonlinear elliptic PDEs, and the p-harmonic equation in particular. Recall from (16.69) that the complex gradient f satisfies $f_{\bar{z}} = \left(\frac{1}{p} - \frac{1}{2}\right)(\mu f_z + \nu \overline{f_z})$ with $\mu = \overline{f}/f = \bar{\nu}$. The ellipticity constant here is $p - 1$, since $k = \left|\frac{1}{p} - \frac{1}{2}\right|(|\mu| + |\nu|) = 1 - \frac{2}{p}$ for $p > 2$, so that

$$K = \frac{1+k}{1-k} = p - 1,$$

which tends to infinity as $p \to \infty$.

16.4. NONLINEAR $\mathcal{A}$-HARMONIC EQUATION

Generally, one cannot expect smoothness (say C^∞) of the gradient of a solution u near critical points. Nevertheless, it appears that some expressions of the form $\Phi(\nabla u)$, where $\Phi : \mathbb{R}^2 \to \mathbb{R}^2$ is a homeomorphism, may have somewhat greater regularity (as in this instance, where $\Phi(z) = |z|^{p-1}z$). This situation is reminiscent of the uniformization of multivalued functions. This suggests itself because the hodographic method asks us to consider, as the dependent variable, the inverse of the complex gradient of a solution to an equation. Inverses are not typically single-valued unless defined so as to be valued in some Riemann surface. Let us illustrate with a particular example. Consider (16.72) with $n = 1$, $A_3 = 1$ and all other $A_l = 0$, $l \neq 3$. In that case we have the solution

$$h(\zeta) = |\zeta|^{\lambda-2}\zeta^2 + \varepsilon|\zeta|^{\lambda-4}\bar\zeta^4 = h(-\zeta),$$

where we have chosen λ so that

$$2\lambda = -p + \sqrt{p^2 + 32p - 32} \quad \text{and} \quad \varepsilon = \frac{\lambda - 2}{\lambda + 4}$$

Now solving the equation $z = h(\zeta)$ for ζ shows that $\zeta = f(z)$ is a two-valued function. Uniformization of $f = u_z$ is achieved by squaring f. Indeed, $g = f^2(z)$ becomes univalent and is obtained by solving the algebraic equation

$$|g|^{\lambda/2-1}g + \varepsilon|g|^{\lambda/2-2}\bar g^2 = z$$

uniquely solvable for g near the origin.

16.4 The Nonlinear $\mathcal{A}$-Harmonic Equation and Its Hodge Conjugate

So far in this chapter we have concentrated on second-order linear differential operators of divergence form, and our first foray into the nonlinear world was with the p-harmonic equation. We now look to extend some of the ideas from quasiconformal mappings in the p-harmonic setting to a more general nonlinear setting. Here we shall consider equations of the form

$$\operatorname{div} \mathcal{A}(z, g + \nabla u) = \operatorname{div} h \tag{16.76}$$

in a domain $\Omega \subset \mathbb{R}^2$. The mapping $\mathcal{A} : \Omega \times \mathbb{R}^2 \to \mathbb{R}^2$ is called a *structural field*. The two vector fields $g, h : \Omega \to \mathbb{R}^2$ are given, and we are supposed to be looking for a real-valued solution $u : \Omega \to \mathbb{R}$.

We first point out that when dealing with an inhomogeneous equation such as (16.76), there is no loss of generality in assuming

$$\mathcal{A}(z, 0) \equiv 0, \qquad z \in \Omega \tag{16.77}$$

We should now look for the *natural setting* of this equation. That is, we seek conditions on $\mathcal{A}$ and regularity assumptions on g and h under which the equation can be given a rigorous formulation and the solution u can be found in an identifiable Sobolev class. However, before doing this, let us make a few fairly general and informal comments so as to facilitate our precise discussion later.

The concept of a stream function that we have used a number of times already in the linear case becomes even more crucial in the nonlinear setting. Since $\mathcal{A}(z, g+\nabla u) - h$ is going to be divergence-free for a solution, we can express it as $-*\nabla v$ for some function v (which is of course the stream function), where $*$ is the Hodge $*$ as defined in (16.7). We therefore have the equation

$$\nabla v = *[\mathcal{A}(z, g + \nabla u) - h], \tag{16.78}$$

which we want to resolve for u. Formally, we see that

$$g + \nabla u = \mathcal{A}^{-1}(z, h - *\nabla v)$$

(here $\mathcal{A}^{-1}$ denotes the inverse of the map $\zeta \mapsto \mathcal{A}(z,\zeta)$). To eliminate u we apply Hodge $*$ and take the divergence, then using the identity $\mathrm{div} * \nabla u = 0$,

$$\mathrm{div}(*g) = \mathrm{div} * \mathcal{A}^{-1}(z, h - *\nabla v)$$

This is the operator we shall call the Hodge $*$ conjugate operator, or simply the Hodge conjugate operator. Let us write it in the same form as in (16.76),

$$\mathrm{div}\mathcal{A}^*(z, g^* + \nabla u^*) = \mathrm{div}\, h^*, \tag{16.79}$$

where

$$\mathcal{A}^*(z,\zeta) = *\mathcal{A}^{-1}(z, -*\zeta) = *\mathcal{A}^{-1}(z, *^t\zeta)$$

and where

$$g^* = *h, \quad \text{and} \quad h^* = *g$$

The solution to this equation $u^* = v$ is the stream function for u. In this notation

$$-\nabla u = *[\mathcal{A}^*(z, g^* + \nabla u^*) - h^*],$$

from which we see that $(u^*)^* = -u$. Here note the clear analogy with the Hilbert transform and the identity $\mathcal{H}^2 = -\mathbf{I}$. Similarly, $(g^*)^* = -g$ and $(h^*)^* = -h$.

We are still looking for the correct Sobolev class to identify our solutions, and so we recall the integral form of the equation,

$$\int_\Omega \langle \mathcal{A}(z, g + \nabla u), \nabla \varphi \rangle = \int_\Omega \langle h, \nabla \varphi \rangle \tag{16.80}$$

for all $\varphi \in C_0^\infty(\Omega)$. Typically, one proceeds from here by trying to use the solution itself as a test function after multiplication by a smooth cutoff function to get a priori estimates. In order to be able to consider anything like this, we would at least require the $\mathcal{A}$-harmonic energy of u to be finite. Thus we define

$$\mathcal{E}_\mathcal{A}[u] = \int_\Omega \langle \mathcal{A}(z, g + \nabla u), \nabla u \rangle \tag{16.81}$$

16.4. NONLINEAR $\mathcal{A}$-HARMONIC EQUATION

and require $\mathcal{E}_\mathcal{A}[u] < \infty$. Looking at this carefully, one sees that a requirement, more or less, is that g and ∇u lie in the same space (since they are added together!) and, moreover, by (16.80) that h should be in the dual space to ∇u and hence in the dual space to g. To get precise results in great generality, we will have to look beyond the usual Sobolev spaces to Orlicz spaces and their duals. These spaces should be determined in a natural way by the structural field $\mathcal{A} : \Omega \times \mathbb{R}^2 \to \mathbb{R}^2$.

16.4.1 δ-Monotonicity of the Structural Field

The usual condition that is imposed on the structural field is that of monotonicity. In order to connect with ellipticity, we have to quantize this notion and hence arrive at the δ-monotonicity as discussed in Section 3.11. For convenience we recall that structural field $\mathcal{A}$ is called δ-*monotone*, $0 < \delta \leq 1$, if for all $z \in \Omega$ and $\zeta, \xi \in \mathbb{R}^2$,

$$\langle \mathcal{A}(z,\zeta) - \mathcal{A}(z,\xi), \zeta - \xi \rangle \geq \delta |\mathcal{A}(z,\zeta) - \mathcal{A}(z,\xi)| \, |\zeta - \xi| \qquad (16.82)$$

In particular, we must have

$$\langle \mathcal{A}(z,\zeta), \zeta \rangle \geq \delta |\mathcal{A}(z,\zeta)||\zeta|$$

Earlier, in (16.2), we quantified, for a symmetric linear matrix function $A(z)$, strong ellipticity by the condition

$$|\zeta|^2 + |A(z)\zeta|^2 \leq \left(K + \frac{1}{K} \right) \langle A(z)\zeta, \zeta \rangle, \qquad \zeta \in \mathbb{R}^2 \qquad (16.83)$$

Since $2|\zeta| \, |A(z)\zeta| \leq |\zeta|^2 + |A(z)\zeta|^2$, we see that such a structural field $\mathcal{A}(z,\zeta) = A(z)\zeta$ is δ-monotone. However, the converse is not true in general since monotonicity is a scale invariant notion, while the condition (16.83) is not. But if we normalize so that $\det A(z) \equiv 1$, then (16.83) holds if and only if

$$2|\zeta| \, |A(z)\zeta| \leq \left(K + \frac{1}{K} \right) \langle A(z)\zeta, \zeta \rangle, \qquad \zeta \in \mathbb{R}^2 \qquad (16.84)$$

Then, in this special case, δ-monotonicity and strong ellipticity are equivalent concepts, with the parameters related by

$$\delta = \frac{2K}{K^2 + 1}$$

The symmetry in the definition (16.82) implies that a structural field $\mathcal{A}$ is δ-monotone if and only if $\mathcal{A}^*$ is δ-monotone. Kovalev's theorem (Section 3.11) shows that δ-monotone mappings are in fact quasisymmetric. Thus there is

$$\eta = \eta_\delta : [0, \infty) \to [0, \infty)$$

such that

$$\frac{|\mathcal{A}(z,\zeta) - \mathcal{A}(z,\xi_1)|}{|\mathcal{A}(z,\zeta) - \mathcal{A}(z,\xi_2)|} \leq \eta \left(\frac{|\zeta - \xi_1|}{|\zeta - \xi_2|} \right) \qquad (16.85)$$

From Theorem 3.11.6, we know that the linear distortion of $\mathcal{A}$ is at most
$$K \leqslant \frac{1 + \sqrt{1-\delta^2}}{1 - \sqrt{1-\delta^2}}$$
Hence we have explicit estimates for η_δ from Corollary 3.10.4 and Theorem 12.6.2. It follows, since $\mathcal{A}(z,0) = 0$, that
$$e^{-\pi K} \min\{|\zeta|^K, |\zeta|^{1/K}\} \leqslant \frac{|\mathcal{A}(z,\zeta)|}{|\mathcal{A}(z,e_1)|} \leqslant e^{\pi K} \max\{|\zeta|^K, |\zeta|^{1/K}\} \qquad (16.86)$$

Because of the considerable technical difficulties and lengthy calculations, which we do not wish to present in this book, we will restrict ourselves by assuming that

the structural field is independent of the space variable.

Many of the key ideas and techniques given in what follows are also valid in the more general setting. We also note that the continuity equations for planar fluid flow, like p-Laplace equation and many other physical divergence-type equations, have this feature.

We then introduce two strictly increasing functions
$$a(t) = \sup_{|\zeta|=t} |\mathcal{A}(\zeta)| \qquad (16.87)$$
$$a^*(t) = \inf_{|\zeta|=t} |\mathcal{A}^*(\zeta)| \qquad (16.88)$$
Thus quasisymmetry gives us the bounds
$$|\mathcal{A}(\zeta)| \leqslant a(|\zeta|) \leqslant \eta(1)|\mathcal{A}(\zeta)| \qquad (16.89)$$
$$\frac{1}{\eta(1)}|\mathcal{A}^*(\zeta)| \leqslant a^*(|\zeta|) \leqslant |\mathcal{A}^*(\zeta)| \qquad (16.90)$$
We note for reference that $a^*(t)$ is the inverse of $a(t)$. This follows simply from topology and can be seen in the following calculation,
$$a^*(a(t)) = \inf_{|\zeta|=a(t)} |\mathcal{A}^*(\zeta)| = \inf_{|\zeta|=a(t)} |\mathcal{A}^{-1}(\zeta)| = \inf_{|\mathcal{A}(\zeta)|=a(t)} |\zeta| = t$$
Note also that for all $k, t \in [0, \infty)$,
$$a(kt) \leqslant \eta(k)\, a(t), \qquad a^*(kt) \leqslant \eta(k)\, a^*(t)$$
From integrating these two functions, we obtain two *convex* functions
$$\Phi(t) = \int_0^t a(s)\, ds \qquad (16.91)$$
$$\Psi(t) = \int_0^t a^*(s)\, ds \qquad (16.92)$$
This setup gives a pair of Young conjugate functions and we shall use them to define the Orlicz spaces of functions necessary for the theory.

16.4. NONLINEAR $\mathcal{A}$-HARMONIC EQUATION

Orlicz Spaces L^Φ and L^Ψ

We set
$$L^\Phi(\Omega) = \{g : \Omega \to \mathbb{R}^2 : \Phi(|g|) \in L^1(\Omega)\}$$
$$L^\Psi(\Omega) = \{g : \Omega \to \mathbb{R}^2 : \Psi(|g|) \in L^1(\Omega)\}$$

and the similarly defined local spaces. These are both Banach spaces when equipped with the Luxemburg norm

$$\|g\|_\Phi = \inf\left\{\lambda : \int_\Omega \Phi\left(\frac{|g|}{\lambda}\right) \leqslant 1\right\}, \tag{16.93}$$

and of course similarly for $\|g\|_\Psi$. Actually, the infimum here is attained at $\lambda = \|g\|_\Phi$ unless $g \equiv 0$. We delve into the theory of these function spaces a little but do not give all the proofs. The functions Φ and Ψ satisfy a doubling condition,

$$\Phi(kt) = \int_0^{kt} a(s)\,ds = k\int_0^t a(ks)\,ds \leqslant k\eta(k)\int_0^t a(s)\,ds = k\eta(k)\Phi(t),$$

so

$$\frac{k}{\eta(1/k)}\Phi(t) \leqslant \Phi(kt) \leqslant k\eta(k)\,\Phi(t), \tag{16.94}$$

and similarly,

$$\frac{k}{\eta(1/k)}\Psi(t) \leqslant \Psi(kt) \leqslant k\eta(k)\,\Psi(t) \tag{16.95}$$

This makes both L^Φ and L^Ψ separable Banach spaces. Because of the estimate in (16.86), we have the following power bounds on the growth of these functions near 0 and ∞,

$$c_\delta\, t\, \min\{t^K, t^{1/K}\} \leqslant \Phi(t), \Psi(t) \leqslant C_\delta\, t\, \max\{t^K, t^{1/K}\} \tag{16.96}$$

In particular, we must have $L^\Phi_{loc}(\Omega), L^\Phi_{loc}(\Omega) \subset L^p_{loc}(\Omega)$ for

$$p = 1 + \frac{1}{K} \geqslant \frac{2}{1 + \sqrt{1-\delta^2}} > 1$$

In view of (16.94) and (16.95), we find that

$$\frac{\|g\|_\Phi}{\eta(1/\|g\|_\Phi)} \leqslant \int_\Omega \Phi(|g|) \leqslant \|g\|_\Phi\, \eta(\|g\|_\Phi) \tag{16.97}$$

$$\frac{\|h\|_\Psi}{\eta(1/\|h\|_\Psi)} \leqslant \int_\Omega \Psi(|h|) \leqslant \|h\|_\Psi\, \eta(\|h\|_\Psi) \tag{16.98}$$

Indeed,

$$\int_\Omega \Phi(|g|) \leqslant \|g\|_\Phi\, \eta(\|g\|_\Phi) \int_\Omega \Phi\left(\frac{|g|}{\|g\|_\Phi}\right),$$

and the latter integral is equal to 1 by the definition of the Luxemburg norm, and so on.

Energy Estimates

The importance of these two spaces L^Φ and L^Ψ for the divergence-type operator we wish to study lies in the following estimates of the energy integrand,

$$c_\delta \Phi(|\zeta|) \leqslant \langle \mathcal{A}(\zeta), \zeta \rangle \leqslant C_\delta \Phi(|\zeta|) \tag{16.99}$$

and

$$c_\delta \Psi(|\zeta|) \leqslant \langle \mathcal{A}^*(\zeta), \zeta \rangle \leqslant C_\delta \Psi(|\zeta|) \tag{16.100}$$

For proving these, in what follows we will use constants $0 < c_\delta \leqslant 1 \leqslant C_\delta < \infty$ that depend only on δ (and are obtained via η) and that may vary from line to line.

Before establishing these estimates we note that

$$\frac{\delta}{\eta(1)} |\zeta| a(|\zeta|) \leqslant \langle \mathcal{A}(\zeta), \zeta \rangle \leqslant |\zeta| a(|\zeta|) \tag{16.101}$$

$$\delta |\zeta| a^*(|\zeta|) \leqslant \langle \mathcal{A}^*(\zeta), \zeta \rangle \leqslant \eta(1) |\zeta| a^*(|\zeta|), \tag{16.102}$$

so in order to establish the estimates in (16.99) and (16.100), we actually only need to compare $t\,a(t)$ with $\Phi(t)$ and $t\,a^*(t)$ with $\Psi(t)$. This is not too difficult. Because a is increasing with $a(0) = 0$,

$$\begin{aligned}\Phi(t) &= \int_0^t a(s)\,ds \leqslant ta(t) \leqslant t\eta(2)a(t/2) \leqslant 2\eta(2) \int_{t/2}^t a(s)\,ds \\ &\leqslant 2\eta(2) \int_0^t a(s)\,ds = 2\eta(2)\Phi(t)\end{aligned}$$

Similarly for Ψ, so we have the two comparisons we want,

$$\Phi(t) \leqslant ta(t) \leqslant 2\eta(2)\Phi(t) \tag{16.103}$$
$$\Psi(s) \leqslant s\,a^*(s) \leqslant 2\eta(2)\Psi(s) \tag{16.104}$$

We therefore have the following corollary to the above discussion.

Corollary 16.4.1. *Suppose the structural field $\mathcal{A}$ is spatially independent and δ-monotone. Then the Sobolev function $u \in W^{1,1}_{loc}(\Omega)$ has finite energy if and only if its gradient lies in the Orlicz space $L^\Phi(\Omega)$. In that case, we have the estimate*

$$c_\delta \int_\Omega \Phi(|\nabla u|) \leqslant \mathcal{E}_\mathcal{A}[u] \leqslant C_\delta \int_\Omega \Phi(|\nabla u|) \tag{16.105}$$

Similarly for the Hodge conjugate (the stream function), for which we have the estimate

$$c_\delta \int_\Omega \Psi(|\nabla v|) \leqslant \mathcal{E}_{\mathcal{A}^*}[v] \leqslant C_\delta \int_\Omega \Psi(|\nabla v|)$$

We next note the inequalities

$$c_\delta\, \Phi(|\zeta|) \leqslant \Psi(|\mathcal{A}(\zeta)|) \leqslant C_\delta \Phi(|\zeta|) \tag{16.106}$$

16.4. NONLINEAR $\mathcal{A}$-HARMONIC EQUATION

and
$$c_\delta \Psi(|\zeta|) \leqslant \Phi(|\mathcal{A}^*(\zeta)|) \leqslant C_\delta \Psi(|\zeta|) \tag{16.107}$$

For instance, with the aid of the estimates in (16.103) and (16.104), we see that
$$\Psi(|\mathcal{A}(\zeta)|) \leqslant \Psi(a(|\zeta|)) \leqslant a(|\zeta|) \cdot a^*(a(|\zeta|)) = |\zeta|a(|\zeta|) \leqslant 2\eta(2)\Phi(|\zeta|),$$
and all the other inequalities follow in a similar fashion.

Next observe that for $g \in L^\Phi(\Omega)$ and $h \in L^\Psi(\Omega)$, we have
$$\mathcal{A}(g) \in L^\Psi(\Omega), \qquad \mathcal{A}(h) \in L^\Phi(\Omega)$$
with the estimates
$$c_\delta \leqslant \left\| \mathcal{A}\left(\frac{g}{\|g\|_\Phi}\right) \right\|_\Psi \leqslant C_\delta \tag{16.108}$$
$$c_\delta \leqslant \left\| \mathcal{A}^*\left(\frac{h}{\|h\|_\Psi}\right) \right\|_\Phi \leqslant C_\delta \tag{16.109}$$

Let us demonstrate the proof of at least one of these inequalities. We may assume that $\|h\|_\Psi = 1$. Set $G = \mathcal{A}^*(h)$, so by (16.107) $\Phi(|G|) \leqslant C_\delta \Psi(|h|)$. Therefore
$$\int_\Omega \Phi(|G|) \leqslant C_\delta \int_\Omega \Psi(|h|) = C_\delta$$

On the other hand, by (16.97), we have
$$\frac{\|G\|_\Phi}{\eta(1/\|G\|_\Phi)} \leqslant \int_\Omega \Phi(|G|) \leqslant C_\delta \tag{16.110}$$

Since the function $t \mapsto t/\eta(1/t)$ is strictly increasing from 0 to ∞, we may apply its inverse function, say $\Gamma(t)$, to estimate $\|G\|_\Phi$ and to conclude (with a possibly new constant) that
$$\left\| \mathcal{A}^*\left(\frac{h}{\|h\|_\Psi}\right) \right\|_\Phi = \|G\|_\Phi \leqslant \Gamma\left(\int_\Omega \Phi(|G|)\right) \leqslant C_\delta, \tag{16.111}$$
which is the estimate we required.

Duality

We now recall that both Φ and Ψ are represented by the integrals in (16.91) and (16.92), where a and a^* are inverse to each other. Thus we have Young's inequality
$$ts \leqslant \Phi(t) + \Psi(s) \tag{16.112}$$

This fact of course plays an important part in establishing the duality between the spaces $L^\Psi(\Omega)$ and $L^\Phi(\Omega)$. Indeed, if we apply (16.112) to $t = |g|/\|g\|_\Phi$ and $s = |h|/\|h\|_\Psi$, upon integration we have the formula
$$\frac{1}{\|g\|_\Phi \|h\|_\Psi} \left| \int_\Omega \langle g, h \rangle \right| \leqslant \int_\Omega \Phi\left(\frac{|g|}{\|g\|_\Phi}\right) + \int_\Omega \Psi\left(\frac{|h|}{\|h\|_\Psi}\right) = 2,$$

and hence Hölder's inequality

$$\left| \int_\Omega \langle g, h \rangle \right| \leqslant 2 \|g\|_\Phi \|h\|_\Psi \qquad (16.113)$$

whenever $g \in L^\Phi(\Omega)$ and $h \in L^\Psi(\Omega)$.

Again, since both Φ and Ψ satisfy a doubling condition, both of these spaces are reflexive and dual to each other. This duality is understood through the Riesz representation theorem.

Theorem 16.4.2. *Every bounded linear functional*

$$T : L^\Phi(\Omega) \to \mathbb{R}$$

is uniquely represented by an $h \in L^\Psi(\Omega)$ via the rule

$$Tg = \int_\Omega \langle g, h \rangle \qquad (16.114)$$

for all $g \in L^\Phi(\Omega)$. Similarly, the bounded linear functionals defined on $L^\Psi(\Omega)$ are represented uniquely by functions $g \in L^\Phi(\Omega)$.

The norm of such a bounded linear functional $T : L^\Phi(\Omega) \to \mathbb{R}$ is comparable with $\|h\|_\Psi$. From (16.100) and (16.114), with $g = \mathcal{A}^*(h/\|h\|_\Psi)$, we see that

$$Tg = \int_\Omega \langle h, \mathcal{A}^*(h/\|h\|_\Psi) \rangle \geqslant c_\delta \|h\|_\Psi \int_\Omega \Psi(h/\|h\|_\Psi) = c_\delta \|h\|_\Psi$$

As $\|g\|_\Phi \leqslant C_\delta$, by (16.109) we have

$$\frac{c_\delta}{C_\delta} \|h\|_\Psi \leqslant \|T\| \leqslant 2 \|h\|_\Psi, \qquad (16.115)$$

where the right hand estimate follows from (16.113).

The Orlicz-Sobolev Spaces $\mathbb{W}^{1,\Phi}(\Omega)$ and $\mathbb{W}_0^{1,\Phi}(\Omega)$

As should be self-evident, Corollary 16.4.1 suggests the spaces we should be working in. We want functions whose gradients are in the appropriate Orlicz class. Thus we define

$$\mathbb{W}^{1,\Phi}(\Omega) = \{ u \in W_{loc}^{1,1}(\Omega) : \nabla u \in L^\Phi(\Omega) \}$$

When functions that differ by a constant are identified, the space $\mathbb{W}^{1,\Phi}(\Omega)$ becomes a separable reflexive Banach space with norm

$$\|u\|_{1,\Phi} = \|\nabla u\|_\Phi$$

With (16.96) we have the inclusion

$$\mathbb{W}^{1,\Phi}(\Omega) \subset W_{loc}^{1,p}(\Omega), \qquad p = \frac{2}{1 + \sqrt{1 - \delta^2}} > 1 \qquad (16.116)$$

16.4. NONLINEAR A-HARMONIC EQUATION

Once again we need to look at those functions vanish on the boundary of Ω and so define $\mathbb{W}_0^{1,\Phi}(\Omega)$ to be the completion of $C_0^\infty(\Omega)$ in $\mathbb{W}^{1,\Phi}(\Omega)$. As a closed subspace, it too is a separable reflexive Banach space. The dual to the space $\mathbb{W}_0^{1,\Phi}(\Omega)$ consists of those Schwarz distributions of the form $T = \operatorname{div} h$, where $h \in L^\Psi(\Omega)$. The action of T on a $\varphi \in \mathbb{W}_0^{1,\Phi}(\Omega)$ does not depend on the choice of h (that is up to divergence-free vector fields) and is defined by the rule

$$T\varphi = -\int_\Omega \langle h, \nabla \varphi \rangle, \tag{16.117}$$

and the norm

$$\|T\|_\Phi = \sup\left\{\int_\Omega \langle h, \nabla \varphi \rangle : \varphi \in \mathbb{W}_0^{1,\Phi}(\Omega), \ \|\nabla \varphi\|_\Phi = 1\right\} \tag{16.118}$$

16.4.2 The Dirichlet Problem for the Nonlinear A-Harmonic Equation

We are now in a position to introduce the natural domain of definition of the A-harmonic operator and present the most general existence and uniqueness result we can for second-order equations of divergence form, Theorem 16.4.3 below, concerning

$$\operatorname{div} \mathcal{A}(g + \nabla u) = \operatorname{div} h \tag{16.119}$$

After settling the Dirichlet problem for this equation, the $C^{1,\alpha}$-regularity of solutions will then be established in Section 16.4.3.

It is quite remarkable that to establish the fundamental properties of solutions we need only to ask the structural field to be δ-monotone. No further assumptions such as the L^p-growth of the structural field are required. On the other hand, within this general setting it quickly becomes evident that a strong use of quasiconformal theory will be unavoidable.

Below the structural field has $\mathcal{A}(0) = 0$ and is δ-monotone. We are also given vector fields $g \in L^\Phi(\Omega)$ and $h \in L^\Psi(\Omega)$ and then seek a solution $u \in \mathbb{W}^{1,\Phi}(\Omega)$. The Dirichlet problem with prescribed boundary values seeks to find a solution

$$u \in u_0 + \mathbb{W}_0^{1,\Phi}(\Omega), \tag{16.120}$$

where $u_0 \in \mathbb{W}^{1,\Phi}(\Omega)$ is referred to as the Dirichlet data. Actually, in our inhomogeneous setting g is arbitrary in $L^\Phi(\Omega)$, so there is no loss of generality in assuming $u_0 \equiv 0$. This is how we shall prove the result, but we will formulate it with nonzero Dirichlet data.

Theorem 16.4.3. *The Dirichlet problem (16.119) with δ-monotone spatially independent structural field $\mathcal{A}$ has a unique solution $u \in u_0 + \mathbb{W}_0^{1,\Phi}(\Omega)$ for any data $u_0 \in L^\Phi(\Omega)$.*

The hypotheses of δ-monotonicity needs a few words. It is clear that some hypothesis on $\mathcal{A}$ is necessary. If $\mathcal{A}$ is a linear map that rotates a region by 90 degrees, then for any $u \in W^{1,1}_{loc}(\Omega)$ the expression $\mathcal{A}(\nabla u)$ is divergence-free. Thus if $\operatorname{div} \mathcal{A}(g) \neq \operatorname{div} h$, there are no solutions at all, while if $\operatorname{div} \mathcal{A}(g) = \operatorname{div} h$, then the equation is identically satisfied and little can be deduced in the way of regularity. The condition of δ-monotonicity precisely bounds us away from this situation. Monotonicity is going to lead us into the class of operators for which there is some fairly classical topological fixed-point theory we can use (Minty-Browder in this case) to find a solution. Monotonicity will also be key in proving uniqueness. On the other hand, δ-monotonicity implies quasisymmetry, which will be crucial to getting estimates—in particular, of the coercivity of our monotone operator.

Let us now prove Theorem 16.4.3.

Proof. Suppose $u_1, u_2 \in \mathbb{W}^{1,\Phi}_0(\Omega)$ are solutions to the $\mathcal{A}$-harmonic equation, that is,

$$\int_\Omega \langle \mathcal{A}(g + \nabla u_j), \nabla \varphi \rangle = \int_\Omega \langle \nabla \varphi, h \rangle \quad \text{for all } \varphi \in C^\infty_0(\Omega),\ j=1,2$$

Since $u_1, u_2 \in \mathbb{W}^{1,\Phi}_0(\Omega)$, we can use the difference as a test function to get

$$\begin{aligned}
0 &= \int_\Omega \langle \mathcal{A}(g+\nabla u_1) - \mathcal{A}(g+\nabla u_2), \nabla u_1 - \nabla u_2 \rangle \\
&\geq \delta \int_\Omega |\mathcal{A}(g+\nabla u_1) - \mathcal{A}(g+\nabla u_2)||\nabla u_1 - \nabla u_2| \geq 0, \quad (16.121)
\end{aligned}$$

and this is possible only if $\nabla u_1 \equiv \nabla u_2$, and so $u_1 = u_2$ as these functions coincide on the boundary.

For existence we have to appeal to the Minty-Browder theory of monotone operators [264, 77]. Let

$$T : \mathbb{X} \to \mathbb{X}^*$$

be a continuous (not necessarily linear) operator of a reflexive Banach space $\mathbb{X}$ into its dual $\mathbb{X}^*$. The operator T is called strictly monotone if whenever $x \neq y$,

$$\langle T(x) - T(y), x - y \rangle > 0$$

Here the scalar field is $\mathbb{R}$, and we interpret the inner product to mean the action of the functional $T(x) - T(y)$ on the point $x - y$. The operator T is coercive if

$$\lim_{\|x\| \to \infty} \frac{\langle Tx, x \rangle}{\|x\|} = \infty \qquad (16.122)$$

Next we recall the Minty-Browder theorem.

Theorem 16.4.4. *Every continuous strictly monotone and coercive operator $T : \mathbb{X} \to \mathbb{X}^*$ of a reflexive Banach space $\mathbb{X}$ is a bijective map.*

16.4. NONLINEAR 𝒜-HARMONIC EQUATION

Of course, we now have to check that we can apply this theorem. We take $\mathbb{X} = \mathbb{W}_0^{1,\Phi}(\Omega)$ and $\mathbb{X}^* = [\mathbb{W}_0^{1,\Phi}(\Omega)]^*$. Our operator is defined by

$$\langle Tu, \varphi \rangle = \int_\Omega \langle \mathcal{A}(g + \nabla u), \nabla \varphi \rangle \tag{16.123}$$

We have already seen that for $h \in L^\Psi(\Omega)$ we have $\operatorname{div} h \in [\mathbb{W}_0^{1,\Phi}(\Omega)]^*$, the dual space. We have already verified monotonicity as in (16.121). We have to argue to achieve coercivity and continuity. By (16.99),

$$\langle \mathcal{A}(\zeta), \zeta \rangle \geqslant c_\delta \, \Phi(|\zeta|)$$

We apply Young's inequality (16.112) and (16.106) to see that

$$\begin{aligned}
\langle \mathcal{A}(\zeta), \xi \rangle &= \varepsilon \langle \mathcal{A}(\zeta), \frac{\xi}{\varepsilon} \rangle \leqslant \varepsilon \Phi\left(\frac{|\xi|}{\varepsilon}\right) + \varepsilon \Psi(|\mathcal{A}(\zeta)|) \\
&\leqslant \varepsilon \, \Phi(|\xi|/\varepsilon) + \varepsilon \, C_\delta \Phi(|\zeta|), \qquad \varepsilon > 0,
\end{aligned}$$

where we choose $\varepsilon = c_\delta/2C_\delta$. These constants c_δ and C_δ are defined above in (16.99) and (16.106), respectively. We have

$$\langle \mathcal{A}(\zeta), \zeta - \xi \rangle \geqslant \frac{1}{2} c_\delta \, \Phi(|\zeta|) - \varepsilon \, \Phi(|\xi|/\varepsilon)$$

Thus

$$\langle Tu, u \rangle \geqslant \frac{1}{2} c_\delta \int_\Omega \Phi(|g + \nabla u|) - \varepsilon \int_\Omega \Phi(|g|/\varepsilon)$$

From (16.97) we obtain

$$\langle Tu, u \rangle \geqslant \frac{c_\delta \|g + \nabla u\|_\Phi}{2\eta(1/\|g + \nabla u\|_\Phi)} - \varepsilon \int_\Omega \Phi(|g|/\varepsilon)$$

We are then able to conclude that

$$\begin{aligned}
\lim_{\|u\| \to \infty} \frac{\langle Tu, u \rangle}{\|u\|} &= \lim_{\|\nabla u\|_\Phi \to \infty} \frac{\langle Tu, u \rangle}{\|g + \nabla u\|_\Phi} \\
&\geqslant \lim_{t \to \infty} \frac{c_\delta}{2\eta(1/t)} - \frac{\varepsilon}{t} \int_\Omega \Phi(|g|/\varepsilon) = \infty,
\end{aligned}$$

as desired.

Next we seek continuity of T. We argue as follows. Let $u \in \mathbb{W}_0^{1,\Phi}(\Omega)$ be fixed. It suffices to show that whenever we have a sequence with $u_j \to u$ in the norm topology of $\mathbb{W}_0^{1,\Phi}(\Omega)$, then there is a subsequence $\{u_{j_\ell}\}$ with $\|Tu_{j_\ell} - Tu\| \to 0$. Now,

$$\begin{aligned}
\|Tu_j - Tu\| &= \sup\left\{ \int_\Omega \langle \mathcal{A}(g + \nabla u_j) - \mathcal{A}(g + \nabla u), \nabla \varphi \rangle \; : \; \varphi \in \mathbb{W}_0^{1,\Phi}(\Omega), \|\nabla \varphi\|_\Phi = 1 \right\} \\
&\leqslant 2\|\mathcal{A}(g + \nabla u_j) - \mathcal{A}(g + \nabla u)\|_\Psi
\end{aligned}$$

by Hölder's inequality (in Orlicz spaces). In view of (16.98) it will suffice if we can show (for a subsequence) that

$$\int_\Omega \Psi(|\mathcal{A}(g+\nabla u_j) - \mathcal{A}(g+\nabla u)|) \to 0 \qquad \text{as } j \to \infty$$

We use Egoroff's theorem and a diagonalization to solve this problem. In view of (16.96) we have $\|\nabla u_j - \nabla u\|_{L^p(\Omega)} \to 0$ for some $p > 1$. Therefore for a subsequence, still denoted by $\{u_j\}$, we have $\nabla u_j \to \nabla u$ almost everywhere. We next write $\Omega = \bigcup_m \Omega_m$, where each Ω_m has finite measure. Then (writing Ω for Ω_j to simplify things) we apply Egoroff's theorem to find a compact set E_n on which $g + \nabla u_j \to g + \nabla u$ uniformly and so that $|\Omega \setminus E_n| < \frac{1}{n}$. Now on E_n we have

$$\int_{E_n} \Psi(|\mathcal{A}(g+\nabla u_j) - \mathcal{A}(g+\nabla u)|) \to 0,$$

and so it remains to estimate this integral on $\Omega \setminus E_n$. From the convexity of the function Ψ and (16.95) we note the following subadditivity-type property of Ψ.

$$\Psi(x+y) \leq \frac{1}{2}(\Psi(2x) + \Psi(2y)) \leq \eta(2)(\Psi(x) + \Psi(y))$$

Thus from (16.106) we have the estimate

$$\int_{\Omega \setminus E_n} \Psi(|\mathcal{A}(g+\nabla u_j) - \mathcal{A}(g+\nabla u)|)$$

$$\leq \eta(2) \left(\int_{\Omega \setminus E_n} \Psi(|\mathcal{A}(g+\nabla u_j)|) + \int_{\Omega \setminus E_n} \Psi(|\mathcal{A}(g+\nabla u)|) \right)$$

$$\leq \eta(2) C_\delta \left(\int_{\Omega \setminus E_n} \Phi(|g+\nabla u_j|) + \int_{\Omega \setminus E_n} \Phi(|g+\nabla u|) \right)$$

$$\leq \eta(2) C_\delta \left(\int_{\Omega \setminus E_n} |\Phi(|g+\nabla u_j|) - \Phi(|g+\nabla u|)| + 2\int_{\Omega \setminus E_n} \Phi(|g+\nabla u|) \right)$$

The last term here causes no difficulty as $\Phi(|g+\nabla u|) \in L^1(\Omega)$ and the set $\Omega \setminus E_n$ has small measure. We have to deal with the first term. We use the easy inequality

$$|\Phi(t_1) - \Phi(t_2)| = \left| \int_{t_1}^{t_2} a(s)\, ds \right| \leq |t_1 - t_2|(a(t_1) + a(t_2))$$

to estimate the first integral using Hölder's inequality as in (16.113),

$$\int_\Omega |\Phi(|g+\nabla u_j|) - \Phi(|g+\nabla u|)| \qquad (16.124)$$

$$\leq 2\|\nabla u_j - \nabla u\|_\Phi \left(\|a(|g+\nabla u_j|)\|_\Psi + \|a(|g+\nabla u|)\|_\Psi \right)$$

16.4. NONLINEAR $\mathcal{A}$-HARMONIC EQUATION

As in (16.111), the norm $\|a(|g + \nabla u|)\|_\Psi$ is bounded by a function of the integral $\int_\Omega \Psi(a(|g + \nabla u|))$, and this in turn is smaller than $\int_\Omega \Phi(|g + \nabla u|)$. Similarly for the term containing u_j, and so we see that the expression (16.124) tends to 0 independently of the set E_n. Finally, we should remove the hypothesis that $\Omega = \Omega_m$ has finite measure. This is done by the standard diagonal selection process, which we leave for the reader to complete. $\square$

16.4.3 Quasiregular Gradient Fields and $C^{1,\alpha}$-Regularity

In our earlier discussion of the p-harmonic operator in Theorem 16.3.2, we found that the complex gradient of a solution is a quasiregular mapping. Here we show this is no accident, and prove that quasiregular gradients are a general feature of nonlinear PDEs.

Theorem 16.4.5. *Any finite energy solution u to the equation*

$$\mathrm{div}\,\mathcal{A}(\nabla u) \equiv 0, \tag{16.125}$$

with δ-monotone spatially independent structural field $\mathcal{A}$, has a quasiregular complex gradient. In fact, $f = u_z$ is K-quasiregular with

$$K = \frac{1 + \sqrt{1 - \delta^2}}{1 - \sqrt{1 - \delta^2}} \tag{16.126}$$

One cannot improve the bound (16.126) in the theorem, as shown by the next simple example. For $0 \leqslant k < 1$ consider the field

$$\mathcal{A}(x,y) = \big((1-k)x, (1+k)y\big), \qquad (x,y) \in \mathbb{R}^2,$$

which is δ-monotone with $\delta = \sqrt{1-k^2}$. Equation (16.125) admits the solution $u(x,y) = x^2 - y^2 + k(x^2 + y^2)$, where $f = u_z = z + k\bar{z}$ is K-quasiconformal with

$$K = \frac{1+k}{1-k} = \frac{1 + \sqrt{1-\delta^2}}{1 - \sqrt{1-\delta^2}}$$

Theorem 16.4.5 has the consequence that for $f = u_z$ the composition $\mathcal{A}(f)$ is a quasiregular mapping and so is in $W^{1,2}_{loc}(\Omega)$. In this case of course, taking the divergence of $\mathcal{A}(f)$ can be understood in the usual sense—that is, not in the distributional sense—via the chain rule.

$C^{1,\alpha}$-Regularity

Another remarkable feature of the result is that for any spatially independent δ-monotone structural field, solutions to (16.125) exhibit $C^{1,\alpha}$-regularity independently of the smoothness of $\mathcal{A}$.

Corollary 16.4.6. *Let $\mathcal{A}$ be a δ-monotone spatially independent structural field. Then every finite energy solution u to (16.125) satisfies*

$$u \in C^{1,\alpha}_{loc}, \qquad \alpha = \frac{1 - \sqrt{1 - \delta^2}}{1 + \sqrt{1 - \delta^2}} \tag{16.127}$$

Every K-quasiregular mapping belongs to $C^{1/K}_{loc}$, which proves the corollary. In fact, as shown by Baernstein and Kovalev [37], all gradient K-quasiregular mappings have even better Hölder regularity. Thus the exponent α in (16.127) can be improved. See [37] for a conjecture on the optimal Hölder exponent for gradient quasiregular mappings. It is possible that their conjectured Hölder exponent is optimal for gradients of solutions to (16.125).

Similarly, we have discreteness of singularities of u, higher integrability for its second derivatives, removability results and so on, all immediate consequences of the theory of quasiregular mappings.

The proof of Theorem 16.4.5 will take us a while; it is covered in the next few subsections. The lack of smoothness in $\mathcal{A}$ forces us to abandon the difference quotient method used in Theorem 16.3.1, but we shall go about deriving a first-order quasilinear equation for f. To do this we initially make an informal calculation deriving a Beltrami equation for f; this calculation will be validated as soon as we prove u_z is quasiregular. We begin with the complex form of the general equation with a spatial variable present in the structural field for the moment, although we shall later return to (16.125),

$$\operatorname{div} \mathcal{A}(z, \nabla u) = 0, \tag{16.128}$$

where $\mathcal{A} = (A_1, A_2) : \Omega \times \mathbb{R}^2 \to \mathbb{R}^2$ is δ-monotone. For the complex gradient $f = u_z$, we have the complex structural field $\mathbf{A}(z, f)$, where

$$\mathbf{A}(z, \zeta) = A_1(z, (2x, -2y)) - iA_2(z, (2x, -2y)), \qquad \zeta = x + iy$$

Then (16.128) can be written in the form

$$\Re\left(\frac{\partial}{\partial \bar{z}} \mathbf{A}(z, f)\right) = 0, \qquad \Im m\left(\frac{\partial}{\partial \bar{z}} f\right) = 0 \tag{16.129}$$

The latter condition tells us that f is a complex gradient. The condition for δ-monotonicity reads as

$$\Re\bigl((\mathbf{A}(z,\zeta) - \mathbf{A}(z,\xi))(\bar{\zeta} - \bar{\xi})\bigr) \geqslant \delta |\mathbf{A}(z,\zeta) - \mathbf{A}(z,\xi)|\, |\zeta - \xi| \tag{16.130}$$

Recall that as a particular consequence of Theorem 3.11.6, every δ-monotone mapping satisfies the distortion inequality

$$|\mathbf{A}_{\bar{\zeta}}| \leqslant \sqrt{1 - \delta^2}\; \Re(\mathbf{A}_\zeta) \tag{16.131}$$

Hence any δ-monotone mapping is K-quasiconformal,

$$K \leqslant \frac{1 + \sqrt{1 - \delta^2}}{1 - \sqrt{1 - \delta^2}},$$

16.4. NONLINEAR $\mathcal{A}$-HARMONIC EQUATION

where the bound is sharp for every δ.

Let us look at an example. The p-harmonic equation $\text{div}(|\nabla u|^{p-2}\nabla u) = 0$ can be expressed in complex notation as

$$\Re \frac{\partial}{\partial \bar{z}}\left(|u_z|^{p-2} u_z\right) = 0 \tag{16.132}$$

Thus $\mathbf{A}(z, \zeta) = |\zeta|^{p-2}\zeta$,

$$\mathbf{A}_\zeta = \frac{p}{2}|\zeta|^{p-2}, \qquad \mathbf{A}_{\bar\zeta} = \frac{p-2}{2}|\zeta|^p \bar\zeta^{-2},$$

and hence

$$|\mathbf{A}_{\bar\zeta}| = \left|1 - \frac{2}{p}\right| \Re e(\mathbf{A}_\zeta) \tag{16.133}$$

The complex structural field $\mathbf{A}$ is therefore K-quasiconformal for

$$K = \max\{p - 1, 1/(p-1)\}$$

In the case of the general δ-monotone structural field, if we assume that the complex gradient $f = u_z$ lies in $W^{1,2}_{loc}(\Omega)$, we may now apply the chain rule to (16.129) to obtain the quasilinear Beltrami system

$$\frac{\partial f}{\partial \bar{z}} = \overline{\alpha(z,f)} \frac{\partial f}{\partial z} + \alpha(z,f)\overline{\frac{\partial f}{\partial z}} + h(z, f), \tag{16.134}$$

where

$$\alpha(z, \zeta) = -\frac{\mathbf{A}_{\bar\zeta}(z,\zeta)}{2\,\Re e(\mathbf{A}_\zeta(z,\zeta))}$$

and

$$h(z, \zeta) = -\frac{\Re e(\mathbf{A}_{\bar z}(z,\zeta))}{\Re e(\mathbf{A}_\zeta(z,\zeta))}$$

Notice that this equation is uniformly elliptic as

$$2|\alpha(z, f)| \leq k = \sqrt{1 - \delta^2} < 1$$

Further note that every solution to (16.134) is the complex gradient of a real-valued function since $\Im m\left(\frac{\partial f}{\partial \bar z}\right) = 0$.

The calculation preceding (16.134) can be reversed, and this process will take us back to (16.128). This suggests a method for solving (16.128) and proving Theorem 16.4.5. We first find a solution to the appropriate Beltrami equation with suitable boundary conditions (thus it is quasiregular) and then demonstrate that it is the complex gradient of some solution u. Now uniqueness will force the issue.

In the case of spatially independent structural field $\mathbf{A}$, the Beltrami equation we wish to solve is

$$\frac{\partial f}{\partial \bar z} = \overline{\alpha(f)}\frac{\partial f}{\partial z} + \alpha(f)\overline{\frac{\partial f}{\partial z}} \tag{16.135}$$

with

$$\alpha(\zeta) = \frac{-\mathbf{A}_{\bar\zeta}(\zeta)}{2\,\Re e(\mathbf{A}_\zeta(\zeta))}$$

Proof of Theorem 16.4.5

It suffices to show that u_z is K-quasiregular in every disk in Ω. We may of course take this arbitrary disk to be the unit disk $\mathbb{D}$. To avoid some technical issues involving the existence of second derivatives, we approximate our given solution by smooth functions. Thus let

$$w^j \to u \quad \text{in } \mathbb{W}^{1,\Phi}(\mathbb{D}),$$

where $w^j \in C^\infty(\mathbb{D})$ is real-valued. These functions will serve as the Dirichlet data. The functions u^j that will approximate u come as solutions to the Dirichlet problem

$$\left. \begin{array}{l} \text{div}\,\mathcal{A}(\nabla u^j) = 0 \\ u^j \in w^j + \mathbb{W}^{1,\Phi}_0(\mathbb{D}) \end{array} \right\}, \tag{16.136}$$

and we aim to prove that $(u^j)_z$ are K-quasiregular mappings in $\mathbb{D}$, with K as in (16.126), converging to u_z in $L^\Phi(\mathbb{D})$, so u_z is K-quasiregular as well.

To see that u^j_z are K-quasiregular we shall solve (16.136) in a different way than above. Recall that the solutions we have chosen to this Dirichlet problem are unique. Instead of (16.136), let us solve in $W^{1,2}(\mathbb{D})$ the associated quasilinear equation

$$f^j_{\bar{z}} = \overline{\alpha(f^j)}\, f^j_z + \alpha(f^j)\, \overline{f^j_{\bar{z}}} \tag{16.137}$$

with

$$\Im m(zf^j - zw^j_z) \in W^{1,2}_0(\mathbb{D}) \tag{16.138}$$

We put aside the difficulty of showing how to do this for a moment; it is the content of Theorem 16.5.1 in the next section. Assuming this, it then follows that in fact $\Im m(f^j_{\bar{z}}) \equiv 0$, and so the K-quasiregular mapping f^j is the complex gradient of a real function,

$$f^j = v^j_z,$$

where v^j is unique up to a constant. In view of (16.138),

$$\Im m(zv^j_z - zw^j_z) \in W^{1,2}_0(\mathbb{D}),$$

which is the same as saying that $v^j = w^j$ on $\partial \mathbb{D}$ up to a constant. We therefore choose v^j so that this constant is equal to zero. That is,

$$v^j \in w^j + W^{1,2}_0(\mathbb{D}) \cap W^{2,2}(\mathbb{D}) \subset w^j + \mathbb{W}^{1,\Phi}_0(\mathbb{D})$$

Now we can get to (16.136) from (16.137), as the v^j have sufficient regularity. Thus v^j and u^j solve the same boundary value problem in (16.136) and by uniqueness are the same. In particular, u^j_z are K-quasiregular.

∇u^j Converge to ∇u

We now subtract (16.125) and (16.136) and test them against $u^j - w^j \in \mathbb{W}_0^{1,\Phi}(\mathbb{D})$. We have

$$\begin{aligned} 0 &= \int_{\mathbb{D}} \langle \mathcal{A}(\nabla u^j) - \mathcal{A}(\nabla u), \nabla u^j - \nabla w^j \rangle \\ &= \int_{\mathbb{D}} \langle \mathcal{A}(\nabla u^j) - \mathcal{A}(\nabla u), \nabla u^j - \nabla u \rangle + \int_{\mathbb{D}} \langle \mathcal{A}(\nabla u^j) - \mathcal{A}(\nabla u), \nabla u - \nabla w^j \rangle \end{aligned}$$

We consider each of these integrals in turn. By Hölder's inequality in (16.113),

$$\int_{\mathbb{D}} \langle \mathcal{A}(\nabla u^j) - \mathcal{A}(\nabla u), \nabla u - \nabla w^j \rangle \leqslant 2\|\nabla u - \nabla w^j\|_\Phi \left(\|\mathcal{A}(\nabla u^j)\|_\Psi + \|\mathcal{A}(\nabla u)\|_\Psi \right)$$

Now $\|\nabla u - \nabla w^j\|_\Phi \to 0$ by our initial approximation of u by w^j. Of course, the term

$$\|\mathcal{A}(\nabla u^j)\|_\Psi + \|\mathcal{A}(\nabla u)\|_\Psi \leqslant C_\delta \left(\Gamma(\|\nabla u^j\|_\Phi) + \Gamma(\|\nabla u\|_\Phi) \right)$$

remains bounded by (16.124). On the other, hand monotonicity gives us the following estimate of the first integral:

$$\int_{\mathbb{D}} \langle \mathcal{A}(\nabla u^j) - \mathcal{A}(\nabla u), \nabla u^j - \nabla u \rangle \geqslant \delta \int_{\Omega} |\mathcal{A}(\nabla u^j) - \mathcal{A}(\nabla u)||\nabla u^j - \nabla u|$$

Thus for a subsequence we have $|\mathcal{A}(\nabla u^j) - \mathcal{A}(\nabla u)||\nabla u^j - \nabla u| \to 0$ pointwise almost everywhere. Since $\mathcal{A}$ is in fact a quasiconformal homeomorphism, we must have $|\nabla u^j - \nabla u| \to 0$ almost everywhere. As $u_{\bar{z}}^j$ now form a sequence of K-quasiregular mappings bounded in $L^\Phi(\mathbb{D}) \subset L^1(\mathbb{D})$ and converging to $u_{\bar{z}}$, we deduce with the Stoilow factorization that $u_{\bar{z}}$ itself is K-quasiregular. This completes the proof of Theorem 16.4.5. □

16.5 Boundary Value Problems

For a discussion on solving the boundary value problem (16.137) and (16.138), we begin again with the classical Dirichlet and Neumann problem for the Laplace equation in a bounded domain $\Omega \subset \mathbb{C}$. The Dirichlet problem asks us to find a harmonic function $u \in W^{1,2}(\Omega)$ coinciding with a given function (the Dirichlet data) $\omega \in W^{1,2}(\Omega)$ on the boundary $\partial\Omega$. We chose to write this as

$$\Delta u = 0, \qquad u \in \omega + W_0^{1,2}(\Omega) \tag{16.139}$$

On the other hand, the Neumann problem prescribes the normal derivative of a harmonic function on $\partial\Omega$. In order to get this problem into the framework of Sobolev functions of class $W^{1,2}(\Omega)$, we observe that for functions sufficiently regular we have div $*\nabla\omega = 0$. Hence on a smooth domain and for every smooth function φ, we have

$$\int_\Omega \langle *\nabla\omega, \nabla\varphi \rangle = \int_{\partial\Omega} \varphi \cdot \frac{\partial \omega}{\partial \mathbf{n}},$$

where **n** is the outer unit normal. Thus we may say that a function $w \in W^{1,2}(\Omega)$ has a *vanishing normal derivative* on $\partial\Omega$ if

$$\int_\Omega \langle *\nabla w, \nabla\varphi\rangle = 0 \qquad (16.140)$$

for every $\varphi \in W^{1,2}(\Omega)$. We denote this class of functions by $W^{1,2}_{\mathbf{n}}(\Omega)$. Now any function $w \in W^{1,2}(\Omega)$ could serve as the Neumann data, and we have the weak formulation

$$\Delta u = 0, \qquad u \in w + W^{1,2}_{\mathbf{n}}(\Omega) \qquad (16.141)$$

The classical formulation of the problems requires some regularity of the boundary, of the data w and also of the solution u. Let us for the moment anyway assume all of them are smooth. Suppose that Ω is a Jordan curve parameterized by arc length

$$z(t) = (x(t), y(t))$$

The tangential derivative along the boundary is

$$\frac{\partial}{\partial \mathbf{t}} = \dot{x}\frac{\partial}{\partial x} + \dot{y}\frac{\partial}{\partial y}$$

So for the Dirichlet problem we have

$$\dot{x}u_x + \dot{y}u_y = \dot{x}w_x + \dot{y}w_y, \qquad (16.142)$$

while for the Neumann problem we find the normal derivative is

$$\frac{\partial}{\partial \mathbf{n}} = \langle \dot{z}(t), *\nabla\rangle = \dot{x}\frac{\partial}{\partial y} - \dot{y}\frac{\partial}{\partial x},$$

so the boundary condition has the form

$$\dot{x}u_y - \dot{y}u_x = \dot{x}w_y - \dot{y}w_x \qquad (16.143)$$

The above simplifies considerably if we adopt complex notation. Put $z = x + iy$ and $f = u_z$, and then the Dirichlet boundary value problem for the Laplace operator reads as follows.

Dirichlet Problem

$$\frac{\partial f}{\partial \bar{z}} = 0 \quad \text{on } \Omega; \qquad \Re(\dot{z}f) = \Re(\dot{z}w_z) \quad \text{on } \partial\Omega$$

Neumann Problem

$$\frac{\partial f}{\partial \bar{z}} = 0 \quad \text{on } \Omega; \qquad \Im(\dot{z}f) = \Im(\dot{z}w_z) \quad \text{on } \partial\Omega,$$

which reveals a quite striking duality. We note here that our normalizations

16.5. BOUNDARY VALUE PROBLEMS

have $|\dot{z}| \equiv 1$.

Let us now suppose that the boundary of Ω is given by the defining equation $\Gamma(z) = 0$, where Γ is a smooth real-valued function defined in a neighborhood of $\partial\Omega$ and has a nonvanishing gradient. Upon differentiating the equation $\Gamma(z(t)) \equiv 0$, we obtain $\dot{z}\Gamma_z + \overline{\dot{z}}\Gamma_{\bar{z}} = 0$. This shows us that

$$\dot{z} = \frac{i\overline{\Gamma_z}}{|\Gamma_z|},$$

and the Dirichlet and Neumann problems take the form $f_{\bar{z}} = 0$, and

$$\Re\left(\frac{i\overline{\Gamma_z}}{|\Gamma_z|}f\right) = \Re\left(\frac{i\overline{\Gamma_z}}{|\Gamma_z|}\omega_z\right) \quad \text{and} \quad \Re\left(\frac{\overline{\Gamma_z}}{|\Gamma_z|}f\right) = \Re\left(\frac{\overline{\Gamma_z}}{|\Gamma_z|}\omega_z\right)$$

We can put both these problems in one general category then,

$$\Re(\rho(z)f(z)) = \gamma(z), \qquad z \in \partial\Omega, \qquad (16.144)$$

where ρ and γ are given continuous functions defined on $\partial\Omega$. The function ρ will be complex-valued and of modulus $|\rho| = 1$, while γ is real-valued. We seek a holomorphic function that satisfies (16.144). This is well known in the literature as the Riemann-Hilbert problem for holomorphic functions. Further generalization comes from the study of the genuinely nonlinear second-order PDE

$$F(x, y, u_x, u_y, u_{xx}, u_{xy}, u_{yy}) = 0$$

in Ω (note that there is no explicit dependence on u). This is because the equation reduces to a first-order system for the complex gradient $f = u_z$. We note

$$\begin{array}{lll} u_{xx} = 2f_{\bar{z}} + f_z + \overline{f_z}, & u_{yy} = 2f_{\bar{z}} - f_z - \overline{f_z}, & u_{xy} = i(f_z - \overline{f_z}) \\ u_{xx} + u_{yy} = 4f_{\bar{z}}, & u_{xx} - u_{yy} - 2iu_{xy} = 4f_z, & \Im m\, f_{\bar{z}} = 0 \end{array}$$

In Section 7.7 we discussed what it means for the nonlinear system

$$\Phi\left(z, f, \frac{\partial f}{\partial z}, \frac{\partial f}{\partial \bar{z}}\right) = 0 \qquad (16.145)$$

to be uniformly elliptic and arrived at the general equation

$$\frac{\partial f}{\partial \bar{z}} = H\left(z, f, \frac{\partial f}{\partial z}\right), \qquad H : \Omega \times \Omega' \times \mathbb{C} \to \mathbb{C}, \qquad (16.146)$$

under the subsequent conditions following (7.43). In our present circumstances, where (16.146) comes from (16.145), we see that H must be a real-valued function as $\Im m\, f_{\bar{z}} = 0$. However, there is no real need to restrict ourselves to this case, and we therefore come to perhaps the most general formulation of the nonlinear Riemann-Hilbert problem.

16.5.1 A Nonlinear Riemann-Hilbert Problem

The problem we now have is the following: Given a nonlinear elliptic system

$$\frac{\partial f}{\partial \bar{z}} = H\left(z, f, \frac{\partial f}{\partial z}\right), \tag{16.147}$$

find a solution (a quasiregular mapping) satisfying the boundary conditions

$$\Re e(\rho(z) f(z)) = \gamma(z), \quad z \in \partial\Omega \tag{16.148}$$

with given real-valued γ and complex-valued ρ, $|\rho| = 1$ both continuous on $\partial\Omega$.

It is this nonlinear problem that we will discuss in the rest of this section, though we will not delve into its very depths. However, for our applications to second-order PDEs we will need to resolve this problem completely in the following setting.

Suppose Ω is a smoothly bounded convex domain in $\mathbb{C}$ defined by $\Gamma \in C^\infty(\bar{\Omega})$,

$$\bar{\Omega} = \{z : \Gamma(z) \leqslant 0, \ \Gamma_z \neq 0\} \quad \text{with } \rho(z) = \frac{i\bar{\Gamma}_z}{|\Gamma_z|}$$

Consider the following quasilinear elliptic system in Ω,

$$\frac{\partial f}{\partial \bar{z}} = \overline{\alpha(f)} \frac{\partial f}{\partial z} + \alpha(f) \overline{\frac{\partial f}{\partial z}}, \tag{16.149}$$

where f is sought in $W^{1,2}(\Omega)$ and where α is a bounded and measurable coefficient satisfying the uniform ellipticity condition

$$2|\alpha(\zeta)| \leqslant k < 1, \quad \text{almost every } \zeta \in \Omega \tag{16.150}$$

We then have the following theorem.

Theorem 16.5.1. *Given a real-valued function* $w \in W^{2,2}(\Omega)$, *the quasilinear equation (16.149) admits a solution of the following weak-type Riemann-Hilbert problem:*

$$\Re e(\rho f) \in \Re e(\rho w_z) + W_0^{1,2}(\Omega) \tag{16.151}$$

Remark. For some range of exponents $2 \leqslant p \leqslant 2 + \varepsilon(k)$, if $w \in W^{1,p}(\Omega)$, then also $f \in W^{1,p}(\Omega)$.

We first assume that the coefficient $\alpha : \mathbb{C} \to \mathbb{C}$ is continuous. In order to solve this Riemann-Hilbert problem we need to recall the integral operator relevant to this boundary value problem. As usual, we look for a solution of the form

$$f(z) = \frac{\partial w}{\partial z} + \mathcal{C}_\Delta \varphi,$$

where $\varphi \in L^2(\Omega)$ is a real-valued function and $\mathcal{C}_\Delta$ is the Cauchy transform for the Dirichlet problem, as described in Sections 4.8.3 and 4.8.3. In particular, by

16.5. BOUNDARY VALUE PROBLEMS

(4.152) we have $\Re e(\rho\mathcal{C}_\Delta\varphi) \in W_0^{1,2}(\Omega)$. This means that f satisfies the boundary value condition in (16.151). The complex partials of our purported solution are

$$\frac{\partial f}{\partial \bar{z}} = \frac{\partial^2 w}{\partial \bar{z}z} + \varphi \quad \text{and} \quad \frac{\partial f}{\partial z} = \frac{\partial^2 w}{\partial zz} + \mathcal{S}_\Delta \varphi$$

We can substitute this back into (16.149) to obtain a quasilinear singular integral equation for φ, namely,

$$\varphi = \bar{\alpha}(f)\mathcal{S}_\Delta\varphi + \alpha(f)\overline{\mathcal{S}_\Delta\varphi} + \bar{\alpha}(f)w_{zz} + \alpha(f)\overline{w_{zz}} - w_{z\bar{z}} \qquad (16.152)$$

Note that there the right hand side is actually real-valued. We solve the integral equation in (16.152) using Schauder's fixed-point theorem. For this we introduce a mapping $\mathbf{T}: L^2(\Omega, \mathbb{R}) \to L^2(\Omega, \mathbb{R})$ by the rule

$$\mathbf{T}\varphi = \Phi,$$

where Φ is the unique solution to the uniformly elliptic inhomogeneous equation

$$\begin{aligned}\Phi &= \bar{\alpha}(w_z + \mathcal{C}_\Delta\varphi)\mathcal{S}_\Delta\Phi + \alpha(w_z + \mathcal{C}_\Delta\varphi)\overline{\mathcal{S}_\Delta\Phi} \\ &\quad + \bar{\alpha}(w_z + \mathcal{C}_\Delta\varphi)w_{zz} + \alpha(w_z + \mathcal{C}_\Delta\varphi)\overline{w_{zz}} - w_{z\bar{z}},\end{aligned} \qquad (16.153)$$

guaranteed by Theorem 14.0.4 (or by considering the Neumann series for this L^2 case). Since $2|\alpha(\zeta)| \leq k < 1$ and $\|\mathcal{S}_\Delta\|_2 = 1$ by Theorem 4.8.3, we obtain a uniform bound on $\|\Phi\|_2$,

$$\|\Phi\|_2 \leq \frac{k}{1-k}\|w_{zz}\| + \frac{1}{1-k}\|w_{z\bar{z}}\|_2 = M \qquad (16.154)$$

We now need to see that $\mathbf{T}$ is continuous and compact. Suppose that $\phi_n \rightharpoonup \phi$ weakly in $L^2(\Omega, \mathbb{R})$ and we need to show that $\mathbf{T}\phi_n \to \mathbf{T}\phi$ strongly in $L^2(\Omega, \mathbb{R})$. The Cauchy transform $\mathcal{C}_\Delta$ defines a bounded linear operator from $L^2(\Omega, \mathbb{R}) \to W^{1,2}(\Omega)$, and as such we must have

$$w_z + \mathcal{C}_\Delta \phi_n \to w_z + \mathcal{C}_\Delta \phi \qquad (16.155)$$

almost everywhere in $\mathbb{C}$, when passing to a suitable subsequence. As the coefficients are supposed continuous for the moment, we see that

$$\alpha_n(z) = \alpha(w_z + \mathcal{C}_\Delta\phi_n) \to \alpha(w_z + \mathcal{C}_\Delta\phi_n) = \alpha(z)$$

almost everywhere in Ω. Now the solutions $\Phi_n = \mathbf{T}\phi_n$ to the equation

$$\Phi_n = \bar{\alpha}_n(z)\mathcal{S}_\Delta\Phi_n + \alpha_n(z)\overline{\mathcal{S}_\Delta\Phi_n} + \bar{\alpha}_n(z)w_{zz} + \alpha_n(z)\overline{w_{zz}} - w_{z\bar{z}}$$

converge strongly in $L^2(\Omega, \mathbb{R})$ to the solution $\Phi = \mathbf{T}\phi$ of the equation

$$\Phi = \bar{\alpha}(z)\mathcal{S}_\Delta\Phi + \alpha(z)\overline{\mathcal{S}_\Delta\Phi} + \bar{\alpha}(z)w_{zz} + \alpha(z)\overline{w_{zz}} - w_{z\bar{z}}$$

Indeed, if we look at the difference here, we have

$$(\Phi_n - \Phi) = \bar{\alpha}_n \mathcal{S}_\Delta(\Phi_n - \Phi) + \alpha_n \overline{\mathcal{S}_\Delta(\Phi_n - \Phi)} + \varepsilon_n(z)$$
$$\varepsilon_n(z) = (\bar{\alpha}_n - \bar{\alpha})\mathcal{S}_\Delta \Phi + (\alpha_n - \alpha)\overline{\mathcal{S}_\Delta \Phi} + (\bar{\alpha}_n - \bar{\alpha})w_{zz} + (\alpha_n - \alpha)\overline{w_{zz}}$$

The last four terms here (denoted by $\varepsilon_n(z)$) clearly converge strongly to 0 in $L^2(\Omega, \mathbb{R})$ by the Lesbegue dominated convergence theorem. But we may use the fact that $\|\mathcal{S}_\Delta\|_2 = 1$ and $2|\alpha_n| \leqslant k < 1$ to get

$$(1-k)\|\Phi_n - \Phi\|_2 \leqslant \|\varepsilon_n\|_2 \to 0$$

This holds no matter which subsequence we choose in (16.155) to obtain the convergence almost everywhere. Hence the whole sequence $\{\Phi_n\}$ converges in $L^2(\mathbb{C})$, proving that **T** is continuous and compact. Its image lies in the closed ball $\{\psi \in L^2(\Omega, \mathbb{R}) : \|\psi\|_2 \leqslant M\}$. Schauder's fixed-point theorem now provides us with the fixed-point $\varphi \in L^2(\Omega, \mathbb{R})$ that we have been seeking, that is, the solution to (16.152).

Our next step in the proof of the theorem must be to remove the hypotheses that α is continuous. Here one could use an approach similar to the proof of Theorem 8.2.1. However, in the present case of quasilinear equations there is less heavy machinery available.

Thus let $2|\alpha| \leqslant k < 1$ and measurable. Using a standard mollifying procedure, we may construct a sequence $\{\alpha_j(\zeta)\}_{j=1}^{\infty}$ that converges to $\alpha(\zeta)$ for almost every $\zeta \in \mathbb{C}$ and for which we retain the bounds $2|\alpha_j(\zeta)| \leqslant k$ for all $\zeta \in \mathbb{C}$. We solve the boundary value problem

$$\frac{\partial f^j}{\partial \bar{z}} = \bar{\alpha}_j(f^j)\frac{\partial f^j}{\partial z} + \alpha_j(f^j)\overline{\frac{\partial f^j}{\partial z}} \qquad (16.156)$$

$$\Re e(\rho f^j) \in \Re e(\rho w_z) + W_0^{1,2}(\Omega)$$

Thus, in particular, the mappings f^j are K-quasiregular, $K = \frac{1+k}{1-k}$, in Ω. As in (16.154), we have a uniform bound on the L^2-norms of the derivatives; $\|f_{\bar{z}}^j\|_2 \leqslant M$, while

$$\|(f^j - w_z)_z\|_2 \leqslant \|(f^j - w_z)_{\bar{z}}\|_2 \leqslant M + \|w_{z\bar{z}}\|_2,$$

which gives us the uniform bounds

$$\|f_{\bar{z}}^j\|_2 + \|f_z^j\|_2 \leqslant C\left(\|w_{zz}\|_2 + \|w_{z\bar{z}}\|_2\right) \qquad (16.157)$$

for $j = 1, 2, \ldots$. We may pass to a subsequence for which

- $f^j \to f$ locally uniformly in Ω and f is quasiregular (or constant).
- $Df^j \to Df$ weakly in $L^2(\Omega)$.

16.5. BOUNDARY VALUE PROBLEMS

As a consequence we have $\Re e(\rho f^j - \rho \omega_z) \to \Re e(\rho f - \rho \omega_z)$ weakly in $W_0^{1,2}(\Omega)$. Thus
$$\Re e(\rho f) \in \Re e(\rho \omega_z) + W_0^{1,2}(\Omega) \qquad (16.158)$$
Of course, if f is constant, then it satisfies the equation
$$\frac{\partial f}{\partial \bar{z}} = \overline{\alpha(f)} \frac{\partial f}{\partial z} + \alpha(f) \overline{\frac{\partial f}{\partial z}} \qquad (16.159)$$
trivially. Thus we suppose otherwise and will show next that f satisfies (16.159) by passing to the weak L^2-limit in (16.156).

Passing to the Limit

The limiting process we need to finish the proof of Theorem 16.5.1 is accomplished by studying the target domains $f^j(\Omega) \subset \mathbb{C}$. As f is a nonconstant quasiregular mapping, it is discrete and open and the branch set of f consists of isolated points (see Section 5.5). It is enough to show that f satisfies the equation (16.159) almost everywhere and so in particular away from these points. Choose such a point z_0 and a neighborhood $\mathbb{D}(z_0, r) \subset \Omega$ of z_0 such that $f|\mathbb{D}(z_0, r)$ is univalent. Such a disk exists simply by the definition of branch points. As the degree is continuous under uniform convergence (see Section 2.8) and as $f_j \to f$ uniformly, we see that for all sufficiently large j each f^j is univalent in $B = \mathbb{D}(z_0, r/2)$. We are now in a situation where the equation coefficients α_j converge and the solutions f^j of
$$\frac{\partial f^j}{\partial \bar{z}} = \overline{\alpha_j(f^j)} \frac{\partial f^j}{\partial z} + \alpha_j(f^j) \overline{\frac{\partial f^j}{\partial z}} \qquad (16.160)$$
are quasiconformal homeomorphisms that also converge to a quasiconformal homeomorphism $f : B \to f(B) = B'$. Choose and then fix an open set $U' \subset B'$ compactly contained in B'. The uniform convergence of f^j to f assures us that $f^j(B)$ contains U' for all sufficiently large j. We show that (16.160) holds on $U = f^{-1}(U')$. Let us consider the inverse mapping $g^j = (f^j)^{-1} : U' \to B$. The quasilinear equation (16.160) is actually a linear equation for g^j,
$$-\frac{\partial g^j}{\partial \bar{w}} = \overline{\alpha_j(w)} \frac{\partial g^j}{\partial w} + \alpha_j(w) \overline{\frac{\partial g^j}{\partial w}}, \qquad w \in U'$$
We can now pass to the weak L^2-limit (and since we must have $g^j \to g = f^{-1}$)
$$-\frac{\partial g}{\partial \bar{w}} = \overline{\alpha(w)} \frac{\partial g}{\partial w} + \alpha(w) \overline{\frac{\partial g}{\partial w}}, \qquad w \in U'$$
We return to the equation for f by once again changing variables to get to
$$\frac{\partial f}{\partial \bar{z}} = \overline{\alpha(f)} \frac{\partial f}{\partial z} + \alpha(f) \overline{\frac{\partial f}{\partial z}}, \qquad z \in U,$$
which is what we wanted. $\square$

We have now completed the proofs of Theorems 16.4.5 and 16.5.1. Actually, the proof shows the power of the hodographic transform (that is, an analysis of the inverse mapping) in these situations. However, if the coefficients were of the form $\alpha(z, f)$, such a trick would not work. Indeed, there are considerable technical difficulties in formulating and proving limit theorems for equations in this and more general situations involving in particular subtleties around convergence in the Lusin sense and so forth. We shall not discuss this further but now turn to consider another aspect of limits of equations.

16.6 G-Compactness of Beltrami Differential Operators

In this section we shall study the compactness properties of *families* of Beltrami differential operators of the form

$$\mathcal{B} = \frac{\partial}{\partial \bar{z}} - \mu(z) \frac{\partial}{\partial z} - \nu(z) \overline{\frac{\partial}{\partial z}} : \mathbb{W}^{1,2}(\mathbb{C}) \to L^2(\mathbb{C}) \tag{16.161}$$

Recall that earlier in Chapter 14 we considered Beltrami operators of the form $\mathcal{B} \circ \bar{\partial}^{-1}$. However, the form (16.161) is customary in the literature on G-convergence, the topic we will study here, and hence we adopt this practice in the present section. We continue to assume uniform ellipticity so that

$$|\mu(z)| + |\nu(z)| \leqslant k < 1 \tag{16.162}$$

almost everywhere in $\mathbb{C}$. Every such operator is a bijection (see Theorem 14.0.4), and so we may speak of the inverse

$$\mathcal{B}^{-1} : L^2(\mathbb{C}) \to \mathbb{W}^{1,2}(\mathbb{C})$$

We now come to the interesting notion of G-convergence. We say that a sequence of Beltrami differential operators

$$\mathcal{B}_j = \frac{\partial}{\partial \bar{z}} - \mu_j(z) \frac{\partial}{\partial z} - \nu_j(z) \overline{\frac{\partial}{\partial z}} \tag{16.163}$$

with $|\mu_j(z)| + |\nu_j(z)| \leqslant k$ will G-*converge* to a Beltrami differential operator $\mathcal{B}$ if for every sequence $\{h_j\}_{j=1}^\infty \subset L^2(\mathbb{C})$ that strongly converges to $h \in L^2(\mathbb{C})$, we have $\mathcal{B}_j^{-1} h_j$ converging weakly to $\mathcal{B}^{-1} h$ in $\mathbb{W}^{1,2}(\mathbb{C})$. That is, $\mathcal{B}_j \to_G \mathcal{B}$ if

$$h_j \to h \Rightarrow \mathcal{B}_j^{-1} h_j \rightharpoonup \mathcal{B}^{-1} h \tag{16.164}$$

Thus, roughly, strong convergence of inhomogeneous data gives weak convergence of solutions to a solution of the limit equation.

We have a particularly good setup here as far as quasiconformal mappings are concerned, as we have proved quite strong normal family results for quasiregular mappings. Virtually any sort of convergence of a family of K-quasiregular

16.6. G-COMPACTNESS

mappings will imply local uniform convergence. Thus we will be able to develop fairly general methods of constructing G-limits of Beltrami operators.

Surprisingly, the notion of G-convergence of operators has little to do with weak convergence of the coefficients. If $\mu_j \rightharpoonup \mu$ and $\nu_j \rightharpoonup \nu$ weakly in L^∞, there is no reason to suspect that the solutions $f^j \rightharpoonup f$ weakly in $\mathbb{W}^{1,2}(\mathbb{C})$ even if we know a priori that the functions $\mathcal{B}_j f^j$ converge strongly in $L^2(\mathbb{C})$. We shall see such examples later, however, the reader may wish to verify that the conclusion will hold should $\mu_j \to \mu$ and $\nu_j \to \nu$ almost everywhere (as discussed for instance in Lemma 5.3.5).

Families of differential operators arise quite naturally in geometric function theory and nonlinear analysis, and usually the only information one has in hand concerns uniform bounds on the ellipticity, as in (16.162). Here the notion of G-compactness, together with our earlier results on regularity and so forth, will help us out. Put briefly, a family $\mathcal{F}$ of Beltrami operators is G-compact if $\mathcal{F}$ contains all limit operators of G-converging sequences $\{\mathcal{B}_j\} \subset \mathcal{F}$. These ideas have a reasonably long and interesting history, particularly studied in the Italian school of PDEs and in relation to second-order equations of divergence-type, see [102, 104, 105, 239, 240, 281, 318] as a good starting place.

Of course, a fundamental precept of our discussion of Beltrami equations in two dimensions is their close relation with second-order equations. Here we shall build up the theory a little further, starting from [69, 139], so as to be able to observe a few surprising results. As a motivational example, let us begin with the most elementary case—Beltrami operators linear over $\mathbb{C}$.

16.6.1 G-Convergence of the Operators $\partial_{\bar{z}} - \mu_j \partial_z$

Let us denote the family of all such operators linear over $\mathbb{C}$ with a fixed ellipticity bound

$$\mathcal{F}_K = \left\{ \frac{\partial}{\partial \bar{z}} - \mu(z)\frac{\partial}{\partial z} : |\mu(z)| \leq k = \frac{K-1}{K+1} < 1 \right\}$$

We have the following theorem.

Theorem 16.6.1. *The family $\mathcal{F}_K$ is G-compact.*

This elegant result actually goes back to S. Spagnolo [332], although he did not formulate it in this manner because he was principally concerned with second-order PDEs. The proof we give for this theorem shows the power of quasiconformal mappings and hints at the way forward for more general equations. With the proof we shall be establishing notation and techniques necessary for the sometimes cumbersome details in the forthcoming treatment of more general Beltrami operators.

Proof of Theorem 16.6.1. Let Φ^j be the unique normalized K-quasiconformal solution to the Beltrami equations

$$\frac{\partial \Phi^j}{\partial \bar{z}} = \mu_j(z) \frac{\partial \Phi^j}{\partial z} \qquad (16.165)$$

Passing to a subsequence if necessary, we may assume that $\Phi^j \to \Phi$ uniformly on compact subsets of $\mathbb{C}$. Let $\mu = \mu(z)$ denote the complex dilatation of Φ, that is,
$$\mu(z) = \frac{\Phi_{\bar{z}}(z)}{\Phi_z(z)} \quad \text{almost everywhere in } \mathbb{C},$$
and we recall that $\Phi_z \neq 0$ almost everywhere. We now claim that the operator $\mathcal{B} = \frac{\partial}{\partial \bar{z}} - \mu(z) \frac{\partial}{\partial z}$ is a G-limit of the sequence $\mathcal{B}_j = \frac{\partial}{\partial \bar{z}} - \mu_j(z) \frac{\partial}{\partial z}$.

To see this we consider a sequence $\{f^j\}_{j=1}^\infty$ weakly converging to f in $\mathbb{W}^{1,2}(\mathbb{C})$ such that the functions $h_j = \mathcal{B}_j f^j$ strongly converge to h in $L^2(\mathbb{C})$. We aim to show that $f_{\bar{z}} - \mu(z) f_z = h$, or equivalently,
$$\frac{\partial \Phi}{\partial z} \frac{\partial f}{\partial \bar{z}} - \frac{\partial \Phi}{\partial \bar{z}} \frac{\partial f}{\partial z} = \frac{\partial \Phi}{\partial z} h \tag{16.166}$$

Those familiar with the subject may recognize that the quadratic expression of the left hand side is a null Lagrangian. We have the following identity:
$$\Phi_z f_{\bar{z}} - \Phi_{\bar{z}} f_z = (\Phi f_{\bar{z}})_z - (\Phi f_z)_{\bar{z}}$$
in the sense of distributions, meaning that
$$\int_\mathbb{C} (\Phi_z f_{\bar{z}} - \Phi_{\bar{z}} f_z) \eta = \int_\mathbb{C} (\eta_{\bar{z}} f_z - \eta_z f_{\bar{z}}) \Phi$$
for every test function $\eta \in C_0^\infty(\mathbb{C})$. Equation (16.166) will then follow if we show that
$$\int_\mathbb{C} (\eta_{\bar{z}} f_z - \eta_z f_{\bar{z}}) \Phi = \int_\mathbb{C} \eta h \Phi_z \tag{16.167}$$

We certainly have these integral formulas for each $j = 1, 2, \ldots$
$$\int_\mathbb{C} (\eta_{\bar{z}} f_z^j - \eta_z f_{\bar{z}}^j) \Phi^j = \int_\mathbb{C} \eta h^j \Phi_z^j \tag{16.168}$$

In order to prove our claim we need only to justify the passage to the limit as $j \to \infty$. For the left hand side of (16.168) we recall that $\Phi^j \to \Phi$ uniformly on compact subsets, while $f_z^j \rightharpoonup f_z$ and $f_{\bar{z}}^j \rightharpoonup f_{\bar{z}}$ weakly in $L^2(\mathbb{C})$. Thus $\Phi^j f_z^j \rightharpoonup \Phi f_z$ and $\Phi^j f_{\bar{z}}^j \rightharpoonup \Phi f_{\bar{z}}$ in the sense of distributions (weakly in $L^2_{loc}(\mathbb{C})$ as well). Therefore, the integrals on the left hand side of (16.168) converge to the corresponding integral in (16.167). As for the integrals on the right hand side of (16.168), we recall that $h^j \to h$ strongly in $L^2(\mathbb{C})$ and $\Phi_z^j \rightharpoonup \Phi_z$ weakly in $L^2(\mathbb{C})$. This implies $h^j \Phi_z^j \rightharpoonup h \Phi_z$ weakly in $L^1(\mathbb{C})$, enough to ensure that these integrals converge to the integral on the right hand side of (16.167). Finally, identity (16.167) shows that $h = f_{\bar{z}} - \mu(z) f_z$ with $\mu = \frac{\Phi_{\bar{z}}}{\Phi_z}$, as desired. $\square$

It is not always easy to predict the G-limits of sequences of Beltrami operators. For instance, the reader may wish to investigate the sequence of operators
$$\mathcal{B}_n = \left\{ \frac{\partial}{\partial \bar{z}} - \mu_n(z) \frac{\partial}{\partial z}, \quad \mu_n(z) = \mu(z/n), \quad n = 1, 2, \ldots \right\},$$

16.6.2 G-Limits and the Weak*-Topology

We discuss here, largely by way of example, the possible relation between G-limits and weak*-convergence when the Beltrami coefficients possess some special algebraic structure. Roughly speaking we assume that μ_j is a nonlinear combination of certain weakly converging terms.

In the rectangular region $\Omega = (a, b) \times (c, d)$ we take real-valued measurable functions

$$\mu_n(z) = \frac{u_n(x) - v_n(y)}{u_n(x) + v_n(y)}, \qquad z = x + iy, \tag{16.169}$$

where

$$\left. \begin{array}{l} \frac{1}{\sqrt{K}} \leqslant u_n(x) \leqslant \sqrt{K} \quad \text{for } a < x < b \\ \frac{1}{\sqrt{K}} \leqslant v_n(y) \leqslant \sqrt{K} \quad \text{for } c < y < d \end{array} \right\} \tag{16.170}$$

Suppose that $u_n \rightharpoonup u$ and $v_n \rightharpoonup v$ and define

$$\mu(z) = \frac{u(x) - v(y)}{u(x) + v(y)} \tag{16.171}$$

We extend μ and each μ_n to the entire complex plane by setting them equal to zero in $\mathbb{C} \setminus \Omega$. Then the G-limit of the Beltrami operators $\frac{\partial}{\partial \bar{z}} - \mu_n(z)\frac{\partial}{\partial z}$ is equal to $\frac{\partial}{\partial \bar{z}} - \mu(z)\frac{\partial}{\partial z}$.

To see this claim, first notice the ellipticity bounds $|\mu_n(z)| \leqslant \frac{K-1}{K+1}$, $n = 1, 2, \ldots$. Let $\Phi^n : \mathbb{C} \to \mathbb{C}$ be the normalized principal solutions of the equations

$$\Phi^n_{\bar{z}} = \mu_n(z)\Phi^n_z \tag{16.172}$$

and let Φ be their local uniform limit where we may pass to a subsequence if necessary. We need only show that Φ is a solution of the equation $\Phi_{\bar{z}} = \mu(z)\Phi_z$ since the requisite behavior near infinity is assured by the convergence of the conformal mappings there. To justify passage to the limit, we examine the complex antiderivatives

$$F^n(z) = \int_a^x u_n(t)\, dt + i \int_c^y v_n(s)\, ds$$

As $u_n \rightharpoonup u$ and $v_n \rightharpoonup v$, we see that $F^n(z)$ converge pointwise everywhere in Ω to

$$F(z) = \int_a^x u(t)\, dt + i \int_c^y v(s)\, ds$$

These mappings are actually Lipschitz-continuous and so uniformly bounded as well. Precisely,

$$|F^n(z_1) - F^n(z_2)| \leqslant K|z_1 - z_2| \quad \text{for } z_1, z_2 \in \Omega \tag{16.173}$$

Their complex derivatives take the form

$$\left.\begin{array}{rcl} 2F^n_{\bar{z}} &=& F^n_x + iF^n_y = u_n(x) - v_n(y) \\ 2F^n_z &=& F^n_x - iF^n_y = u_n(x) + v_n(y) \geqslant \frac{2}{\sqrt{K}} \end{array}\right\} \tag{16.174}$$

Thus $\mu_n(z)$ is in fact the Beltrami coefficient of $F^n(z)$, $\mu_n(z) = F^n_{\bar{z}}(z)/F^n_z(z)$, $n = 1, 2, \ldots$ and $z \in \Omega$. Now equations (16.172) become $F^n_z \Phi^n_{\bar{z}} - F^n_{\bar{z}} \Phi^n_z = 0$. The integral form of these equations is

$$\int_\Omega (\eta_z \Phi^n_{\bar{z}} - \eta_{\bar{z}} \Phi^n_z) F^n = 0$$

for every test function $\eta \in C^\infty_0(\Omega)$, and so it is legitimate to pass to the limit as $n \to \infty$,

$$\int_\Omega (\eta_z \Phi_{\bar{z}} - \eta_{\bar{z}} \Phi_z) F = 0,$$

which is the same as

$$\Phi_{\bar{z}} = \frac{F_{\bar{z}}}{F_z} \Phi_z = \mu(z) \Phi_z$$

almost everywhere in Ω. Finally, the equation $\Phi_{\bar{z}} = \mu(z)\Phi_z$ holds in $\mathbb{C} \setminus \Omega$ by trivial means since $\mu(z) \equiv 0$ and $\Phi_{\bar{z}} \equiv 0$ outside Ω. This establishes our claim.

One particular case merits additional note. Take $v_n(y) \equiv 1$ for $c < y < d$ and all $n = 1, 2, \ldots$. Define each $u_n(x)$ in the closed interval $[0, 2]$ by the rule

$$u_n(x) = \begin{cases} a & \text{if } \frac{2k-2}{n} < x < \frac{2k-1}{n} \\ b & \text{if } \frac{2k-1}{n} < x < \frac{2k}{n} \end{cases}$$

for $k = 1, 2, \ldots, n$. This sequence converges weakly to the constant function,

$$u_n(x) \rightharpoonup u(x) = \frac{a+b}{2}$$

From our example (16.169) above, the G-limit of μ_n is also a constant function in $\Omega = (0, 2) \times (c, d)$,

$$G\text{-}\lim \mu_n = \mu(z) = \frac{a+b-2}{a+b+2}$$

On the other hand, the Beltrami coefficients

$$\mu_n(z) = \frac{u_n(x) - 1}{u_n(x) + 1} = \begin{cases} \frac{a-1}{a+1} & \text{if } \frac{2k-2}{n} < x < \frac{2k-1}{n} \\ \frac{b-1}{b+1} & \text{if } \frac{2k-1}{n} < x < \frac{2k}{n} \end{cases} \tag{16.175}$$

converge in the weak*-topology of $L^\infty(\Omega)$ to the constant function

$$\mu_\infty(z) = \frac{ab-1}{(a+1)(b+1)},$$

which is different from $\mu(z)$.

We next discuss another interesting outcome.

16.6.3 The Jump from $\partial_{\bar z} - \nu\overline{\partial_z}$ to $\partial_{\bar z} - \mu\partial_z$

It is somewhat surprising that the G-limit of the operators $\partial_{\bar z} - \nu_n\overline{\partial_z}$ with $|\nu_n(z)| \equiv k < 1$ may upon passing to G-limits wind up in the class of operators $\partial_{\bar z} - \mu\partial_z$ with $\mu(z) \equiv k^2$. Even more surprisingly, this jump cannot be reversed by Theorem 16.6.1. In light of such examples considerable caution is required in guessing whether a given class of linear PDEs is closed under G-convergence.

We follow an idea of Marcellini [239] to illustrate this phenomenon. Consider a sequence of operators

$$\frac{\partial}{\partial \bar z} - \nu^j(z)\overline{\frac{\partial}{\partial z}}$$

where the measurable coefficients $\nu^j(z)$ depend only on the real part of z. The point here is that $\left(\frac{\partial}{\partial \bar z} - \frac{\partial}{\partial z}\right)\nu^j \equiv 0$ in the sense of Schwartz distributions. Assume, in addition, that all $\nu^j(z)$ have the same constant modulus, say $|\nu^j(z)| \equiv k < 1$, and converge to zero in weak*-topology, for example $\nu_n(z) = ke^{inx}$, $n = 1, 2, \ldots$.

Theorem 16.6.2. *Under the above hypotheses we have*

$$\left(\frac{\partial}{\partial \bar z} - \nu^j(z)\overline{\frac{\partial}{\partial z}}\right) \to_G \left(\frac{\partial}{\partial \bar z} - k^2\frac{\partial}{\partial z}\right) \qquad (16.176)$$

Proof. Suppose we are given a sequence of function f^j weakly converging to f in $\mathbb{W}^{1,2}(\mathbb{C})$ such that the functions $h^j = f^j_{\bar z} - \nu^j(z)\overline{f^j_z}$ converge strongly in $L^2(\mathbb{C})$ to a function h. We begin with the identity

$$\nu^j\left(\frac{\partial}{\partial \bar z} - \frac{\partial}{\partial z}\right)\overline{f^j} = \left(\frac{\partial}{\partial \bar z} - \frac{\partial}{\partial z}\right)\left(\nu^j\,\overline{f^j}\right) \to 0 \qquad (16.177)$$

in the sense of distribution. Note that with Theorem 4.3.14 we can assume that f^j converges strongly in $L^2_{loc}(\mathbb{C})$.

However, the left hand side can be expressed as

$$\begin{aligned}
\nu^j\,\overline{f^j_z} - \nu^j\,\overline{f^j_{\bar z}} &= (f^j_{\bar z} - h^j) - \nu^j\left(\overline{\nu^j f^j_z} + \overline{h^j}\right)\\
&= \left(f^j_{\bar z} - k^2 f^j_z - h^j - \nu^j\overline{h^j}\right) \rightharpoonup (f_{\bar z} - k^2 f_z - h)
\end{aligned}$$

weakly in $L^2(\mathbb{C})$. Hence $f_{\bar z} - k^2 f_z = h$, as desired. $\square$

For more generality, suffice it to note that when $|\nu^j(z)| \to k$ almost everywhere and $\nu^j \rightharpoonup 0$, then the G-limit will be $\frac{\partial}{\partial \bar{z}} - k^2 \frac{\partial}{\partial z}$.

Another interesting example of these sorts of phenomena is contained in the observation that

$$\mathcal{B}_n = \frac{\partial}{\partial \bar{z}} - \sqrt{k}\, \frac{z^n}{\bar{z}^n} \overline{\frac{\partial}{\partial z}} \to_G \frac{\partial}{\partial \bar{z}} - k\, \frac{z}{\bar{z}} \frac{\partial}{\partial z}$$

for $k = \frac{1}{2}, \frac{2}{3}, \ldots, \frac{n-1}{n}, \ldots$ as $n \to \infty$. See [139]. In this example note that the coefficients $\sqrt{k}\, \frac{z^n}{\bar{z}^n}$ converge weakly to 0.

16.6.4 The Adjacent Operator's Two Primary Solutions

A somewhat more sophisticated approach is needed to handle general Beltrami operators. As these operators are linear only over $\mathbb{R}$, two independent quasiconformal solutions will have a part to play in G-convergence.

Associated with each operator

$$\mathcal{B} = \frac{\partial}{\partial \bar{z}} - \mu(z) \frac{\partial}{\partial z} - \nu(z)\, \overline{\frac{\partial}{\partial z}}$$

is its adjacent operator

$$\mathcal{B}^* = \frac{\partial}{\partial \bar{z}} - \mu(z) \frac{\partial}{\partial z} + \overline{\nu(z)}\, \overline{\frac{\partial}{\partial z}}$$

The game is to establish a linear class of quasiconformal mappings $F : \mathbb{C} \to \mathbb{C}$ that solve the same homogeneous equation $\mathcal{B}^* F = 0$. Consider the two-dimensional space

$$\text{span}(\Phi, \Psi) = \{ F = \alpha \Phi + \beta \Psi;\ \alpha, \beta \in \mathbb{R}\}, \qquad (16.178)$$

where Φ and Ψ, called *primary solutions*, are quasiconformal homeomorphisms satisfying

$$\left. \begin{array}{ll} \Phi_{\bar{z}} - \mu\, \Phi_z + \bar{\nu}\, \overline{\Phi_z} = 0, & \Phi(0) = 0,\ \Phi(1) = 1 \\ \Psi_{\bar{z}} - \mu\, \Psi_z + \bar{\nu}\, \overline{\Psi_z} = 0, & \Psi(0) = 0,\ \Psi(1) = i \end{array} \right\} \qquad (16.179)$$

We point out that each $F = \alpha \Phi + \beta \Psi$, apart from the trivial case $\alpha = \beta = 0$, will be a K-quasiconformal normalized solution to the equation

$$\mathcal{B}^* F = 0, \qquad F(0) = 0 \text{ and } F(1) = \alpha + i\beta \neq 0$$

From Corollary 6.2.5 we have the following theorem.

Theorem 16.6.3. *The adjacent operator $\mathcal{B}^*$ has a linear family of K-quasiconformal solutions spanned by the pair (Φ, Ψ) of primary solutions.*

We now go about establishing some preliminary results before discussing the G-compactness of general families of Beltrami operators.

16.6.5 The Independence of $\Phi_z(z)$ and $\Psi_z(z)$

If we fix real parameters $\alpha, \beta \in \mathbb{R}$, $\alpha^2 + \beta^2 \neq 0$, then the linear combination $\alpha \Phi + \beta \Psi$ of the primary solutions is a quasiconformal mapping. Hence for every nonzero pair of real numbers, there is a corresponding set $E \subset \mathbb{C}$ of full measure such that

$$\alpha \Phi_z(z) + \beta \Psi_z(z) \neq 0 \quad \text{for } z \in E \qquad (16.180)$$

This suggests that for almost every $z \in \mathbb{C}$ the complex numbers $\Phi_z(z)$ and $\Psi_z(z)$ might be linearly independent over the field $\mathbb{R}$. In other words, we want to find a set E of full measure in which (16.180) holds for all $\alpha, \beta \in \mathbb{R}$ apart from the trivial case $\alpha = \beta = 0$, of course. This is not an easy task since the set of parameters α, β is uncountable. In fact, the problem is equivalent to the following condition:

$$\Im m(\Phi_z \overline{\Psi_z}) \neq 0 \quad \text{almost everywhere in } \mathbb{C} \qquad (16.181)$$

Thus the $\mathbb{R}$-linear independence of $\Phi_z(z)$ and $\Psi_z(z)$ is a consequence of the following result.

Theorem 16.6.4. *Let Φ and Ψ be primary solutions. Then for almost every $z \in \mathbb{C}$, we have*

$$\Im m\left(\frac{\partial \Phi}{\partial z} \overline{\frac{\partial \Psi}{\partial z}}\right) < 0 \qquad (16.182)$$

Proof. From the general factorization Theorem 6.1.1, we know that

$$\Psi = F \circ \Phi,$$

where F satisfies the reduced Beltrami equation (6.6) with $F(0) = 0$ and $F(1) = i$. Using the chain rule and the identities (2.49) and (2.50), we have

$$\begin{aligned} J(z, \Phi) F_w(w) &= \Psi_z(z) \overline{\Phi_z(z)} - \Psi_{\bar{z}}(z) \overline{\Phi_{\bar{z}}(z)} \\ &= (1 - |\mu|^2) \Psi_z \overline{\Phi_z} - |\nu|^2 \Phi_z \overline{\Psi_z} + 2 \Re e(\mu \bar{\nu} \Phi_z \Psi_z), \qquad w = \Phi(z) \end{aligned}$$

Thus

$$J(z, \Phi) \Im m(F_w \circ \Phi) = (-1 + |\mu|^2 - |\nu|^2) \Im m(\Phi_z \overline{\Psi_z}), \qquad (16.183)$$

and since Φ preserves sets of zero measure, the result follows from Theorems 6.3.2 and 6.4.1. $\square$

A special case here deserves attention. The distortion inequality

$$\left|\frac{\partial \Phi}{\partial \bar{z}}\right| \leqslant k \left|\Re e\left(\frac{\partial \Phi}{\partial z}\right)\right|, \qquad 0 \leqslant k < 1, \qquad (16.184)$$

defines a class of K-quasiconformal homeomorphisms.

Corollary 16.6.5. *Every homeomorphic solution $\Phi : \mathbb{C} \to \mathbb{C}$ to (16.184) normalized by $\Phi(0) = 0$ and $\Phi(1) = 1$ has $\Re e(\Phi_z) > 0$ almost everywhere in $\mathbb{C}$. In particular, we have $|\Phi_{\bar{z}}| \leqslant k \Re e(\Phi_z)$ almost everywhere.*

Proof. The distortion inequality can be expressed as a Beltrami equation of the form
$$2\frac{\partial \Phi}{\partial \bar{z}} - \mu \frac{\partial \Phi}{\partial z} - \mu \overline{\frac{\partial \Phi}{\partial z}} = 0, \qquad |\mu(z)| \leqslant k < 1 \tag{16.185}$$

The point here is that the complex linear function $\Psi(z) = iz$ solves this equation (with $\Psi(0) = 0$, $\Psi(1) = i$) regardless of the coefficient μ. Since
$$\Re e \frac{\partial \Phi}{\partial z} = -\Im m \left(\frac{\partial \Phi}{\partial z} \overline{\frac{\partial \Psi}{\partial z}} \right)$$

the claim follows from Theorem 16.6.4. □

16.6.6 Linear Families of Quasiregular Mappings

In general a linear combination of quasiregular mappings need not remain quasiregular. However, there is one obvious special case where quasiregularity is preserved under linear combinations. Namely, where f and g both satisfy the same Beltrami equation such as (16.179).

Our task here is to complete the discussion in Section 6.2 and show that the converse holds for any (two-dimensional) linear family of quasiregular mappings. That is, such families are always associated to a linear equation. In fact for linear families of homeomorphisms, families of quasiconformal mappings, we further establish the uniqueness of this Beltrami equation. These arguments will play a key role in the proof of the G-compactness for general Beltrami operators.

Theorem 16.6.6. *Suppose $f, g : \mathbb{C} \to \mathbb{C}$ generate a linear family of K-quasiconformal mappings, that is, every mapping in the family*
$$\mathcal{F} = \{ \alpha f + \beta g : \mathbb{C} \to \mathbb{C}, \quad \alpha^2 + \beta^2 \neq 0 \} \tag{16.186}$$

is K-quasiconformal. Then there are unique Beltrami coefficients μ and ν such that
$$|\mu(z)| + |\nu(z)| \leqslant k = \frac{K-1}{K+1} < 1 \tag{16.187}$$

and each $\phi \in \mathcal{F}$ satisfies the equation
$$\frac{\partial \phi}{\partial \bar{z}} = \mu \frac{\partial \phi}{\partial z} + \nu \overline{\frac{\partial \phi}{\partial z}} \qquad \text{almost everywhere in } \mathbb{C} \tag{16.188}$$

Proof. We need to find coefficients μ and ν such that
$$\left. \begin{array}{rcl} f_{\bar{z}} &=& \mu f_z + \nu \overline{f_z} \\ g_{\bar{z}} &=& \mu g_z + \nu \overline{g_z} \end{array} \right\} \tag{16.189}$$

16.6. G-COMPACTNESS

almost everywhere in $\mathbb{C}$. In the *regular set* $\mathcal{R}_{\mathcal{F}}$ of $\mathcal{F}$, the set of those points $\zeta \in \mathbb{C}$ where the matrix

$$M = M(\zeta) = \begin{bmatrix} f_z(\zeta) & \overline{f_z(\zeta)} \\ g_z(\zeta) & \overline{g_z(\zeta)} \end{bmatrix}$$

is invertible, the values $\mu(\zeta)$ and $\nu(\zeta)$ are uniquely determined by (16.189). They are given by the following identities,

$$\left.\begin{aligned} \mu(\zeta) &= i\left(g_{\bar{z}}\,\overline{f_z} - \overline{g_z}\,f_{\bar{z}}\right)/\left(2\,\Im m(f_z\,\overline{g_z})\right) \\ \nu(\zeta) &= i\left(f_{\bar{z}}\,g_z - f_z\,g_{\bar{z}}\right)/\left(2\,\Im m(f_z\,\overline{g_z})\right) \end{aligned}\right\} \qquad (16.190)$$

A change of generators of $\mathcal{F}$ corresponds to multiplying $M(\zeta)$ by an invertible constant matrix. Thus the regular set and its complement, the *singular set* $\mathcal{S}_{\mathcal{F}}$, depend only on the family $\mathcal{F}$ and not on the particular choice of the generators.

On the singular set

$$\mathcal{S}_{\mathcal{F}} = \{\zeta \in \mathbb{C} : \ 2i\,\Im m\left(f_z(\zeta)\,\overline{g_z(\zeta)}\right) = \det M(\zeta) = 0\},$$

we are to show that for almost every point ζ the vector $(f_{\bar{z}}(\zeta), g_{\bar{z}}(\zeta))$ lies in the range of the linear operator $M(\zeta) : \mathbb{C}^2 \to \mathbb{C}^2$. We need to use the assumption that the linear family $\mathcal{F}$ consists entirely of quasiconformal mappings, so that for every $\alpha, \beta \in \mathbb{R}$,

$$|\alpha\,f_{\bar{z}}(\zeta) + \beta\,g_{\bar{z}}(\zeta)| \leqslant k\,|\alpha\,f_z(\zeta) + \beta\,g_z(\zeta)|, \qquad \text{for almost every } \zeta \in \mathbb{C} \quad (16.191)$$

The technical difficulty here is that the set in which this inequality holds generally depends on the parameters α and β, and so getting (16.191) on the same set of full measure for all reals requires an argument.

As the intersection of a countable family of sets of full measure is again a set of full measure, inequality (16.191) holds on a set of full measure $E \subset \mathbb{C}$ for rational α, β. We may further require that $f_z(\zeta) \neq 0$ and $g_z(\zeta) \neq 0$ for all $\zeta \in E$. An important observation is that on this particular set E the estimate (16.191) remains valid for all real coefficients, as follows from a rational approximation of the numbers α, β while keeping ζ fixed.

Now, if $\zeta \in E \cap \mathcal{S}_{\mathcal{F}}$, then $\det M(\zeta) = 0$ or equivalently, $f_z(\zeta)$ and $g_z(\zeta)$ are linearly dependent over the real numbers. Thus $\alpha\,f_z(\zeta) + \beta\,g_z(\zeta) = 0$ for some real numbers α and β different from zero. From (16.191) we now have $\alpha\,f_{\bar{z}}(\zeta) + \beta\,g_{\bar{z}}(\zeta) = 0$, so that

$$\frac{f_{\bar{z}}(\zeta)}{f_z(\zeta)} = \frac{g_{\bar{z}}(\zeta)}{g_z(\zeta)}$$

Hence for $\zeta \in E \cap \mathcal{S}_{\mathcal{F}}$ may define the value $\mu(\zeta)$ to be this common ratio, and set $\nu(\zeta) = 0$. With these choices the equations (16.189) then hold for almost every $\zeta \in \mathbb{C}$.

It remains to prove the ellipticity bounds (16.187). For points in the singular set this follows immediately from the definition of $\mu(\zeta)$ and $\nu(\zeta)$, as f and g are K-quasiregular. For the regular set the argument is more difficult. We consider the points
$$\zeta \in E \cap \mathcal{R}_\mathcal{F}$$
For such points we can test the inequality (16.191) by real-valued measurable functions in place of the parameters α and β. Given a real-valued measurable function $\theta = \theta(\zeta)$ defined in $E \cap \mathcal{R}_\mathcal{F}$, we may solve the equation
$$\alpha(\zeta) f_z + \beta(\zeta) g_z = e^{i\theta(\zeta)}, \qquad \zeta \in E \cap \mathcal{R}_\mathcal{F}, \qquad (16.192)$$
for the measurable real coefficients
$$\alpha(\zeta) = \frac{\Im m(e^{i\theta} \overline{g_z})}{\Im m(f_z \overline{g_z})}, \qquad \beta(\zeta) = \frac{\Im m(e^{i\theta} \overline{f_z})}{\Im m(f_z \overline{g_z})}$$
Note that $\alpha(\zeta)$ and $\beta(\zeta)$ are well defined functions since $\Im m(f_z \overline{g_z}) \neq 0$ on the regular set.

Next, we can certainly select a measurable function $\theta = \theta(\zeta)$ in such a way that $|\mu(\zeta)| + |\nu(\zeta)| = |\mu e^{i\theta} + \nu e^{-i\theta}|$ almost everywhere in $E \cap \mathcal{R}_\mathcal{F}$. In view of (16.189) and (16.192),
$$\alpha(\zeta) f_{\bar{z}} + \beta(\zeta) g_{\bar{z}} = \mu e^{i\theta} + \nu e^{-i\theta}$$
and thus for each given point $\zeta \in E \cap \mathcal{R}_\mathcal{F}$,
$$|\mu(\zeta)| + |\nu(\zeta)| = |\alpha(\zeta) f_{\bar{z}} + \beta(\zeta) g_{\bar{z}}| \leqslant k |\alpha(\zeta) f_z + \beta(\zeta) g_z| = k,$$
as claimed. We have thus established the ellipticity bounds (16.187) almost everywhere on the regular set.

Up to this point only local properties of the mappings have been used, and therefore the argument applies to any linear family of K-quasiregular mappings, defined in a domain $\Omega \subset \mathbb{C}$. However, to establish the uniqueness of the Beltrami coefficients μ and ν global properties must be employed. Indeed, we saw in (16.190) that the coefficients are uniquely determined on the regular set, while Theorem 16.6.4 proves that for a linear family of homeomorphic solutions to an elliptic equation, defined in the entire plane, the singular set is empty. Thus the uniqueness follows, completing the proof of Theorem 16.6.6. $\square$

Remark 16.6.7. *As we noted during the proof of Theorem 16.6.6, the argument above shows that for any (two dimensional) linear family $\mathcal{F}$ of quasiregular mappings defined in a domain Ω there corresponds a Beltrami equation*
$$\frac{\partial \phi}{\partial \bar{z}} = \mu \frac{\partial \phi}{\partial z} + \nu \overline{\frac{\partial \phi}{\partial z}}, \qquad \textit{almost everywhere in } \Omega,$$
satisfied by every element ϕ of the family.

16.6.7 G-Compactness for Beltrami Operators

As a consequence of the results of the previous section, we shall now establish the G-compactness of the family of all Beltrami operators with a given bound on the ellipticity constant.

Theorem 16.6.8. *The family* $\mathcal{F}_K(\mathbb{C})$, $1 \leqslant K < \infty$, *of Beltrami differential operators*

$$\frac{\partial}{\partial \bar{z}} - \mu(z)\frac{\partial}{\partial z} - \nu(z)\overline{\frac{\partial}{\partial z}}$$

with

$$|\mu(z)| + |\nu(z)| \leqslant k = \frac{K-1}{K+1} < 1$$

is G-compact.

In spite of the close analogy with Theorem 16.6.1, the general setting developed in [139] is really different and, as we have seen in the preliminaries, the proof requires more involved methods. Essential for the complete solution are also the results from Section 6.4.

Proof of Theorem 16.6.8. Given a sequence of general Beltrami operators defined in $\mathbb{C}$

$$\mathcal{B}_j = \frac{\partial}{\partial \bar{z}} - \mu_j(z)\frac{\partial}{\partial z} - \nu_j(z)\overline{\frac{\partial}{\partial z}}, \qquad j=1,2,\ldots,$$

with $|\mu_j| + |\nu_j| \leqslant k < 1$, we identify a subsequence and its G-limit by solving the adjacent equations

$$\Phi^j_{\bar{z}} - \mu_j \Phi^j_z + \overline{\nu_j}\,\overline{\Phi^j_z} = 0, \quad \Phi^j(0)=0, \quad \Phi^j(1)=1$$

$$\Psi^j_{\bar{z}} - \mu_j \Psi^j_z + \overline{\nu_j}\,\overline{\Psi^j_z} = 0, \quad \Psi^j(0)=0, \quad \Psi^j(1)=i$$

to get a pair of primary solutions, namely, K-quasiconformal homeomorphisms $\Phi^j, \Psi^j : \mathbb{C} \to \mathbb{C}$. Theorem 16.6.4 ensures that $\Im m(\Phi^j_z \overline{\Psi^j_z}) < 0$ almost everywhere. We can uniquely express the coefficients μ_j and $\overline{\nu_j}$ in terms of the complex derivatives of Φ^j and Ψ^j, namely,

$$\left.\begin{array}{rcl}\mu_j(z) &=& i\left(\Psi^j_{\bar{z}}\,\overline{\Phi^j_z} - \overline{\Psi^j_z}\,\Phi^j_{\bar{z}}\right)/\left(2\,\Im m(\Phi^j_z \overline{\Psi^j_z})\right) \\[4pt] \overline{\nu_j(z)} &=& i\left(\Psi^j_{\bar{z}}\,\Phi^j_z - \Psi^j_z\,\Phi^j_{\bar{z}}\right)/\left(2\,\Im m(\Phi^j_z \overline{\Psi^j_z})\right)\end{array}\right\} \qquad (16.193)$$

We may assume, passing to a subsequence if necessary, that the nonconstant K-quasiconformal limits

$$\Phi(z) = \lim_{j\to\infty} \Phi^j(z), \qquad \Psi(z) = \lim_{j\to\infty} \Psi^j(z)$$

exist in both cases, the convergence being locally uniform. Similarly, every non-trivial linear combination

$$f = \alpha\Phi + \beta\Psi = \lim_{j\to\infty}\left(\alpha\Phi^j + \beta\Psi^j\right)$$

is a K-quasiconformal homeomorphism with $f(0) = 0$ and $f(1) = \alpha + i\beta \neq 0$. From Theorems 16.6.4 and 16.6.6 we infer that $\Im m(\Phi_z\overline{\Psi_z}) \neq 0$ almost everywhere in $\mathbb{C}$. Next, we mimic the formulas (16.193) to define the following coefficients for the limit operator. Set

$$\left.\begin{aligned}\mu(z) &= i\left(\Psi_{\bar{z}}\overline{\Phi_z} - \overline{\Psi_z}\,\Phi_{\bar{z}}\right)/\left(2\,\Im m(\Phi_z\overline{\Psi_z})\right) \\ \overline{\nu(z)} &= i\left(\Psi_{\bar{z}}\Phi_z - \Psi_z\,\Phi_{\bar{z}}\right)/\left(2\,\Im m(\Phi_z\overline{\Psi_z})\right)\end{aligned}\right\} \quad (16.194)$$

so that

$$\left.\begin{aligned}\Phi_{\bar{z}} - \mu(z)\Phi_z + \overline{\nu(z)}\,\overline{\Phi_z} &= 0 \\ \Psi_{\bar{z}} - \mu(z)\Psi_z + \overline{\nu(z)}\,\overline{\Psi_z} &= 0\end{aligned}\right\} \quad (16.195)$$

The uniform ellipticity bound for $\mu(z)$ and $\nu(z)$ are established in (16.187). Therefore the next result will complete the proof of Theorem 16.6.8.

Lemma 16.6.9. *The differential operators $\mathcal{B}_j$ G-converge to*

$$\mathcal{B} = \frac{\partial}{\partial\bar{z}} - \mu(z)\frac{\partial}{\partial z} - \nu(z)\overline{\frac{\partial}{\partial z}}$$

Proof. Suppose we are given a sequence $\{f^j\}_{j=1}^{\infty}$ weakly converging to f in the Sobolev space $W_{loc}^{1,2}(\mathbb{C})$ such that $\mathcal{B}_j f^j$ converge strongly in $L_{loc}^2(\mathbb{C})$. We must show that

$$\lim_{j\to\infty}\mathcal{B}_j f^j = \mathcal{B}f$$

We need a few identities. Let us begin with elementary, though lengthy, differentiation in the sense of distributions

$$\begin{aligned}&[(\Psi_z - i\Phi_z)f + \overline{(\Psi_{\bar{z}} + i\Phi_{\bar{z}})}\,\overline{f}\,]_{\bar{z}} - [(\Psi_{\bar{z}} - i\Phi_{\bar{z}})f + \overline{(\Psi_z + i\Phi_z)}\,\overline{f}\,]_z \\ &= [(\Psi_z - i\Phi_z)f_{\bar{z}} + \overline{(\Psi_{\bar{z}} + i\Phi_{\bar{z}})}\,\overline{f_z}\,] - [(\Psi_{\bar{z}} - i\Phi_{\bar{z}})f_z + \overline{(\Psi_z + i\Phi_z)}\,\overline{f_{\bar{z}}}\,] \\ &= (\Psi_z - i\Phi_z)f_{\bar{z}} + (\overline{\mu}\Psi_z - \nu\Psi_z - i\,\overline{\mu}\Phi_z + i\,\nu\Phi_z)\overline{f_z} \\ &\quad - (\mu\Psi_z - \overline{\nu}\Psi_z - i\,\mu\Phi_z + i\,\overline{\nu}\Phi_z)f_z - \overline{(\Psi_z + i\Phi_z)}\,\overline{f_{\bar{z}}} \\ &= (\Psi_z - i\Phi_z)(f_{\bar{z}} - \mu f_z - \nu\overline{f_z}) - \overline{(\Psi_z + i\Phi_z)}\,\overline{(f_{\bar{z}} - \mu f_z - \nu\overline{f_z})}\end{aligned}$$

We can apply this identity to f^j, Φ^j and Ψ^j to obtain

$$\begin{aligned}&[(\Psi_z^j - i\Phi_z^j)f^j + \overline{(\Psi_{\bar{z}}^j + i\Phi_{\bar{z}}^j)}\,\overline{f^j}\,]_{\bar{z}} - [(\Psi_{\bar{z}}^j - i\Phi_{\bar{z}}^j)f^j + \overline{(\Psi_z^j + i\Phi_z^j)}\,\overline{f^j}\,]_z \\ &= (\Psi_z^j - i\Phi_z^j)\mathcal{B}_j f^j - \overline{(\Psi_z^j + i\Phi_z^j)}\,\overline{\mathcal{B}_j f^j}\end{aligned}$$

16.6. G-COMPACTNESS

We denote by h the strong $L^2_{loc}(\mathbb{C})$ limit of the sequence $\{\mathcal{B}_j f^j\}_{j=1}^\infty$. Since we have $(\Phi^j_z, \Phi^j_{\bar z}, \Psi^j_z, \Psi^j_{\bar z}) \rightharpoonup (\Phi_z, \Phi_{\bar z}, \Psi_z, \Psi_{\bar z})$ weakly in $L^2(\mathbb{C})$ and $f^j \to f$ strongly in $L^2_{loc}(\mathbb{C})$, we are in a position to pass to the limit in the last equation. The formula in the limit is then

$$[(\Psi_z - i\,\Phi_z)f + \overline{(\Psi_{\bar z} + i\,\Phi_{\bar z})}\,\overline{f}\,]_{\bar z} - [(\Psi_{\bar z} - i\,\Phi_{\bar z})f + \overline{(\Psi_z + i\,\Phi_z)}\,\overline{f}\,]_z$$
$$= (\Psi_z - i\,\Phi_z)h - \overline{(\Psi_{\bar z} + i\,\Phi_{\bar z})}\,\overline{h}$$

After we cancel the second-order derivatives of Φ and Ψ, this formula simplifies to

$$(\Psi_z - i\,\Phi_z)(f_{\bar z} - h) + \overline{(\Psi_{\bar z} + i\,\Phi_{\bar z})}\,\overline{f_z}$$
$$= (\Psi_{\bar z} - i\,\Phi_{\bar z})f_z + \overline{(\Psi_z + i\,\Phi_z)}\,\overline{(f_{\bar z} - h)}$$

The rest of the proof is simply a matter of elementary algebraic manipulation. We first express $f_{\bar z} - h$ as a linear combination of f_z and $\overline{f_z}$, the detailed verification of this computation being left to the reader.

$$2\,\Im m(\Phi_z \overline{\Psi_z})\,(f_{\bar z} - h)$$
$$= i\,(\Psi_{\bar z}\overline{\Phi_z} - \overline{\Psi_z}\Phi_{\bar z})f_z - i\,\overline{(\Psi_{\bar z}\Phi_z - \Phi_{\bar z}\Psi_z)}\,\overline{f_z}$$

To continue we need to know that $\Im m(\Phi_z \overline{\Psi_z}) \ne 0$ almost everywhere. This follows from Theorems 16.6.4 and 16.6.6, and thus in view of (16.194) we arrive at the equation

$$h = f_{\bar z} - \mu f_z - \nu \overline{f_z} = \mathcal{B}f,$$

completing the proof of Lemma 16.6.9. This also proves Theorem 16.6.8. $\square$

There is one very interesting observation we want to make in respect to the proof above. For this we note that the numerators of the quotients defining the coefficients $\bar\nu_j(z)$ and $\bar\nu(z)$ are null Lagrangians:

$$\mathcal{J}(\Phi, \Psi) = \Phi_z \Psi_{\bar z} - \Phi_{\bar z}\Psi_z \in L^1_{loc}(\mathbb{C})$$

Being a null Lagrangian (for more information see Chapter 19) simply means (via integration by parts) that

$$\int_{\mathbb{C}} \eta(z)\,\mathcal{J}(\Phi, \Psi) = -\int_{\mathbb{C}} \Phi\,\mathcal{J}(\eta, \Psi)$$

for every $\eta \in C_0^\infty(\mathbb{C})$. Of course, the same applies for each $j = 1, 2, \ldots$,

$$\int_{\mathbb{C}} \eta(z)\,\mathcal{J}(\Phi^j, \Psi^j) = -\int_{\mathbb{C}} \Phi^j\,\mathcal{J}(\eta, \Psi^j)$$

These two identities immediately imply the weak continuity of the bilinear differential form $\mathcal{J} : \mathbb{W}^{1,2}(\mathbb{C}) \times \mathbb{W}^{1,2}(\mathbb{C}) \to L^1(\mathbb{C})$, an analog of the celebrated div-curl lemma of F. Murat and L. Tatar [281].

Lemma 16.6.10. *Suppose that $\Phi^j \rightharpoonup \Phi$ and $\Psi^j \rightharpoonup \Psi$ weakly in $\mathbb{W}^{1,2}(\mathbb{C})$. Then for every $\eta \in C_0^\infty(\mathbb{C})$ we have*

$$\int_{\mathbb{C}} \eta(z)\,\mathcal{J}(\Phi^j, \Psi^j) \to \int_{\mathbb{C}} \eta(z)\,\mathcal{J}(\Phi, \Psi) \tag{16.196}$$

Of course, in our application Φ^j and Ψ^j will be quasiconformal, and so (16.196) holds for $\eta \in L^\infty(\mathbb{C})$ with compact support by an approximation of η (where we note that the derivatives of Φ^j and Ψ^j stay uniformly bounded in $L^p(\mathbb{C})$ for some $p > 2$ by virtue of the uniform quasiconformality of these functions and their uniform convergence).

Now let us suppose that, for $z \in \mathbb{C}$ and some measurable function λ, the ν_j-coefficients of the Beltrami operators satisfy the relation

$$\Re\mathfrak{e}\bigl(\lambda(z)\nu_j(z)\bigr) = 0, \qquad j = 1, 2, \ldots \tag{16.197}$$

Then the same relation holds in any G-limit; that is,

$$\Re\mathfrak{e}\bigl(\lambda(z)\nu(z)\bigr) = 0 \tag{16.198}$$

Actually, (16.197) is equivalent to the condition $\Im\mathfrak{m}\bigl(\bar\lambda(z)\mathcal{J}(\Phi^j, \Psi^j)\bigr) \equiv 0$, as is seen from (16.194). To see that the identity remains true in the G-limit, let η be any real-valued non-negative compactly supported bounded function. We have

$$\Im\mathfrak{m}\left(\int_{\mathbb{C}} \eta(z)\frac{\bar\lambda(z)}{1+|\lambda|}\mathcal{J}(\Phi, \Psi)\right) = \lim_{j\to\infty} \Im\mathfrak{m}\left(\int_{\mathbb{C}} \eta(z)\frac{\bar\lambda(z)}{1+|\lambda|}\mathcal{J}(\Phi^j, \Psi^j)\right) = 0,$$

so that $\Im\mathfrak{m}\bigl(\bar\lambda(z)\mathcal{J}(\Phi, \Psi)\bigr) \equiv 0$, as claimed. Similarly, if we have instead

$$\Re\mathfrak{e}\bigl(\lambda(z)\nu_j(z)\bigr) \geqslant 0, \qquad j = 1, 2, \ldots,$$

then the same relation again holds in the G-limit because the denominators of the quotients defining ν_j and ν do not change sign (they are strictly negative almost everywhere). What is rather peculiar here is that the same sorts of relationships for the μ_j-coefficients are lost when passing to a G-limit. For example, in Theorem 16.6.2 we have $\mu_j(z) \equiv 0$ for $j = 1, 2, \ldots$, and yet $\mu(z) \equiv k^2 \neq 0$.

There is one subclass of Beltrami operators of special interest, namely, the Beltrami operators for gradient quasiregular mappings. These take the form

$$\frac{\partial f}{\partial \bar z} = \mu\frac{\partial f}{\partial z} + \bar\mu\,\overline{\frac{\partial f}{\partial z}}, \qquad 2|\mu(z)| \leqslant k < 1 \tag{16.199}$$

This particular form is clear since it implies that $\Im\mathfrak{m}(f_{\bar z}) = 0$, which is a necessary and sufficient condition for f to be the complex gradient of a real-valued function. We are thus motivated to consider the following operators:

$$\mathcal{B}_j = \frac{\partial}{\partial \bar z} - \mu_j\frac{\partial}{\partial z} - \bar\mu_j\,\overline{\frac{\partial}{\partial z}}, \qquad 2|\mu(z)| \leqslant k < 1 \tag{16.200}$$

Our construction of the G-limit via primary solutions of the adjacent equations yields the following result.

16.6. G-COMPACTNESS

Theorem 16.6.11. *The class of Beltrami operators for gradient mappings is G-compact for every $k < 1$.*

Proof. The adjacent equations take the form

$$\left.\begin{array}{r}\Phi^j_{\bar{z}} - \mu_j(z)\Phi^j_z + \mu_j(z)\overline{\Phi^j_z} = 0 \\ \Psi^j_{\bar{z}} - \mu_j(z)\Psi^j_z + \mu_j(z)\overline{\Psi^j_z} = 0\end{array}\right\} \qquad (16.201)$$

An important point to notice here is that $\Phi(z) = z$ is a primary solution regardless of the coefficients. A subsequence of the other primary solutions converge to a quasiconformal Ψ locally uniformly in $\mathbb{C}$. As above in (16.194), the coefficients of the G-limit are expressed by the formula

$$\mu(z) = i\frac{\Psi_{\bar{z}}}{2\,\Im m\,\overline{\Psi_z}}, \qquad \overline{\nu(z)} = i\frac{\Psi_z}{2\,\Im m\,\overline{\Psi_z}}, \qquad (16.202)$$

hence $\mu = \bar{\nu}$, as desired. $\square$

Chapter 17

PDEs Not of Divergence Type: Pucci's Conjecture

In the previous chapter we made a fairly thorough study of second-order elliptic PDEs of divergence-type. However, in two dimensions the methods and techniques developed there actually apply as well to equations that are not of divergence-type. Such equations arise naturally in many different contexts, for instance in stochastics, as Monge Ampere–type equations (for transport and related problems) and also in the linearization of nonlinear elliptic PDEs.

The inequality of Alexandrov, Bakel'man and Pucci is a basic tool in the theory of linear elliptic partial differential equations that are not in divergence form, as well as in the more general theory of nonlinear elliptic PDEs (see [84, 143]). In its usual formulation this inequality is concerned with uniformly elliptic second-order differential operators of the form

$$(\mathcal{L}_A u)(x) = \sum_{i,j=1}^{n} a_{i,j}(x)\, u_{x_i x_j} \qquad (17.1)$$

defined for functions u of the Sobolev class $W^{2,p}_{loc}(\Omega)$ for a bounded subdomain $\Omega \subset \mathbb{R}^n$. Here the coefficient matrix

$$A = A(x) = [a_{i,j}(x)]_{i,j=1}^{n}$$

is a symmetric positive definite matrix function with real-valued measurable entries $a_{i,j}(x)$. Thus

$$\langle A(x)\zeta, \zeta \rangle > 0, \qquad \zeta \in \mathbb{R}^n \setminus \{0\} \qquad (17.2)$$

The operator (17.1) can also be written as

$$\mathcal{L}_A u = \langle A(x), \mathcal{H}u \rangle,$$

where $\mathcal{H}u = \left(u_{x_i x_j}\right)_{i,j=1}^n$ is the *Hessian matrix*. For rather obvious reasons the natural domain of definition of the operator $\mathcal{L}_A$ is the Sobolev class $W^{2,n}_{loc}(\Omega)$, for in this class the determinant of the Hessian matrix is locally integrable,

$$\det \mathcal{H}u \in L^1_{loc}(\Omega) \tag{17.3}$$

As a matter of fact, the condition (17.3) is less restrictive than the assumption that the Sobolev class is $W^{2,n}_{loc}(\Omega)$ when uniform ellipticity bounds are lost. As far as degenerate equations are concerned, for $n=2$, they will be briefly discussed in Chapter 20 where we shall see that (17.3) is a natural restriction.

A fact worth noting at this point is that

$$W^{2,s}_{loc}(\Omega) \subset C(\Omega), \quad \text{whenever } s > n/2$$

We point this out, as we will be interested in studying the operator $\mathcal{L}_A$ in weaker domains of definition than $W^{2,n}_{loc}$, mainly the Sobolev classes $W^{2,s}_{loc}(\Omega)$ with $s < n$.

The classical maximum principle of Alexandrov [143, Theorem 9.1] is as follows:

Theorem 17.0.12. *There is a constant C_n such that, for every $u \in W^{2,n}(\Omega)$ and every relatively compact subdomain $\Omega' \subset \Omega$, we have the estimate*

$$\max_{\Omega'} |u| \leqslant \max_{\partial \Omega'} |u| + C_n \operatorname{diam}(\Omega) \left(\int_\Omega |\mathcal{L}_A u(x)|^n \det A^{-1}(x)\, dx \right)^{1/n} \tag{17.4}$$

Of course, as a special case of this result when $A(x) = \mathbf{I}$ and $\mathcal{L}_A$ is the Laplacian, one knows that the estimate remains valid for $u \in W^{2,s}(\Omega)$, for all $s > n/2$, where it is

$$\max_{\Omega'} |u| \leqslant \max_{\partial \Omega'} |u| + C(s,n) \operatorname{diam}^{2-n/s}(\Omega) \left(\int_\Omega |\Delta u(x)|^s\, dx \right)^{1/s} \tag{17.5}$$

We shall see below in Section 17.6 that an estimate cannot be achieved at the critical exponent $n/2$.

This chapter is concerned with interpolation in two-dimensions between the estimates (17.4) (for general elliptic operators) and (17.5) (for the Laplacian). Other recent extensions of this maximum principle, in a different direction and for all dimensions, can be found in [222]. It is only natural from the point of view of applications and the theory of weak solutions to PDEs to try to extend the range of exponents for which the estimate (17.4) holds when one has control of the ellipticity constant. In particular, one would seek to replace the $L^n(\Omega)$ norm of $\mathcal{L}_A u$ by some $L^p(\Omega)$ norm, $p < n$. In general, this is not possible. It was shown by Alexandrov [15] and Pucci [304] that without any restriction on the ellipticity constant, the $L^n(\Omega)$ estimate in (17.4) is the best possible.

However, if one fixes the ellipticity constant, say K, for the equation, then at least in two dimensions one can obtain estimates in $L^p(\Omega)$ for some p below 2.

Here once again we meet the question of parameterizing the bounds of uniform ellipticity. In the case (17.1) it turns out that the correct choice is

$$\frac{|\zeta|^2}{\sqrt{K}} \leqslant \langle A(x)\zeta, \zeta \rangle \leqslant \sqrt{K}\, |\zeta|^2 \qquad (17.6)$$

In 1966 Pucci [305] made a conjecture on the precise range of values for p for which there is an L^p-version of (17.4). The theme of this chapter is the following theorem from [28], proving the conjecture in dimension $n = 2$.

Theorem 17.0.13. *Let $u \in W^{2,p}(\Omega)$, $p > \frac{2K}{K+1}$. Then for every relative compact subdomain $\Omega' \subset \Omega$, we have*

$$\max_{\Omega'} |u| \leqslant \max_{\partial \Omega'} |u| + C_p(\Omega) \|\mathcal{L}_A u\|_{L^p(\Omega)}, \qquad (17.7)$$

where the constant depends only on p and Ω.

In his paper Pucci gave an example, which we recall in Section 17.6, showing that the inequality fails for $p = \frac{2K}{K+1}$. Thus the theorem gives the sharp generalization and determines the values of p for which an estimate of the form (17.4) holds.

The theorem has for instance a natural interpretation in terms of the related Green's operator. Namely, given a function $h \in L^p(\Omega)$, the operator

$$\mathcal{G}_A(h) = u$$

associates with h the solution, if such exists, to

$$\mathcal{L}_A u = h, \qquad u|\partial\Omega = 0$$

According to (17.4), for any A uniformly elliptic, the operator $\mathcal{G}_A : L^2(\Omega) \to L^\infty(\Omega)$ is well defined and bounded,

$$\|\mathcal{G}_A(h)\|_\infty \leqslant C \|h\|_2, \qquad h \in L^2(\Omega)$$

A natural question is whether the operator $\mathcal{G}_A$ extends to any larger space $L^p(\Omega)$, $p < 2$. From Theorem 17.0.13 we immediately have the following corollary.

Corollary 17.0.14. *Let $K \geqslant 1$ and suppose $A(z)$ satisfies (17.6) almost everywhere in a bounded domain Ω. Then for every $p > \frac{2K}{K+1}$, the Green's operator $\mathcal{G}_A$ is continuous $L^p(\Omega) \to L^\infty(\Omega)$,*

$$\|\mathcal{G}_A(h)\|_\infty \leqslant C \|h\|_p, \qquad h \in L^p(\Omega)$$

Moreover, for every $p \leqslant \frac{2K}{K+1}$ the result fails for some operators $\mathcal{G}_A$, with $A(z)$ satisfying (17.6).

We will give further consequences and discuss related topics in subsequent sections.

17.1 Reduction to a First-Order System

We begin with a few elementary remarks and the algebraic computations that will reduce the equations $\mathcal{L}_A u = h$ to a system of first-order PDEs where we can use the results concerning quasiregular mappings that we have developed in this book.

Let $u \in W^{2,1}_{loc}(\Omega)$. Motivated by the results in Chapter 16, we consider the complex gradient $f = u_z$ of u. The second-order complex derivatives are

$$\left. \begin{array}{rcl} u_{zz} & = & \frac{1}{4}(u_{xx} - u_{yy} - 2iu_{xy}) \\ u_{z\bar{z}} & = & \frac{1}{4}(u_{xx} + u_{yy}) = \frac{1}{4}\Delta u \end{array} \right\} \quad (17.8)$$

We rearrange these to obtain the equivalent

$$\left. \begin{array}{rcl} u_{xx} & = & 2u_{z\bar{z}} + u_{zz} + \overline{u_{zz}} \\ u_{yy} & = & 2u_{z\bar{z}} - u_{zz} - \overline{u_{zz}} \\ u_{xy} & = & i(u_{zz} - \overline{u_{zz}}) \end{array} \right\} \quad (17.9)$$

We now wish to express the operator $\mathcal{L}_A$ in complex notation. Suppose that

$$A = \begin{bmatrix} a_{11}(z) & a_{12}(z) \\ a_{12}(z) & a_{22}(z) \end{bmatrix}$$

Then in terms of the complex derivatives,

$$\begin{array}{rcl} \mathcal{L}_A u & = & a_{11} u_{xx} + 2a_{12} u_{xy} + a_{22} u_{yy} \\ & = & 2(a_{11} + a_{22}) u_{z\bar{z}} + (a_{11} - a_{22} + 2ia_{12}) u_{zz} + (a_{11} - a_{22} - 2ia_{12}) \overline{u_{zz}} \\ & = & \operatorname{tr} A(z) \left[2 u_{z\bar{z}} + \mu\, u_{zz} + \bar{\mu}\, \overline{u_{zz}} \right], \end{array}$$

where

$$\mu(z) = \frac{a_{11} - a_{22} + 2ia_{12}}{a_{11} + a_{22}} \quad (17.10)$$

Hence,

$$|\mu(z)|^2 = 1 - \frac{4 \det A}{\operatorname{tr}^2 A} < 1$$

While in terms of the eigenvalues $0 < \lambda_1 \leqslant \lambda_2$ of the matrix A, we see that $|\mu(z)| = \left| \frac{\lambda_2 - \lambda_1}{\lambda_2 + \lambda_1} \right|$ and hence

$$|\mu(z)| = \frac{K_A(z) - 1}{K_A(z) + 1}, \quad K_A(z) = \lambda_2/\lambda_1 \leqslant K \quad (17.11)$$

This inequality also explains our choice (17.6) for the ellipticity parameters.

We summarize the above discussion in the next lemma.

Lemma 17.1.1. *For every* $u \in W^{2,1}_{loc}(\Omega)$ *we have*

$$\frac{1}{\operatorname{tr} A} \mathcal{L}_A u = 2 u_{z\bar{z}} + \mu(z) u_{zz} + \bar{\mu}(z) \overline{u_{zz}}, \qquad (17.12)$$

where the complex coefficient $\mu = \mu(z)$ *satisfies*

$$|\mu(z)| \leqslant \frac{K-1}{K+1} \qquad (17.13)$$

17.2 Second-Order Caccioppoli Estimates

We see from Lemma 17.1.1 that if $u \in W^{2,2}_{loc}(\Omega)$ is a solution to the equation $\mathcal{L}_A u = 0$, then the complex gradient $f = u_z = \frac{1}{2}(u_x - i u_y)$ is a quasiregular mapping. This fact makes it necessary for us to have a brief discussion of quasiregular gradient mappings. Let

$$f = \nabla V = (V_{x_1}, V_{x_2}) \in W^{1,2}_{loc}(\Omega, \mathbb{C}),$$

where V, called the potential function for f, lies in the Sobolev space $W^{2,2}_{loc}(\Omega)$. If f is K-quasiregular, we show how one can control $|f|$ and $|Df|$ by integral averages of the potential V alone. We begin with the classical Caccioppoli inequality (5.27),

$$\int_\Omega |\phi D f|^2 \leqslant K^2 \int_\Omega |\nabla \phi|^2 |f|^2, \qquad \phi \in C_0^\infty(\Omega) \qquad (17.14)$$

Next, given any Lipschitz function η with compact support in Ω, we use integration by parts to get

$$\int |\eta f|^2 = \int |\eta|^2 \langle \nabla V, f \rangle \leqslant 2 \int |\eta| |f| |\nabla \eta| |V| + \int |\eta|^2 |Df| |V|$$

With the aid of Young's inequality and after arranging the terms, we obtain

$$\int |\eta f|^2 \leqslant C_0 \int |\nabla \eta|^2 |V|^2 + C_0 \int |\eta|^2 |Df| |V| \qquad (17.15)$$

We choose $\eta = |\phi \nabla \phi|$ and combine this estimate with the Caccioppoli inequality (17.14), where ϕ is replaced by ϕ^2,

$$\begin{aligned}\int |\phi|^4 |Df|^2 &\leqslant C_K \int |\phi \nabla \phi|^2 |f|^2 \\ &\leqslant C \int \left(|\nabla \phi|^2 + |\phi| |\nabla^2 \phi| \right)^2 |V|^2 + C \int |\phi|^2 |Df| |\nabla \phi|^2 |V|\end{aligned}$$

By again using Young's inequality, we reduce the estimate to

$$\int |\phi|^4 |Df|^2 \leqslant C_K \int \left(|\nabla \phi|^2 + |\phi \nabla^2 \phi| \right)^2 |V|^2,$$

17.2. CACCIOPPOLI ESTIMATES

or equivalently,
$$\|\phi^2 Df\|_2 \leqslant C_K \|(|\nabla\phi|^2 + |\phi\,\nabla^2\phi|)\,V\|_2 \tag{17.16}$$

This is what we call the *second Cacciopoli inequality*.

We now combine the second Cacciopoli inequality with the estimate in (17.15) and use Hölder's inequality to obtain

$$\int_\Omega |\eta f|^2 \leqslant C \int_\Omega |\nabla\eta|^2 |V|^2 \tag{17.17}$$
$$+ C_K \Big[\int_\Omega |V|^2\Big]^{1/2} \Big[\int_\Omega (|\nabla\eta|^2 + |\eta\,\nabla^2\eta|)^2 |V|^2\Big]^{1/2}$$

Next, we restrict these estimates to concentric balls $B = \mathbb{D}(a, R)$ and $\mathbb{D}(a, sR) = sB$, where $s > 1$. With the appropriate choice of a test function $\eta \in C_0^\infty(sB)$, we infer from (17.17) that

$$\Big(\frac{1}{|B|}\int_B |f|^2\Big)^{1/2} \leqslant \frac{sC(K)}{(s-1)R}\Big(\frac{1}{|sB|}\int_{sB} |V|^2\Big)^{1/2} \tag{17.18}$$

We also appeal to the Sobolev-Poincaré-inequality (A.6.3) for V to write (17.18) as

$$\Big(\frac{1}{|B|}\int_B |V - V_B|^4\Big)^{1/4} \leqslant CR\Big(\frac{1}{|B|}\int_B |\nabla V|^2\Big)^{1/2} \leqslant \frac{sC(K)}{(s-1)}\Big(\frac{1}{|sB|}\int_{sB} |V|^2\Big)^{1/2},$$

where the integral mean can be estimated as

$$|V_B| \leqslant \Big(\frac{1}{|B|}\int_B |V|^2\Big)^{1/2} \leqslant s\Big(\frac{1}{|sB|}\int_{sB} |V|^2\Big)^{1/2}$$

Hence we arrive at the weak reverse Hölder inequality for the potential function,

$$\Big(\frac{1}{|B|}\int_B |V|^4\Big)^{1/4} \leqslant \frac{sC(K)}{(s-1)}\Big(\frac{1}{|sB|}\int_{sB} |V|^2\Big)^{1/2} \tag{17.19}$$

It is well known that reverse Hölder inequalities imply improved integrability properties; here we use the fact that by an iteration process the exponent 2 on the right hand side of (17.19) can be arbitrarily decreased, see Theorem A.9.1. In particular, we obtain

$$\Big(\frac{1}{|B|}\int_B |V|^2\Big)^{1/2} \leqslant C(K)\frac{1}{|sB|}\int_{sB}|V| \tag{17.20}$$

With this information in hand we observe that the inequality (17.18) and the second Cacciopoli estimate (17.16) together take the elegant form

$$\Big(\frac{1}{|B|}\int_B |f|^2\Big)^{1/2} \leqslant \frac{C(K)}{R}\frac{1}{|2B|}\int_{2B}|V| \tag{17.21}$$

and

$$\Big(\frac{1}{|2B|}\int_B |Df|^2\Big)^{1/2} \leqslant \frac{C(K)}{R^2}\frac{1}{|2B|}\int_{2B}|V| \tag{17.22}$$

Actually, via the second-order Poincaré inequalities, Lemma A.6.5, these estimates in turn yield better inequalities for the potential function. We have

$$\max_{B} |V| \leqslant C(K) \frac{1}{|2B|} \int_{2B} |V| \qquad (17.23)$$

for every disk B with $2B \subset \Omega$.

As one should immediately observe, the estimates (17.21) and (17.23) hold for any function V whose gradient is a quasiconformal mapping, whether it is the usual gradient $V_{\bar{z}}$ or the complex gradient V_z.

In studying solutions to elliptic equations, we have the latter case. Combined with the results from Chapter 14 we obtain the following corollary.

Corollary 17.2.1. *Let*

$$\frac{2K}{K+1} \leqslant q \leqslant p < \frac{2K}{K-1} \qquad (17.24)$$

and assume that the complex gradient $f = u_z$ of a function $u \in W^{2,q}_{loc}(\Omega)$ is weakly K-quasiregular. Then $D^2 u \in L^p_{loc}(\Omega)$, and we have the estimate

$$\|D^2 u\|_{L^p(\Omega')} \leqslant C \|u\|_{L^1(\Omega')}$$

for any relatively compact subdomain $\Omega' \subset \Omega$. Here, $C = C(p, q, \Omega, \Omega')$.

Proof. According to Theorem 14.4.3, f is quasiregular and $f = u_z \in W^{1,2}_{loc}(\Omega)$. Therefore the Caccioppoli inequalities from Corollary 14.3.3 give

$$\|D^2 u\|_{L^p(\Omega')} = \|f_z\|_{L^p(\Omega')} + \|f_{\bar{z}}\|_{L^p(\Omega')} \leqslant C \|f_z\|_{L^2(\Omega')} \leqslant C \|u\|_{L^1(\Omega')}$$

as long as (17.24) is satisfied. Here, the last inequality was based on (17.22). □

17.3 The Maximum Principle and Pucci's Conjecture

This section is devoted to proving the sharp form of the maximum principle, Theorem 17.0.13, conjectured by Pucci. The key tool will be the invertibility of Beltrami operators discussed in Chapter 14. The corresponding questions for degenerate elliptic equations are briefly discussed in Chapter 20.

We present the proof of Theorem 17.0.13 in a form that gives an explicit bound on the constant $C_p(\Omega)$.

Theorem 17.3.1. *Let $u \in W^{2,p}(\Omega)$, $p > \frac{2K}{K+1}$. Then for every relative compact subdomain $\Omega' \subset \Omega$ we have*

$$\max_{\Omega'} |u| \leqslant \max_{\partial \Omega'} |u| + C_p(\Omega) \|\mathcal{L}_A u\|_{L^p(\Omega)}, \qquad (17.25)$$

17.3. PUCCI'S CONJECTURE

where

$$C_p(\Omega) \leq C(p)|\Omega|^{1-1/p} \log \frac{(\operatorname{diam}\Omega)^2}{|\Omega|} \leq C(p)(\operatorname{diam}\Omega)^{2-2/p} \qquad (17.26)$$

Proof. Let us denote by $h \in L^p(\Omega)$ the function

$$h = \frac{1}{\operatorname{tr} A} \mathcal{L}_A u = 2u_{z\bar{z}} + \mu(z)u_{zz} + \bar{\mu}(z)\overline{u_{zz}}, \qquad (17.27)$$

where μ is given in (17.10). We extend μ and h to the entire complex plane $\mathbb{C}$ by defining them to be 0 outside of Ω. We then use the Beltrami operator

$$\mathcal{B} = 2\mathbf{I} + \mu \mathcal{S} + \bar{\mu}\bar{\mathcal{S}} : L^p(\mathbb{C}) \to L^p(\mathbb{C}), \qquad (17.28)$$

which enables us to study the equation

$$2U_{z\bar{z}} + \mu(z)U_{zz} + \bar{\mu}(z)\overline{U_{zz}} = H(z), \qquad z \in \mathbb{C}, \qquad (17.29)$$

auxiliary to the equation $\mathcal{L}_A u = h$.

We start with the general case where $H \in L^p(\mathbb{C})$, with p in the critical interval $2K/(K+1) < p < 2K/(K-1)$, and we will look for solutions U to (17.29) for which $D^2 U \in L^p_{loc}(\mathbb{C})$. Since the result concerns local properties, to simplify matters we shall assume that both μ and H are compactly supported in a bounded domain $\Omega \subset \mathbb{C}$.

A solution U should be of the form

$$U(z) = \frac{2}{\pi} \int_\Omega \log \frac{|z-\tau|}{\operatorname{diam}(\Omega)} \omega(\tau)\, d\tau \qquad (17.30)$$

for some function $\omega \in L^p(\mathbb{C})$ supported in Ω. The complex gradient of such a solution will be given by the Cauchy formula

$$\Phi(z) = \frac{\partial U}{\partial z} = \frac{1}{\pi} \int_\mathbb{C} \frac{\omega(\tau)d\tau}{z-\tau} = \mathcal{C}\omega, \qquad (17.31)$$

which upon further differentiation yields

$$U_{z\bar{z}} = \frac{\partial \Phi}{\partial \bar{z}} = \omega, \qquad U_{zz} = \frac{\partial \Phi}{\partial z} = -\frac{1}{\pi} \int_\mathbb{C} \frac{\omega(\tau)d\tau}{(z-\tau)^2} = \mathcal{S}\omega,$$

and so $\omega, \mathcal{S}\omega \in L^p(\mathbb{C})$. Now of course, (17.29) takes the form $\mathcal{B}\omega = H$. The key idea is to use the invertibility Beltrami operators, Theorem 14.0.4. We can rephrase this as

$$\|U_{z\bar{z}}\|_{L^p(\mathbb{C})} + \|U_{zz}\|_{L^p(\mathbb{C})} \leq C_p \|H\|_{L^p(\mathbb{C})}, \qquad (17.32)$$

so that in particular, using (17.30),

$$\|U\|_\infty \leq C_\Omega \|\omega\|_{L^p(\Omega)} \leq C_p \|H\|_{L^p(\Omega)} \qquad (17.33)$$

Let us then apply this general argument to the specific situation of the theorem and solve (in $\mathbb{C}$) the auxiliary equation in (17.29) with $H = h\chi_\Omega$. For the theorem we may assume $\frac{2K}{K+1} < p \leqslant 2$. From (17.32) we see that $U \in W^{2,p}_{loc}(\mathbb{C})$. The difference $U - u \in W^{2,p}(\Omega)$ solves the homogeneous equation

$$\mathcal{L}_A(U - u) = 0, \quad \text{in } \Omega \tag{17.34}$$

Therefore, by Lemma 17.1.1, its complex gradient $F = U_z - u_z$ satisfies $F_{\bar{z}} = -\frac{1}{2}(\mu F_z + \bar{\mu}\bar{F_z})$ and so is a weakly K-quasiregular mapping defined in Ω. Corollary 17.2.1 tells us that in fact $F \in W^{1,2}_{loc}(\Omega)$.

We are now in the situation where we may apply Alexandrov's maximum principle, Theorem 17.0.12. Accordingly,

$$\max_{\Omega'} |U - u| \leqslant \max_{\partial\Omega'} |U - u|,$$

from which the triangle inequality yields

$$\max_{\Omega'} |u| \leqslant \max_{\partial\Omega'} |u| + \max_{\partial\Omega'} |U| + \max_{\Omega'} |U|$$
$$\leqslant \max_{\partial\Omega'} |u| + 2\max_{\Omega} |U|$$

Of course, what remains is to estimate the last term here by quantities involving $\mathcal{L}_A u$. Recall that U is defined by (17.30). From Hölder's inequality, with $q = p/(p-1)$, we have

$$\max_\Omega |U| \leqslant \frac{2}{\pi} \|\omega\|_{L^p(\Omega)} \max_{z \in \Omega} \left\| \log \frac{|z - \tau|}{\operatorname{diam}\Omega} \right\|_{L^q(\Omega)},$$

where

$$\|\omega\|_{L^p(\Omega)} \leqslant C_p \|H\|_{L^p(\Omega)} = \|h\|_{L^p(\Omega)}$$
$$= C_p \|\mathcal{L}_A u\|_{L^p(\Omega)}$$

We therefore arrive at the inequality in (17.25) with the constant

$$C_p(\Omega) = \max_{z \in \Omega} \left(\int_\Omega \left| \log \frac{|z - \tau|}{\operatorname{diam}\Omega} \right|^q d\tau \right)^{1/q} \tag{17.35}$$

Analysis of this constant is of independent interest. Because of the homogeneity, we can certainly assume that $|\Omega| = 1$ by rescaling z and τ above. The isodiametric inequality shows that

$$\operatorname{diam}^2(\Omega) \geqslant \frac{4}{\pi} > 1,$$

and thus $\log(\operatorname{diam}(\Omega)/|z - \tau|)$ increases when $|z - \tau|$ decreases. We can apply the method of circular symmetrization to determine that the integral in (17.35)

17.4. INTERIOR REGULARITY

assumes its largest value when Ω is a disk of area 1. Then

$$C_p(\Omega) \leq \left(\int_{\pi|\zeta|^2 < 1} \left(\log \frac{\operatorname{diam}(\Omega)}{|\zeta|} \right)^q d\zeta \right)^{1/q}$$

$$\leq \left(\int_{\pi|\zeta|^2 < 1} \log^q \operatorname{diam}(\Omega) \right)^{1/q} + \left(\int_{\pi|\zeta|^2 < 1} \log^q 1/|\zeta| \right)^{1/q}$$

$$\leq \log \operatorname{diam}(\Omega) + C_q \leq C \log \operatorname{diam}(\Omega) \leq C \log \left(\frac{\operatorname{diam}^2(\Omega)}{|\Omega|} \right)$$

which, when the normalization of the area is removed, gives us the constant in the statement of the theorem. $\square$

We are unsure as to whether the logarithmic factor in (17.26) is in fact necessary. It is shown in [82] that for $p = 2$ one can do without this term.

17.4 Interior Regularity

The above proof for the maximum principle gives further regularity estimates which we now consider. The first concerns the interior regularity of solutions to $\mathcal{L}_A u = h$ in the complex plane.

Theorem 17.4.1. *Suppose that* $u \in W^{2,q}(\Omega)$, *whereas* $\mathcal{L}_A \in L^p(\Omega)$ *for some pair of exponents*

$$\frac{2K}{K+1} < q \leq p < \frac{2K}{K-1} \tag{17.36}$$

Then $u \in W^{2,p}_{loc}(\Omega)$. *Moreover, for every relatively compact subdomain* $\Omega' \subset \Omega$, *we have the uniform estimate*

$$\|D^2 u\|_{L^p(\Omega')} \leq C \|u\|_{L^1(\Omega)} + C \|\mathcal{L}_A u\|_{L^p(\Omega)}, \tag{17.37}$$

where $C = C(p, q, \Omega', \Omega)$.

Proof. Denote by $h \in L^p(\Omega)$ the function

$$h = \frac{1}{\operatorname{tr} A} \mathcal{L}_A u = 2u_{z\bar{z}} + \mu(z) u_{zz} + \bar{\mu}(z) \overline{u_{zz}}, \tag{17.38}$$

where μ is given in (17.10), and extend as before μ and h to $\mathbb{C}$ by defining them to be 0 outside of Ω. For p in the critical interval (17.36), we can then solve the auxiliary equation (17.29) in the entire plane $\mathbb{C}$, with $H = h\chi_\Omega$. From (17.32) we have $U \in W^{2,p}_{loc}(\mathbb{C})$. Now the function $U - u \in W^{2,q}(\Omega)$ solves in Ω the homogeneous equation

$$\mathcal{L}_A(U - u) = 0, \tag{17.39}$$

and thus by Lemma 17.1.1 its complex gradient $F = U_z - u_z$ is a weakly K-quasiregular mapping in Ω. Again, Corollary 17.2.1 tells us that $F \in W^{1,2}_{loc}(\Omega)$ and further gives the estimate

$$\|U_{z\bar z} - u_{z\bar z}\|_{L^p(\Omega')} + \|U_{zz} - u_{zz}\|_{L^p(\Omega')} \leq C\|U - u\|_{L^1(\Omega)}$$

Hence the triangle inequality implies that

$$\|D^2 u\|_{L^p(\Omega')} \leq C(\|u\|_{L^1(\Omega)} + \|U\|_{L^1(\Omega)} + \|D^2 U\|_{L^p(\Omega)})$$

Finally, we use (17.32) and (17.33) to control the last two terms by

$$\|H\|_{L^p(\mathbb{C})} = \|h\|_{L^p(\Omega)} \leq 2K\|\mathcal{L}_A u\|_{L^p(\Omega)}$$

This completes the proof. $\square$

The example of Section 17.6 demonstrates that the estimate (17.37) fails in the borderline case $q = \frac{2K}{K+1}$. Recent work by Astala, Faraco and Székelyhidi [24] shows that also in the other borderline case $p = \frac{2K}{K-1}$ the L^p regularity for the operator

$$2\frac{\partial}{\partial \bar z} + \mu \frac{\partial}{\partial z} + \bar\mu \overline{\frac{\partial}{\partial z}}$$

fails. Therefore (compare with Lemma 17.1.1) we see that (17.37) fails at $p = \frac{2K}{K-1}$.

We remark at this point that the Sobolev embedding theorem implies that solution $u \in W^{2,2}_{loc}(\Omega)$ of the homogeneous equation $\mathcal{L}_A u = 0$ have Hölder continuous first derivatives, $u \in C^{1,\alpha}_{loc}(\Omega)$, for every $0 < \alpha < \frac{1}{K}$. As a matter of fact, the borderline exponent is also right, and we have sharp energy estimates over measurable sets.

Theorem 17.4.2. *Let $u \in W^{2,2}_{loc}(\Omega)$ solve $\mathcal{L}_A u = 0$. Then $u \in C^{1,1/K}_{loc}(\Omega)$. Moreover, for every compact set $\Omega' \subset \Omega$ there corresponds a constant C, depending only on K, Ω, Ω' and $\|u\|_{L^1(\Omega)}$, such that the energy on a measurable set $E \subset \Omega'$ is estimated as*

$$\int_E |D^2 u|^2 \leq C|E|^{1/K}$$

Proof. With Lemma 17.1.1 the result reduces to Corollary 13.2.5. $\square$

We now turn to a discussion of the removability of singularities of solutions to the equation $\mathcal{L}_A u = h$.

Theorem 17.4.3. *Let $E \subset \mathbb{C}$ be a closed set of α-Hausdorff measure zero,*

$$\mathcal{H}^\alpha(E) = 0, \qquad \alpha = \frac{2}{K+1}, \tag{17.40}$$

and consider a function of the Sobolev class

$$u \in W^{1,\infty}(\Omega) \cap W^{2,p}_{loc}(\Omega \setminus E), \qquad 2 \leq p < \frac{2K}{K-1}$$

If $\mathcal{L}_A u \in L^p_{loc}(\Omega)$, then $u \in W^{2,p}_{loc}(\Omega)$.

17.5. EQUATIONS WITH LOWER-ORDER TERMS

Proof. We continue with the notation of the previous section. What we need to do here is to show that
$$U - u \in W^{2,p}_{loc}(\Omega)$$

We are able to assume that the complex gradient $F = U_z - u_z$ is of class $W^{1,p}_{loc}(\Omega \setminus E)$ where $2 \leqslant p < 2K/(K-1)$. Hence F is K-quasiregular in $\Omega \setminus E$. Moreover, as $U_{z\bar{z}}$ is a compactly supported function in $L^2(\mathbb{C})$ and $u \in W^{1,\infty}(\Omega)$, the complex gradient F belongs to $BMO(\Omega)$. It was proven in Corollary 13.5.12 that for K-quasiregular mappings in this class, E is removable if its Hausdorff measure $\mathcal{H}^{2/(K+1)}(E) = 0$. Thus F is K-quasiregular in Ω, and in particular $F \in W^{1,p}_{loc}(\Omega)$, which in turn implies $U - u \in W^{2,p}_{loc}(\Omega)$, as desired. □

Example 13.6.1 shows that in general sets $E \subset \mathbb{C}$ of Hausdorff dimension equal to $2/(K+1)$ are not removable. Further, the above result can be generalized by weakening the requirement that the gradient of u is bounded. We need only assume a certain degree of integrability of the gradient. However, under these circumstances we must assume that the dimension of E is smaller.

Theorem 17.4.4. *Let $E \subset \mathbb{C}$ be a closed set of Hausdorff dimension α,*
$$\dim_{\mathcal{H}}(E) = \alpha < 2\left(1 + \frac{sK}{s - 2K}\right)^{-1}, \qquad s > 2K, \qquad (17.41)$$

and consider a function of Sobolev class
$$u \in W^{1,s}_{loc}(\Omega) \cap W^{2,p}_{loc}(\Omega \setminus E), \qquad \frac{2s}{2+s} < p < \frac{2K}{K-1}$$

If $\mathcal{L}_A u \in L^p_{loc}(\Omega)$, then $u \in W^{2,p}_{loc}(\Omega)$.

The proof can be found by using the same lines of reasoning as above. We need only establish that the weaker assumption $F \in L^s_{loc}(\Omega)$, $s > 2K$, and the stronger condition in (17.41) are enough to force the removability of E. This is shown in [191, Corollary 17.3.1].

17.5 Equations with Lower-Order Terms

Alexandrov's estimate in (17.4) actually holds when lower-order terms are added to the operator $\mathcal{L}_A$. In particular, let us consider the operator
$$\mathcal{Q}_A u = \mathcal{L}_A u + \langle b, \nabla u \rangle \qquad (17.42)$$

defined for $u \in W^{2,p}(\Omega)$, where $p > 2K/(K+1)$. Here the vector-valued function $b(z) = (b_1(z), b_2(z))$ is assumed to lie in the space $L^s(\Omega)$. For the classical case $p \geqslant 2$ and $s = 2$, the generalization of (17.4) is
$$\max_{\Omega'} |u| \leqslant \max_{\partial \Omega'} |u| + C \operatorname{diam}(\Omega) \|\mathcal{Q}_A u\|_{L^2(\Omega)} \qquad (17.43)$$

(see [143]). We shall then consider the case $2K/(K+1) < p < 2$. Note that by the embedding theorem $\nabla u \in L_{loc}^{2p/(2-p)}(\Omega)$, so that in fact

$$\langle b, \nabla u\rangle \in L_{loc}^r(\Omega), \qquad r = \frac{2sp}{2p+2s-sp},$$

as $\frac{1}{s} + \frac{2-p}{2p} = \frac{1}{r}$. As we are going to consider the function $\mathcal{L}_A u$ in the $L^p(\mathbb{C})$ space, it is natural to asssume that $s \geqslant 2$. We shall see in a moment that the case $s = 2$ is not sufficient for our arguments below. Thus we shall assume henceforth that

$$s > 2 \quad \text{(which implies } r > p\text{)} \tag{17.44}$$

We now have the following version of the Pucci conjecture.

Theorem 17.5.1. *Let $u \in W^{2,p}(\Omega)$, $p > \frac{2K}{K+1}$. Then for every relative compact subdomain $\Omega' \subset \Omega$, we have*

$$\max_{\Omega'} |u| \leqslant \max_{\partial \Omega'} |u| + C_p(\Omega)\|\mathcal{Q}_A u\|_{L^p(\Omega)}, \tag{17.45}$$

where $\mathcal{Q}_A$ is defined in (17.42) and

$$C_p(\Omega) \leqslant C(p)|\Omega|^{1-1/p} \log \frac{\mathrm{diam}^2(\Omega)}{|\Omega|} \leqslant C(p)\mathrm{diam}^{2-2/p}(\Omega).$$

Proof. There is no loss of generality in assuming that $\mathrm{tr}\, A = 1$, see Lemma 17.1.1. Then we are led to consider the equation

$$h = 2u_{z\bar{z}} + \mu u_{zz} + \bar{\mu}\overline{u_{zz}} + bu_z + \bar{b}\overline{u_z},$$

where we have chosen to write $b = b_1 + ib_2$. As before, we consider h, μ and b to be defined in $\mathbb{C}$ by setting them to zero outside Ω. We now solve the auxiliary equation

$$H = 2U_{z\bar{z}} + \mu\, U_{zz} + \bar{\mu}\,\overline{U_{zz}} + b\, U_z + \bar{b}\,\overline{U_z} \tag{17.46}$$

in the plane $\mathbb{C}$ for a function $U \in W_{loc}^{2,p}(\mathbb{C})$ of the form

$$U(z) = \frac{2}{\pi}\int_\Omega \log \frac{|z-\tau|}{\mathrm{diam}(\Omega)}\, \omega(\tau)d\tau,$$

where the function $\omega \in L^p(\Omega)$ is to be found, and $H = h\chi_\Omega$. We shall see in a moment that this is indeed possible. Meanwhile, consider the solution $U - u \in W^{2,p}(\Omega)$ of the homogeneous equation $\mathcal{Q}_A(U-u) = 0$. Its complex gradient $F = (U-u)_z$ satisfies the equation

$$2F_{\bar{z}} + \mu F_z + \bar{\mu}\overline{F_z} + bF + \bar{b}\bar{F} = 0,$$

and so $F \in W_{loc}^{1,2}(\Omega)$. To see this we may assume that $2 < s < 2K/(K-1)$. We proceed by induction to show that

$$F \in W_{loc}^{1,p_n}(\Omega) \tag{17.47}$$

17.5. EQUATIONS WITH LOWER-ORDER TERMS

with $p_0 = p$ and
$$p_{n+1} = \frac{2sp_n}{2s + 2p_n - sp_n}$$
for $n = 0, 1, 2, \ldots, N$, where N is chosen so that
$$\frac{(2-p)s}{(s-2)p} - 1 \leqslant N < \frac{(2-p)s}{(s-2)p}$$

Note that $p_0 < p_1 < \cdots < p_N < 2 \leqslant p_{N+1} < 2K/(K-1)$. We need to explain the induction step. Suppose $F \in W^{1,p_n}_{loc}(\Omega) \subset L^{2p_n/(2-p_n)}(\Omega)$. We may use Hölder's inequality $bF \in L^{p_{n+1}}_{loc}(\Omega)$, while by Theorem 17.4.1 we have $F \in W^{1,p_{n+1}}_{loc}(\Omega)$, as desired.

Now we may use Alexandrov's theorem in this setting for $U - u$ to get
$$\max_{\Omega'} |U - u| \leqslant \max_{\partial \Omega'} |U - u|$$

As before, the triangle inequality gives
$$\begin{aligned}\max_{\Omega'} |u| &\leqslant \max_{\partial \Omega'} |u| + 2 \max_{\Omega} |U| \\ &\leqslant \max_{\partial \Omega'} |u| + C_p(\Omega) \|w\|_{L^p(\Omega)},\end{aligned}$$
as desired.

We now return to discussing the issue of solving (17.46). The question reduces to the invertibility for the operator
$$\mathcal{F} = 2\mathbf{I} + \mu \mathcal{S} + \bar{\mu}\bar{\mathcal{S}} + b\mathcal{C} + \bar{b}\bar{\mathcal{C}},$$
where $\mathcal{C}$ denotes the usual Cauchy transform. It is important to note that the operator
$$b\mathcal{C} : L^p(\mathbb{C}) \to L^p(\mathbb{C})$$
is compact because b vanishes outside the bounded set Ω. Since $\mathcal{B}$ as defined in (17.28) is invertible, we see that $\mathcal{F}$ is Fredholm of index zero. What we need to ensure the invertibility of $\mathcal{F} : L^p(\mathbb{C}) \to L^p(\mathbb{C})$ is to show that the kernel of $\mathcal{F}$ is trivial, $\mathrm{Ker}(\mathcal{F}) = 0$. Equivalently, we must show that the equation
$$2g_{\bar{z}} + \mu g_z + \bar{\mu}\bar{g}_z + bg + \bar{b}\bar{g} = 0, \tag{17.48}$$
for $g \in W^{1,p}_{loc}(\mathbb{C}) \cap C_0(\hat{\mathbb{C}})$, admits only the trivial solution $g \equiv 0$. Just as we established (17.47), we can see that
$$g \in W^{1,2}_{loc}(\mathbb{C}) \cap C_0(\hat{\mathbb{C}})$$

We also have the differential inequality
$$|g_{\bar{z}}| \leqslant \frac{K-1}{K+1}|g_z| + \sigma(z)|g(z)|,$$

where $\sigma(z) \leqslant |b(z)| \in L^{2+\varepsilon}(\mathbb{C}) \cap L^{2-\varepsilon}(\mathbb{C})$, for some $0 < \varepsilon < 1$, since b has compact support. Finally, from the Liouville-type Theorem 8.5.1 we conclude that $g = 0$, as desired. □

The reader may wish to verify that the $W^{1,2}_{loc}(\Omega)$ regularity for the equation $F_{\bar{z}} + \sigma F = 0$ with $\sigma \in L^2_{loc}(\Omega)$ fails even if $F \in W^{1,p}_{loc}(\Omega)$ for all $1 < p < 2$. Indeed, consider

$$F(z) = \sqrt{\log(e|z|^{-2})} \quad \text{and} \quad \sigma(z) = \frac{1}{2\bar{z}\log(e|z|^{-2})}$$

in the unit disk $\mathbb{D}$. We find that both $F_z = -1/(2zF)$ and $F_{\bar{z}} = -1/(2\bar{z}F)$ lie in every $L^p(\mathbb{D})$, $1 < p < 2$, but not in $L^2(\mathbb{D})$.

17.5.1 The Dirichlet Problem

We illustrate for the unit disk $\mathbb{D} = \{z : |z| < 1\}$ how our estimates yield the solution of the Dirichlet problem in $W^{2,p}(\mathbb{D})$.

Theorem 17.5.2. Let $b \in L^s(\mathbb{D}, \mathbb{R}^2)$ and $h \in L^p(\mathbb{D})$ for some $s > 2$ and $\frac{2K}{K+1} < p < \frac{2K}{K-1}$. Then the equation

$$\operatorname{tr}(A\, D^2 u) + \langle b, \nabla u \rangle = h$$

has a unique solution $u \in W^{2,p}(\mathbb{D})$ vanishing on $\partial \mathbb{D}$. Moreover, we have the uniform estimate

$$\|D^2 u\|_{L^p(\mathbb{D})} \leqslant C \|h\|_{L^p(\mathbb{D})},$$

where C does not depend on h.

Proof. Suppose we are given a function $u \in W^{2,p}(\mathbb{D}) \subset C(\overline{\mathbb{D}})$ that vanishes on the boundary $\partial \mathbb{D}$. We consider the differential operator

$$\mathcal{L}_A u = 2 u_{z\bar{z}} + \mu\, u_{zz} + \overline{\mu}\, \overline{u_{zz}} + b u_z + \overline{b} \overline{u_z} \stackrel{\text{def}}{=} h \qquad (17.49)$$

defining the function h. We look for estimates of $D^2 u$ in terms of h.

We make the following assumptions about the coefficients,

$$|\mu(z)| \leqslant k < 1 \quad \text{almost everywhere on } \mathbb{D} \qquad (17.50)$$
$$b \in L^s(\mathbb{D}) \quad \text{for some } s > 2 \qquad (17.51)$$

To make use of the global estimates developed in the previous sections, we extend u outside $\mathbb{D}$ by reflection and set

$$U(z) = u(z) \quad \text{for } |z| \leqslant 1 \quad \text{and} \quad U(z) = -u(\bar{z}^{-1}) \quad \text{for } |z| \geqslant 1$$

Clearly, $U \in W^{2,p}_{loc}(\mathbb{C})$. A calculation shows that the extended function U solves the following equation in the entire complex plane $\mathbb{C}$,

$$2U_{z\bar{z}} + \mu^* U_{zz} + \overline{\mu^* U_{zz}} + b^* U_z + \overline{b^* U_z} = h^*,$$

17.5. EQUATIONS WITH LOWER-ORDER TERMS

where $\mu^*(z) = \mu(z)$ for $|z| \leqslant 1$ and $\mu^*(z) = z^2 \bar{z}^{-2} \bar{\mu}(\bar{z}^{-1})$ for $|z| \geqslant 1$. Thus in particular,
$$|\mu^*(z)| \leqslant k < 1 \quad \text{almost everywhere on } \mathbb{C}$$

The function b^* is defined by
$$b^*(z) = b(z) \quad \text{for } |z| \leqslant 1 \quad \text{and} \quad b^*(z) = (2z/\bar{z}^2)\bar{\mu}(1/\bar{z}) - \bar{z}^{-2} \quad \text{for } |z| \geqslant 1,$$

while the inhomogeneous part attains the form
$$h^*(z) = h(z) \quad \text{for } |z| \leqslant 1 \quad \text{and} \quad h^*(z) = -|z|^{-4} h(\bar{z}^{-1}) \quad \text{for } |z| \geqslant 1$$

The point to make here is that we have uniform control of the norms of U over the double disk $2\mathbb{D}$ in terms of $u(z), z \in \mathbb{D}$.

The first-order equation for the complex gradient $F(z) = U_z(z)$ now takes the form
$$2F_{\bar{z}} + \mu^* F_z + \overline{\mu^*}\, \overline{F}_z + b^* F + \overline{b^*}\, \overline{F} = h^* \tag{17.52}$$

To use the global estimates here we fix a non-negative bump function $\phi \in C_0^\infty(2\mathbb{D})$ such that $\phi \equiv 1$ on $\mathbb{D}$. Then multiply (17.52) by ϕ to obtain
$$2(\phi F)_{\bar{z}} + \mu^*(\phi F)_z + \overline{\mu^*}\,\overline{(\phi F)_z} + b^* \phi F + \overline{b^*}\,\overline{(\phi F)}$$
$$= \phi h^* + 2\phi_{\bar{z}} F + \mu^* \phi_z F + \overline{\mu^*}\,\overline{\phi_z F}$$

It makes no difference if we redefine b^* as being equal to zero outside $2\mathbb{D}$. As shown in the proof of Theorem 17.5.1, the operator
$$\mathcal{F} = 2\mathbf{I} + \mu^* S + \overline{\mu^*}\, \overline{S} + b^* \mathcal{C} + \overline{b^*}\, \overline{\mathcal{C}} : L^p(\mathbb{C}) \to L^p(\mathbb{C})$$

is then invertible, and we have the L^p-estimate
$$\|D(\phi F)\|_{L^p(\mathbb{C})} \leqslant C\|\phi h^*\|_{L^p(\mathbb{C})} + C\| |\nabla \phi| |F| \|_{L^p(\mathbb{C})}$$

for all p in the critical interval. Hence $\|DF\|_{L^p(\mathbb{D})} \leqslant C\|h^*\|_{L^p(2\mathbb{D})} + C\|F\|_{L^p(2\mathbb{D})}$, which in turn reads in terms of u as
$$\|D^2 u\|_{L^p(\mathbb{D})} \leqslant C\|h\|_{L^p(\mathbb{D})} + C\|\nabla u\|_{L^p(\mathbb{D})} \tag{17.53}$$

We now invoke the well-known interpolation inequality
$$\|\nabla u\|_{L^p(\mathbb{D})} \leqslant \varepsilon \|D^2 u\|_{L^p(\mathbb{D})} + C_\varepsilon \|u\|_{L^p(\mathbb{D})}$$

Choosing ε sufficiently small, the term with the second derivatives is absorbed by the left hand side of (17.53), and we arrive at the estimate
$$\|D^2 u\|_{L^p(\mathbb{D})} \leqslant C\|h\|_{L^p(\mathbb{D})} + C\|u\|_{L^p(\mathbb{D})},$$

Finally, by using Theorem 17.5.1 we see that $\|u\|_{L^p(\mathbb{D})}$ is dominated by $\|h\|_{L^p(\mathbb{D})}$. We conclude with the a priori estimate
$$\|D^2 u\|_{L^p(\mathbb{D})} \leqslant C\|h\|_{L^p(\mathbb{D})} \tag{17.54}$$

This result has an interesting interpretation. To this effect, given $w \in L^p(\mathbb{D})$, consider the logarithmic potential

$$u(z) = \frac{2}{\pi} \int_{\mathbb{D}} \log\left(\frac{|z-\tau|}{|1-z\overline{\tau}|}\right) w(\tau) d\tau$$

Then $u \in C(\overline{\mathbb{D}})$, and u vanishes on the boundary $\partial \mathbb{D}$. The complex gradient is given by

$$u_z(z) = (\mathcal{C}_{\mathbb{D}} w)(z) \stackrel{\text{def}}{=} \frac{1}{\pi} \int_{\mathbb{D}} \left[\frac{1}{z-\tau} + \frac{\overline{\tau}}{1-z\overline{\tau}}\right] w(\tau) d\tau,$$

while the second derivatives are $u_{z\overline{z}} = w$ with

$$u_{zz} = (\mathcal{S}_{\mathbb{D}} w)(z) \stackrel{\text{def}}{=} -\frac{1}{\pi} \int_{\mathbb{D}} \left[\frac{1}{(z-\tau)^2} - \frac{\overline{\tau}^2}{(1-z\overline{\tau})^2}\right] w(\tau) d\tau$$

Estimate (17.54) tells us that $\|w\|_{L^p(\mathbb{D})} \leqslant C\|h\|_{L^p(\mathbb{D})}$ when h is defined through (17.49). In other words, the operator

$$\mathcal{F}_{\mathbb{D}} = 2\mathbf{I} + \mu \mathcal{S}_{\mathbb{D}} + \overline{\mu} \overline{\mathcal{S}_{\mathbb{D}}} + b\mathcal{C}_{\mathbb{D}} + \overline{b\mathcal{C}_{\mathbb{D}}}$$

is invertible in $L^p(\mathbb{D})$ for all p in the critical interval $\frac{2K}{K+1} < p < \frac{2K}{K-1}$. Of course, Theorem 17.5.2 is an immediate consequence of this fact. $\square$

At this point the reader may wish to go on and look at questions of removable singularities and so forth in this more general setting with lower-order terms, using the methods and ideas outlined above. We will finish this chapter by presenting Pucci's example.

17.6 Pucci's Example

In this section we give the example of Pucci [304], which exhibits sharpness in all of the theorems we have stated with regard to the critical exponents. Our domain will be the unit disk $\mathbb{D}$. To begin we put

$$A(z) = \frac{z \otimes z}{|z|^2}\left(\sqrt{K} - \frac{1}{\sqrt{K}}\right) + \frac{1}{\sqrt{K}}\mathbf{I}$$

An elementary calculation reveals the ellipticity bounds

$$\frac{1}{\sqrt{K}} \leqslant \langle A(z)\zeta, \zeta\rangle \leqslant \sqrt{K}, \qquad |\zeta| = 1$$

Let $u(z) = \varphi(|z|)$ be any radial function of Sobolev class $W^{2,2}_{loc}(\Omega)$. Then

$$\mathcal{L}_A u = \text{tr}(A(z)\mathcal{H}u) = \sqrt{K}\varphi''(|z|) + \frac{1}{\sqrt{K}}\frac{1}{|z|}\varphi'(|z|)$$

17.6. PUCCI'S EXAMPLE

We now make a careful choice of function φ. When $K > 1$, let

$$\varphi(r) = \begin{cases} (\log r) r^{1-1/K} - \left(\log \nu - \frac{K}{K-1}\right)(1 - r^{1-1/K}) & \text{if } \frac{1}{\nu} \leq r < 1 \\ -\log(\nu) + \frac{K}{K-1}(1 - \nu^{-1+1/K}) & \text{if } 0 \leq r < \frac{1}{\nu} \end{cases}$$

We compute that

$$\varphi'(r) = \begin{cases} (1 - 1/K) \log(\nu r) r^{-1/K} & \text{if } \frac{1}{\nu} \leq r < 1 \\ 0 & \text{if } 0 \leq r < \frac{1}{\nu} \end{cases}$$

Thus our function $u(z) = \varphi(|z|)$ has

- $u \in W^{2,\infty}(\mathbb{D})$.
- $u|\partial \mathbb{D} = 0$.
- $\mathcal{L}_A u = h$, where $h = \left(\sqrt{K} - \frac{1}{\sqrt{K}}\right) |z|^{-1-1/K} \chi_{\{1/\nu < |z| < 1\}}(z)$.
- $\|h\|_{L^p(\mathbb{D})} = \left(\sqrt{K} - \frac{1}{\sqrt{K}}\right) (2\pi \log \nu)^{(K+1)/2K}$ when $p = 2K/(K+1)$.
- $\frac{1}{2} \log \nu \leq \|u\|_{L^\infty(\mathbb{D})} \leq 2 \log \nu$ for all sufficiently large ν.
- $\|D^2 u\|_{L^p(\mathbb{D})} \geq C(K) (\log \nu)^{1+(K+1)/2K}$.

It follows that

$$\frac{\|u\|_\infty}{\|h\|_{L^p}} \geq c(K)(\log \nu)^{1-(K+1)/2K} = C(\log \nu)^{(K-1)/2K} \to \infty$$

as $\nu \to \infty$. Hence (17.7) fails for $p = 2K/(K+1)$. Similarly,

$$\frac{\|D^2 u\|_{L^p}}{\|u\|_\infty} \geq c(K)(\log \nu)^{(K+1)/2K} \to \infty$$

as $\nu \to \infty$ so that (17.37) fails when $q = p = 2K/(K+1)$.

When $K = 1$, let

$$\varphi(r) = \begin{cases} \log^2(\nu r) - \log^2 \nu & \text{if } \frac{1}{\nu} \leq r < 1 \\ \log^2 \nu & \text{if } 0 \leq r \leq \frac{1}{\nu} \end{cases}$$

Then $\mathcal{L}_A u = h = 2|z|^{-2} \chi_{\{1/\nu < |z| < 1\}}(z)$, and again the ratio $\|u\|_\infty / \|h\|_{L^1} = (4\pi)^{-1} \log \nu \to \infty$ as $\nu \to \infty$.

Chapter 18

Quasiconformal Methods in Impedance Tomography: Calderón's Problem

In impedance tomography one aims to determine the internal structure of a body from electrical measurements on its surface. Such methods have a variety of different applications for instance in engineering and medical diagnostics. For a general expository presentation for medical applications, see [106].

Medical imaging in two dimensions

In 1980 A.P. Calderón showed that the impedance tomography problem admits a clear and precise mathematical formulation. Indeed, suppose that $\Omega \subset \mathbb{R}^n$ is a bounded domain with connected complement and let $\sigma : \Omega \to (0, \infty)$ be a measurable function that is bounded away from zero and infinity. According to Theorem 16.1.1, the Dirichlet problem

$$\nabla \cdot \sigma \nabla u = 0 \quad \text{in } \Omega, \tag{18.1}$$

$$u\big|_{\partial \Omega} = \phi \in W^{1,1/2}(\partial \Omega) \tag{18.2}$$

admits a unique solution $u \in W^{1,2}(\Omega)$. Here $W^{1,1/2}(\partial\Omega) = W^{1,2}(\Omega)/W_0^{1,2}(\Omega)$ stands for the space of elements $\phi + W_0^{1,2}(\Omega)$, where $\phi \in W^{1,2}(\Omega)$. This is the most general space of functions that can possibly arise as Dirichlet boundary values or traces of general $W^{1,2}(\Omega)$-functions in a bounded domain Ω.

In physical terms, if we charge the "body" Ω with an electric current, then $\phi = u|\partial\Omega$ represents the potential difference from $\partial\Omega$ to ∞. Furthermore, the electric current on the boundary is equal to $(\sigma\nabla u)|\partial\Omega$. In practice, one can measure only the normal component of the current, $\sigma\partial u/\partial\nu$, with ν the unit outer normal to the boundary. For smooth σ this quantity is well defined pointwise, while for general bounded measurable σ we need to use the (equivalent) definition

$$\langle \sigma \frac{\partial u}{\partial \nu}, \psi \rangle = \int_\Omega \sigma \nabla u \cdot \nabla \psi, \qquad \psi \in W^{1,2}(\Omega), \tag{18.3}$$

as an element of the dual of $W^{1,1/2}(\partial\Omega)$.

The inverse conductivity problem of Calderón asks if we can recover the pointwise conductivity $\sigma(x)$ *inside* the domain Ω from voltage/current measurements on the boundary $\partial\Omega$. In mathematical terms, the question is if the Dirichlet-to-Neumann boundary map

$$\Lambda_\sigma : \phi \mapsto \sigma \frac{\partial u}{\partial \nu}\bigg|_{\partial\Omega} \tag{18.4}$$

determines the coefficient $\sigma(x)$ in (18.1) for all $x \in \Omega$.

The positive answer to this problem in two dimensions was recently given in [35], making strong use of the techniques and quasiconformal methods developed in this book. Therefore we present this chapter as an example of those techniques applied in a real setting, and we will explain the answer to Calderón's question in the plane. Earlier results on this problem, assuming greater regularity, have been obtained by R. Brown, J. Sylvester, G. Uhlmann and A. Nachmann; see for instance [79, 282, 346]. Calderón's problem remains open in higher dimensions. For the largest class of potentials where the uniqueness has been shown so far see [291]. In higher dimensions the usual method is to reduce, by substituting $v = \sigma^{1/2}u$, the conductivity equation (18.1) to the Schrödinger equation and then to apply the methods of scattering theory. Indeed, after such a substitution v satisfies

$$\Delta v - qv = 0,$$

where $q = \sigma^{-1/2}\Delta\sigma^{1/2}$. This substitution is possible only if σ has some smoothness. In the case $\sigma \in L^\infty$, relevant for practical applications, in general there is no smoothness and the reduction to the Schrödinger equation fails. Therefore one must turn to complex analytic tools.

To avoid some of the technical complications, in this book we shall assume that the domain $\Omega = \mathbb{D}$, the unit disk. In fact, the reduction of general Ω to this case is not difficult; see [35].

We will prove the following theorem.

Theorem 18.0.1. *Let $\sigma_j \in L^\infty(\mathbb{D})$, $j = 1, 2$. Suppose that there is a constant $c > 0$ such that $c^{-1} \leqslant \sigma_j \leqslant c$. If*
$$\Lambda_{\sigma_1} = \Lambda_{\sigma_2},$$
then $\sigma_1 = \sigma_2$ almost everywhere. Here Λ_{σ_i}, $i = 1, 2$, are defined by (18.4).

From Theorem 16.1.6 we see that if $u \in W^{1,2}(\mathbb{D})$ is a real-valued solution of (18.1), then it has the σ-harmonic conjugate $v \in W^{1,2}(\mathbb{D})$ such that

$$\partial_x v = -\sigma \partial_y u \qquad (18.5)$$
$$\partial_y v = \sigma \partial_x u \qquad (18.6)$$

Equivalently (see (16.26)), the function $f = u + iv$ satisfies the $\mathbb{R}$-linear Beltrami equation

$$\frac{\partial f}{\partial \bar{z}} = \mu(z) \overline{\frac{\partial f}{\partial z}}, \qquad (18.7)$$

where
$$\mu = \frac{1 - \sigma}{1 + \sigma}$$

In particular, note that μ is real-valued and that the assumptions on σ in Theorem 18.0.1 imply $\|\mu\|_{L^\infty} \leqslant k < 1$. This reduction to the Beltrami equation and the complex analytic methods it provides will be the main tools in our analysis of the Dirichlet-to-Neumann map and the solutions to (18.1).

On the other hand, the basic concepts from the scattering theory approach and, in particular, ideas from the inverse scattering method in KdV-type equations, do lie behind the argument we are going to present. With the help of the above Beltrami equation, we will first construct global solutions to the conductivity equation (18.1) that grow exponentially,

$$u(z, \xi) = e^{i\xi z}\left(1 + \mathcal{O}\left(\frac{1}{|z|}\right)\right) \quad \text{as } |z| \to \infty$$

Studying the ξ-dependence of these solutions then gives rise to the concept of the *nonlinear Fourier transform* $\tau_\sigma(\xi)$, see Section 18.4. It is not difficult to show that the Dirichlet-to-Neumann boundary operator Λ_σ determines the nonlinear Fourier transforms $\tau_\sigma(\xi)$ for all $\xi \in \mathbb{C}$. Therefore the main difficulty, and our main strategy, is to show that the nonlinear Fourier transform $\tau_\sigma(\xi)$ determines $\sigma(z)$ for almost all z.

The properties of the nonlinear Fourier transform depend on the underlying differential equation. In one dimension the basic properties of the transform are fairly well understood, while deeper results such as analogs of Carleson's L^2-converge theorem remain open. The reader should consult the excellent lecture notes of Tao and Thiele [347] for an introduction to the one-dimensional theory.

For (18.1) with nonsmooth σ, many basic questions concerning the nonlinear Fourier transform, such as the Plancherel formula, remain open. For results on related equations, see [78].

18.1 Complex Geometric Optics Solutions

We will use the following convenient notation

$$e_\xi(z) = e^{i(z\xi + \bar{z}\bar{\xi})}, \qquad z, \xi \in \mathbb{C} \tag{18.8}$$

We will also extend σ to the entire plane $\mathbb{C}$ by requiring $\sigma(z) \equiv 1$ when $|z| \geq 1$. Clearly, this keeps σ and $1/\sigma$ in L^∞. Moreover, then

$$\mu(z) = \frac{1 - \sigma(z)}{1 + \sigma(z)} \equiv 0, \qquad |z| \geq 1$$

As a first step toward Theorem 18.0.1, we establish the existence of a family of special solutions to (18.7). These, called complex geometric optics solutions, are specified by having the asymptotics

$$f_\mu(z, \xi) = e^{i\xi z} M_\mu(z, \xi), \tag{18.9}$$

where

$$M_\mu(z, \xi) - 1 = \mathcal{O}\left(\frac{1}{z}\right) \quad \text{as } |z| \to \infty \tag{18.10}$$

Theorem 18.1.1. *For each parameter $\xi \in \mathbb{C}$ and for each $2 \leq p < 1 + 1/k$, the equation*

$$\frac{\partial f}{\partial \bar{z}} = \mu(z) \overline{\frac{\partial f}{\partial z}} \tag{18.11}$$

admits a unique solution $f = f_\mu \in W^{1,p}_{loc}(\mathbb{C})$ that has the form (18.9) with (18.10) holding. In particular, $f(z, 0) \equiv 1$.

Proof. Any solution to (18.11) is quasiregular. If $\xi = 0$, (18.9) and (18.10) imply that f is bounded, hence constant by the Liouville theorem.

If $\xi \neq 0$, look for a solution $f = f_\mu(z, \xi)$ in the form

$$f_\mu(z, \xi) = e^{i\xi \psi_\xi(z)}, \qquad \psi_\xi(z) = z + \mathcal{O}\left(\frac{1}{z}\right) \quad \text{as } |z| \to \infty \tag{18.12}$$

Substituting (18.12) into (18.11) indicates that ψ_ξ is the principal solution to the quasilinear equation

$$\frac{\partial}{\partial \bar{z}} \psi_\xi(z) = -\frac{\bar{\xi}}{\xi} e_{-\xi} \circ \psi_\xi(z) \, \mu(z) \, \overline{\frac{\partial}{\partial z} \psi_\xi(z)} \tag{18.13}$$

The function $H(z, w, \zeta) = -(\bar{\xi}/\xi) \mu(z) e_\xi(w) \bar{\zeta}$ satisfies requirements 1–4 of Theorem 8.2.1 (see Section 8.2), as well as the Lipschitz condition (8.39). From Theorem 8.6.1 we obtain the existence and uniqueness of the principal solution ψ_ξ in $W^{1,2}_{loc}(\mathbb{C})$. Equation (18.13) together with Theorem 5.4.2 yields $\psi_\xi \in W^{1,p}_{loc}(\mathbb{C})$ for all $p < 1 + 1/k$ since $|\mu(z)| \leq k$ and e_ξ is unimodular.

Finally, to see the uniqueness of the complex geometric optics solution f_μ, let $f \in W^{1,2}_{loc}(\mathbb{C})$ satisfy (18.9) and (18.10). Denote

$$\mu_1(z) = \mu(z) \frac{\overline{\partial_z f(z)}}{\partial_z f(z)}$$

and let φ be the unique principal solution to

$$\frac{\partial \varphi}{\partial \bar{z}} = \mu_1 \frac{\partial \varphi}{\partial z} \qquad (18.14)$$

Then the Stoilow factorization, Theorem 5.5.1, gives $f = h \circ \varphi$, where $h : \mathbb{C} \to \mathbb{C}$ is an entire analytic function. But (18.9) and (18.10) show that

$$\frac{h \circ \varphi(z)}{\exp(i\xi\varphi(z))} = \frac{f(z)}{\exp(i\xi\varphi(z))}$$

has the limit 1 when the variable $z \to \infty$. Thus

$$h(z) \equiv e^{i\xi z}$$

Therefore $f(z) = \exp(i\xi\varphi(z))$ has the form (18.12). In particular, φ satisfies (18.13), and thus $\varphi = \psi_\xi$ by Theorem 8.6.1. $\square$

It is useful to note that if a function f satisfies (18.11), then if satisfies not the same equation but the equation where μ is replaced by $-\mu$. In terms of the real and imaginary parts of $f = u + iv$, we see that

$$f_{\bar{z}} = \mu \overline{f_z} \quad \Leftrightarrow \quad \nabla \cdot \sigma \nabla u = 0 \quad \text{and} \quad \nabla \cdot \frac{1}{\sigma} \nabla v = 0, \qquad (18.15)$$

where $\mu = (1 - \sigma)/(1 + \sigma)$. From these identities we obtain the complex geometric optics solutions also for the conductivity equation (18.1).

Corollary 18.1.2. *Suppose that $\sigma, 1/\sigma \in L^\infty(\mathbb{D})$ and that $\sigma(z) \equiv 1$ for $|z| \geqslant 1$. Then the equation $\nabla \cdot \sigma(z)\nabla u(z) = 0$ admits a unique weak solution $u = u_\xi \in W^{1,2}_{loc}(\mathbb{C})$ such that*

$$u(z, \xi) = e^{i\xi z}\left(1 + \mathcal{O}\left(\frac{1}{|z|}\right)\right) \quad \text{as } |z| \to \infty \qquad (18.16)$$

Proof. For existence, in view of (18.15) the function $u_\xi = \Re e\, f_\mu + i\Im m\, f_{-\mu}$ is precisely what we are looking for.

When it comes to uniqueness, if the function u satisfies the divergence equation $\nabla \cdot \sigma(z)\nabla u(z) = 0$, then using Theorem 16.1.6 for the real and imaginary parts of u, we can write it as

$$u = \Re e\, f_+ + i\Im m\, f_- = \frac{1}{2}(f_+ + f_- + \overline{f_+} - \overline{f_-}),$$

18.2. THE HILBERT TRANSFORM

where $f_\pm$ are quasiregular mappings with

$$\frac{\partial f_+}{\partial \bar{z}} = \mu \overline{\frac{\partial f_+}{\partial z}} \quad \text{and} \quad \frac{\partial f_-}{\partial \bar{z}} = -\mu \overline{\frac{\partial f_-}{\partial z}}$$

and where $\mu = (1-\sigma)/(1+\sigma)$. Given the asymptotics (18.16), it is not hard to see that both f_+ and f_- satisfy (18.9) with (18.10). Therefore $f_+ = f_\mu$ and $f_- = f_{-\mu}$. $\square$

The exponentially growing solutions of Corollary 18.1.2 can be considered σ-harmonic counterparts of the usual exponential functions $e^{i\xi z}$. They are the building blocks of the *nonlinear Fourier transform* to be discussed in more detail in Section 18.4.

18.2 The Hilbert Transform $\mathcal{H}_\sigma$

Assume that $u \in W^{1,2}(\mathbb{D})$ is a weak solution to $\nabla \cdot \sigma(z)\nabla u(z) = 0$. Then, by Theorem 16.1.6, u admits a conjugate function $v \in W^{1,2}(\mathbb{D})$ such that

$$\partial_x v = -\sigma \partial_y u$$
$$\partial_y v = \sigma \partial_x u$$

Let us now elaborate on the relationship between u and v. Since the function v is defined only up to a constant, we will normalize it by assuming

$$\int_{\partial \mathbb{D}} v \, ds = 0 \qquad (18.17)$$

This way we obtain a unique map $\mathcal{H}_\mu : W^{1,1/2}(\partial \mathbb{D}) \to W^{1,1/2}(\partial \mathbb{D})$ by setting

$$\mathcal{H}_\mu : u\big|_{\partial \mathbb{D}} \mapsto v\big|_{\partial \mathbb{D}} \qquad (18.18)$$

In other words, $v = \mathcal{H}_\mu(u)$ if and only if $\int_{\partial \mathbb{D}} v \, ds = 0$, and $u + iv$ has a $W^{1,2}$-extension f to the disk $\mathbb{D}$ satisfying $f_{\bar{z}} = \mu \overline{f_z}$. We call $\mathcal{H}_\mu$ the *Hilbert transform* corresponding to (18.11).

Since the function $g = -if = v - iu$ satisfies $g_{\bar{z}} = -\mu \overline{g_z}$, we have

$$\mathcal{H}_\mu \circ \mathcal{H}_{-\mu} u = \mathcal{H}_{-\mu} \circ \mathcal{H}_\mu u = -u + \frac{1}{2\pi}\int_{\partial \mathbb{D}} u \, ds \qquad (18.19)$$

So far we have defined $\mathcal{H}_\mu(u)$ only for real-valued functions u. By setting

$$\mathcal{H}_\mu(iu) = i\mathcal{H}_{-\mu}(u),$$

we extend the definition of $\mathcal{H}_\mu(\cdot)$ to all $\mathbb{C}$-valued functions in $W^{1,1/2}(\partial \mathbb{D})$. Note, however, that $\mathcal{H}_\mu$ still remains only $\mathbb{R}$-linear.

As in the case of analytic functions, the Hilbert transform defines a projection, now on the "μ-analytic" functions. That is, we define $Q_\mu : W^{1,1/2}(\partial \mathbb{D}) \to W^{1,1/2}(\partial \mathbb{D})$ by

$$Q_\mu(g) = \frac{1}{2}(g - i\mathcal{H}_\mu g) + \frac{1}{4\pi} \int_{\partial \mathbb{D}} g \, ds \qquad (18.20)$$

Then it follows that $Q_\mu^2 = Q_\mu$. Furthermore, we have the following lemma.

Lemma 18.2.1. *If $g \in W^{1,1/2}(\partial \mathbb{D})$, the following conditions are equivalent:*

(a) $g = f|_{\partial \mathbb{D}}$, where $f \in W^{1,2}(\mathbb{D})$ satisfies $f_{\bar{z}} = \mu \overline{f_z}$

(b) $Q_\mu(g)$ is a constant

Proof. Condition (a) holds if and only if $g = u + i\mathcal{H}_\mu u + ic$ for some real-valued $u \in W^{1,1/2}(\partial \mathbb{D})$ and real constant c. If g has this representation, then $Q_\mu(g) = \frac{1}{4\pi} \int_{\partial \mathbb{D}} u \, ds + ic$. On the other hand, if $Q_\mu(g)$ is a constant, then we put $g = u + iw$ into (18.20) and use (18.19) to show that $w = \mathcal{H}_\mu u + $ constant. This shows that (a) holds. $\square$

The Dirichlet-to-Neumann map (18.4) and the Hilbert transform (18.18) are closely related, as the next lemma shows.

Theorem 18.2.2. *Choose the counterclockwise orientation for $\partial \mathbb{D}$ and denote by ∂_T the tangential (distributional) derivative on $\partial \mathbb{D}$ corresponding to this orientation. We then have*

$$\partial_T \mathcal{H}_\mu(u) = \Lambda_\sigma(u) \qquad (18.21)$$

In particular, the Dirichlet-to-Neumann map Λ_σ uniquely determines $\mathcal{H}_\mu$, $\mathcal{H}_{-\mu}$ and $\Lambda_{1/\sigma}$.

Proof. By the definition of Λ_σ we have

$$\int_{\partial \mathbb{D}} \varphi \Lambda_\sigma u \, ds = \int_{\mathbb{D}} \nabla \varphi \cdot \sigma \nabla u, \qquad \varphi \in C^\infty(\overline{\mathbb{D}})$$

Thus, by (18.5) and (18.6) and integration by parts, we get

$$\int_{\partial \mathbb{D}} \varphi \Lambda_\sigma u \, ds = \int_{\mathbb{D}} (\partial_x \varphi \, \partial_y v - \partial_y \varphi \, \partial_x v) = -\int_{\partial \mathbb{D}} v \, \partial_T \varphi \, ds,$$

and (18.21) follows. Next,

$$-\mu = (1 - 1/\sigma)/(1 + 1/\sigma),$$

and so $\Lambda_{1/\sigma}(u) = \partial_T \mathcal{H}_{-\mu}(u)$. Since by (18.19) $\mathcal{H}_\mu$ uniquely determines $\mathcal{H}_{-\mu}$, the proof is complete. $\square$

With these identities we can now show that, for the points z that lie outside $\mathbb{D}$, the values of the complex geometric optics solutions $f_\mu(z, \xi)$ and $f_{-\mu}(z, \xi)$ are determined by the Dirichlet-to-Neumann operator Λ_σ.

18.3. DEPENDENCE ON PARAMETERS

Theorem 18.2.3. *Let σ and $\widetilde{\sigma}$ be two conductivities satisfying the assumptions of Theorem 18.0.1 and assume $\Lambda_\sigma = \Lambda_{\widetilde\sigma}$. Then if μ and $\widetilde\mu$ are the corresponding Beltrami coefficients, we have*

$$f_\mu(z,\xi) = f_{\widetilde\mu}(z,\xi) \quad \text{and} \quad f_{-\mu}(z,\xi) = f_{-\widetilde\mu}(z,\xi) \qquad (18.22)$$

for all $z \in \mathbb{C} \setminus \overline{\mathbb{D}}$ and $\xi \in \mathbb{C}$.

Proof. By Theorem 18.2.2 the condition $\Lambda_\sigma = \Lambda_{\widetilde\sigma}$ implies that $\mathcal{H}_\mu = \mathcal{H}_{\widetilde\mu}$. In the same way Λ_σ determines $\Lambda_{\sigma^{-1}}$, and so it is enough to prove the first claim of (18.22).

Fix the value of the parameter $\xi \in \mathbb{C}$. From (18.20) we see that the projections $Q_\mu = Q_{\widetilde\mu}$, and thus by Lemma 18.2.1

$$Q_\mu(\widetilde f) = Q_{\widetilde\mu}(\widetilde f) \quad \text{is constant}$$

Here we have written

$$\widetilde f = (f_{\widetilde\mu})\big|_{\partial \mathbb{D}}$$

Using Lemma 18.2.1 again, we see that there exists a function $G \in W^{1,2}(\mathbb{D})$ such that $G_{\overline z} = \mu \overline{G_z}$ in $\mathbb{D}$ and

$$G\big|_{\partial \mathbb{D}} = \widetilde f$$

We then define $G(z) = f_{\widetilde\mu}(z,\xi)$ for z outside $\mathbb{D}$. Now $G \in W^{1,2}_{loc}(\mathbb{C})$, and it satisfies $G_{\overline z} = \mu \overline{G_z}$ in the whole plane. Thus it is quasiregular, and so $G \in W^{1,p}_{loc}(\mathbb{C})$ for all $2 \leqslant p < 2 + 1/k$, $k = \|\mu\|_\infty$. But now G is a solution to (18.9) and (18.10). By the uniqueness part of Theorem 18.1.1, we obtain $G(z) \equiv f_\mu(z,\xi)$. $\square$

Similarly, the Dirichlet-to-Neumann operator determines the complex geometric optics solutions to the conductivity equation at every point z outside the disk $\mathbb{D}$.

Corollary 18.2.4. *Let σ and $\widetilde\sigma$ be two conductivities satisfying the assumptions of Theorem 18.0.1 and assume $\Lambda_\sigma = \Lambda_{\widetilde\sigma}$.*
Then

$$u_\sigma(z,\xi) = u_{\widetilde\sigma}(z,\xi) \quad \text{for all} \quad z \in \mathbb{C} \setminus \overline{\mathbb{D}} \quad \text{and} \quad \xi \in \mathbb{C}$$

Proof. The claim follows immediately from the previous theorem and the representation $u_\sigma(z,\xi) = \Re e\, f_\mu(z,\xi) + i\Im m\, f_{-\mu}(z,\xi)$. $\square$

18.3 Dependence on Parameters

Our strategy will be to extend the identities $f_\mu(z,\xi) = f_{\widetilde\mu}(z,\xi)$ and $u_\sigma(z,\xi) = u_{\widetilde\sigma}(z,\xi)$ from outside the disk to points z inside $\mathbb{D}$. Once we do that, Theorem 18.0.1 follows via the equation $f_{\overline z} = \mu \overline{f_z}$.

For this purpose we need to understand the ξ-dependence in $f_\mu(z,\xi)$ and the quantities controlling it. In particular, we will derive equations relating the solutions and their derivatives with respect to the ξ-variable. For this purpose we prove the following theorem.

Theorem 18.3.1. *The complex geometric optics solutions $u_\sigma(z, \xi)$ and $f_\mu(z, \xi)$ are (Hölder)-continuous in z and C^∞-smooth in the parameter ξ.*

The continuity in the z-variable is of course clear since f_μ is a quasiregular function of z. However, for analyzing the ξ-dependence we need to realize the solutions in a different manner, by identities involving linear operators that depend smoothly on the variable ξ.

Let $f_\mu(z, \xi) = e^{i\xi z} M_\mu(z, \xi)$ and $f_{-\mu}(z, \xi) = e^{i\xi z} M_{-\mu}(z, \xi)$ be the solutions of Theorem 18.1.1 corresponding to conductivities σ and σ^{-1}, respectively. We can write (18.7), (18.9) and (18.10) in the form

$$\frac{\partial}{\partial \bar{z}} M_\mu = \mu(z) \overline{\frac{\partial}{\partial z} (e_\xi M_\mu)}, \qquad M_\mu - 1 \in W^{1,p}(\mathbb{C}) \tag{18.23}$$

when $2 < p < 1 + 1/k$. By taking the Cauchy transform and introducing a $\mathbb{R}$-linear operator L_μ,

$$L_\mu g = \mathcal{C} \left(\mu \overline{\frac{\partial}{\partial \bar{z}} (e_{-\xi} \bar{g})} \right), \tag{18.24}$$

we see that (18.23) is equivalent to

$$(\mathbf{I} - L_\mu) M_\mu = 1 \tag{18.25}$$

Theorem 18.3.2. *Assume that $\xi \in \mathbb{C}$ and $\mu \in L^\infty(\mathbb{C})$ is compactly supported with $\|\mu\|_\infty \leqslant k < 1$. Then for $2 < p < 1 + 1/k$ the operator*

$$\mathbf{I} - L_\mu : W^{1,p}(\mathbb{C}) \oplus \mathbb{C} \to W^{1,p}(\mathbb{C}) \oplus \mathbb{C}$$

is bounded and invertible.

Here we denote by $W^{1,p}(\mathbb{C}) \oplus \mathbb{C}$ the Banach space consisting of functions of the form $f = \text{constant} + f_0$, where $f_0 \in W^{1,p}(\mathbb{C})$.

Proof. We write $L_\mu(g)$ as

$$L_\mu(g) = \mathcal{C} \left(\mu e_{-\xi} \overline{g_z} - i \bar{\xi} \mu e_{-\xi} \bar{g} \right) \tag{18.26}$$

Then Theorem 4.3.12 shows that

$$L_\mu : W^{1,p}(\mathbb{C}) \oplus \mathbb{C} \to W^{1,p}(\mathbb{C}) \tag{18.27}$$

is bounded. Thus we need only establish invertibility.

To this end let us assume $h \in W^{1,p}(\mathbb{C})$. Consider the equation

$$(\mathbf{I} - L_\mu)(g + C_0) = h + C_1, \tag{18.28}$$

where $g \in W^{1,p}(\mathbb{C})$ and C_0, C_1 are constants. Then

$$C_0 - C_1 = g - h - L_\mu(g + C_0),$$

18.4. NONLINEAR FOURIER TRANSFORM

which by (18.27) gives $C_0 = C_1$. By differentiating and rearranging we see that (18.28) is equivalent to $g_{\bar{z}} - \mu(e_{-\xi}\bar{g})_z = h_{\bar{z}} + \mu(\overline{C_0}e_{-\xi})_{\bar{z}}$, or in other words, to

$$g_{\bar{z}} - (\mathbf{I} - \mu e_{-\xi}\overline{S})^{-1}\big(\mu(e_{-\xi})_{\bar{z}}\bar{g}\big) = (\mathbf{I} - \mu e_{-\xi}\overline{S})^{-1}\big(h_{\bar{z}} + \mu(\overline{C_0}e_{-\xi})_{\bar{z}}\big) \quad (18.29)$$

We are now faced with the operator R defined by

$$R(g) = \mathcal{C}\left(\mathbf{I} - \nu\overline{S}\right)^{-1}(\alpha\bar{g}),$$

where $\nu(z) = \mu e_{-\xi}$ satisfies $|\nu(z)| \leqslant k\chi_{\mathbb{D}}(z)$ and α is defined by $\alpha = \mu(e_{-\xi})_{\bar{z}} = -i\bar{\xi}\mu e_{-\xi}$. According to Theorem 14.0.4, $\mathbf{I} - \nu\overline{S}$ is invertible in $L^p(\mathbb{C})$ when $1 + k < p < 1 + 1/k$, while the Cauchy transform requires $p > 2$. Therefore R is a well-defined and bounded operator on $L^p(\mathbb{C})$ for $2 < p < 1 + 1/k$. Moreover, the right hand side of (18.29) belongs to $L^p(\mathbb{C})$ for each $h \in W^{1,p}(\mathbb{C})$. Hence this equation admits a unique solution $g \in W^{1,p}(\mathbb{C})$ if and only if the operator $\mathbf{I} - R$ is invertible in $L^p(\mathbb{C})$, $2 < p < 1 + 1/k$.

To get this we will use Fredholm theory. First, Theorem 4.3.14 shows that R is a compact operator on $L^p(\mathbb{C})$ when $2 < p < 1 + 1/k$. Therefore it suffices to show that $\mathbf{I} - R$ is injective. Suppose now that $g \in L^p(\mathbb{C})$ satisfies

$$g = Rg = \mathcal{C}\left(\mathbf{I} - \nu\overline{S}\right)^{-1}(\alpha\bar{g})$$

Then $g \in W^{1,p}(\mathbb{C})$ by Theorem 4.3.12 and $g_{\bar{z}} = \left(\mathbf{I} - \nu\overline{S}\right)^{-1}(\alpha\bar{g})$. Equivalently

$$g_{\bar{z}} - \nu\overline{g_z} = \alpha\bar{g} \quad (18.30)$$

Thus the assumptions of Theorem 8.5.1 are fulfilled, and we must have $g \equiv 0$. Therefore $\mathbf{I} - R$ is indeed injective on $L^p(\mathbb{C})$. As a Fredholm operator, it therefore is invertible in $L^p(\mathbb{C})$. Therefore the operator $\mathbf{I} - L_\mu$ is invertible in $W^{1,p}(\mathbb{C})$, $2 < p < 1 + 1/k$. □

A glance at (18.24) shows that $\xi \to L_\mu$ is an infinitely differentiable family of operators. Therefore, with Theorem 18.3.2, we see that $M_\mu = (\mathbf{I} - L_\mu)^{-1}1$ is C^∞-smooth in the parameter ξ. Thus we have obtained Theorem 18.3.1.

18.4 Nonlinear Fourier Transform

The idea of studying the $\bar{\xi}$-dependence of operators associated with complex geometric optics solutions was introduced by Beals and Coifman [47] in connection with the inverse scattering approach to KdV-equations. Here we will apply this method to the solutions u_σ to the conductivity equation (18.1) and show that they satisfy a simple $\bar{\partial}$-equation with respect to the parameter ξ.

We start with the representation $u_\sigma(z,\xi) = \Re f_\mu(z,\xi) + i\Im f_{-\mu}(z,\xi)$, where $f_{\pm\mu}$ are the solutions to the corresponding Beltrami equations; in particular, they are analytic outside the unit disk. Hence with the asymptotics (18.10) they admit the following power series development,

$$f_{\pm\mu}(z,\xi) = e^{i\xi z}\bigg(1 + \sum_{n=1}^{\infty} b_n^{\pm}(\xi)z^{-n}\bigg), \qquad |z| > 1, \quad (18.31)$$

where $b_n^+(\xi)$ and $b_n^-(\xi)$ are the coefficients of the series, depending on the parameter ξ. For the solutions to the conductivity equation, this gives

$$u_\sigma(z,\xi) = e^{i\xi z} + \frac{b_1^+(\xi) + b_1^-(\xi)}{2z} e^{i\xi z} + \frac{\overline{b_1^+(\xi)} - \overline{b_1^-(\xi)}}{2\bar{z}} e^{-i\bar{\xi}\bar{z}} + e^{i\xi z}\,\mathcal{O}\!\left(\frac{1}{|z|^2}\right)$$

as $z \to \infty$. Fixing the z-variable, we take the $\partial_{\bar{\xi}}$-derivative of $u_\sigma(z,\xi)$ and get

$$\partial_{\bar{\xi}} u_\sigma(z,\xi) = -i\tau_\sigma(\xi)\, e^{-i\bar{\xi}\bar{z}}\left(1 + \mathcal{O}\!\left(\frac{1}{|z|}\right)\right), \tag{18.32}$$

where the coefficient

$$\tau_\sigma(\xi) = \frac{1}{2}\left(\overline{b_1^+(\xi)} - \overline{b_1^-(\xi)}\right) \tag{18.33}$$

However, the derivative $\partial_{\bar{\xi}} u_\sigma(z,\xi)$ is another solution to the conductivity equation! From the uniqueness of the complex geometric optics solutions under the given exponential asymptotics, Corollary 18.1.2, we therefore have the simple but important relation

$$\partial_{\bar{\xi}} u_\sigma(z,\xi) = -i\,\tau_\sigma(\xi)\,\overline{u_\sigma(z,\xi)} \quad \text{for all } \xi, z \in \mathbb{C} \tag{18.34}$$

The remarkable feature of this relation is that the coefficient τ_σ does not depend on the space variable z. Later, this phenomenon will become of crucial importance in the solution to the Calderón problem.

In analogy with the one-dimensional scattering theory of integrable systems and associated inverse problems (see [47, 79, 282]), we call τ_σ the *nonlinear Fourier transform* of σ.

To understand the basic properties of the nonlinear Fourier transform, we need to return to the Beltrami equation. We will first show that the Dirichlet-to-Neumann data determines τ_σ. This is straightforward. Then the later sections are devoted to showing that the nonlinear Fourier transform τ_σ determines the coefficient σ almost everywhere. There does not seem to be any direct method for this, rather we will have to show that from τ_σ we can determine the exponentially growing solutions $f_{\pm\mu}$ defined in the entire plane. From this information the coefficient μ, and hence σ, can be found.

In any case it seems that most properties of τ_σ are important and interesting in and of themselves. Not much is known concerning questions such as the possibility of a Plancherel formula. However, in the first instance simple bounds can be achieved, and for this we need the following result. Here let $f_\mu(z,\xi) = e^{i\xi z} M_\mu(z,\xi)$ and $f_{-\mu}(z,\xi) = e^{i\xi z} M_{-\mu}(z,\xi)$ be the solutions of Theorem 18.1.1 corresponding to conductivities σ and σ^{-1}, respectively.

Theorem 18.4.1. *For every $\xi, z \in \mathbb{C}$ we have $M_{\pm\mu}(z,\xi) \neq 0$. Moreover,*

$$\mathfrak{Re}\left(\frac{M_\mu(z,\xi)}{M_{-\mu}(z,\xi)}\right) > 0 \tag{18.35}$$

18.4. NONLINEAR FOURIER TRANSFORM

Proof. First, note that (18.7) implies, for $M_{\pm\mu}$,

$$\frac{\partial}{\partial \bar{z}} M_{\pm\mu} \mp \mu e_{-\xi} \overline{\frac{\partial}{\partial z} M_{\pm\mu}} = \mp i\bar{\xi}\mu e_{-\xi}\overline{M_{\pm\mu}} \qquad (18.36)$$

Thus we may apply Theorem 8.5.1 to get

$$M_{\pm\mu}(z) = \exp(\eta_{\pm}(z)) \neq 0, \qquad (18.37)$$

and consequently $M_\mu/M_{-\mu}$ is well defined. Second, if (18.35) is not true, the continuity of $M_{\pm\mu}$ and the fact $\lim_{z\to\infty} M_{\pm\mu}(z,\xi) = 1$ imply the existence of $z_0 \in \mathbb{C}$ such that

$$M_\mu(z_0,\xi) = it M_{-\mu}(z_0,\xi)$$

for some $t \in \mathbb{R} \setminus \{0\}$ and $\xi \in \mathbb{C}$. But then, $g = M_\mu - it M_{-\mu}$ satisfies

$$\frac{\partial}{\partial \bar{z}} g = \mu(z)\overline{\frac{\partial}{\partial z}(e_\xi g)},$$

$$g(z) = 1 - it + \mathcal{O}\left(\frac{1}{z}\right), \quad \text{as } z \to \infty$$

According to Theorem 8.5.1, this implies

$$g(z) = (1-it)\exp(\eta(z)) \neq 0,$$

contradicting the assumption $g(z_0) = 0$. □

The boundedness of the nonlinear Fourier transform is now a simple corollary of Schwarz's lemma.

Theorem 18.4.2. *The functions $f_{\pm\mu}(z,\xi) = e^{i\xi z} M_{\pm\mu}(z,\xi)$ satisfy, for $|z| > 1$ and for all $\xi \in \mathbb{C}$,*

$$\left|\frac{M_\mu(z,\xi) - M_{-\mu}(z,\xi)}{M_\mu(z,\xi) + M_{-\mu}(z,\xi)}\right| \leq \frac{1}{|z|} \qquad (18.38)$$

Moreover, for the nonlinear Fourier transform τ_σ, we have

$$|\tau_\sigma(\xi)| \leq 1 \quad \text{for all } \xi \in \mathbb{C} \qquad (18.39)$$

Proof. Fix the parameter $\xi \in \mathbb{C}$ and denote

$$m(z) = \frac{M_\mu(z,\xi) - M_{-\mu}(z,\xi)}{M_\mu(z,\xi) + M_{-\mu}(z,\xi)}$$

Then by Theorem 18.4.1, $|m(z)| < 1$ for all $z \in \mathbb{C}$. Moreover, m is holomorphic for $z \in \mathbb{C} \setminus \overline{\mathbb{D}}$, $m(\infty) = 0$, and thus by Schwarz's lemma we have $|m(z)| \leq 1/|z|$ for all $z \in \mathbb{C} \setminus \overline{\mathbb{D}}$.

On the other hand, from the development (18.31),

$$M_\mu(z,\xi) = 1 + \sum_{n=1}^{\infty} b_n(\xi) z^{-n} \quad \text{for } |z| > 1,$$

and similarly for $M_{-\mu}(z,\xi)$. We see that

$$\tau_\sigma(\xi) = \frac{1}{2}\left(\overline{b_1^+(\xi)} - \overline{b_1^-(\xi)}\right) = \lim_{z\to\infty} \overline{z\,m(z)}$$

Therefore the second claim also follows. □

With these results the Calderón problem reduces to the question whether we can invert the nonlinear Fourier transform.

Theorem 18.4.3. *The operator Λ_σ uniquely determines the nonlinear Fourier transform τ_σ.*

Proof. The claim follows immediately from Theorem 18.2.3, from the development (18.31) and from the definition (18.33) of τ_σ. □

From the relations $-\mu = (1 - 1/\sigma)/(1 + 1/\sigma)$, we have the symmetry

$$\tau_\sigma(\xi) = -\tau_{1/\sigma}(\xi)$$

It follows that the functions

$$u_1 = \Re e\, f_\mu + i\,\Im m\, f_{-\mu} = u_\sigma \quad \text{and} \quad u_2 = i\,\Re e\, f_{-\mu} - \Im m\, f_\mu = iu_{1/\sigma} \quad (18.40)$$

form a "primary pair" of complex geometric optics solutions.

Corollary 18.4.4. *The functions $u_1 = u_\sigma$ and $u_2 = iu_{1/\sigma}$ are complex-valued $W^{1,2}_{loc}(\mathbb{C})$-solutions to the conductivity equations*

$$\nabla \cdot \sigma \nabla u_1 = 0 \quad \text{and} \quad \nabla \cdot \frac{1}{\sigma}\nabla u_2 = 0, \quad (18.41)$$

respectively. In the ξ-variable they are solutions to the same $\partial_{\bar{\xi}}$-equation,

$$\frac{\partial}{\partial\bar{\xi}}\,u_j(z,\xi) = -i\,\tau_\sigma(\xi)\,\overline{u_j}(z,\xi), \quad j = 1,2, \quad (18.42)$$

and their asymptotics, as $|z| \to \infty$, are

$$u_\sigma(z,\xi) = e^{i\xi z}\left(1 + \mathcal{O}\left(\frac{1}{|z|}\right)\right), \qquad u_{1/\sigma}(z,\xi) = e^{i\xi z}\left(i + \mathcal{O}\left(\frac{1}{|z|}\right)\right)$$

18.5 Argument Principle

The solution to the Calderón problem combines analysis with topological arguments that are specific to two dimensions. For instance, we need a version of the argument principle, which we consider next. The following theorem is a refinement of the Liouville-type theorems discussed in Section 8.5.

18.5. ARGUMENT PRINCIPLE

Theorem 18.5.1. *Let $F \in W^{1,p}_{loc}(\mathbb{C})$ and $\gamma \in L^p_{loc}(\mathbb{C})$ for some $p > 2$. Suppose that, for some constant $0 \leqslant k < 1$, the differential inequality*

$$\left|\frac{\partial F}{\partial \bar{z}}\right| \leqslant k \left|\frac{\partial F}{\partial z}\right| + \gamma(z)\,|F(z)| \tag{18.43}$$

holds for almost every $z \in \mathbb{C}$ and assume that, for large z, $F(z) = \lambda z + \varepsilon(z)z$, where the constant $\lambda \neq 0$ and $\varepsilon(z) \to 0$ as $|z| \to \infty$.

Then $F(z) = 0$ at exactly one point, $z = z_0 \in \mathbb{C}$.

Proof. The continuity of $F(z) = \lambda z + \varepsilon(z)z$ and an elementary topological argument show that F is surjective, and consequently there exists at least one point $z_0 \in \mathbb{C}$ such that $F(z_0) = 0$.

To show that F cannot have more zeros, let $z_1 \in \mathbb{C}$ and choose a large disk $B = \mathbb{D}(0, R)$ containing both z_1 and z_0. If R is so large that $\varepsilon(z) < \lambda/2$ for $|z| = R$, then $F|_{\{|z|=R\}}$ is homotopic to the identity relative to $\mathbb{C} \setminus \{0\}$. Next, we express (18.43) in the form

$$\frac{\partial F}{\partial \bar{z}} = \nu(z)\frac{\partial F}{\partial z} + A(z)\,F, \tag{18.44}$$

where $|\nu(z)| \leqslant k < 1$ and $|A(z)| \leqslant \gamma(z)$ for almost every $z \in \mathbb{C}$. Now $A\chi_B \in L^r(\mathbb{C})$ for all $2 \leqslant r \leqslant p' = \min\{p, 1 + 1/k\}$, and we obtain from Theorem 14.0.4 that $(\mathbf{I} - \nu\mathcal{S})^{-1}(A\chi_B) \in L^r$ for all $p'/(p'-1) < r < p'$.

Next, we define $\eta = \mathcal{C}\big((\mathbf{I} - \nu\mathcal{S})^{-1}(A\chi_B)\big)$. By Theorem 4.3.11, $\eta \in C_0(\mathbb{C})$, and we also have

$$\frac{\partial \eta}{\partial \bar{z}} - \nu\frac{\partial \eta}{\partial z} = A(z), \qquad z \in B \tag{18.45}$$

Therefore simply by differentation we see that the function

$$g = e^{-\eta}F \tag{18.46}$$

satisfies

$$\frac{\partial g}{\partial \bar{z}} - \nu\frac{\partial g}{\partial z} = 0, \qquad z \in B \tag{18.47}$$

Since η has derivatives in $L^r(\mathbb{C})$, we have $g \in W^{1,r}_{loc}(\mathbb{C})$. As $r \geqslant 2$, the mapping g is quasiregular in B. The Stoilow factorization theorem with the formulation of Corollary 5.5.3 gives $g = h \circ \psi$, where $\psi : B \to B$ is a quasiconformal homeomorphism and h is holomorphic, both continuous up to the boundary.

Since η is continuous, (18.46) shows that $g|_{|z|=R}$ is homotopic to the identity relative to $\mathbb{C} \setminus \{0\}$, as is the holomorphic function h. Therefore the argument principle, Theorem 2.9.15, shows that h has precisely one zero in $B = \mathbb{D}(0, R)$. Already, $h(\psi(z_0)) = e^{-\eta(z_0)}F(z_0) = 0$, and there can be no further zeros for F either. This finishes the proof. □

18.6 Subexponential Growth

A basic obstacle in the solution to Calderón's problem is to find methods to control the asymptotic behavior in the parameter ξ for complex geometric optics solutions. If we knew that the assumptions of Theorem 8.5.1 were valid in (18.42), then the equation, hence the Dirichlet-to-Neumann map, would uniquely determine $u_\sigma(z,\xi)$ with $u_{1/\sigma}(z,\xi)$, and we would be finished. However, we only know from Theorem 18.4.2 that $\tau_\sigma(\xi)$ is bounded in ξ. Hence some further considerations are needed.

On the other hand, it is clear that some control of the parameter ξ is needed for $u_\sigma(z,\xi)$. Within the category of conductivity equations with L^∞-coefficients σ, the complex analytic and quasiconformal methods provide by far the most powerful methods. Therefore we return to the Beltrami equation. The purpose of this section is to study the ξ-behavior in the functions $f_\mu(z,\xi) = e^{i\xi z} M_\mu(z,\xi)$ and to show that for a fixed z, $M_\mu(z,\xi)$ grows at most subexponentially in ξ as $\xi \to \infty$. Subsequently, the result will be applied to $u_j(z,\xi)$.

For some later purposes we will also need to generalize the situation a bit by considering complex Beltrami coefficients μ_λ of the form $\mu_\lambda = \lambda\mu$, where the constant $\lambda \in \partial \mathbb{D}$ and μ is as before. Exactly as in Theorem 18.1.1, we can show the existence and uniqueness of $f_{\lambda\mu} \in W^{1,p}_{loc}(\mathbb{C})$ satisfying

$$\frac{\partial}{\partial \bar{z}} f_{\lambda\mu} = \lambda\mu \overline{\frac{\partial}{\partial z} f_{\lambda\mu}} \quad \text{and} \tag{18.48}$$

$$f_{\lambda\mu}(z,\xi) = e^{i\xi z}\left(1 + \mathcal{O}\left(\frac{1}{z}\right)\right) \quad \text{as } |z| \to \infty \tag{18.49}$$

In fact, we have that the function $f_{\lambda\mu}$ admits a representation of the form

$$f_{\lambda\mu}(z,\xi) = e^{i\xi \varphi_\lambda(z,\xi)}, \tag{18.50}$$

where for each fixed $\xi \in \mathbb{C} \setminus \{0\}$ and $\lambda \in \partial\mathbb{D}$, $\varphi_\lambda(z,\xi) = z + \mathcal{O}\left(\frac{1}{z}\right)$ for $z \to \infty$. The principal solution $\varphi = \varphi_\lambda(z,\xi)$ satisfies the nonlinear equation

$$\frac{\partial}{\partial \bar{z}} \varphi(z) = \kappa_{\lambda,\xi}\, e_{-\xi}\big(\varphi(z)\big)\, \mu(z) \overline{\frac{\partial}{\partial z} \varphi(z)} \tag{18.51}$$

where $\kappa = \kappa_{\lambda,\xi} = -\lambda \bar{\xi}^2 |\xi|^{-2}$ is constant with $|\kappa_{\lambda,\xi}| = 1$.

The main goal of this section is to show the following theorem.

Theorem 18.6.1. *If $\varphi = \varphi_\lambda$ and $f_{\lambda\mu}$ are as in (18.48)–(18.51), then*

$$\varphi_\lambda(z,\xi) \to z$$

uniformly in $z \in \mathbb{C}$ and $\lambda \in \partial\mathbb{D}$ as $\xi \to \infty$.

We shall split the proof of Theorem 18.6.1 up into several lemmas.

18.6. SUBEXPONENTIAL GROWTH

Lemma 18.6.2. *Suppose $\varepsilon > 0$ is given. Suppose also that for $\mu_\lambda(z) = \lambda\mu(z)$, we have*
$$f_n = \mu_\lambda S_n \mu_\lambda S_{n-1} \mu_\lambda \cdots \mu_\lambda S_1 \mu_\lambda, \tag{18.52}$$
where $S_j : L^2(\mathbb{C}) \to L^2(\mathbb{C})$ are Fourier multiplier operators, each with a unimodular symbol. Then there is a number $R_n = R_n(k,\varepsilon)$ depending only on $k = \|\mu\|_\infty$, n and ε such that
$$|\widehat{f_n}(\eta)| < \varepsilon \quad \text{for } |\eta| > R_n \tag{18.53}$$

Proof. It is enough to prove the claim for $\lambda = 1$. By assumption,
$$\widehat{S_j g}(\eta) = m_j(\eta)\widehat{g}(\eta),$$
where $|m_j(\eta)| = 1$ for $\eta \in \mathbb{C}$. We have by (18.52),
$$\|f_n\|_{L^2} \leqslant \|\mu\|_{L^\infty}^n \|\mu\|_{L^2} \leqslant \sqrt{\pi} k^{n+1} \tag{18.54}$$
since $\operatorname{supp}(\mu) \subset \mathbb{D}$. Choose ρ_n so that
$$\int_{|\eta|>\rho_n} |\widehat{\mu}(\eta)|^2 \, d\eta < \varepsilon^2 \tag{18.55}$$
After this, choose $\rho_{n-1}, \rho_{n-2}, \ldots, \rho_1$ inductively so that for $l = n-1, \ldots, 1$,
$$\pi \int_{|\eta|>\rho_l} |\widehat{\mu}(\eta)|^2 \, d\eta \leqslant \varepsilon^2 \left(\prod_{j=l+1}^n \pi \rho_j\right)^{-2} \tag{18.56}$$
Finally, choose ρ_0 so that
$$|\widehat{\mu}(\eta)| < \varepsilon \pi^{-n} \left(\prod_{j=1}^n \rho_j\right)^{-1} \quad \text{when} \quad |\eta| > \rho_0 \tag{18.57}$$

All these choices are possible since $\mu \in L^1 \cap L^2$.

Now, we set $R_n = \sum_{j=0}^n \rho_j$ and claim that (18.53) holds for this choice of R_n. Hence assume that $|\eta| > \sum_{j=0}^n \rho_j$. We have
$$\begin{aligned}|\widehat{f_n}(\eta)| &\leqslant \int_{|\eta-\zeta|\leqslant\rho_n} |\widehat{\mu}(\eta-\zeta)||\widehat{f_{n-1}}(\zeta)| \, d\zeta \\ &\quad + \int_{|\eta-\zeta|\geqslant\rho_n} |\widehat{\mu}(\eta-\zeta)||\widehat{f_{n-1}}(\zeta)| \, d\zeta\end{aligned} \tag{18.58}$$
But if $|\eta - \zeta| \leqslant \rho_n$, then $|\zeta| > \sum_{j=0}^{n-1} \rho_j$. Thus, if we denote
$$\Delta_n = \sup\left\{|\widehat{f_n}(\eta)| : |\eta| > \sum_{j=0}^n \rho_j\right\},$$

it follows from (18.58) and (18.54) that

$$\Delta_n \leqslant \Delta_{n-1}(\pi\rho_n^2)^{1/2}\|\mu\|_{L^2} + \left(\int_{|\zeta|\geqslant\rho_n} |\widehat{\mu}(\zeta)|^2 \, d\zeta\right)^{1/2} \|\widehat{f}_{n-1}\|_{L^2}$$

$$\leqslant \pi\rho_n k \, \Delta_{n-1} + k^n \left(\pi \int_{|\zeta|\geqslant\rho_n} |\widehat{\mu}(\zeta)|^2 \, d\zeta\right)^{1/2}$$

for $n \geqslant 2$. Moreover, the same argument shows that

$$\Delta_1 \leqslant \pi\rho_1 \, k \, \sup\{|\widehat{\mu}(\eta)| : |\eta| > \rho_0\} + k \left(\pi \int_{|\zeta|>\rho_1} |\widehat{\mu}(\zeta)|^2 \, d\zeta\right)^{1/2}$$

In conclusion, after iteration we will have

$$\Delta_n \leqslant (k\pi)^n \left(\prod_{j=1}^n \rho_j\right) \sup\{|\widehat{\mu}(\eta)| : |\eta| > \rho_0\}$$

$$+ k^n \sum_{l=1}^n \left(\prod_{j=l+1}^n \pi\rho_j\right) \left(\pi \int_{|\zeta|>\rho_l} |\widehat{\mu}(\zeta)|^2 \, d\zeta\right)^{1/2}$$

With the choices (18.55)–(18.57), this leads to

$$\Delta_n \leqslant (n+1)k^n \varepsilon \leqslant \frac{\varepsilon}{1-k},$$

which proves the claim. □

Our next goal is to use Lemma 18.6.2 to prove the asymptotic result required in Theorem 18.6.1 for the solution of a closely related linear equation.

Theorem 18.6.3. *Suppose $\psi \in W^{1,2}_{loc}(\mathbb{C})$ satisfies*

$$\frac{\partial \psi}{\partial \bar{z}} = \kappa \, \mu(z) \, e_{-\xi}(z) \, \frac{\partial \psi}{\partial z} \quad \text{and} \quad (18.59)$$

$$\psi(z) = z + \mathcal{O}\left(\frac{1}{z}\right) \quad \text{as } z \to \infty, \quad (18.60)$$

where κ is a constant with $|\kappa| = 1$.
Then $\psi(z, \xi) \to z$, uniformly in $z \in \mathbb{C}$ and $\kappa \in \partial \mathbb{D}$, as $\xi \to \infty$.

To prove Theorem 18.6.3 we need some preparation. First, since the L^p-norm of the Beurling transform $\mathbf{S}_p \to 1$ when $p \to 2$, we can choose a $\delta_k > 0$ so that $k\mathbf{S}_p < 1$ whenever $2 - \delta_k \leqslant p \leqslant 2 + \delta_k$. With this notation we then have the following lemma.

Lemma 18.6.4. *Let $\psi = \psi(\cdot, \xi)$ be the solution of (18.59) and let $\varepsilon > 0$. Then $\psi_{\bar{z}}$ can be decomposed as $\psi_{\bar{z}} = g + h$, where*

18.6. SUBEXPONENTIAL GROWTH

1. $\|h(\cdot,\xi)\|_{L^p} < \varepsilon$ for $2 - \delta_k \leqslant p \leqslant 2 + \delta_k$ uniformly in ξ.
2. $\|g(\cdot,\xi)\|_{L^p} \leqslant C_0 = C_0(k)$ uniformly in ξ.
3. $\widehat{g}(\eta,\xi) \to 0$ as $\xi \to \infty$.

In statement 3 convergence is uniform on compact subsets of the η-plane and also uniform in $\kappa \in \partial\mathbb{D}$. The Fourier transform is with respect to the first variable only.

Proof. We may solve (18.59) using a Neumann series, which will converge in L^p,

$$\frac{\partial \psi}{\partial \bar{z}} = \sum_{n=0}^{\infty} (\kappa \mu e_{-\xi} \mathcal{S})^n (\kappa \mu e_{-\xi})$$

Let

$$h = \sum_{n=n_0}^{\infty} (\kappa \mu e_{-\xi} \mathcal{S})^n (\kappa \mu e_{-\xi})$$

Then

$$\|h\|_{L^p} \leqslant \pi^{1/p} \frac{k^{n_0+1} \mathbf{S}_p^{n_0}}{1 - k\mathbf{S}_p}$$

We obtain the first statement by choosing n_0 large enough.

The remaining part clearly satisfies the second statement with a constant C_0 that is independent of ξ and λ. To prove statement 3 we first note that

$$\mathcal{S}(e_{-\xi}\phi) = e_{-\xi} S_\xi \phi,$$

where $\widehat{(S_\xi \phi)}(\eta) = m(\eta - \xi)\,\widehat{\phi}(\eta)$ and $m(\eta) = \eta/\bar{\eta}$. Consequently,

$$(\mu e_{-\xi} \mathcal{S})^n \mu e_{-\xi} = e_{-(n+1)\xi}\, \mu S_{n\xi}\, \mu S_{(n-1)\xi} \cdots \mu S_\xi \mu,$$

and so

$$g = \sum_{j=1}^{n_0} \kappa^j\, e_{-j\xi}\, \mu S_{(j-1)\xi}\, \mu \cdots \mu S_\xi \mu$$

Therefore

$$g = \sum_{j=1}^{n_0} e_{-j\xi} G_j,$$

where by Lemma 18.6.2, $|\widehat{G}_j(\eta)| < \widetilde{\varepsilon}$ whenever $|\eta| > R = \max_{j \leqslant n_0} R_j$. As $\widehat{(e_{j\xi} G_j)}(\eta) = \widehat{G}_j(\eta + j\xi)$, for any fixed compact set K_0, we can take ξ so large that $j\xi + K_0 \subset \mathbb{C} \setminus \mathbb{D}(0,R)$ for each $1 \leqslant j \leqslant n_0$. Then

$$\sup_{\eta \in K_0} |\widehat{g}(\eta,\xi)| \leqslant n_0 \widetilde{\varepsilon}$$

This proves the lemma. $\square$

Proof of Theorem 18.6.3. We show first that when $\xi \to \infty$, $\psi_{\bar{z}} \to 0$ weakly in L^p, $2 - \delta_k \leqslant p \leqslant 2 + \delta_k$. For this suppose that $f_0 \in L^q$, $q = p/(p-1)$, is fixed and choose $\varepsilon > 0$. Then there exists $f \in C_0^\infty(\mathbb{C})$ such that $\|f_0 - f\|_{L^q} < \varepsilon$, and so by Lemma 18.6.4,

$$|\langle f_0, \psi_{\bar{z}} \rangle| \leqslant \varepsilon C_1 + \left| \int \widehat{f}(\eta) \widehat{g}(\eta, \xi) \, d\eta \right|,$$

First choose R so large that

$$\int_{\mathbb{C} \setminus \mathbb{D}(0,R)} |\widehat{f}(\eta)|^2 \, d\eta \leqslant \varepsilon^2$$

and then $|\xi|$ so large that $|\widehat{g}(\eta, \xi)| \leqslant \varepsilon/(\sqrt{\pi}R)$ for all $\eta \in \mathbb{D}(R)$. Now,

$$\left| \int \widehat{f}(\eta) \widehat{g}(\eta, \xi) \, d\eta \right| \leqslant \int_{\mathbb{D}(R)} \widehat{f}(\eta) \widehat{g}(\eta, \xi) \, d\eta + \int_{\mathbb{C} \setminus \mathbb{D}(R)} \widehat{f}(\eta) \widehat{g}(\eta, \xi) \, d\eta$$
$$\leqslant \varepsilon (\|f\|_{L^2} + \|g\|_{L^2}) \leqslant C_2(f) \varepsilon \qquad (18.61)$$

The bound is the same for all κ, hence

$$\sup_{\kappa \in \partial \mathbb{D}} |\langle f_0, \psi_{\bar{z}} \rangle| \to 0 \qquad (18.62)$$

as $|\xi| \to \infty$.

To prove the uniform convergence of ψ itself, we write

$$\psi(z, \xi) = z - \frac{1}{\pi} \int_{\mathbb{D}} \frac{1}{\zeta - z} \frac{\partial}{\partial \bar{\zeta}} \psi(\zeta, \xi) \qquad (18.63)$$

Here note that $\mathrm{supp}(\psi_{\bar{z}}) \subset \mathbb{D}$ and $\chi_{\mathbb{D}}(\zeta)/(\zeta - z) \in L^q$ for all $q < 2$. Thus by the weak convergence we have

$$\psi(z, \xi) \to z \quad \text{as } \xi \to \infty \qquad (18.64)$$

for each fixed $z \in \mathbb{C}$, but uniformly in $\kappa \in \partial \mathbb{D}$. On the other hand, as

$$\sup_\xi \left\| \frac{\partial \psi}{\partial \bar{z}} \right\|_{L^p} \leqslant C_0 = C_0(p, \|\mu\|_\infty) < \infty$$

for all z sufficiently large, $|\psi(z, \xi) - z| < \varepsilon$, uniformly in $\xi \in \mathbb{C}$ and $\kappa \in \partial \mathbb{D}$. Moreover, (18.63) shows also that the family $\{\psi(\cdot, \xi) : \xi \in \mathbb{C}, \kappa \in \partial \mathbb{D}\}$ is equicontinuous. Combining all these observations shows that the convergence in (18.64) is uniform in $z \in \mathbb{C}$ and $\kappa \in \partial \mathbb{D}$. □

Finally, we proceed to the nonlinear case: Assume that φ_λ satisfies (18.48) and (18.50). Since φ is a (quasiconformal) homeomorphism, we may consider its inverse $\psi_\lambda : \mathbb{C} \to \mathbb{C}$,

$$\psi_\lambda \circ \varphi_\lambda(z) = z, \qquad (18.65)$$

18.6. SUBEXPONENTIAL GROWTH

which also is quasiconformal. By differentiating (18.65) with respect to z and $\bar{z}$ we find that ψ satisfies

$$\frac{\partial}{\partial \bar{z}}\psi_\lambda = -\frac{\bar{\xi}}{\xi}\lambda(\mu \circ \psi_\lambda)e_{-\xi}\frac{\partial}{\partial z}\psi_\lambda \quad \text{and} \qquad (18.66)$$

$$\psi_\lambda(z,\xi) = z + \mathcal{O}\left(\frac{1}{z}\right) \quad \text{as } z \to \infty \qquad (18.67)$$

Proof of Theorem 18.6.1. It is enough to show that

$$\psi_\lambda(z,\xi) \to z \qquad (18.68)$$

uniformly in z and λ as $\xi \to \infty$. For this we introduce the notation

$$\Sigma_k = \{g \in W^{1,2}_{loc}(\mathbb{C}) : g_{\bar{z}} = \nu g_z, |\nu| \leqslant k\chi_{\mathbb{D}(2)} \quad \text{and} \quad g = z + \mathcal{O}\left(\frac{1}{z}\right) \text{ as } z \to \infty\} \qquad (18.69)$$

Note that all mappings $g \in \Sigma_k$ are principal solutions and hence, according to Theorem 5.3.2, homeomorphisms.

The support of the coefficient $\mu \circ \psi_\lambda$ in (18.66) need no longer be contained in $\mathbb{D}$. However, by Theorem 2.10.4, $\varphi_\lambda(\mathbb{D}) \subset \mathbb{D}(0,2)$ and thus $\text{supp}(\mu \circ \psi_\lambda) \subset \mathbb{D}(0,2)$. Accordingly, $\psi_\lambda \in \Sigma_k$.

Corollary 3.9.2 now shows us that the family Σ_k is compact in the topology of uniform convergence. Given sequences $\xi_n \to \infty$ and $\lambda_n \in \partial \mathbb{D}$, we may pass to a subsequence and assume that $\kappa_{\lambda_n, \xi_n} = -\lambda_n \bar{\xi_n}^2 |\xi_n|^{-2} \to \kappa \in \partial \mathbb{D}$ and that the corresponding mapping $\psi_{\lambda_n}(\cdot, \xi_n) \to \psi_\infty$ uniformly, with $\psi_\infty \in \Sigma_k$. To prove Theorem 18.6.1 it is enough to show that for any such sequence $\psi_\infty(z) \equiv z$.

Hence we assume that there is such a limit function ψ_∞. We consider the $W^{1,2}_{loc}$-solution $\Phi(z) = \Phi_\lambda(z,\xi)$ of

$$\frac{\partial \Phi}{\partial \bar{z}} = \kappa(\mu \circ \psi_\infty)e_{-\xi}\frac{\partial \Phi}{\partial z}$$

$$\Phi(z) = z + \mathcal{O}\left(\frac{1}{z}\right) \quad \text{as } z \to \infty$$

This is now a linear Beltrami equation which, by Theorem 5.3.2 has a unique solution $\Phi \in \Sigma_k$ for each $\xi \in \mathbb{C}$ and $|\lambda| = 1$. According to Theorem 18.6.3,

$$\Phi_\lambda(z,\xi) \to z \quad \text{as } \xi \to \infty \qquad (18.70)$$

Further, when $2 < p < 1 + 1/k$, by Lemma 5.3.1,

$$|\psi_{\lambda_n}(z,\xi_n) - \Phi_\lambda(z,\xi_n)|$$

$$= \frac{1}{\pi}\left|\int_\mathbb{D} \frac{1}{\zeta - z}\frac{\partial}{\partial \bar{z}}(\psi_{\lambda_n}(\zeta,\xi_n) - \Phi_\lambda(\zeta,\xi_n))\,d\zeta\right|$$

$$\leqslant C_1 \left\|\frac{\partial}{\partial \bar{z}}(\psi_{\lambda_n}(\zeta,\xi_n) - \Phi_\lambda(\zeta,\xi_n))\right\|_{L^p}$$

$$\leqslant C_2|\kappa_{\lambda_n,\xi_n} - \kappa|$$

$$+ C_2 \left(\int_{2\mathbb{D}} |\mu(\psi_{\lambda_n}(\zeta,\xi_n)) - \mu(\psi_\infty(\zeta))|^{\frac{p(1+\varepsilon)}{\varepsilon}}\,d\zeta\right)^{\frac{\varepsilon}{p(1+\varepsilon)}} \qquad (18.71)$$

Finally, we apply our higher-integrability results, such as Corollary 13.2.4. Thus for all $2 < p < 1 + 1/k$ and for all $g = \psi^{-1}$, $\psi \in \Sigma_k$, we have the estimate for the Jacobian $J(z,g)$:

$$\int_{\mathbb{D}} J(z,g)^{p/2} \leqslant \int_{\mathbb{D}} \left|\frac{\partial g}{\partial z}\right|^p \leqslant C(k) < \infty, \qquad (18.72)$$

where $C(k)$ depends only on k. We use this estimate in the cases $\psi(z) = \psi_{\lambda_n}(z, \xi_n)$ and $\psi = \psi_\infty$. Namely, we have for each $\gamma \in C_0^\infty(\mathbb{D})$ that

$$\int_{2\mathbb{D}} |\mu(\psi) - \gamma(\psi)|^{\frac{p(1+\varepsilon)}{\varepsilon}} = \int_{\mathbb{D}} |\mu - \gamma|^{\frac{p(1+\varepsilon)}{\varepsilon}} J_g$$
$$\leqslant \left(\int_{\mathbb{D}} |\mu - \gamma|^{\frac{p^2(1+\varepsilon)}{\varepsilon(p-2)}}\right)^{(p-2)/p} \left(\int_{\mathbb{D}} J_g^{p/2}\right)^{2/p}$$

Since μ can be approximated in the mean by smooth γ, the last term can be made arbitrarily small. By uniform convergence $\gamma(\psi_{\lambda_n}(z, \xi_n)) \to \gamma(\psi_\infty(z))$, and so we see that the last bound in (18.71) converges to zero as $\kappa_{\lambda_n, \xi_n} \to \kappa$. In view of (18.70) and (18.71), we have established that

$$\psi_{\lambda_n}(z, \xi_n) \to z$$

and that $\psi_\infty(z) \equiv z$. The theorem is proved. □

18.7 The Solution to Calderón's Problem

The Jacobian $J(z, f)$ of a quasiregular map can vanish only on a set of Lebesque measure zero, see Corollary 5.5.2. Since $J(z,f) = |f_z|^2 - |f_{\bar z}|^2 \leqslant |f_z|^2$, this implies that once we know the values $f_\mu(z, \xi)$ for every $z \in \mathbb{C}$, then we can recover the values $\mu(z)$ and hence $\sigma(z)$ almost everywhere, from f_μ by the formulas

$$\frac{\partial f_\mu}{\partial \bar z} = \mu(z) \overline{\frac{\partial f_\mu}{\partial z}} \quad \text{and} \quad \sigma = \frac{1 - \mu}{1 + \mu} \qquad (18.73)$$

On the other hand, considering the functions

$$u_1 = u_\sigma = \Re\mathfrak{e}\, f_\mu + i \Im\mathfrak{m}\, f_{-\mu} \quad \text{and} \quad u_2 = iu_{1/\sigma} = i\Re\mathfrak{e}\, f_{-\mu} - \Im\mathfrak{m}\, f_\mu$$

that were described in Corollary 18.4.4, it is clear that the pair $\{u_1(z, \xi), u_2(z, \xi)\}$ determines the pair $\{f_\mu(z, \xi), f_{-\mu}(z, \xi)\}$, and vice versa. Therefore to prove Theorem 18.0.1 it will suffice to establish the following result.

Theorem 18.7.1. *Assume that $\Lambda_\sigma = \Lambda_{\tilde\sigma}$ for two L^∞-conductivities σ and $\tilde\sigma$. Then for all $z, \xi \in \mathbb{C}$,*

$$u_\sigma(z, \xi) = u_{\tilde\sigma}(z, \xi) \quad \text{and} \quad u_{1/\sigma}(z, \xi) = u_{1/\tilde\sigma}(z, \xi)$$

18.7. THE SOLUTION

For the proof of the theorem, our first task it to determine the asymptotic behavior of $u_\sigma(z,\xi)$. We state this as a separate result.

Lemma 18.7.2. *We have $u_\sigma(z,\xi) \neq 0$ for every $(z,\xi) \in \mathbb{C} \times \mathbb{C}$. Furthermore, for each fixed $\xi \neq 0$, we have with respect to z*

$$u_\sigma(z,\xi) = \exp(i\xi z + v(z)),$$

where $v = v_\xi \in L^\infty(\mathbb{C})$. On the other hand, for each fixed z we have with respect to ξ

$$u_\sigma(z,\xi) = \exp(i\xi z + \xi\varepsilon(\xi)), \tag{18.74}$$

where $\varepsilon(\xi) \to 0$ as $\xi \to \infty$.

Proof. For the first claim we write

$$u_\sigma = \frac{1}{2}\left(f_\mu + f_{-\mu} + \overline{f_\mu} - \overline{f_{-\mu}}\right)$$

$$= f_\mu \left(1 + \frac{f_\mu - f_{-\mu}}{f_\mu + f_{-\mu}}\right)^{-1}\left(1 + \frac{\overline{f_\mu} - \overline{f_{-\mu}}}{f_\mu + f_{-\mu}}\right)$$

Each factor in the product is continuous and nonvanishing in z by Theorem 18.4.1. Taking the logarithm and using $f_{\pm\mu}(z,\xi) = e^{i\xi z}(1 + \mathcal{O}_\xi(1/z))$ leads to

$$u_\sigma(z,\xi) = \exp\left(i\xi z + \mathcal{O}_\xi\left(\frac{1}{z}\right)\right)$$

For the ξ-asymptotics we apply Theorem 18.6.1, which governs the growth of the functions f_μ for $\xi \to \infty$. We see that for (18.74) it is enough to show that

$$\inf_t \left|\frac{f_\mu - f_{-\mu}}{f_\mu + f_{-\mu}} + e^{it}\right| \geq e^{-|\xi|\varepsilon(\xi)} \tag{18.75}$$

For this, define

$$\Phi_t = e^{-it/2}(f_\mu \cos t/2 + if_{-\mu} \sin t/2)$$

Then for each fixed ξ,

$$\Phi_t(z,\xi) = e^{i\xi z}\left(1 + \mathcal{O}_\xi\left(\frac{1}{z}\right)\right) \quad \text{as } z \to \infty,$$

and

$$\frac{\partial}{\partial \bar{z}}\Phi_t = \mu e^{-it} \overline{\frac{\partial}{\partial z}\Phi_t}$$

Thus for $\lambda = e^{-it}$, the mapping $\Phi_t = f_{\lambda\mu}$ is precisely the exponentially growing solution from (18.48) and (18.49). But

$$\frac{f_\mu - f_{-\mu}}{f_\mu + f_{-\mu}} + e^{it} = \frac{2e^{it}\Phi_t}{f_\mu + f_{-\mu}} = \frac{f_{\lambda\mu}}{f_\mu}\frac{2e^{it}}{1 + M_{-\mu}/M_\mu} \tag{18.76}$$

By Theorem 18.6.1,
$$e^{-|\xi|\varepsilon_1(\xi)} \leqslant |M_{\pm\mu}(z,\xi)| \leqslant e^{|\xi|\varepsilon_1(\xi)} \tag{18.77}$$
and
$$e^{-|\xi|\varepsilon_2(\xi)} \leqslant \inf_{\lambda\in\partial\mathbb{D}}\left|\frac{f_{\lambda\mu}(z,\xi)}{f_{\mu}(z,\xi)}\right| \leqslant \sup_{\lambda\in\partial\mathbb{D}}\left|\frac{f_{\lambda\mu}(z,\xi)}{f_{\mu}(z,\xi)}\right| \leqslant e^{|\xi|\varepsilon_2(\xi)}, \tag{18.78}$$
where $\varepsilon_j(\xi) \to 0$ as $\xi \to \infty$. Since $\Re e(M_{-\mu}/M_\mu) > 0$, the inequality (18.75) follows, completing the proof of the lemma. $\square$

As discussed earlier, the functions $u_1 = u_\sigma$ and $u_2 = iu_{1/\sigma}$ satisfy a $\partial_{\bar\xi}$-equation as a function of the parameter ξ, but it is clear that for a fixed z the asymptotics in (18.74) are not strong enough to determine the individual solution $u_j(z,\xi)$. However, if we consider the entire family $\{u_j(z,\xi) : z \in \mathbb{C}\}$, then, somewhat surprisingly, uniqueness properties do arise.

To prove this assume that the Dirichlet-to-Neumann operators are equal for the conductivities σ and $\tilde\sigma$. By Lemma 18.7.2, we have $u_\sigma(z,\xi), u_{\tilde\sigma}(z,\xi) \neq 0$ at every point (z,ξ). Therefore we can take their logarithms δ_σ and $\delta_{\tilde\sigma}$, respectively, where for each fixed $z \in \mathbb{C}$,

$$\delta_\sigma(z,\xi) = \log u_\sigma(z,\xi) = i\xi z + \xi\varepsilon_1(\xi) \tag{18.79}$$
$$\delta_{\tilde\sigma}(z,\xi) = \log u_{\tilde\sigma}(z,\xi) = i\xi z + \xi\varepsilon_2(\xi) \tag{18.80}$$

Here, for $|\xi| \to \infty$, $\varepsilon_j(\xi) \to 0$. Moreover, by Theorem 18.1.1,
$$\delta_\sigma(z,0) \equiv \delta_{\tilde\sigma}(z,0) \equiv 0$$
for all $z \in \mathbb{C}$.

In addition, $z \mapsto \delta_\sigma(z,\xi)$ is continuous, and we have
$$\delta_\sigma(z,\xi) = i\xi z\left(1 + \frac{v_\xi(z)}{i\xi z}\right), \qquad \xi \neq 0, \tag{18.81}$$
where by Lemma 18.7.2, $v_\xi \in L^\infty(\mathbb{C})$ for each fixed $\xi \in \mathbb{C}$. Since δ_σ is close to a multiple of the identity for $|z|$ large, an elementary topological argument shows that $z \mapsto \delta_\sigma(z,\xi)$ is surjective $\mathbb{C} \to \mathbb{C}$.

To prove the theorem it suffices to show that, if $\Lambda_\sigma = \Lambda_{\tilde\sigma}$, then
$$\delta_{\tilde\sigma}(z,\xi) \neq \delta_\sigma(w,\xi) \quad \text{for } z \neq w \text{ and } \xi \neq 0 \tag{18.82}$$

If this property is established, then (18.82) and the surjectivity of $z \mapsto \delta_\sigma(z,\xi)$ show that we necessarily have $\delta_\sigma(z,\xi) = \delta_{\tilde\sigma}(z,\xi)$ for all $\xi, z \in \mathbb{C}$. Hence $u_{\tilde\sigma}(z,\xi) = u_\sigma(z,\xi)$.

We are now at a point where the $\partial_{\bar\xi}$-method and (18.42) can be applied. Substituting $u_\sigma = \exp(\delta_\sigma)$ in this identity shows that $\xi \to \delta_\sigma(z,\xi)$ and $\xi \to \delta_{\tilde\sigma}(w,\xi)$ both satisfy the $\partial_{\bar\xi}$-equation
$$\frac{\partial \delta}{\partial \bar\xi} = -i\tau(\xi)e^{(\overline{\delta}-\delta)}, \qquad \xi \in \mathbb{C}, \tag{18.83}$$

18.7. THE SOLUTION

where by Theorem 18.2.3 and the assumption $\Lambda_\sigma = \Lambda_{\tilde{\sigma}}$, the coefficient $\tau(\xi)$ is the same for both functions δ_σ and $\delta_{\tilde{\sigma}}$. The difference

$$g(\xi) := \delta_{\tilde{\sigma}}(w,\xi) - \delta_\sigma(z,\xi)$$

thus satisfies the identity

$$\frac{\partial g}{\partial \overline{\xi}} = -i\tau(\xi)\, e^{(\overline{\delta} - \delta)} \left[e^{(\overline{g} - g)} - 1 \right]$$

In particular,

$$\left| \frac{\partial g}{\partial \overline{\xi}} \right| \leqslant |\overline{g} - g| \leqslant 2|g| \tag{18.84}$$

From (18.79) we have $g(\xi) = i(w-z)\xi + \xi\varepsilon(\xi)$. Now we only need to apply Theorem 18.5.1 (with respect to ξ) to see that for $w \neq z$ the function g vanishes only at $\xi = 0$. This establishes (18.82).

According to Theorem 18.2.2 (or by the identity $\tau_\sigma = -\tau_{1/\sigma}$), if $\Lambda_\sigma = \Lambda_{\tilde{\sigma}}$, the same argument works to show that $u_{1/\tilde{\sigma}}(z,\xi) = u_{1/\sigma}(z,\xi)$ as well. Theorem 18.7.1 is thus proved. As the pair $\{u_1(z,\xi), u_2(z,\xi)\}$ pointwise determines the pair $\{f_\mu(z,\xi), f_{-\mu}(z,\xi)\}$, we find via (18.73) that $\sigma \equiv \tilde{\sigma}$. Therefore the proof of Theorem 18.0.1 is complete. □

Chapter 19

Integral Estimates for the Jacobian

In this chapter, besides providing some fundamental estimates for the Jacobian of a quasiconformal mapping, we aim to connect the theory of quasiconformal mappings with central problems in the calculus of variations and in particular the notions of quasiconvexity and rank-one convexity due to Morrey [273]. We show how conjectures relating these ideas would provide the answers to important questions in the theory of quasiconformal mappings such as for instance the precise form of the p-norms of the Beurling transform. Actually, the natural place for this discussion is in $\mathbb{R}^n$, and so for this chapter we will frame some of our discussion there, giving proofs in two dimensions where applicable.

19.1 The Fundamental Inequality for the Jacobian

We begin with the following elementary observation from Lemma 2.9.2. Suppose that Ω is a domain in $\mathbb{R}^n$ and that we have two mappings $f, g \in W^{1,n}(\Omega, \mathbb{R}^n)$ that coincide on the boundary $\partial\Omega$, meaning that $f - g \in W_0^{1,n}(\Omega, \mathbb{R}^n)$. Then we have

$$\int_\Omega J(x, f)\, dx = \int_\Omega J(x, g)\, dx, \qquad (19.1)$$

and since f and g are otherwise arbitrary, there must be some quite subtle cancellation in these integrals of Jacobian determinants to achieve this identity. Indeed, it is precisely this nonlinear cancellation phenomena that leads to many of the interesting consequences in the L^p-theory of mappings of finite distortion. In many ways therefore, L^p-estimates for Jacobians provide nonlinear counterparts for the oscillatory cancellation phenomena that underpin the Calderón-Zygmund theory of singular integral operators [87, 86]. This point of view has quite profound implications in the higher-dimensional theory of quasi-

19.1. THE FUNDAMENTAL INEQUALITY

conformal mappings, see [191] for an overview. In $\mathbb{R}^n$ most L^p-estimates for the Jacobian are based on the Hodge decomposition of differential forms and nonlinear commutators. Our first estimate is known as the fundamental inequality for the Jacobian, and we will give an elementary proof in two dimensions using only the Beurling transform.

Theorem 19.1.1. *There exists a number $M = M_p \geqslant 1$ such that if $1 < p < \infty$, then*

$$\int_{\mathbb{C}} |Df(z)|^{p-2} J(z,f) \, dz \leqslant \frac{M-1}{M+1} \int_{\mathbb{C}} |Df(z)|^p \, dz, \qquad (19.2)$$

or equivalently,

$$\int_{\mathbb{C}} (M|f_{\bar{z}}| - |f_z|)(|f_{\bar{z}}| + |f_z|)^{p-1} \geqslant 0 \qquad (19.3)$$

for every function $f \in \mathbb{W}^{1,p}(\mathbb{C})$.

We are concerned here with the operator norm $|Df(z)| = |f_z(z)| + |f_{\bar{z}}(z)|$. We shall prove this inequality with the constant

$$M = \mathbf{S}_p^p \geqslant \mathbf{S}_p \geqslant \max\{p - 1, 1/(p-1)\}$$

where $\mathbf{S}_p$ stands for the p-norm of the Beurling transform from Theorem 4.5.3. It is important to remark here that the constant on the right hand side

$$\frac{M-1}{M+1} < 1$$

The inequality (19.2) is of course trivial with any constant greater than or equal to 1.

We begin our proof of Theorem 19.1.1 with the following elementary lemma.

Lemma 19.1.2. *For every $M \geqslant \max\{p - 1, 1/(p-1)\}$ and $x, y \geqslant 0$, we have*

$$Mx^p - y^p \leqslant (Mx - y)(x + y)^{p-1} \qquad (19.4)$$

Proof. Consider the function

$$\varphi(x) = [pM + (M - p + 1)x](1 + x)^{p-2} - pM$$

We calculate for $x \geqslant 0$,

$$\varphi'(x) = (p-1)(1+x)^{p-3}\big[(p-1)M + (M - p + 1)x - 1\big] \geqslant 0$$

since $M \geqslant p - 1$ and $M \geqslant 1/(p-1)$. Thus $\varphi(x) \geqslant \varphi(0) = 0$. Now consider

$$\psi(x) = (Mx - 1)(1 + x)^{p-1} - (Mx^p - 1), \qquad \psi(0) = 0,$$

and compute that $\psi'(x) = x^{p-1}\varphi(1/x) \geqslant 0$, and so $\psi(x) \geqslant 0$. Then of course

$$(Mx - y)(x + y)^{p-1} - (Mx^p - y^p) = y^p \psi\Big(\frac{x}{y}\Big) \geqslant 0$$

We note here that the bound on M is the best possible. □

We now apply (19.4) with $M = \mathbf{S}_p^p$. We know M satisfies the hypothesis of the lemma from (4.87) and

$$\|f_z\|_p^p = \|\mathcal{S} f_{\bar{z}}\|_p^p \leqslant \mathbf{S}_p^p \|f_{\bar{z}}\|_p^p = M \|f_{\bar{z}}\|_p^p$$

So

$$0 \leqslant M\|f_{\bar{z}}\|_p^p - \|f_z\|_p^p = \int_{\mathbb{C}} (M|f_{\bar{z}}|^p - |f_z|^p) \leqslant \int_{\mathbb{C}} (M|f_{\bar{z}}| - |f_z|)(|f_{\bar{z}}| + |f_z|)^{p-1}$$

We recall that $J(z, f) = |f_z|^2 - |f_{\bar{z}}|^2$ and see that this is equivalent to

$$\int_{\mathbb{C}} |Df(z)|^{p-2} J(z, f)\, dz \leqslant \frac{M-1}{M+1} \int_{\mathbb{C}} |Df(z)|^p\, dz \qquad (19.5)$$

as claimed.

The best possible constant on the right hand side here is unknown. Finding this is a challenging open problem. We offer the following conjecture.

Conjecture 19.1.3. *Let $M = \max\{p-1, 1/(p-1)\}$. Then*

$$\int_{\mathbb{C}} |Df(z)|^{p-2} J(z, f)\, dz \leqslant \frac{M-1}{M+1} \int_{\mathbb{C}} |Df(z)|^p\, dz, \qquad (19.6)$$

or equivalently,

$$\int_{\mathbb{C}} |Df(z)|^{p-2} J(z, f)\, dz \leqslant \left| 1 - \frac{2}{p} \right| \int_{\mathbb{C}} |Df(z)|^p\, dz \qquad (19.7)$$

for every function $f \in \mathbb{W}^{1,p}(\mathbb{C})$.

We have included the equivalent formulation in the above to point out the behavior of the constant as $p \to 2$. It is conjectured that, appropriately formulated, (19.7) holds in all dimensions. More precisely,

$$\int_{\mathbb{R}^n} |Df(x)|^{p-n} J(x, f)\, dx \leqslant \left| 1 - \frac{n}{p} \right| \int_{\mathbb{R}^n} |Df(x)|^p\, dx \qquad (19.8)$$

for all $f \in \mathbb{W}^{1,p}(\mathbb{R}^n, \mathbb{R}^n)$ and $p > \frac{n}{2}$.

There is a proven bound here in [175] that has the correct behavior.

Theorem 19.1.4. *For each $n \geqslant 2$ there is $p_0 \in [\frac{n}{2}, n)$ such that for all $p > p_0$ there is $\lambda_p < 1$ with*

$$\int_{\mathbb{R}^n} |Df(x)|^{p-n} J(x, f)\, dx \leqslant \lambda_p \int_{\mathbb{R}^n} |Df(x)|^p\, dx \qquad (19.9)$$

for all $f \in \mathbb{W}^{1,p}(\mathbb{R}^n, \mathbb{R}^n)$. Moreover, $\lambda_p = \mathcal{O}(|p-n|)$ as $p \nearrow n$.

19.1. THE FUNDAMENTAL INEQUALITY

The lower bound p_0 for the exponents in even dimensions is known to be $n/2$ [189], where there is an analog of the planar Beurling transform available. The fact that $p_0 < n$ plays a crucial role in the higher-dimensional theory of weakly quasiregular mappings with regularity below their natural domain of definition, that is, in spaces larger that $\mathbb{W}^{1,n}(\mathbb{R}^n, \mathbb{R}^n)$. This leads to estimates for the Painlevé problem of removable singularities. We circumvent the considerable difficulties in proving sharp Painlevé-type results in the plane via this route through use of the area distortion theorem and its consequences. Unfortunately, nothing like this is known in higher dimensions, and estimates such as those in (19.9) provide our only hope.

Returning to the complex plane, any of the conjectures above would imply the conjecturally correct bound for the norm of the Beurling transform $\mathbf{S}_p = p - 1$, $p \geqslant 2$. This can be seen from the following inequality.

Lemma 19.1.5. *For all $M \geqslant \max\{p-1, 1/(p-1)\}$ and $x, y \geqslant 0$, we have*

$$(1+M)^{p-1}M^{1-p}(M^p x^p - y^p) \geqslant p(Mx-y)(x+y)^{p-1} \tag{19.10}$$

Proof. There is no loss of generality in assuming that $x+y=1$ (divide through by $(x+y)^p$ and make the obvious variable substitution), and so $0 \leqslant x \leqslant 1$. We examine the function

$$\gamma(x) = M^p x^p - (1-x)^p - pM^{p-1}(1+M)^{1-p}(Mx+x-1)$$

on the interval $[0,1]$. If $p=2$, then $\gamma(x) = \frac{M-1}{M+1}(Mx+x-1)^2 \geqslant 0$, and our claim follows. Thus we suppose $p \neq 2$, so that $M > 1$. We compute

$$\gamma''(x) = p(p-1)\big(M^p x^{p-2} - (1-x)^{p-2}\big),$$

which shows that γ has exactly one inflection point and, consequently, at most one local minimum. Since $x_0 = (1+M)^{-1}$ is such a local minimum where $\gamma'(x_0) = \gamma(x_0) = 0$, we need only check the end points. There we must have $\gamma(0) = pM^{p-1}(1+M)^{1-p} - 1 \geqslant 0$ and $\gamma(1) = M^p - pM^p(1+M)^{1-p} \geqslant 0$. Both of these occur with our choice of M. $\square$

Returning to our discussion of the values of $\mathbf{S}_p$, we apply the inequality in (19.10) to $x = |f_{\bar{z}}|$ and $y = |f_z|$ to find the first pointwise almost everywhere inequality,

$$(1+M)^{p-1}M^{1-p}(M^p|f_{\bar{z}}|^p - |f_z|^p) \geqslant p(M|f_{\bar{z}}| - |f_z|)(|f_{\bar{z}}| + |f_z|)^{p-1},$$

which we integrate to get

$$\int_{\mathbb{C}} M^p|f_{\bar{z}}|^p - |f_z|^p \geqslant \frac{pM^{p-1}}{(1+M)^{p-1}} \int_{\mathbb{C}} (M|f_{\bar{z}}| - |f_z|)(|f_{\bar{z}}| + |f_z|)^{p-1} \tag{19.11}$$

The right hand side here is positive by (19.3). Hence

$$\int_{\mathbb{C}} |f_z|^p \leqslant M^p \int_{\mathbb{C}} |f_{\bar{z}}|^p, \tag{19.12}$$

so that $\mathbf{S}_p \leqslant M$.

A point to make here is that the left hand side of (19.11) does not have good convexity properties (as we will discuss in a moment), but the right hand side is rank-one convex. In order to study the p-norms $\mathbf{S}_p$ it seems that a study of (19.5) might be more profitable than a direct attack on the inequality (19.12). This leads us directly to discuss another fundamentally important problem in modern analysis and the calculus of variations.

19.2 Rank-One Convexity and Quasiconvexity

The subjects that we discuss in this section have evolved from attempts, many and varied, to extend the direct method of the calculus of variations from convex to nonconvex energy integrals. Initially, they sprang out of the pioneering work of C.B. Morrey [273], who introduced the notions of rank-one convexity and quasiconvexity, and were brought into the modern world by J. Ball [39] with the notions of polyconvexity and null Lagrangians. These theories were developed largely for applications in nonlinear elasticity, a subject concerned with elastic deformations $f : \Omega \to \mathbb{R}^n$ of a body $\Omega \subset \mathbb{R}^n$ that minimize a given energy integral

$$\mathcal{E}[f] = \int_\Omega \mathbf{E}(Df)\, dx \qquad (19.13)$$

usually subject to some boundary constraints. Here the integrand $\mathbf{E} : \mathbb{R}^{n \times n} \to \mathbb{R}$ is a continuous function defined on matrices.

As an example, let us consider Ω smooth and bounded and $\mathbf{E}$ smooth and convex. We seek to minimize the energy integral (19.13) among all deformations (mappings) that coincide with a given linear deformation on the boundary, say $A : \mathbb{R}^n \to \mathbb{R}^n$. Thus

$$f(x) = Ax + \phi(x), \qquad \phi(x) = 0,\ x \in \partial\Omega$$

As $\mathbf{E}$ is supposed convex, the graph of $\mathbf{E}$ lies above its tangent hyperplanes. This means that

$$\mathbf{E}(X) \geqslant \mathbf{E}(A) + \langle \mathbf{E}'(A), X - A \rangle, \qquad (19.14)$$

where $\mathbf{E}' : \mathbb{R}^{n \times n} \to \mathbb{R}^{n \times n}$ is the derivative of $\mathbf{E}$. Integrating this inequality yields

$$\int_\Omega \mathbf{E}(Df) \geqslant \int_\Omega \mathbf{E}(A) + \langle \mathbf{E}'(A),\ \int_\Omega D\phi \rangle = \int_\Omega \mathbf{E}(A) \qquad (19.15)$$

Thus convex integrands possess the following property.

The energy of any affine deformation $A : \Omega \to \mathbb{R}^n$ is an absolute minimum among the energies of all other (nonlinear) deformations $f : \Omega \to \mathbb{R}^n$ that coincide with A on the boundary.

It turns out that, roughly speaking, a property such as that above allows one to minimize the energy integrals subject to almost any boundary data—not

19.2. QUASICONVEXITY

only linear data. This observation pertains to nonconvex functionals as well. More precisely it yields lower semicontinuity of the energy functional, and this is of course one of the major prerequisites in the calculus of variations. However, we would stray too far in discussing this any further in such a general manner. The reader is referred to [101, 273]. It is the statement above that leads to the far-reaching generalization of convexity proposed by Morrey.

Quasiconvexity

Definition 19.2.1. *A continuous function* $\mathbf{E} : \mathbb{R}^{n \times n} \to \mathbb{R}$ *is said to be* quasiconvex *if for every* $A \in \mathbb{R}^{n \times n}$ *and* $\phi \in C_0^\infty(\mathbb{R}^n, \mathbb{R}^n)$ *we have*

$$\int_{\mathbb{R}^n} [\mathbf{E}(A + D\phi) - \mathbf{E}(A)] \geqslant 0 \qquad (19.16)$$

In the quest for examples of quasiconvex functionals, other than those that are convex, the intermediate concept of polyconvexity emerged in the work of Ball [39].

Polyconvexity

Definition 19.2.2. *A continuous function* $\mathbf{E} : \mathbb{R}^{n \times n} \to \mathbb{R}$ *is said to be* polyconvex *if, for each* $X \in \mathbb{R}^{n \times n}$, $\mathbf{E}(X)$ *can be expressed as a convex function of the minors (subdeterminants) of* X. *In particular, this includes the entries of* X *and* $\det(X)$.

We wish to point out that a key ingredient in our very first computation for convex functionals was the identity

$$\int_\Omega Df = \int_\Omega Dg$$

for all deformations $f, g : \Omega \to \mathbb{R}^n$ that agree on the boundary $\partial \Omega$. There are many such (nonlinear) differential expressions that possess this property. They are called *null Lagrangians*. The identity in (19.1) shows us that the determinant is a null Lagrangian. All other Jacobian subdeterminants are also null Lagrangians. Let us go back to the plane to illustrate the utility of null Lagrangians.

Suppose $\mathbf{E} : \mathbb{R}^{2 \times 2} \to \mathbb{R}$ can be written as

$$\mathbf{E}(X) = P(X, \det X), \qquad X \in \mathbb{R}^{2 \times 2}, \qquad (19.17)$$

where $P : \mathbb{R}^{2 \times 2} \times \mathbb{R} \to \mathbb{R}$ is convex. Such a P is going to be Lipschitz-regular, so we have the subgradient estimate

$$P(X, \det X) \geqslant P(A, \det A) + \langle P'(A, \det A), X - A \rangle + P_t(A, \det A)(\det X - \det A),$$

where
$$P_t(X,t) = \frac{\partial}{\partial t} P(X,t)$$
From this we obtain the desired property of quasiconvexity as follows,
$$\int_\Omega \mathbf{E}(A + D\phi) \geq \int_\Omega \mathbf{E}(A) + \langle P'(A, \det A), \int_\Omega D\phi \rangle$$
$$+ P_t(A, \det A) \int_\Omega \big(\det(A + D\phi) - \det(A)\big)$$
$$= \int_\Omega \mathbf{E}(A)$$
And more generally polyconvex functionals are quasiconvex—Ball's theorem.

We have already met a number of polyconvex functionals in this book. For example, the distortion function of a matrix $A \in \mathbb{R}^{2\times 2}$, defined on matrices of positive determinant by the rule
$$K(A) = \frac{|A|^2}{\det A},$$
is polyconvex (an excercise!). A still weaker but nevertheless very interesting notion is that of rank-one convexity.

Rank-one Convexity

Definition 19.2.3. *A continuous function* $\mathbf{E} : \mathbb{R}^{n\times n} \to \mathbb{R}$ *is said to be* rank-one convex *if it is convex in the direction of every rank-one matrix, that is, if the function* $t \mapsto \mathbf{E}(A + tX)$ *is a convex function of the real variable* t *for every* $A \in \mathbb{R}^{n\times n}$ *and for all* $X \in \mathbb{R}^{n\times n}$ *with* $\text{rank}(X) = 1$.

The following chain of implications actually holds,

Convexity $\Rightarrow$ *Polyconvexity* $\Rightarrow$ *Quasiconvexity* $\Rightarrow$ *Rank-one Convexity*

Furthermore, none of the implications can be reversed in dimensions $n \geq 3$, see [16, 101] and the well-known paper of V. Šverák [344]. In two dimensions, however, one of the most outstanding and important problems is that of Morrey from 1952.

Problem. Does rank-one convexity of functionals of 2×2 matrices imply quasiconvexity?

If we return to Theorem 19.1.4, and its associated conjecture in all dimensions $n \geq 2$, and reformulate it in the terminology introduced above, we are simply asking for which numbers λ_p the matrix function
$$\mathbf{E}(X) = \lambda_p |X|^p - |X|^{p-n} \det(X) \qquad (19.18)$$

19.2. QUASICONVEXITY

is quasiconvex at the origin, meaning that (19.16) holds with $A = 0$. Of course, this problem cannot be so simple! Although the question remains open for the conjectural sharp value $\lambda_p = \lambda_p(n) = |1 - \frac{n}{p}|$, the rank-one convexity of $\mathbf{E}(X)$ with this λ_p was established in [179], and, moreover, this is the smallest possible value for which rank-one convexity of the functional in (19.18) actually holds. It may very well be the case that $\mathbf{E}$ defined in (19.18) is quasiconvex. If so, we would have a very sharp tool to explore the L^p-theory of quasiconformal mappings in space.

We now return to two dimensions and recall an important result of Burkholder [80] reformulated in our present terminology.

Theorem 19.2.4. *The matrix function*

$$\mathbf{E}(X) = \mathbf{E}_{p,\lambda}(X) = \lambda |X|^p - |X|^{p-2}\det(X), \qquad X \in \mathbb{R}^{2\times 2},$$

is rank-one convex for all $1 < p < \infty$ and for all $\lambda \geqslant |1 - \frac{2}{p}|$.

In fact, the function $\mathbf{E}_{p,\lambda}$ is convex in the direction of matrices with nonpositive determinant. In terms of the complex derivatives and for $f \in W^{1,p}(\mathbb{C})$, we have

$$\mathcal{E}_{p,\lambda}[f] := \int_{\mathbb{C}} \mathbf{E}_{p,\lambda}(Df) = (1-\lambda)\int_{\mathbb{C}}(\gamma|f_{\bar{z}}| - |f_z|)(|f_{\bar{z}}| + |f_z|)^{p-1}, \qquad \gamma = \frac{1+\lambda}{1-\lambda},$$

and in fact more general energy integrands of the form

$$\mathbf{B}(\zeta, \xi) = (\gamma|\zeta| - |\xi|)(|\zeta| + |\xi|)^{p-1}, \qquad \zeta, \xi \in \mathbb{C}, \tag{19.19}$$

have emerged in Burkholder's theory of stochastic integrals and martingales, see [80, 81]. Related variational forms have been introduced by Šverák [345, 343]. We refer the reader to the article by A. Baernstein and S. Montgomery-Smith [38] for a greater and deeper discussion of these topics.

When $p = 2$ and $\gamma = 1$, the form in (19.19) reduces to the Jacobian

$$\mathbf{B}(\zeta, \xi) = |\zeta|^2 - |\xi|^2$$

And considering convexity, we are led to the quadratic polynomial in t,

$$\begin{aligned}\mathbf{B}(\zeta + t\zeta_0, \xi + t\xi_0) &= |\zeta + t\zeta_0|^2 - |\xi + t\xi_0|^2 \\ &= |\zeta|^2 - |\xi|^2 + 2t\,\Re(\zeta_0\bar{\zeta} - \xi_0\bar{\xi}) + t^2(|\zeta_0|^2 - |\xi_0|^2),\end{aligned}$$

and this polynomial is convex as long as $|\zeta_0|^2 - |\xi_0|^2 \geqslant 0$, or equivalently, we see that $\mathbf{B}$ is convex in the direction of matrices with nonpositive determinant. Let us discuss this statement for a larger range of p.

19.2.1 Burkholder's Theorem

For proving Burkholder's Theorem 19.2.4 let $A, X \in \mathbb{R}^{2\times 2}$ and $\det(X) \leqslant 0$. Let us consider the function φ of one variable t defined by

$$\varphi(t) = \lambda |A + tX|^p - |A + tX|^{p-2}\det(A + tX),$$

and we will now show that this function is convex for all $p > 1$ and $\lambda \geqslant |1 - \frac{2}{p}|$. It is clear that the borderline case $\lambda = |1 - \frac{2}{p}|$ is the worst because $(\lambda - |1 - \frac{2}{p}|)|A + tX|^p$ is convex in t. Thus we consider this case only. First, we make some observations to simplify the problem. Given a nonzero matrix X, then for almost every other matrix A the straight line $\{A + tX : t \in \mathbb{R}\} \subset \mathbb{R}^{2 \times 2}$ contains no matrix with double singular values, and thus $|A + tX|$ is a smooth function of t. Of course, we need only establish convexity for matrices with this property. It suffices to show that $\varphi''(0) \geqslant 0$ since we can write

$$A + tX = (A + t_0 X) + (t - t_0)X$$

and replace A by $A + t_0 X$ and t by $t - t_0$. Also, homogeneity assures us that we need consider only those matrices A for which $|A| = 1$, the case $A = 0$ being trivial. We can diagonalize A with two orthogonal transformations $O_1, O_2 \in SO(2)$ and so come to the formula

$$A + tX = O_1 \left(\begin{bmatrix} a & 0 \\ 0 & 1 \end{bmatrix} + tO_1^* X O_2^* \right) O_2,$$

where $|a| < 1$. Note that $\det(O_1^* X O_2^*) = \det(X) \leqslant 0$. We have therefore shown that there is no loss of generality in assuming that

$$A = \begin{bmatrix} a & 0 \\ 0 & 1 \end{bmatrix}, \quad |a| < 1, \quad X = \begin{bmatrix} x_{11} & x_{12} \\ x_{21} & x_{22} \end{bmatrix}, \quad \det(X) \leqslant 0$$

and $\lambda = |1 - \frac{2}{p}|$, $1 < p < \infty$. We simplify notation and write $N(t) = |A + tX|$ and $D(t) = \det(A + tX)$, both functions of t. Differentiation of the function φ shows

$$\begin{aligned}\varphi''(0) =\ & N^{p-3}[|p - 2|N^2 - (p - 2)D]\ddot{N} - N^{p-2}\ddot{D} \\ & + N^{p-4}[|p - 2|(p - 1)N^2 \dot{N}^2 - (p - 2)(p - 3)\dot{N}^2 D - 2(p - 2)N\dot{N}\dot{D}]\end{aligned}$$

where all the functions are evaluated at 0. We find $D = a$, $\dot{D} = x_{11} + ax_{22}$ and $\ddot{D} = 2 \det(X) \leqslant 0$. Unfortunately, computation of the derivatives of $|A + tX|$ is somewhat more involved. The formula for the operator norm of a matrix Y is given by

$$|Y| = \frac{1}{\sqrt{2}} \left(\|Y\|^2 + \sqrt{\|Y\|^4 - 4\det^2(Y)} \right)^{1/2}, \tag{19.20}$$

where $\|Y\|^2 = \operatorname{tr}(Y^t Y)$ is the Hilbert-Schmidt norm. We put $Y = A + tX$ in this equation, and simplify to see that

$$N = 1, \qquad \dot{N} = x_{22}, \qquad \ddot{N} = \frac{1}{1 - a^2}(x_{12}^2 + x_{21}^2 + 2\,a\,x_{12}x_{21})$$

We substitute this back into the formula for $\varphi''(0)$ above and obtain

$$\begin{aligned}\varphi''(0) =\ & [|p - 2| - (p - 2)a]\ddot{N} - 2(x_{11}x_{22} - x_{21}x_{12}) \\ & + |p - 2|(p - 1)x_{22}^2 - (p - 2)(p - 3)ax_{22}^2 - 2(p - 2)x_{22}(x_{11} + ax_{22})\end{aligned}$$

19.3. INTEGRABILITY OF THE JACOBIAN

Since $|a| < 1$, the terms involving x_{22}^2 add up to a non-negative number,

$$[|p-2|(p-1) - (p-2)(p-3)a - 2(p-2)a]x_{22}^2 = (p-1)[|p-2| - a(p-2)]x_{22}^2 \geqslant 0$$

The terms involving $x_{11}x_{22}$ sum up to

$$-2(p-1)x_{11}x_{22} \geqslant -2(p-1)x_{12}x_{21}$$

as $\det(X) \leqslant 0$. Thus

$$\varphi''(0) \geqslant |p-2|\ddot{N} - (p-2)a\ddot{N} - 2(p-2)x_{12}x_{21}$$

We simplify this by breaking it up into two cases with $p \geqslant 2$ and $p < 2$. In the following calculation the $\pm$ signs correspond to the first and second cases, respectively,

$$\begin{aligned}
\varphi''(0) &\geqslant |p-2|\left((1 \mp a)\ddot{N} \mp 2x_{12}x_{21}\right) \\
&= |p-2|\left[\frac{x_{12}^2 + x_{21}^2 + 2ax_{12}x_{21}}{1 \pm a} \mp 2x_{12}x_{21}\right] \\
&= \frac{|p-2|}{1 \pm a}(x_{12} \mp x_{21})^2 \geqslant 0
\end{aligned}$$

This last fact then proves Theorem 19.2.4.

19.3 L^1-Integrability of the Jacobian

We recall that Theorem 19.1.1 gave us the estimate

$$\int_{\mathbb{C}} |Df(z)|^{p-2} J(z, f)\, dz \leqslant \frac{\mathbf{S}_p^p - 1}{\mathbf{S}_p^p + 1} \int_{\mathbb{C}} |Df(z)|^p\, dz \tag{19.21}$$

for every function $f \in \mathbb{W}^{1,p}(\mathbb{C})$. We want to discuss an application of (19.21) in examining the L^1-integrability properties of the Jacobian determinant of an orientation-preserving mapping under minimal regularity hypotheses. Obviously, if $f = u + iv \in W^{1,2}_{loc}(\Omega)$, then the Jacobian $J(z, f) \in L^1_{loc}(\Omega)$. Further, the mapping f obeys the rule for integration by parts,

$$\int_\Omega \varphi\, (u_x v_y - u_y v_x) = \int_\Omega u(v_x \varphi_y - v_y \varphi_x) = \int_\Omega v(\varphi_x u_y - \varphi_y u_x) \tag{19.22}$$

for every $\varphi \in C_0^\infty(\Omega)$. In general, if $f \in W^{1,1}_{loc}(\Omega)$ is a homeomorphism, then $J(z, f) \in L^1_{loc}(\Omega)$, but the formula for integration by parts may fail. However, if f is orientation-preserving, that is, $J(z, f) \geqslant 0$, then less regularity than $f \in W^{1,2}_{loc}(\Omega)$ suffices.

Theorem 19.3.1. Let $f = u + iv$ be an orientation-preserving mapping of the Orlicz-Sobolev (or Zygmund) class $\mathbb{W}^{1,Q}(\Omega)$ with

$$Q(t) = t^2 \log^{-1}(1 + t) \tag{19.23}$$

That is,
$$\int_\Omega \frac{|Df|^2}{\log(1+|Df|)} < \infty \tag{19.24}$$

Then the Jacobian determinant is locally integrable and obeys the rule of integration by parts.

There are examples to show that (19.24) fails for orientation-preserving mappings in *weak*-$W^{1,2}(\Omega)$. Take, for instance, the homeomorphism of the punctured disk onto an annulus, defined by

$$f(z) = z + \frac{z}{|z|}, \qquad 0 < |z| < 1, \ 1 < |f(z)| < 2 \tag{19.25}$$

Here
$$|Df(z)| = |f_z| + |f_{\bar z}| = 1 + \frac{1}{|z|} \in weak - L^2(\mathbb{D})$$

but does not satisfy (19.24). Being an orientation-preserving map in *weak*-$W^{1,2}_{loc}(\Omega)$ still implies integrability of the Jacobian but not a change of variables in general. For instance, if we view the function f in (19.25) as being defined in the disk (so no longer a homeomorphism), integration by parts produces a Dirac mass at the origin: If $\varphi \in C_0^\infty(\mathbb{D})$ is defined over 0, then

$$\int_\Omega \varphi(z) J(z,f) = \int_\Omega u(v_x \varphi_y - v_y \varphi_x) + \pi \varphi(0)$$

Readers are urged to cook up more exotic examples for themselves as an exercise; see also [191].

Proof of Theorem 19.3.1. Let F be any compactly supported mapping of Orlicz-Sobolev class $\mathbb{W}^{1,Q}(\Omega)$ with Q as in (19.23). We have the following pointwise estimate based on the inequality $\varepsilon t^{-\varepsilon} \leqslant \log^{-1}(t)$ for $t \geqslant 1$.

$$\varepsilon |DF|^{2-\varepsilon} \leqslant \varepsilon(1+|DF|)^{2-\varepsilon} \leqslant \frac{(1+|DF|)^2}{\log(1+|DF|)}$$

Now, the Lebesgue dominated convergence theorem shows us that

$$\lim_{\varepsilon \to 0} \varepsilon \int_\Omega |DF|^{2-\varepsilon} = 0 \tag{19.26}$$

Although we do not pursue the matter here, the reader may wish to verify that if $F \in weak$-$W^{1,2}(\Omega)$ with distribution $|\{z : |DF| > t\}| \leqslant At^{-2}$, then $\limsup_{\varepsilon \to 0} \varepsilon \int_\Omega |DF|^{2-\varepsilon} \leqslant 2A$. This is why we would end up with integrability of the Jacobian but not integration by parts. We now apply the estimate in (19.21) that certainly holds for $F \in \mathbb{W}^{1,Q}(\Omega)$ and is compactly supported. Our earlier estimate of the p-norms $\mathbf{S}_p$ in (4.89) shows $\mathbf{S}_p \leqslant 1 + 3(p-2)$ for $p \geqslant 2$, so that $\mathbf{S}_{2-\varepsilon} \leqslant 1 + \frac{3\varepsilon}{1-\varepsilon}$ for $p = 2 - \varepsilon < 2$. Thus

$$\frac{\mathbf{S}_p^p - 1}{\mathbf{S}_p^p + 1} < 3\varepsilon,$$

19.3. INTEGRABILITY OF THE JACOBIAN

and hence
$$\int_\Omega |DF(z)|^{-\varepsilon} J(z,F) \leqslant 3\varepsilon \int_\Omega |DF(z)|^{2-\varepsilon} \tag{19.27}$$

We can therefore let $\varepsilon \to 0$ to see that, for any compactly supported $F \in \mathbb{W}^{1,Q}(\Omega)$,
$$\lim_{\varepsilon \to 0} \int_\Omega |DF(z)|^{-\varepsilon} J(z,F) = 0 \tag{19.28}$$

We may not be able to pass the limit through the integral here though.

Suppose now that $f = u + iv \in \mathbb{W}^{1,Q}(\Omega)$ has a non-negative Jacobian and that $\varphi \in C_0^\infty(\Omega)$ is a fixed test function. We apply the above argument to $F = f\varphi$, which has compact support and hence extends to a function of $\mathbb{W}^{1,Q}(\mathbb{C})$. For technical reasons we shall also need another such function, say $\eta \in C_0^\infty(\Omega)$, which is equal to 1 in the support of φ. Both these functions are supposed to be real-valued, and we will assume at some point that φ is non-negative. Consider the compactly supported map
$$F = (\varphi u, \eta v) \in \mathbb{W}^{1,Q}(\Omega)$$

We note that
$$DF = \begin{bmatrix} \varphi u_x & \varphi u_y \\ \eta v_x & \eta v_y \end{bmatrix} + \begin{bmatrix} \varphi_x u & \varphi_y u \\ \eta_x v & \eta_y v \end{bmatrix} = A + B$$

Then we integrate
$$\begin{aligned}\int_\Omega |A|^{-\varepsilon} \det(A) &= \int_\Omega \left(|A|^{-\varepsilon} \det A - |A+B|^{-\varepsilon} \det(A+B)\right) \\ &\quad + \int_\Omega |DF|^{-\varepsilon} J(z,F)\end{aligned} \tag{19.29}$$

and let $\varepsilon \to 0$. The last integral tends to 0 by (19.28). Because of the estimate
$$\left||A+B|^{-\varepsilon} \det(A+B) - |A|^{-\varepsilon} \det(A)\right| \leqslant C|B|(|A|+|B|)^{1-\varepsilon},$$

the first integral remains bounded. Precisely, we have
$$\begin{aligned}\limsup_{\varepsilon \to 0} \int_\Omega &\left||A|^{-\varepsilon} \det A - |A+B|^{-\varepsilon} \det(A+B)\right| \\ &\leqslant C(\|\nabla\varphi\|_\infty + \|\nabla\eta\|_\infty) \int_\Omega |f|(|f|+|Df|) < \infty\end{aligned}$$

Together they yield
$$\limsup_{\varepsilon \to 0} \left|\int_\Omega |A|^{-\varepsilon} \det(A)\right| < \infty \tag{19.30}$$

On the other hand, $\det(A) = \varphi\eta \det(Df) = \varphi \det(Df) \geqslant 0$ (here we assume $\varphi \geqslant 0$). We now apply the monotone convergence theorem, which allows us to pass the limit under the integral sign and conclude that $\det(A)$ is integrable.

It then follows that $J(z, F) = \det(DF)$ is integrable and hence the family $|DF(z)|^{-\varepsilon} J(z, F)$ has a L^1-majorant. Thus we can pass the limit under the integral sign in (19.28), so
$$\int_\Omega J(z, F) = 0$$

Finally, note that
$$\det(DF) = \varphi J(z, f) + u(\varphi_x v_y - \varphi_y u_x),$$
which gives us the local integrability of $J(z, f)$ and the desired integration by parts,
$$\int_\Omega \varphi J(z, f) = \int_\Omega u \left(\varphi_y v_x - \varphi_x v_y \right),$$
completing the proof. $\square$

The integrability of the Jacobian has strong consequences, such as the following.

Theorem 19.3.2. *Suppose $f \in \mathbb{W}^{1,Q}(\Omega)$ is a homeomorphism, where $Q(t) = t^2 \log^{-1}(1+t)$. Then f has the Lusin property $\mathcal{N}$.*

The proof of this theorem can be found by following the lines of argument from the same result in the $W^{1,2}$-setting, given in Theorem 3.3.7, now that we have the rule of integration by parts, Theorem 19.3.1.

Chapter 20

Solving the Beltrami Equation: Degenerate Elliptic Case

This chapter is focused on the study of the degenerate elliptic equation

$$\frac{\partial f}{\partial \bar{z}} = \mu(z)\frac{\partial f}{\partial z}, \tag{20.1}$$

where $|\mu(z)| < 1$ almost everywhere, but one might have

$$\|\mu\|_\infty = 1 \tag{20.2}$$

These equations arise naturally in hydrodynamics, nonlinear elasticity, holomorphic dynamics and several other related areas. Let us briefly consider two examples:

In two-dimensional hydrodynamics we have seen that the fluid velocity (the complex gradient of the potential function (see (16.51)) satisfies a Beltrami equation that degenerates as the flow approaches a critical value, the local speed of sound (see (16.54)). What happens to these equations and their solutions as we approach or break the speed of sound? What sort of degeneracies might occur? We should expect analytic degeneracy—the solution failing to lie in a certain Sobolev class (say $W^{1,2}$ in this case)—but what about topological degeneracy? Do the streamlines change their nature, for instance, exhibiting discontinuities?

In discussing holomorphic dynamics, Chapter 12, we saw that the Julia set of $\lambda z + z^2$ is a $\frac{1+|\lambda|}{1-|\lambda|}$-quasicircle if $|\lambda| < 1$ (equivalently for $z^2 + c$, when $c = \lambda/2 - \lambda^2/4$ lies in the primary component of the Mandelbrot set). The flow of these quasicircles is determined by a family of Beltrami equations whose ellipticity constant is controlled by λ. What happens as $|\lambda| \to 1$? Does the Julia set remain a topological circle? We know the answer is generically no, so these equations must degenerate in quite a bad way.

Similar questions arise naturally in the study of Kleinian groups in degenerating sequences of quasi-Fuchsian groups and materials science. The aim of this chapter is to develop some tools to explore these questions. In fact, it is essentially from the study of (20.1) in the degenerate case and the related higher-dimensional equations that the theory of mappings of finite distortion, which we shall spend considerable time discussing below, was born.

Our main goal is to see how to relax the assumptions on the Beltrami equation (20.1) in the degenerate setting (20.2), away from the uniformly elliptic case, and yet save as much as possible of the theory. In the degenerate case it is not clear if the Beltrami equation (20.1) has continuous or even nonconstant solutions at all. And even if such solutions exist, their topological properties may depend drastically on a specific situation. Consider in the first instance the equation

$$\frac{\partial f}{\partial \bar{z}} = \mu(z) \frac{\partial f}{\partial z}, \qquad \mu = -\frac{z}{2z - \bar{z}}$$

Notice that $|\mu| < 1$ away from the real axis and $\|\mu\|_\infty = 1$. This equation has the particularly degenerate solution $f(z) = z(z-\bar{z})$, mapping the real axis to a point. The distortion function is $K(z,f) = \frac{1+|\mu|}{1-|\mu|} \sim \frac{x}{y}$ for y small and x away from 0. Thus $K(z,f)$ is not integrable. Stoilow factorization must fail, and so forth. Yet the partial derivatives $f_z = 2z - \bar{z}$ and $f_{\bar{z}} = -z$ are exceptionally regular. The moral here is that we should not seek higher degrees of differentiablity but should try to control the distortion function.

To begin with we should recall that the Beltrami equation (20.1) is equivalent to the apparently quite different distortion inequality for a mapping f, $|Df(z)|^2 = K(z,f)J(z,f)$, since

$$K(z,f) = \frac{|Df(z)|^2}{J(z,f)} = \frac{(|f_z| + |f_{\bar{z}}|)^2}{|f_z|^2 - |f_{\bar{z}}|^2} = \frac{1 + |\mu(z)|}{1 - |\mu(z)|} \qquad (20.3)$$

In particular, at a given point z we have $K(z,f) < \infty$ if and only if $|\mu(z)| < 1$.

The most general class of mappings where at least some reasonable theory is to be expected is the following.

Mappings of Finite Distortion

Definition 20.0.3. *We say that a mapping $f = u + iv$ defined in a domain $\Omega \subset \mathbb{C}$ is a mapping of finite distortion if*

1. $f \in W^{1,1}_{loc}(\Omega)$.

2. $J(\cdot, f) = u_x v_y - u_y v_x \in L^1_{loc}(\Omega)$.

3. *There is a measurable function $K(z)$, finite almost everywhere, such that*

$$|Df(z)|^2 \leq K(z) J(z,f) \quad \text{almost everywhere in } \Omega$$

The smallest such function is denoted $K(z, f)$ and is called the *distortion function* of f.

Let us discuss this definition for a moment. The first requirement says that f has locally integrable distributional partial derivatives. This is the smallest degree of smoothness where one can begin to discuss what it means for f to be a (weak) solution of an equation such as (20.1). The second condition is a (weak) regularity property that is automatically satisfied by all local homeomorphisms f (or more generally, mappings that are locally finite to 1) defined on Ω and satisfying the first condition. The last condition, that the distortion function $1 \leqslant K(z, f) < \infty$, merely asks that the pointwise Jacobian $J(z, f) \geqslant 0$ almost everywhere and that the gradient $Df(z)$ vanishes at those points z where $J(z, f) = 0$. This seems to be a minimal requirement for a mapping to carry any geometric information.

In the case of bounded distortion, that is, $1 \leqslant K(z, f) \leqslant K < \infty$, the notion of the principal solution has been fundamental to our study, and we required in the previous chapters that the principal solution be in the class $W^{1,2}_{loc}(\mathbb{C})$. However, for degenerate elliptic equations this requirement is too strong. We therefore state the following definition.

Definition 20.0.4. *A solution f to the degenerate Beltrami equation (20.1), (20.2) is called a* principal solution *if*

- *f is a homeomorphism of $\mathbb{C}$.*
- *For some discrete set $E \subset \mathbb{C}$, f lies in the Sobolev space*

$$f \in W^{1,1}_{loc}(\mathbb{C} \setminus E), \qquad (20.4)$$

  and finally

- *At infinity f has the Taylor expansion*

$$f(z) = z + \frac{a_1}{z} + \frac{a_2}{z^2} + \cdots \qquad (20.5)$$

Thus to balance the very weak regularity (20.4), the principal solution is assumed to be a homeomorphism. Actually, of course, the monodromy theorem implies that a local homeomorphism satisfying the normalization (20.5) is a homeomorphism. In the setting of bounded distortion and $W^{1,2}_{loc}(\mathbb{C})$-regularity, this property was a basic consequence; see for instance Theorem 5.3.2.

Naturally, the notion of a principal solution is relevant only when $\mu(z)$ is compactly supported, or equivalently, $K(z) \equiv 1$ for $|z|$ large enough.

20.1 Mappings of Finite Distortion; Continuity

The first fundamental property of mappings of finite distortion we discuss is their continuity. Simple examples show that not all of these mappings are continuous.

Take for example the following example of Ball [39],

$$f(z) = z + \frac{z}{|z|}, \quad \text{for } 0 < |z| < 1, \quad f(0) = 0$$

Then f maps the punctured disk $\mathbb{D} \setminus \{0\}$ onto the annulus $2\mathbb{D} \setminus \mathbb{D}$, so that continuity at $z = 0$ certainly fails; elsewhere f is smooth. A straightforward calculation shows that

$$|Df(z)| = |f_z(z)| + |f_{\bar{z}}(z)| = 1 + \frac{1}{|z|} = J(z, f), \quad z \in \mathbb{D}$$

Hence $Df \in \text{weak-}L^2(\mathbb{D})$ and $K(z, f) = 1 + \frac{1}{|z|} < \infty$ for $z \neq 0$.

This example shows that continuity requires an extra regularity assumption. Indeed, it turns out that $f \in W^{1,2}_{loc}$ will suffice. This was shown first by Goldstein and Vodop'yanov [145]. This regularity requirement can be even further weakened by using the language of Orlicz-Sobolev spaces; for the optimal result, see [182].

Here we will give a proof for the case $f \in W^{1,2}_{loc}$ by using the notion of weak monotonicity introduced by Manfredi [238].

Theorem 20.1.1. *Assume that f is a mapping of finite distortion in a domain $\Omega \subset \mathbb{C}$ and assume that*

$$f \in W^{1,2}_{loc}(\Omega)$$

Then f is continuous in Ω and, moreover, we have the modulus of continuity estimate

$$|f(a) - f(b)|^2 \leqslant \frac{2\pi \int_{2B} \|Df\|^2}{\log\left(e + \frac{\text{diam}(B)}{|a - b|}\right)}$$

for every disk B with $a, b \in B \subset 2B \subset \Omega$.

Before we give the proof of this theorem, we need to discuss some different concepts of monotonicity.

20.1.1 Topological Monotonicity

We have already discussed monotonicity in a geometric setting in Section 3.11. Unfortunately, the term "monotonicity" has many different meanings in the extant literature and we now have to discuss another term, namely, "topological monotonicity". Recall that a function $u : \mathbb{R} \to \mathbb{R}$ is nondecreasing if $u(x) \leqslant u(y)$ whenever $x < y$, and u is monotone if either u or $-u$ is nondecreasing. This familiar notion from real analysis can be expressed in many equivalent ways, leading to different generalizations in more general spaces. For instance, a real-valued function u is monotone if it satisfies the maximum and minimum principles; that is, $\sup_I u = \sup_{\partial I} u$ and $\inf_I u = \inf_{\partial I} u$ for each interval $I \subset \mathbb{R}$, and a natural generalization is to replace the interval by an arbitrary open subset when functions $u : \Omega \to \mathbb{R}$ are considered. This point of view was introduced

20.1. CONTINUITY

by Lebesgue in his studies on the Dirichlet problem, and we shall return to this view a bit later.

A more topological aspect is the observation that the function $u : \mathbb{R} \to \mathbb{R}$ is monotone if and only if the inverse image of each point $u^{-1}(\{x\})$ is connected, that is, an interval. We therefore follow Morrey [270] and state the following definition.

Definition 20.1.2. *A mapping $f : X \to Y$ between two compact metric spaces is (topologically)* monotone *iff f is continuous, onto and*

$$f^{-1}(\{a\}) \text{ is connected in } X \tag{20.6}$$

for each singleton $\{a\} \subset Y$.

In the next sections we will approach the degenerate elliptic equations via approximations and limiting processes. Then we are faced with the problem that a uniform limit of homeomorphisms need not be a homeomorphism. In general, such limits are only monotone, that is, satisfy condition (20.6). It is for this reason, among others, that the concept of a monotone mapping is of fundamental importance in many questions in analysis and topology.

We shall prove the stability of monotone mappings under uniform limits only in the special case where $X = Y = \hat{\mathbb{C}}$, and $\hat{\mathbb{C}}$ is equipped with the spherical metric. For the general case of locally connected compact metric spaces see [254].

We first have the following theorem of G.T. Whyburn [371].

Lemma 20.1.3. *Suppose $f : X \to Y$ is a monotone mapping between compact metric spaces. Then for each closed and connected subset (continuum) $A \subset Y$, the inverse image $f^{-1}(A)$ is closed and connected (so another continuum).*

Proof. By definition f is onto, and so $A = f\left(f^{-1}(A)\right) = \bigcup_{m \in \mathcal{J}} f(U_m)$, where U_m, $m \in \mathcal{J}$, are the connected components of $f^{-1}(A)$. Here the sets $f(U_m)$, $m \in \mathcal{J}$, are disjoint. Namely, if the intersection $f(U_m) \cap f(U_n)$ contains a point $a \in A$, then the connected set $f^{-1}(\{a\})$ intersects both U_m and U_n. Hence the set $U_m \cup f^{-1}(\{a\}) \cup U_n \subset f^{-1}(A)$ is connected and thus U_m, U_n are contained in the same component, that is, $m = n$. Furthermore, since f is continuous and A compact, $f^{-1}(A)$ and hence each U_m and $f(U_m)$ is compact. Now $A = \bigcup_{m \in \mathcal{J}} f(U_m)$ is a disjoint union of compact subsets. But since A is connected, this is not possible unless we have only one component. Therefore $f^{-1}(A)$ is connected. □

Theorem 20.1.4. *Assume that f_j, $j = 1, 2, \ldots$, are continuous mappings from the compact metric space X onto the compact metric space Y that are monotone in the sense of (20.6). Suppose that the sequence f_j converges uniformly on X to a mapping $f : X \to Y$. Then f is monotone.*

Proof. As the uniform limit of continuous surjective mappings $f : \hat{\mathbb{C}} \to \hat{\mathbb{C}}$ is clearly continuous and surjective, it remains to show (20.6). Let $a \in Y$. Assume that we can express $f^{-1}(\{a\})$ as a union of two disjoint nonempty (but not necessarily connected) compact subsets A and B,

$$f^{-1}(\{a\}) = A \cup B$$

By the uniform convergence, the distances $\text{dist}_Y(a, f_j(A))$, $\text{dist}_Y(a, f_j(B)) \to 0$ when $j \to \infty$. We can therefore assume that the closed ball $B_j = B_Y(a, 1/j)$ contains both $f_j(A)$ and $f_j(B)$ for each j. Since f_j are monotone mappings, by Lemma 20.1.3, the sets $f_j^{-1}(B_j)$ are connected, and they all intersect both A and B. Hence there must be a point $x_j \in f_j^{-1}(B_j)$ with $\text{dist}(x_j, A) = \text{dist}(x_j, B)$. (Apply the mean value theorem to the continuous function $\text{dist}(x, A) - \text{dist}(x, B)$ on the connected space $f_j^{-1}(B_j)$). But taking a subsequence, again denoted x_j, we obtain a point $x = \lim_j x_j \in X$ with $\text{dist}(x, A) = \text{dist}(x, B)$ and $f(x) = a$. This gives $x \in f^{-1}(\{a\}) = A \cup B$ and $\text{dist}(A, B) = 0$, which is a contradiction. It follows that $f^{-1}(\{a\})$ is connected. □

We now return to our analysis and to the Lebesgue view of monotone functions. We will need a counterpart of monotonicity for real-valued functions, defined in a domain $\Omega \subset \mathbb{C}$, which are only in the class $W_{loc}^{1,p}(\Omega)$ and thus a priori not necessarily continuous. The following notion, introduced by Manfredi [238], will serve our purposes perfectly.

We say that a function $u \in W_{loc}^{1,p}(\Omega)$ is *weakly monotone* if, for each constant $b \in \mathbb{R}$ and each relatively compact subdomain $\Omega' \subset \Omega$, we have

1. $(u - b)^+ \in W_0^{1,p}(\Omega') \Rightarrow u(z) \leqslant b$ for almost all $z \in \Omega'$, and
2. $(u - b)^- \in W_0^{1,p}(\Omega') \Rightarrow u(z) \geqslant b$ for almost all $z \in \Omega'$

Note that, to be precise, weak monotonicity may depend on the Sobolev exponent p. However, in what follows the precise Sobolev smoothness we require will be evident from the setup.

There is a particularly elegant geometric approach to obtaining continuity estimates of weakly monotone functions. The idea goes back to Gehring [133] and his study of the Liouville theorem in space. While many other interesting applications of Gehring's oscillation lemma have been discussed in the literature, its use with weakly monotone functions seems less familiar. In this setting the oscillation Lemma 3.5.1 has the following variant. Its proof combines the earlier Lemma 3.5.1 with a convolution approximation and is left to the reader. The details can also be found in [191, p. 152].

Lemma 20.1.5. *Let $u \in W^{1,1}(\mathbb{D})$ be weakly monotone. Then for all Lebesgue points $a, b \in \mathbb{D}(r) \subset \mathbb{D}$, we have*

$$|u(a) - u(b)| \leqslant \frac{1}{2} \int_{\partial \mathbb{D}(t)} |\nabla u| |dz| \tag{20.7}$$

for almost every $t \in [r, 1]$.

20.1. CONTINUITY

The numbers $t \in [r, 1]$ for which (20.7) holds can be identified as the Lebesgue points of the function $t \mapsto \int_{\partial \mathbb{D}(t)} |\nabla u|$, which by Fubini's theorem is an integrable function defined on the interval $[r, 1]$.

Theorem 20.1.6. *Let $u \in W^{1,2}(2\mathbb{D})$ be weakly monotone. Then for all Lebesgue points $a, b \in \mathbb{D}$, we have*

$$|u(a) - u(b)|^2 \leqslant \frac{\pi \int_{2\mathbb{D}} |\nabla u|^2}{\log\left(e + \frac{1}{|a-b|}\right)} \qquad (20.8)$$

In particular, u has a continuous representative for which (20.8) holds for all a and b in $\mathbb{D}$.

Proof. Consider the two concentric disks $\mathbb{D}(z, r)$ and $\mathbb{D}(z, 1)$, where $z = \frac{1}{2}(a+b)$ and $r = \frac{1}{2}|a - b|$. Note that all disks $\mathbb{D}(z, t)$ with $t \leqslant \rho_0 = \max\{1, r\}$ lie in $2\mathbb{D}$. From Lemma 20.1.5 we have

$$\frac{|u(a) - u(b)|}{\pi t} \leqslant \frac{1}{2\pi t} \int_{\partial \mathbb{D}(z,t)} |\nabla u| |dw|$$

for almost all $r < t < 1$. Jensen's inequality gives, for any convex function P,

$$P\left(\frac{|u(a) - u(b)|}{\pi t}\right) \leqslant \frac{1}{2\pi t} \int_{\partial \mathbb{D}(z,t)} P(|\nabla u|) |dw| \qquad (20.9)$$

We choose $P(t) = t^2$, multiply by t and integrate with respect to t, $r/2 < t < \rho_0$, to get

$$|u(a) - u(b)|^2 \log(2\rho_0/r) \leqslant \frac{\pi}{2} \int_{\mathbb{D}(z, \rho_0)} |\nabla u|^2 \leqslant \frac{\pi}{2} \int_{2\mathbb{D}} |\nabla u|^2$$

The estimate (20.8) is a quick consequence. $\square$

Remark 20.1.7. *It is clear from the above discussion that for weakly monotone mappings the continuity holds in strictly larger classes than $W^{1,2}$. Indeed, from (20.9) we see that all one needs is that $P(|\nabla u|) \in L^1(\mathbb{D})$ for a convex function with*

$$\int_0^1 t P\left(\frac{1}{t}\right) dt = \int_1^\infty P(s) \frac{ds}{s^3} = \infty \qquad (20.10)$$

For instance,

$$P(s) = \frac{s^2}{\log^\alpha(e+s)}, \qquad 0 \leqslant \alpha \leqslant 1,$$

suffices, and this leads to the modulus of continuity in Zygmund classes and their close relations [191].

20.1.2 Proof of Continuity in $W^{1,2}$

To prove Theorem 20.1.1, it remains to be shown that for any mapping of finite distortion in the class $W^{1,2}_{loc}(\Omega)$, the coordinate functions are weakly monotone. This will follow easily from the next lemma.

Lemma 20.1.8. *Assume that $u \in W^{1,2}_0(\mathbb{D})$ and $v \in W^{1,2}(\mathbb{D})$ are real-valued. If the function $h = u + iv$ satisfies*

$$J(z,h) \equiv u_x(z)v_y(z) - u_y(z)v_x(z) \geq 0 \quad \text{for almost all } z \in \mathbb{D},$$

then $J(z,h) = 0$ for almost all $z \in \mathbb{D}$.

Proof. Let $g(z) = 0 + iv(z)$. Then $h - g \in W^{1,2}_0(\mathbb{D})$, and we see from Corollary 2.9.2 that

$$\int_\mathbb{D} J(z,h)dz = \int_\mathbb{D} J(z,g)dz = 0$$

Since $J(z,h) \geq 0$, the Jacobian $J(z,h)$ vanishes almost everywhere. □

Proof of Theorem 20.1.1. The proof of the theorem is now straightforward. In fact, let $f = u + iv \in W^{1,2}_{loc}(\Omega)$ be a mapping of finite distortion. We can assume $\Omega = \mathbb{D}$ and that $f \in W^{1,2}(\mathbb{D})$. To show the weak monotonicity of f, we assume that $b \in \mathbb{R}$ and that $\Omega' \subset \mathbb{D}$ is a subdomain for which $(u-b)^+ \in W^{1,2}_0(\Omega')$. Write $u_0 = (u-b)^+$. Outside the domain Ω' define u_0 to be zero so that $u_0 \in W^{1,2}_0(\mathbb{D})$.

Now we put $h = u_0 + iv$ and $E = \{z \in \Omega' : u(z) > b\}$. By Lemma A.4.1, $\nabla u_0 = 0$ for almost all points of $\mathbb{D} \setminus E$. On the set E, $J(z,h) = J(z,f)$, while $J(z,h) = 0$ on $\mathbb{D} \setminus E$. Since the Jacobian $J(z,f) \geq 0$, we must also have $J(z,h) \geq 0$ almost everywhere. But then Lemma 20.1.8 shows that $J(z,h) \equiv 0$. We obtain almost everywhere on E that $J(z,f) = 0$, and since f is a mapping of finite distortion, for almost all points of E we have $|Dh(z)|^2 = |Df(z)|^2 = J(z,f)K(z,f) = 0$. It follows that $\nabla u_0(z) = 0$ for almost all points $z \in \mathbb{D}$, thus $u_0 \in W^{1,2}_0(\Omega')$ is a constant, and hence it vanishes. We have shown that $u(z) \leq b$ almost everywhere on Ω'.

This argument proves that the real part of f is weakly monotone. Similarly, the imaginary part will be weakly monotone. Since $f \in W^{1,2}(\mathbb{D})$, Theorem 20.1.6 proves that f is continuous, and hence the proof of Theorem 20.1.1 is complete. □

The reader should note that the proof of Theorem 20.1.1 given here works for any mapping of finite distortion as soon as one obtains counterparts to Theorem 20.1.6 and Lemma 20.1.8. We shall return to this question later in Section 20.4.8 when studying mappings of exponentially integrable distortion.

20.2 Integrable Distortion; $W^{1,2}$-Solutions and Their Properties

Naturally, a basic question that the theory of mappings of finite distortion should help us to understand is when do we have the existence of *principal* solutions

20.2. INTEGRABLE DISTORTION

to (20.1) for a given complex dilatation $\mu(z)$. To address this question we begin with the study of the case where we assume a priori that we are given one fairly regular nonconstant solution to the Beltrami equation

$$F_{\bar{z}} = \mu(z) F_z \tag{20.11}$$

From this assumption we will draw conclusions about F, which will then lead to the existence of the principal solution as well as to understanding the properties of more general solutions to (20.11).

Aside from the assumption $F \in W^{1,2}(\Omega)$ in the previous section, we shall make a (weak) restriction on the class of distortion functions under consideration and assume that $K \in L^1(\Omega)$. It turns out that with these assumptions many of the classical properties of solutions to the (uniformly elliptic) Beltrami equation do hold in more generality.

To be precise, we normalize $\Omega = \mathbb{D}$ and assume that

1. μ is compactly supported in the unit disk $\mathbb{D}$

2. $K(z) = \frac{1+|\mu(z)|}{1-|\mu(z)|} \in L^1(\mathbb{D})$

3. The Beltrami equation admits a nonconstant solution $F \in W^{1,2}(\mathbb{D})$,

$$F_{\bar{z}} = \mu F_z \qquad \text{for almost every } z \in \mathbb{D} \tag{20.12}$$

We wish to stress here that these are not unnatural assumptions. The planar theory of nonlinear elasticity deals with mappings $F \in W^{1,2}(\mathbb{D})$ with integrable distortion, see for instance J. Ball [40]. Such mappings are usually obtained as minimizers of certain energy integrals. Thus in fairly natural circumstances the existence of at least one $W^{1,2}(\mathbb{D})$-solution to (20.12) is possible. On the other hand, in general the existence of a solution in $W^{1,2}(\mathbb{D})$ is quite a strong assumption. (However, one could jump ahead for a moment to consider this result in light of Theorem 20.4.7.)

Theorem 20.2.1. *Under the hypotheses 1–3 above, there exists a principal solution to the equation*

$$\frac{\partial f}{\partial \bar{z}} = \mu(z) \frac{\partial f}{\partial z} \qquad \text{for almost every } z \in \mathbb{C},$$

which has the following additional properties:

- *Partial regularity:* $f \in W^{1,1}_{loc}(\mathbb{C})$, *and there is a finite set E such that*

$$f \in W^{1,2}_{loc}(\mathbb{C} \setminus E)$$

- *Factorization: each solution $h \in W^{1,2}_{loc}(\Omega)$ to the equation*

$$h_{\bar{z}}(z) = \mu(z) h_z(z) \qquad \text{for almost every } z \in \Omega,$$

admits a Stoilow factorization

$$h(z) = (\Phi \circ f)(z),$$

where Φ is holomorphic in $f(\Omega)$. In particular, all nonconstant solutions in $W^{1,2}_{loc}(\Omega)$, including F, are open and discrete. Here Ω is an arbitrary domain in $\mathbb{C}$.

- Uniqueness: Let E' be a finite set and let $f_0 \in W^{1,2}_{loc}(\mathbb{C} \setminus E')$ be a solution to the Beltrami equation, continuous in $\mathbb{C}$, with $f_0(z) = z + o(1)$ near ∞. Then $f_0 = f$, the principal solution.

- Inverse: Denote inverse mapping of f by $g = g(w)$. Then g has no singularities; $g(w) - w \in W^{1,2}(\mathbb{C})$ with the precise estimate

$$\int_{\mathbb{C}} (|g_{\overline{w}}|^2 + |g_w - 1|^2) \, dw \leq C \int_{\mathbb{D}} K(z) \, dz \qquad (20.13)$$

- Modulus of continuity: We have the modulus of continuity estimate

$$|g(a) - g(b)|^2 \leq C \, \frac{1 + |a|^2 + |b|^2}{\log\left(e + \frac{1}{|a-b|}\right)} \int_{\mathbb{D}} K(z) \, dz$$

In the last two estimates C is an absolute constant.

It is not difficult to find examples showing that in general the exceptional set E for the $W^{1,2}_{loc}$-regularity of f is not empty. We will point out later one such situation, at the end of Section 20.4.1.

For the proof of this theorem we will need a few auxiliary results. First, we establish the uniform bound (20.13) for the L^2-derivatives of the inverse of the principal solution to the (uniformly elliptic) Beltrami equation.

Lemma 20.2.2. Let $f \in W^{1,2}_{loc}(\mathbb{C})$ be the principal solution to $f_{\overline{z}}(z) = \mu(z) f_z(z)$, $z \in \mathbb{C}$, where μ is supported in the unit disk $\mathbb{D}$ with $\|\mu\|_\infty < 1$. Let $g = f^{-1}$ be the inverse of f. Then

$$\int_{\mathbb{C}} (|g_{\overline{w}}|^2 + |g_w - 1|^2) \, dw \leq 2 \int_{\mathbb{D}} K(z) \, dz$$

Proof. We proved in (5.9) that $g_w - 1 = S g_{\overline{w}}$. Since the Beurling operator is an isometry on $L^2(\mathbb{C})$ and $g_{\overline{w}}$ vanishes outside $f(\mathbb{D})$,

$$\int_{\mathbb{C}} (|g_{\overline{w}}|^2 + |g_w - 1|^2) \, dw = 2 \int_{\mathbb{C}} |g_{\overline{w}}|^2 \, dw \leq 2 \int_{g^{-1}(\mathbb{D})} (|g_{\overline{w}}| + |g_w|)^2 \, dw$$

From (2.49) and (2.50) we have

$$-f_{\overline{z}} = J(z, f) \, g_{\overline{w}}(f), \qquad f_z = J(z, f) \, \overline{g_w(f)} \qquad (20.14)$$

20.2. INTEGRABLE DISTORTION

Hence a change of variables, allowed by Theorem 3.8.1, gives

$$\int_{g^{-1}(\mathbb{D})} (|g_{\bar{w}}| + |g_w|)^2 \, dw = \int_{\mathbb{D}} (|f_{\bar{z}}| + |f_z|)^2 J(z,f)^{-1} \, dz$$

$$= \int_{\mathbb{D}} K(z,f) \, dz, \quad (20.15)$$

and we are done. □

Also, for the modulus of continuity of the inverse mapping, we have a uniform estimate. The key fact here and in Lemma 20.2.2 is that the bounds do not depend on the value of $\|\mu\|_\infty$.

Lemma 20.2.3. *Let f and $g = f^{-1}$ be as in Lemma 20.2.2. Then*

$$|g(a) - g(b)|^2 \leqslant \frac{16\pi^2 \left(R^2 + \int_{\mathbb{D}} K(z,f) \, dz\right)}{\log\left(e + \frac{1}{|a-b|}\right)} \quad (20.16)$$

whenever $a, b \in \mathbb{D}_R$. Moreover, $|g(w) - w| < 1$ for all $|w| > 3$.

Proof. From the bound (2.61) we have $|f(z) - z| < 1$ whenever $|z| > 1.3$. By the maximum principle, this implies the last claim of the lemma. Similarly, $|g(w)| \leqslant 2 + |w|$ for every $w \in \mathbb{C}$. Arguing as in (20.14) and (20.15), we can estimate

$$\pi \int_{\mathbb{D}_{2R}} |Dg(w)|^2 \, dw \leqslant \pi \int_{g(\mathbb{D}_{2R})} K(z,f) \, dz$$

Since $K(z,f) \equiv 1$ outside the unit disk and $g(\mathbb{D}_{2R}) \subset \mathbb{D}_{2R+2}$, for $R \geqslant 1$, the last integral above is therefore less than $\pi \int_{\mathbb{D}} K(z,f) \, dz + 15\pi^2 R^2$. Hence for $R \leqslant 1$, the integral in question is less than $16\pi \int_{\mathbb{D}} K(z,f) \, dz$.

Now let $a, b \in \mathbb{D}_R$. If $R \leqslant 1$, the claim (20.16) follows from Theorem 20.1.6 since a homeomorphism is weakly monotone. For $R > 1$, apply Theorem 20.1.6 to $u(z) = g(Rz), |z| < 2$, to obtain

$$|g(a) - g(b)|^2 \leqslant \frac{\pi \int_{\mathbb{D}_{2R}} |Dg(w)|^2}{\log\left(e + \frac{R}{|a-b|}\right)},$$

proving the claim (20.16) in this case also. □

Proof of Theorem 20.2.1. We begin with the given nonconstant function $F \in W^{1,2}(\mathbb{D})$ and the complex dilatation μ satisfying $F_{\bar{z}} = \mu F_z$ at almost every point $z \in \mathbb{D}$. First, note that F, being a mapping of finite distortion in the Sobolev class $W^{1,2}(\mathbb{D})$, is continuous by Theorem 20.1.1. Consider then the truncated Beltrami equation

$$f_{\bar{z}}^\varepsilon(z) = \mu_\varepsilon(z) f_z^\varepsilon(z), \quad \mu_\varepsilon(z) = \begin{cases} \mu(z) & \text{if } |\mu(z)| \leqslant 1 - \varepsilon \\ (1-\varepsilon) \frac{\mu(z)}{|\mu(z)|} & \text{otherwise} \end{cases}$$

$$(20.17)$$

with its principal solution f_ε. We obtain f_ε conveniently from the given μ, but it is hard to operate on it directly. Instead, we operate on the inverse mappings $g_\varepsilon = f_\varepsilon^{-1}$ for which Lemma 20.2.3 applies.

From Lemma 20.2.3 we have first for the spherical distance $d(g_\varepsilon(z), \infty) = 2/(1 + |g_\varepsilon(z)|^2)^{1/2} \leqslant 2d(z, \infty)$ when $|z| > 3$. Since

$$K(z, f_\varepsilon) = \frac{1 + |\mu_\varepsilon(z)|}{1 - |\mu_\varepsilon(z)|} \leqslant \frac{1 + |\mu(z)|}{1 - |\mu(z)|} = K(z, F), \tag{20.18}$$

Lemma 20.2.3 gives in $\hat{\mathbb{C}}$ a uniform estimate for the modulus of continuity of g_ε. Thus we can let $\varepsilon \to 0$ and use the Arzela-Ascoli theorem. We obtain a subsequence of the inverse mappings $g_{\varepsilon_j} = f_{\varepsilon_j}^{-1}$, converging uniformly in $\hat{\mathbb{C}}$ to a mapping g.

Theorem 20.1.4 says that g is a monotone mapping of $\hat{\mathbb{C}}$. In fact, each g_ε has a Taylor expansion of the form

$$g_\varepsilon(w) = w + \frac{a_1^\varepsilon}{w} + \frac{a_2^\varepsilon}{w^2} + \cdots, \tag{20.19}$$

where the area formula (2.59) gives bounds for the coefficients. Therefore the limit g is actually conformal near ∞. The inverse image $\Omega' = g^{-1}(\mathbb{D})$ is the nested union of the sets $g^{-1}(\overline{\mathbb{D}_r})$, $0 < r < 1$, and hence connected by Lemma 20.1.3. We have just deduced that Ω' is a domain in $\mathbb{C}$. Furthermore, according to Lemma 20.2.2 and (20.18), the functions $g_{\varepsilon_j}(z) - z$ have uniformly bounded L^2-derivatives in $\mathbb{C}$. Hence, using standard properties of Sobolev functions and in particular weak compactness, we may assume that the mappings g_{ε_j} converge weakly in $W^{1,2}(\mathbb{D}_R)$ for each $R > 1$ and that $g \in W^{1,2}_{loc}(\mathbb{C})$.

Next, let us define $\phi: \Omega' \to \mathbb{D}$ by the rule

$$\phi(w) = F(g(w)), \quad w \in \Omega' = g^{-1}(\mathbb{D}) \tag{20.20}$$

We claim that ϕ is a holomorphic function in Ω'. To see this we approximate ϕ by the mappings $\phi_\varepsilon(w) = F(g_\varepsilon(w))$ and therefore need estimates for ϕ_ε. Applying the chain rule gives

$$\begin{aligned}(\phi_\varepsilon)_{\overline{w}} &= (F_z \circ g_\varepsilon)(g_\varepsilon)_{\overline{w}} + (F_{\bar{z}} \circ g_\varepsilon)\overline{(g_\varepsilon)_w} \\ &= (F_z \circ g_\varepsilon)\overline{(g_\varepsilon)_w}\Big(\mu(g_\varepsilon) - \mu_\varepsilon(g_\varepsilon)\Big),\end{aligned} \tag{20.21}$$

where we have used (20.12) and the identity $(g_\varepsilon)_{\overline{w}} = -\mu_\varepsilon(g_\varepsilon)\overline{(g_\varepsilon)_w}$ from Lemma 10.3.1. Similarly,

$$\begin{aligned}(\phi_\varepsilon)_w &= (F_z \circ g_\varepsilon)(g_\varepsilon)_w + (F_{\bar{z}} \circ g_\varepsilon)\overline{(g_\varepsilon)_{\overline{w}}} \\ &= (F_z \circ g_\varepsilon)(g_\varepsilon)_w\Big(1 - \mu(g_\varepsilon)\overline{\mu_\varepsilon(g_\varepsilon)}\Big)\end{aligned} \tag{20.22}$$

Since (20.17) shows that

$$|\mu - \mu_\varepsilon|^2 \leqslant \varepsilon(1 - |\mu_\varepsilon|^2)$$

20.2. INTEGRABLE DISTORTION

and we have $J(w, g_\varepsilon) = |(g_\varepsilon)_w|^2(1 - |\mu_\varepsilon(g_\varepsilon)|^2)$, we can estimate

$$\int_{\Omega'} |(\phi_\varepsilon)_{\overline{w}}|^2 \, dw \leqslant \varepsilon \int_{\Omega'} J(w, g_\varepsilon)|F_z \circ g_\varepsilon|^2 \, dw = \varepsilon \int_{\mathbb{D}} |F_z|^2 \, dz < \infty \quad (20.23)$$

To get bounds for ϕ_w note that $|1 - \mu(z)\overline{\mu_\varepsilon(z)}|^2 \leqslant 1 - |\mu_\varepsilon(z)|^2$. Thus (20.22) yields

$$\int_{\Omega'} |(\phi_\varepsilon)_w|^2 \, dw \leqslant \int_{\Omega'} J(w, g_\varepsilon)|F_z \circ g_\varepsilon|^2 \, dw = \int_{\mathbb{D}} |F_z|^2 \, dz \quad (20.24)$$

We now let $\varepsilon \to 0$. Weak compactness in the Sobolev space $W^{1,2}(\Omega')$ with the two estimates (20.23) and (20.24) implies that a further subsequence $\phi_{\varepsilon_j} = F(g_{\varepsilon_j})$ converges to the mapping $\phi = F(g)$ not only pointwise (as F is continuous by Theorem 20.1.1) but also weakly in $W^{1,2}(\mathbb{D})$. Furthermore, from (20.23) we see that the weak derivative $\phi_{\overline{w}} = 0$ in $\mathbb{D}$. By Weyl's Lemma A.6.10 we see that ϕ is indeed a holomorphic function in Ω'.

It is important to observe that $\phi \not\equiv \text{const}$. To see this consider points $a, b \in \mathbb{D}$ such that $F(a) \neq F(b)$. As g is monotone, and hence a surjection, we may pick up points $A \in g^{-1}\{a\} \subset \Omega'$ and $B \in g^{-1}\{b\} \subset \Omega'$ to obtain $\phi(A) = F(a) \neq F(b) = \phi(B)$.

We now use this to prove that g is a homeomorphism of $\hat{\mathbb{C}}$. Explicitly, note that g was obtained as a local uniform limit of homeomorphisms $g_\varepsilon : \hat{\mathbb{C}} \to \hat{\mathbb{C}}$, which are conformal in $g_\varepsilon^{-1}(\mathbb{C} \setminus A)$, where A is a relatively compact subset of $\mathbb{D}$, namely, the support of μ. Thus g is conformal in $g^{-1}(\mathbb{C} \setminus A)$. Since by the monotonicity each $g^{-1}(\{a\})$ is connected, we see that $g^{-1}(\{a\})$ are singletons for each $a \in \mathbb{C} \setminus A$. On the other hand, for each $a \in \mathbb{D}$, the holomorphic function ϕ is a constant on $\Omega' \cap g^{-1}(\{a\})$. Therefore in this case, too, the connected set $g^{-1}(\{a\})$ is a singleton. We have shown that g is injective, and as a monotone mapping g is onto. Therefore it is in fact a homeomorphism of $\mathbb{C}$.

Now we may define f by setting $f = g^{-1} : \hat{\mathbb{C}} \to \hat{\mathbb{C}}$. This is the principal solution we seek.

Clearly, f is holomorphic in $\mathbb{C} \setminus A \supset \mathbb{C} \setminus \mathbb{D}$. Concerning the regularity in $\mathbb{D}$, we use the factorization

$$F(z) = \phi(f(z)),$$

where F is our given nonconstant solution in $W^{1,2}_{loc}(\mathbb{D})$. We have shown that ϕ is holomorphic and nonconstant in $\Omega' = f(\mathbb{D})$ and therefore has only a finite number of critical points on any relatively compact subdomain. Away from these critical points ϕ has a holomorphic inverse. Thus $f \in W^{1,2}_{loc}(\mathbb{C} \setminus E)$ outside a finite set E. Moreover, $f \in W^{1,2}(\mathbb{D})$,

$$\int_{\mathbb{D}} |Df(z)| dz \leqslant \int_{\mathbb{D}} \sqrt{K(z) J(z, f)} dz \leqslant \left(\int_{\mathbb{D}} K(z) dz \right)^{1/2} \left(\int_{\mathbb{D}} J(z, f) dz \right)^{1/2}$$

$$\leqslant \left(\int_{\mathbb{D}} K(z) dz \right)^{1/2} |f(\mathbb{D})|^{1/2} < \infty$$

In the last step we have used inequality (3.10) from Corollary 3.3.6. Therefore $f \in W^{1,1}_{loc}(\mathbb{C})$, and there are no exceptional sets for $W^{1,1}_{loc}$-regularity.

It remains to prove the Stoilow factorization and the uniqueness of the principal solution. For the first, if $\Omega \subset \mathbb{C}$ is a subdomain, let μ be the Beltrami coefficient from our starting point (20.12) and consider an arbitrary $W^{1,2}(\Omega)$-solution

$$\frac{\partial h}{\partial \bar{z}} = \mu(z)\frac{\partial h}{\partial z}, \quad z \in \Omega.$$

As in (20.20), we define an auxiliary function $\Phi(z) = h(g(z))$, $z \in g^{-1}(\Omega)$. Then h is continuous by Theorem 20.1.1, and arguing precisely as in (20.21)–(20.24), we see that Φ is holomorphic. That is, $h = \Phi \circ f$ admits the Stoilow factorization.

Last, concerning uniqueness, let $f_0 \in W^{1,2}_{loc}(\mathbb{C} \setminus E')$ be another principal solution continuous in $\mathbb{C}$. In particular, using the Stoilow factorization, we may write

$$f_0(z) = \Phi(f(z)),$$

where Φ is holomorphic in $f(\mathbb{C}\setminus(E\cup E'))$. The point here is that the singular set $f(E\cup E')$ is countable (having finite linear measure would suffice) and therefore removable for the locally bounded Φ. Since Φ is now linear and near ∞, we have $\Phi(w) = w + o(1)$ and it follows that $f_0 = f$. □

20.3 A Critical Example

We now turn to establishing the existence of a principal solution without assuming the existence of a nonconstant $W^{1,2}$-solution. We begin by presenting a simple but dramatic example that implies that the mere assumption $K \in L^1(\mathbb{D})$ is much too weak. This of course means that obtaining principal solutions f with reasonable properties requires a higher degree of integrability of the distortion function K.

The example will also hint at the optimal limits of requirements on K or on the regularity of the solution f. It will become evident that a full understanding requires the notion of Orlicz-Sobolev spaces.

Theorem 20.3.1. *Let* $\mathcal{A}: [1, \infty) \to [1, \infty)$ *be a* C^∞-*smooth strictly increasing function with* $\mathcal{A}(1) = 1$ *and such that*

$$\int_1^\infty \frac{\mathcal{A}(t)}{t^2} dt < \infty \tag{20.25}$$

Then there is a Beltrami coefficient μ *compactly supported in the unit disk,* $|\mu(z)| < 1$, *with the following properties:*

1. The distortion function $K(z) = \frac{1+|\mu(z)|}{1-|\mu(z)|}$ satisfies

$$\int_{\mathbb{D}} e^{\mathcal{A}(K(z))} dz < \infty \tag{20.26}$$

20.3. A CRITICAL EXAMPLE

2. If f is a continuous $W^{1,1}_{loc}(\mathbb{D})$-solution to the Beltrami equation

$$f_{\bar{z}} = \mu f_z \quad \text{for almost every } z \in \mathbb{D}, \tag{20.27}$$

then f is constant.

3. There is a bounded solution $w = f(z)$ to the Beltrami equation in the space weak-$W^{1,2}(\mathbb{D}) \subset \bigcap_{1 \leqslant q < 2} W^{1,q}(\mathbb{D})$ that homeomorphically maps the punctured disk $\mathbb{D} \setminus \{0\}$ onto the annulus $1 < |w| < R$.

Note that the integrability condition (20.25) implies that $\mathcal{A}$ is slightly less than sublinear. For instance, the function defined for large values of t by

$$\mathcal{A}(t) = \frac{t}{(\log t)^{1+\varepsilon}}$$

satisfies (20.25) for all $\varepsilon > 0$ but not for $\varepsilon = 0$. Condition 3 in the theorem is no accident. We shall show in the next sections that if $e^K \in L^p(\mathbb{D})$ for some $p > 0$, then there is a homeomorphic (hence continuous) solution in $W^{1,q}(\mathbb{D})$ for all $q < 2$.

Proof of Theorem 20.3.1. Given a function $\mathcal{A}$ satisfying (20.25), we define a function $K(s)$, $0 < s \leqslant 1$, via the functional relation

$$e^{\mathcal{A}(K(s))} = \frac{e}{s}, \quad 0 < s \leqslant 1 \tag{20.28}$$

As the map $K \mapsto e^{\mathcal{A}(K)}$ is monotone increasing, we see that the solution $K(s)$ is well defined and decreasing from ∞ to 1, $K(1) = 1$.

We compute the integral $\int_0^1 \frac{ds}{sK(s)}$ by making the substitution $t = K(s)$. From (20.28), we find that $\mathcal{A}'(t)\,dt = -ds/s$, so that

$$\int_0^1 \frac{ds}{sK(s)} = \int_1^\infty \mathcal{A}'(t)\frac{dt}{t} = \left.\frac{\mathcal{A}(t)}{t}\right|_1^\infty + \int_1^\infty \mathcal{A}(t)\frac{dt}{t^2}$$

$$= -1 + \int_1^\infty \mathcal{A}(t)\frac{dt}{t^2}$$

The last identity is justified by

$$\frac{\mathcal{A}(R)}{R} \leqslant 2\int_R^{2R} \mathcal{A}(t)\frac{dt}{t^2} \to 0 \quad \text{as } R \to \infty \tag{20.29}$$

Now set

$$f(z) = \frac{z}{|z|}\rho(|z|), \tag{20.30}$$

where

$$\rho(t) = \exp\left(\int_0^t \frac{ds}{sK(s)}\right), \quad 0 \leqslant t < 1$$

Then f is a radial stretching, defined on $\mathbb{D} \setminus \{0\}$, and as in (2.40) and (2.41) we compute (away from the origin)

$$\dot{\rho}(t) = \frac{\rho(t)}{tK(t)}$$

$$|Df(z)| = \frac{\rho(|z|)}{|z|}$$

$$J(z,f) = \frac{\rho(|z|)\dot{\rho}(|z|)}{|z|} = \frac{1}{K(|z|)} \frac{\rho^2(|z|)}{|z|^2}$$

$$K(z,f) = \frac{\rho}{|z|\dot{\rho}} = K(|z|)$$

$$\mu_f(z) = -\frac{z}{\bar{z}} \frac{K(z)-1}{K(z)+1}$$

It is clear by construction that $\mathcal{A} \circ K$ is exponentially integrable. Indeed,

$$\int_{\mathbb{D}} e^{\mathcal{A}(K(z,f))} = 2\pi \int_0^1 e^{\mathcal{A}(K(s))} \, s \, ds = 2\pi e$$

Notice here that the function f is not continuous at the origin. However, near the origin we have $\rho(|z|) \approx 1$, and therefore the formula for $|Df|$ gives us the bounds

$$\frac{1}{|z|} \leqslant |Df(z)| \leqslant \frac{1}{|z|} \exp\left(\int_1^\infty \mathcal{A}(t) \frac{dt}{t^2}\right),$$

which uncovers our bounded solution in weak-$W^{1,2}(\mathbb{D})$.

We also observe that $w = f(z)$ is a C^∞-diffeomorphism of the punctured disk $\mathbb{D} \setminus \{0\}$ onto the annulus $\{w : 1 < |w| < \rho(1)\}$; and we have "cavitation" at the origin.

Finally, we want to show that the only solutions in $W^{1,1}_{loc}(\mathbb{D})$ that are continuous at the origin are the constants. To this end let $\varepsilon > 0$ and note that $\rho(\varepsilon) \to 1$ as $\varepsilon \to 0$. Let g be a $W^{1,1}_{loc}(\mathbb{D})$-solution to (20.27) with $\mu = \mu_f$ as above and which is continuous at 0. Set

$$\varphi(z) = g \circ f^{-1} : \{\rho(\varepsilon) < |w| < \rho(1)\} \to \mathbb{C}$$

As $\varphi \in W^{1,1}(\{\rho(\varepsilon) < |w| < \rho(1)\})$ and f is a C^∞-smooth diffeomorphism away from the origin, a simple computation shows $\bar{\partial}\varphi = 0$, and the Weyl's Lemma A.6.10 gives φ holomorphic. This is true for every $\varepsilon > 0$, and thus we are provided with an analytic function $\varphi : \{1 < |w| < \rho(1)\} \to \mathbb{C}$, which has the limit $g(0)$ as $|w| \to 1$.

This argument shows that both of the functions φ and g are constant, and hence completes the proof. □

20.4 Distortion in the Exponential Class

When aiming at a generalization of the measurable Riemann mapping theorem 5.3.2, valid in the degenerate case, the most natural candidate for an interesting and cohesive theory is the class of mappings with exponentially integrable distortion. Thus we should be prepared to assume

$$e^K \in L^p_{loc}(\mathbb{C}) \quad \text{for some } 0 < p < \infty \tag{20.31}$$

As before, $K(z) = \frac{1+|\mu(z)|}{1-|\mu(z)|}$ and $1 \leqslant K(z) < \infty$ almost everywhere, with $K(z) = 1$ outside the support of μ.

Theorem 20.3.1 provides strong motivation for the study of this class, and indeed we will show that under the assumption (20.31) one always has a principal solution f, as well as counterparts to the other basic properties such as the Stoilow factorization. We will also describe the optimal regularity properties of the solutions f discovered recently in [25]. In brief, our purpose here is to systematically uncover the properties of the class of mappings of finite distortion for which we have (20.31).

As we have repeatedly stressed and as is again evident from the previous sections, the Sobolev space $W^{1,2}_{loc}(\mathbb{C})$ is the most natural space in which to look for a solution to the Beltrami equation. Theorems 20.1.1 and 20.2.1 promote the importance of $W^{1,2}$-regularity even further. And above all, the $W^{1,2}$-regularity automatically quarantees that the Jacobian determinant $J(z, f) = |f_z|^2 - |f_{\bar{z}}|^2$ is locally integrable, a property one cannot dispense with.

On the other hand, if we are given a solution to the Beltrami equation $f_{\bar{z}} = \mu f_z$, even in the degenerate case with $|\mu(z)| < 1$ almost everywhere, we will always have $J(z, f) \geqslant 0$ by virtue of the distortion inequality. Theorem 19.3.1 shows that there is a chance of working with regularity slightly below $W^{1,2}$.

To uncover and explain the basic phenomena, we will begin with an explicit example concerning the regularity of mappings with exponentially integrable dilatation, close in spirit to that of the previous section. However, this time we will have continuity. This example serves as a guide for the results one might obtain in the exponential class. In particular, we wish to understand what the expected Sobolev regularity is for solutions to $f_{\bar{z}} = \mu f_z$ when (20.31) holds.

20.4.1 Example: Regularity in Exponential Distortion

Given a positive number θ, we set

$$f(z) = f_\theta(z) = \begin{cases} \frac{z}{|z|} \left(1 + \frac{1}{\theta} \log \frac{1}{|z|}\right)^{-\theta}, & |z| \leqslant 1 \\ z, & |z| > 1 \end{cases} \tag{20.32}$$

From the computations in Section 2.6, with $\rho(t) = (1 + \frac{1}{\theta} \log \frac{1}{t})^{-\theta}$, we find the Beltrami coefficient of f_θ to be

$$\mu = \mu_\theta(z) = \frac{z}{\bar{z}} \frac{\log|z|}{2\theta - \log|z|} \chi_\mathbb{D}(z),$$

544   CHAPTER 20. DEGENERATE BELTRAMI EQUATIONS

where as usual $\chi_\mathbb{D}$ denotes the characteristic function of the unit disk. This shows f to be a principal solution to the Beltrami equation

$$(f_\theta)_{\bar{z}} = \mu_\theta \, (f_\theta)_z$$

The distortion function is finite except at the origin. We have

$$K(z) = K_\theta(z) = \left(1 + \frac{1}{\theta} \log \frac{1}{|z|}\right) \chi_\mathbb{D}(z),$$

and the norm of the differential is given by

$$|Df_\theta(z)| = \begin{cases} \frac{1}{|z|} \left(1 + \frac{1}{\theta} \log \frac{1}{|z|}\right)^{-\theta}, & |z| \leq 1 \\ 1, & |z| > 1 \end{cases} \qquad (20.33)$$

From this we see that

The function $e^{K(z)} \in L^p_{loc}(\mathbb{C})$ if and only if $p < 2\theta$.

Indeed,

$$\int_\mathbb{D} e^{pK(z)} \, dz = \frac{2\theta e^p \pi}{2\theta - p}, \qquad p < 2\theta$$

On the other hand, we find that f_θ belongs to the natural Sobolev space; that is,

$$f_\theta \in W^{1,2}_{loc}(\mathbb{C})$$

if and only if $1 < 2\theta$. Indeed, we have

$$\int_\mathbb{D} |Df_\theta(z)|^2 \, dz = \frac{2\theta \pi}{2\theta - 1}$$

This leads us to the following observation.

In order for a principal solution to exist in the natural Sobolev space $W^{1,2}_{loc}(\mathbb{C})$, it is necessary that the exponent p in (20.31) be large, at least $p \geq 1$.

Indeed, one of the main goals of this chapter is to show that exponential integrabilty (20.31) with any $p > 1$ suffices for the $W^{1,2}_{loc}(\mathbb{C})$-regularity.

Remark. In Theorem 20.2.1 we discussed mappings of integrable distortion and the exceptional sets E for the $W^{1,2}_{loc}$-regularity of the principal solutions. If above $2\theta < 1 < 4\theta$, then $F(z) = f_\theta(z)^2 \in W^{1,2}(\mathbb{D})$ and F has integrable distortion $K_\theta(z) \in L^1(\mathbb{D})$. However, the corresponding principal solution $f_\theta \in W^{1,2}_{loc}(\mathbb{D} \setminus \{0\})$ only. In particular, in this case the exceptional set $E = \{0\}$ is not empty.

20.4.2 Beltrami Operators for Degenerate Equations

The results from Chapter 14 show that the use of Beltrami operators gives a systematic and flexible approach to uniformly elliptic PDEs and systems in the plane. Therefore it must be expected that these operators also play a significant a role in the degenerate case. On the other hand, as we saw in Chapter 14, Theorem 14.0.4, the general Beltrami operator $\mathbf{I} - \mu \mathcal{S}$ is invertible in $L^p(\mathbb{C})$ only for $1 + \|\mu\|_\infty < p < 1 + 1/\|\mu\|_\infty$. Thus in the degenerate case with $\|\mu\|_\infty = 1$, invertibility may occur only in L^2, and even there one must be careful in finding the correct formulation.

To advance these ideas we first look for a BMO-characterization of functions that satisfy the condition of exponential integrability (20.31).

Theorem 20.4.1. *A necessary and sufficient condition that a function $K(z)$ defined on a domain $\Omega \subset \mathbb{C}$ should satisfy*

$$|K(z)| \leq K^*(z) \quad \text{for almost every } z \in \Omega \tag{20.34}$$

for some $K^* \in BMO(\mathbb{C})$ is that the integral

$$\int_\Omega \frac{e^{\alpha |K(z)|}\, dz}{(1 + |z|)^3} < \infty \tag{20.35}$$

for some numbers $\alpha > 0$.

Of course, the denominator term in the integrand in (20.35) is relevant only if Ω is unbounded.

Proof. The exponential integrability of any function $u \in BMO$ follows from the John-Nirenberg Lemma A.8.3. The global exponential integrability of any function $K^* \in BMO(\mathbb{C})$ as in (20.35) follows from the global version of this theorem, see Theorem 4.6.1.

On the other hand, if (20.35) holds, we may construct a BMO-majorant $K^*(z)$ by using the Hardy-Littlewood maximal operator. We regard $K = K(z)$ as being defined everywhere in $\mathbb{C}$ by defining it to be zero outside of Ω.

Assuming $e^{\alpha K}(1 + |z|)^{-3} \in L^1(\mathbb{C})$, the Hardy-Littlewood maximal function $\mathcal{M}[e^{\alpha K}(1 + |z|)^{-3}]$ lies in weak-$L^1(\mathbb{C})$ and so is finite almost everywhere. Theorem 4.4.9 therefore applies, and we see that the logarithm of this maximal function belongs to $BMO(\mathbb{C})$. We therefore define

$$K^*(z) = \frac{1}{\alpha} \log \mathcal{M}[e^{\alpha K}(1 + |z|)^{-3}] + \frac{3}{\alpha} \log(1 + |z|) \tag{20.36}$$

which we have observed belongs to $BMO(\mathbb{C})$ and has

$$\|K^*\|_* \leq \frac{C}{\alpha},$$

where C is an absolute constant. Moreover,

$$K^*(z) \geq \frac{1}{\alpha} \log\left(e^{\alpha K}(1 + |z|)^{-3}\right) + \frac{3}{\alpha} \log(1 + |z|) = K(z)$$

almost everywhere in Ω. □

We will call any function $K^* \in BMO$ satisfying (20.34) a *BMO-majorant* of K. In particular, by Theorem 20.4.1, a function admits a BMO-majorant if and only if it is exponentially integrable.

Corollary 20.4.2. *The BMO-majorant of K can be chosen to satisfy*

$$\|K^*\|_{BMO} \leqslant \frac{C_2}{\alpha}$$

with C_2 a universal constant, and α is such that we have (20.35).

Proof. The estimate follows immediately from (20.36) and Theorem 4.4.9. □

With this terminology we can give, following [180, 191], the correct formulation of the invertibility of the Beltrami operators $\mathbf{I} - \mu \mathcal{S}$ in the degenerate setting.

Theorem 20.4.3. *There exists a number $p_0 \geqslant 1$ with the following property: Suppose that the distortion function $K(z) = \frac{1+|\mu(z)|}{1-|\mu(z)|}$ is exponentially integrable, so that*

$$\int_{\mathbb{D}} e^{pK(z)}\, dz < \infty \quad \text{for some } p > p_0 \geqslant 1, \tag{20.37}$$

and that $\mu(z) = 0$ for $|z| > 1$. Then the equation

$$\omega - \mu \mathcal{S}\omega = h \tag{20.38}$$

has a unique solution $\omega \in L^2(\mathbb{C})$ for every h such that $K^*(z)h(z) \in L^2(\mathbb{C})$, where K^* is a BMO-majorant of K as in (20.36). Furthermore, for such functions h,

$$\|\omega\|_2 = \|(\mathbf{I} - \mu\mathcal{S})^{-1} h\|_2 \leqslant 4\|K^* h\|_2 \tag{20.39}$$

The regularity results from Section 20.4.6 strongly suggest that $p_0 = 1$ should suffice. For the proof of Theorem 20.4.3 we need to understand the action of BMO functions on Jacobian determinants. Namely, if $\omega = \phi_{\bar{z}}$, then the assumption (20.38) may be formulated as

$$|D\phi(z)|^2 \leqslant 2K^*(z)J(z,\phi) + 4\left(K^*(z)\right)^2 |h(z)|^2;$$

see the forthcoming inequality (20.42). Hence we have to estimate the integral of $K^*(z)J(z,\phi)$, and for this we recall Fefferman's duality theorem [125], which asserts $(H^1)^* = BMO$. Thus what we need is that the Jacobian $J(z,\phi)$ should belong to the Hardy space H^1. This is guaranteed by the famous theorem of R. Coifman, P. Lions, Y. Meyer and S. Semmes that (in $\mathbb{R}^n$) the Jacobian of a $W^{1,n}$-mapping ϕ is contained in the Hardy space $H^1(\mathbb{R}^n)$ [95]. We see that Jacobians of $W^{1,2}$-mappings are the natural class for us to work with.

20.4. EXPONENTIAL DISTORTION

We will use the commutator results from Section 4.6.5 to present a refined approach to these crucial BMO bounds. Bounded functions will suffice for our purposes, and with them we avoid the subtle questions of integrability. Given a proper interpretation, though, one could introduce general BMO functions b here.

Theorem 20.4.4. *Let $b \in L^\infty(\mathbb{C})$ and assume that $\phi \in W^{1,1}_{loc}(\mathbb{C})$ with derivative $D\phi \in L^2(\mathbb{C})$. Then*

$$\int_\mathbb{C} b(z) J(z, \phi) \, dz \leqslant C_1 \|b\|_{BMO(\mathbb{C})} \int_\mathbb{C} |D\phi|^2 \, dz \qquad (20.40)$$

for an absolute constant C_1.

Proof. Let $\omega = \phi_{\bar z}$. According to our assumptions, $\omega, \mathcal{S}\omega \in L^2(\mathbb{C})$. Therefore the integral

$$\int_\mathbb{C} b(z) J(x, \phi) dz = \int_\mathbb{C} b \left(|S\omega|^2 - |\omega|^2 \right) dz$$

converges absolutely. The Beurling transform S is symmetric, and $S(\overline{S\omega}) = \overline{\omega}$ since $\overline{S}$ is the L^2-adjoint of S; see (4.22). We hence arrive at the estimate of the commutator,

$$\int_\mathbb{C} b \left(|S\omega|^2 - |\omega|^2 \right) dz = \int_\mathbb{C} \omega (Sb - bS)(\overline{S\omega}) dz$$
$$\leqslant \|\omega\|_{L^2(\mathbb{C})} \|(Sb - bS)(\overline{S\omega})\|_{L^2(\mathbb{C})} \leqslant C_1 \|\omega\|^2_{L^2(\mathbb{C})} \|b\|_{BMO(\mathbb{C})},$$

where the last estimate is based on Theorem 4.6.13. $\square$

After these preparations we now return to the invertibility properties of the Beltrami operators.

Proof of Theorem 20.4.3. We go directly to the main point of the theorem, the proof of (20.39) under the assumption that the solution $\omega \in L^2(\mathbb{C})$ exists. We start with the elementary inequality

$$(|u| + |v|)^2 \leqslant 2K(|u|^2 - |v|^2) + 4K^2 |v - w|^2 \qquad (20.41)$$

valid for all complex numbers u, v, w such that $|w| \leqslant \frac{K-1}{K+1}|u|$ and $K \geqslant 1$. Indeed, $|v - w| \geqslant |v| - |w| \geqslant |v| - \frac{K-1}{K+1}|u|$, or equivalently,

$$|u| + |v| \leqslant K(|u| - |v|) + (K + 1)|v - w|$$

We multiply both sides of this inequality by $2(|u| + |v|)$ to find

$$\begin{aligned} 2(|u| + |v|)^2 &\leqslant 2K(|u|^2 - |v|^2) + 2(K+1)|v - w|(|u| + |v|) \\ &\leqslant 2K(|u|^2 - |v|^2) + (K+1)^2 |v - w|^2 + (|u| + |v|)^2, \end{aligned}$$

from which inequality (20.41) is straightforward.

Now let h be a measurable function, assume $\omega \in L^2(\mathbb{C})$ is a solution to (20.38) and apply the inequality (20.41) pointwise with

$$u = \mathcal{S}\omega(z), \qquad v = \omega(z), \qquad w = (\mu\mathcal{S}\omega)(z), \qquad K = K^*(z),$$

where $K^*(z)$ is the BMO-majorant of the distortion function $K(z)$ as defined in (20.36).

We end up with the estimate

$$(|\mathcal{S}\omega| + |\omega|)^2 \leqslant 2K^*(|\mathcal{S}\omega|^2 - |\omega|^2) + 4(K^*)^2|h|^2 \tag{20.42}$$

Upon integration one expects strong cancellation in the first term on the right hand side. However, there is the technical complication that the integral of this term may not be absolutely convergent. For this reason, prior to integration we divide both sides by the factor $1 + \delta K^*(z)$, where the positive parameter δ is let go to 0 in the end. In this way we are reduced to considering the inequality

$$\frac{|\omega|^2 + |\mathcal{S}\omega|^2}{1 + \delta K^*} \leqslant 2\mathcal{K}(|\mathcal{S}\omega|^2 - |\omega|^2) + 4(K^*)^2|h|^2 \tag{20.43}$$

The new factor $\mathcal{K} = \frac{K^*}{1+\delta K^*}$ is bounded. Moreover, its BMO-norm has an upper bound that is independent of δ,

$$\|\mathcal{K}\|_{BMO} \leqslant 2\|K^*\|_{BMO} \leqslant \frac{2C_2}{\alpha} \tag{20.44}$$

Here we have used Lemma 4.4.8 and Corollary 20.4.2.

Integrating (20.43) is now justified. If ϕ is the Cauchy transform of ω, Theorem 20.4.4 with $b = 2\mathcal{K}$ applies, and hence

$$\int_{\mathbb{C}} \frac{|\omega|^2 + |\mathcal{S}\omega|^2}{1 + \delta K^*} dz \leqslant \frac{4C_1 C_2}{\alpha} \int_{\mathbb{C}} (|\omega|^2 + |\mathcal{S}\omega|^2) \, dz + \int_{\mathbb{C}} 4(K^*)^2|h|^2 \, dz \tag{20.45}$$

Using Fatou's lemma allows us to let $\delta \to 0$ to find, via the identity $\int_{\mathbb{C}} |\mathcal{S}\omega|^2 = \int_{\mathbb{C}} |\omega|^2$, that

$$2\int_{\mathbb{C}} |\omega|^2 \, dz \leqslant \frac{8C_1 C_2}{\alpha} \int_{\mathbb{C}} |\omega|^2 \, dz + \int_{\mathbb{C}} 4(K^*)^2|h|^2 \, dz$$

for every measurable function h and every solution $\omega \in L^2(\mathbb{C})$ to (20.38).

We now choose the parameter α to be greater than $8C_1 C_2$. In view of Theorem 20.4.1 this also determines the critical exponent; that is,

$$p_0 = 8C_1 C_2$$

satisfies the requirements of the theorem. The reader may note that this value for the critical exponent depends only on the constants arising in Theorem 20.4.4 and Corollary 20.4.2, or Theorems 4.4.9 and 4.6.13, respectively.

With this choice of α we obtain the inequality (20.39). The inequality of course proves the uniqueness claim as well.

20.4. EXPONENTIAL DISTORTION

We are left with the problem of proving the existence of an L^2-solution to (20.38) whenever $K^*h \in L^2(\mathbb{C})$. We will approach this via an approximation. As in (20.17), let us define a "good" approximation of μ by the sequence

$$\mu_m(z) = \begin{cases} \mu(z) & \text{if } |\mu(z)| \leqslant 1 - \frac{1}{m} \\ (1 - \frac{1}{m}) \frac{\mu(z)}{|\mu(z)|} & \text{otherwise,} \end{cases} \quad (20.46)$$

defined for $m = 2, 3, \ldots$. Of course, $|\mu_m(z)| \leqslant 1 - \frac{1}{m}$. We also have the bound

$$|\mu_m(z)| \leqslant \frac{K(z) - 1}{K(z) + 1},$$

which is independent of m.

Since $\|\mu_m\|_\infty < 1$ strictly and $\mathcal{S}$ is an L^2-isometry, the operator $\mathbf{I} - \mu_m \mathcal{S}$ is invertible in $L^2(\mathbb{C})$. The function $|h| \leqslant K^*|h| \in L^2(\mathbb{C})$, and thus we can solve the Beltrami equation

$$\omega_m - \mu_m \mathcal{S}\omega_m = h$$

for $\omega_m \in L^2(\mathbb{C})$. Inequality (20.39) applies to ω_m as well, giving the uniform L^2-bound

$$\|\omega_m\|_2 \leqslant 4\|K^*h\|_2 \quad (20.47)$$

There is no loss of generality in assuming that in fact ω_m converges weakly to an element $\omega \in L^2(\mathbb{C})$. Since the operator $\mathcal{S}$ is bounded on $L^2(\mathbb{C})$, the sequence $\mathcal{S}\omega_m$ will converge weakly to $\mathcal{S}\omega$.

In addition,

$$\mu_m(z) \to \mu(z) \quad \text{for almost every } z \in \Omega,$$

and $\|\mathcal{S}\omega_m\|_2 = \|\omega_m\|_2 \leqslant 4\|K^*h\|_2 < \infty$ for every m. This quickly implies that

$$\omega_m - \mu_m \mathcal{S}\omega_m \to \omega - \mu \mathcal{S}\omega \quad (20.48)$$

weakly in the space $L^2(\mathbb{C})$ and yields

$$\omega - \mu \mathcal{S}\omega = h,$$

which completes the proof of the theorem. □

20.4.3 Decay of the Neumann Series

The previous section gave one method for establishing the L^2-bounds unavoidable in solving the degenerate equations. Here we discuss an alternative route to the L^2-bounds, further developing the approach of G. David [103]. Recall from Chapter 5, and Section 5.1 in particular, that for the uniformly elliptic case the Beltrami equation was initially solved by using the Neumann series (5.3),

$$(\mathbf{I} - \mu\mathcal{S})^{-1}\mu = \mu + \mu\mathcal{S}\mu + \mu\mathcal{S}\mu\mathcal{S}\mu + \mu\mathcal{S}\mu\mathcal{S}\mu\mathcal{S}\mu + \cdots \quad (20.49)$$

It is of course natural to study how far this method carries over to the present situation of degenerate equations. The main question here is the rate of decay of the L^2-norms of the terms in the corresponding series.

The next result from [25] gives the optimal answer.

Theorem 20.4.5. *Suppose $|\mu(z)| < 1$ almost everywhere, with $\mu(z) \equiv 0$ for $|z| > 1$. If the distortion function $K(z) = \frac{1+|\mu(z)|}{1-|\mu(z)|}$ satisfies*

$$e^K \in L^p(\mathbb{D}), \qquad p > 0, \tag{20.50}$$

then we have for every $0 < \beta < p$,

$$\int_{\mathbb{C}} |(\mu \mathcal{S})^n \mu|^2 \leqslant C_0 \, n^{-\beta}, \qquad n \in \mathbb{N}, \tag{20.51}$$

where $C_0 \leqslant C_{p,\beta} \cdot \int_{\mathbb{D}} e^{pK}$ with $C_{p,\beta}$ depending only on β and p.

Hereafter $(\mu \mathcal{S})^n$ stands for the nth iterate of the operator $\mu \mathcal{S}$, that is, $(\mu \mathcal{S})^n = (\mu \mathcal{S}) \circ \cdots \circ (\mu \mathcal{S})$.

Proof. Since

$$K(z) + 1 = \frac{2}{1 - |\mu(z)|},$$

with Chebychev's inequality we have the measure estimates

$$\left|\left\{z \in \mathbb{D} : |\mu(z)| \geqslant 1 - \frac{1}{t}\right\}\right| \leqslant e^{-2pt} \int_{\mathbb{D}} e^{p(K+1)} = Ce^{-2pt} \tag{20.52}$$

for each $t > 1$. With this control on the Beltrami coefficient, we will iteratively estimate the terms $(\mu \mathcal{S})^n \mu$ of the Neumann series. For this purpose we first fix the parameter $0 < \beta < p$, and then for each $n \in \mathbb{N}$ divide the unit disk into "bad" and "good" points,

$$B_n = \left\{z \in \mathbb{D} : |\mu(z)| > 1 - \frac{\beta}{2n + \beta}\right\} \quad \text{and} \quad G_n = \mathbb{D} \setminus B_n$$

For $n \in \mathbb{N}$ the above bounds on area are now

$$|B_n| \leqslant C_1 \, e^{-4n\, p/\beta}, \qquad C_1 \leqslant \int_{\mathbb{D}} e^{pK} \tag{20.53}$$

Next, let us consider the terms $\psi_n = (\mu \mathcal{S})^n \mu$ of the Neumann series (20.49), obtained inductively by

$$\psi_n = \mu \mathcal{S}(\psi_{n-1}), \qquad \psi_0 = \mu \tag{20.54}$$

It is helpful to first look at the following auxiliary terms:

$$g_n = \chi_{G_n} \mu \mathcal{S}(g_{n-1}), \qquad g_0 = \mu; \tag{20.55}$$

20.4. EXPONENTIAL DISTORTION

that is, at each iterative step we restrict μ to the corresponding good part of the disk.

The terms g_n are easy to estimate,

$$\|g_n\|_{L^2(\mathbb{C})}^2 = \int_{G_n} |\mu \mathcal{S}(g_{n-1})|^2 \leqslant \left(1 - \frac{\beta}{2n+\beta}\right)^2 \|g_{n-1}\|_{L^2(\mathbb{C})}^2$$

and therefore

$$\|g_n\|_{L^2(\mathbb{C})} \leqslant \prod_{j=1}^n \left(1 - \frac{\beta}{2j+\beta}\right) \|\mu\|_{L^2(\mathbb{C})} \leqslant \left(1+\frac{\beta}{2}\right)^{\beta/2} n^{-\beta/2}$$

In other words, the terms g_n have the correct decay. Thus it suffices only to prove that the differences $\psi_n - g_n$ decay at the same rate.

For this purpose note that

$$\psi_n - g_n = \chi_{G_n} \mu \mathcal{S}(\psi_{n-1} - g_{n-1}) + \chi_{B_n} \mu \mathcal{S}(\psi_{n-1}),$$

which yields the estimate

$$\|\psi_n - g_n\|_{L^2(\mathbb{C})}^2 \leqslant \left(1 - \frac{\beta}{2n+\beta}\right)^2 \|\psi_{n-1} - g_{n-1}\|_{L^2(\mathbb{C})}^2 + R(n), \qquad (20.56)$$

where

$$R(n) = \|\chi_{B_n} \mu \mathcal{S}(\psi_{n-1})\|_{L^2(\mathbb{C})}^2 = \int_{B_n} |(\mu \mathcal{S})^n \mu|^2 \qquad (20.57)$$

The point of the proof is to bound this last integral term. A natural possibility is to apply Hölder's inequality and estimate the norms $\|(\mu \mathcal{S})^n \mu\|_p$ for $p > 2$. On the other hand, the best bounds for the norms known at the moment are all based on the area distortion results from Section 13.1. To reveal more clearly the heart of the argument, we will therefore use the area distortion directly. This can be done with a holomorphic representation as follows.

Recall from Section 5.7 that the principal solution $f = f^\lambda$ to the equation

$$f_{\bar{z}}(z) = \lambda \mu(z) f_z(z), \qquad \lambda \in \mathbb{D},$$

depends holomorphically on the parameter $\lambda \in \mathbb{D}$. This fact is based on the power series representation for the derivative,

$$f_{\bar{z}}^\lambda = \lambda \mu + \lambda^2 \mu \mathcal{S} \mu + \cdots + \lambda^n (\mu \mathcal{S})^{n-1} \mu + \cdots$$

The series converges absolutely in $L^2(\mathbb{C})$ since $\|\lambda \mu\|_\infty \leqslant |\lambda| < 1$ and $\mathcal{S}$ is an L^2-isometry.

The $L^2(\mathbb{C})$-valued holomorphic function $\lambda \to f_{\bar{z}}^\lambda$ can as well be represented by the Cauchy integral. In fact, in this way we obtain for any measurable set $E \subset \mathbb{D}$ the following integral representation,

$$\chi_E (\mu \mathcal{S})^n \mu = \frac{1}{2\pi i} \int_{|\lambda|=\rho} \frac{1}{\lambda^{n+1}} (f^\lambda)_{\bar{z}} \chi_E \, d\lambda, \qquad E \subset \mathbb{D}, \qquad (20.58)$$

valid for any $0 < \rho < 1$. Observe from the representation that what one now needs is an estimate for the norms

$$\|(f^\lambda)_{\bar z}\, \chi_E\|^2_{L^2} = \int_E |(f^\lambda)_{\bar z}|^2 \leqslant \frac{|\lambda|^2}{1-|\lambda|^2} \int_E J(z, f^\lambda) = \frac{|\lambda|^2}{1-|\lambda|^2} |f^\lambda(E)|$$

It is here that we see the need for control on the quasiconformal distortion of area. We have from Theorem 13.1.4

$$|f^\lambda(E)| \leqslant \pi M |E|^{1/M}, \qquad |\lambda| = \frac{M-1}{M+1}$$

If we now take $\rho = \frac{M-1}{M+1}$ in (20.58) and combine it with the above estimates, we end up with

$$\|\chi_E (\mu \mathcal{S})^n \mu\|_2 \leqslant \frac{\sqrt{\pi}}{2} \frac{(M+1)^n}{(M-1)^{n-1}} |E|^{1/2M}$$
$$\leqslant M e^{2n/(M-1)} |E|^{1/2M} \qquad (20.59)$$

This bound is valid for every $M > 1$ and for any Beltrami coefficient with $|\mu| \leqslant \chi_{\mathbb{D}}$ almost everywhere.

For later purposes we recall the similar power series representation of the ∂_z-derivative,

$$f^\lambda_z - 1 = \mathcal{S} f^\lambda_{\bar z} = \lambda \mathcal{S} \mu + \lambda^2 \mathcal{S}(\mu \mathcal{S}) + \cdots + \lambda^n \mathcal{S}(\mu \mathcal{S})^{n-1} \mu + \cdots$$

In much the same way we obtain

$$\|\chi_E \mathcal{S}(\mu \mathcal{S})^n \mu\|_2 \leqslant \sqrt{\pi} \left(\frac{M+1}{M-1}\right)^n \frac{(M+1)^2}{2M} |E|^{1/2M}$$
$$\leqslant 4 M e^{2n/(M-1)} |E|^{1/2M}, \qquad M > 1 \qquad (20.60)$$

Let us then return to estimating the $L^2(\mathbb{C})$-norms of the terms $\psi_n - g_n$. Unwinding the iteration in (20.56) gives

$$\|\psi_n - g_n\|^2_{L^2(\mathbb{C})} \leqslant \sum_{j=1}^n R(j) \prod_{k=j+1}^n \left(1 - \frac{\beta}{2j+\beta}\right)^2$$
$$\leqslant \left(1 + \frac{\beta}{2}\right)^\beta n^{-\beta} \sum_{j=1}^n j^\beta R(j) \qquad (20.61)$$

It remains to show that the reminder term $R(n)$ decays exponentially. But here we may use the bound (20.59) for the set $E = B_n$. This has area $|B_n| \leqslant e^{-4n\rho/\beta} \int_{\mathbb{D}} e^{pK}$, and we obtain for every $M > 1$,

$$R(n) \leqslant M^2 e^{4n/(M-1)} |B_n|^{1/M} \leqslant M^2 \left(\int_{\mathbb{D}} e^{pK}\right)^{1/M} e^{\frac{4n}{M}\left(\frac{M}{M-1} - \frac{p}{\beta}\right)}$$

20.4. EXPONENTIAL DISTORTION

Given $\beta < p$, we choose $M = \frac{p+\beta}{p-\beta}$ so that

$$R(n) \leqslant C e^{-\delta n} \quad \text{for } n \in \mathbb{N},$$

where $C = (\frac{p+\beta}{p-\beta})^2 \int_{\mathbb{D}} e^{pK}$ and $\delta = (\frac{p}{\beta} - 1)^2 > 0$.

With the exponential decay, (20.61) gives

$$\|\psi_n - g_n\|_{L^2(\mathbb{C})}^2 \leqslant C \left(1 + \frac{\beta}{2}\right)^\beta n^{-\beta} \sum_{j=1}^\infty j^\beta e^{-\delta j} \leqslant C_{p,\beta}\, n^{-\beta} \int_{\mathbb{D}} e^{pK}$$

In particular, we have obtained the required bounds (20.51),

$$\int_{\mathbb{C}} |(\mu \mathcal{S})^n \mu|^2 \leqslant 2\|g_n\|_{L^2(\mathbb{C})}^2 + 2\|\psi_n - g_n\|_{L^2(\mathbb{C})}^2 \leqslant C_0\, n^{-\beta}$$

The proof is complete. $\square$

According to Theorem 20.4.5, the terms of the Neumann series decay with the rate $\|(\mu\mathcal{S})^n\mu\|_2 \leqslant C\, n^{-\beta/2}$ for any $\beta < p$. If the decay were a little better, say of the order $n^{-\beta}$, then for any $p > 1$ the series would be norm-convergent in $L^2(\mathbb{C})$. This would immediately show $p_0 = 1$ as the critical exponent for the $W^{1,2}$-regularity. However, [25] provides examples showing that the order of decay from Theorem 20.4.5 cannot be improved. Hence further means are required for optimal regularity; see Section 20.4.6.

If μ and $0 < \beta < p$ are as in Theorem 20.4.5, then at least for $\beta > 2$ the series

$$\sigma_\mu := \sum_{n=0}^\infty (\mu\mathcal{S})^n \mu \tag{20.62}$$

is absolutely convergent in $L^2(\mathbb{C})$. In this case we have the next auxiliary result, which the reader may recognize as a distortion of area for mappings in the exponential class. The result here is rather weak but is a necessary step toward the optimal measure distortion bounds, which will be established in Section 20.4.6.

Corollary 20.4.6. *If $\beta > 2$, then*

$$\int_E \left(|\sigma_\mu|^2 + |\mathcal{S}\sigma_\mu|^2\right) \leqslant C \log^{2-\beta}\left(e + \frac{1}{|E|}\right), \qquad E \subset \mathbb{D},$$

where $C = C_0\, C_\beta < \infty$ with C_β depending only on β.

Proof. Theorem 20.4.5 gives us two different ways to estimate the terms. First,

$$\|\chi_E(\mu\mathcal{S})^n\mu\|_2 + \|\chi_E \mathcal{S}(\mu\mathcal{S})^n\mu\|_2 \leqslant 2\|(\mu\mathcal{S})^n\mu\|_2 \leqslant 2\sqrt{C_0}\, n^{-\beta/2}$$

Summing this up gives

$$\sum_{n=m+1}^\infty \|\chi_E(\mu\mathcal{S})^n\mu\|_2 + \|\chi_E\mathcal{S}(\mu\mathcal{S})^n\mu\|_2 \leqslant \frac{4\sqrt{C_0}}{\beta-2}\, m^{1-\beta/2}, \qquad m \in \mathbb{N}$$

Second, let us choose, say, $M = 3$ in (20.59) and (20.60). Then for every $n \in \mathbb{N}$ we have the estimates $\|\chi_E (\mu\mathcal{S})^n \mu\|_2 + \|\chi_E \mathcal{S}(\mu\mathcal{S})^n \mu\|_2 \leqslant 15 \cdot e^n |E|^{1/6}$. This in turn we can sum to

$$\sum_{n=0}^{m} \|\chi_E(\mu\mathcal{S})^n\mu\|_2 + \|\chi_E \mathcal{S}(\mu\mathcal{S})^n\mu\|_2 \leqslant 15 \cdot e^{m+1}|E|^{1/6}$$

Combining, we arrive at

$$\|\chi_E \sigma_\mu\|_2 + \|\chi_E \mathcal{S}\sigma_\mu\|_2 \leqslant \frac{4\sqrt{C_0}}{\beta - 2} m^{1-\beta/2} + 15 \cdot e^{m+1}|E|^{1/6} \qquad (20.63)$$

The bound holds for any integer $m \in \mathbb{N}$ but is of course at its best when the two terms on the right hand side are roughly equal. We let

$$\frac{1}{10} \log\left(e + \frac{1}{|E|}\right) \leqslant m < 1 + \frac{1}{10} \log\left(e + \frac{1}{|E|}\right),$$

and with this choice the terms on the right hand side of (20.63) are both smaller that $\sqrt{C_0} C_\beta \log^{1-\beta/2}(e + \frac{1}{|E|})$, where the constant C_β depends only on the parameter $\beta > 2$. □

20.4.4 Existence Above the Critical Exponent

In this subsection we prove a generalization of the measurable Riemann mapping theorem for exponentially integrable distortion when p is above the critical exponent, that is, in the situation where $e^K \in L^p_{loc}$ for $p > p_0$. The key idea here is to use the a priori $W^{1,2}$-regularity established in the previous sections and the general equicontinuity properties this brings. The case of low exponential integrability requires further considerations and is postponed to Section 20.4.5.

David [103] was the first to prove the existence of solutions in the case of exponentially integrable distortion functions. However, the methods developed here substantially refine this approach.

Theorem 20.4.7. *There exists a number $p_0 \geqslant 1$ with the following property: If μ is a Beltrami coefficient such that*

$$|\mu(z)| \leqslant \frac{K(z) - 1}{K(z) + 1} \chi_\mathbb{D}$$

and

$$e^K \in L^p(\mathbb{D}) \quad for\ some\ p > p_0, \qquad (20.64)$$

then the Beltrami equation

$$\frac{\partial f}{\partial \bar{z}} = \mu(z) \frac{\partial f}{\partial z}, \quad for\ almost\ every\ z \in \mathbb{C},$$

admits a unique principal solution $f \in W^{1,2}_{loc}(\mathbb{C})$.

20.4. EXPONENTIAL DISTORTION

In Section 20.4.6 we will show that the *critical exponent*, the smallest number p_0 for which above $W^{1,2}$-regularity holds, is precisely $p_0 = 1$.

We have already seen the uniqueness established: The basic properties of the principal solution are listed in Theorem 20.2.1, which now applies. Note in particular that we obtain a Stoilow factorization for all $W^{1,2}$-solutions as soon as Theorem 20.4.7 is proven.

Alternative proofs for uniqueness and Stoilow factorization will be given in Sections 20.4.7 and 20.4.8, respectively.

Proof of Theorem 20.4.7. Since only existence needs to be established, we use the good approximations $\mu_m(z)$ as defined in (20.46). These give us corresponding distortion functions

$$K_m(z) = \frac{1 + |\mu_m(z)|}{1 - |\mu_m(z)|} \leqslant \frac{1 + |\mu(z)|}{1 - |\mu(z)|} = K(z), \qquad m = 1, 2, \ldots$$

Note that each $K_m(z) \in L^\infty(\mathbb{C})$.

With the measurable Riemann mapping Theorem 5.3.2 we have unique principal solutions $f^m : \mathbb{C} \to \mathbb{C}$ to the Beltrami equation

$$(f^m)_{\bar{z}} = \mu_m (f^m)_z, \qquad m = 1, 2, \ldots$$

Moreover,

$$f^m = z + \mathcal{C}(\omega_m),$$

where $\omega_m = (f^m)_{\bar{z}}$ satisfies the identity

$$\omega_m(z) = \mu_m(z)(\mathcal{S}\omega_m)(z) + \mu_m(z) \qquad \textit{for almost every } z \in \mathbb{C}$$

In order for these approximate solutions f^m to converge, we need uniform L^2-bounds for their derivatives; once these have been established, the proof follows quickly. Actually we have two different ways to achieve the a priori estimates, using either the Beltrami operators from Section 20.4.2 or the Neumann series discussed in Section 20.4.3.

Within the first method we find BMO-majorants for the functions K_m. Constructing them with the procedure (20.36) shows that we can take $(K_m)^* \leqslant K^*$ for each $m = 1, 2, \ldots$ Now all is set for the Beltrami operators and Theorem 20.4.3, which also gives us one estimate for the critical exponent p_0. Assuming the exponential integrability (20.37), we obtain from (20.39) the uniform bounds

$$\|(f^m)_z - 1\|_2 = \|(f^m)_{\bar{z}}\|_2 = \|\omega_m\|_2 \leqslant 4\|(K_m)^* \mu_m\|_2 \leqslant 4\|K^*\|_{L^2(\mathbb{D})} \quad (20.65)$$

Another possible approach is to use Theorem 20.4.5. This gives for any $2 < \beta < p$ the estimates

$$\|(\mu_m \mathcal{S})^k \mu_m\|_2 \leqslant C_m \, k^{-\beta/2}, \qquad k \in \mathbb{N}$$

The theorem also gives bounds for the constant term C_m. Since the approximate solutions have distortion $K_m \leqslant K$ pointwise, we have

$$C_m \leqslant C_{p,\beta} \left(\int_{\mathbb{D}} e^{pK_m} \right)^{1/2} \leqslant C_{p,\beta} \left(\int_{\mathbb{D}} e^{pK} \right)^{1/2}$$

with $C_{p,\beta}$ depending only on p and β. Hence if $e^K \in L^p(\mathbb{D})$ for some $p > 2$, we obtain L^2-estimates for the sums (5.3),

$$\|(f^m)_{\bar{z}}\|_2 \leqslant \sum_{k=0}^{\infty} \|(\mu_m \mathcal{S})^k \mu_m\|_2 \leqslant C_m \sum_{k=0}^{\infty} k^{-\beta/2} < \infty \qquad (20.66)$$

Again we end up with uniform estimates, now of the form

$$\|(f^m)_z - 1\|_2 = \|(f^m)_{\bar{z}}\|_2 \leqslant C_{p,\beta} \|e^K\|_p^{p/2}$$

This approach gives the estimate $1 \leqslant p_0 \leqslant 2$ for the critical exponent.

Whichever method we choose, we arrive at

$$\int_{\mathbb{D}_R} |Df^m|^2 \leqslant 4\pi R^2 + M, \qquad R > 1$$

Now Theorem 20.1.6 applies and shows that on compact subsets of $\mathbb{C}$ the sequence $\{f^m\}$ has a uniform modulus of continuity. Considering a subsequence, if necessary, we may assume that $f^m(z) \to f(z)$ locally uniformly with the derivatives Df^m converging weakly in $L^2_{loc}(\mathbb{C})$. Their weak limit necessarily equals Df, and arguing as in (5.16), we obtain

$$f_{\bar{z}} = \mu \, f_z \quad \text{for almost every } z \in \mathbb{C}$$

On the other hand, we can use Lemma 20.2.3 for the inverse mappings $g_m = (f^m)^{-1}$ and see that $g^m(z) \to g(z)$ uniformly on compact subsets of the plane, where $g : \mathbb{C} \to \mathbb{C}$ is continuous. From $g^m(f^m(z)) = z$ we have $g \circ f(z) = z$. That is, f is a homeomorphism, hence the principal solution we were looking for. □

Corollary 20.4.8. *If μ, p_0 and the principal solution $f \in W^{1,2}_{loc}(\mathbb{C})$ are as in Theorem 20.4.7, then f has the properties $\mathcal{N}$ and $\mathcal{N}^{-1}$, meaning that*

$$|f(E)| = 0 \quad \Leftrightarrow \quad |E| = 0, \qquad E \subset \mathbb{C}$$

Proof. We need to establish only the Neumann series representation

$$f_{\bar{z}} = \sum_{n=0}^{\infty} (\mu \mathcal{S})^n \mu \qquad (20.67)$$

Note that the sum is absolutely convergent in $L^2(\mathbb{C})$. Since approximate coefficients $\mu_m \to \mu$ pointwise, the terms $(\mu_m \mathcal{S})^k \mu_m \to (\mu \mathcal{S})^k \mu$ weakly and hence

20.4. EXPONENTIAL DISTORTION

the identity (20.67) follows with the help of the uniform bounds (20.66). That f preserves sets of Lebesgue measure zero follows now from Corollary 20.4.6. On the other hand, since $K(\cdot, f) \in L^1(\mathbb{D})$, Lemma 20.2.2 puts the inverse map $g = f^{-1}$ into $W^{1,2}_{loc}(\mathbb{C})$. Thus the inverse map preserves Lebesgue null sets by Theorem 3.3.7. □

A little later, in Theorem 20.4.21, we attribute both properties $\mathcal{N}$ and $\mathcal{N}^{-1}$ to all mappings of exponentially integrable distortion.

20.4.5 Exponential Distortion: Existence of Solutions

The situation is rather different in the general case, where the integrability exponent of e^K is smaller than the critical exponent p_0. Here we see from Section 20.4.1 that the corresponding principal solution need not be in $W^{1,2}_{loc}(\mathbb{C})$. Instead, the example puts the principal solutions f_θ in $W^{1,Q}_{loc}(\mathbb{C})$, where

$$Q(t) = \frac{t^2}{\log(e+t)} \qquad (20.68)$$

This is not an exception: Suppose $f \in W^{1,1}_{loc}(\mathbb{C})$ is an orientation-preserving mapping (that is, $J(z, f) \geq 0$) whose distortion function $K(z)$ satisfies $e^K \in L^p_{loc}(\mathbb{C})$ for some $0 < p < \infty$. Then we have the inequality

$$|Df(z)|^2 \leq K(z)J(z,f), \qquad z \in \mathbb{C}$$

If for some reason the Jacobian function $J = J(z, f)$ is locally integrable (for instance if $f \in W^{1,1}_{loc}(\mathbb{C})$ is a homeomorphism), we may use the elementary inequality

$$ab \leq a\log(1+a) + e^b - 1$$

to find that

$$\frac{|Df|^2}{\log(e+|Dh|^2)} \leq \frac{KJ}{\log(e+KJ)}$$
$$\leq \frac{1}{p}\frac{J}{\log(e+J)}\,pK \leq \frac{1}{p}\left(J + e^{pK} - 1\right)$$

for all $p > 0$. Thus

$$\int_\Omega \frac{|Df|^2}{\log(e+|Df|)} \leq \frac{2}{p}\int_\Omega J(z,f) + \frac{2}{p}\int_\Omega [e^{pK(z)} - 1]\,dz \qquad (20.69)$$

for any bounded domain. This shows that f belongs to the Orlicz-Sobolev class $W^{1,Q}_{loc}(\mathbb{C})$

Conversely, in Theorem 19.3.1 it is shown that for an orientation-preserving mapping f the $L^2 \log^{-1} L$-integrability of the differential implies that $J(z,f) \in L^1_{loc}(\Omega)$. This slight gain in the regularity of the Jacobian determinant is precisely why it will be possible to study solutions to the Beltrami equation using the space $W^{1,Q}_{loc}$ as a starting point.

In this setup we can now establish one of the main results of this chapter, the generalizations of the measurable Riemann mapping theorem and Stoilow factorization to the mappings with exponentially integrable distortion.

Existence of Solutions for Exponential Distortion

Theorem 20.4.9. *Suppose the distortion function $K = K(z)$ is such that*

$$e^K \in L^p(\mathbb{D}) \quad \text{for some } p > 0$$

Assume also that $\mu(z) = 0$ for $|z| > 1$. Then the Beltrami equation $f_{\bar{z}}(z) = \mu(z) f_z(z)$ admits a unique principal solution f for which

$$f \in W^{1,Q}_{loc}(\mathbb{C}), \qquad Q(t) = t^2 \log^{-1}(e+t) \tag{20.70}$$

Moreover, every other $W^{1,Q}_{loc}$-solution h to this Beltrami equation in a domain $\Omega \subset \mathbb{C}$ admits the factorization

$$h = \phi \circ f,$$

where ϕ is a holomorphic function in the domain $f(\Omega)$.

Proof. There are different ways to approach the theorem. For instance, it is possible to establish estimates analogous to those leading to Theorem 20.4.7. In fact, we will describe such a line of argument later in a more general context when discussing distortion bounds via general Orlicz functions. Naturally, these results apply to the particular gauge $\mathcal{A}(t) = pt$. However, we find it useful to apply a different and more direct route here, which quickly reduces the proof to Theorem 20.4.7.

Let $0 < p < p_0$, where $p_0 \geqslant 1$ is the constant from Theorem 20.4.7. Consider

$$K(z) = \frac{p_0}{p} \cdot \frac{pK(z)}{p_0} = K_1(z) K_2(z),$$

where $K_1(z) = \frac{pK(z)}{p_0}$ satisfies the hypotheses of Theorem 20.4.7 and the constant factor $K_2(z) = \frac{p_0}{p} \geqslant 1$ could be viewed as representing the distortion of a quasiconformal mapping. There is one problem here in that at points where $K(z)$ is already finite and perhaps small, $K_1(z)$ might be less than 1 and so cannot be a distortion function. We have to argue around this point and construct the related Beltrami coefficients using hyperbolic geometry as per Section 2.3.

Let $M > 0$. For each $z \in \mathbb{D}$, choose a point $\nu = \nu(z)$ on the radial segment determined by $\mu(z)$ so that

$$\rho_\mathbb{D}(0, \nu) + \rho_\mathbb{D}(\nu, \mu) = \rho_\mathbb{D}(0, \mu) = \log \frac{1 + |\mu|}{1 - |\mu|} \tag{20.71}$$

If $\rho_\mathbb{D}(0, \mu) > \log M$, we require $\rho_\mathbb{D}(\nu, \mu) = \log M$, and otherwise we set $\nu = 0$. In any case we always have

$$MK_\nu = e^{\log M} e^{\rho_\mathbb{D}(0,\nu)} \leqslant K_\mu + M$$

20.4. EXPONENTIAL DISTORTION

It follows that
$$\int_{\mathbb{D}} e^{pMK_\nu} \leqslant e^{pM} \int_{\mathbb{D}} e^{pK_\mu} < \infty, \qquad (20.72)$$

so that ν satisfies the hypotheses of Theorem 20.4.7 as soon as we put $M = \frac{p_0}{p}$. We can therefore solve the Beltrami equation for ν to get a principal mapping F of class $W^{1,2}(\mathbb{D})$. Now, for $z \in \mathbb{D}$ set

$$\kappa(w) = \frac{\mu(z) - \nu(z)}{1 - \mu(z)\overline{\nu(z)}} \left(\frac{F_z}{|F_z|}\right)^2, \qquad w = F(z) \qquad (20.73)$$

According to Corollary 20.4.8, the coefficient κ is well defined almost everywhere. We also see that

$$\frac{1 + |\kappa|}{1 - |\kappa|} = \frac{1 + \left|\frac{\mu - \nu}{1 - \mu\bar\nu}\right|}{1 - \left|\frac{\mu - \nu}{1 - \mu\bar\nu}\right|} = e^{\rho_\mathbb{D}(\nu,\mu)} \leqslant M = \frac{p_0}{p} < \infty \qquad (20.74)$$

Thus we may solve the Beltrami equation for κ to obtain a M-quasiconformal principal mapping g.

We next put $f = g \circ F$. This is differentiable almost everywhere with

$$|Df(z)|^2 \leqslant MJ(w,g)K(z,F)J(z,F) = MK_\nu(z)J(z,f), \qquad w = F(z)$$

Arguing as in (20.69), we see that

$$\frac{|Df|^2}{\log(e + |Df|^2)} \leqslant \frac{1}{p}\left[J(\cdot,f) + e^{pMK_\nu}\right]$$

is locally integrable. Thus f is a Sobolev mapping in the class $W^{1,Q}_{loc}$,

$$Q(t) = \frac{t^2}{\log(e+t)}$$

However, note that although the quasiconformal map g lies in the space $W^{1,s}_{loc}(\mathbb{C})$ for all

$$2 \leqslant s < \frac{2p_0}{p_0 - p},$$

by Theorem 13.2.3, the composition $f = g \circ F$ does not lie in $W^{1,2}_{loc}(\mathbb{C})$ in general.

We defined the Beltrami coefficient $\kappa = \mu_g$ through (20.73), but we may identify $\mu_g = \mu_{f \circ F^{-1}}$ as well via (5.45). Comparing the expressions shows that $\mu_f = \mu$, hence f in fact solves the required Beltrami equation $f_{\bar z} = \mu f_z$. Clearly, f is a principal solution.

Uniqueness and Stoilow factorization also follow once we have established Theorems 20.4.15 and 20.4.19, see Sections 20.4.7 and 20.4.8, respectively. □

From the above proof and (20.72), we distill a powerful factorization, one of the key facts in obtaining the sharp regularity.

Corollary 20.4.10. *Suppose the distortion function $K = K(z)$ satisfies $e^K \in L^p(\mathbb{D})$ for some $p > 0$. Then for any $M \geqslant 1$ the principal solution to $f_{\bar{z}}(z) = \mu(z) f_z(z)$ admits a factorization*

$$f = g \circ F,$$

where both g and F are principal mappings, g is M-quasiconformal and F satisfies

$$\int_{\mathbb{D}} e^{pMK(z,F)} \leqslant C_0 < \infty$$

Note that Theorem 20.4.9 actually allows us to improve the Stoilow factorization theorem even in the favorable situation of Theorem 20.4.7 where the principal solution is $W_{loc}^{1,2}$-smooth; the solution h needs only to belong to the class $W_{loc}^{1,Q}(\mathbb{C})$. On the other hand, we want to stress that the regularity cannot be weakened much further: In Section 20.4.9 we will construct a mapping $f \in \bigcap_{1 \leqslant q < 2} W^{1,q}(\mathbb{D})$ with exponentially integrable distortion such that f is neither open nor continuous. In particular, f does not admit a Stoilow factorization.

Consequently, the example shows that there is no reasonable theory of mappings of exponentially integrable distortion in the classes $W^{1,q}$, $q < 2$!

20.4.6 Optimal Regularity

In Example 20.4.1 we presented a family of mappings f_θ that have exponential distortion with $e^{K(z,f_\theta)} \in L^p$ if and only if $2\theta > p$, while $f_\theta \in W_{loc}^{1,2}$ precisely when $2\theta > 1$. A similar example by Kovalev,

$$g_p(z) = \frac{z}{|z|} \left[\log\left(e + \frac{1}{|z|}\right) \log\log\left(e + \frac{1}{|z|}\right) \right]^{-p/2}, \qquad |z| < 1, \qquad (20.75)$$

and $g_p(z) = c_0 z$, $|z| > 1$, satisfies

$$e^{K(z,g_p)} \in L^p(\mathbb{D}), \qquad p > 0$$

However, $g_1 \notin W_{loc}^{1,2}(\mathbb{C})$ so that at the borderline $p = 1$ the integrability of $e^{K(z)}$ is in general *not* sufficient for the $W_{loc}^{1,2}$-regularity.

These examples and the results from the previous sections lead to natural conjectures [180, 191, 196] on the optimal regularity within the mappings of exponential distortion. We will next describe recent work [25] establishing these conjectures. The main result of this section is the following theorem.

Theorem 20.4.11. *Suppose that f is a principal solution to $f_{\bar{z}} = \mu(z) f_z$, where $|\mu(z)| < 1$ almost everywhere with $\mu(z) \equiv 0$ for $|z| > 1$. If*

$$e^{K(z,f)} \in L_{loc}^p(\mathbb{C}) \quad \text{for some } p > 1,$$

then $f \in W_{loc}^{1,2}(\mathbb{C})$.

20.4. EXPONENTIAL DISTORTION

It is natural to inquire if improved exponential integrability of the distortion function would give any extra regularity for the corresponding principal solutions—that is, beyond L^2-integrability of the derivatives. This turns out to be the case, and we shall actually prove the following refined version of Theorem 20.4.11 that gives the correct regularity at every scale of exponential distortion.

Theorem 20.4.12. *Suppose that f is a principal solution to $f_{\bar{z}} = \mu(z) f_z$, where $|\mu(z)| < 1$ almost everywhere and $\mu(z) \equiv 0$ for $|z| > 1$. If*

$$e^{K(z,f)} \in L^p_{loc}(\mathbb{C}) \quad \text{for some } p > 0, \tag{20.76}$$

then we have for every $0 < \beta < p$,

$$J(z,f) \log^{\beta}(e + J(z,f)) \in L^1_{loc}(\mathbb{C}) \quad \text{and} \tag{20.77}$$

$$|Df|^2 \log^{\beta-1}(e + |Df|) \in L^1_{loc}(\mathbb{C}) \tag{20.78}$$

Moreover, for every $p > 0$, there are examples satisfying (20.76) yet failing (20.77) and (20.78) for $\beta = p$.

With the Stoilow factorization as described in Section 20.4.8, the regularity extends to mappings f of exponential distortion defined in domains $\Omega \subset \mathbb{C}$.

We start the proof of Theorem 20.4.12 (and hence also of Theorem 20.4.11) by establishing general area distortion bounds. With the factorization method described in Corollary 20.4.10 we may in fact improve the distortion exponent of Corollary 20.4.6.

Theorem 20.4.13. *Let $|\mu(z)| < 1$ almost everywhere and $\mu(z) \equiv 0$ outside the unit disk. Suppose f is a principal solution to $f_{\bar{z}} = \mu f_z$. If*

$$e^{K(z,f)} \in L^p(\mathbb{D}) \quad \text{for some } p > 0,$$

then for any $0 < \beta < p$ we have

$$|f(E)| \leq C \log^{-\beta}\left(e + \frac{1}{|E|}\right), \quad E \subset \mathbb{D} \tag{20.79}$$

The constant depends on β, p and $\|e^{K(z,f)}\|_p$ only.

Proof. Choose $\beta < \beta_0 < p$ and $M \geq 1$ so that

$$\frac{2}{M} < \beta < \beta_0 - \frac{2}{M}$$

We will then use the factorization $f = g \circ F$ from Corollary 20.4.10. Since $pM > \beta_0 M > 2$, Corollary 20.4.6 applies to $\sigma_\mu = F_{\bar{z}}$ and $F_z = 1 + \mathcal{S} F_{\bar{z}}$,

$$|F(E)| \leq \int_E |F_z|^2 \leq 2|E| + 2\int_E |F_z - 1|^2 \leq C \log^{-\beta_0 M + 2}\left(e + \frac{1}{|E|}\right)$$

On the other hand, since g is a M-quasiconformal principal mapping, we can use Theorem 13.1.4. This gives

$$|f(E)| = |g \circ F(E)| \leqslant \pi M |F(E)|^{1/M} \leqslant \pi C \left[\log \left(e + \frac{1}{|E|} \right) \right]^{-\beta_0 + 2/M}$$

Since $\beta < \beta_0 - \frac{2}{M}$, the result follows. $\square$

Once we have the area distortion, the improved integrability quickly follows, in fact by quite general arguments on higher integrability.

Theorem 20.4.14. *Let $\mathcal{J} \subset L^1(\Omega)$ be a non-negative function such that*

$$\int_E \mathcal{J}(z) \leqslant \Phi(|E|) \leqslant 1 \quad \text{for every } E \subset \Omega,$$

where Φ is a given increasing function in $\mathbb{R}_+$.

Let Ψ be continuously differentiable on $[0, \infty)$ with $\Psi(0) = 0$. Then

$$\int_\Omega \mathcal{J} \Psi(\mathcal{J}) \leqslant \int_0^\infty |\Psi'(t)| \, \Phi\left(\frac{1}{t}\right) dt$$

Proof. Let $E_t = \{z \in \Omega : \mathcal{J}(z) > t\}$ for $t > 0$. From Chebyshev's inequality,

$$|E_t| \leqslant \frac{1}{t} \int_{E_t} \mathcal{J} \leqslant \frac{1}{t} \Phi(|E_t|) \leqslant \frac{1}{t}$$

Therefore

$$\int_{E_t} \mathcal{J} \leqslant \Phi(|E_t|) \leqslant \Phi\left(\frac{1}{t}\right)$$

It remains only to apply Fubini's theorem,

$$\int_\Omega \mathcal{J} \Psi(\mathcal{J}) = \int_\Omega \mathcal{J} \int_0^{\mathcal{J}} \Psi'(t) \, dt$$

$$= \int_0^\infty \Psi'(t) \left(\int_{\{\mathcal{J} > t\}} \mathcal{J} \right) dt \leqslant \int_0^\infty |\Psi'(t)| \, \Phi\left(\frac{1}{t}\right),$$

and the claim follows. $\square$

As an example illustrating the above theorem, suppose

$$\int_E \mathcal{J} \leqslant \log^{-\alpha} \left(e + \frac{1}{|E|} \right), \quad E \subset \Omega \tag{20.80}$$

Letting $\Psi(t) = -1 + \log^\beta(e + t)$, we obtain for $\beta < \alpha$

$$\int_\Omega \mathcal{J} \log^\beta(e + \mathcal{J}) \leqslant \int_\Omega \mathcal{J} + \beta \int_0^\infty \frac{\log^{\beta-1}(e + t)}{(e + t) \log^\alpha(e + t)} \leqslant \frac{\alpha}{\alpha - \beta} \tag{20.81}$$

20.4. EXPONENTIAL DISTORTION

As another example, with the same assumption (20.80), we may choose

$$\Psi(t) = \frac{\log^\alpha(e^2 + t)}{[\log\log(e^2 + t)]^2} - \frac{2^\alpha}{\log 2}$$

to get

$$\int_\Omega \mathcal{J} \frac{\log^\alpha(e^2 + \mathcal{J})}{[\log\log(e^2 + \mathcal{J})]^2} < \infty$$

Proof of Theorem 20.4.12. Suppose f is a principal solution to $f_{\bar z} = \mu f_z$ where

$$e^{K(z,f)} \in L^p(\mathbb{D}) \quad \text{for some } p > 0$$

If $0 < \beta < p$, choose a number $\beta < \alpha < p$. From Theorem 20.4.13 we have

$$|f(E)| = \int_E J(z,f) \leqslant C \log^{-\alpha}\left(e + \frac{1}{|E|}\right), \qquad E \subset \mathbb{D}$$

Therefore (20.80) and (20.81) show that

$$J(z,f) \log^\beta(e + J(z,f)) \in L^1_{loc}(\mathbb{C}),$$

proving the first claim (20.77) in Theorem 20.4.12.

The second claim (20.78) can be deduced from (20.77) by observing that for every $\beta, p > 0$ there are positive constants C_1 and C_2 such that

$$xy \log^{\beta-1}(e + \sqrt{xy}) \leqslant C_1 x \log^\beta(e + \sqrt{x}) + C_2 e^{py} \quad \text{for } x, y > 0$$

For the inequality, we consider separately the cases where $x < e^{py/2}$ and where $x \geqslant e^{py/2}$. Since $s \mapsto s^2 \log^{\beta-1}(e + s)$ is increasing, we conclude that

$$|Df(z)|^2 \log^{\beta-1}(e + |Df(z)|) \leqslant K(z,f) J(z,f) \log^{\beta-1}\left(e + \sqrt{K(z,f) J(z,f)}\right)$$

$$\leqslant C_1 J(z,f) \log^\beta(e + J(z,f)) + C_2 e^{pK(z,f)} \in L^1_{loc}(\mathbb{D})$$

The family g_p from (20.75) shows that (20.77) and (20.78) may fail at the borderline $\beta = p$. $\square$

20.4.7 Uniqueness of Principal Solutions

We have seen several instances where the basic integral bounds on the Jacobian have turned out to be surprisingly powerful tools in the theory of elliptic differential equations. Here we will apply a line of argument already used in Lemma 20.1.8 to quickly establish the uniqueness in Theorem 20.4.9.

Theorem 20.4.15. *Suppose the distortion function $K(z) = \frac{1+|\mu(z)|}{1-|\mu(z)|}$ satisfies $e^K \in L^p(\mathbb{D})$ for some positive p, with $\mu(z)$ vanishing outside $\mathbb{D}$. Then the Beltrami equation*

$$\frac{\partial f}{\partial \bar{z}} = \mu(z) \frac{\partial f}{\partial z} \quad \text{for almost every } z \in \mathbb{C},$$

admits at most one principal solution with $f \in W^{1,1}_{loc}(\mathbb{C})$.

Proof. We saw in (20.69) that for the exponentially integrable distortion functions their principal solutions lie in $W^{1,Q}_{loc}(\mathbb{C})$, $Q(t) = t^2 \log^{-1}(e+t)$. Assuming that we are given two principal solutions f_1 and f_2, both in this class, then the difference

$$h = f_1 - f_2$$

is an orientation-preserving mapping. Indeed, $h_{\bar{z}} = \mu(z) h_z$, so $J(z,h) = (1 - |\mu|^2)|h_z|^2 \geq 0$. In addition, $|Dh(z)| = \mathcal{O}(1/|z|^2)$ near ∞. Theorem 19.3.1 applies and in particular we have the formula (19.22), which yields

$$\int_{\mathbb{C}} \varphi J(z,h) \leq \int_{\mathbb{C}} |h| \, |Dh| \, |\nabla \varphi|$$

for non-negative test functions $\varphi \in C_0^\infty(\mathbb{C})$. We can certainly take $\varphi(z) = 1$ for $|z| \leq R$ and $|\nabla \varphi| \leq 1$ in $\mathbb{C}$. This gives

$$\int_{|z| \leq R} J(z,h) \leq C \int_{|z| \geq R} |z|^{-3},$$

and letting $R \to \infty$, we conclude that $J(z,h) \equiv 0$. As $J(z,h) = (1-|\mu|^2)|h_z|^2$ and $|\mu(z)| < 1$ almost everywhere, Dh vanishes almost everywhere. Since $h(\infty) = 0$, we obtain $f_1 = f_2$. □

20.4.8 Stoilow Factorization

For quasiregular mappings in $W^{1,2}_{loc}(\Omega)$ the Stoilow factorization was achieved basically by a change of variables. This was possible because the principal solutions of bounded distortion, that is, quasiconformal mappings, preserve the class $W^{1,2}_{loc}$. In the case of degenerate equations we cannot use such an approach. However, the methods developed in Section 20.1 will save the day.

We will consider the Stoilow factorization of solutions to the Beltrami equation

$$\frac{\partial f}{\partial \bar{z}} = \mu(z) \frac{\partial f}{\partial z} \tag{20.82}$$

for degenerate cases, where $|\mu(z)| < 1$ almost everywhere. The remarkable phenomenon we will face here is that the integrability properties of the distortion function $K(z) = \frac{1+|\mu(z)|}{1-|\mu(z)|}$ will play no role in these results, *assuming there is a homeomorphic solution* to (20.82).

20.4. EXPONENTIAL DISTORTION

Beyond this assumption the point in achieving Stoilow factorization is to get by with a minimal regularity. We will assume that $f \in W^{1,Q}_{loc}(\Omega)$ where, as in the previous sections,

$$Q(t) = \frac{t^2}{\log(e+t)}$$

Using general Orlicz spaces, this regularity can be slightly relaxed; for details see Section 20.5. However, as we will see in Section 20.4.9, no $W^{1,q}$-regularity with $q < 2$ will work even for the case of exponentially integrable distortion.

We begin with a variant of Lemma 20.1.8.

Lemma 20.4.16. *Suppose $f = u + iv \in W^{1,Q}_{loc}(\Omega)$ is an orientation-preserving mapping such that*

$$u \in W^{1,Q}_0(\Omega')$$

for some relatively compact subdomain $\Omega' \subset \Omega$. Then

$$J(z, f) = 0 \quad \text{for almost every } z \in \Omega'$$

Proof. This is immediate from the formula for integration by parts (19.22) in Theorem 19.3.1, by applying it with $\varphi \in C^\infty_0(\Omega)$, which is equal to 1 on Ω'. □

Imitating the argument in the proof of Theorem 20.1.1 and replacing the earlier Lemma 20.1.8 by Lemma 20.4.16, we now have the following corollary.

Corollary 20.4.17. *Suppose $f = u + iv \in W^{1,Q}_{loc}(\Omega)$ is a solution to the Beltrami equation $f_{\bar{z}} = \mu f_z$, where $|\mu(z)| < 1$ for almost every $z \in \Omega$. Then the coordinate functions u and v are weakly monotone.*

On the other hand, the gauge Q satisfies the inequality (20.10), and therefore the weakly monotone functions in $W^{1,Q}_{loc}(\Omega)$ are continuous; see Remark 20.1.7.

Corollary 20.4.18. *Let $f = u + iv \in W^{1,Q}_{loc}(\Omega)$ be a solution to the Beltrami equation as in the previous corollary. Then f is continuous.*

We now have the general Stoilow factorization result.

Theorem 20.4.19. *Suppose we are given a homeomorphic solution $f \in W^{1,Q}_{loc}(\Omega)$ to the Beltrami equation*

$$\frac{\partial f}{\partial \bar{z}} = \mu(z) \frac{\partial f}{\partial z}, \qquad z \in \Omega, \tag{20.83}$$

where $|\mu(z)| < 1$ almost everywhere. Then every other solution $h \in W^{1,Q}_{loc}(\Omega)$ to (20.83) takes the form

$$h(z) = \phi(f(z)), \qquad z \in \Omega,$$

where $\phi : f(\Omega) \to \mathbb{C}$ is holomorphic.

Proof. By Corollary 20.4.18, h is continuous. Fix an arbitrary closed disk $\overline{B} \subset f(\Omega)$. The function $h \circ f^{-1} : \overline{B} \to \mathbb{C}$ is continuous, and the solution to the classical Dirichlet problem tells us that there is a function u continuous on $\overline{B}$ that is harmonic in B and equal to $\Re(h \circ f^{-1})$ on ∂B. Let v be the harmonic conjugate of u on B. Then the function $\phi = u + iv$ is holomorphic on B (the reader is cautioned about the continuity of ϕ on ∂B). Consider the two functions $\phi \circ f$ and h defined and continuous on $f^{-1}(B)$. By construction, these two functions have the same real part on $\partial(f^{-1}B)$. Moreover, both of these functions lie in $W_{loc}^{1,Q}(f^{-1}B)$ and satisfy the same Beltrami equation. Therefore their difference, $g = h - \phi \circ f$, is a mapping of finite distortion. Corollaries 20.4.17 and 20.4.18 show that the real part of g is weakly monotone, is continuous in $f^{-1}(\overline{B})$ and vanishes on the boundary of this set. We deduce from the definition of weak monotonicity that $\Re(g) \equiv 0$ in $f^{-1}(B)$. In particular, $J(z,g) \equiv 0$ in $f^{-1}(B)$, and as g is a mapping of finite distortion, $Dg(z) \equiv 0$ in $f^{-1}(B)$. From this we deduce that $g(z) \equiv c$, an imaginary constant.

We therefore have the factorization

$$h(z) = \phi(f(z)) + c, \qquad z \in B$$

Last, recall that B was an arbitrary disk compactly contained in $f(\Omega)$. As f is continuous in Ω, the principle of analytic continuation gives us a unique extension of $\phi + c$ to a holomorphic function; call it ϕ again, defined in $f(\Omega)$ and such that $h = \phi \circ f$. $\square$

As a particular consequence, we have the optimal smoothness for mappings of exponential distortion. We see a jump in regularity inside the class $W_{loc}^{1,Q}(\mathbb{C})$, $Q(t) = \frac{t^2}{\log(e+t)}$. To be precise, the following result is an immediate corollary of Theorems 20.4.12 and 20.4.19.

Optimal Smoothness for Mappings of Exponential Distortion

Theorem 20.4.20. *Suppose $f \in W_{loc}^{1,Q}(\mathbb{C})$ is a mapping of finite distortion with*

$$e^{K(z,f)} \in L_{loc}^p(\Omega), \qquad p > 0$$

Then for every $0 < \beta < p$,

$$|Df|^2 \log^{\beta-1}(e + |Df|) \in L_{loc}^1(\Omega) \tag{20.84}$$

On the other hand, for the family g_β from (20.75), the regularity (20.84) fails at $\beta = p$.

With the Stoilow factorization in hand, we immediately obtain a large spectrum of different further properties for the class of mappings with exponentially integrable distortion. For instance, those concerning removability, modulus of continuity and so on are all modeled on the topics discussed in earlier chapters

20.4. EXPONENTIAL DISTORTION

for mappings of bounded distortion, that is, quasiconformal mappings. We leave it for the reader to further contemplate these aspects of the theory. Here we shall prove only the following.

Theorem 20.4.21. *Suppose* $\phi \in W^{1,Q}_{loc}(\mathbb{C})$ *is a mapping of finite distortion with distortion* $e^{K(z,\phi)} \in L^p_{loc}(\Omega)$ *for some* $p > 0$. *Then* ϕ *has the properties* $\mathcal{N}$ *and* $\mathcal{N}^{-1}$, *that is,*

$$|\phi(E)| = 0 \quad \Leftrightarrow \quad |E| = 0$$

Proof. We may factor $\phi = h \circ f$, where h is holomorphic and f is the principal solution from Theorem 20.4.9. Applying Corollary 20.4.10, we can make a further factorization,

$$f = g \circ F,$$

where g is quasiconformal and F has exponential distortion $e^K \in L^p$ with $p > p_0$. According to Corollary 20.4.8, each factor in $\phi = h \circ g \circ F$ has the properties $\mathcal{N}$ and $\mathcal{N}^{-1}$. $\square$

20.4.9 Failure of Factorization in $W^{1,q}$ When $q < 2$

Here we present a mapping $g \in \bigcap_{q<2} W^{1,q}_{loc}(\Omega)$ that has exponentially integrable distortion,

$$\int_\Omega \left(e^{pK(z,g)} - 1\right) dz < \infty$$

for a $p > 0$ (we can take p as large as we wish) but such that g is neither continuous nor open or discrete. Thus no Stoilow factorization is possible for such a mapping g. The example in particular emphasizes the role of the regularity assumption $W^{1,Q}_{loc}$ in the theory of mappings of exponentially integrable distortion.

The map is constructed by piecing together suitable "quasi-inversions" in carefully arranged circles. Given a domain $\Omega \in \mathbb{C}$, an *exact packing* of Ω by balls is an infinite family $\mathcal{F} = \{\mathbb{D}_j\}_{j=1}^\infty$ of disjoint open disks $\mathbb{D}_j \subset \Omega$ such that

$$\left|\Omega \setminus \bigcup_{j=1}^\infty \mathbb{D}_j\right| = 0$$

The existence of such an exact packing for any domain Ω follows from Vitali's covering lemma. Indeed, one can require that the disks all have radius bounded by 1, say.

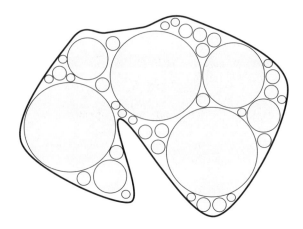

An exact packing of a domain

Theorem 20.4.22. *Let $\Omega \subset \mathbb{C}$ be a domain. Suppose*

$$0 < p < \alpha - 1$$

Then there is a Beltrami coefficient μ that has the following properties. First, the distortion function $K(z) = \frac{1+|\mu(z)|}{1-|\mu(z)|}$ is locally exponentially integrable,

$$\int_E e^{pK(z)} dz < \infty$$

for all E relatively compact in Ω. Second, the Beltrami equation

$$g_{\bar{z}} = \mu(z)\, g_z \qquad (20.85)$$

admits a solution $g \in W^{1,1}_{loc}(\Omega)$ with

$$\frac{|Dg|^2}{\log^\alpha(e+|Dg|)} \in L^1_{loc}(\Omega)$$

In particular, $g \in W^{1,q}_{loc}(\Omega)$ for all $q < 2$. Third, g is neither continuous nor open or discrete. Hence it cannot be factored as a composition of a holomorphic map and the principal solution to (20.85).

Proof. Assume for convenience that $|\Omega| < \infty$.

Recall the family of examples $f = f_\theta$ described in (20.32). We consider an analogous family of maps,

$$\Psi_{a,r}(z) = a + r\, \frac{z-a}{|z-a|}\left(1 + \frac{1}{\theta}\log\frac{r}{|z-a|}\right)^\theta, \qquad |z-a| < r, \qquad (20.86)$$

20.4. EXPONENTIAL DISTORTION

where we leave the θ-dependence unmarked. Notice that in fact

$$\Psi_{a,r}(z) = a + \frac{r}{f\left(\frac{\bar{z}-\bar{a}}{r}\right)}$$

Let $\mathcal{F} = \{\mathbb{D}_j\}_{j=1}^\infty$ be an exact packing of Ω by disks $\mathbb{D}_j = \mathbb{D}(a_j, r_j)$ of radius at most 1. We define a mapping g of Ω in the following piecewise fashion by requiring that its complex conjugate

$$\overline{g(z)} = \begin{cases} \Psi_{a_j,r_j}(z) & \text{if } z \in \mathbb{D}_j \\ z & \text{otherwise} \end{cases} \qquad (20.87)$$

Following the argument in [191, pp. 293–297], we see that g has integrable distributional derivates and is hence a Sobolev mapping. With (20.33) the differential

$$|D\Psi_{a_j,r_j}(z)| = \frac{r_j}{|z-a_j|}\left(1 + \frac{1}{\theta}\log\frac{r_j}{|z-a_j|}\right)^\theta, \qquad z \in \mathbb{D}_j$$

Therefore

$$\int_\Omega \frac{|Dg|^2}{\log^\alpha(e+|Dg|)} < \infty$$

for all $\alpha > 2\theta + 1$. Since g is continuous in the punctured disks $\mathbb{D}(a_j, r_j) \setminus \{a_j\}$ with $g(z) \to \infty$ as $z \to a_j$, we see that g is not continuous, nor is it open or discrete in Ω.

From the construction we see that g satisfies the Beltrami equation with coefficient given

$$\mu_g(z) = \frac{z-a_j}{\bar{z}-\bar{a}_j} \frac{\log r_j - \log|z-a_j|}{2\theta + \log r_j - \log|z-a_j|}, \qquad z \in \mathbb{D}(a_j, r_j)$$

The corresponding distortion function is

$$K(z) = \frac{1+|\mu(z)|}{1-|\mu(z)|} = 1 + \frac{1}{\theta}\log\frac{r_j}{|z-a_j|}, \qquad z \in \mathbb{D}(a_j, r_j)$$

Therefore, arguing as in Section 20.4.1, we have

$$\int_\Omega e^{pK(z)}\,dz = \sum_{j=1}^\infty \int_{\mathbb{D}(a_j,r_j)} e^{pK(z)}\,dz = \sum_{j=1}^\infty \frac{2\theta e^p}{2\theta-p}|\mathbb{D}(a_j,r_j)| = \frac{2\theta e^p}{2\theta-p}|\Omega| < \infty$$

whenever $p < 2\theta$. Hence g has the required properties once we choose the parameter θ so that $p < 2\theta < \alpha - 1$. □

20.5 Optimal Orlicz Conditions for the Distortion Function

For clarity of presentation, in the previous section we concentrated on mappings with exponentially integrable distortion and the regularity class $W^{1,Q}_{loc}$. However, reflecting back on Theorem 20.3.1, we see that the results achieved and the methods used indicate that slightly further steps will be possible using general Orlicz space conditions such as

$$\int_{\mathbb{D}} e^{\mathcal{A}(K(z))} dz < \infty, \quad \text{where} \quad \int_1^\infty \frac{\mathcal{A}(t)}{t^2} dt = \infty \tag{20.88}$$

Such further refinements not only turn out to be true but also are quite useful. Therefore we give a brief description of the optimal conditions that enable the fullest possible extension of the measurable Riemann mapping theorem and the Stoilow factorization theorem.

Naturally, both the distortion and the degree of Sobolev regularity must be measured in terms of Orlicz gauges. Let us recall the notation

$$W^{1,P}(\Omega) = \{f \in W^{1,1}_{loc}(\Omega) : \int_\Omega P(|Df|) < \infty\}$$

According to Theorem 20.3.1, a necessary condition to have the required generalization of the measurable Riemann mapping theorem is that, for the gauge $\mathcal{A}$ controlling the distortion function, the integral of $\mathcal{A}(t)t^{-2}$ diverges, as in (20.88). Up to natural regularity assumptions on $\mathcal{A}$, this requirement also turns out to be sufficient. More precisely, we suppose that $\mathcal{A}$ has the following properties:

1. $\mathcal{A}: [1, \infty) \to [0, \infty)$ is a smooth increasing function with $\mathcal{A}(1) = 0$.
2.
$$\int_1^\infty \frac{\mathcal{A}(t)}{t^2} dt = \infty$$

We will establish the following theorem.

Theorem 20.5.1. *Let $f : \Omega \to \mathbb{C}$ be a mapping of finite distortion as in Definition 20.0.3. Assume that in the distortion inequality*

$$|Df(z)|^2 \leq K(z) J(z, f) \tag{20.89}$$

the function $K = K(z)$ satisfies

$$\int_\Omega e^{\mathcal{A}(K(z))} dz < \infty \tag{20.90}$$

Then f is contained in the Orlicz-Sobolev class $W^{1,P}_{loc}(\Omega)$, where

$$P(t) = \begin{cases} t^2, & 0 \leq t \leq 1 \\ \frac{t^2}{\mathcal{A}^{-1}(\log t^2)}, & t \geq 1 \end{cases} \tag{20.91}$$

20.5. OPTIMAL ORLICZ CONDITIONS

Proof. We recall that mappings of finite distortion belong to $W^{1,1}_{loc}(\Omega)$ and have a locally integrable Jacobian. Thus the theorem is immediate once we establish the pointwise inequality

$$P(|Df(z)|) \leqslant e^{\mathcal{A}(K(z))} + J(z, f) \tag{20.92}$$

To see this inequality consider two cases: If $|Df(z)|^2 \leqslant e^{\mathcal{A}(K(z))}$, then

$$P(|Df(z)|) \leqslant e^{\mathcal{A}(K(z))}$$

If, however, $|Df(z)|^2 > e^{\mathcal{A}(K(z))}$, then from the distortion inequality (20.89) we have

$$P(|Df(z)|) \leqslant \frac{K(z)J(z,f)}{\mathcal{A}^{-1}(\log e^{\mathcal{A}(K(z))})} = J(z,f),$$

as desired. $\square$

As an example, note that if $\mathcal{A}$ is affine, $\mathcal{A}(t) = pt - p$ for some number $p > 0$, then the condition (20.90) takes us back to the exponentially integrable distortion of Section 20.4, while $P(t) = t^2 \left(1 + \frac{1}{p}\log^+ t^2\right)^{-1}$ is equivalent to the gauge $Q(t)$ used in Theorem 20.4.9.

The divergence of the integral $\int_1^\infty \frac{\mathcal{A}(t)}{t^2} dt$ is equivalent to

$$\int_1^\infty \frac{P(t)}{t^3} dt = \frac{1}{2}\int_1^\infty \frac{\mathcal{A}'(t)}{t} dt = \frac{1}{2}\int_1^\infty \frac{\mathcal{A}(t)}{t^2} dt = \infty \tag{20.93}$$

by substituting $\mathcal{A}(s) = \log t^2$ and using (20.29). The condition (20.93) is critical for various phenomena concerning homeomorphisms in $W^{1,P}_{loc}(\Omega)$, such as regularity of Jacobians and modulus of continuity, as well as properties of weakly monotone functions. However, for further computations we will need to make an additional technical assumption about $\mathcal{A}$. We require

3. $\quad\quad\quad t\,\mathcal{A}'(t) \geqslant 5 \quad$ for large values of t

In view of the integral condition (20.93), this easily verifiable condition involves an insignificant loss of generality as the function $t\mathcal{A}'(t)$ behaves more or less like $\mathcal{A}(t)$, at least in the typical examples we have in mind for applications.

Theorem 20.5.2. (Existence and Uniqueness) *Let $\mathcal{A} = \mathcal{A}(t)$ satisfy the above conditions 1 – 3. Suppose the Beltrami coefficient, with $|\mu(z)| < 1$ almost everywhere, is compactly supported and the associated distortion function $K(z) = \frac{1+|\mu(z)|}{1-|\mu(z)|}$ satisfies*

$$e^{\mathcal{A}(K(z))} \in L^1_{loc}(\mathbb{C}) \tag{20.94}$$

Then the Beltrami equation $f_{\bar z}(z) = \mu(z)\,f_z(z)$ admits a unique principal solution $f \in W^{1,P}_{loc}(\mathbb{C})$ with $P(t)$ as in (20.91). Moreover, any solution $h \in W^{1,P}_{loc}(\Omega)$ to this Beltrami equation in a domain $\Omega \subset \mathbb{C}$ admits a factorization

$$h = \phi \circ f,$$

where ϕ is holomorphic in $f(\Omega)$.

Proof. For the proof we require a sequence of auxiliary results, Lemmas 20.5.3–20.5.8, which we first present. We begin with auxiliary Orlicz gauges closely related to P. The first of them, $\Psi : [0, \infty) \to [0, \infty)$, is defined by the rule

$$\Psi(s) = s, \quad 0 \leqslant s \leqslant 1 \tag{20.95}$$

$$\Psi'\left(e^{\mathcal{A}(s)}\right) = \frac{1}{s}, \quad 1 \leqslant s \tag{20.96}$$

These identities determine Ψ uniquely on $[0, \infty)$ and make Ψ a continuously differentiable function. The second gauge is given by

$$\Phi(t) = \Psi(t^2) \tag{20.97}$$

Lemma 20.5.3. *For all $t \geqslant 0$ we have*

$$P(t) \leqslant \Phi(t) \leqslant 2P(t)$$

Proof. We need to consider only the case $t \geqslant 1$ and show that

$$\frac{\tau}{\mathcal{A}^{-1}(\log \tau)} \leqslant \Psi(\tau) \leqslant \frac{2\tau}{\mathcal{A}^{-1}(\log \tau)}, \quad \tau > 1 \tag{20.98}$$

The computation

$$\Psi(\tau) - 1 = \int_1^\tau \Psi'(\tau) d\tau = \int_1^{\mathcal{A}^{-1}(\log \tau)} \Psi'\left(e^{\mathcal{A}(s)}\right) de^{\mathcal{A}(s)}$$

$$= \int_1^{\mathcal{A}^{-1}(\log \tau)} \frac{1}{s} de^{\mathcal{A}(s)} = \frac{\tau}{\mathcal{A}^{-1}(\log \tau)} - 1 + \int_1^{\mathcal{A}^{-1}(\log \tau)} \frac{e^{\mathcal{A}(s)}}{s^2} ds$$

proves the left-hand-side inequality in (20.98). For the right hand side we argue as follows:

$$\left[s^{-2} e^{\mathcal{A}(s)}\right]' = s^{-3} e^{\mathcal{A}(s)} \left(s\mathcal{A}'(s) - 2\right) \geqslant 0$$

Hence

$$\int_1^{\mathcal{A}^{-1}(\log \tau)} \frac{e^{\mathcal{A}(s)}}{s^2} ds \leqslant \frac{\tau}{\left(\mathcal{A}^{-1}(\log \tau)\right)^2} \int_1^{\mathcal{A}^{-1}(\log \tau)} ds \leqslant \frac{\tau}{\mathcal{A}^{-1}(\log \tau)},$$

which proves the required estimate. $\square$

The advantage of using $\Phi(t)$ instead of $P(t)$ is that the former has useful convexity properties.

Lemma 20.5.4. *The function $t \mapsto t^{1/4} \Psi'(t^{5/4})$ is increasing for large values of t. In particular, $t \mapsto \Psi(t^{5/4})$ and $t \mapsto \Phi(t^{5/8})$ are convex on some interval $[t_0, \infty)$.*

20.5. OPTIMAL ORLICZ CONDITIONS

Proof. For $t \geqslant 1$ we can define an auxiliary parameter $s \geqslant 1$ by setting $e^{\mathcal{A}(s)} = t^{5/4}$. Thus

$$t^{1/4}\Psi'(t^{5/4}) = e^{\mathcal{A}(s)/5}\Psi'\left(e^{\mathcal{A}(s)}\right) = e^{\mathcal{A}(s)/5}\frac{1}{s} = e^{(\mathcal{A}(s) - 5\log(s))/5},$$

which is increasing for large values of s and thus for large values of t as well. □

Next, we require a counterpart to the estimate (20.69). This is provided by the following lemma.

Lemma 20.5.5. *For each $\tau \geqslant 0$ and $K \geqslant 1$, we have*

$$\Psi(\tau) \leqslant \frac{2\tau}{\mathcal{A}^{-1}(\log \tau)} \leqslant 2\, e^{\mathcal{A}(K)} + \frac{2\tau}{K}$$

In particular,

$$\lim_{t \to \infty} \frac{\Psi(t)}{t} = 0$$

Proof. Simply consider the two cases, either $\tau \leqslant e^{\mathcal{A}(K)}$ or $\tau > e^{\mathcal{A}(K)}$. For the limit argue by

$$\limsup_{t \to \infty} \frac{\Psi(t)}{t} \leqslant \frac{2}{K}$$

for all $K \geqslant 1$. □

In establishing the existence of the principal solution for degenerate equations, one naturally aims to find uniform continuity estimates, independent of the distortion, for the approximating functions solving the elliptic equations. For this we require a variant of the inequality (20.8). An argument similar to that used for (20.8) also applies to the Orlicz-Sobolev classes $\mathbb{W}^{1,\Phi}$.

Lemma 20.5.6. *Suppose $\Phi : [1, \infty) \to [1, \infty)$ is an increasing function satisfying the divergence condition*

$$\int_1^\infty \frac{\Phi(t)}{t^3}\, dt = \infty \tag{20.99}$$

Assume further that $t \mapsto \Phi(t^{5/8})$ is convex for large values of t. Let $f \in \mathbb{W}^{1,\Phi}(\Omega)$ be a mapping of finite distortion. Then f is continuous and satisfies the following uniform modulus of continuity,

$$|f(z) - f(w)| \leqslant \rho(|z - w|) \quad \text{whenever } z, w \in \mathbb{D}(a, R) \subset \mathbb{D}(a, 2R) \subset \Omega$$

Here the modulus of continuity $\rho : [0, \infty) \to [0, \infty)$ is continuously increasing with $\rho(0) = 0$. Furthermore, ρ depends only on the gauge Φ, the radius R and the integral $\int_\Omega \Phi(|Df|)$.

Proof. We omit the details of the proof and refer the reader to [191, Theorems 7.3.1 and 7.5.1]. □

Recall from Theorem 20.4.15 that for degenerate equations the uniqueness of the principal solutions is most conveniently established by integrating by the Jacobian determinant by parts whenever this is possible. It turns out that under the above assumptions on $\Phi(t)$ the fundamental Jacobian identities also remain true for the sense-preserving functions in $\mathbb{W}^{1,\Phi}(\Omega)$. For details we refer the reader to [191, Theorem 7.2.1], where a complete argument is presented.

Lemma 20.5.7. *Suppose* $\Phi : [1, \infty) \to [1, \infty)$ *and* $f = u + iv \in \mathbb{W}^{1,\Phi}(\Omega)$ *are as in Lemma 20.5.6, so that in particular* $J(z, f) \geq 0$. *Then* $J(z, f) \in L^1_{loc}(\Omega)$, *and for every* $\phi \in C_0^\infty(\Omega)$ *we have*

$$\int_\Omega \phi(z) J(z, f) = \int_\Omega u\left(v_x \phi_y - v_y \phi_x\right) = -\int_\Omega v\left(u_x \phi_y - u_y \phi_x\right)$$

There are a few remaining ingredients required for the Stoilow factorization, proving the counterpart to Corollaries 20.4.17 and 20.4.18. We refer the reader directly to [191], in particular to Corollary 7.2.1 there, for this.

Lemma 20.5.8. *Suppose* $\Phi(t)$ *and* $f = u + iv$ *are as in Lemma 20.5.6. If* f *is a solution to a Beltrami equation* $f_{\bar{z}} = \mu f_z$, *where* $|\mu(z)| < 1$ *for almost every* $z \in \Omega$, *then the coordinate functions* u *and* v *are weakly monotone and the mapping* f *is continuous.*

We can now continue the proof of Theorem 20.5.2 with the above auxiliary results and establish a general measurable Riemann mapping theorem with the minimal assumptions about the distortion $K(z)$.

The proof of existence is very similar to that described in Theorems 20.4.7. We use the approximations $\mu_m(z)$ to the Beltrami coefficient defined in (20.46). The distortion functions

$$K_m(z) = \frac{1 + |\mu_m(z)|}{1 - |\mu_m(z)|} \leq \frac{1 + |\mu(z)|}{1 - |\mu(z)|} = K(z), \quad m = 1, 2, \ldots$$

each satisfy $K_m(z) \in L^\infty(\mathbb{C})$, and with the measurable Riemann mapping Theorem 5.3.2 we have unique principal solutions $f^m(z) = z + \mathcal{O}(1/z)$ to the Beltrami equations

$$(f^m)_{\bar{z}} = \mu_m\, (f^m)_z \quad \text{almost everywhere in } \mathbb{C}; \ m = 1, 2, \ldots$$

To establish the equicontinuity of the sequence $\{f^m\}$ first use Lemma 20.5.5 for the pointwise estimate

$$\begin{aligned}\Phi(|Df^m(z)|) &= \Psi\bigl(|Df^m(z)|^2\bigr) \leq \Psi\bigl(K_m(z) J(z, f^m)\bigr) \\ &\leq 2e^{\mathcal{A}(K_m(z))} + 2J(z, f^m)\end{aligned} \quad (20.100)$$

20.5. OPTIMAL ORLICZ CONDITIONS

The first term on the right is locally uniformly integrable by assumption (20.94), while the last terms are locally uniformly integrable as f^m are principal quasiconformal mappings with

$$\int_{\mathbb{D}(0,R)} J(z, f^m) \leqslant \pi R^2$$

by Theorem 2.10.1.

On the other hand, according to Lemmas 20.5.4 and 20.5.3, Φ satisfies the assumptions of Lemma 20.5.6, the convexity of $\Phi(t^{5/8})$ and the divergence condition (20.99). Thus Lemma 20.5.6 applies, and we have the equicontinuity of $\{f^m\}$. For the inverse mappings $g^m = (f^m)^{-1}$, the equicontinuity follows immediately from Lemma 20.2.3.

Therefore we have for subsequences

$$f^{m_j}(z) \to f(z), \qquad g^{m_j}(z) \to g(z)$$

locally uniformly, with $f^m \circ g^m(z) = g^m \circ f^m(z) \equiv z = f \circ g(z) = g \circ f(z)$. In particular, f is a homeomorphism. Lemma 20.5.4 shows that $\Phi(t) = \Psi(t^2)$ is convex for large values of t and therefore for the limit map $\Phi(|Df|) = \Psi(|Df|^2) \in L^1_{loc}(\mathbb{C})$.

As in the proof of Theorem 20.4.7, we see that f satisfies the Beltrami equation $f_{\bar{z}} = \mu f_z$ almost everywhere in $\mathbb{C}$. Hence we have the existence of a principal solution. In view of Lemma 20.5.7, repeating the reasoning of Theorem 20.4.15 word for word gives the uniqueness of the principal solutions. Similarly, using the argument of Corollaries 20.4.17 and 20.4.18, we see that Lemma 20.5.7 implies that the coordinates u, v of a solution $h = u + iv \in W^{1,P}_{loc}(\Omega)$ are continuous and weakly monotone. These were the properties needed for the proof of Theorem 20.4.19. Therefore the argument of Theorem 20.4.19 implies the last remaining fact to be proven, the Stoilow factorization for solutions h with $\int_\Omega P(|Dh|) < \infty$. The proof is complete. □

For further recent developments concerning the study of optimal Orlicz-type conditions on the distortion functions we refer the reader to [26, 180, 181, 182, 191]. However, there are still many interesting open questions. For instance, is it true that Theorem 20.2.1 holds under the assumptions

$$K \in L^1(\mathbb{D}), \qquad F \in \mathbb{W}^{1,Q}(\mathbb{D}),$$

where $Q(t) = t^2 \log^{-1}(e+t)$? Or more generally, does Theorem 20.2.1 hold if $\mathcal{A}(K) \in L^1(\mathbb{D})$ and $F \in \mathbb{W}^{1,P}(\mathbb{D})$, where

$$\int_1^\infty \frac{\mathcal{A}(t)}{t^2} dt = \int_1^\infty P(s) \frac{ds}{s^3} = \infty?$$

20.6 Global Solutions

In this section we give fairly general existence and uniqueness results for solutions to the Beltrami equation without the assumption that μ is compactly supported. We point out, as should be obvious, that without any condition on the distortion function at ∞, we cannot hope to guarantee that a homeomorphic solution in the entire plane $\mathbb{C}$ extends continuously to the Riemann sphere $\hat{\mathbb{C}}$.

20.6.1 Solutions on $\mathbb{C}$

Theorem 20.6.1. *Let $\mu : \mathbb{C} \to \mathbb{B}$ be a measurable function valued in the unit disk and suppose that the distortion function of the associated Beltrami equation $h_{\bar{z}} = \mu h_z$ has the property that $\exp(K) \in L^p_{loc}(\mathbb{C})$ for some $p > 0$. Then there is a continuous solution $f : \mathbb{C} \to \mathbb{C}$ to the Beltrami equation with the following properties:*

- *f is injective but not necessarily onto.*
- *$f \in W^{1,Q}_{loc}(\mathbb{C})$ with $Q(t) = \frac{t^2}{\log(e+t)}$.*
- *If $g \in W^{1,Q}_{loc}(\mathbb{C})$ is any other solution, then there is a holomorphic function $\phi : f(\mathbb{C}) \to \mathbb{C}$ such that $g(z) = \phi \circ f(z)$.*

It is to be noted that locally, say on a bounded open set U, the solution we prove the existence of enjoys the same regularity properties as the principal solution of the equation whose Beltrami coefficient is $\mu \chi_U$.

Proof. For every $n = 1, 2, \ldots$, we set

$$\mu_n(z) = \begin{cases} \mu(z), & |z| \leq n \\ 0, & |z| > n \end{cases}$$

As μ_n has a distortion function $K_n(z) \leq K(z)$ with $\exp(K_n) \in L^p(\mathbb{D}_n)$, there is a unique principal solution $h_n : \mathbb{C} \to \mathbb{C}$ to the Beltrami equation

$$\frac{\partial}{\partial \bar{z}} h_n = \mu_n(z) \frac{\partial}{\partial z} h_n, \qquad h_n(z) = z + o(1)$$

We normalize this solution by setting

$$f_n(z) = \frac{h_n(z) - h_n(0)}{h_n(1) - h_n(0)}$$

We want to extract from this sequence a subsequence converging locally uniformly to a mapping $f : \mathbb{C} \to \mathbb{C}$ with $f(0) = 0$ and $f(1) = 1$. It suffices to show that for each disk $\mathbb{D}(0, R)$, $R \geq 2$, there is a subsequence converging locally uniformly on it, for then the usual diagonal selection argument will apply to generate the sequence we seek. Fix such a disk $B = \mathbb{D}(0, R)$ and let $h : \mathbb{C} \to \mathbb{C}$ be the homeomorphic solution to the equation

$$\bar{\partial} h = \mu(z) \chi_\mathbb{D}(z) \partial h$$

20.6. GLOBAL SOLUTIONS

normalized so that $h(0) = 0$ and $h(1) = 1$. This is unique by Theorem 5.3.4. The factorization theorem tells us that for each n, with $n \geq R$, the map f_n can be written as
$$f_n(z) = \phi_n(h(z)),$$
where $\phi_n : h(B) \to \mathbb{C}$ form a family of conformal mappings with $\phi_n(0) = 0$ and $\phi_n(1) = 1$. The family $\{\phi_n\}$ is normal on $h(B)$, for instance, by Theorems 3.9.1 and 2.10.9, with every limit being a conformal map.

This process provides us with our limit map on B, namely,
$$\lim_{k \to \infty} f_{n_k} = \lim_{k \to \infty} \phi_{n_k}(h(z)) = \phi(h(z)) = f(z),$$
where $\phi_{n_k} \to \phi$, a conformal mapping locally uniformly on $h(B)$.

Now we need to find uniform bounds on the integrals of the Jacobians. Since each f_{n_k} is a homeomorphism of Sobolev class $W^{1,1}_{loc}(B)$, we have
$$\int_U J(z, f_{n_k}) \, dz \leq |f_{n_k}(U)| \leq C_U$$
for every relatively compact subset U of B. The constant C_U should not depend on n_k, and this choice can be made as $f_{n_k} \to f$ uniformly on compact subsets of B. With the aid of the elementary inequality
$$\frac{|Df_{n_k}|^2}{\log(e + |Df_{n_k}|^2)} \leq \frac{K(z)J(z, f_{n_k})}{\log(e + K(z)J(z, f_{n_k}))}$$
$$\leq C_p J(z, f_{n_k}) + C_p \exp(pK(z)),$$
we conclude that the sequence $\{Df_{n_k}\}$ is bounded in $L^Q(U)$ for every relatively compact subdomain U of $\mathbb{D}$. Thus $f_{n_k} \to f$ weakly in $W^{1,Q}(U)$. In particular, $f \in W^{1,Q}_{loc}(\mathbb{D})$, $Q(t) = t^2 \log^{-1}(e + t^2)$. Moreover,
$$f_{\bar{z}} - \mu f_z = \lim_{k \to \infty} (\bar{\partial} f_{n_k} - \mu \partial f_{n_k}) = 0,$$
the limit being the weak limit in $W^{1,Q}_{loc}(B)$. In particular, f solves the Beltrami equation almost everywhere, and this last observation completes the proof of the theorem. $\square$

Note here that although each f_{n_k} is a homeomorphism of $\mathbb{C}$ onto $\mathbb{C}$, the limit map f need not be. As an example, consider the map f constructed in Theorem 20.3.1 from a weight function $\mathcal{A}$. More precisely, for a smooth strictly increasing function $\mathcal{A}(s)$, we set
$$f(z) = \frac{z}{|z|} \rho(|z|), \quad \text{where } \rho(t) = \exp\left(\int_0^t \frac{ds}{sK(s)}\right), \quad 0 \leq t \leq 1,$$
and the weight $\mathcal{A}(s)$ and $K(s) = K_{\mathcal{A}}(s)$ were related by
$$e^{\mathcal{A}(K(s))} = \frac{e}{s}$$

For $|z| > 1$ we can set $f(z) = \rho(1)z$, so that f is conformal outside the unit disk and continuous at ∞ with the definition $f(\infty) = \infty$. Now the mapping of finite distortion

$$g(z) = \frac{1}{f(\frac{1}{z})}, \qquad g(0) = 0,$$

is defined and continuous on $\mathbb{C}$, but it is bounded, $g : \mathbb{C} \to \mathbb{D}$. However, the distortion $K(z, g) = K(\frac{1}{z}, f) = K_A(\frac{1}{z})$ is locally bounded, thus certainly in $L^p_{loc}(\mathbb{C})$.

In fact, it is probable that this example describes the precise limits for the existence of bounded entire mappings of finite distortion (or, if one prefers, solutions to the Beltrami equation).

20.6.2 Solutions on $\hat{\mathbb{C}}$

What we want to do now is to give a condition that implies that the solution defined above can be extended continuously, and therefore homeomorphically, to $\hat{\mathbb{C}}$ by setting $f(\infty) = \infty$. One could achieve this by demanding uniform asymptotic estimates on the integrability properties of K to give modulus of continuity bounds, as is done in [74, 315]. However, this approach does not provide the regularity estimates we want. We will simply assume exponential integrability of the distortion in a spherical sense.

Theorem 20.6.2. *Let $\mu : \hat{\mathbb{C}} \to \mathbb{D}$ be a measurable function and suppose that the distortion function of the associated Beltrami equation $h_{\bar{z}} = \mu h_z$ has the property that*

$$\int_{\mathbb{C}} \exp(pK) \frac{dz}{1 + |z|^4} < \infty \qquad (20.101)$$

for some $p > 0$. Then there is a homeomorphic solution $f : \hat{\mathbb{C}} \to \hat{\mathbb{C}}$ to the Beltrami equation such that

$$\int_{\mathbb{C}} \frac{|\Psi(z)|^2}{\log(e + \Psi(z))} \frac{dz}{1 + |z|^4} < \infty, \qquad (20.102)$$

where $\Psi(z) = \frac{1 + |z|^2}{1 + |f(z)|^2} |Df(z)|$.

Of course, it follows from our factorization results that this solution is unique up to the normalization $f(0) = 0$, $f(1) = 1$ and $f(\infty) = \infty$. Further note that if

$$\limsup_{z \to \infty} \frac{|z|}{|f(z)|} < \infty,$$

then (20.102) reduces to the weighted L^Q-estimate for $|Df|$ with weight given by the spherical measure $\frac{d\zeta}{1+|\zeta|^4}$. In general, the weight depends on the growth rate of f near ∞. Bounds on the rate can be provided by local Hölder or modulus of continuity estimates at ∞. However, when formulated in these terms, the result appears to be quite technical and not to be sharp.

20.7. A LIOUVILLE THEOREM

Proof. Let f be the normalized injective solution given by Theorem 20.6.1. We first want to show that f has a well-defined limit at ∞. To this end set

$$g(z) = \frac{1}{f(\frac{1}{z})}, \qquad z \in \mathbb{C} \setminus \{0\}$$

We compute that

$$\mu_g(z) = \mu\left(\frac{1}{z}\right)\left(\frac{z}{|z|}\right)^4$$

The distortion functions are therefore related by the formula $K_g(z) = K_f(\frac{1}{z})$. We then have

$$\int_{\mathbb{D}} \exp(pK_g)\, dz = \int_{\mathbb{C}\setminus\mathbb{D}} \exp(pK_f)\frac{d\zeta}{|\zeta|^4}$$

$$\leq 2\int_{\mathbb{C}} \exp(pK_f)\frac{d\zeta}{1+|\zeta|^4} < \infty$$

Therefore we can find a unique principal solution to the equation

$$h_{\bar{z}} = \mu_g(z)\chi_{\mathbb{D}}(z)h_z$$

defined on $\mathbb{C}$. Now on the domain $\mathbb{D}\setminus\{0\}$, both g and h are embeddings satisfying the same Beltrami equation. We apply the factorization theorem to see that

$$g(z) = (\varphi \circ h)(z), \qquad z \in \mathbb{D} \setminus \{0\}$$

Now

$$\varphi : h(\mathbb{D}\setminus\{0\}) \to g(\mathbb{D}\setminus\{0\}) \qquad (20.103)$$

is conformal. Therefore φ admits a continuous extension over the point $h(0)$. This value must be finite since $f(0) = 0$ and f is a homeomorphism onto its image. Hence g has a well-defined limit at 0, and therefore we are given a well-defined limit to f at ∞. As f is a homeomorphism on $\mathbb{C}$, it must be the case that $f(\infty) = \infty$. Furthermore, $g \in W^{1,Q}_{loc}(\mathbb{D})$ for gauge Q as above, and as $\frac{1}{f(z)} = g(\frac{1}{z})$, we have

$$|Dg(z)| = |Df(1/z)|\frac{1}{|z|^2|f(\frac{1}{z})|^2}$$

The result claimed follows from the change-of-variables formula. $\square$

20.7 A Liouville Theorem

Earlier in Theorem 14.4.7 we discussed the consequences of the integral bound

$$\frac{1}{\pi R^2}\int_{\mathbb{D}(0,R)} K(z,f) \leq K_\infty < \infty \qquad (20.104)$$

required for all *sufficiently large* disks $\mathbb{D}(0,R)$ centered at the origin. The constant K_∞ is of course supposed to be independent of the disk. Combining the isoperimetric inequality with appropriate differential inequalities, we proved the following result, which is restated here in terms of mappings of finite distortion.

Theorem 20.7.1. *Let $f : \mathbb{C} \to \mathbb{C}$ be of finite distortion and assume that the distortion function K of f satisfies the bound (20.104) for all $R > R_0 > 1$. If, moreover, $J(z,f) \in L^q(\mathbb{C})$ for some $1 < q \leq \frac{K_\infty}{K_\infty - 1}$, then f is constant.*

20.8 Applications to Degenerate PDEs

The theory of mappings of finite distortion has of course immediate consequences in the theory of degenerate elliptic PDEs. We will not cover this theme systematically but only recall the first few immediate consequences where we combine the method of $\mathcal{A}$-harmonic fields described in Section 16.1.6 with the results of the present chapter.

Let $\Omega \subset \mathbb{C}$ be a bounded domain and suppose that

$$\mathcal{A} : \Omega \times \mathbb{R}^2 \to \mathbb{R}^2$$

is measurable in $z \in \Omega$, is continuous in $\zeta \in \mathbb{R}^2$ and satisfies, for $(z,\zeta) \in \Omega \times \mathbb{R}^2$, the ellipticity condition

$$|\zeta|^2 + |\mathcal{A}(z,\zeta)|^2 \leq \left(K(z) + \frac{1}{K(z)}\right) \langle \zeta, \mathcal{A}(z,\zeta) \rangle, \qquad (20.105)$$

where $K(z) \geq 1$, $z \in \Omega$.

Theorem 20.8.1. *Suppose $u \in W^{1,1}_{loc}(\Omega)$ is a solution to the nonlinear equation*

$$\operatorname{div}\mathcal{A}(z,\nabla u) = 0 \qquad (20.106)$$

with finite energy

$$\int_\Omega \langle \mathcal{A}(z,\nabla u), \nabla u \rangle < \infty$$

If the ellipticity bound $K(z)$ satisfies

$$e^{K(z)} \in L^p(\Omega) \qquad (20.107)$$

for some $p > 1$, then the solution u has the regularity

$$|\nabla u|^2 + |\mathcal{A}(z,\nabla u)|^2 \in L^1_{loc}(\Omega) \qquad (20.108)$$

Proof. We consider the conjugate equation

$$\mathcal{A}(z,\nabla u) = *\nabla v$$

20.9. LEHTO'S CONDITION

and the associated map $f = u + iv$. In view of (16.33) the ellipticity bounds (20.105) read as
$$|Df(z)|^2 \leqslant K(z)J(z,f)$$
Theorem 20.4.20 now proves that $|Df| \in L^2_{loc}(\Omega)$, which is equivalent to the claim (20.108). $\square$

We can also ask for a finer scale of regularity estimates when (20.107) holds for some $p > 0$. In this case the above results show that
$$|\nabla u|^2 \log^{\beta-1}(e + |\nabla u|) \in L^1_{loc}(\Omega), \qquad 0 < \beta < p,$$
and similarly
$$|\mathcal{A}(z, \nabla u)|^2 \log^{\beta-1}(e + |\mathcal{A}(z, \nabla u)|) \in L^1_{loc}(\Omega), \qquad 0 < \beta < p$$

Moreover, analogs of basic results from Section 16.2 follow in a similar fashion. For instance, under the assumptions of Theorem 20.8.1, we can make use of Stoilow factorization and apply Theorem 20.4.19. We see that
$$u = h(f(z)),$$
where f is a principal mapping, hence a homeomorphism, of exponentially integrable distortion with Beltrami coefficient supported on Ω, while h is harmonic on the domain $\Omega' = f(\Omega)$. Thus, for instance, u shares all the topological properties of harmonic functions.

20.9 Lehto's Condition

Our approach to existence, uniqueness and regularity of mappings of finite distortion solving a, possibly degenerate, Beltrami equation has so far been almost entirely analytically based. This is because existence and regularity go hand in hand when approaching these problems analytically. However, in this section we depart from that approach to give a proof of a theorem of Lehto [227] that shows how to obtain homeomorphic solutions to the Beltrami equation under very weak assumptions on the distortion function. The price we must pay is that we see very little regularity in the solution – but it is a mapping of finite distortion. The proof is based around moduli estimates, which themselves provide one of the many possible approaches to the theory of quasiconformal mappings. We shall not go here into great detail here since this approach is covered in greater depth elsewhere, see for instance [8, 18, 229, 367, 365]. The key idea is the classical length area method, [8]. Some of the ideas in Lehto's proof have been developed into a theory of mappings which distort moduli by a bounded amount in some integral sense. These are the Q-homeomorphisms, see [246, 247]. This approach has been used earlier to study existence of homeomorphic solutions to degenerate Beltrami equations, see eg. [74].

Geometric bounds on the moduli of (topological) annuli, provide powerful methods in the geometric study of mappings and Lehto's theorem illustrates this, which is why we have chosen to present it here. The reader will see the main idea is simply that the bounds on moduli and an associated distortion estimate, even in an average sense, provide equicontinuity estimates. From these we can produce a solution to the equation via an approximation process. Further, the ideas around this theorem are the key in understanding the weldings of random homeomorphisms of the real line.

We begin with the following lemma.

Lemma 20.9.1. *Consider two annuli*

$$\mathbb{A} = \{z : r < |z| < R\} \quad \text{and} \quad \mathbb{A}' = \{\zeta : r' < |\zeta| < R'\}$$

For every homeomorphism $f : \mathbb{A} \to \mathbb{A}'$ of finite distortion, we have

$$4\pi^2 \int_r^R \frac{1}{\int_0^{2\pi} K(\rho e^{i\theta}, f)\, d\theta} \frac{d\rho}{\rho} \leqslant 2\pi \log \frac{R'}{r'} \tag{20.109}$$

The estimate holds as an equality for the radial stretchings $f(z) = z|z|^{\frac{1}{K}-1}$.

Proof. The proof of this lemma uses some elementary geometric facts which we later develop more systematically in our study of free Lagrangians, see eg. Lemma 21.3.9.

First, from the increment of the argument,

$$\int_0^{2\pi} \frac{f_\theta(\rho e^{i\theta})}{f(\rho e^{i\theta})}\, d\theta = 2\pi i \quad \text{for almost every } \rho \in (r, R)$$

Writing the complex derivatives in polar coordinates, see (21.51), shows that $\rho^{-2}|f_\theta|^2 \leqslant (|f_z| + |f_{\bar z}|)^2 \leqslant K(z, f)J(z, f)$. Hence

$$2\pi \leqslant \rho \int_0^{2\pi} \frac{\sqrt{K(\rho e^{i\theta}, f)J(\rho e^{i\theta}, f)}}{|f(\rho e^{i\theta})|}\, d\theta \tag{20.110}$$

Subsequently Hölder's inequality gives us

$$\frac{4\pi^2}{\rho \int_0^{2\pi} K(\rho e^{i\theta})\, d\theta} \leqslant \rho \int_0^{2\pi} \frac{J(\rho e^{i\theta}, f)}{|f(\rho e^{i\theta})|^2}\, d\theta$$

Integrating this estimate with respect to the radius ρ and observing that

$$\int_r^R \int_0^{2\pi} \frac{J(\rho e^{i\theta}, f)}{|f(\rho e^{i\theta})|^2} \rho\, d\theta\, d\rho = \int_\mathbb{A} \frac{J(z,f)}{|f(z)|^2} \leqslant \int_{\mathbb{A}'} \frac{1}{|z|^2} = 2\pi \log \frac{R'}{r}$$

gives us the result we seek. The inequality here, arising from the change of variables, is based on Corollary 3.3.6. (For a more precise discussion see (21.59) below.)

That the estimate is sharp is easy to establish as the power mappings have constant distortion. □

20.9. LEHTO'S CONDITION

The inequality at (20.109) is written in this fashion to identify the right hand side as the modulus of the annulus $\mathbb{A}'$,

$$\mathrm{mod}(\mathbb{A}') = 2\pi \log \frac{R'}{r'}$$

Of course, any topological annulus, a doubly connected domain Ω, is conformally equivalent to a round annulus $\mathbb{A}$, and we may define the conformal invariant $\mathrm{mod}(\Omega) = \mathrm{mod}(\mathbb{A})$. If we do this, then both sides of (20.109) become conformally invariant, and we obtain the following corollary.

Corollary 20.9.2. *Suppose $f : \mathbb{A} \to \Omega$ is a homeomorphism of finite distortion. Then*

$$4\pi^2 \int_r^R \frac{1}{\int_0^{2\pi} K(\rho e^{i\theta}, f)\, d\theta} \frac{d\rho}{\rho} \leqslant \mathrm{mod}(\Omega)$$

In particular, if f is quasiconformal, we see from the corollary that the mapping f preserves the moduli of all annuli, up to the multiplicative factor $K(f)$. The mappings $z \mapsto z|z|^\alpha$, $\alpha > 0$, show such an estimate to be optimal. Conversely [229], a homeomorphism that preserves the moduli of annuli up to a uniform multiplicative constant is quasiconformal. Thus estimates that connect the geometric bounds of the annuli with the modulus provide one possible way to develop the theory of quasiconformal mappings. We illustrate this approach with the following key estimate, often coined as the Loewner property of the complex plane.

Theorem 20.9.3. *Suppose $\Omega = \mathbb{C} \setminus (E \cup F)$ is a topological annulus with connected components of the complement E bounded, and F unbounded. Then*

$$\mathrm{mod}(\Omega) \leqslant \Phi\left(\frac{\mathrm{dist}(E, F)}{\mathrm{diam}(E)}\right), \qquad (20.111)$$

where $\Phi : [0, \infty] \to [0, \infty]$ is an increasing homeomorphism, independent of the continua.

Proof. The estimate is invariant under similarities—rotation and scaling—and thus we may assume that $0, 1 \in E$ with $\mathrm{diam}(E) = 1$.

Assume first that $\mathrm{dist}(E, F) = r > 1$. Let $\phi : \Omega \to \mathbb{A} = \{z : \rho < |z| < 1\}$ be a conformal mapping to a round annulus. Then the function

$$u(z) = \frac{\log |\phi(z)|}{\log \rho}, \qquad z \in \Omega,$$

extends to $u \in W^{1,2}_{loc}(\mathbb{C})$ with $u|F \equiv 0$ and $u|E \equiv 1$. In particular, the conditions 1. – 4. in the proof of Theorem 3.5.3 are now satisfied. Since

$$\int_{\mathbb{C}} |\nabla u|^2 = (\log(1/\rho))^{-2} \int_\Omega \frac{|\phi'(z)|^2}{|\phi(z)|^2} = \frac{4\pi^2}{\mathrm{mod}(\Omega)},$$

(3.25) gives the required estimate (20.111), with the function

$$\Phi(t) = \frac{16\pi^3}{\log\left(1 + \frac{2}{(1+t)^2}\right)}, \quad t \geq 1$$

(For our subsequent application this suffices.) In the remaining case $\text{dist}(E, F) = r \leq 1$ one needs to compare Ω with the Teichmüller ring domain [18, 229, 367] which, by a symmetrization argument, has the largest modulus among all annuli with the given geometric data. The modulus of the Teicmüller ring can be explicitly calculated, and in this way we find the required function $\Phi(R)$ for $0 \leq R \leq 1$. □

After these preparations we are ready to state and prove Lehto's theorem.

Theorem 20.9.4. *Suppose μ is measurable, compactly supported and $|\mu(z)| < 1$ for almost all $z \in \mathbb{C}$. Assume that distortion function*

$$K(z) = \frac{1 + |\mu(z)|}{1 - |\mu(z)|}$$

is locally integrable, $K \in L^1_{loc}(\mathbb{C})$, and

$$\lim_{r \to 0} \int_r^1 \frac{1}{\int_0^{2\pi} K(z + \rho e^{i\theta})\, d\theta} \frac{d\rho}{\rho} = \infty, \quad z \in \mathbb{C} \qquad (20.112)$$

Then the Beltrami equation

$$\frac{\partial f}{\partial \bar{z}}(z) = \mu(z)\frac{\partial f}{\partial z}(z) \quad \text{for almost every } z \in \mathbb{C} \qquad (20.113)$$

admits a homeomorphic $W^{1,1}_{loc}$-solution $f : \mathbb{C} \to \mathbb{C}$.

Note that the requirements on the distortion function K are closely related but weaker than those used for Theorem 20.5.2. However, the regularity of the solution is now much weaker, and in particular too weak to guarantee the uniqueness of homeomorphic solutions, let alone the Stoilow factorization or the classification of all solutions to (20.113).

Lehto [227] originally applied this theorem for the welding of homeomorphisms of the real line (as discussed earlier in Section 5.10). In this case, the uniqueness of the solution to (20.113) is reduced to the uniqueness of the welding mapping where methods and techniques from conformal mappings are applicable.

Proof of Theorem 20.9.4. We may assume that μ is supported in the unit disk. We then construct good approximations μ_n as defined in (20.46). Let f_n be the (quasiconformal) principal solutions to the Beltrami equation

$$f_{\bar{z}} = \mu_n(z) f_z$$

20.9. LEHTO'S CONDITION

We would like to show the equicontinuity of the family $\{f_n\}$. Given $z_0 \in \mathbb{D}$ consider the annulus $\mathbb{A}(r) = \{z : r < |z - z_0| < 1\}$ and apply Theorem 20.9.3 with $E = \overline{f_n \mathbb{D}(z_0, r)}$ and $F = \mathbb{C} \setminus f_n \mathbb{D}(z_0, 1)$. By Theorem 2.10.4 we have

$$\operatorname{dist}(E, F) \leqslant \operatorname{diam}(f_n \mathbb{D}(z_0, 1)) \leqslant 8,$$

so that Corollary 20.9.2 gives

$$\int_r^1 \frac{1}{\rho \int_0^{2\pi} K(z + \rho e^{i\theta}) \, d\theta} \, d\rho \leqslant \int_r^1 \frac{1}{\rho \int_0^{2\pi} K(z + \rho e^{i\theta}, f_n) \, d\theta} \, d\rho$$

$$\leqslant \frac{1}{4\pi^2} \Phi\left(\frac{8}{\operatorname{diam}(f_n \mathbb{D}(z_0, r))}\right)$$

Our hypothesis (20.112) shows that, as $r \searrow 0$, the left hand side tends to ∞. This allows us to see that $\operatorname{diam}(f_n \mathbb{D}(z_0, r)) \to 0$ with a rate independent of n which proves the equicontinuity of the sequence $\{f_n\}$. We conclude that there is a continuous mapping f so that for a subsequence $f_n \to f$ locally uniformly in $\mathbb{C}$. To see that f is a homeomorphism, note that by Lemma 20.2.3 the family $\{f_n^{-1}\}$ of inverse mappings is equicontinuous and hence converges uniformly with limit f^{-1}. Next, since

$$|Df_n(z)| \leqslant \sqrt{K(z, f_n) J(z, f_n)} \leqslant \sqrt{K(z) J(z, f_n)},$$

an elementary application of Hölder's inequality implies that the sequence $\{|Df_n|\}$ is uniformly integrable and thus for a subsequence, still denoted by $\{f_n\}$, the derivatives converge weakly in $L^1_{loc}(\mathbb{C})$ to the corresponding derivatives of f. Consequently $f \in W^{1,1}_{loc}(\mathbb{C})$. It remains to show that f solves (20.113). For this let $\phi \in C_0^\infty(\mathbb{C})$ and consider the identity

$$\int \phi \left(\frac{\partial f_n}{\partial \bar{z}} - \mu \frac{\partial f_n}{\partial z}\right) = \int \phi (\mu_n - \mu) \frac{\partial f_n}{\partial z} \qquad (20.114)$$

As the derivatives of f_n converge weakly, the left hand side tends to $\int \phi(f_{\bar{z}} - \mu f_z)$. The right hand side can be estimated by

$$\int |\phi| |\mu_n - \mu| |Df_n| \leqslant \int |\phi| |\mu_n - \mu| \sqrt{K(z) J(z, f_n)}$$

$$\leqslant \left(\int_\mathbb{D} |\mu_n - \mu| K\right)^{1/2} \left(\int_\mathbb{D} J(z, f_n)\right)^{1/2}$$

With the area formula (2.58) the last integrals are uniformly bounded, so that by Lebesgue's Dominated convergence theorem the right hand side of (20.114) tends to zero. Thus f is a solution to the Beltrami equation and, in particular, is a mapping of finite distortion. □

Chapter 21

Aspects of the Calculus of Variations

The theory of mappings of finite distortion arose out of a need to extend the ideas and applications of the classical theory of quasiconformal mappings to the degenerate elliptic setting. There one finds concrete applications in materials science, particularly nonlinear elasticity and critical phase phenomena, and in the calculus of variations. In this chapter we consider applications of mappings of finite distortion to some interesting problems in the calculus of variations.

21.1 Minimizing Mean Distortion

Here we discuss recent advances in the study of existence and uniqueness properties for mappings between planar domains whose boundary values are prescribed and have the smallest mean distortion. For closed Riemann surfaces, existence and uniqueness within a homotopy class follows, more or less, from known results in the theory of harmonic mappings based on the work of Shoen and Yau [323]; see also Wolf [372]. The relevance of this work comes out of a surprising connection discovered in [27] between the minimizers of the mean distortion functional and harmonic mappings in the plane, and this carries over into the setting of Riemann surfaces.

It is very important in applications, such as the deformation theory of conformal dynamical systems or low-dimensional topology and geometry, to study these mappings in degenerating sequences. This quickly leads one to consider extremal mappings of mean distortion between annuli with flat metrics, a classical and well-understood problem for extremal quasiconformal mapping (that is, mappings that minimize the L^∞-norm of the mean distortion). This is known classically as the Grötsch problem. Of course, annuli are where one first observes nontrivial conformal invariants such as moduli. Here we point out a close relationship between the existence and uniqueness problems with the well-known conjecture of Nitsche concerning harmonic homeomorphisms. P. Duren's book

21.1. MINIMIZING MEAN DISTORTION

[114] is a good reference for the theory of harmonic mappings. We recall that, according to the fundamental theorem of Radó, Kneser and Choquet, if $\Omega \subset \mathbb{R}^2$ is a bounded convex domain, then for each homeomorphism $h_o : \partial\Omega' \to \partial\Omega$ the harmonic extension $h : \overline{\Omega'} \to \overline{\Omega}$ maps Ω' univalently onto Ω. By a theorem of Lewy the univalent harmonic map has a nonvanishing Jacobian. Its inverse is therefore a real analytic diffeomorphism.

A second reason for our studies is that we find many new and unexpected phenomena concerning existence, uniqueness and regularity for these extremal problems where the functionals are polyconvex but typically not convex. These seem to differ markedly from phenomena observed when studying multi-well-type functionals in the calculus of variations. The phenomena observed concerning mappings between annuli present another case in point.

An important observation is the following. The principal advantage of minimizing over families of homeomorphisms in the variational problems we shall study below lies in the fact that the inverse maps are also extremal for their own variational integrals. As we shall see, sometimes this associated problem is easier to solve than the original one, as it may involve minimizing a convex functional [27].

Thus our basic aim in this section is to extend the theory of extremal quasiconformal mappings to the consideration of minimizers of integral averages of the distortion function instead of its maximum, that is, the L^1-norm as opposed to the L^∞-norm. For L^p, $p < 1$, we show that there is no viable theory, though there is a chance for certain Zygmund spaces around L^1. Near L^∞, we have mappings with exponentially integrable distortion and a viable theory for minimal distortion, though we do not discuss that in this book. See [27] for a discussion of this subject. It remains an interesting open question as to whether there is a good theory for any L^p with $1 < p < \infty$. In particular, is there a good theory of mappings minimizing the L^2-norm of the distortion function?

Definition of the Distortion $\mathbb{K}(z, f)$

The most common notion measuring the distortion of a mapping f is the *linear distortion*, given at points of differentiability by $K(z,f) = |Df(z)|^2 J(z,f)^{-1}$. Here $|Df(z)| = |f_z| + |f_{\bar{z}}|$ is the operator norm of the derivative. This norm has the disadvantage of being insufficiently regular to deal with variational equations. Therefore in the study of extremal mean distortion we replace the operator norm by the Hilbert-Schmidt norm of the derivative,

$$\|Df(z)\|^2 = \operatorname{tr}(Df^t Df) = 2|f_z|^2 + 2|f_{\bar{z}}|^2$$

and in the entire chapter use the distortion function $\mathbb{K}(z, f)$ defined by

$$\mathbb{K}(z,f) = \frac{|f_z|^2 + |f_{\bar{z}}|^2}{|f_z|^2 - |f_{\bar{z}}|^2} = \frac{1}{2}\frac{\|Df(z)\|^2}{J(z,f)} \qquad (21.1)$$

This has better convexity properties than $K(z, f)$ and behaves better in the variational questions concerning the distortion, see for instance Theorems 21.1.1 and 21.1.2 below. However, notice that

$$\mathbb{K}(z, f) = \frac{1}{2}\left(K(z, f) + \frac{1}{K(z, f)}\right)$$

is a convex function of $K(z, f)$, so the L^∞-minimizers are the same.

21.1.1 Formulation of the General Problem

We can now formulate the major minimization problem we shall address.

Let $\mathcal{F}$ consist of homeomorphisms $f : \Omega \to \Omega'$ of finite distortion such that

$$\int_\Omega \mathbb{K}(z, f) \, dz < \infty \tag{21.2}$$

Problem: *Given $f_o \in \mathcal{F}$, find a mapping $f \in \mathcal{F}$ that coincides with f_o on $\partial\Omega$ and minimizes the integral at (21.2).*

The lack of convexity for $K(z, f)$ motivating the definition given in (21.1) is easily seen by considering the classical Grötzsch problem in a new setting, in the next results from [27].

21.1.2 The L^1-Grötzsch Problem

In this section we present an L^1-variant of the celebrated Grötzsch extremal problem for mappings between rectangles [149]. Let $\mathbb{Q}$ be the unit square in the plane,

$$\mathbb{Q} = [0, 1] \times [0, 1] \subset \mathbb{R}^2,$$

and let $\mathbb{Q}'$ be a rectangle,

$$\mathbb{Q}' = [0, 2] \times [0, 1] \subset \mathbb{R}^2$$

We shall show that

$$\min_{f \in \mathcal{F}} \int_\mathbb{Q} \mathbb{K}(z, f) \, dz = \frac{5}{4}, \tag{21.3}$$

when $\mathcal{F}$ consists of homeomorphisms $f : \overline{\mathbb{Q}} \to \overline{\mathbb{Q}'}$ of integrable distortion, contained in the Sobolev class $W^{1,1}_{loc}(\mathbb{Q}, \mathbb{R}^2)$ and taking vertices into vertices. Our goal is to show that this free boundary value problem has a unique extremal. Before proving this result, let us first demonstrate that uniqueness is lost for the usual distortion function $K(z, f)$.

Theorem 21.1.1. *The minimization problem*

$$\min_{f \in \mathcal{F}} \int_\mathbb{Q} K(z, f) \, dz \tag{21.4}$$

has infinitely many extremals.

21.1. MINIMIZING MEAN DISTORTION

Proof. Suppose $f \in \mathcal{F}$. We first show that

$$2 \leqslant \int_{\mathbb{Q}} K(z, f)\, dz \qquad (21.5)$$

To see this we note that

$$2 \leqslant \int_0^1 |Df(x+iy)|\, dx \quad \text{for almost all } y \in [0,1],$$

and then after integrating over y, we find that

$$2 \leqslant \int_{\mathbb{Q}} |Df(z)|\, dz = \int_{\mathbb{Q}} \sqrt{K(z,f)}\,\sqrt{J(z,f)}\, dz$$

Upon squaring this we find that Hölder's inequality further implies

$$4 \leqslant \int_{\mathbb{Q}} K(z,f)\, dz \cdot \int_{\mathbb{Q}} J(z,f)\, dz \qquad (21.6)$$

Since $f : \mathbb{Q} \to \mathbb{Q}'$ is a homeomorphism of the Sobolev class $W^{1,1}_{loc}(\mathbb{Q}, \mathbb{R}^2)$, it follows that

$$\int_{\mathbb{Q}} J(z,f)\, dz \leqslant |\mathbb{Q}'|,$$

where equality holds if f maps sets of zero measure into sets of zero measure. The claim (21.5) follows.

For every $0 \leqslant a < 1$, the piecewise linear map

$$g(z) = \begin{cases} x + iy & \text{if } z \in [0, a] \times [0, 1] \\ \frac{2-a}{1-a}(x - a) + a + iy & \text{if } z \in [a, 1] \times [0, 1] \end{cases}$$

is a minimizer. Indeed, g is the identity if $0 \leqslant x \leqslant a$, while for $x \leqslant a < 1$ we have

$$Dg(z) = \begin{pmatrix} \frac{2-a}{1-a} & 0 \\ 0 & 1 \end{pmatrix} \quad |Dg| \equiv \frac{2-a}{1-a} \quad \text{hence } K(z,g) \equiv \frac{2-a}{1-a}$$

The integral of $K(z, g)$ does not depend on a; indeed,

$$\int_{\mathbb{Q}} K(z,g)\, dz = a \cdot 1 + (1-a)\frac{2-a}{1-a} = 2$$

This completes the proof of Theorem 21.1.1. $\square$

The situation is dramatically different for the distortion function $\mathbb{K}$.

Theorem 21.1.2. *The minimization problem (21.3) has a unique extremal.*

Proof. Let $f : \overline{\mathbb{Q}} \to \overline{\mathbb{Q}'}$ be any admissible mapping, that is, $f \in \mathcal{F}$, and write
$$f(z) = u(x,y) + iv(x,y), \qquad z = x + iy$$
We observe that for almost every $0 < y < 1$,
$$\int_0^1 u_x(x+iy)\,dx = u(1+iy) - u(iy) = 2,$$
while for almost every $0 < x < 1$,
$$\int_0^1 v_y(x+iy)\,dx = v(x+i) - v(x) = 1$$
Further integration yields
$$\int_{\mathbb{Q}} u_x\,dx\,dy = 2 \quad \text{and} \quad \int_{\mathbb{Q}} v_y\,dx\,dy = 1$$
We combine these equations in one weighted sum and use the Schwarz inequality to obtain
$$5 = \int_{\mathbb{Q}} (2u_x + v_y)\,dx\,dy \leqslant \sqrt{5}\int_{\mathbb{Q}} \sqrt{u_x^2 + v_y^2}\,dx\,dy \qquad (21.7)$$
$$\leqslant \sqrt{5}\int_{\mathbb{Q}} \|Df\| = \sqrt{10}\int_{\mathbb{Q}} \sqrt{\mathbb{K}(z,f)}\sqrt{J(z,f)}\,dz \qquad (21.8)$$
Upon squaring both sides, Hölder's inequality implies
$$25 \leqslant 10\,|\mathbb{Q}'|\int_{\mathbb{Q}} \mathbb{K}(z,f)\,dz = 20\int_{\mathbb{Q}} \mathbb{K}(z,f)\,dz$$
In other words,
$$\int_{\mathbb{Q}} \mathbb{K}(z,f)\,dz \geqslant \frac{5}{4} \quad \text{for all } f \in \mathcal{F}$$
As expected, equality occurs when $f(z) = 2x + iy$. We need only show that the equation
$$\int_{\mathbb{Q}} \mathbb{K}(z,f)\,dz = \frac{5}{4} \quad \text{for } f \in \mathcal{F} \text{ yields } f = 2x + iy$$
To this end, we must examine when equalities occur in the chain of the above estimates. First, (21.8) holds as an equality only if
$$u_y \equiv v_x \equiv 0,$$
meaning that u depends only on x and v depends only on y. Recall that (21.7) came from using the Schwartz inequality. This forces the following relation
$$2v_y = u_x \quad \text{almost everywhere in } \mathbb{Q}$$
As u_x and v_y depend on different variables, we infer that u_x and v_y are constant functions. This leaves only one possibility,
$$u(x,y) = 2x \quad \text{and} \quad v(x,y) = y,$$
as claimed. $\square$

21.1.3 Sublinear Growth: Failure of Minimization

Given a mapping $f_o : \overline{\mathbb{D}} \to \mathbb{C}$, we now study minima of the variational integral

$$\min \int_{\mathbb{D}} \Psi\left[\mathbb{K}(z, f)\right] dz \qquad (21.9)$$

in the class $\mathcal{F}_o = \mathcal{F}(f_o)$ of all $W^{1,2}(\mathbb{D}, \mathbb{C})$-homeomorphisms $f : \overline{\mathbb{D}} \to \mathbb{C}$ that coincide with f_o on $\partial \mathbb{D}$. We demonstrate that these integrals almost never attain their minimum value if Ψ exhibits *sublinear growth*, meaning that

$$\lim_{t \to \infty} \frac{\Psi(t)}{t} = 0$$

In many ways this situation is reminiscent of the well-known Lavrentiev phenomenon in the calculus of variations [224]. For simplicity we consider quasiconformal boundary data. It is easy to generalize this situation.

Theorem 21.1.3. *Let $\Psi \in C([1, \infty))$ be a positive strictly increasing function of sublinear growth. Given a quasiconformal map $f_o : \overline{\mathbb{D}} \to \mathbb{C}$, we have*

$$\inf_{f \in \mathcal{F}_o} \int_{\mathbb{D}} \Psi\left[\mathbb{K}(z, f)\right] dz = \pi \Psi(1) \qquad (21.10)$$

In particular, the minimization problem (21.9) has no solution in $\mathcal{F}_o$ if the boundary values $f_o : \partial \mathbb{D} \to \mathbb{C}$ are not those of a conformal mapping of $\mathbb{D}$.

Proof. Let $a \in \mathbb{D}$ be a Lebesgue point for both $Df(z)$ and $\Psi\left[\mathbb{K}(z, f_o)\right]$ such that $\det Df_o(a) > 0$. We need to be able to modify f_o without changing its boundary values and while retaining quasiconformality. We could use the deformation result in Theorem 12.7.1 but shall instead make use of Theorem A.10.1. This result deals with deformations of arbitrary Sobolev functions. With this modified map, still denoted by f_o, both Df_o and $\Psi\left[\mathbb{K}(z, f_o)\right]$ have the origin as a Lebesgue point. What we have gained here is that f_o is an isometry near the origin; we have $Df_o(0) = \mathbf{I}$ and $\Psi\left[\mathbb{K}(0, f_o)\right] = \Psi(1)$. For every $0 < \varepsilon < r < 1$ we consider the radial mapping $g : \overline{\mathbb{D}} \to \overline{\mathbb{D}}$ defined by the rule

$$g(z) = z \, \rho(|z|),$$

where

$$\rho(|z|) = \begin{cases} \varepsilon, & |z| \leqslant r \\ \frac{\varepsilon - r}{1 - r} + \frac{1 - \varepsilon}{1 - r}|z|, & r \leqslant |z| \leqslant 1 \end{cases}$$

The distortion function of g can be easily computed using the formulas in Section 2.6 (see (2.42)), giving the following estimate.

$$\mathbb{K}(z, g) \leqslant \begin{cases} 1, & |z| \leqslant r \\ \frac{2}{1-r}, & r \leqslant |z| \leqslant 1 \end{cases}$$

Note that $g(z) = \varepsilon z$ for $|z| \leqslant r$. Now the composition

$$f(z) = f_o(g(z))$$

is a legitimate competitor for our variational integral. The chain rule is valid for quasiconformal mappings, so we can write

$$Df(z) = Df_o(g(z)) \cdot Dg(z) = \varepsilon Df_o(\varepsilon z), \qquad |z| \leqslant r$$

Hence we obtain the following estimate of the inner distortion function of f

$$\mathbb{K}(z,f) \leqslant \begin{cases} \mathbb{K}(\varepsilon z, f_o), & |z| \leqslant r \\ \frac{2K}{1-r}, & r \leqslant |z| \leqslant 1 \end{cases}$$

where $K = \|\mathbb{K}(z, f_o)\|_\infty$. We may now evaluate the variational integral

$$\int_\mathbb{D} \Psi\left[\mathbb{K}(z,f)\right] dz \leqslant \int_{|z|\leqslant r} \Psi\left[\mathbb{K}(\varepsilon z, f_o)\right] dz + \int_{r\leqslant |z|<1} \Psi\left(\frac{2K}{1-r}\right) dz$$

Hence

$$\int_\mathbb{D} \Psi\left[\mathbb{K}(z,f)\right] dz \leqslant \varepsilon^{-2} \int_{|z|\leqslant \varepsilon} \Psi\left[\mathbb{K}(z,f_o)\right] dz + \pi(1-r^2)\Psi\left(\frac{2K}{1-r}\right)$$

Using the supposed sublinear growth of Ψ, we find that the latter term goes to zero as $r \to 1$. In the first term, we let ε go to zero. Since the origin is the Lebesgue point of $\Psi[K(z, f_o)]$, the integral mean converges to $\pi \Psi[K(0, f_o)] = \pi \Psi(1)$. In conclusion, the infimum in (21.10) does not exceed $\pi \Psi(1)$. On the other hand, the integrand is always greater than or equal to $\pi \Psi(1)$. Thus (21.10) holds.

Finally, any minimizer of finite distortion must satisfy the pointwise equation

$$\Psi[K(z,f)] = \Psi(1) \quad \text{for almost every } z \in \mathbb{D},$$

or equivalently, $K(z, f) = 1$ for almost every $z \in \mathbb{D}$ because Ψ is strictly increasing. Thus f is conformal. This shows that for arbitrary boundary values, other than those of a conformal mapping of course, there can be no minimizers to the problem (21.9). $\square$

21.1.4 Inverses of Homeomorphisms of Integrable Distortion

In this section we state and give applications of an interesting and quite striking result of Hencl, Koskela and Onninen [166] generalizing an earlier result of ours [27]. We obtain a surprising connection between the problems of minimizing the mean distortion and minimizing the Dirichlet energy of homeomorphisms. Throughout this section Ω and Ω' will be bounded domains in the complex plane.

Theorem 21.1.4. *Let $f \in W^{1,1}_{loc}(\Omega, \Omega')$ be a homeomorphism of finite distortion with*

$$\int_\Omega \mathbb{K}(z,f) \, dz < \infty$$

21.1. MINIMIZING MEAN DISTORTION

Then the inverse map $h : \Omega' \to \Omega$ belongs to $W^{1,2}(\Omega', \Omega)$ and

$$\int_{\Omega'} \|Dh(\zeta)\|^2 \, d\zeta = 2 \int_{\Omega} \mathbb{K}(z, f) \, dz \qquad (21.11)$$

Proof. Complete details with the minimal $W^{1,1}_{loc}$ regularity would lead us too far astray at this late stage in the book. Therefore we give a proof of the theorem only under the stronger assumption $f \in W^{1,2}_{loc}(\Omega, \Omega')$.

Now the homeomorphism f satisfies a Beltrami equation

$$\frac{\partial f}{\partial \bar{z}} = \mu(z) \frac{\partial f}{\partial z}, \qquad |\mu(z)|^2 \leqslant \frac{\mathbb{K}(z, f) - 1}{\mathbb{K}(z, f) + 1}$$

Since f is assumed to have locally square integrable derivatives, according to Theorem 20.2.1 we may represent f as a composition of a conformal mapping and a principal solution to this Beltrami equation. Theorem 20.2.1 puts the inverse of the principal solution in $W^{1,2}_{loc}$ and as conformal mappings preserve this regularity, we see that $h = f^{-1} \in W^{1,2}_{loc}(\Omega')$.

Now the identity (21.11) quickly follows. Indeed, by Theorem 3.3.7 both f and h satisfy Lusin's condition $\mathcal{N}$, and so with formulae (2.49) and (2.50) we have

$$-\frac{\partial f}{\partial \bar{z}} = J(z, f) \frac{\partial h}{\partial \bar{w}}(f), \quad \frac{\partial f}{\partial z} = J(z, f) \overline{\frac{\partial h}{\partial w}}(f) \qquad \text{almost everywhere}$$

In other words, $\|Df\|^2 = J(z, f)^2 \|Dh(f)\|^2$. Moreover, the Lusin properties imply that we can perform the change of variables with f and that the Jacobian $J(z, f) > 0$ almost everywhere. Thus

$$\int_{\Omega'} \|Dh(\zeta)\|^2 = \int_{\Omega} \|Dh(f)\|^2 \, J(z, f) = 2 \int_{\Omega} \mathbb{K}(z, f)$$

With the identity $h \in W^{1,2}(\Omega', \Omega)$, and this completes the proof. $\square$

To apply Theorem 21.1.4 to our variational problem, we need to recall some of the considerable literature on the existence and topological properties of harmonic maps between planar domains. We refer the reader to the recent book of Duren [114]. In particular, according to the fundamental theorem of Radó [306], Kneser [212] and Choquet [93], if $\Omega \subset \mathbb{R}^2$ is a bounded convex domain, then each homeomorphism $h_o : \partial \Omega' \to \partial \Omega$ has a unique continuous extension $h : \overline{\Omega'} \to \overline{\Omega}$, which is univalent and maps Ω' harmonically to Ω. Then, by a theorem of Lewy [231] the univalent harmonic map has a nonvanishing Jacobian. Its inverse is therefore a real analytic diffeomorphism.

We now let $\mathcal{F} = \mathcal{F}(\Omega, \Omega')$ stand for the class of all $W^{1,1}_{loc}(\Omega, \mathbb{R}^2)$-regular homeomorphisms $f : \overline{\Omega} \to \overline{\Omega'}$ of finite distortion for which $\mathbb{K}(z, f)$ is integrable in Ω.

Theorem 21.1.5. *Let $\Omega \subset \mathbb{R}^2$ be a convex domain and $f_o \in \mathcal{F}(\Omega, \Omega')$. Then the minimization problem*

$$\min_{f \in \mathcal{F}} \int_\Omega \mathbb{K}(z, f)\, dz, \qquad f = f_0 \text{ on } \partial\Omega,$$

has a unique solution. This extremal map is a C^∞-diffeomorphism whose inverse is harmonic in Ω'.

Proof. Let $\mathcal{H} = \mathcal{H}(\Omega', \Omega)$ denote the class of inverse mappings $h = f^{-1} : \overline{\Omega'} \to \overline{\Omega}$, where $f \in \mathcal{F}(\Omega, \Omega')$. Thus, in particular, $h_0 = f_0^{-1} \in \mathcal{H}(\Omega', \Omega)$. In light of Theorem 21.1.4 we are reduced, equivalently, to the Dirichlet problem

$$\min_{h \in \mathcal{H}} \int_{\Omega'} \|Dh\|^2, \qquad h = h_0 \text{ on } \partial\Omega'$$

The existence and uniqueness of the minimizer in the appropriate Sobolev class $h_o + W_0^{1,2}(\Omega', \mathbb{R}^2)$ is well known. The only point to make is that such a minimizer lies in $\mathcal{H}(\Omega', \Omega)$ by Radó-Kneser-Choquet and Lewy theorems. □

Corollary 21.1.6. *The extremal map $f : \Omega \to \Omega'$ solves the quasilinear Beltrami equation*

$$\frac{\partial f}{\partial \bar{z}} = \mu(f(z)) \frac{\partial f}{\partial z}, \tag{21.12}$$

where $\mu : \Omega' \to \mathbb{D}$ is an antianalytic function valued in the unit disk.

Proof. As f is a diffeomorphism with positive Jacobian,

$$J(z, f) = |f_z|^2 - |f_{\bar{z}}|^2 > 0,$$

and thus $|f_{\bar{z}}/f_z| < 1$. Let us define $\mu : \Omega' \to \mathbb{D}$ by (21.12). We need only show that μ is antianalytic. For this reason we consider the inverse map $h(\xi) = f^{-1}(\xi)$ and write

$$\mu(\xi) = -\frac{\overline{h_{\bar{\xi}}}}{\overline{h_\xi}}$$

Hence

$$\frac{\partial \mu}{\partial \xi} = -\frac{\partial}{\partial \xi}\left(\frac{\overline{h_{\bar{\xi}}}}{\overline{h_\xi}}\right) = \frac{-\overline{h_{\xi\bar{\xi}}}\,\overline{h_\xi} + \overline{h_{\bar{\xi}}}\,\overline{h_{\xi\xi}}}{(\overline{h_\xi})^2} = 0,$$

as claimed.

Now the following is immediate.

Corollary 21.1.7 (Maximum Principle). *Let $\mathbb{K}(z, f)$ denote the inner distortion of the extremal map. Then*

$$\max_\mathbb{U} \mathbb{K}(z, f) \leqslant \max_{\partial \mathbb{U}} \mathbb{K}(z, f)$$

for every $\mathbb{U} \subset \Omega$.

21.1.5 The Traces of Mappings with Integrable Distortion

Theorem 21.1.5 demands, for the sake of completeness, that we give necessary and sufficient conditions for a homeomorphism $f_0 : \partial\Omega \to \partial\Omega'$ to admit an extension $f : \overline{\Omega} \to \overline{\Omega'}$ that lies in $\mathcal{F}(\Omega, \Omega')$; the class of $W^{1,1}_{loc}(\Omega, \mathbb{R}^2)$-regular homeomorphisms $f : \overline{\Omega} \to \overline{\Omega'}$ of finite distortion for which $\mathbb{K}(z, f)$ is integrable in Ω. Recall that for the L^∞-minimization problem the requisite notion is that of quasisymmetry, though in this case there is a surprise; see Theorem 21.1.10.

Theorem 21.1.8. *Suppose $\Gamma = \partial\Omega$ and $\Gamma' = \partial\Omega'$ are C^1-regular Jordan curves. A necessary and sufficient condition that a homeomorphism $f_o : \Gamma \to \Gamma'$ should extend to an $f : \overline{\Omega} \to \overline{\Omega'}$, $f \in \mathcal{F}(\Omega, \Omega')$, is that the double integral*

$$L_\Gamma(f_o) = -\frac{1}{\pi} \int_{\Gamma \times \Gamma} \log |f_o(z) - f_o(w)| \, dz \, d\overline{w} \qquad (21.13)$$

converges absolutely. That is,

$$\int_{\Gamma \times \Gamma} \big| \log |f_o(z) - f_o(w)| \big| \, |dz| \, |dw| < \infty$$

Among all such extensions of f_o there is one that maps Ω diffeomorphically onto Ω'.

Using the Riemann mapping theorem, we reduce to the case when both Ω and Ω' are disks in $\mathbb{C}$. Because of the C^1-regularity of $\Gamma = \partial\Omega$ and $\Gamma' = \partial\Omega'$, neither the hypotheses nor the assertion of this theorem will be affected by such a change of variables. Therefore we shall have established Theorem 21.1.8 once we prove the following more precise special case of it.

Theorem 21.1.9. *Theorem 21.1.8 holds when Ω' is a disk and Ω has the additional property of being convex. In this case, for every extension of f_0 to an $f \in \mathcal{F}(\Omega, \Omega')$, we have*

$$-\frac{1}{\pi} \int_{\Gamma \times \Gamma} \log |f_o(z) - f_o(w)| \, dz \, d\overline{w} \leqslant 2 \int_\Omega \mathbb{K}(z, f) \qquad (21.14)$$

Equality occurs only when $f = h^{-1}$, where $h : \overline{\Omega'} \to \overline{\Omega}$ is the unique harmonic extension of $f_0^{-1} : \Gamma' \to \Gamma$. This extremal extension turns out to be a diffeomorphism.

Remark. Our results are reminiscent of the ideas of Douglas [112], characterizing boundary functions whose harmonic extensions have finite Dirichlet energy. Douglas' condition for $h_o = f_o^{-1}$ reads as

$$\int_{\Gamma' \times \Gamma'} \left| \frac{h_o(\xi) - h_o(\zeta)}{\xi - \zeta} \right|^2 |d\xi||d\zeta| < \infty$$

Proof. Observe that (21.14) is invariant under translation and rescaling of Ω', so we may assume that Ω' is the unit disk, $\Omega' = \mathbb{D} \subset \mathbb{C}$. Consider the inverse homeomorphism $h_o : \partial \mathbb{D} \to \partial \Omega$. As shown in the previous section, f_o admits an extension to $f \in \mathcal{F}(\Omega, \Omega')$ if and only if the Poisson extension $h : \overline{\mathbb{D}} \to \overline{\Omega}$ of h_o has finite energy. Moreover, in this case the inverse map $f = h^{-1}$ provides us with one of the desired extensions of f_o to $f \in \mathcal{F}(\Omega, \Omega')$.

We begin with integral representation formulas of the complex derivatives of the harmonic map h in terms of $f_o : \Gamma \to \partial \mathbb{D}$:

$$\frac{\partial h}{\partial a} = \frac{1}{2\pi i} \int_\Gamma \frac{dz}{f_o(z) - a} \tag{21.15}$$

$$\frac{\partial h}{\partial \overline{a}} = \frac{1}{2\pi i} \int_\Gamma \frac{dz}{\overline{a} - \overline{f_o(z)}} \tag{21.16}$$

for every $a \in \mathbb{D}$. We give the proof only for (21.15), and the second identity follows in much the same way. Consider an exhaustion of Ω by smooth domains $\Omega_1 \subset \Omega_2 \subset \cdots \subset \Omega$, with Ω_j relatively compact in Ω_{j+1}, such that

$$a \in f(\Omega_1) \subset f(\Omega_2) \subset \cdots \subset \mathbb{D}$$

We have

$$\int_\Gamma \frac{dz}{f_o(z) - a} = \lim_{n \to \infty} \int_{\partial \Omega_n} \frac{dz}{f(z) - a} = \lim_{n \to \infty} \int_{C_n} \frac{dh(\xi)}{\xi - a}$$

$$= \lim_{n \to \infty} \int_{C_n} \frac{h_\xi \, d\xi}{\xi - a} + \lim_{n \to \infty} \int_{C_n} \frac{h_{\overline{\xi}} \, d\overline{\xi}}{\xi - a}, \tag{21.17}$$

where the curves $C_n = f(\partial \Omega_n)$ approach $\partial \mathbb{D}$ uniformly as $n \to \infty$. We do not claim here that the lengths of C_n stay bounded. Since h_ξ is an analytic function in $\mathbb{D}$, the first integral is independent of the curve C_n and equals $2\pi i \frac{\partial h}{\partial a}$ by Cauchy's formula. Concerning the second integral, it would be equal to zero if C_n were a circle. Indeed, we would have

$$\int_{|\xi|=\rho} \frac{h_{\overline{\xi}} \, d\overline{\xi}}{\xi - a} = \int_{|\xi|=\rho} \frac{\overline{\xi} h_{\overline{\xi}} \, d\overline{\xi}}{\rho^2 - a\overline{\xi}} = 0$$

by Green's formula (2.52). The above arguments suggest embedding every $f(\Omega_n)$ in a disk, say $f(\Omega_n) \subset \mathbb{D}_n \subset \mathbb{D}$. We can now express the curve integral by the area integral by using Stokes' formula

$$\int_{C_n} \frac{h_{\overline{\xi}} \, d\overline{\xi}}{\xi - a} = \int_{C_n} \frac{h_{\overline{\xi}} \, d\overline{\xi}}{\xi - a} - \int_{\partial \mathbb{D}_n} \frac{h_{\overline{\xi}} \, d\overline{\xi}}{\xi - a} = \int_{\mathbb{D}_n \setminus f(\Omega_n)} d \left[\frac{h_{\overline{\xi}} \, d\overline{\xi}}{\xi - a} \right]$$

$$= - \int_{\mathbb{D}_n \setminus f(\Omega_n)} \frac{h_{\overline{\xi}}}{(\xi - a)^2} \tag{21.18}$$

21.1. MINIMIZING MEAN DISTORTION

Hölder's inequality yields

$$\left| \int_{C_n} \frac{h_{\bar\xi}\, d\bar\xi}{\xi - a} \right| \leqslant C \left(\int_{\mathbb{D}\setminus f(\Omega_n)} |Dh|^2 \right)^{1/2} \to 0 \quad \text{as } n \to \infty,$$

completing the proof of (21.15).

Having disposed of formulas (21.15) and (21.16), since $|Dh| \in L^2(\mathbb{D})$ we can compute the Dirichlet integral of h over an arbitrary disk $\mathbb{D}_r = \{\xi : |\xi| < r\}$ with $0 < r < 1$,

$$\int_{\mathbb{D}_r} \|Dh\|^2 = 2 \int_{|a| \leqslant r} \left| \frac{\partial h}{\partial a} \right|^2 + 2 \int_{|a| \leqslant r} \left| \frac{\partial h}{\partial \bar a} \right|^2$$

The computation of the first integral is as follows.

$$\left| \frac{\partial h}{\partial a} \right|^2 = \frac{1}{4\pi^2} \int_{\Gamma \times \Gamma} \frac{dz\, d\bar w}{[f_o(z) - a][\overline{f_o(w)} - \bar a]}$$

Hence, by Fubini's theorem,

$$\int_{|a| \leqslant r} \left| \frac{\partial h}{\partial a} \right|^2 = \frac{1}{4\pi^2} \int_{\Gamma \times \Gamma} \left(\int_{|a| \leqslant r} \frac{da}{[f_o(z) - a][\overline{f_o(w)} - \bar a]} \right) dz\, d\bar w$$

A tedious (but elementary) computation, developing the integrand as a power series, yields an explicit expression for the area integral

$$\int_{|a| \leqslant r} \frac{da}{(\xi - a)(\bar\zeta - \bar a)} = -\pi \log(1 - r^2 \bar\xi \zeta),$$

where $|\xi| = |\zeta| = 1$. We substitute this value into the latter formula to obtain

$$2 \int_{|a| \leqslant r} \left| \frac{\partial h}{\partial a} \right|^2 = -\frac{1}{2\pi} \int_{\Gamma \times \Gamma} \log\left(1 - r^2 f_o(z)\overline{f_o(w)}\right) dz\, d\bar w$$

Arguments similar to those above show that

$$2 \int_{|a| \leqslant r} \left| \frac{\partial h}{\partial \bar a} \right|^2 = -\frac{1}{2\pi} \int_{\Gamma \times \Gamma} \log\left(1 - r^2 \overline{f_o(z)} f_o(w)\right) dz\, d\bar w$$

These two equations add up to

$$\int_{\mathbb{D}_r} \|Dh\|^2 = -\frac{1}{2\pi} \int_{\Gamma \times \Gamma} \log \left| 1 - r^2 f_o(z) \overline{f_o(w)} \right|^2 dz\, d\bar w,$$

which, in view of the identity (21.11), can be stated as

$$-\frac{1}{\pi} \int_{\Gamma \times \Gamma} \log \left| r^2 - f_o(z)\overline{f_o(w)} \right| dz\, d\bar w = \int_{\mathbb{D}_r} \|Dh\|^2 = 2 \int_{f^{-1}(\mathbb{D}_r)} \mathbb{K}(z, f)\, dz$$

It is now clear that the integral on the left hand side increases with r. Letting r go to 1, we see that the limit exists if and only if $\mathbb{K}(z, f)$ is integrable. The only point remaining concerns the equivalence of the following two properties of the boundary map $f_o : \Gamma \to \partial \mathbb{D}$: the existence of this limit and the absolute convergence of the integral

$$\int_{\Gamma \times \Gamma} \left| \log |f_o(z) - f_o(w)| \right| \, |dz| \, |dw| \tag{21.19}$$

It is clear that, regardless of the regularity of Γ, the absolute convergence of the integral in (21.19) implies

$$\lim_{r \nearrow 1} \int_{\Gamma \times \Gamma} \log \left| r^2 - f_o(z) \overline{f_o(w)} \right| \, dz \, d\overline{w} = \int_{\Gamma \times \Gamma} \log |f_o(z) - f_o(w)| \, dz \, d\overline{w} \tag{21.20}$$

by the Lebesgue dominated convergence theorem. For the converse, we need C^1-regularity of Γ. Suppose that the above limit exists. Since the integrand is invariant under the interchange of variable z and w, we may replace the complex area element $dz \, d\overline{w}$ by the real one $\Re(dz \, d\overline{w})$. However, on C^1-regular curves the latter element is comparable with $|dz| \, |dw|$ when z is sufficiently close to w, say

$$\frac{1}{2} |dz| \, |dw| \leqslant \Re(dz \, d\overline{w}) \leqslant 2 |dz| \, |dw| \tag{21.21}$$

provided $|z - w| \leqslant \varepsilon$. The interested reader may wish to observe that this estimate fails for the cube $\Omega = [0, 1] \times [0, 1]$. Indeed, near the corner $(0, 0)$ we may take $z = x$ and $w = iy$ to obtain

$$\Re(dz \, d\overline{w}) = -\Re(i \, dx \, dy) = 0$$

On the other hand, (21.21), with 2 replaced by some positive number, remains valid, e.g., for polygons with obtuse angles. Now, for C^1-regular curves, in view of (21.21), the existence of the limit and equality in (21.20) becomes equivalent to the absolute convergence of the integral in (21.13), as desired. $\square$

We must now consider the following example. Let f_o denote the homeomorphism of the unit circle onto itself $f_o : \partial \mathbb{D} \to \partial \mathbb{D}$ given by

$$f_o(e^{i\theta}) = e^{i\Phi(\theta)} \quad \text{for} \quad -\pi < \theta \leqslant \pi,$$

where

$$\Phi(\theta) = \pi \, \text{sgn} \, \theta \, e^{1 - \pi^2/\theta^2}$$

Then the double integral

$$\int_0^\pi \int_{-\pi}^0 \log \left| e^{i\Phi(\alpha)} - e^{i\Phi(\beta)} \right| \, d\alpha \, d\beta$$

diverges (we leave this as an exercise).

As a corollary, we see that f_o has no homeomorphic extension into the unit disk with integrable distortion.

Note that in the proof of Theorem 21.1.9 we did not really have to use the C^1-regularity of the convex domain Ω; we could have used the limit formula in (21.20) instead of the integral in (21.14). As Ω is convex, its boundary $\Gamma = \partial\Omega$ is Lipschitz and therefore a rectifiable Jordan curve. We leave the details of such an extension of Theorem 21.1.9 to arbitrary convex domains to the reader.

Finally, in this section, we wish to observe that given quasiconformal boundary data, even for the disk $\mathbb{D}$, the minimizer of the L^1-problem is seldom quasiconformal.

Theorem 21.1.10. *Let $f_o : \mathbb{D} \to \mathbb{D}$ be quasiconformal and $\mathcal{F}_o$ as in Theorem 21.1.5. Then the unique minimizer of the problem*

$$\min_{f \in \mathcal{F}} \int_{\mathbb{D}} \mathbb{K}(z, f), \qquad f = f_0 \text{ on } \partial\mathbb{D},$$

is quasiconformal if and only if f_o is bilipschitz.

Proof. Set $g_o = f_o^{-1}$. We know that the minimizer f exists and that its inverse h is the unique harmonic extension of $g_o|\partial\mathbb{D}$. If f is quasiconformal, then so too is h. However, a theorem of Pavlović [292] states that the Poisson extension of the boundary values g_o of a quasiconformal mapping is quasiconformal if and only if the map g_o is bilipschitz. Thus f_o is bilipschitz. □

Actually, [292] points out that the quasiconformality of the Poisson extension is equivalent to the boundary values themselves being bilipschitz or that the Hilbert transform of their derivative lies in L^∞.

21.2 Variational Equations

This chapter gives a quite developed picture of the theory of minimal mean distortion of mappings. If we replace the mean by more general functionals Ψ then much less is known, for instance on the properties of the extremals minimizing

$$\int_\Omega \Psi(\mathbb{K}(z, f)) \, dz$$

subject to given boundary values. Suggested by many problems in the calculus of variations, we strongly believe that if $\Psi \in C^\infty[1, \infty)$, then the extremals are continuously differentiable, as they are in the case $\Psi(t) = t$.

Conjecture 21.2.1. *Suppose Ψ is C^∞-smooth, convex and strictly increasing. Then every homeomorphism of finite distortion on a domain Ω, which minimizes the variational integral*

$$\int_\Omega \Psi(\mathbb{K}(z, f)) \, dz \qquad (21.22)$$

subject to given boundary values, is a $C^{1,\alpha}$-diffeomorphism in Ω.

Unfortunately, we do not even know whether the minimizers enjoy partial regularity as in the quasiconvex calculus of variations [119].

Assuming the conjecture, we are free to make a few general remarks and computations on the minimizers for (21.22). We begin with a variation of the complex Beltrami coefficient

$$\mu(z,f) = \frac{f_{\bar{z}}}{f_z},$$

where we assume that $f : \overline{\Omega} \to \overline{\Omega'}$ is an orientation-preserving diffeomorphism. Let $\eta \in C_0^\infty(\Omega)$ be a complex-valued test function. For all sufficiently small complex parameters λ, we still have $J(z, f + \lambda\eta) > 0$ and $f + \lambda\eta$ enjoys the same boundary values as f. The complex differential of $\mu(z,f)$, denoted by $\dot{\mu} = \dot{\mu}(z,f)$, is a $\mathbb{C}$-linear operator on $C_0^\infty(\Omega)$. It acts on a test function $\eta \in C_0^\infty(\Omega)$ by the rule

$$\dot{\mu}[\eta] = \dot{\mu}(z,f)[\eta] = \left.\frac{\partial \mu(z, f + \lambda\eta)}{\partial \lambda}\right|_{\lambda=0} = \frac{f_z \eta_{\bar{z}} - f_{\bar{z}} \eta_z}{(f_z)^2} \qquad (21.23)$$

Now consider the function

$$\kappa = \kappa(z,f) = |\mu(z,f)|^2 = \frac{|f_{\bar{z}}|^2}{|f_z|^2}$$

Its complex differential is computed by using the chain rule:

$$\dot{\kappa} = \dot{\kappa}(z,f) = \overline{\mu}\,\dot{\mu}$$

More explicitly, for each $\eta \in C_0^\infty(\Omega)$,

$$\dot{\kappa}[\eta] = \dot{\kappa}(z,f)[\eta] = \kappa \left(\frac{\eta_{\bar{z}}}{f_{\bar{z}}} - \frac{\eta_z}{f_z} \right)$$

Next, recall the distortion function

$$\mathbb{K}(z,f) = \frac{|f_z|^2 + |f_{\bar{z}}|^2}{|f_z|^2 - |f_{\bar{z}}|^2} = \frac{\|Df(z)\|^2}{2J(z,f)} = \frac{1+\kappa(z,f)}{1-\kappa(z,f)} \qquad (21.24)$$

Again by the chain rule, we find that

$$\dot{\mathbb{K}} = \dot{\mathbb{K}}(z,f) = \frac{2\dot{\kappa}(z,f)}{(1-\kappa)^2},$$

that is,

$$\dot{\mathbb{K}}[\eta] = \dot{\mathbb{K}}(z,f)[\eta] = \frac{2\kappa}{(1-\kappa)^2} \left(\frac{\eta_{\bar{z}}}{f_{\bar{z}}} - \frac{\eta_z}{f_z} \right)$$

21.2. VARIATIONAL EQUATIONS

More generally, for every convex $\Psi : [1, \infty) \to [1, \infty)$, we have the corresponding distortion function

$$\mathbb{K}_\Psi = \mathbb{K}_\Psi(z, f) = \Psi\big(\mathbb{K}(z, f)\big) = \Psi\left(\frac{|f_z|^2 + |f_{\bar{z}}|^2}{|f_z|^2 - |f_{\bar{z}}|^2}\right)$$

whose complex differential equals

$$\begin{aligned}
\dot{\mathbb{K}}_\Psi[\eta] &= \dot{\mathbb{K}}_\Psi(z, f)[\eta] = \Psi'\left(\frac{1+\kappa}{1-\kappa}\right) \frac{2\kappa}{(1-\kappa)^2} \left(\frac{\eta_{\bar{z}}}{f_{\bar{z}}} - \frac{\eta_z}{f_z}\right) \\
&= \frac{2}{(1-\kappa)^2} \Psi'\left(\frac{1+\kappa}{1-\kappa}\right) \left(\frac{\overline{f_{\bar{z}}}\,\eta_{\bar{z}} - \kappa \overline{f_z}\,\eta_z}{|f_z|^2}\right) \quad (21.25)
\end{aligned}$$

21.2.1 The Lagrange-Euler Equations

We now want to discuss the minimizers f of the general variational integrals

$$\int_\Omega \mathbb{K}_\Psi(z, f)\, dz \quad (21.26)$$

If f is also a $C^1(\Omega)$-diffeomorphism, then

$$\frac{\partial}{\partial \lambda} \int_\Omega \mathbb{K}_\Psi(z, f + \lambda \eta)\, dz = 0 \quad \text{at } \lambda = 0, \quad (21.27)$$

for every test function $\eta \in C_0^\infty(\Omega)$. This gives

$$\left[\frac{1}{(1-\kappa)^2} \Psi'\left(\frac{1+\kappa}{1-\kappa}\right) \frac{\overline{f_{\bar{z}}}}{|f_z|^2}\right]_{\bar{z}} = \left[\frac{\kappa}{f_z(1-\kappa)^2} \Psi'\left(\frac{1+\kappa}{1-\kappa}\right)\right]_z$$

or equivalently,

$$\frac{\partial}{\partial \bar{z}}\left(\frac{A(\kappa)}{f_{\bar{z}}}\right) = \frac{\partial}{\partial z}\left(\frac{A(\kappa)}{f_z}\right), \quad (21.28)$$

where

$$A(\kappa) = \Psi'\left(\frac{1+\kappa}{1-\kappa}\right) \frac{2\kappa}{(1-\kappa)^2} = \Psi'\left(\frac{|f_z|^2 + |f_{\bar{z}}|^2}{|f_z|^2 - |f_{\bar{z}}|^2}\right) \frac{2|f_z|^2 |f_{\bar{z}}|^2}{\big(|f_z|^2 - |f_{\bar{z}}|^2\big)^2}$$

Let us now introduce the conjugate stationary solution $g = g(z)$ in order to express (21.28) as a first-order system. From now on we need to assume that Ω is a simply connected domain.

Note that $\frac{A(\kappa)}{f_{\bar{z}}}$ and $\frac{A(\kappa)}{f_z}$ are continuous in Ω. Therefore there is $g \in C^1(\Omega)$, unique up to a constant, such that

$$\frac{\partial g}{\partial z} = \frac{A(\kappa)}{f_{\bar{z}}}, \qquad \frac{\partial g}{\partial \bar{z}} = \frac{A(\kappa)}{f_z}$$

Notice that g need not be a homeomorphism even supposing that f is. However, we do have the following.

Corollary 21.2.2. *The minimizer f and its conjugate stationary function g have the same complex Beltrami coefficient,*

$$\frac{\partial g}{\partial \bar{z}} = \mu(z) \frac{\partial g}{\partial z} \quad \text{with} \quad \frac{\partial g}{\partial z} \frac{\partial f}{\partial \bar{z}} = \frac{\partial g}{\partial \bar{z}} \frac{\partial f}{\partial z} = A(\kappa) > 0,$$

for almost every $z \in \Omega$.

To every extremal mapping $f : \Omega \to \Omega'$, there corresponds a holomorphic function $F : \Omega' \to \mathbb{C}$ defined by

$$F(\xi) = g\left(f^{-1}(\xi)\right),$$

where g denotes the conjugate function to f. Indeed, since f is a homeomorphism, we can always express $g(z) = F(f(z))$ for some mapping $F : \Omega' \to \mathbb{C}$. Now, F is holomorphic because f and g have the same Beltrami coefficient and both are C^1-smooth. The chain rule gives the following relations

$$g_z = F'(f(z))\, f_z, \qquad g_{\bar{z}} = F'(f(z))\, f_{\bar{z}}$$

Hence

$$F'(f) = \frac{g_z}{f_z} = \frac{g_{\bar{z}}}{f_{\bar{z}}} = \frac{A(\kappa)}{f_{\bar{z}} f_z}$$

We recall the derivatives of the inverse map $h(\xi) = f^{-1}(\xi)$,

$$h_{\bar{\xi}}(\xi) = -f_{\bar{z}}(z)\, J(\xi, h), \qquad h_\xi(\xi) = \overline{f_z(z)}\, J(\xi, h)$$

and compute

$$k\frac{|F'(\xi)|}{F'(\xi)} = k \frac{f_{\bar{z}} f_z}{|f_{\bar{z}}|\,|f_z|} = \frac{f_{\bar{z}}}{f_z} = -\frac{h_{\bar{\xi}}}{\overline{h_\xi}},$$

where

$$k = k(z, f) = |\mu(z, f)| = |\mu(\xi, h)| = k(\xi, h) \tag{21.29}$$

Corollary 21.2.3. *Let $f : \Omega \to \Omega'$ be a C^1-diffeomorphism that is a minimizer for the variational integral (21.26) subject to given boundary values. Then its inverse $h : \Omega' \to \Omega$ satisfies the Beltrami equation*

$$\frac{\partial h}{\partial \bar{\xi}} = k \frac{\overline{\varphi(\xi)}}{|\varphi(\xi)|} \overline{\frac{\partial h}{\partial \xi}}, \tag{21.30}$$

where $\varphi(\xi) = -F'(\xi)$ is a holomorphic function in Ω' and $k = k(\xi, h)$ is as in (21.29).

It is appropriate at this stage to recall that mappings satisfying (21.30) with a constant $0 \leqslant k < 1$ are referred to as Teichmüller mappings [228, p.231]. For this reason we shall call h a *pseudo-Teichmüller mapping*. What is so special about the Beltrami coefficient of h is that its argument is a harmonic function.

Note that in the case where $\Psi(t) = t$ and h is harmonic, as in Theorem 21.1.4, we find that

$$\varphi(\xi) = 2\, h_\xi\, \overline{h_{\bar{\xi}}},$$

21.2. VARIATIONAL EQUATIONS

where both h_ξ and $\overline{h_{\bar\xi}}$ are analytic functions.

The conjugate stationary solutions reduced (21.28) to a first-order equation for the inverse. As an alternative development, we note that nondivergence forms of (21.28) are also interesting. These are actually systems of second-order PDEs for the real and imaginary parts of f. Since the integrand in (21.26) is polyconvex, such systems must satisfy the Legendre-Hadamard ellipticity condition; see [101, 151] for more details. Let us consider the simplest case of the distortion function

$$\mathbb{K}(z,f) = \frac{\|Df(z)\|^2}{2J(z,f)} = \frac{|f_z|^2 + |f_{\bar z}|^2}{|f_z|^2 - |f_{\bar z}|^2}$$

In this case $\Psi(t) = t$, and hence

$$A(\kappa) = \frac{2|f_z|^2 |f_{\bar z}|^2}{\left(|f_z|^2 - |f_{\bar z}|^2\right)^2} \tag{21.31}$$

The Euler-Lagrange equation reads as

$$\frac{\partial}{\partial z}\left(\frac{|f_z|^2 f_{\bar z}}{J(z,f)^2}\right) = \frac{\partial}{\partial \bar z}\left(\frac{|f_{\bar z}|^2 f_z}{J(z,f)^2}\right) \tag{21.32}$$

Lengthy computation, see [27], reduces (21.32) to an elegant nondivergence equation

$$f_{z\bar z} = \alpha f_{zz} + \beta f_{\bar z\bar z}, \tag{21.33}$$

where

$$\alpha = \frac{\mu(z,f)}{1+|\mu(z,f)|^2} = \frac{\overline{f_z} f_{\bar z}}{|f_z|^2 + |f_{\bar z}|^2} = \overline{\beta}$$

This quasilinear system is elliptic, meaning that

$$|\alpha(z)| + |\beta(z)| = \frac{2k}{1+k^2} < 1$$

It is somewhat peculiar that (21.33) turns out to be $\mathbb{C}$-linear with respect to the second-order derivatives.

We have seen systems of type (21.33) with $\beta = \bar\alpha$ before in Section 16.4.3. See, for instance, (16.135). We end this subsection with an interesting discussion concerning an application of the continuity properties of mappings of finite distortion to such systems.

Theorem 21.2.4. *Let $f \in W^{2,2}_{loc}(\Omega)$ be a solution of the linear (over $\mathbb{C}$) elliptic, but not necessarily uniformly elliptic, equation*

$$f_{z\bar z} = \alpha(z) f_{zz} + \beta(z) f_{\bar z\bar z}, \tag{21.34}$$

where $\beta = \bar\alpha$ and

$$|\alpha(z)| + |\beta(z)| < 1 \quad \text{almost everywhere in } \Omega$$

Then $f \in C^1(\Omega)$.

Proof. First observe that if we conjugate both sides of (21.34), we obtain the same equation for $\bar{f}$. Therefore the real and imaginary parts of f,

$$u = \frac{1}{2}(f + \bar{f}), \qquad v = \frac{1}{2i}(f - \bar{f}),$$

also solve the same equation $w_{z\bar{z}} = \alpha(z) w_{zz} + \beta(z) w_{\bar{z}\bar{z}}$. Now if w is any real-valued solution, we have $w_{\bar{z}\bar{z}} = \overline{w_{zz}}$. This gives us a first-order elliptic equation for the complex gradient $g = w_z$, namely,

$$g_{\bar{z}} = \alpha(z) g_z + \beta(z) \overline{g_z}, \qquad |\alpha(z)| + |\beta(z)| < 1$$

Since $g \in W^{1,2}_{loc}(\Omega)$, it is a mapping of finite distortion, and we infer from Theorem 20.1.1 that g is continuous. Thus $f = u + iv$ is continuously differentiable as claimed since both u and v are. $\square$

Note that $f_z = u_z + iv_z$ itself need not be a mapping of finite distortion. Of course, the C^1-regularity of $W^{2,2}_{loc}$-solutions of (21.34) is a consequence of the structure of the equations themselves and not a general fact. In particular, this result remains valid (with the same proof) for quasilinear systems of the form

$$f_{z\bar{z}} = \alpha(z, f, f_z, f_{\bar{z}}) f_{zz} + \beta(z, f, f_z, f_{\bar{z}}) f_{\bar{z}\bar{z}} \tag{21.35}$$

when $\alpha = \bar{\beta}$, with the only proviso being that we must have $\alpha(z, f, f_z, f_{\bar{z}})$ a measurable function.

21.2.2 Equations for the Inverse Map

Suppose that a C^1-diffeomorphism $f : \Omega \to \Omega'$ is a minimizer of the variational integral

$$\int_\Omega \mathbb{K}_\Psi(z, f) \, dz = \int_\Omega \Psi\left(\frac{|f_z|^2 + |f_{\bar{z}}|^2}{|f_z|^2 - |f_{\bar{z}}|^2}\right) dz$$

This just amounts to saying that the inverse map $h : \Omega' \to \Omega$ minimizes the integral

$$\int_{\Omega'} J(\xi, h) \mathbb{K}_\Psi(\xi, h) \, d\xi = \int_{\Omega'} \left(|h_\xi|^2 - |h_{\bar{\xi}}|^2\right) \Psi\left(\frac{|h_\xi|^2 + |h_{\bar{\xi}}|^2}{|h_\xi|^2 - |h_{\bar{\xi}}|^2}\right) d\xi$$

This time the variation of the integrand reads as

$$\left(\dot{J}\Psi + J\Psi'\dot{\mathbb{K}}\right)[\eta] = \left(\overline{h_\xi}\,\eta_\xi - \overline{h_{\bar{\xi}}}\,\eta_{\bar{\xi}}\right)\Psi + \frac{2\kappa J}{(1-\kappa)^2}\left(\frac{\eta_{\bar{\xi}}}{h_{\bar{\xi}}} - \frac{\eta_\xi}{h_\xi}\right)\Psi'$$

whose complex conjugate gives the following divergence form of the Euler-Lagrange equation:

$$\frac{\partial}{\partial\bar{\xi}}\left(\left[(\mathbb{K}+1)\Psi' - \Psi\right]h_{\bar{\xi}}\right) + \frac{\partial}{\partial\xi}\left(\left[\Psi - (\mathbb{K}-1)\Psi'\right]h_\xi\right) = 0 \tag{21.36}$$

21.2. VARIATIONAL EQUATIONS

First, we view the square brackets as given measurable coefficients

$$\frac{\partial}{\partial \xi}\left(M\, h_{\bar{\xi}}\right) + \frac{\partial}{\partial \bar{\xi}}\left(N\, h_{\xi}\right) = 0, \tag{21.37}$$

where

$$M(\xi) = (\mathbb{K}+1)\Psi'(\mathbb{K}) - \Psi(\mathbb{K}) \quad \text{and} \quad N(\xi) = \Psi(\mathbb{K}) - (\mathbb{K}-1)\Psi'(\mathbb{K}),$$

$\mathbb{K} = \mathbb{K}(\xi, h)$. From this point of view, if $M = M(\xi)$ and $N = N(\xi)$ happen to be smooth, then (21.37) is elliptic. Indeed,

$$(M+N)\,h_{\xi\bar{\xi}} + M_\xi h_{\bar{\xi}} + N_{\bar{\xi}} h_\xi = 0, \quad \text{where } M+N = 2\,\Psi'(\mathbb{K}) > 0 \tag{21.38}$$

On the other hand, we may consider (21.37) the Lagrange-Euler equation of a quadratic energy integrand

$$\mathcal{E}[h] = \int_{\Omega'} \left(M\, |h_{\bar{\xi}}|^2 + N\, |h_\xi|^2 \right) d\xi$$

However, inequality $M + N > 0$ is insufficient for this functional to be convex. We must assume that both coefficients $M(\xi)$ and $N(\xi)$ are non-negative. This happens if and only if

$$\mathbb{K} - 1 \leqslant \frac{\Psi(\mathbb{K})}{\Psi'(\mathbb{K})} \leqslant \mathbb{K} + 1 \tag{21.39}$$

The case $\Psi(\mathbb{K}) = \mathbb{K}$ has already been investigated in Theorem 21.1.5, where we showed that h then satisfies the Laplace equation. Practically, there are no other examples since (21.39) forces almost linear growth of Ψ.

That is why the variational approach has to be abandoned when Ψ has overlinear growth. There is, however, an interesting and promising nondivergence form of (21.36). A tedious computation leads to a second-order equation

$$h_{\xi\bar{\xi}} = \alpha\, h_{\xi\xi} + \beta\, h_{\bar{\xi}\bar{\xi}} + \gamma\, \overline{h_{\xi\xi}} + \delta\, \overline{h_{\bar{\xi}\bar{\xi}}}, \tag{21.40}$$

where the complex coefficients α, β, γ and δ depend in a rather explicit way, by means of Ψ, only on the first-order derivatives h_ξ and $h_{\bar{\xi}}$.

We shall not bother with these explicit formulas here but refer the interested reader to [27]. What is perhaps more interesting is the ellipticity condition

$$\left|1 + \alpha(\xi)\lambda + \beta(\xi)\bar{\lambda}\right| > \left|\gamma(\xi)\bar{\lambda} + \delta(\xi)\lambda\right| \tag{21.41}$$

for every complex number λ of modulus 1. This means that our second-order equation (21.40) not only is elliptic but also lies in the same homotopy class as the Laplacian [68]. Verification of the ellipticity condition in (21.41) is again challenging, and we refer the reader to the original source [27].

21.3 Mean Distortion, Annuli and the Nitsche Conjecture

The L^1-variant of the classical extremal Grötsch problem was considered earlier in Section 21.1.2. Now we examine the first instance where there are nontrivial conformal invariants. Thus we next investigate the case of minimizing mean distortion in doubly connected domains.

Given two annuli

$$\mathbb{A} = \{z : r < |z| < R\} \quad \text{and} \quad \mathbb{A}' = \{\zeta : r' < |\zeta| < R'\},$$

we shall consider homeomorphisms $f : \mathbb{A} \to \mathbb{A}'$. Here, note that $|f|$ extends continuously to $\overline{\mathbb{A}}$, with values r' and R' on the boundary of $\mathbb{A}$. We shall normalize our mappings so that

$$|f(z)| = r' \text{ for } |z| = r \quad \text{and} \quad |f(z)| = R' \text{ for } |z| = R$$

Throughout this and the remaining sections we let $\mathcal{F}(\mathbb{A}, \mathbb{A}')$ denote the family of all such homeomorphisms $f : \mathbb{A} \to \mathbb{A}'$ of finite distortion.

There are two integral means of the distortion function $\mathbb{K}(z, f)$ that concern us in this section: first, the *average distortion*,

$$\mathcal{K}_f := \frac{1}{|\mathbb{A}|} \int_{\mathbb{A}} \mathbb{K}(z, f) \, dz$$

and second the *weighted average*,

$$\mathcal{K}_f^* := \frac{1}{\mu(\mathbb{A})} \int_{\mathbb{A}} \mathbb{K}(z, f) \, d\mu(z),$$

where we have used the notations $d\mu(z) = dz/|z|^2$, $dz = dx\, dy$.

The minimization problems we now address are to evaluate the following infima:

$$\inf\{\mathcal{K}_f : f \in \mathcal{F}(\mathbb{A}, \mathbb{A}')\} \tag{21.42}$$

and

$$\inf\{\mathcal{K}_f^* : f \in \mathcal{F}(\mathbb{A}, \mathbb{A}')\}$$

Further, decide if the infimum is attained and prove uniqueness up to rotation of the annuli.

The concept of conformal modulus will prove useful in formulating our results. For the annulus $\mathbb{A}$ we have the equivalent definitions

$$\text{mod}(\mathbb{A}) = 2\pi \log \frac{R}{r} = \int_{\mathbb{A}} \frac{dz}{|z|^2} = \mu(\mathbb{A})$$

Every topological annulus is conformally equivalent to a round annulus $\mathbb{A}$, and we can set $\text{mod}(\Omega) = \text{mod}(\mathbb{A})$.

For weighted averages we can now prove the following theorem.

21.3. MEAN DISTORTION AND ANNULI

Theorem 21.3.1. *Among all mappings $f \in \mathcal{F}(\mathbb{A}, \mathbb{A}')$, the infimum of*

$$\mathcal{K}_f^* = \frac{1}{\mu(\mathbb{A})} \int_{\mathbb{A}} \mathbb{K}(z, f) \, d\mu(z)$$

is attained for the power function

$$f^\alpha(z) = r' r^{-\alpha} |z|^{\alpha-1} z, \quad \text{where } \alpha = \frac{\operatorname{mod}(\mathbb{A}')}{\operatorname{mod}(\mathbb{A})}$$

Furthermore, we have

$$\frac{1}{\mu(\mathbb{A})} \int_{\mathbb{A}} \mathbb{K}(z, f^\alpha) \, d\mu(z) = \frac{1}{2} \left(\frac{\operatorname{mod}(\mathbb{A}')}{\operatorname{mod}(\mathbb{A})} + \frac{\operatorname{mod}(\mathbb{A})}{\operatorname{mod}(\mathbb{A}')} \right) = \frac{1}{2} \left(\frac{1}{\alpha} + \alpha \right) \quad (21.43)$$

This extremal map is unique up to rotations of the annuli.

Minimization of the average $\mathcal{K}_f$ for mappings $f : \mathbb{A} \to \mathbb{A}'$ of finite distortion is more subtle. The infimum is attained only when the annulus $\mathbb{A}'$ is not too fat when compared with $\mathbb{A}$. This special restriction on the moduli of the annuli is explicitly described by the Nitsche bound, which in terms of the radii, reads as

$$\frac{R'}{r'} \leqslant \frac{R}{r} + \sqrt{\frac{R^2}{r^2} - 1}, \quad \text{or equivalently,} \quad \frac{R}{r} \geqslant \frac{1}{2} \left(\frac{R'}{r'} + \frac{r'}{R'} \right) \quad (21.44)$$

These relations between the annuli $\mathbb{A}$ and $\mathbb{A}'$ arise naturally in the following conjecture due to Nitsche.

Conjecture 21.3.2. *If*

$$\frac{R}{r} < \frac{1}{2} \left(\frac{R'}{r'} + \frac{r'}{R'} \right)$$

then there are no homeomorphic harmonic mappings $h : \mathbb{A}' \to \mathbb{A}$.

Under the condition (21.44) harmonic homeomorphisms are easy to construct. Namely, assuming that this condition holds, we can find (uniquely) a real parameter $\omega \geqslant -r^2$ such that

$$\frac{R'}{r'} = \frac{R + \sqrt{R^2 + \omega}}{r + \sqrt{r^2 + \omega}} \quad (21.45)$$

Because of scale invariance in the target, it involves no loss of generality to assume that, in fact,

$$r' = r + \sqrt{r^2 + \omega} \quad \text{and} \quad R' = R + \sqrt{R^2 + \omega}$$

and we then have the complex harmonic Nitsche maps

$$h^\omega : \mathbb{A}' \to \mathbb{A}, \quad h^\omega(\zeta) = \frac{1}{2} \left(\zeta - \frac{\omega}{\bar{\zeta}} \right)$$

and their inverses

$$f^\omega : \mathbb{A} \to \mathbb{A}', \quad f^\omega(z) = z + z\left(1 + \frac{\omega}{|z|^2}\right)^{1/2} \tag{21.46}$$

By taking inverses, minimizing mean distortion in $\mathcal{F}(\mathbb{A}, \mathbb{A}')$ is equivalent to minimizing the Dirichlet energy among all homeomorphisms $h : \mathbb{A}' \to \mathbb{A}$. Of course, homeomorphisms of minimal energy, if they exist, are necessarily harmonic. We will later see that, in fact, if the Nitsche condition (21.44) fails, then the minimal energy is not attained by any homeomorphism between the annuli. However, under the Nitsche condition we have the following result.

Theorem 21.3.3. *If the Nitsche condition (21.44) holds, then we have for each homeomorphism $f \in \mathcal{F}(\mathbb{A}, \mathbb{A}')$ the estimate*

$$\frac{1}{|\mathbb{A}|}\int_\mathbb{A} \mathbb{K}(z,f)\,dz \geqslant \frac{1}{|\mathbb{A}|}\int_\mathbb{A} \mathbb{K}(z,f^\omega)\,dz = \frac{R^2 + r^2 + \omega}{R\sqrt{R^2 + \omega} + r\sqrt{r^2 + \omega}} \tag{21.47}$$

The extremal map $f = f^\omega$ is unique up to rotations of the annuli.

We are also interested in the extremal problem outside the Nitsche bound (21.44). It is important to look more closely at the critical inverse Nitsche map corresponding to $\omega = -r^2$, which we denote by

$$f^{\#}(z) = z + z\left(1 - \frac{r^2}{|z|^2}\right)^{1/2}, \quad r \leqslant |z| \leqslant R \tag{21.48}$$

This maps $\mathbb{A}$ onto the critical annulus

$$\mathbb{A}^{\#} = \{\zeta : r < |\zeta| < R + \sqrt{R^2 - r^2}\} \tag{21.49}$$

We easily calculate its mean distortion

$$\mathcal{K}_{f^{\#}} = \frac{R}{\sqrt{R^2 - r^2}},$$

where by Theorem 21.3.3,

$$\mathcal{K}_{f^{\#}} = \inf\{\mathcal{K}_f : f \in \mathcal{F}(\mathbb{A}, \mathbb{A}^{\#})\}$$

The inverse of $f^{\#}$ is the critical Nitsche harmonic map

$$h^{\#}(\zeta) = \frac{1}{2}\left(\zeta + \frac{r^2}{\bar{\zeta}}\right)$$

Its Jacobian determinant vanishes identically on the inner boundary of $\mathbb{A}^{\#}$,

$$|h^{\#}_\zeta|^2 - |h^{\#}_{\bar\zeta}|^2 \equiv \frac{1}{4}\left(1 - \frac{r^4}{|\zeta|^4}\right) = 0 \quad \text{for } |\zeta| = r$$

21.3. MEAN DISTORTION AND ANNULI

Further,
$$|Df^{\#}| = |f_z^{\#}| + |f_{\bar{z}}^{\#}| = \frac{|f^{\#}|}{\sqrt{|z|^2 - r^2}} \in L^2_{weak}(\mathbb{A}),$$

but $|Df^{\#}| \notin L^2(\mathbb{D})$. Thus we see various degenerations of our extremals in the critical case, suggesting that there are no extremal homeomorphisms beyond this range of the moduli.

Suppose now that the target annulus is fatter than the critical one. In other words, let
$$\mathbb{A}' = \{\zeta : r' < |\zeta| < R'\} \supsetneq \mathbb{A}^{\#},$$
where $0 < r' < r < R \leqslant R' = R + \sqrt{R^2 - r^2}$. Thus

$$\frac{R'}{r'} > \frac{R}{r} + \sqrt{\frac{R^2}{r^2} - 1} \tag{21.50}$$

Theorem 21.3.4. *Under the fatness condition (21.50), the infimum in (21.42) is not attained by any homeomorphism $f \in \mathcal{F}(\mathbb{A}, \mathbb{A}')$.*

The sequences $\{f_n\} \subset \mathcal{F}(\mathbb{A}, \mathbb{A}')$ minimizing the mean distortion $\int_{\mathbb{A}} \mathbb{K}(z, f)$ will be bounded in $W^{1,1}(\mathbb{A}, \mathbb{A}')$, but the derivatives will lack equi-integrability. Consequently, $\{f_n\}$ will contain no subsequence that converges weakly in the space $W^{1,1}(\mathbb{A}, \mathbb{A}')$. However, the derivatives of f_n will converge in the sense of measures to a limit. The precise description of the limit measure is given in Section 21.3.7.

The minimization of the integral means of the distortion functions of homeomorphisms $f : \mathbb{A} \to \mathbb{A}'$ turns out to be equivalent to the Dirichlet-type problems for the inverse mappings $h = f^{-1} : \mathbb{A}' \to \mathbb{A}$. If a homeomorphism $f \in W^{1,1}_{loc}(\mathbb{A}, \mathbb{A}')$ has integrable distortion, then $h \in W^{1,2}(\mathbb{A}', \mathbb{A})$ and we can consider the two energy functionals

$$E[h] := \int_{\mathbb{A}'} \|Dh(\zeta)\|^2 \, d\zeta = 2 \int_{\mathbb{A}} \mathbb{K}(z, f) \, dz < \infty$$

$$F[h] := \int_{\mathbb{A}'} \frac{\|Dh(\zeta)\|^2}{|h|^2} \, d\zeta = 2 \int_{\mathbb{A}} \frac{\mathbb{K}(z, f)}{|z|^2} \, dz < \infty$$

In general, the converse is not true simply because the inverse of a homeomorphism $h \in W^{1,2}(\mathbb{A}', \mathbb{A})$ need not even belong to the Sobolev class $W^{1,1}_{loc}(\mathbb{A}', \mathbb{A})$. It has bounded variation but fails to be absolutely continuous on lines; see [159, 165] for related questions and results.

We overcome such subtleties by proving a correction lemma. Accordingly, for every homeomorphism $h \in W^{1,2}(\mathbb{A}', \mathbb{A})$, we can construct a homeomorphism $\tilde{h} \in W^{1,2}(\mathbb{A}', \mathbb{A})$, with $E[\tilde{h}] \leqslant E[h]$ and $F[\tilde{h}] \leqslant F[h]$, whose inverse lies in $\mathcal{F}(\mathbb{A}, \mathbb{A}')$. For details, see Lemma 21.3.15 below. As a consequence, the minimization problems for $\mathcal{K}_f$ and $\mathcal{K}_f^*$ are equivalent to the corresponding minimization problems for $E[h]$ and $F[h]$ with $h = f^{-1}$.

Corollary 21.3.5. *The minimum of the homogeneous energy $F[h]$ is attained (uniquely up to a rotation of the annuli) by a power stretching.*

Corollary 21.3.6. *Within the Nitsche range (21.44) for the annuli $\mathbb{A}$ and $\mathbb{A}'$, the minimum of the harmonic energy $E[h]$ is obtained (uniquely up to rotation) by a Nitsche map*

$$h(\zeta) = \lambda\left(\zeta - \frac{\omega}{\bar{\zeta}}\right), \qquad \omega \in \mathbb{R}, \; \lambda > 0$$

Corollary 21.3.7. *Outside the Nitsche range (21.44) for the annuli $\mathbb{A}$ and $\mathbb{A}'$, the infimum of $E[h]$ is not attained by any homeomorphism $h \in W^{1,2}(\mathbb{A}', \mathbb{A})$.*

In studying $E[h]$ outside the Nitsche range, it is interesting to look at the minimizing sequence $\{h^n\} \subset W^{1,2}(\mathbb{A}', \mathbb{A})$. Its limit exists in the Sobolev space $W^{1,2}(\mathbb{A}', \mathbb{A})$. Moreover, although the limit fails to be a homeomorphism, no cavity or crack occurs within the solid body of ring $\mathbb{A}$ under such deformation of $\mathbb{A}'$. For this reason, we shall accept such a limit as a legitimate minimizer of the Dirichlet integral.

Both energy integrals $E[h]$ and $F[h]$ are invariant under a conformal change of the independent variable. Thus, given any topological annulus $\Omega \subset \mathbb{C}$ of the same modulus as $\mathbb{A}'$, we may consider homeomorphisms $g : \Omega \to \mathbb{A}$ in the Sobolev class $W^{1,2}(\Omega, \mathbb{A})$. To each such g there corresponds a mapping $h = g \circ \chi : \mathbb{A}' \to \mathbb{A}$, where $\chi : \mathbb{A}' \to \Omega$ is a conformal transformation. The corresponding energy integrals for $g \in W^{1,2}(\Omega, \mathbb{A})$ reduce to

$$E[g] = \int_\Omega \|Dg\|^2 = \int_{\mathbb{A}'} \|Dh\|^2 = E[h]$$

and

$$F[g] = \int_\Omega \frac{\|Dg\|^2}{|g|^2} = \int_{\mathbb{A}'} \frac{\|Dh\|^2}{|h|^2} = F[h]$$

Of particular interest in elasticity theory is to know how to deform, with minimal energy, a given topological annulus onto a given round annulus. Corollaries 21.3.5–21.3.7 provide us with the solution to this problem. For example, as a result of our computation, we find the minimum energy required to deform a given annulus onto a punctured disk.

Corollary 21.3.8. *Let $h \in W^{1,2}(\Omega, \mathbb{D} \setminus \{0\})$ be a homeomorphism from a topological annulus Ω onto the punctured unit disk $\mathbb{D} \setminus \{0\}$. Then*

$$E[h] = \int_{\mathbb{A}'} \|Dh\|^2 \geqslant \pi \frac{e^{\mathrm{mod}(\Omega)/\pi} + 1}{e^{\mathrm{mod}(\Omega)/\pi} - 1}$$

The infimum is attained by the harmonic map

$$h(\zeta) = C \frac{|\chi(\zeta)|^2 - 1}{\overline{\chi(\zeta)}},$$

21.3. MEAN DISTORTION AND ANNULI

where $\chi : \Omega \to \mathbb{A}' = \{w : 1 < |w| < e^{\mathrm{mod}(\Omega)/2\pi}\}$ is a conformal map of Ω onto a round annulus and the constant

$$C = \left(e^{\mathrm{mod}(\Omega)/\pi} + e^{-\mathrm{mod}(\Omega)/\pi} - 2\right)^{-1/2}$$

The conformally invariant nature of the minimization problems described above strongly suggests natural generalizations also within the theory of Teichmüller spaces of Riemann surfaces. The study of minimal deformations of annuli described here will indeed prove an important tool, and we refer the reader to [29] for such further developments.

The rest of this chapter is devoted to the proofs of Theorems 21.3.1 - 21.3.4.

21.3.1 Polar Coordinates

We frequently view functions $f = f(z)$ of one complex variable $z = \rho e^{i\theta}$ as functions of the polar coordinates $0 < \rho < \infty$, $0 \leqslant \theta < 2\pi$. Then the Cauchy-Riemann derivatives of f are

$$f_z = \frac{1}{2} e^{-i\theta} \left(f_\rho - \frac{i}{\rho} f_\theta\right), \qquad f_{\bar{z}} = \frac{1}{2} e^{i\theta} \left(f_\rho + \frac{i}{\rho} f_\theta\right) \tag{21.51}$$

Hence

$$|f_z|^2 + |f_{\bar{z}}|^2 = \frac{1}{2}\left(|f_\rho|^2 + \rho^{-2}|f_\theta|^2\right),$$

$$J(z,f) = |f_z|^2 - |f_{\bar{z}}|^2 = \frac{1}{\rho}\Im m(f_\theta \overline{f_\rho}), \tag{21.52}$$

which together yield

$$\mathbb{K}(z,f) = \frac{\rho|f_\rho|^2 + \rho^{-1}|f_\theta|^2}{2\,\Im m(f_\theta \overline{f_\rho})} \tag{21.53}$$

We expect extremal mappings between annuli to be radial stretchings of the form

$$f(\rho e^{i\theta}) = F(\rho)e^{i\theta},$$

with $F : [r, R] \to [r', R']$ increasing, $0 < r < R$ and $0 < r' < R'$. A remarkable feature of radial mappings is that

$$\frac{f_\theta}{\rho f} = \frac{i}{\rho} \in i\mathbb{R}, \qquad \frac{f_\rho}{\rho f} = \frac{\dot{F}}{\rho F} \in \mathbb{R}$$

At points where F is differentiable, we calculate the distortion function as

$$\mathbb{K}(z,f) = \frac{1}{2}\left(\frac{\rho \dot{F}}{F} + \frac{F}{\rho \dot{F}}\right) \tag{21.54}$$

21.3.2 Free Lagrangians

For certain nonlinear differential expressions, defined for diffeomorphisms $f : \mathbb{A} \to \mathbb{A}'$, their integral mean does not depend on the specific choice of the diffeomorphism f. We call them *free Lagrangians*. They play a fundamental role in our computations in a way similar to that played by the integral means of null Lagrangians in the polyconvex calculus of variations [39]. As a typical example, consider the expression $J(z,f)|f|^{-2}$ with integral mean

$$\int_{\mathbb{A}} \frac{J(z,f)dz}{|f|^2} = \int_{\mathbb{A}'} \frac{d\zeta}{|\zeta|^2} = \mathrm{mod}(\mathbb{A}')$$

depending only on $\mathbb{A}'$.

This identity generalizes, as an inequality, to all homeomorphisms $f : \mathbb{A} \to \mathbb{A}'$ in the Sobolev class $W^{1,1}_{loc}(\mathbb{A}, \mathbb{A}')$, namely,

$$\int_{\mathbb{A}} \frac{J(z,f)dz}{|f|^2} \leqslant \int_{\mathbb{A}'} \frac{d\zeta}{|\zeta|^2} = \mathrm{mod}(\mathbb{A}') \tag{21.55}$$

Precisely, there is a set $\mathbb{E} \subset \mathbb{A}$ of measure zero such that

$$\int_{\mathbb{A}} \frac{J(z,f)dz}{|f|^2} = \int_{f(\mathbb{A}\setminus\mathbb{E})} \frac{d\zeta}{|\zeta|^2}$$

The reader is referred to Sections 3.1.4, 3.1.8 and 3.2.5 of [124] for a proof. For other and more explicit statements, see for instance, [142, 150] and Section 6.3 of [191]. We shall introduce the following nonlinear differential expressions defined for homeomorphisms $f : \mathbb{A} \to \mathbb{A}'$ in the Sobolev class $W^{1,1}_{loc}(\mathbb{A}, \mathbb{A}')$,

$$\mathcal{J}_f = \frac{J(z,f)}{|f|^2}, \qquad \text{where } \mathcal{J}_f = \frac{\dot{F}}{\rho F} \text{ for } f(z) = F(|z|)\frac{z}{|z|}$$

$$\mathcal{A}_f = \Re e \frac{f_\rho(z)}{|z|f(z)}, \qquad \text{where } \mathcal{A}_f = \frac{\dot{F}}{\rho F} \text{ for } f(z) = F(|z|)\frac{z}{|z|}$$

$$\mathcal{B}_f = \Im m \frac{f_\theta(z)}{|z|^2 f(z)}, \qquad \text{where } \mathcal{B}_f = \frac{1}{\rho^2} \text{ for } f(z) = F(|z|)\frac{z}{|z|}$$

With the next lemma we show that these give rise to free Lagrangians.

Lemma 21.3.9. *The following integrals are independent of the mapping $f \in \mathcal{F}(\mathbb{A}, \mathbb{A}')$:*

$$\int_{\mathbb{A}} \mathcal{A}_f \, dz = \mathrm{mod}(\mathbb{A}') \tag{21.56}$$

$$\int_{\mathbb{A}} \mathcal{B}_f \, dz = \mathrm{mod}(\mathbb{A}) \tag{21.57}$$

More generally, for every continuous function $u : [r, R] \to \mathbb{R}$,

$$\int_{\mathbb{A}} \mathcal{B}_f(z)u(|z|)dz = \int_{\mathbb{A}} \frac{u(|z|)}{|z|^2}dz = 2\pi \int_r^R \frac{u(\rho)d\rho}{\rho} \tag{21.58}$$

21.3. MEAN DISTORTION AND ANNULI

In addition, for every $f \in \mathcal{F}(\mathbb{A}, \mathbb{A}')$, we have the estimate

$$\int_\mathbb{A} \frac{J(z,f)dz}{|f|^2} \leqslant \int_{\mathbb{A}'} \frac{d\zeta}{|\zeta|^2} = \mathrm{mod}(\mathbb{A}'), \tag{21.59}$$

where equality holds whenever f belongs to the class $W^{1,2}_{loc}(\mathbb{A}, \mathbb{A}')$.

Proof. Note first that

$$|f|_\rho = \frac{1}{2|f|}\left(|f|^2\right)_\rho = |f|\frac{f_\rho \overline{f} + f \overline{f_\rho}}{2 f \overline{f}} = |f|\Re\mathrm{e}\left(\frac{f_\rho}{f}\right)$$

Therefore

$$\mathcal{A}_f(z) = \frac{|f|_\rho}{\rho|f|} \tag{21.60}$$

and integration gives

$$\int_\mathbb{A} \mathcal{A}_f(z) dz = \int_0^{2\pi}\left(\int_r^R \frac{|f|_\rho}{|f|} d\rho\right) d\theta = \int_0^{2\pi}(\log R' - \log r')\, d\theta = \mathrm{mod}(\mathbb{A}')$$

For (21.57) and (21.58), from the increment of the argument, we have

$$\int_0^{2\pi} \frac{f_\theta(\rho e^{i\theta})}{f(\rho e^{i\theta})}\, d\theta = 2\pi i \quad \text{for almost every } \rho \in (r, R) \tag{21.61}$$

Hence by Fubini's theorem,

$$\int_\mathbb{A} \mathcal{B}_f(z) u(|z|) dz = \Im m \int_r^R \frac{u(\rho)}{\rho} \int_0^{2\pi} \frac{f_\theta(\rho e^{i\theta})}{f(\rho e^{i\theta})}\, d\theta\, d\rho = 2\pi \int_r^R \frac{u(\rho)}{\rho}\, d\rho$$

Finally, we remark that a homeomorphism $f \in W^{1,2}_{loc}(\mathbb{A}, \mathbb{A}')$ satisfies the Lusin condition $\mathcal{N}$, and therefore (21.59) holds as an equality in this case. □

21.3.3 Lower Bounds by Free Lagrangians

The free Lagrangians provide powerful tools for solving extremal problems on mappings of finite distortion. As a particular example, in this section we will show how to obtain optimal integral mean estimates for the distortion functions.

For this purpose, let $f : \mathbb{A} \to \mathbb{A}'$ be a given mapping of finite distortion, where $\mathbb{A} = \{z : r < |z| < R\}$ and $\mathbb{A}' = \{z : r' < |z| < R'\}$. We begin with pointwise lower bounds for the distortion function $\mathbb{K}(z, f)$.

Lemma 21.3.10. *For every parameter $\lambda \geqslant -r^2$, we have*

$$\mathbb{K}(z, f) \geqslant \frac{-\lambda |f_\rho(z)|^2}{2|z|^2 J(z, f)} + \frac{\sqrt{|z|^2 + \lambda}}{|z|}, \tag{21.62}$$

or equivalently,

$$\mathbb{K}(z, f) \geqslant \frac{\lambda |f_\theta(z)|^2}{2(|z|^2 + \lambda)|z|^2 J(z, f)} + \frac{|z|}{\sqrt{|z|^2 + \lambda}}, \tag{21.63}$$

almost everywhere on the set $\mathbb{A}^+ = \{z \in \mathbb{A} : J(z,f) > 0\}$. For $z \in \mathbb{A}^+$, equality holds if and only if
$$f_\theta(z) = i\sqrt{|z|^2 + \lambda}\, f_\rho(z) \qquad (21.64)$$

Proof. These estimates are equivalent to the obvious inequality
$$|\sqrt{|z|^2 + \lambda}\, f_\rho + i f_\theta|^2 \geq 0, \qquad \rho^2 \geq -\lambda, \qquad (21.65)$$
once we take into account the identities (21.52) and (21.53). $\square$

For later purposes, we note that for every mapping of finite distortion,
$$f_\rho = f_\theta = 0 \quad \text{outside } \mathbb{A}^+$$
This observation is useful in integrating the pointwise bounds.

Lemma 21.3.11. *For every $f \in \mathcal{F}(\mathbb{A}, \mathbb{A}')$, we have*
$$\int_{\mathbb{A}^+} \frac{|f_\rho(z)|^2}{|z|^2 J(z,f)} \geq \mathrm{mod}(\mathbb{A}') \qquad (21.66)$$
The equality occurs if and only if almost everywhere on $\mathbb{A}^+$ we have
$$\Re e\left(\frac{f_\rho}{f}\right) = \left|\frac{f_\rho}{f}\right| \qquad (21.67)$$
and
$$\Im m\left(\frac{f_\theta}{f}\right) = k \qquad (21.68)$$
for some constant $k > 0$.

Proof. We start with (21.56) and obtain
$$\mathrm{mod}(\mathbb{A}') = \int_{\mathbb{A}} \Re e\, \frac{f_\rho(z)}{|z| f(z)} \leq \int_{\mathbb{A}^+} \frac{|f_\rho(z)|}{|z||f(z)|} \qquad (21.69)$$
Combining the estimate with Hölder's inequality gives
$$\mathrm{mod}(\mathbb{A}') \leq \int_{\mathbb{A}^+} \frac{|f_\rho|}{|z||f|} \leq \left(\int_{\mathbb{A}^+} \frac{|f_\rho|^2}{|z|^2 J(z,f)}\right)^{1/2} \left(\int_{\mathbb{A}^+} \frac{J(z,f)}{|f|^2}\right)^{1/2} \qquad (21.70)$$

A glance at (21.55) shows that we have proved the claim (21.66).
To achieve equality it is necessary that, almost everywhere on $\mathbb{A}^+$,
$$\Re e\left(\frac{f_\rho}{f}\right) = \left|\frac{f_\rho}{f}\right|,$$
and since we have used Hölder's inequality, that almost everywhere on $\mathbb{A}^+$,
$$\frac{J(z,f)}{|f(z)|^2} = k\, \frac{|f_\rho(z)|^2}{|z|^2 J(z,f)}$$

21.3. MEAN DISTORTION AND ANNULI

for some constant $k > 0$. The latter condition takes the form

$$k|f||f_\rho| = |z|J(z,f) = \Im\left(f_\theta \overline{f_\rho}\right),$$

which together with (21.67) reduces to (21.68). Conversely, under the conditions (21.67) and (21.68), we have equality in (21.66). □

Lemma 21.3.12. *For every $f \in \mathcal{F}(\mathbb{A}, \mathbb{A}')$ and for every parameter $\lambda \geqslant -r^2$, we have*

$$\int_{\mathbb{A}^+} \frac{|f_\theta(z)|^2}{|z|^2(|z|^2 + \lambda)J(z,f)} \geqslant \frac{4\pi^2}{\mathrm{mod}(\mathbb{A}')} \log^2 \frac{R + \sqrt{R^2 + \lambda}}{r + \sqrt{r^2 + \lambda}} \qquad (21.71)$$

Equality occurs if and only if, almost everywhere on $\mathbb{A}^+$, we have

$$\Im\left(\frac{f_\theta}{f}\right) = \left|\frac{f_\theta}{f}\right| \qquad (21.72)$$

and

$$\Re\left(\frac{f_\rho}{f}\right) = \frac{\alpha}{\sqrt{|z|^2 + \lambda}} \qquad (21.73)$$

for some constant $\alpha > 0$.

Proof. Let us first apply (21.58) and choose as the auxiliary function $u(\rho) = \rho\left(\rho^2 + \lambda\right)^{-1/2}$. That yields

$$2\pi \int_r^R \frac{d\rho}{\sqrt{\rho^2 + \lambda}} = \int_{\mathbb{A}} \Im\left(\frac{f_\theta(z)}{f(z)}\right) \frac{dz}{|z|\sqrt{|z|^2 + \lambda}} \qquad (21.74)$$

Clearly, this quantity is less than or equal to

$$\int_{\mathbb{A}^+} \frac{|f_\theta|}{|z|\sqrt{|z|^2 + \lambda}|f|} \leqslant \left(\int_{\mathbb{A}^+} \frac{|f_\theta|^2}{|z|^2(|z|^2 + \lambda)J(z,f)}\right)^{1/2} \left(\int_{\mathbb{A}} \frac{J(z,f)}{|f(z)|^2}\right)^{1/2}$$

According to (21.59), the last integral is bounded by $\mathrm{mod}(\mathbb{A}')$. On the other hand,

$$2\pi \int_r^R \frac{d\rho}{\sqrt{\rho^2 + \lambda}} = 2\pi \log \frac{R + \sqrt{R^2 + \lambda}}{r + \sqrt{r^2 + \lambda}}$$

Hence we have the inequality (21.71). Finally, for the equality to hold there, in complete analogy with the proof of Lemma 21.3.11, we see that it occurs if and only if both (21.72) and (21.73) are true. □

21.3.4 Weighted Mean Distortion

The idea in proving Theorem 21.3.1 is to find combinations of invariant integrals involving $|z|^{-2}\mathbb{K}$ and then apply the integral estimates of the previous section. In fact, using (21.52) and (21.53), we may write

$$\frac{\mathbb{K}(z,f)}{|z|^2} = \frac{|f_\rho|^2}{2|z|^2 J(z,f)} + \frac{|f_\theta|^2}{2|z|^4 J(z,f)} \qquad (21.75)$$

on the set $\mathbb{A}^+ = \{z \in \mathbb{A} : J(z,f) > 0\}$. As we observed before, outside $\mathbb{A}^+$ we have $f_\rho = f_\theta = 0$. Hence,

$$2\int_\mathbb{A} \frac{\mathbb{K}(z,f)}{|z|^2} \geqslant \int_{\mathbb{A}^+} \frac{|f_\rho|^2}{|z|^2 J(z,f)} + \int_{\mathbb{A}^+} \frac{|f_\theta|^2}{|z|^4 J(z,f)} \tag{21.76}$$

For the first integral on the right hand side we use Lemma 21.3.11, while for the second we apply Lemma 21.3.12 with the parameter $\lambda = 0$. Consequently,

$$2\int_\mathbb{A} \frac{\mathbb{K}(z,f)}{|z|^2} \geqslant \mathrm{mod}(\mathbb{A}') + \frac{\mathrm{mod}(\mathbb{A})^2}{\mathrm{mod}(\mathbb{A}')} \tag{21.77}$$

This is precisely the desired estimate (21.43). Furthermore, it is a quick calculation to verify that the equality occurs for the power function

$$f^\alpha(z) = \frac{r'}{r^\alpha}|z|^{\alpha-1}z, \qquad \alpha = \frac{\mathrm{mod}(\mathbb{A}')}{\mathrm{mod}(\mathbb{A})}$$

As for the uniqueness of the extremal mappings, if for $f \in \mathcal{F}(\mathbb{A}, \mathbb{A}')$ the equality holds in the lower bound (21.77), then by the above proof f satisfies (21.67) and (21.68) with (21.72) and (21.73), with $\lambda = 0$. The combination of these identities is simply

$$f_\theta = ikf \quad \text{and} \quad \rho f_\rho = \alpha f, \tag{21.78}$$

where α and k are positive constants.

If we integrate f_θ/f over the circle $|z| = \rho$ in $\mathbb{A}$, we find that $k = 1$; see (21.61). Thus the general solution to the system of PDEs (21.78) is

$$f(z) = Cz|z|^{\alpha-1},$$

which is unique up to the rotations of the annuli, as claimed. The proof of Theorem 21.3.1 is complete. $\square$

21.3.5 Minimizers within the Nitsche Range

The proof of Theorem 21.3.3 is similar to that of Theorem 21.3.1, but now we make use of Lemma 21.3.10 and integrate the pointwise bounds it provides for the distortion function.

We first consider the case $-r^2 \leqslant \lambda < 0$. Since $\mathbb{K}(z,f) \geqslant 1$, we can write

$$\int_\mathbb{A} \mathbb{K}(z,f) \geqslant \int_{\mathbb{A}^+} \mathbb{K}(z,f) + \int_{\mathbb{A}\setminus\mathbb{A}^+} \frac{\sqrt{|z|^2 + \lambda}}{|z|}$$

$$\geqslant \frac{-\lambda}{2}\int_{\mathbb{A}^+} \frac{|f_\rho(z)|^2}{|z|^2 J(z,f)} + \int_\mathbb{A} \frac{\sqrt{|z|^2 + \lambda}}{|z|}$$

Here we may appeal to Lemma 21.3.11 for the first integral and compute the second explicitly. This argument gives the first part of the following lemma.

21.3. MEAN DISTORTION AND ANNULI

Lemma 21.3.13. *Suppose $f \in \mathcal{F}(\mathbb{A}, \mathbb{A}')$ and $-r^2 \leqslant \lambda < 0$. Then*

$$\int_{\mathbb{A}} \mathbb{K}(z, f) \geqslant \frac{-\lambda}{2} \mathrm{mod}(\mathbb{A}') + \pi \left[R\sqrt{R^2 + \lambda} - r\sqrt{r^2 + \lambda} \right] + \pi \lambda \log \frac{R + \sqrt{R^2 + \lambda}}{r + \sqrt{r^2 + \lambda}}$$

Furthermore, the equality occurs in this estimate if and only if

$$f(z) = C \frac{z}{|z|} \left(|z| + \sqrt{|z|^2 + \lambda} \right),$$

where C is any nonzero complex number.

Proof. It remains to prove the uniqueness part of the claim. For the equality in the above integral estimate, we must have $\mathbb{A} = \mathbb{A}^+$ and $f_\theta(z) = i\sqrt{|z|^2 + \lambda}\, f_\rho(z)$. In addition, since we used Lemma 21.3.11, it is necessary to add the conditions (21.67) and (21.68). Together these requirements reduce to the simple necessary and sufficient condition

$$\frac{f_\theta}{f} = ik \quad \text{and} \quad \frac{f_\rho}{f} = \frac{k}{\sqrt{\rho^2 + \lambda}} \qquad (21.79)$$

for some constant $k > 0$. Integrating over a circle, as before, gives $k = 1$. Thus the general solution to the system of PDEs (21.79) is

$$f(z) = C \frac{z}{|z|} \left(|z| + \sqrt{|z|^2 + \lambda} \right),$$

where C is any nonzero complex number. □

Next consider the case $\lambda > 0$ and proceed by using the lower bound (21.63) in Lemma 21.3.10, obtaining

$$\int_{\mathbb{A}} \mathbb{K}(z, f) \geqslant \frac{\lambda}{2} \int_{\mathbb{A}^+} \frac{|f_\theta(z)|^2}{|z|^2(|z|^2 + \lambda)\, J(z, f)} + \int_{\mathbb{A}} \frac{|z|}{\sqrt{|z|^2 + \lambda}} \qquad (21.80)$$

By analogy to Lemma 21.3.13, one obtains the following estimate.

Lemma 21.3.14. *For every $f \in \mathcal{F}(\mathbb{A}, \mathbb{A}')$ and for every $\lambda > 0$, we have*

$$\int_{\mathbb{A}} \mathbb{K}(z, f) \geqslant \frac{2\lambda \pi^2}{\mathrm{mod}(\mathbb{A}')} \log^2 \frac{R + \sqrt{R^2 + \lambda}}{r + \sqrt{r^2 + \lambda}}$$
$$+ \pi \left[R\sqrt{R^2 + \lambda} - r\sqrt{r^2 + \lambda} \right] - \pi \lambda \log \frac{R + \sqrt{R^2 + \lambda}}{r + \sqrt{r^2 + \lambda}}$$

The equality holds if and only if

$$f(z) = C \frac{z}{|z|} \left(|z| + \sqrt{|z|^2 + \lambda} \right), \qquad (21.81)$$

where C is any nonzero complex number.

Proof. We apply Lemma 21.3.12 to the first integral in (21.80) and compute the second integral explicitly. This proves the lower bound. For the equality, as before, we must have $\mathbb{A} = \mathbb{A}^+$ and $f_\theta(z) = i\sqrt{|z|^2 + \lambda}\, f_\rho(z)$. Further, this time we must add the conditions (21.72) and (21.73). These requirements reduce to the same set of equations as in (21.79). Thus the equality occurs if and only if f takes the form (21.81). □

We are now ready for the proof of Theorem 21.3.3. Assuming that the target annulus $\mathbb{A}' = \{z : r' < |z| < R'\}$ satisfies the relative Nitsche bounds (21.44), we found in (21.45) a unique parameter $\omega \geqslant -r^2$ such that

$$r' = r + \sqrt{r^2 + \omega} < R' = R + \sqrt{R^2 + \omega}$$

We will then use the above integral estimates with the special choice $\lambda = \omega$.

If $-r^2 \leqslant \omega < 0$, then the lower bound in Lemma 21.3.13 reads as

$$\int_{\mathbb{A}} \mathbb{K}(z, f) \geqslant \pi\left(R\sqrt{R^2 + \omega} - r\sqrt{r^2 + \omega}\right) \qquad (21.82)$$

For the case $\omega > 0$, we use the lower bound in Lemma 21.3.14. However, a manipulation shows that the lower bound now also attains precisely the same form (21.82). In other words we have shown that under the Nitsche bound (21.44),

$$\int_{\mathbb{A}} \mathbb{K}(z, f) \geqslant \int_{\mathbb{A}} \mathbb{K}(z, f^\omega) = \pi\left(R\sqrt{R^2 + \omega} - r\sqrt{r^2 + \omega}\right),$$

where

$$f^\omega(z) = \frac{z}{|z|}\left(|z| + \sqrt{|z|^2 + \omega}\right)$$

is the Nitsche map. Moreover, according to Lemmas 21.3.13 and 21.3.14, the mappings f^ω are the only minimizers up to rotation of the rings. Hence we have completed the proof of Theorem 21.3.3. □

21.3.6 Beyond the Nitsche Bound

We are now given two round annuli $\mathbb{A}$ and $\mathbb{A}'$ with inner and outer radii r, R and r', R', respectively. Moreover, we assume that the relative fatness condition (21.50) is satisfied.

Here it will be convenient to use the normalization

$$r = 1, \qquad r' < 1 < R' = R + \sqrt{R^2 - 1} \qquad (21.83)$$

Suppose $f : \mathbb{A} \to \mathbb{A}'$ is a homeomorphism of finite distortion. Then we may use Lemma 21.3.13, this time with the choice $\lambda = -1$. As a result, we get

$$2\int_{\mathbb{A}} \mathbb{K}(z, f) \geqslant \mathrm{mod}(\mathbb{A}') + 2\pi R\sqrt{R^2 - 1} - 2\pi \log\left(R + \sqrt{R^2 - 1}\right), \qquad (21.84)$$

21.3. MEAN DISTORTION AND ANNULI

or equivalently,
$$\int_{\mathbb{A}} \mathbb{K}(z,f) \geqslant \int_{\mathbb{A}} \mathbb{K}(z,f^{\#}) + \pi \log \frac{1}{r'}, \qquad (21.85)$$
where
$$f^{\#}(z) = \frac{z}{|z|}\left(|z| + \sqrt{|z|^2 - 1}\right)$$
is the critical inverse Nitsche map from (21.48).

In the next section we show that the lower bound (21.84) and (21.85) is optimal. Namely, we will construct a minimizing sequence that attains the bound in the limit. However, no single homeomorphism $f: \mathbb{A} \to \mathbb{A}'$ of finite distortion can achieve the bound. Otherwise, if $f \in \mathcal{F}(\mathbb{A}, \mathbb{A}')$ could satisfy (21.84) with an equality, then Lemma 21.3.13 would show that such an f has to be of the form
$$f(z) = C\frac{z}{|z|}\left(|z| + \sqrt{|z|^2 - 1}\right) = Cf^{\#}(z)$$
for some rotation C. But then $\mathbb{A}' = f^{\#}(\mathbb{A})$, which is in contradiction to the fatness condition (21.50). Hence Theorem 21.3.4 follows once we have constructed the minimizing sequence.

We will complete this section with a remark suggesting how to construct a minimizing sequence: The target annulus $\mathbb{A}'$ splits into a union of two annuli,
$$\mathbb{A}'_0 = \{\zeta : r' < |\zeta| < 1\} \quad \text{and} \quad \mathbb{A}'_{\#} = \{\zeta : 1 < |\zeta| < R'\},$$
where $\mathbb{A}'_{\#}$ is the critical Nitsche annulus with $f^{\#}(\mathbb{A}) = \mathbb{A}'_{\#}$. The additional term in (21.85) is nothing other than the infimum of the integrals
$$\int_{\mathcal{A}} \mathbb{K}(z, \phi)\, dz$$
over all annuli $\mathcal{A} = \{z : 1 < |z| < \sigma\}$ and over all $\phi \in \mathcal{F}(\mathcal{A}, \mathbb{A}'_0)$ with the fixed target $\mathbb{A}'_0 = \{\zeta : r' < |\zeta| < 1\}$. Indeed, by Theorem 21.3.1, we see that
$$\int_{\mathcal{A}} \frac{\mathbb{K}(z,\phi)}{|z|^2} = \left(\frac{\mathrm{mod}(\mathbb{A}'_0)}{2\,\mathrm{mod}(\mathcal{A})} + \frac{\mathrm{mod}(\mathcal{A})}{2\,\mathrm{mod}(\mathbb{A}'_0)}\right)\mathrm{mod}(\mathcal{A}) > \frac{\mathrm{mod}(\mathbb{A}'_0)}{2} = \pi \log \frac{1}{r'}$$

This suggests that in order to reach the right-hand side of (21.85), we must shrink $\mathcal{A}$ to the unit circle and, by Theorem 21.3.1, for ϕ we may take the power functions.

21.3.7 The Minimizing Sequence and Its BV-limit

The Sobolev space $W^{1,1}(\mathbb{A}, \mathbb{A}')$ appears to be the natural domain of definition for homeomorphisms $f: \mathbb{A} \to \mathbb{A}'$ of finite distortion. However, when studying the extremal problem beyond the Nitsche bound, we found that the lack of weak compactness in $W^{1,1}(\mathbb{A}, \mathbb{A}')$ prevents the minimizing sequences to converge to a mapping with a minimal integral mean distortion. The weak*-closure of this Sobolev space hence plays a role here.

Let us briefly recall the space $BV(\Omega, \mathbb{C})$ of functions $f : \Omega \to \mathbb{C}$ of bounded variation; the interested reader may wish to look at [17] for a complete BV theory. The Banach space $BV(\Omega, \mathbb{C})$ consists of those integrable functions defined in a domain $\Omega \subset \mathbb{C}$ whose distributional derivatives f_z and $f_{\bar{z}}$ are complex Radon measures, bounded linear functionals on $C_0(\Omega)$. The BV-norm is given by

$$\|f\|_{BV(\Omega)} = \|f\|_{L^1(\Omega)} + [f_z]_\Omega + [f_{\bar{z}}]_\Omega,$$

where the latter two symbols denote the total variation of the measures f_z and $f_{\bar{z}}$, respectively.

Every bounded sequence $\{f^n\}_{n=1}^\infty$ in $BV(\Omega, \mathbb{C})$ contains a subsequence $\{f^{n_j}\}$ converging to an $f \in BV(\Omega, \mathbb{C})$ in the weak*-topology. This means that $f^{n_j} \to f$ in $L^1(\Omega)$, and for every test function $\phi \in C_0(\Omega)$, we have

$$\langle f_z^{n_j}, \phi \rangle \to \langle f_z, \phi \rangle, \qquad \langle f_{\bar{z}}^{n_j}, \phi \rangle \to \langle f_{\bar{z}}, \phi \rangle$$

Here the angular brackets stand for the duality action of measures on $C_0(\Omega)$.

Let us now return to the problem of minimizing the integral mean distortion

$$\inf \left\{ \frac{1}{|\mathbb{A}|} \int_{\mathbb{A}} \mathbb{K}(z, f) \, dz \; : \; f \in \mathcal{F}(\mathbb{A}, \mathbb{A}') \right\} \qquad (21.86)$$

in the case of a fat target (21.50). Suppose $\{f^n\} \subset \mathcal{F}(\mathbb{A}, \mathbb{A}')$ is a minimizing sequence. By Hölder's inequality, we find that

$$\int_{\mathbb{A}} |Df^n| \leqslant \int_{\mathbb{A}} \sqrt{\mathbb{K}(z, f^n)} \sqrt{J(z, f^n)} \qquad (21.87)$$

$$\leqslant \left[\int_{\mathbb{A}} \mathbb{K}(z, f^n) \right]^{1/2} \left[\int_{\mathbb{A}} J(z, f^n) \right]^{1/2}, \qquad (21.88)$$

which shows that $\{f^n\}$ is a bounded sequence in $W^{1,1}(\mathbb{A}, \mathbb{A}')$. Thus $\{f^n\}$ contains a subsequence, again denoted by $\{f^n\}$, converging in the weak*-topology to a mapping f of bounded variation in $\mathbb{A}$.

We shall discuss this phenomenon in more detail for the mappings $f^n : \mathbb{C} \to \mathbb{C}$ defined by the rule

$$f^n(z) = \begin{cases} \frac{z}{|z|} \left(|z| + \sqrt{|z|^2 - 1} \right), & |z| \geqslant r_n \\ r' z |z|^{n-1}, & 1 \leqslant |z| \leqslant r_n \\ r' z, & |z| \leqslant 1 \end{cases} \qquad (21.89)$$

Here $0 < r' < 1$ is fixed, and the radii $r_n > 1$ are uniquely determined from the equation

$$r_n + \sqrt{r_n^2 - 1} = r' (r_n)^n \qquad (21.90)$$

Elementary analysis shows that

$$\lim_{n \to \infty} (r_n)^n = \frac{1}{r'} \quad \text{and} \quad \lim_{n \to \infty} n(r_n - 1) = \log \frac{1}{r'} \qquad (21.91)$$

21.3. MEAN DISTORTION AND ANNULI

The pointwise limit of $\{f^n(z)\}$ exhibits a discontinuity along the unit circle,

$$f^n(z) \to f(z) = \begin{cases} \frac{z}{|z|}\left(|z| + \sqrt{|z|^2 - 1}\right), & |z| > 1 \\ r'z, & |z| \leqslant 1 \end{cases}$$

Outside the unit disk f is the critical inverse Nitsche map.

Computing the derivatives of f^n, one has

$$f_z^n = \begin{cases} 1 + \frac{1}{2}\left(\frac{|z|}{\sqrt{|z|^2-1}} + \frac{\sqrt{|z|^2-1}}{|z|}\right), & |z| > r_n \\ \frac{n+1}{2}r'|z|^{n-1}, & 1 < |z| < r_n \\ r', & |z| < 1 \end{cases} \quad (21.92)$$

and

$$f_{\bar{z}}^n = \begin{cases} \frac{z^2}{2|z|^3\sqrt{|z|^2-1}}, & |z| > r_n \\ \frac{n-1}{2}r'|z|^{n-3}z^2, & 1 < |z| < r_n \\ 0, & |z| < 1 \end{cases} \quad (21.93)$$

The formulas show that the sequence $\{f^n\}$ is bounded in $W^{1,1}(\Omega, \mathbb{C})$ for every bounded domain $\Omega \subset \mathbb{C}$. Thus f is its weak*-limit.

Let us then examine the singular part of the measures f_z and $f_{\bar{z}}$. With the aid of (21.91), we compute

$$f_z^{\text{sing}} = \pi(1 - r')\, d\nu \quad (21.94)$$
$$f_{\bar{z}}^{\text{sing}} = \pi(1 - r')\frac{z}{\bar{z}}\, d\nu, \quad (21.95)$$

where $d\nu$ stands for the unit mass uniformly distributed along the unit circle.

Next we fix a radius $R > 1$ and define $R' = R + \sqrt{R^2 - 1}$. Each f^n maps the annulus $\mathbb{A} = \{z : 1 < |z| < R\}$ homeomorphically onto the annulus

$$\mathbb{A}' = \{\zeta : r' < |\zeta| < R'\}$$

From the above formulas on the derivatives we may easily determine the distortion function of f^n. This turns out to be

$$\mathbb{K}(z, f^n) = \begin{cases} \frac{1}{2}\left(\frac{|z|}{\sqrt{|z|^2-1}} + \frac{\sqrt{|z|^2-1}}{|z|}\right) & \text{for } |z| > r_n \\ \frac{1}{2}(n + 1/n) & \text{for } 1 < |z| < r_n \\ 1 & \text{for } |z| < 1 \end{cases} \quad (21.96)$$

The limit of the integral means of the distortion functions is now easy to determine,

$$\lim_{n\to\infty} \int_{\mathbb{A}} \mathbb{K}(z, f^n)\, dz = \int_{\mathbb{A}} \mathbb{K}(z, f)\, dz + \lim_{n\to\infty} \int_{1<|z|<r_n} \mathbb{K}(z, f^n)\, dz$$
$$= \int_{\mathbb{A}} \mathbb{K}(z, f)\, dz + \pi \log \frac{1}{r'},$$

where the last equality follows from (21.91). The right hand side coincides with that in inequality (21.85), as $f(z) = f^\#(z)$ for all $z \in \mathbb{A}$. In other words, the sequence $\{f^n\}$ is a minimizing sequence for the problem (21.86).

21.3.8 Correction Lemma

In order to apply the above results to energy minimization problems for the inverse mappings and to Corollaries 21.3.5–21.3.7, we need to establish the following lemma.

Lemma 21.3.15. *Let $h : \Omega' \to \Omega$ be a homeomorphism of finite Dirichlet energy,*

$$E[h] = \int_\Omega \|Dh(\zeta)\|^2\, d\zeta = \int_\Omega \left(|h_\zeta|^2 + |h_{\bar\zeta}|^2\right) d\zeta < \infty$$

Then there exists a homeomorphism $\tilde{h} : \Omega' \to \Omega$ such that

- $E[\tilde{h}] \leqslant E[h]$.

- *The inverse $\tilde{f} = \tilde{h}^{-1}$ belongs to $W^{1,1}(\Omega, \Omega')$ and has finite distortion.*

- *We have the identity*

$$\int_\Omega \mathbb{K}(z, \tilde{f})dz = E[\tilde{h}]$$

Proof. We express Ω as a locally finite countable union of closed convex domains $\overline{\Omega_j} \subset \Omega$, $j = 1, 2, \ldots$, with pairwise disjoint interiors. Through the homeomorphism h we have a decomposition

$$\Omega' = \bigcup_{j=1}^\infty \overline{\Omega'_j}$$

into a countable union of closed Jordan domains $\overline{\Omega'_j} = h^{-1}(\overline{\Omega_j})$. In each Ω'_j we solve the Dirichlet problem for $h_j \in C(\overline{\Omega'_j})$,

$$\begin{aligned} \Delta h_j &= 0 & &\text{in } \Omega'_j \\ h_j(\zeta) &= h(\zeta) & &\text{on } \partial \Omega'_j \end{aligned}$$

Since h lies in the Royden algebra $C(\overline{\Omega'_j}) \cap W^{1,2}(\Omega'_j)$ and Ω'_j is a Jordan domain, we are reduced via the Riemann mapping to the unit disk and the Poisson

21.3. MEAN DISTORTION AND ANNULI

integral. This results in the well-known energy estimates

$$\int_{\Omega'_j} \|Dh_j\|^2 \leq \int_{\Omega'_j} \|Dh\|^2$$

Next we observe that $h_j : \partial \Omega'_j \to \partial \Omega_j$ is a homeomorphism and the domain Ω_j is convex. Thus by the Rado-Kneser-Choquet theorem (see [114]), each h_j is a homeomorphism of $\overline{\Omega'_j}$ onto $\overline{\Omega_j}$. Hence we may consider $f_j = h_j^{-1} : \overline{\Omega_j} \to \overline{\Omega'_j}$ and define $\tilde{f} : \Omega \to \Omega'$ and $\tilde{h} : \Omega' \to \Omega$ by the rules

$$\tilde{f}(z) = f_j(z), \qquad z \in \overline{\Omega_j}$$
$$\tilde{h}(\zeta) = h_j(\zeta), \qquad z \in \overline{\Omega'_j}$$

It follows that $\tilde{f}$ and $\tilde{h}$ are well-defined homeomorphisms inverse to each other.

We then prove that $\tilde{f} \in W^{1,1}(\Omega, \Omega')$. For this, note that we may argue as in (21.87) to locally estimate the L^1-norms of the derivative. Summing up the estimates and using the Schwarz inequality yields

$$\int_\Omega \|D\tilde{f}\| \leq \left[\sum_{j=1}^\infty \int_{\Omega_j} \mathbb{K}(z, \tilde{f}) \right]^{1/2} \left[\sum_{j=1}^\infty \int_{\Omega_j} J(z, \tilde{f}) \right]^{1/2}$$

On the other hand, since f_j and h_j are smooth, we may change variables as in [27] to obtain

$$\int_{\Omega_j} \mathbb{K}(z, f_j) dz = \int_{\Omega'_j} \|Dh_j\|^2 \leq \int_{\Omega'_j} \|Dh\|^2$$

Thus

$$\int_\Omega \|Df_j\| \leq \left[\sum_{j=1}^\infty \int_{\Omega_j} \|Dh\|^2 \right]^{1/2} \left[\sum_{j=1}^\infty |\Omega'_j| \right]^{1/2} \leq \sqrt{E[h]} \sqrt{|\Omega'|}$$

Since $\tilde{f}$ is continuous on Ω, the above proves $\tilde{f} \in W^{1,1}(\Omega, \Omega')$. This map has integrable distortion since

$$\int_\Omega \mathbb{K}(z, \tilde{f}) = \sum_{j=1}^\infty \int_{\Omega_j} \mathbb{K}(z, f_j) \leq \int_{\Omega'} \|Dh\|^2$$

Finally, we appeal to Theorem 21.1.4, which tells us that the homeomorphism $\tilde{f} : \Omega \to \Omega'$ of integrable distortion in the Sobolev class $W^{1,1}(\Omega, \Omega')$ has its inverse $\tilde{h} \in W^{1,2}(\Omega', \Omega)$ and satisfies the identity

$$\int_\Omega \mathbb{K}(z, \tilde{f}) dz = \int_{\Omega'} \|D\tilde{h}\|^2$$

This completes the proof of the lemma. □

Other minimization problems for mean distortion, such as the Teichmüller problem where one seeks a minimizer which holds the boundary of a domain fixed and moves one internal point to another, are discussed in [242].

Appendix: Elements of Sobolev Theory and Function Spaces

This appendix contains a brief review of a number of facts concerning Sobolev spaces that are repeatedly used throughout the text. Most of the unreferenced results in this section can be found in basic real analysis texts, see for instance [376]. We largely restrict ourselves to the case of the complex plane, though the reader will often see that generalizations of both definitions and results to $\mathbb{R}^n$ are straightforward.

A.1 Schwartz Distributions

The theory of distributions, or generalized functions, used in a somewhat non-rigorous fashion by physicists A. M. Dirac and E. Schrödinger since 1933 was first put in the explicit and presently accepted form by S. L. Sobolev in 1936. In 1950–1951 Laurent Schwartz's monograph *Théorie des Distributions* [324] appeared, in which he systematized the early approaches. The principal idea is to view distributions as linear functionals on the space of test functions.

For an arbitrary set $\Omega \subset \mathbb{C}$, we denote by $C_0^\infty(\Omega)$ the algebra of all infinitely differentiable functions $\varphi : \mathbb{C} \to \mathbb{C}$ with compact support contained in Ω. The differential monomial of order α is the operator

$$\partial^\alpha = \frac{\partial^{\alpha_1 + \alpha_2}}{\partial x^{\alpha_1} \partial y^{\alpha_2}},$$

which can be applied to sufficiently smooth functions. By convention, 0 is the zero multi-index, and ∂^0 stands for the identity operator. There are the related complex differentials and define the complex derivatives

$$\partial = \frac{\partial}{\partial z} = \frac{1}{2}\left(\frac{\partial}{\partial x} - i\frac{\partial}{\partial y}\right), \quad \bar\partial = \frac{\partial}{\partial \bar z} = \frac{1}{2}\left(\frac{\partial}{\partial x} + i\frac{\partial}{\partial y}\right)$$

with $\partial + \bar\partial = \frac{\partial}{\partial x}$ and $\partial - \bar\partial = i\frac{\partial}{\partial y}$. Let Ω be an open subset of $\mathbb{C}$ and $\mathbb{V}$ a finite-dimensional inner-product space. A *distribution* f in Ω with values in $\mathbb{V}$

A.1. SCHWARTZ DISTRIBUTIONS

is a linear form
$$f : C_0^\infty(\Omega) \to \mathbb{V}$$
such that, for every compact set $X \subset \Omega$ and any test function $\varphi \in C_0^\infty(X)$, we have the estimate
$$|f[\varphi]| \leq C(X) \sum_{|\alpha| \leq k} \sup |\partial^\alpha \varphi| \tag{97}$$

In general, the integer k may also depend on the compact set X. If not, we say that f has finite order in Ω, and the smallest such integer k is called the order of f in Ω. The space of all distributions will be denoted by $\mathcal{D}'(\Omega, \mathbb{V})$. As a matter of fact, $\mathcal{D}'(\Omega, \mathbb{V})$ is a module over the algebra $C^\infty(\Omega)$. The multiplication of $f \in \mathcal{D}'(\Omega, \mathbb{V})$ by $\lambda \in C^\infty(\Omega)$ is given by $(\lambda f)[\varphi] = f[\lambda \varphi]$.

Distributions can be locally defined; that is, they are uniquely determined by their values on test functions supported in an arbitrary neighborhood of every point. It is immediate from the definition that the space of distributions in Ω is complete under pointwise convergence. Specifically, given a sequence $\{f_j\}$ of distributions in Ω such that $\lim_{j \to \infty} f_j[\varphi]$ exists for every test function $\varphi \in C_0^\infty(\Omega)$, we define $f : C_0^\infty(\Omega) \to \mathbb{V}$ by $f[\varphi] = \lim_{j \to \infty} f_j[\varphi]$. Then it is not too difficult to see that $f \in \mathcal{D}'(\Omega, \mathbb{V})$. We then say that
$$f = \lim_{j \to \infty} f_j \tag{98}$$
in the sense of distributions. This notion of convergence will often be used. Every locally integrable function $f \in L^1_{loc}(\Omega, \mathbb{V})$ can be viewed as a distribution (of order zero) defined by the rule
$$\varphi \mapsto \int_\Omega \varphi(z) f(z) \, dz \quad \text{for } \varphi \in C_0^\infty(\Omega) \tag{99}$$

It is customary to use the same letter f for the distribution associated with the locally integrable function f. It is in this sense that we understand the inclusion
$$L^1_{loc}(\Omega, \mathbb{V}) \subset \mathcal{D}'(\Omega, \mathbb{V}) \tag{100}$$

Quite often, locally integrable functions are referred to as regular distributions. Although it is not apparent at this point, it is true that regular distributions are dense in $\mathcal{D}'(\Omega, \mathbb{V})$; see Lemma A.1.1. Of fundamental importance is the *Dirac delta* $\delta_a \in \mathcal{D}'(\Omega)$ at the point $a \in \Omega$ that assigns to each $\varphi \in C_0^\infty(\Omega)$ its value at a. That is, $\delta_a[\varphi] = \varphi(a)$. This distribution has order zero but is not a regular distribution. We next discuss the approximation of δ_a by smooth functions because this approximation lies at the heart of the regularization techniques in this book.

Fix a nonnegative function $\Phi \in C_0^\infty(\mathbb{C})$ supported in the closed unit disk and having integral 1. For example,
$$\Phi(z) = C \exp\left(\frac{1}{|z|^2 - 1}\right) \quad \text{if } |z| < 1 \tag{101}$$
$$\Phi(z) = 0 \quad \text{if } |z| \geq 1,$$

where the constant C is chosen so that $\int_{\mathbb{D}} \Phi(z) = 1$.

Next, define a one-parameter family of *mollifiers*

$$\Phi_t(z) = t^{-2}\Phi(t^{-1}z), \qquad t > 0 \tag{102}$$

It is easy to see that

$$\lim_{t \to 0} \Phi_t(a - z) = \delta_a \tag{103}$$

in the sense of distributions; that is,

$$\lim_{t \to 0} \int \Phi_t(a - z)\varphi(z) = \varphi(a) = \delta_a[\varphi] \tag{104}$$

for every test function φ.

The integral on the left hand side is called the *convolution* of φ with the mollifier Φ_t, usually denoted by $(\varphi * \Phi_t)(a)$. Less obvious but of crucial importance is that, for every $f \in L^1_{loc}(\mathbb{C}, \mathbb{V})$,

$$\lim_{t \to 0}(f * \Phi_t)(a) = \lim_{t \to 0} \int \Phi_t(a - z)f(z)\,dz = f(a) \tag{105}$$

for almost every $a \in \mathbb{C}$. For an arbitrary distribution $f \in \mathcal{D}'(\Omega, \mathbb{V})$, we cannot speak of the integral in (105), but its convolution with Φ_t can nevertheless be formally defined. It is a function on the set $\Omega_t = \{a \in \Omega : \mathrm{dist}(a, \partial\Omega) > t\}$ given by

$$f_t(a) = (f * \Phi_t)(a) = f[\Phi_t(a - \cdot)], \tag{106}$$

where we notice that the function $z \mapsto \Phi_t(a - z)$ belongs to $C_0^\infty(\Omega)$.

It should be reasonably evident that $f_t \in C^\infty(\Omega_t, \mathbb{V})$. The stage is now almost set for the following lemma.

Lemma A.1.1. *The space $C_0^\infty(\Omega, \mathbb{V})$ is dense in $\mathcal{D}'(\Omega, \mathbb{V})$.*

That is to say, for every $f \in \mathcal{D}'(\Omega, \mathbb{V})$, there exists a sequence $\{f_j\}$ of functions in $C_0^\infty(\Omega, \mathbb{V})$ such that

$$f[\varphi] = \lim_{j \to \infty} \int_\Omega \varphi(z) f_j(z)\,dz \tag{107}$$

for all test functions $\varphi \in C_0^\infty(\Omega)$.

Distributions of order zero, such as the Dirac delta function, are all represented by integration with respect to a suitable $\mathbb{V}$-valued Radon measure on Ω. This fundamental fact is usually referred to as the Riesz representation theorem. It asserts that each $f \in \mathcal{D}'(\Omega, \mathbb{V})$ can be written as

$$f[\varphi] = \int_\Omega \varphi(z)\,d\mu(z) \qquad \text{for all } \varphi \in C_0^\infty(\Omega) \tag{108}$$

We recall that a Radon measure μ on Ω is a signed Borel measure that is finite on compact subsets. It is in this way that we identify the space of

A.2. DEFINITIONS OF SOBOLEV SPACES

Radon measures with the corresponding distributions of order zero. The regular distributions are the absolutely continuous ones, having no singular part with respect to Lebesgue measure.

A distribution $f \in \mathcal{D}'(\Omega, \mathbb{R})$ is said to be *positive* if $f[\varphi] \geq 0$ whenever $\varphi \geq 0$. Positive distributions have order zero and therefore are represented by Borel measures.

The theory of distributions mimics the calculus of differentiable functions very cleverly. As a starting point, recall the formula of integration by parts. For $f \in C^\infty(\Omega, \mathbb{V})$, we have

$$\int_\Omega \varphi(\partial^\alpha f) = (-1)^{|\alpha|} \int_\Omega (\partial^\alpha \varphi) f, \qquad \varphi \in C_0^\infty(\Omega) \tag{109}$$

This procedure can be extended to all $f \in \mathcal{D}'(\Omega, \mathbb{V})$ by setting

$$(\partial^\alpha f)[\varphi] = (-1)^{|\alpha|} f[\partial^\alpha \varphi] \tag{110}$$

Of course, differentiation increases the order of f by $|\alpha|$. Locally, every distribution can be written as $\partial^\alpha f$ for some $\alpha = (\alpha_1, \alpha_2)$ and $f \in L^1(\Omega, \mathbb{V})$. Let us point out that the original purpose of the theory of distributions was to make it possible to differentiate locally integrable functions. From this point of view, Schwartz distributions offer us the most economical extension of the space $L^1(\Omega, \mathbb{V})$ suitable for carrying out this task.

It makes sense to say that a given distribution $f \in \mathcal{D}'(\Omega, \mathbb{V})$ vanishes on an open subset $U \subset \Omega$. This simply means that $f[\varphi] = 0$ whenever $\varphi \in C_0^\infty(U)$. The union of such sets is the largest open subset of Ω on which f vanishes. Its complement is called the *support* of f and is denoted by $\operatorname{supp} f$. Every distribution $f \in \mathcal{D}'$ supported at a single point $a \in \Omega$ is of the form

$$f = \sum_{|\alpha| \leq k} c_\alpha \partial^\alpha(\delta_a)$$

As a note of warning, no reasonable multiplication of distributions can be defined.

A.2 Definitions of Sobolev Spaces

While distributions are elegant and effective tools for dealing with linear PDEs, their use in nonlinear problems is quite limited. To be able to make the pointwise estimates, we shall need to study distributions that together with partial derivatives up to a certain order are represented by locally integrable functions. Such objects are called Sobolev functions. The principal feature of these functions and their use in the theory of nonlinear PDEs lies in the pointwise behavior of their derivatives, see Lemma A.4.1. We next present a few basic definitions. Let Ω be a domain in $\mathbb{C}$ and $\mathbb{V}$ a finite-dimensional inner-product space.

For $1 \leqslant p \leqslant \infty$ and $k = 1, 2, \ldots$, the *Sobolev space* $W^{k,p}(\Omega, \mathbb{V})$ consists of all $f \in \mathcal{D}'(\Omega, \mathbb{V})$ for which the distributional derivatives $\partial^\alpha f$ are represented by functions in $L^p(\Omega, \mathbb{V})$, with $|\alpha| \leqslant k$. That is, there exist $g_\alpha \in L^p(\Omega), |\alpha| \leqslant k$, such that

$$(-1)^{|\alpha|} f[\partial^\alpha \varphi] = \partial^\alpha f[\varphi] = \int g_\alpha(z) \, \varphi(z) \quad \text{for all } \varphi \in C_0^\infty(\Omega)$$

As in (99), we identify $\partial^\alpha f = g_\alpha$. When $\mathbb{V} = \mathbb{C}$, we simplify notation and write $W^{k,p}(\Omega)$.

The space $W^{k,p}(\Omega, \mathbb{V})$ is equipped with the norm

$$\|f\|_{k,p} = \left(\sum_{|\alpha| \leqslant k} \int_\Omega |\partial^\alpha f(\zeta)|^p d\zeta \right)^{1/p} \tag{111}$$

if $p \neq \infty$ and

$$\|f\|_{k,\infty} = \max_{|\alpha| \leqslant k} \|\partial^\alpha f\|_\infty \tag{112}$$

These norms give $W^{k,p}(\Omega, \mathbb{V})$ a Banach space structure if $1 \leqslant p \leqslant \infty$. The local Sobolev spaces $W^{k,p}_{loc}(\Omega, \mathbb{V})$ consist of those functions whose partials of order k or less are in $L^p_{loc}(\Omega, \mathbb{V})$. We say that the sequence $\{f_j\}$ converges to f in $W^{k,p}_{loc}(\Omega, \mathbb{V})$ if

$$\lim_{j \to \infty} \|\partial^\alpha f_j - \partial^\alpha f\|_{L^p(X)} = 0 \tag{113}$$

for every compact set $X \subset \Omega$ and every $|\alpha| \leqslant k$.

On many occasions in the theory of PDEs, such as the variational formulations for the Dirichlet problem in unbounded domains, it is natural to require p-integrability only of the derivatives, not of the function itself. This leads to the notion of the homogeneous Sobolev space,

$$\mathbb{W}^{1,p}(\Omega) = \{f \in \mathcal{D}'(\Omega, \mathbb{V}) : Df \in L^p(\Omega)\},$$

equipped with the seminorm,

$$\|f\|_{\mathbb{W}^{1,p}} = \left(\int_\Omega |Df(\zeta)|^p d\zeta \right)^{1/p}$$

A.3 Mollification

Given $f \in L^1_{loc}(\Omega, \mathbb{V})$, the mollification of f is the family of functions f_t, $t > 0$, defined on the set $\Omega_t = \{z \in \Omega : \text{dist}(z, \partial\Omega) > t\}$ by the convolution formula

$$f_t(z) = \int_\Omega \Phi_t(z - \zeta) f(\zeta), \tag{114}$$

where Φ is defined in (101). If $f \in L^1(\Omega, \mathbb{V})$, this formula is valid for all $z \in \mathbb{C}$ and the functions f_t belong to the space $C^\infty(\mathbb{C}, \mathbb{V})$, as one can differentiate under the

A.3. MOLLIFICATION

integral sign in (114). However, various bounds for f can be transferred to f_t only on the set Ω_t. We now come to a central, although elementary, regularization result.

Theorem A.3.1. *Let $f \in L^1_{loc}(\Omega, \mathbb{V})$. Then*

1. $\lim_{t \to 0} f_t(z) = f(z)$ *for almost all $z \in \Omega$*

2. *If f is continuous, then the convergence in statement 1 is locally uniform.*

3. *Mollification does not increase the norms. More specifically, for each compact $X \subset \Omega$ and $0 \leqslant t < \operatorname{dist}(X, \partial\Omega)$, we have*

$$\|\partial^\alpha f_t\|_{L^p(X)} \leqslant \|\partial^\alpha f\|_{L^p(\Omega)}$$

whenever $f \in W^{k,p}(\Omega, \mathbb{V})$ with $1 \leqslant p \leqslant \infty$ and $|\alpha| \leqslant k$. Also, on X we have

$$\partial^\alpha f_t = (\partial^\alpha f)_t$$

4. *For $1 \leqslant p < \infty$, we have*

$$\lim_{t \to 0} \|\partial^\alpha f_t - \partial^\alpha f\|_{L^p(X)} = 0$$

5. *When $\Omega = \mathbb{C}$, statements 3 and 4 remain valid for $X = \mathbb{C}$.*

This theorem quickly implies the following corollary.

Corollary A.3.2. $C^\infty(\mathbb{C}, \mathbb{V})$ *is dense in $W^{k,p}_{loc}(\Omega, \mathbb{V})$ for $1 \leqslant p \leqslant \infty$.*

Next, the space $W^{k,p}_0(\Omega, \mathbb{V})$ is defined as the closure of $C^\infty_0(\Omega, \mathbb{V})$, the compactly supported smooth functions, with respect to the norms defined in (111) and (112). Functions in the space $W^{k,p}_0(\Omega, \mathbb{V})$ are quite often said to vanish on $\partial\Omega$ in the Sobolev sense.

Theorem A.3.1 will suffice for most of our forthcoming approximation arguments. However, it is not too difficult to derive from it the following global variant we attribute to Friedrichs [129] and Meyers and Serrin [260]. The reader familiar with partitions of unity will have no difficulty in proving this result given Theorem A.3.1 above.

Theorem A.3.3. *Let $f \in W^{k,p}_{loc}(\Omega, \mathbb{V})$, $1 \leqslant p < \infty$, $k = 1, 2, \ldots$. Then there is a sequence of smooth functions $f_j \in C^\infty(\Omega, \mathbb{V})$ such that*

1. $f - f_j \in W^{k,p}_0(\Omega, \mathbb{V})$ *and*

2. $\lim_{j \to \infty} \|f - f_j\|_{k,p} = 0$

It should be noted that for regular domains, such as disks or squares in $\mathbb{C}$, one can approximate any $f \in W^{k,p}(\Omega, \mathbb{V})$ by functions $f_j \in C^\infty_0(\mathbb{C}, \mathbb{V})$. This is not in general possible for irregular domains.

A.4 Pointwise Coincidence of Sobolev Functions

Various approximation techniques will be repeatedly used throughout this text. In this section we have selected an example to explicitly illustrate how this technique works.

Lemma A.4.1. *Suppose two functions $f, g \in W_{loc}^{k,p}(\Omega, \mathbb{V})$ agree on a set $E \subset \Omega$. Then $\partial^\alpha f(x) = \partial^\alpha g(x)$ for almost every $x \in E$ and all $|\alpha| \leq k$.*

Proof. We begin by assuming that $p = 1$. We need only consider the scalar case $\mathbb{V} = \mathbb{R}$ and by linearity of the operators ∂^α, $g \equiv 0$ in Ω. Let f_j be smooth functions converging almost everywhere to f and locally in $W_{loc}^{k,1}(\Omega)$. We observe the elementary identity

$$\frac{\partial}{\partial x_i}\left(\frac{f_j^3}{\varepsilon + f_j^2}\right) = -\frac{f_j^4 - 3\varepsilon f_j^2}{(\varepsilon + f_j^2)^2}\frac{\partial f_j}{\partial x_i},$$

which holds for every $\varepsilon > 0$ and $i = 1, 2$. If we multiply this equation by a test function $\phi \in C_0^\infty(\Omega)$ and integrate by parts, we find that

$$\int_\Omega \frac{f_j^3}{\varepsilon + f_j^2}\frac{\partial \phi}{\partial x_i} = \int_\Omega \phi \frac{3\varepsilon f_j^2 - f_j^4}{(\varepsilon + f_j^2)^2}\frac{\partial f_j}{\partial x_i}$$

Now by the dominated convergence theorem, we see at once that this identity remains valid for the limit function f in place of f_j. After passing to the limit as $\varepsilon \to 0$ and again using the dominated convergence theorem, we arrive at the identity

$$\int_\Omega f\frac{\partial \phi}{\partial x_i} = -\int_\Omega \left(\chi_F \frac{\partial f}{\partial x_i}\right)\phi,$$

where χ_F as usual denotes the characteristic function of the set $F = \Omega \setminus E$. Using the definition of the distributional derivative, we see that

$$\frac{\partial f}{\partial x_i} = \chi_F \frac{\partial f}{\partial x_i} \quad \text{for almost every } z \in \Omega$$

Hence $\frac{\partial f}{\partial x_i} = 0$ almost everywhere in E and all $i = 1, 2$.

We may repeat this procedure k times to conclude that $\partial^\alpha f(x) = 0$ for almost every $x \in E$ and all $|\alpha| \leq k$, completing the proof of the lemma. $\square$

A.5 Alternate Characterizations

One identity that lies at the heart of Sobolev theory is the formula for integration by parts.

$$\int_\Omega \langle f, \partial^\alpha g \rangle = (-1)^{|\alpha|} \int_\Omega \langle g, \partial^\alpha f \rangle \tag{115}$$

whenever $f \in W^{p,k}(\Omega, \mathbb{V})$ and $g \in W_0^{q,k}(\Omega, \mathbb{V})$ with $1 \leq p, q \leq \infty$, $\frac{1}{p} + \frac{1}{q} = 1$ and $|\alpha| \leq k$.

A.5. ALTERNATE CHARACTERIZATIONS

The formula for integration by parts implies, in particular, the following corollary.

Corollary A.5.1. *For each $f \in W^{k,p}(\Omega, \mathbb{V})$, there are constants $C_\alpha = C_\alpha(f)$, $|\alpha| \leqslant k$, such that*

$$\int_\Omega \langle f, \partial^\alpha \phi \rangle \leqslant C_\alpha \|\phi\|_{L^q(\Omega)}, \tag{116}$$

where $\phi \in C_0^\infty(\Omega, \mathbb{V})$ and q is the Hölder conjugate exponent to p.

We actually find that, conversely, for a $f \in \mathcal{D}'(\Omega, \mathbb{V})$ these inequalities imply that $f \in W^{k,p}(\Omega, \mathbb{V})$ if $1 < p < \infty$. We also obtain the estimates $\|\partial^\alpha f\|_p \leqslant C_\alpha$ for all $|\alpha| \leqslant k$.

There is another, more geometric, characterization of Sobolev spaces based on the absolute continuity on lines (ACL) property. A function $f : [a,b] \to \mathbb{R}^m$ is said to be *absolutely continuous* if for each $\varepsilon > 0$ there exists a $\delta > 0$ such that

$$\sum_{i=1}^k |f(b_i) - f(a_i)| < \varepsilon \tag{117}$$

whenever $a = a_1 < b_1 \leqslant \cdots \leqslant a_k < b_k = b$ and

$$\sum_{i=1}^k (b_i - a_i) < \delta$$

According to the fundamental theorem of calculus, every such f is differentiable almost everywhere, and we have the integration by parts formula

$$\int_a^b \langle \varphi(t), f'(t) \rangle dt = -\int_a^b \langle \varphi'(t), f(t) \rangle dt \tag{118}$$

for all smooth functions $\varphi : [a,b] \to \mathbb{R}^m$ that vanish at the end points, $\varphi(a) = \varphi(b) = 0$.

Next, for $\nu = 1, 2, \ldots, n$, let $I_\nu = [a_\nu, b_\nu]$ be closed intervals such that the associated rectangular box $Q = I_1 \times I_2 \times \cdots \times I_n$ lies in Ω. Consider the νth face

$$Q_\nu = I_1 \times \cdots \times I_{\nu-1} \times \{a_\nu\} \times I_{\nu+1} \times \cdots \times I_n$$

We say a continuous function $f : \Omega \to \mathbb{R}^m$ is $ACL(\Omega, \mathbb{R}^m)$, *absolutely continuous on lines*, if for all $\nu = 1, 2, \ldots, n$ and almost every $a \in Q_\nu$, with respect to $(n-1)$-measure, the function

$$t \mapsto f(a + te_\nu), \qquad 0 \leqslant t \leqslant b_\nu - a_\nu$$

is absolutely continuous. Here of course e_ν denotes the νth unit basis vector.

Clearly, such functions have first-order partials almost everywhere in Ω, and these partials are Borel functions. Indeed, for an *ACL* function f for almost every $x \in \Omega$ we can define $Df(x)$ as the following linear operator $Df(x) : \mathbb{R}^m \to \mathbb{R}^n$,

$$Df(x)h = \sum_{j=1}^{m} h_j \frac{\partial f}{\partial x_j}(x), \qquad h = (h_j)_{j=1}^m \in \mathbb{R}^m$$

We have the next important, and often useful, characterization of Sobolev functions.

Lemma A.5.2. *A function $f \in W^{1,p}_{loc}(\Omega, \mathbb{R}^m)$, $1 \leqslant p \leqslant \infty$, if and only if $f \in ACL(\Omega, \mathbb{R}^n)$ and $Df \in L^p_{loc}(\Omega, \mathbb{R}^{m \times n})$.*

In particular, the existence of pointwise partial derivatives almost everywhere (that is, convergence of partial difference quotients at almost every point) holds for every Sobolev function $f \in W^{1,p}_{loc}$, $p \geqslant 1$.

There is yet another important technique used for dealing with weak solutions of nonlinear equations we want to point out. Formally, one can differentiate the equations to deduce various properties of solutions. To implement this formal process, where in fact the functions concerned may not be differentiable, one works with the difference quotients

$$\frac{|f(z+h) - f(z)|}{|h|},$$

instead of derivatives, for $f \in L^1_{loc}(\Omega, \mathbb{V})$ and $0 < |h| < \text{dist}(z, \partial \Omega)$.

Uniform integral bounds on these quotients imply that the derivatives exist and lie in the desired Sobolev spaces with corresponding estimates. A particular instance of this technique when $k = 1$ is to replace the partials $\partial/\partial x$ and $\partial/\partial y$ in (116) by the corresponding difference quotients, leading to a very useful characterization of $W^{1,p}(\Omega, \mathbb{V})$, which we state in two lemmas.

Lemma A.5.3. *Let $f \in W^{1,p}_{loc}(\Omega, \mathbb{V})$ with $1 \leqslant p \leqslant \infty$. Then for each non-negative $\omega \in C_0^\infty(\Omega)$, we have*

$$\limsup_{h \to 0} |h|^{-p} \int |f(z+h) - f(z)|^p \omega(z)\, dz \leqslant \int |\nabla f(z)|^p \omega(z)\, dz,$$

where

$$\nabla f(z) = \left(\frac{\partial f}{\partial x}, \frac{\partial f}{\partial y} \right) \in \mathbb{V}^2$$

The converse statement reads as

Lemma A.5.4. *Let $f \in L^p_{loc}(\Omega, \mathbb{V})$, $1 < p \leqslant \infty$. Suppose that there is a function $F \in L^p_{loc}(\Omega)$ for which*

$$\limsup_{h \to 0} |h|^{-p} \int |f(z+h) - f(z)|^p \omega(z)\, dz \leqslant \int |F(z)|^p \omega(z)\, dz$$

whenever $\omega \in C_0^\infty(\Omega)$ is non-negative. Then, $f \in W^{1,p}_{loc}(\Omega, \mathbb{V})$ and

$$|\nabla f(z)| \leqslant |F(z)| \quad \text{for almost every } z \in \Omega$$

A.6. EMBEDDING THEOREMS

We point out here that this characterisation of Sobolev functions fails if $p = 1$. To see this, simply consider the characteristic function of a disk. On the other hand, for $p = \infty$ we find the space $W^{1,\infty}_{loc}(\Omega, \mathbb{V})$ consists precisely of those locally Lipschitz functions valued in $\mathbb{V}$. We denote this space by $Lip_{loc}(\Omega, \mathbb{V})$. It follows from these two lemmas that Sobolev functions do not change their nature when composed with Lipschitz functions. We record this as the chain rule.

Lemma A.5.5. (Chain Rule) *Let $\mathbb{V}$ and $\mathbb{W}$ be finite-dimensional inner-product spaces and $T \in Lip(\mathbb{V}, \mathbb{W})$. Suppose that both $f \in W^{1,p}_{loc}(\Omega, \mathbb{V})$ and that $T \circ f \in L^p_{loc}(\Omega, \mathbb{W})$ for some p with $1 \leqslant p \leqslant \infty$. Then, $T \circ f \in W^{1,p}_{loc}(\Omega, \mathbb{W})$ and*

$$D(T \circ f)(z) = DT(f(z)) \circ Df(z) \quad \text{for almost every } z \in \Omega$$

A.6 Embedding Theorems

Embedding theorems are fundamental to any method aimed at establishing the regularity of solutions to elliptic equations with nonsmooth coefficients. We will not strive here for the utmost generality, instead we recall a few main examples needed in our study.

Theorem A.6.1. (Sobolev Inequality) *Let $\Omega \subset \mathbb{R}^n$ be a cube and $u \in W^{1,p}(\Omega)$ with $1 \leqslant p < n$. Then, $u \in L^{\frac{np}{n-p}}(\Omega)$, and we have the estimate*

$$\|u\|_{\frac{np}{n-p}} \leqslant \frac{pn-p}{n-p} \|\nabla u\|_p + |\Omega|^{-1/n} \|u\|_p$$

With a little work, Theorem A.6.1 can be used to show that for an arbitrary domain Ω, if $u \in W^{1,p}_0(\Omega)$, then

$$\|u\|_{\frac{np}{n-p}} \leqslant \frac{np-p}{n-p} \|\nabla u\|_p$$

Theorem A.6.2. (Poincaré Inequality) *Let Ω be a disk or cube in $\mathbb{R}^n$ and $u \in W^{1,p}(\Omega)$ with $1 \leqslant p < \infty$. Then for every measurable set $E \subset \Omega$ of positive measure, we have*

$$\int_\Omega |u(x) - u_E|^p dx \leqslant 2^n (\operatorname{diam}(\Omega))^p \frac{|\Omega|}{|E|} \int_\Omega |\nabla u(x)|^p \, dx,$$

where $u_E = \frac{1}{|E|} \int_E u(x) dx$ is the integral mean of u on E.

Combining this result with Theorem A.6.1 gives the following theorem.

Theorem A.6.3. (Poincaré-Sobolev Inequality) *Let Ω be a disk or cube in $\mathbb{R}^n$, and E a subset of Ω with positive measure. Then, for each $u \in W^{1,p}(\Omega)$ with $1 \leqslant p < n$, we have*

$$\|u - u_E\|_{\frac{np}{n-p}} \leqslant \frac{C(n)}{n-p} \left(\frac{|\Omega|}{|E|}\right)^{1/p} \|\nabla u\|_p$$

We will also need estimates in terms of the second derivatives.

Lemma A.6.4. *Let $u \in W^{2,1}(\Omega)$, where $\Omega = [-a,a] \times [-b,b] \subset \mathbb{C}$. Then*

$$|u(z) - u_\Omega| \leq \frac{1}{2b}\int_\Omega |u_x| + \frac{1}{2a}\int_\Omega |u_y| + \int_\Omega |u_{xy}|$$

Proof. By approximation we assume that u is smooth. Fix two points $(x_0, y_0) \in \Omega$ and $(x,y) \in \Omega$. Then

$$u(x_0, y_0) - u(x,y) = \int_x^{x_0} u_s(s,y)ds + \int_y^{y_0} u_t(x,t)dt + \int_x^{x_0}\int_y^{y_0} u_{st}(s,t)ds\,dt$$

We integrate this with respect to $(x,y) \in \Omega$ and obtain

$$\begin{aligned}
u(x_0, y_0) - u_\Omega &= \frac{1}{4ab}\int_\Omega \Big(\int_x^{x_0} u_s(s,y)ds\Big)dx\,dy \\
&+ \frac{1}{4ab}\int_\Omega \Big(\int_y^{y_0} u_t(x,t)dt\Big)dx\,dy \\
&+ \frac{1}{4ab}\int_\Omega \Big(\int_x^{x_0}\int_y^{y_0} u_{st}(s,t)ds\,dt\Big)dx\,dy
\end{aligned}$$

We then estimate $|u(x_0, y_0) - u_\Omega|$ with triangle inequality and take in the above representation the absolute values inside the integrals. The claim follows. □

When Ω is a cube, say, $\Omega = [-a,a] \times [-a,a]$, then $\operatorname{diam}\Omega = 2a\sqrt{2}$, and as $|u_x| + |u_y| \leq \sqrt{2}|\nabla u|$, we have by Lemma A.6.4,

$$\begin{aligned}
|u(z) - u(0)| &\leq |u(z) - u_\Omega| + |u(0) - u_\Omega| \\
&\leq \frac{4}{\operatorname{diam}\Omega}\int_\Omega |\nabla u| + 2\int_\Omega |D^2 u| \qquad (119)
\end{aligned}$$

If instead of a cube we consider a disk $B = \mathbb{D}(0,R)$, if $z \in B$, let $\Omega \subset B$ be a cube centered at the origin and of diameter $2R$ and rotate it so that $z \in \Omega$. This gives

$$|u(z) - u(0)| \leq \frac{2}{R}\int_\Omega |\nabla u| + 2\int_\Omega |D^2 u| \leq \frac{2}{R}\int_B |\nabla u| + 2\int_B |D^2 u|$$

We now have the following useful theorem.

Theorem A.6.5. *Let $u \in W^{2,1}_{loc}(\Omega)$, where $\Omega \subset \mathbb{C}$ is any domain in the plane. Then, if $B = \mathbb{D}(z_0, R) \subset \Omega$, we have*

$$|u(z)| \leq |u_B| + 4\int_B |D^2 u| + \frac{8}{\operatorname{diam}(B)}\int_B |\nabla u|, \qquad z \in B$$

A.6. EMBEDDING THEOREMS

Proof. We need only to apply the above estimate for

$$u - u_B = u(z) - u(0) - \frac{1}{|B|} \int_B [u(\zeta) - u(0)]$$

to obtain the desired inequality. □

Let us now consider the situation where $p = n$. The relevant, though not the strongest possible, conclusion for functions in $W^{1,2}$ shows that they are exponentially integrable. Precisely, combining the inequalities of John-Nirenberg and Poincaré, in other words, Theorems A.8.3 and A.6.2, we have

Theorem A.6.6. *Let $u \in W^{1,2}(\mathbb{D})$ be a nonconstant function. Then*

$$\int_{\mathbb{D}} \exp\left(\frac{\theta |u(z) - u_{\mathbb{D}}|}{\|\nabla u\|_2}\right) dz \leqslant 2|\mathbb{D}|,$$

where θ is a positive constant.

With the following well-known inequality of Trudinger [355] the integrability can be further improved.

Theorem A.6.7. *There are positive constants α and C such that if $u \in W^{1,2}(\mathbb{D})$ satisfies $\int_{\mathbb{D}} |\nabla u|^2 \leqslant 1$, then*

$$\int_{\mathbb{D}} \exp\left(\alpha |u(z) - u_{\mathbb{D}}|^2\right) \leqslant C|\mathbb{D}|$$

Let us next invoke the case $p > n$ to obtain the Hölder continuity of the Sobolev functions in $W^{1,p}(\Omega)$. First, we recall that a function $f : \Omega \to \mathbb{V}$ is said to be Hölder-continuous with exponent $0 < \alpha \leqslant 1$ if there is a constant C such that

$$|f(x) - f(y)| \leqslant C|x - y|^\alpha \quad \text{for all } x, y \in \Omega$$

We denote the space of all such functions as $C^\alpha(\Omega, \mathbb{V})$. This space becomes a Banach space when supplied with the norm

$$\|f\|_{C^\alpha(\Omega, \mathbb{V})} = \sup_{x \in \Omega} |f(x)| + \sup_{x \neq y} \frac{|f(x) - f(y)|}{|x - y|^\alpha}$$

Theorem A.6.8. *Let Ω be a cube in $\mathbb{R}^n$ and $p > n$. Then, $W^{1,p}(\Omega) \subset C^{1-n/p}(\Omega)$, and for each $u \in W^{1,p}(\Omega)$ we have the estimate*

$$|u(x) - u(y)| \leqslant \frac{2pn}{p-n} |x-y|^{1-n/p} \|\nabla u\|_p$$

Perhaps one of the most important features of functions in the Sobolev space $W^{1,p}(\Omega)$ for $p > n$ is their pointwise almost everywhere differentiability.

Let $\mathbb{W}$ and $\mathbb{V}$ be normed vector spaces and $\Omega \subset \mathbb{W}$. A function $f : \Omega \to \mathbb{V}$ is differentiable at $x \in \Omega$ if there is a continuous linear map $Df(x) : \mathbb{W} \to \mathbb{V}$ such that

$$\lim_{h \to 0} \frac{|f(x+h) - f(x) - Df(x)h|}{|h|} = 0$$

Theorem A.6.9. *Let $\Omega \subset \mathbb{R}^n$ be a domain and $\mathbb{V}$ a finite-dimensional inner-product space. Every function $f \in W^{1,p}_{loc}(\Omega, \mathbb{V})$, with $p > n$, is differentiable at almost every $x \in \Omega$. At such points its, differential $Df(x) : \mathbb{R}^n \to \mathbb{V}$ is represented by the matrix of partial derivatives*

$$Df(x)h = \sum_{i=1}^{n} h_i \frac{\partial f}{\partial x_i}$$

for $h = (h_1, h_2, \ldots, h_n) \in \mathbb{R}^n$.

As a basic example of an elliptic regularity result, in the case of smooth coefficients, let us recall the classical Weyl lemma.

Lemma A.6.10. (Weyl) *Every distribution $h \in \mathcal{D}'(\Omega)$ satisfying the Laplace equation $\Delta h = 0$ is represented by a $C^\infty(\Omega)$ function.*

In particular, suppose $f \in L^1_{loc}(\Omega)$ with the distributional derivative $f_{\bar{z}} = 0$. Then f is analytic in Ω.

As the example $f(z) = 1/z$ shows, for function $f \in L^1_{loc}(\Omega)$ the condition

$$\frac{\partial f}{\partial \bar{z}}(z) = 0 \quad \text{almost everywhere in } \Omega \tag{120}$$

is not enough to conclude that f is analytic. However, if we assume that f has locally integrable distributional derivatives, that is, $f \in W^{1,1}_{loc}(\Omega)$, then (120) implies that f is analytic in Ω.

A.7 Duals and Compact Embeddings

We recall that a sequence $\{x_i\}$ in a normed linear space X is said to *converge weakly* to $x \in X$ if $\phi(x_i) \to \phi(x)$ for every bounded linear functional $\phi : X \to \mathbb{R}$.

The Hahn-Banach theorem implies that

$$\|x\| \leqslant \liminf_{i \to \infty} \|x_i\|$$

A Banach space X is reflexive if and only if the closed unit ball in X is weakly compact. In particular, a separable Banach space is reflexive if and only if every bounded sequence has a weakly convergent subsequence. An important fact concerning the Sobolev spaces $W^{k,p}(\Omega, \mathbb{V})$ with $1 < p < \infty$ is that they are reflexive. As $W^{k,p}_0(\Omega, \mathbb{V})$ is a closed subspace of $W^{k,p}(\Omega, \mathbb{V})$, it is also reflexive.

It is not difficult to see that every bounded linear functional on $W^{k,p}(\Omega, \mathbb{V})$, $1 \leqslant p < \infty$, can be written as

$$u \mapsto \int_\Omega \sum_{|\alpha|=k} \langle v_\alpha, \partial^\alpha u \rangle \tag{121}$$

A.8. HARDY SPACES AND BMO

for some collection of functions $v_\alpha \in L^q(\Omega, \mathbb{V})$, $\frac{1}{p} + \frac{1}{q} = 1$. For $1 \leqslant p < \infty$, the dual of the space $W_0^{k,p}(\Omega, \mathbb{V})$ is commonly denoted by $W^{-k,q}(\Omega, \mathbb{V})$, with p, q again a Hölder conjugate pair. It consists of distributions of the form

$$f = \sum_{|\alpha| \leqslant k} \partial^\alpha v_\alpha, \qquad v_\alpha \in L^q(\Omega, \mathbb{V})$$

We also want to recall here the theorem of Rellich-Kondrachov. First, a few definitions are needed. Let X and Y be Banach spaces. An injective bounded operator from X into Y is called an *embedding* and is written $X \xhookrightarrow{i} Y$. The embedding is said to be *compact* if the image of any bounded subset of X is relatively compact in Y, that is, its closure in Y is compact.

Theorem A.7.1. (Rellich-Kondrachov) *Let Ω be a ball or a cube in $\mathbb{R}^n$ or any bounded Lipschitz domain. Then the following inclusions are compact embeddings:*

$$W^{1,p}(\Omega) \xhookrightarrow{i} L^q(\Omega)$$

if $1 \leqslant p < n$ and $1 \leqslant q < \frac{np}{n-p}$, and

$$W^{1,p}(\Omega) \xhookrightarrow{i} C^\beta(\overline{\Omega})$$

if $p > \frac{n}{1-\beta}$, for some $0 < \beta < 1$.

A.8 Hardy Spaces and BMO

Delicate cancellation properties of various nonlinear objects such as Jacobians cannot be discussed without introducing Hardy spaces. It is our objective here to give a brief account of this, restricting the discussion to the classical Hardy spaces $H^p(\Omega)$. For the more general notions such as Hardy-Orlicz spaces $H^P(\Omega)$, we refer the reader to the monograph [191].

We are concerned with the Hardy $H^p(\Omega)$, $0 < p < \infty$, on domains $\Omega \subset \mathbb{C}$. Here the most convenient definition is based on the maximal characterization of the Hardy spaces. That is, we first define a maximal function of a distribution and then assume that this function belongs to the appropriate L^p-space.

The Hardy Space

We shall rely on the mollifiers and convolutions constructed in (101) – (106). Which family of mollifiers we choose to fix is quite immaterial to our results. A good general reference here is the book by E. Stein [335].

Given any distribution $f \in \mathcal{D}'(\Omega, \mathbb{V})$, it is legitimate to write

$$f_t(x) = f * \Phi_t(x) \tag{122}$$

for $x \in \Omega$ whenever $0 < t < \text{dist}(x, \partial\Omega)$. Then we can define the mollified maximal function of f as

$$(\mathcal{M}f)(x) = (\mathcal{M}_\Omega f)(x) = \sup\{|f_t(x)|; \quad 0 < t < \text{dist}(x, \partial\Omega)\} \qquad (123)$$

for all $x \in \Omega$. Most often we shall ignore the subscript Ω when the dependence of $\mathcal{M}$ on the domain need not be emphasized. For the Dirac delta, an easy computation shows that

$$(\mathcal{M}\delta)(x) = \frac{C}{|x|^2} \quad \text{for } x \in \mathbb{C} \setminus \{0\}$$

Now, the Hardy spaces $H^p(\Omega, \mathbb{V})$, $0 < p < \infty$, are made up of Schwartz distributions $f \in \mathcal{D}'(\Omega, \mathbb{V})$ such that

$$\|f\|_{H^p(\Omega)} = \|\mathcal{M}_\Omega f\|_{L^p(\Omega)} < \infty \qquad (124)$$

Clearly, $H^p(\Omega, \mathbb{V})$ is a complete linear metric space with respect to the distance

$$\text{dist}(f, g) = \inf\left\{\frac{1}{\lambda} > 0; \ \lambda \int_\Omega (\mathcal{M}(\lambda f - \lambda g))^p < 1\right\}$$

If, moreover, $1 \leq p < \infty$, the nonlinear functional in (124) defines a norm, which makes $H^p(\Omega)$ a Banach space. For $\Omega = \mathbb{C}$, $\mathbb{V} = \mathbb{R}$ and p greater than 1, our definition results [336] in the usual Lebesgue spaces, that is, $H^p(\mathbb{C}) = L^p(\mathbb{C})$, for $1 < p < \infty$. However, note that $H^1(\Omega) \subset L^1(\Omega)$ is a proper subspace with $\|f\|_{L^1(\Omega)} \leq \|f\|_{H^1(\Omega)}$.

The next well-known result of E. Stein [334] characterizes the growth properties of distributions in the local Hardy spaces $H^1_{loc}(\Omega)$, that is, on relatively compact subdomains.

Theorem A.8.1. *A non-negative function f belongs to $H^1_{loc}(\Omega)$ if and only if $f \log f \in L^1_{loc}(\Omega)$.*

However, let us emphasize that the nature of a distribution in $H^p(\Omega)$ is determined not only by its size but also by its internal cancellation properties. These properties are perfectly visible in the atomic decompositions that we shall now turn to. We discuss only the case $p = 1$.

A measurable function $a(z)$ supported in some disk $\mathbb{D}$ is called an $\mathcal{H}^1$-*atom* if it satisfies both the conditions

$$|a(z)| \leq \frac{1}{|\mathbb{D}|} \quad \text{for almost every } z \in \mathbb{C}$$

$$a_\mathbb{D} = \frac{1}{|\mathbb{D}|} \int_\mathbb{D} a(z) dz = 0$$

A function $f \in L^1(\mathbb{C})$ belongs to $\mathcal{H}^1(\mathbb{C})$ if and only if it can be written as a (possibly infinite) linear combination of $\mathcal{H}^1$-atoms, $f = \sum_{k=1}^\infty \lambda_k a_k$, with $\sum_{k=1}^\infty |\lambda_k| < \infty$. The norm is then defined by

$$\|f\|_{\mathcal{H}^1} = \inf\left\{\sum_{k=1}^\infty |\lambda_k| : f = \sum_{k=1}^\infty \lambda_k a_k\right\},$$

A.8. HARDY SPACES AND BMO

where the infimum is taken over all atomic decompositions of f. It is important to notice that such an f satisfies the *moment condition*

$$\int_{\mathbb{C}} f(z)dz = 0 \tag{125}$$

The space BMO

Next, for a measurable function $g : \Omega \to \mathbb{V}$ and a disk $\mathbb{D} \subset \Omega$, we define the average of g on $\mathbb{D}$ as

$$g_{\mathbb{D}} = \frac{1}{|\mathbb{D}|} \int_{\mathbb{D}} g(z)dz$$

If $g \in L^1_{loc}(\Omega, \mathbb{V})$ and if the norm

$$\|g\|_{BMO} = \sup_{\mathbb{D}} \frac{1}{|\mathbb{D}|} \int_{\mathbb{D}} |g(z) - g_{\mathbb{D}}|dz < \infty,$$

then we say g is of *bounded mean oscillation*, $g \in BMO(\Omega, \mathbb{V})$.

There are two central facts to be noted here. The first is the duality theorem of Fefferman, which states that $BMO(\mathbb{C})$ is the dual space of $\mathcal{H}^1(\mathbb{C})$, and also a result of Sarason, which states that $\mathcal{H}^1(\mathbb{C})$ is the dual space of $VMO(\mathbb{C})$, this latter space being the completion of $C_0^\infty(\mathbb{C})$ in $BMO(\mathbb{C})$. In particular, we note the following [125].

Theorem A.8.2. *There is a constant C such that if $f \in \mathcal{H}^1(\mathbb{C})$ and $g \in BMO(\mathbb{C})$, then*

$$\left| \int_{\mathbb{C}} f(x)g(z)dz \right| \leqslant C\|f\|_{\mathcal{H}^1} \|g\|_{BMO} \tag{126}$$

In general, the integral (126) does not converge, however, there are a number of ways to give meaning to it.

Finally, the fundamental integrability properties of BMO-functions are found in the well-known John-Nirenberg lemma [201].

Theorem A.8.3. *There exists a constant $\Theta > 0$ such that, for every $h \in BMO(\Omega, \mathbb{V})$ and every disk or cube $\mathbb{D} \subset \mathbb{C}$, we have*

$$\frac{1}{|\mathbb{D}|} \int_{\mathbb{D}} \exp\left(\frac{\Theta |h(z) - h_{\mathbb{D}}|}{\|h\|_{BMO}}\right) dz \leqslant 2$$

A.9 Reverse Hölder Inequalities

Reverse Hölder estimates are among the basic tools in harmonic analysis. We will make use of the following version of this method, described in [193].

Theorem A.9.1. *Let $f \in L^p_{loc}(\Omega)$ and suppose there is an exponent $t < p$ such that*

$$\left(\frac{1}{|B|}\int_B |f|^p\right)^{1/p} \leqslant C(p,t) \left(\frac{1}{|2B|}\int_{2B} |f|^t\right)^{1/t}$$

whenever the disk $B = \mathbb{D}(z_0, R)$ satisfies $2B = \mathbb{D}(z_0, 2R) \subset \Omega$. Then for each $\sigma > 0$ and $r > 0$ and for each disk B with $\sigma B \subset \Omega$, we have

$$\left(\frac{1}{|B|}\int_B |f|^p\right)^{1/p} \leqslant C \left(\frac{1}{|\sigma B|}\int_{\sigma B} |f|^r\right)^{1/r}$$

A.10 Variations of Sobolev Mappings

Variations of the identity needed in Chapter 21 on the L^1-minimization problem are fairly well known in geometric function theory; see [70, 217]. As we are largely concerned with Sobolev classes of homeomorphisms, we shall need to establish the existence of suitable variations of such mappings. These variations may be required to preserve quasiconformality, as well as some other natural properties of mappings. The following theorem captures all these needs.

Theorem A.10.1. *Let $f : \mathbb{D} \to \mathbb{C}$ be a homeomorphism (onto its image) of the Sobolev class $W^{1,1}_{loc}(\mathbb{D})$ and let $a \in \mathbb{D}$ be a Lebesgue point of the differential Df such that $J(a, f) > 0$. Then there exists a diffeomorphism $h : \mathbb{D} \to \mathbb{D}$, referred to as a change of variables, such that the composite mapping $\tilde{f}(z) = f(h(z))$ satisfies.*

(i) $\tilde{f}(z) = f(z)$ near $\partial \mathbb{D}$.

(ii) *The origin is a Lebesgue point of $D\tilde{f}$.*

(iii) $D\tilde{f}(0) = \mathbf{I}$.

The proof for the construction of this change of variables consists of three lemmas that are of independent interest.

Lemma A.10.2. *Given a point $a \in \mathbb{D}$ and $r > |a|$, there exists a diffeomorphism $\Phi : \mathbb{C} \to \mathbb{C}$ such that*

$$\Phi(0) = a, \qquad D\Phi(0) = \mathbf{I}, \quad \text{and} \quad \Phi(z) = z \text{ for } |z| \geqslant r$$

A.10. VARIATIONS OF SOBOLEV MAPPINGS

Proof. Such a diffeomorphism Φ will be a perturbation of the identity map, namely,
$$\Phi(z) = z + \eta(z)a,$$
where $\eta \in C_0^\infty(\mathbb{D}_r)$ satisfies
$$0 \leqslant \eta \leqslant 1, \qquad \eta(0) = 1, \qquad \text{and} \quad \|\nabla\eta\|_\infty < \frac{1}{|a|}$$

Its differential is a matrix of the form
$$D\Phi(z) = \mathbf{I} + a \otimes \nabla\eta$$

Since η assumes its largest value 1 at the origin, we infer that $\nabla\eta(0) = 0$. Hence $D\Phi(0) = \mathbf{I}$, as claimed. The Jacobian determinant of Φ is computed in $\mathbb{D}_r$ as
$$\det D\Phi = 1 + \langle a, \nabla\eta \rangle \geqslant 1 - |a|\,|\nabla\eta| \geqslant 1 - |a|\,\|\nabla\eta\|_\infty > 0 \qquad (127)$$
and $\det D\Phi = 1$ outside $\mathbb{D}_r$. In particular, Φ is a local diffeomorphism of $\mathbb{C}$ into itself. Since $\Phi(z) = z$ outside the disk $\mathbb{D}_r$, by topological arguments we conclude that $\Phi : \mathbb{C} \to \mathbb{C}$ is one to one. We next have a lemma showing how to decompose matrices.

Lemma A.10.3. *Given $A \in \mathbb{R}_+^{2\times 2}$ and $\varepsilon > 0$, there exist $A_1, A_2, \ldots, A_k \in \mathbb{R}_+^{2\times 2}$ such that*
$$|\mathbf{I} - A_i| \leqslant \varepsilon \quad \text{for } i = 1, 2, \ldots, k$$
and
$$A = A_1 \cdot A_2 \cdots A_k$$

Proof. As $\mathbb{R}_+^{2\times 2}$ is a connected matrix Lie group, we can decompose A as
$$A = e^{X_1} e^{X_2} \cdots e^{X_m},$$
where $X_1, X_2, \ldots, X_m \in \mathbb{R}^{2\times 2}$; see [152, Corollary 2.31]. Next choose $\delta = \frac{1}{N}$, where N is a large positive integer such that $e^{\delta|X_k|} - 1 \leqslant \varepsilon$ for $k = 1, 2, \ldots, m$. Hence
$$\begin{aligned} A &= \underbrace{e^{\delta X_1} \cdots e^{\delta X_1}}_{N \text{ times}} \cdot \underbrace{e^{\delta X_2} \cdots e^{\delta X_2}}_{N \text{ times}} \cdots \underbrace{e^{\delta X_m} \cdots e^{\delta X_m}}_{N \text{ times}} \\ &= A_1 \cdot A_2 \cdots A_k, \qquad \text{where } k = mN \end{aligned} \qquad (128)$$

For every A_i we have
$$|\mathbf{I} - A_i| \leqslant \max_{1 \leqslant k \leqslant m} \left|1 - e^{\delta X_k}\right| \leqslant \max_{1 \leqslant k \leqslant m} \left|1 - e^{\delta|X_k|}\right| \leqslant \varepsilon,$$
as desired.

Lemma A.10.4. *Given $A \in \mathbb{R}_+^{2\times 2}$ and $r > 0$, there exists a diffeomorphism $\Psi : \mathbb{C} \to \mathbb{C}$ such that*
$$\Psi(0) = 0, \qquad D\Psi(0) = A, \qquad \text{and} \quad \Psi(x) = x \text{ for } |x| \geqslant r$$

Proof. First assume that A is sufficiently close to the identity matrix, say,

$$|\mathbf{I} - A| < \frac{1}{4} \tag{129}$$

We construct Ψ as a perturbation of the identity $\Psi(x) = x - (x - Ax)\eta$, where $\eta \in C_0^\infty(\mathbb{D}_r)$, $0 \leqslant \eta(x) \leqslant 1$, $\eta(0) = 1$ and $\|\nabla\eta\|_\infty < \frac{2}{r}$. We find the differential of Ψ as follows

$$D\Psi(x) = \mathbf{I} - (\mathbf{I} - A)\eta - (x - Ax) \otimes \nabla\eta$$

Clearly, $D\Psi(0) = A$. In order to see that $\det D\Psi(x) \neq 0$, we view $D\Psi(x)$ as a small perturbation of $\mathbf{I}$. For $|x| \leqslant r$, we have the following estimate of the perturbation term

$$\begin{aligned}|\eta(x)(\mathbf{I} - A) + (x - Ax) \otimes \nabla\eta| &\leqslant \eta(x)|\mathbf{I} - A| + |x - Ax||\nabla\eta(x)| \\ &\leqslant |\mathbf{I} - A| + r|\mathbf{I} - A|\|\nabla\eta\|_\infty \\ &\leqslant 3|\mathbf{I} - A| < \frac{3}{4}\end{aligned} \tag{130}$$

If follows that $\det D\Psi(x) > 4^{-n}$. As in Lemma A.10.2, we conclude that Ψ is a diffeomorphism of $\mathbb{C}$ onto itself. We can now free ourselves from assumption (129) by using the decomposition at Lemma A.10.3. Accordingly we put $A = A_1 \cdot A_2 \cdots A_k$, where $|\mathbf{I} - A_i| < \frac{1}{4}$ for $i = 1, 2, ..., k$. Let $\Psi_1, \Psi_2, ..., \Psi_k : \mathbb{C} \to \mathbb{C}$ be the diffeomorphisms constructed above, that is,

$$\Psi_i(0) = 0, \qquad \Psi_i(x) = x \text{ for } |x| \geqslant r \quad \text{and} \quad D\Psi_i(0) = A_i \text{ for } i = 1, 2, ..., k$$

The composition $\Psi = \Psi_1 \circ \Psi_2 \circ \cdots \circ \Psi_k : \mathbb{C} \to \mathbb{C}$ satisfies all the assertions of Lemma A.10.4.

Proof of Theorem A.10.1. We define h as composition of Φ from Lemma A.10.2 and Ψ from Lemma A.10.4. This latter map Ψ is determined by taking $A = [Df(a)]^{-1} \in \mathbb{R}_+^{2\times 2}$. Actually, the composite map

$$h = \Phi \circ \Psi : \mathbb{C} \to \mathbb{C}, \qquad h(0) = a,$$

is a diffeomorphism of the entire space $\mathbb{C}$. It maps the unit disk onto itself and is the identity near $\partial \mathbb{D}$. The chain rule applies to $\tilde{f}$ as follows:

$$D\tilde{f}(x) = Df(h(x))\, Df(x) \qquad \text{for almost every } x \in \mathbb{D}$$

In particular, the origin is a Lebesgue point of $D\tilde{f}$. Moreover,

$$D\tilde{f}(0) = Df(a)\, Dh(0) = \mathbf{I}$$

because

$$Dh(0) = D\Phi(0)\, D\Psi(0) = D\Psi(0) = [Df(a)]^{-1}$$

Basic Notation

Here we have collected together some of the standard notation used throughout the text.

- $\mathbb{C}$, the complex plane
- $\hat{\mathbb{C}} = \mathbb{C} \cup \{\infty\}$, the Riemann sphere
- $\mathbb{D}(a, r) = \{z \in \mathbb{C} : |z - a| < r\}$, the open disk about a of radius r
- $\overline{\mathbb{D}}(a, r) = \{z \in \mathbb{C} : |z - a| \leq r\}$, the closed disk about a of radius r
- $\mathbb{D}_r = \mathbb{D}(0, r)$, the open disk of radius r and center 0
- $\mathbb{D} = \mathbb{D}(0, 1)$, the unit disk
- $\operatorname{diam}(E)$, the diameter of the set $E \subset \mathbb{C}$
- $|E|$, the Lebesgue measure of the set E
- $\operatorname{dist}(E, F)$, the distance between the sets E and F,
$$\operatorname{dist}(E, F) = \inf_{z \in E, w \in F} |z - w|$$
- $\mathcal{H}^s(E)$, the s-dimensional Hausdorff measure of a set E
- $\mathcal{M}^s(E)$, the s-dimensional content of a set E
- $\chi_F(z)$, the characteristic function of the set F,
$$\chi_F(z) = \begin{cases} 1, & z \in F \\ 0, & z \notin F \end{cases}$$
- $\chi_{\mathbb{D}(0,R)}$, the characteristic function of the disk $\mathbb{D}(0, R)$
- $GL(n, \mathbb{R})$, the general linear group, that is, the space of invertible $n \times n$ matrices with real entries
- $SL(n, \mathbb{R})$, those matrices $A \in GL(n, \mathbb{R})$ with determinant equal to 1, $\det(A) = 1$

- $SO(n, \mathbb{R})$, the orthogonal matrices in $SL(n, \mathbb{R})$
- $|A|$, the operator norm of $A \in GL(n, \mathbb{R})$,
$$|A| = \max_{|\zeta|=1} |A\zeta|$$
- $\|A\|$, the Hilbert-Schmidt norm of $A \in GL(n, \mathbb{R})$,
$$\|A\| = \left(\sum_{i,j=1}^{n} a_{ij}^2 \right)^{1/2}$$
- Df, the differential matrix of the function $f(z) = u(z) + iv(z)$,
$$Df(z) = \begin{bmatrix} u_x & u_y \\ v_x & v_y \end{bmatrix}$$
- $D^t f = (Df)^t$, the transpose differential matrix
- $\frac{\partial f}{\partial \bar{z}} = f_{\bar{z}}$, the $\bar{z}$-derivative of the function f,
$$\frac{\partial f}{\partial \bar{z}} = \frac{1}{2} \left(\frac{\partial f}{\partial x} + i \frac{\partial f}{\partial y} \right)$$
- $\frac{\partial f}{\partial z} = f_z$, the complex or z-derivative of the function f,
$$\frac{\partial f}{\partial z} = \frac{1}{2} \left(\frac{\partial f}{\partial x} - i \frac{\partial f}{\partial y} \right)$$
- $\partial_\alpha f$, the directional derivative of f in the direction $\alpha \in \mathbb{C} \setminus \{0\}$
- $f|E$ the function f restricted to the set E
- $L_f(z)$, the maximal derivative of a function f
- $(\mathcal{M}f)(z)$, the Hardy-Littlewood maximal operator
- $\mathcal{M}_\Omega(f)$, the mollified maximal operator of a function f
- $f^\flat(z)$, the spherical maximal function of f
- $f^\#(z)$, the Fefferman-Stein sharp maximal function of f
- $(\mathcal{L}f)(z)$, the logarithmic potential of a function f
- $(\mathcal{C}f)(z)$, the Cauchy Transform of a function f
- $(\mathcal{C}_\mathbb{D}f)(z)$, the Cauchy transform in the unit disk
- $(\mathcal{C}_\Delta f)(z)$, the Cauchy transform for the Dirichlet problem

BASIC NOTATION

- $(\mathcal{S}f)(z)$, the Beurling transform of a function f
- $(\mathcal{S}_\Omega f)(z)$, the Beurling transform in a domain Ω
- $(\mathcal{S}_\Delta f)(z)$, the Beurling transform for the Dirichlet problem
- dz, the area element, $dz = dx\, dy$, in a two-dimensional Lebesgue integral
- dz, the complex line-element in integrating over a curve
- $|dz|$, ds, line-elements in integrating with respect to arc length
- $\operatorname{supp}(f)$, the support of the function f, $\operatorname{supp}(f) = \overline{\{z : f(z) \neq 0\}}$
- $C(\Omega)$, the space of continuous real-valued functions defined on an open set Ω
- $C_0(\Omega)$, those functions in $C(\Omega)$ whose support is compactly contained in Ω
- $C_0(\hat{\mathbb{C}})$, those functions $\phi \in C(\mathbb{C})$ for which $\lim_{z \to \infty} \phi(z) = 0$
- $C^\infty(\Omega)$, the space of infinitely differentiable real-valued functions defined on an open set Ω
- $C^\infty(\Omega, \mathbb{V})$, the space of infinitely differentiable functions defined on an open set Ω and valued in the vector space $\mathbb{V}$
- $C_0^\infty(\Omega)$, those functions in $C^\infty(\Omega)$ whose support is compactly contained in Ω
- $C^\alpha(\Omega)$, those functions that satisfy a Hölder estimate with exponent α
- $C^{\ell,\alpha}(\Omega)$, those functions for which the ℓth derivatives exist and satisfy a Hölder estimate with exponent α
- $L^p(\Omega)$, the Banach space ($p \geqslant 1$) of functions f with $|f|^p$ integrable in Ω
- $L^p_{loc}(\Omega)$, the Banach space ($p \geqslant 1$) of functions f with $|f|^p$ locally integrable in Ω, that is, integrable in each compact subset of Ω
- $L^\infty(\Omega)$, the Banach space of essentially bounded measurable functions
- $L^Q(\Omega)$, the Orlicz-Banach space of functions with $\int_\Omega Q(|f|) < \infty$
- $\mathcal{D}'(\Omega, \mathbb{V})$, the space of Schwartz distributions on Ω with values in the vector space $\mathbb{V}$; note that the space $C_0^\infty(\Omega, \mathbb{V})$ is a dense subset

- $W^{k,p}(\Omega, \mathbb{V})$, $1 \leqslant p \leqslant \infty$, $k \in \mathbb{N}$, the *Sobolev space* of all distributions $f \in D'(\Omega, \mathbb{V})$ whose derivatives up to the kth order are represented by functions in $L^p(\Omega, \mathbb{V})$ and equipped with the norm

$$\|f\|_{k,p} = \left(\sum_{|\alpha+\beta| \leqslant k} \int_\Omega \left| \frac{\partial^{\alpha+\beta} f(\zeta)}{\partial z^\alpha \, \partial \bar{z}^\beta} \right|^p \right)^{1/p}$$

for $p < \infty$ and

$$\|f\|_{k,\infty} = \operatorname{essup}_{|\alpha+\beta| \leqslant k} \left| \frac{\partial^{\alpha+\beta} f(\zeta)}{\partial z^\alpha \, \partial \bar{z}^\beta} \right|;$$

when $\mathbb{V} = \mathbb{C}$, we denote these spaces by $W^{k,p}(\Omega)$. The space $C^\infty(\Omega, \mathbb{V})$ is a dense subspace of $W^{k,p}(\Omega, \mathbb{V})$ for all k and p, $1 \leqslant p < \infty$

- $W^{k,p}(\Omega, \Omega')$, the space of functions $f \in W^{k,p}(\Omega)$ with $f(\Omega) \subset \Omega'$

- $W^{k,p}_{loc}(\Omega)$, the space of functions f with $f \in W^{k,p}(\Omega')$ for every relatively compact subdomain with $\overline{\Omega'} \subset \Omega$

- $\mathbb{W}^{1,p}(\Omega)$, the homogeneous Sobolev space, the Banach space (modulo constants) of functions whose gradient lies in $L^p(\Omega)$

- $\mathbb{W}^{1,Q}(\Omega)$, the Orlicz-Banach space (modulo constants) of functions whose gradient satisfies $\int_\Omega Q(|Df|) < \infty$

- $BV(\Omega, \mathbb{C})$, the Banach space of functions $f : \Omega \to \mathbb{C}$ of bounded variation

- $BMO(\Omega)$, the space of functions of bounded mean oscillation, that is, those $f \in L^1_{loc}(\Omega, \mathbb{V})$ for which the norm

$$\|f\|_{BMO} = \sup_D \frac{1}{|D|} \int_D |f(\zeta) - f_D| < \infty, \qquad f_D = \frac{1}{|D|} \int_D f(\zeta),$$

where the supremum is taken over all disks $D \subset \Omega$

- $BMO_2(\Omega)$ the space $BMO(\Omega)$ equipped with the norm

$$\|\phi\|_{BMO_2} = \sup \left(\frac{1}{|D|} \int_D |\phi(\tau) - \phi_D|^2 \, d\tau \right)^{1/2}$$

- $VMO(\Omega)$, those functions $f \in BMO(\Omega)$ with vanishing mean oscillation, the completion of $C_0^\infty(\Omega)$ in $BMO(\Omega)$

- $H^p(\Omega)$, the Hardy space consisting of those distributions for which the mollified maximal function $\mathcal{M}_\Omega(f) \in L^p(\Omega)$

Bibliography

[1] L.V. Ahlfors, *Zur Theorie der Überlagerungsflächen*, Acta Math., **65** (1935).

[2] L.V. Ahlfors, *Commentary on: "Zur Theorie der Überlagerungsflächen" (1935)*, Fields Medallists' Lectures, 8–9, World Sci. Ser. 20th Century Math., 5, World Science. Publishing, River Edge, NJ, 1997.

[3] L.V. Ahlfors, *On quasiconformal mappings*, J. Anal. Math., **3** (1953/54), 1–58.

[4] L.V. Ahlfors, *Quasiconformal reflections*, Acta Math., **109** (1963), 291–301.

[5] L.V. Ahlfors, *Finitely generated Kleinian groups*, Amer. J. Math., **86** (1964), 413–429.

[6] L.V. Ahlfors, *Complex analysis*, McGraw-Hill, New York, 1966.

[7] L.V. Ahlfors, *Conformal invariants: Topics in geometric function theory*, McGraw-Hill Series in Higher Mathematics, McGraw-Hill, New York-Dusseldorf-Johannesburg, 1973.

[8] L.V. Ahlfors, *Lectures on quasiconformal mappings*, Van Nostrand, Princeton, NJ, 1966; reprinted by Wadsworth, Belmont, CA, 1987.

[9] L.V. Ahlfors and L. Bers, *Riemann's mapping theorem for variable metrics*, Ann. of Math., **72** (1960), 385–404.

[10] L.V. Ahlfors and A. Beurling, *Conformal invariants and function theoretic null sets*, Acta Math., **83** (1950), 101–129.

[11] L.V. Ahlfors and A. Beurling, *The boundary correspondence under quasiconformal mappings*, Acta Math., **96** (1956), 125–142.

[12] L.V. Ahlfors and L. Sario, *Riemann surfaces*, Princeton Mathematical Series, 26, Princeton University Press, Princeton, NJ, 1960.

[13] L.V. Ahlfors and G. Weill, *A uniqueness theorem for Beltrami equations*, Proc. Amer. Math. Soc., **13** (1962), 975–978.

[14] G. Alessandrini and V. Nesi, *Beltrami operators, non-symmetric elliptic equations and quantitative Jacobian bounds*, Ann. Acad. Sci Fenn. Math., **34** (2009), 47–67.

[15] A.D. Alexandrov, *The impossibility of general estimates for solutions and of uniqueness for linear equations with norms weaker than in L^n*, Vestnik Leningrad Univ. Math., **21** (1966), 5–10; Amer. Math. Soc. Transl. **68** (1968), 162–168.

[16] J.-J. Alibert and B. Dacorogna, *An example of a quasiconvex function that is not polyconvex in two dimensions*, Arch. Ration. Mech. Anal., **117** (1992), 155–166.

[17] L. Ambrosio, N. Fusco and D. Pallara, *Functions of bounded variation and free discontinuity problems*, Oxford Mathematical Monographs, Clarendon Press, Oxford University Press, New York, 2000.

[18] G. Anderson, M. Vamanamurthy and M. Vuorinen, *Conformal invariants and quasiconformal maps*, Canadian Mathematical Society Monographs, John Wiley, 1997.

[19] S.S. Antman, *Fundamental mathematical problems in the theory of nonlinear elasticity*, Numerical solution of partial differential equations, III, Academic Press, New York, 1976, pp. 35–54.

[20] K. Astala, *A remark on quasiconformal mappings and BMO-functions*, Michigan Math. J., **30** (1983), 209–212.

[21] K. Astala, *Area distortion of quasiconformal mappings*, Acta Math., **173** (1994), 37–60.

[22] K. Astala, *Planar quasiconformal mappings; deformations and interactions*, Quasiconformal Mappings and Analysis: A Collection of Papers Honoring F.W. Gehring, Springer-Verlag, Berlin-New York, 1998, pp. 33–54.

[23] K. Astala, A. Clop, J. Mateu, J. Orobitg, I. Uriarte-Tuero, *Distortion of Hausdorff measures and improved Painlevé removability for quasiregular mappings*, Duke Math. J., **141** (2008), 539–571.

[24] K. Astala, D. Faraco and L. Székelyhidi, *Convex integration and the L^p-theory of elliptic equations*, Ann. Sc. Norm. Super. Pisa Cl. Sci. (5), **7** (2008), 1–50.

[25] K. Astala, J.T. Gill, E. Saksman and S. Rohde, *Optimal regularity for planar mappings of finite distortion*, Ann. Inst. H. Poincaré Anal. Non linéaire, To appear.

[26] K. Astala, T. Iwaniec, P. Koskela and G.J. Martin, *Mappings of BMO-bounded distortion*, Math. Ann., **317** (2000), 703–726.

[27] K. Astala, T. Iwaniec, G.J. Martin and J. Onninen, *Extremal mappings of finite distortion*, Proc. London Math. Soc., **91** (2005), 655–702.

[28] K. Astala, T. Iwaniec and G.J. Martin, *Pucci's conjecture and the Alexandrov inequality for elliptic PDEs in the plane*, J. Reine Angew. Math., **591** (2006), 49–74.

[29] K. Astala, T. Iwaniec and G.J. Martin, *Deformations of annuli with smallest mean distortion*, Preprint.

[30] K. Astala, T. Iwaniec and G.J. Martin, *Monotone maps of $\mathbb{R}^n$ are quasiconformal*, Methods and Applications of Analysis; Honoring the 65th birthday of Neil Trudinger, To appear.

[31] K. Astala, T. Iwaniec and E. Saksman, *Beltrami operators*, Duke Math. J., **107** (2001), 27–56.

[32] K. Astala and J. Jääskeläinen, *Homeomorphic solutions to the reduced Beltrami equations*, Preprint (2008).

[33] K. Astala and G.J. Martin, *Holomorphic motions*, in Papers on Analysis: A Volume Dedicated to Olli Martio on the Occasion of His 60th Birthday, Rep. Univ. Jyväskylä, **83**, 2001, pp. 27–40.

[34] K. Astala and V. Nesi, *Composites and quasiconformal mappings: New optimal bounds in two dimensions,* Calculus of Variations and Partial Differential Equations, **18** (2003), 335–355.

[35] K. Astala and L. Päivärinta, *Calderón's inverse conductivity problem in the plane*, Ann. of Math., **163** (2006), 265–299.

[36] K. Astala, S. Rohde and O. Schramm, *Dimension of quasicircles*, Preprint.

[37] A. Baernstein II and L.V. Kovalev, *On Hölder regularity for elliptic equations of non-divergence type in the plane*, Ann. Sc. Norm. Super. Pisa Cl. Sci. **4**, (2005), 295–317.

[38] A. Baernstein II and S.J. Montgomery-Smith, *Some conjectures about integral means of ∂f and $\overline{\partial} f$*, Complex Analysis and Differential Equations, Uppsala University, 1999, pp. 92–109.

[39] J.M. Ball, *Convexity conditions and existence theorems in non-linear elasticity*, Arch. Ration. Mech. Anal., **63** (1977), 337–403.

[40] J.M. Ball, *The calculus of variations and materials science*, Current and Future Challenges in the Applications of Mathematics, (Providence, RI, 1997). Quart. Appl. Math., **56** (1998), 719–740.

[41] J.M. Ball, *Singularities and computation of minimizers for variational problems*, Foundations of Computational Mathematics (Oxford, 1999), London Math. Soc. Lecture Notes Ser., 284, Cambridge University Press, Cambridge, 2001, pp. 1–20.

[42] J.M. Ball and K.W. Zhang, *Lower semicontinuity of multiple integrals and the biting lemma*, Proc. Roy. Soc. Edinburgh Sect. A, **114** (1990), 367–379.

[43] Z.M. Balogh and P. Koskela, *Quasiconformality, quasisymmetry and removability in Loewner spaces*, Duke Math. J., **101** (2000), 555–577.

[44] R. Bañuelos and P. Janakiraman, *L^p-Bounds for the Beurling-Ahlfors transform*, Trans. Amer. Math. Soc., **360** (2008), 3603–3612.

[45] R. Bañuelos and A. Lindeman II, *A Martingale study of the Beurling-Ahlfors transform in $\mathbb{R}^n$*, J. Funct. Anal., **145** (1997), 224–265.

[46] R. Bañuelos and G. Wang, *Orthogonal Martingales under differential subordination and applications to Riesz transforms*, Illinois J. Math., **40** (1996), 678–691.

[47] R. Beals and R. Coifman, *The spectral problem for the Davey-Stewartson and Ishimori hierarchies*, Nonlinear Evolution Equations: Integrability and Spectral Methods, Manchester University Press, 1988, pp. 15–23.

[48] A. Beardon, *The geometry of discrete groups*, Graduate Texts in Math., 91 Springer-Verlag, Berlin-New York, 1983.

[49] J. Becker and Ch. Pommerenke, *On the Hausdorff dimension of quasicircles*, Ann. Acad. Sci., Ser. A. I. Math., **12** (1987), 329–333.

[50] W. Beckner, *Inequalities in Fourier analysis*, Ann. of Math., **102** (1975), 159–182.

[51] E.T. Bell, *The development of mathematics*, 2nd ed., McGraw-Hill, New York, 1945.

[52] E. Beltrami, *Saggio di interpretazione della geometria non euclidea*, Giornale di Mathematica, **6** (1867).

[53] L. Bers, *Uniformization by Beltrami equations*, Comm. Pure Appl. Math., **14** (1961), 215–228.

[54] L. Bers, *Mathematical aspects of subsonic and transonic gas dynamics*, Surveys in Applied Math., 3, John Wiley, New York; Chapman & Hall, London, 1958.

[55] L. Bers, *Theory of pseudo-analytic functions*. Institute for Mathematics and Mechanics, New York University, New York, 1953.

[56] L. Bers and L. Nirenberg, *On a representation theorem for linear elliptic systems with discontinuous coefficients and its applications*, Convegno Internazionale sulle Equazioni Lineari alle Derivate Parziali, Trieste, 1954, pp. 111–140. Edizioni Cremonese, Rome, 1955, pp. 111–140.

[57] L. Bers and L. Nirenberg, *On linear and non-linear elliptic boundary value problems in the plane*, Convegno Internazionale sulle Equazioni Lineari alle Derivate Parziali, Trieste, 1954, Edizioni Cremonese, Rome, 1955, pp. 141–167.

[58] L. Bers and H. Royden, *Holomorphic families of injections*, Acta Math., **157** (1986), 259–286.

[59] A. Beurling and L. Ahlfors, *The boundary correspondence under quasiconformal mappings*, Acta Math., **96** (1956), 125–142.

[60] L. Bieberbach, *Über die Koeffizienten derjenigen Potenzreihen, welche eine schlichte Abbildung des Einheitskreises vermitteln*, Sitz Ber. Preuss. Akad. Wiss., **138** (1916), 940–955.

[61] C.J. Bishop, *Some homeomorphisms of the sphere conformal off a curve*, Ann. Acad. Sci. Fenn., Ser. A. I. Math., **19** (1994), 323–338.

[62] C.J. Bishop, *Conformal welding and Koebe's theorem*, Ann. of Math., **166** (2007), 613–656.

[63] A. Bloch, *Les théorèmes de M. Valiron sur les fonctions entières et la théorie de l'uniformisation*. Ann. Fac. Sci. Toulouse Sci. Math. Sci. Phys., **17** (1925), 1–22.

[64] B. Bojarski, *Homeomorphic solutions of Beltrami systems*, Dokl. Akad. Nauk. SSSR, **102** (1955), 661–664.

[65] B. Bojarski, *Subsonic flow of compressible fluid*, Arch. Mech. Stos., **18** (1966), 497–520.

[66] B. Bojarski, *Generalised solutions of PDE system of the first order and elliptic type with discontinuous coefficients*, Mat. Sb., **43** (1957), 451–503.

[67] B. Bojarski, *Remarks on stability of reverse Hölder inequalities and quasiconformal mappings*, Ann. Acad. Sci. Fenn. Ser. A. I. Math., **10** (1985), 89–94.

[68] B. Bojarski, *On the first boundary value problem for elliptic systems of second order in the plane*, Bull. Acad. Polon. Sci. Ser. Sci. Math. Astr. Phys., **7** (1959), 565–570.

[69] B. Bojarski, L. D'Onofrio, T. Iwaniec and C. Sbordone, *G-closed classes of elliptic operators in the complex plane*, Ricerche Mat., **54** (2005), 403–432.

[70] B. Bojarski and T. Iwaniec, *Another approach to Liouville theorem*, Math. Nachr., **107** (1982), 253–262.

[71] B. Bojarski and T. Iwaniec, *Analytical foundations of the theory of quasiconformal mappings in $\mathbb{R}^n$*, Ann. Acad. Sci. Fenn. Ser. A. I. Math., **8** (1983), 257–324.

[72] M. Bourdon and H. Pajot, *Poincaré inequalities and quasiconformal structure on the boundaries of some hyperbolic buildings*, Proc. Amer. Math. Soc., **127** (1999), 2315–2324.

[73] J. Bourgain, *Averages in the plane over convex curves and maximal operators*, J. Anal. Math., **47** (1986), 69–85.

[74] M.A. Brakalova and J.A. Jenkins, *On solutions of the Beltrami equation*, J. Anal. Math., **76** (1998), 67–92.

[75] H. Brezis, N. Fusco and C. Sbordone, *Integrability of the Jacobian of orientation preserving mappings*, J. Funct. Anal., **115** (1993), 425–431.

[76] J.K. Brooks and R.V. Chacon, *Continuity and compactness of measures*, Adv. in Math., **37** (1980), 16–26.

[77] F.E. Browder, *Nonlinear elliptic boundary value problems*, Bull. Amer. Math. Soc., **69** (1963), 862–874.

[78] R. Brown, *Estimates for the scattering map associated with a two-dimensional first-order system*, J. Nonlinear Sci., **11** (2001), 459–471.

[79] R. Brown and G. Uhlmann, *Uniqueness in the inverse conductivity problem for nonsmooth conductivities in two dimensions*, Comm. Partial Differential Equations, **22** (1997), 1009–1027.

[80] D.L. Burkholder, *Sharp inequalities for martingales and stochastic integrals*, Astérisque, **157-158** (1988), 75–94.

[81] D.L. Burkholder, *A proof of Pełczyński's conjecture for the Haar system*, Studia Math., **91** (1988), 79–83.

[82] X. Cabré, *On the Alexandroff-Bakel'man-Pucci estimate and the reversed Hölder inequality for solutions of elliptic and parabolic equations*, Comm. Pure Appl. Math., **48** (1995), 539–570.

[83] R. Caccioppoli, *Limitazioni integrali per le soluzioni di un'equazione lineare ellitica a derivate parziali*, Giorn. Mat. Battaglini, **4 (80)** (1951), 186–212.

[84] L.A. Caffarelli and X. Cabré, *Fully nonlinear elliptic equations*, Amer. Math. Soc. Colloq. Publ., **43** (1995).

[85] A.P. Calderón, *On an inverse boundary value problem*, Seminar on Numerical Analysis and Its Applications to Continuum Physics, Soc. Brasil. Mat., Rio de Janeiro, 1980, pp. 65–73.

[86] A.P. Calderón and A. Zygmund, *On the existence of certain singular integrals*, Acta Math., **88** (1952), 88–139.

[87] A.P. Calderón and A. Zygmund, *On singular integrals*, Amer. J. Math., **78** (1956), 289–309.

[88] C. Carathéodory, *Über das lineare Mass von Punktmengen, eine Verallgemeinerung des Längenbegriffs*, Nachr. Ges. Wiss. Göttingen (1914), 406–426.

[89] C. Carathéodory, *Theory of functions of a complex variable*, Vol. II, Chelsea, 1954.

[90] L. Carleson and T.W. Gamelin, *Complex dynamics*, Springer-Verlag, Berlin-New York, 1993.

[91] M. Cheney, D. Isaacson and J.C. Newell, *Electrical impedance tomography*, SIAM Rev., **41** (1999), 85–101.

[92] E.M. Chirka, *On the propagation of holomorphic motions*, Dokl. Akad. Nauk **397** (2004), 37–40.

[93] G. Choquet, *Sur un type de transformation analytique généralisant la représentation conforme et définie au moyen de fonctions harmoniques*, Bull. Sci. Math., **69** (1945), 156–165.

[94] R.R. Coifman and C. Fefferman, *Weighted norm inequalities for maximal functions and singular integrals*, Studia Math., **51** (1974), 241–250.

[95] R.R. Coifman, P.L. Lions, Y. Meyer and S. Semmes, *Compensated compactness and Hardy spaces*, J. Math. Pures Appl., **72** (1993), 247–286.

[96] R.R. Coifman and R. Rochberg, *Another characterization of BMO*, Proc. Amer. Math. Soc., **79** (1980), 249–254.

[97] R.R. Coifman, R. Rochberg and G. Weiss, *Factorization theorems for Hardy spaces in several variables*, Ann. of Math., **103** (1978), 569–645.

[98] R. Courant, *Dirichlet's principle, conformal mapping, and minimal surfaces*, with an appendix by M. Schiffer. Reprint of the 1950 original, Springer-Verlag, Berlin-New York-Heidelberg, 1977.

[99] R. Courant, B. Manel and M. Shiffman, *A general theorem on conformal mapping of multiply connected domains*, Proc. Natl. Acad. Sci. U.S.A., **26** (1940), 503–507.

[100] J.W. Craggs, *The breakdown of the hodograph transformation for irrotational compressible fluid flow in two dimensions*, Proc. Cambridge Philos. Soc., **44** (1948), 360–379.

[101] B. Dacorogna, *Direct methods in the calculus of variations*, Appl. Math. Sci., 78, Springer-Verlag, Berlin-New York, 1989.

[102] G. Dal Maso. *An introduction to Γ-convergence*, Birkhäuser, Boston, 1993.

[103] G. David, *Solutions de l'equation de Beltrami avec $\|\mu\|_\infty = 1$*, Ann. Acad. Sci. Fenn. Ser. A. I. Math., **13** (1988), 25–70.

[104] E. De Giorgi, Γ-*convergenza e G-convergenza*, Boll. Un. Mat. Ital. A (5), **14** (1977), 213–220.

[105] E. De Giorgi and S. Spagnolo, *Sulla convergenza degli integrali dell'energia per operatori ellittici del secondo ordine*, Boll. Unione Mat. Ital., **8** (1973), 391–411.

[106] A. Dijkstra, B. Brown, N. Harris, D. Barber and D. Endbrooke, *Review: Clinical applications of electrical impedance tomography*, J. Med. Eng. Technol., **17** (1993), 89–98.

[107] A. Dold, *Lectures on algebraic topology*, Springer-Verlag, Berlin-New York, 1980.

[108] A. Douady, *Systèmes dynamiques holomorphes*, Astérisque, **105–106** (1983), 39–63.

[109] A. Douady, *Prolongement de mouvements holomorphes (d'après Slodkowski et autres)*, Astérisque, **227** (1995), 7–20.

[110] A. Douady and C.J. Earle, *Conformally natural extension of homeomorphisms of the circle*, Acta Math., **157** (1986), 23–48.

[111] A. Douady and J.H. Hubbard, *On the dynamics of polynomial-like mappings*, Ann. Sci. École Norm. Sup., **18** (1985), 287–343.

[112] J. Douglas, *Solution of the problem of Plateau*, Trans. Amer. Math. Soc., **33** (1931), 231–321.

[113] P. Duren, Theory of H^p-spaces, Academic Press, New York, 1970.

[114] P. Duren, Harmonic mappings, Cambridge University Press, New York, 2004.

[115] P. Duren, J. Heinonen, B. Osgood and B. Palka, *Quasiconformal mappings and analysis*, A Collection of Papers Honoring F.W. Gehring (Ann Arbor 1995), Springer-Verlag, Berlin-New York, 1998.

[116] C. Earle, I. Kra and S. Krushkal, *Holomorphic motions and Teichmüller spaces*, Trans. Amer. Math. Soc., **343** (1994), 927–948.

[117] A. Elcrat and N. Meyers, *Some results on regularity for solutions of nonlinear elliptic systems and quasiregular functions*, Duke Math. J., **42** (1975), 121–136.

[118] A. Eremenko and D.H. Hamilton, *On the area distortion by quasiconformal mappings*, Proc. Amer. Math. Soc., **123** (1995), 2793–2797.

[119] L.C. Evans, *Quasiconvexity and partial regularity in the calculus of variations*, Arch. Ration. Mech. Anal., **95** (1986), 227–252.

[120] L.C. Evans and R.F. Gariepy, *Measure theory and fine properties of functions*, CRC Press, Boca Raton, FL, 1992.

[121] E.B. Fabes and D.W. Stroock, *The L^p-integrability of Green's functions and fundamental solutions for elliptic and parabolic equations*, Duke Math. J., **51** (1984), 997–1016.

[122] H.M. Farkas and I. Kra, *Riemann surfaces*, 2nd ed. Graduate Texts in Math., 71, Springer-Verlag, Berlin-New York, 1992.

[123] P. Fatou, *Sur les quations fonctionnelles*, Bull. Soc. Math. France, **47** (1919), 161–271.

[124] H. Federer, *Geometric measure theory*, Springer-Verlag, Berlin-New York, 1969.

[125] C. Fefferman, *Characterisations of bounded mean oscillation*, Bull. Amer. Math. Soc., **77** (1971), 587–588.

[126] R. Finn, *Isolated singularities of solutions of non-linear partial differential equations*, Trans. Amer. Math. Soc., **75** (1953), 385–404.

[127] R. Finn, *On a problem of type, with application to elliptic partial differential equations*, Arch. Ration. Mech. Anal., **3** (1954), 789–799.

[128] F. Forstnerič, *Polynomial hulls of sets fibered over the circle*, Indiana Univ. Math. J., **37** (1988), 869–889.

[129] K.O. Friedrichs, *The identity of weak and strong extensions of differential operators*, Trans. Amer. Math. Soc., **55** (1944), 132–151.

[130] F.P. Gardiner, *Teichmüller theory and quadratic differentials*, Pure App. Math., John Wiley, New York, 1987.

[131] J. Garnett, *Bounded analytic functions*, Academic Press, New York, 1972.

[132] J. Garnett, *Analytic capacity and measure*, Lecture Notes in Math., 297, Springer-Verlag, Berlin-New York, 1972.

[133] F.W. Gehring, *Rings and quasiconformal mappings in space*, Trans. Amer. Math. Soc., **103** (1962), 353–393.

[134] F.W. Gehring, *The L^p-integrability of the partial derivatives of a quasiconformal mapping*, Acta Math., **130** (1973), 265–277.

[135] F.W. Gehring, *Characteristic properties of quasidisks*, Séminaire de Mathmatiques Supérieures, 84, Presses de l'Université de Montréal, Montreal, 1982.

[136] F.W. Gehring and K. Hag, *Quasidisks*, To appear.

[137] F.W. Gehring and O. Lehto, *On the total differentiability of functions of a complex variable*, Ann. Acad. Sci. Fenn. A. I. Math., **272** (1959), 1–9.

[138] F.W. Gehring and E. Reich, *Area distortion under quasiconformal mappings*, Ann. Acad. Sci. Fenn. Ser. A. I. Math., **388** (1966), 3–15.

[139] F. Giannetti, T. Iwaniec, L. Kovalev, G. Moscariello, and C. Sbordone, *On G-compactness of the Beltrami operators*, Nonlinear Homogenization and Its Applications to Composites, Polycrystals and Smart Materials, Kluwer Academic, Norwell, MA, 2004, pp. 107–138.

[140] M. Giaquinta, *Multiple integrals in the calculus of variations and nonlinear elliptic systems*, Ann. of Math. Stud., 105, Princeton University Press, Princeton, NJ, 1983.

[141] M. Giaquinta and E. Giusti, *On the regularity of the minima of variational integrals*, Acta Math., **148** (1982), 31–46.

[142] M. Giaquinta, G. Modica and J. Souček, *Area and the area formula*, Rend. Sem. Mat. Fis. Milano, **62** (1992), 53–87.

[143] D. Gilbarg and N.S. Trudinger, *Elliptic Partial Differential Equations of Second Order*, Springer-Verlag, Berlin-New York, 1983.

[144] S. Goldstein and M.J. Lighthill, *A note on the hodograph transformation for the two-dimensional vortex flow of an incompressible fluid*, Quart. J. Mech. Appl. Math., **3** (1950), 297–302.

[145] V.M. Goldstein and S.K. Vodop'yanov, *Quasiconformal mappings and spaces of functions with generalised first derivatives*, Sb. Mat. Z., **17** (1976), 515–531.

[146] L. Greco, T. Iwaniec and G. Moscariello, *Limits of the improved integrability of the volume forms*, Indiana Math. J., **44** (1995), 305–339.

[147] M.J. Greenberg and J.R. Harper, *Algebraic topology: A first course*, Math. Lecture Note Series, 58, Benjamin/Cummings, Reading, MA, 1981.

[148] T.H. Gronwall, *Some remarks on conformal representation*, Ann. of Math., **16** (1914/15), 72–76.

[149] H. Grötzsch, *Über die Verzerrung bei schlichten nichtkonformen Abbildungen und über eine damit zusammenhängende Erweiterung des Picardschen Satzes*, Ber. Verh. Sächs. Akad. Wiss. Leipzig, **80** (1928), 503–507.

[150] S.S. Habre, *The Fredholm alternative for second-order linear elliptic systems with VMO coefficients*, Houston J. Math., **22** (1996), 417–433.

[151] J. Hadamard, *Leçons sur la propagation des ondes et les équations de l'hydrodynamique*, Hermann, Paris, 1903.

BIBLIOGRAPHY

[152] B.C. Hall, *Lie groups, Lie algebras, and representations. An elementary introduction,* Graduate Texts in Math., 222, Springer-Verlag, Berlin-New York, 2003.

[153] D.H. Hamilton, *Generalised conformal welding,* Ann. Acad. Sci. Fenn., Ser. A. I. Math., **16** (1991), 333–343.

[154] D.H. Hamilton, *Conformal welding,* The Handbook of Geometric Function Theory, North Holland, Amsterdam, 2002.

[155] P. Haïssinsky, *Chirurgie parabolique,* C.R. Acad. Sci. Paris, **327** (1998), 195–198.

[156] G.H. Hardy and J.E. Littlewood, *Collected papers of G.H. Hardy (including joint papers with J.E. Littlewood and others),* vols. I–III, edited by a committee appointed by the London Mathematical Society, Clarendon Press, Oxford, 1967.

[157] A. Harnack, *Die Grundlagen der Theorie des logarithmischen Potentiales und die eindeutiger Potentialfunction in der Ebene,* Leipziger Beruchter, 1887.

[158] F. Hausdorff, *Dimension und äusseres Mass,* Math. Ann., **79** (1919), 157–179.

[159] J. Heinonen and P. Koskela, *Sobolev mappings with integrable dilatation,* Arch. Ration. Mech. Anal., **125** (1993), 81–97.

[160] J. Heinonen and P. Koskela, *Definitions of quasiconformality,* Invent. Math., **120** (1995), 61–79.

[161] J. Heinonen and P. Koskela, *Quasiconformal maps in metric spaces with controlled geometry,* Acta Math., **181** (1998), 1–61.

[162] S. Helgason, *Differential geometry, Lie groups and symmetric spaces,* Academic Press, New York, 1978.

[163] J.A. Hempel, *The Poincaré metric of the twice punctured plane and the theorems of Landau and Schottky,* J. London Math. Soc., **2** (1979), 435–445.

[164] J.A. Hempel, *Precise bounds in the theorems of Schottky and Picard,* J. London Math. Soc., **21** (1980), 279–286.

[165] S. Hencl and P. Koskela, *Regularity of the inverse of a planar Sobolev homeomorphism,* Arch. Ration. Mech. Anal., **180** (2006), 75–95.

[166] S. Hencl, P. Koskela and J. Onninen, *A note on extremal mappings of finite distortion,* Math. Res. Lett., **12** (2005), 231–237.

[167] A. Hinkkanen, *Uniformly quasiregular semigroups in two dimensions,* Ann. Acad. Sci. Fenn., **21** (1996), 205–222.

[168] J.E. Hutchinson. *Fractals and self-similarity*, Indiana Univ. Math. J., **30** (1981), 713–747.

[169] T. Iwaniec, *Green's function of multiply connected domain and Dirichlet problem for systems of second order in the plane*, Lecture Notes in Math., 561, Springer-Verlag, Berlin-New York, 1976, pp. 261–276.

[170] T. Iwaniec, *Quasiconformal mapping problem for general nonlinear systems of partial differential equations*, Symposia Mathematica, vol. XVIII, Academic Press, London, 1976, pp. 501–517.

[171] T. Iwaniec, *Extremal inequalities in Sobolev spaces and quasiconformal mappings*, Z. Anal. Anwen, **1** no. 6, (1982), 1–16.

[172] T. Iwaniec, *Some aspects of PDEs and quasiregular mappings*, Proc. of ICM, Warsaw, (1983), 1193–1208.

[173] T. Iwaniec, *Projections onto gradient fields and L^p-estimates for degenerated elliptic operators*, Studia Math., **75** (1983), 293–312.

[174] T. Iwaniec, *The best constant in a BMO-inequality for the Beurling–Ahlfors transform*, Michigan Math. J., **34** (1987), 407–434.

[175] T. Iwaniec, *p-Harmonic tensors and quasiregular mappings*, Ann. of Math., **136** (1992), 589–624.

[176] T. Iwaniec, *L^p-theory of quasiregular mappings*, Quasiconformal Space Mappings, Lecture Notes in Math., 1508, Springer-Verlag, Berlin-New York, 1992, pp. 39–64.

[177] T. Iwaniec, *Integrability theory of the Jacobians*, Lipschitz Lectures, Univ. Bonn Sonderforschungsberiech, 256, Bonn, 1995.

[178] T. Iwaniec, *The Gehring lemma*, Quasiconformal Mappings and Analysis, P. Duren, J. Heinonen, B. Osgood and B. Palka eds, Springer-Verlag, Berlin-New York, 1998.

[179] T. Iwaniec, *Nonlinear Cauchy-Riemann operators in $\mathbb{R}^n$*, Trans. Amer. Math. Soc., **354** (2002), 1961–1995.

[180] T. Iwaniec, P. Koskela and G.J. Martin, *Mappings of BMO–bounded distortion*, J. Anal. Math., **88** (2002), 337–381.

[181] T. Iwaniec, P. Koskela, G.J. Martin and C. Sbordonne, *Mappings of finite distortion: $L^n \log^\alpha$-integrability*, Proc. London Math. Soc., **67** (2003), 123–136.

[182] T. Iwaniec, P. Koskela and J. Onninen, *Mappings of finite distortion: monotonicity and continuity*, Invent. Math., **144** (2001), 507–531.

[183] T. Iwaniec and A. Lutoborski, *Integral estimates for null Lagrangians*, Arch. Ration. Mech. Anal., **125** (1993), 25–79.

[184] T. Iwaniec and A. Lutoborski, *Polyconvex functionals for nearly conformal deformations*, SIAM J. Math. Anal., **27** (1996), 609–619.

[185] T. Iwaniec and J. Manfredi, *Regularity of p-harmonic functions on the plane*, Rev. Mat. Iberoamericana, **5** (1989), 1–19.

[186] T. Iwaniec, L. Migliaccio, G. Moscariello, P. Gioconda and A. di Napoli, *A priori estimates for nonlinear elliptic complexes*, Adv. Differential Equations, **8** (2003), 513–546.

[187] T. Iwaniec, L. Migliaccio, L. Nania and C. Sbordone, *Integrability and removability results for quasiregular mappings in high dimensions*, Math. Scand., **75** (1994), 263–279.

[188] T. Iwaniec and G.J. Martin, *Quasiconformal mappings and capacity*, Indiana Math. J., **40** (1991), 101–122.

[189] T. Iwaniec and G.J. Martin, *Quasiregular mappings in even dimensions*, Acta Math., **170** (1993), 29–81.

[190] T. Iwaniec and G.J. Martin, *Riesz transforms and related singular integrals*, J. Reine Angew. Math., **473** (1993), 29–81.

[191] T. Iwaniec and G.J. Martin, *Geometric function theory and nonlinear analysis*, Oxford University Press, New York, 2001.

[192] T. Iwaniec and G.J. Martin, *The Beltrami Equation*, Memoirs of the Amer. Math. Soc., **191** (2008), 1–92.

[193] T. Iwaniec and C.A. Nolder, *Hardy-Littlewood inequality for quasiregular mappings in certain domains in $\mathbb{R}^n$*, Ann. Acad. Sci. Fenn. Ser. A. I. Math., **10** (1985), 267–282.

[194] T. Iwaniec and C. Sbordone, *On the integrability of the Jacobian under minimal hypotheses*, Arch. Ration. Mech. Anal., **119** (1992), 129–143.

[195] T. Iwaniec and C. Sbordone, *Weak minima of variational integrals*, J. Reine Angew. Math., **454** (1994), 143–161.

[196] T. Iwaniec and C. Sbordone, *Quasiharmonic fields*, Ann. Inst. H. Poincaré Anal. Non linéaire, **18** (2001), 519–572.

[197] T. Iwaniec and V. Šverák, *On mappings with integrable dilatation*, Proc. Amer. Math. Soc., **118** (1993), 181–188.

[198] T. Iwaniec and A. Verde, *A study of Jacobians in Hardy-Orlicz spaces*, Proc. Roy. Soc. Edinburgh, Sect. A, **129** (1999), 539–570.

[199] S. Janson, *Generalizations of Lipschitz spaces and an application to Hardy spaces and bounded mean oscillation*, Duke Math. J., **47** (1980), 959–982.

[200] D.S. Jerison and C.E. Kenig, *Hardy spaces, A_∞, and singular integrals on chord-arc domains*, Math. Scand., **50** (1982), 221–247.

[201] F. John and L. Nirenberg, *On functions of bounded mean oscillation*, Comm. Pure Appl. Math., **14** (1961), 415–426.

[202] T. Jørgensen, *On discrete groups of Möbius transformations*, Amer. J. Math., **92** (1976), 739–749.

[203] G. Julia, *Mémoire sur l'itération des fonctions rationnelles*, J. Math. Pures Appl., **8** (1918), 47–245.

[204] J. Kaipio, V. Kolehmainen, E. Somersalo and M. Vauhkonen, *Statistical inversion and Monte Carlo sampling methods in electrical impedance tomography*, Inverse Problems, **16** (2000), 1487–1522.

[205] S. Kallunki and P. Koskela, *Exceptional sets for the definition of quasiconformality*, Amer. J. Math., **122** (2000), 735–743.

[206] N. Kalton, *Non-linear commutators in interpolation theory*, Mem. Amer. Math. Soc., **385** (1998), 1–85.

[207] T. von Kármán, *The similarity law of transonic flow*, J. Math. Phys. Mass. Inst. Tech., **26** (1947), 182–190.

[208] R. Kaufman, *Hausdorff measure, BMO, and analytic functions*, Pacific J. Math., **102** (1982), 369–371.

[209] J. Kauhanen, P. Koskela and J. Malý, *On functions with a derivative in a Lorentz space*, Manuscripta Math., **100** (1999), 87–101.

[210] J. Kauhanen, P. Koskela and J. Malý, *Mappings of finite distortion: discreteness and openness*, Arch. Ration. Mech. Anal., **160** (2001), 135–151.

[211] J. Kauhanen, P. Koskela and J. Malý, *Mappings of finite distortion: Condition N*, Michigan Math. J., **49** (2001), 169–181.

[212] H. Kneser, *Lösung der Aufgabe 41*, Jahresber. Deutsch. Math.-Verein., **35** (1926), 123–124.

[213] R.V. Kohn and M. Vogelius, *Determining conductivity by boundary measurements*, Comm. Pure Appl. Math., **37** (1984), 289–298.

[214] R.V. Kohn and M. Vogelius, *Determining conductivity by boundary measurements. II. Interior results.* Comm. Pure Appl. Math., **38** (1985), 643–667.

[215] P. Koskela, *Notes on quasisymmetric mappings*, University of Jyväskylä.

[216] P. Koskela and J. Malý, *Mappings of finite distortion: the zero set of the Jacobian*, J. Eur. Math. Soc., **5** (2003), 95–105.

[217] P. Koskela and J. Onninen, *Mappings of finite distortion: the sharp modulus of continuity*, Trans. Amer. Math. Soc., **355** (2003), 1905–1920.

[218] L.V. Kovalev, *Quasiconformal geometry of monotone mappings*, J. Lond. Math. Soc., **75** (2007), 391–408.

[219] L.V. Kovalev, *Conformal dimension does not assume values between zero and one*, Duke Math. J., **134** (2006), 1–13.

[220] J. Král, *Analytic capacity*, Elliptische differentialgleichungen, Wilhelm-Pieck-University, Rostock, 1978, pp. 133–142.

[221] S. Krushkal and R. Kühnau, *Quasikonforme Abbildungen-neue Methoden und Anwendungen*, Teubner-Texte zur Mathematik, 54, BSB, B. G. Teubner Verlagsgesellschaft, Leipzig, 1983.

[222] H.-J. Kuo and N.S. Trudinger, *New maximum principles for linear elliptic equations*, Indiana Univ. Math. J., **56** (2007), 2439–2452.

[223] M.T. Lacey, E.T. Sawyer and I. Uriarte-Tuero, *Astala's conjecture on distortion of Hausdorff measures under quasiconformal maps in the plane*, Preprint, arXiv:0805.4711 (2008).

[224] M.A. Lavrentiev, *Sur une critère differentiel des transformations homéomorphes des domains à trois dimensions*, Dokl. Acad. Nauk. SSSR, **20** (1938), 241–242.

[225] H. Lebesgue, *Oeuvres scientifiques*, vols. I–V, Sous la rédaction de Francois Châtelet et Gustave Choquet, Institut de Mathématiques de l'Université de Genève, Geneva, 1973.

[226] O. Lehto, *Homeomorphisms with a given dilatation*, Lecture Notes in Math., 118, Springer-Verlag, Berlin-New York, 1970, pp. 58–73.

[227] O. Lehto, *Quasiconformal mappings and singular integrals*, Symposia Math., 18, Academic Press, London, 1976, pp. 423–429.

[228] O. Lehto, *Univalent functions and Teichmüller spaces*, Springer-Verlag, Berlin-New York, 1987.

[229] O. Lehto and K. Virtanen, *Quasiconformal mappings in the plane*, Springer-Verlag, Berlin-New York, 1971.

[230] L.G. Lewis, *Quasiconformal mappings and Royden algebras in space*, Trans. Amer. Math. Soc., **158** (1971), 481–492.

[231] H. Lewy, *On the non-vanishing of the Jacobian in certain one-to-one mappings*, Bull. Amer. Math. Soc., **42** (1936), 689–692.

[232] S. Lie, *Vorlesungen über continuierliche Gruppen mit geometrischen und anderen Anwendungen*, B. G. Teubner, Liepzig, 1893.

[233] J. Liouville, *Théorèm sur l'équation $dx^2 + dy^2 + dz^2 = \lambda(d\alpha^2 + d\beta^2 + d\gamma^2)$*, J. Math. Pures Appl., **1** (15) (1850), 103.

[234] N.G. Lloyd, *Degree theory*, Cambridge University Press, New York, 1978.

[235] A. Lyzzaik, *The modulus of the image annuli under univalent harmonic mappings and a conjecture of J.C.C. Nitsche*, J. London Math. Soc., **64** (2001), 369–384.

[236] J. Malý and O. Martio, *Lusin's condition (N) and mappings of the class $W^{1,n}$*, J. Reine Angew. Math., **458** (1995), 19–36.

[237] R. Mañé, P. Sad and D. Sullivan, *On the dynamics of rational maps*, Ann. Sci. École Norm. Sup., **16** (1983), 193–217.

[238] J. Manfredi, *Weakly monotone functions*, J. Geometric Anal., **3** (1994), 393–402.

[239] P. Marcellini, *Convergence of second order linear elliptic operators*, Boll. Unione Mat. Ital. Sez. B (5), **16** (1979), 278–290.

[240] A. Marino and S. Spagnolo, *Un tipo di approssimazione dell'operatore $\sum_{i,j=1}^{n} D_i(a_{ij}(x)D_j)$ con operatori $\sum_{j=1}^{n} D_j(\beta(x)D_j)$*, Ann. Scuola Norm. Sup. Pisa, **23** (1969), 657–673.

[241] G.J. Martin, *The distortion theorem for quasiconformal mappings, Schottky's theorem and holomorphic motions*, Proc. Amer. Math. Soc., **125** (1997), 1095–1103.

[242] G.J. Martin, *The Teichmüller problem for mean distortion*, Ann. Acad. Sci. Fenn. Math., **34** (2009), 1–15.

[243] O. Martio, S. Rickman and J. Väisälä, *Definitions for quasiregular mappings*, Ann. Acad. Sci. Fenn. Ser. A.I. Math., **448** (1969), 1–40.

[244] O. Martio, S. Rickman and J. Väisälä, *Distortion and Singularities of Quasiregular Mappings*, Ann. Acad. Sci. Fenn. Ser. A. I. Math., **465** (1970), 1–13.

[245] O. Martio, S. Rickman and J. Väisälä, *Topological and metric properties of quasiregular mappings*, Ann. Acad. Sci. Fenn. Ser. A. I. Math., **488** (1971), 1–31.

[246] O. Martio, V. Ryazanov, U. Srebro, and E. Yakubov, *Q-homeomorphisms*, Contemp. Math., **364** (2004), 193–203.

[247] O. Martio, V. Ryazanov, U. Srebro, and E. Yakubov, *On Q-homeomorphisms*, Ann. Acad. Sci. Fenn. Math., **30** (2005), 49–69.

BIBLIOGRAPHY

[248] O. Martio and J. Sarvas, *Injectivity theorems in plane and space*, Ann. Acad. Sci. Fenn. Ser. A. I. Math., **4** (1978/79), 383–401.

[249] B. Maskit, *The conformal group of a plane domain*, Amer. J. Math., **90** (1968), 718–722.

[250] J. Mateu and J. Verdera, L^p *and weak* L^1 *estimates for the maximal Riesz transform and the maximal Beurling transform*, Math. Res. Lett., **13** (2006), 957–966.

[251] P. Mattila, *Geometry of sets and measures in Euclidean spaces*, Cambridge University Press, New York, 1995.

[252] P. Mattila, *On the analytic capacity and curvature of some Cantor sets with non σ-finite length*, Pub. Mat., **40** (1996), 195–204.

[253] G.V. Maz'ja, *Sobolev spaces*, Springer-Verlag, Berlin-New York, 1985.

[254] L.F. McAuley, *Monotone mappings-some milestones*, General topology and modern analysis, Academic Press, New York-London, 1981, pp. 117–141.

[255] C.T. McMullen, *Complex dynamics and renormalization*, Ann. of Math. Stud., 135, Princeton University Press, Princeton, NJ, 1994.

[256] C.T. McMullen, *Renormalization and 3-manifolds which fiber over the circle*, Ann. of Math. Stud., 142, Princeton University Press, Princeton, NJ, 1996.

[257] C.T. McMullen and D. Sullivan, *Quasiconformal homeomorphisms and dynamics. III. The Teichmller space of a holomorphic dynamical system*, Adv. Math., **135** (1998), 351–395.

[258] D. Menchoff, *Sur les differentielles totales des fonctions univalentes*, Math. Ann., **105** (1931), 75–85.

[259] P.A. Meyer, *Transformations de Riesz pour les lois gaussiennes*. Seminaire de Prob. XVIII, Lecture Notes in Math., 1059, Springer-Verlag, Berlin-New York, 1984, 179–193.

[260] N. Meyers and J. Serrin, $H = W$, Proc. Natl. Acad. Sci. USA, **51** (1964), 1055–1056.

[261] L. Migliaccio and G. Moscariello, *Higher integrability of the gradient of solutions of certain P.D.E.'s*, Rend. Accad. Sci. Fis. Mat. Napoli, **65** (1998), 7–10.

[262] J. Milnor, *Dynamics in one complex variable*, 3rd ed., Ann. of Math. Stud., 160, Princeton University Press, Princeton, NJ, 2006.

[263] Y. Minsky, *End invariants and the classification of hyperbolic 3-manifolds*, Current Developments in Mathematics, International Press, Somerville, MA, 2003, pp.181–217.

[264] G.J. Minty, *Monotone (nonlinear) operators in Hilbert space*, Duke Math. J., **29** (1962), 341–346.

[265] D.S. Mitrinovic and P.M. Vasic, *Analytic inequalities*, Springer-Verlag, Berlin-New York, 1970.

[266] D. Montgomery and L. Zippin, *Topological transformation groups*, Interscience, New York, 1955.

[267] F. Morgan, *Geometric measure theory: A beginner's guide*, 2nd ed., Academic Press, San Diego, CA, 1995.

[268] A. Mori, *An absolute constant in the theory of quasiconformal mappings*, J. Math. Soc. Japan, **8** (1956), 156–166.

[269] A. Mori, *On quasiconformality and pseudo-analyticity*, Trans. Amer. Math. Soc., **84** (1957), 56–77.

[270] C.B. Morrey, *The topology of (path) surfaces*, Amer. J. Math., **57** (1935), 17–50.

[271] C.B. Morrey, *On the solutions of quasi-linear elliptic partial differential quations*, Trans. Amer. Math. Soc., **43** (1938), 126–166.

[272] C.B. Morrey, *Quasi-convexity and the lower semicontinuity of multiple integrals*, Pacific J. Math., **2** (1952), 25–53.

[273] C.B. Morrey, *Multiple integrals in the calculus of variations*, Springer-Verlag, Berlin-New York, 1966.

[274] G. Moscariello, *On the integrability of the Jacobian in Orlicz spaces*, Math. Japonica, **40** (1992), 323–329.

[275] J. Moser, *On Harnack's theorem for elliptic differential equations*, Commun. on Pure Appl. Math., **14** (1961), 577–591.

[276] B. Muckenhoupt, *Weighted norm inequalities for the Hardy maximal function*, Trans. Amer. Math. Soc., **165** (1972), 207–226.

[277] B. Muckenhoupt and R.L. Wheeden, *Weighted norm inequalities for singular and fractional integrals*, Trans. Amer. Math. Soc., **161** (1971), 249–258.

[278] S. Müller, *A surprising higher integrability property of mappings with positive determinant*, Bull. Amer. Math. Soc., **21** (1989), 245–248.

[279] S. Müller, T. Qi and B.S. Yan, *On a new class of elastic deformations not allowing for cavitation*, Ann. Inst. H. Poincaré Anal. Non Linéaire, **11** (1994), 217–243.

[280] F. Murat, *Compacité par compensation*, Ann. Sc. Pisa, **5** (1978), 489–507.

[281] F. Murat and L. Tartar, *H-convergence, Topics in the mathematical modelling of composite materials*, 31, Progr. Nonlinear Differential Equations Appl., Birkhäuser, Boston, MA, 1997, pp. 21–43.

[282] A. Nachman, *Global uniqueness for a two-dimensional inverse boundary value problem*. Ann. of Math., **143** (1996), 71–96.

[283] A. Nachman, *Reconstructions from boundary measurements*, Ann. of Math., **128** (1988), 531–576.

[284] A. Nachman, J. Sylvester and G. Uhlmann, *An n-dimensional Borg-Levinson theorem*. Comm. Math. Phys., **115** (1988), 595–605.

[285] R. Nevanlinna, *Analytic functions*, Grundlehren der mathematischen Wissenshaften, 162, Springer-Verlag, Berlin-New York, 1970.

[286] L. Nirenberg, *On nonlinear elliptic partial differential operators and Hölder continuity*, Comm. Pure Appl. Math., **6** (1953), 103–156.

[287] J.C.C. Nitsche, *On the modulus of doubly connected regions under harmonic mappings*, Amer. Math. Monthly, **69** (1962), 781–782.

[288] E. Noether and W. Schmeidler, *Modul in nichtkommutativen Bereichen, insbesondere aus Differential- und Differenzenausdrcken*. Math. Z., **1** (1918), 1–35.

[289] K. Oikawa, *Welding of polygons and the type of Riemann surfaces*, Kodai Math. Sem. Rep., **13** (1961), 37–52.

[290] B. Palka, *An introduction to complex function theory*, Undergraduate Texts in Math. Springer-Verlag, Berlin-New York, 1991.

[291] L. Päivärinta, A. Panchenko and G. Uhlmann, *Complex geometrical optics solutions for Lipschitz conductivities*, Rev. Math. Iberoamericana, **19** (2003), 57–72.

[292] M. Pavlović, *Boundary correspondence under harmonic quasiconformal homeomorphisms of the unit disk*, Ann. Acad. Sci. Fenn. Math., **27** (2002), 365–372.

[293] H.O. Peitgen and P.H. Richter, *The beauty of fractals; Images of complex dynamical systems*, Springer-Verlag, Berlin-New York, 1986.

[294] S. Petermichl and A. Volberg, *Heating of the Ahlfors-Beurling operator: Weakly quasiregular maps on the plane are quasiregular*, Duke Math. J., **112** (2002), 281–305.

[295] A. Pfluger, *Une propriété métrique de la représentation quasi conforme*. C. R. Acad. Sci., Paris, **226** (1948), 623–625.

[296] A. Pfluger, *Sur une propriété de l'application quasi conforme d'une surface de Riemann ouverte*, C. R. Acad. Sci., Paris, **227** (1948), 25–26.

[297] A. Pfluger, *Über die Konstruction Riemannischer Flächen durch Verheftung*, J. Indian Math. Soc., **24** (1961), 401–412.

[298] É. Picard, *Sur une propriete des fonctions entiéres*, C.R. Acad. Sci. Paris, **88** (1879), 1024–1027.

[299] S.K. Pichorides, *On the best values of the constants in the theorem of M. Riesz, Zygmund and Kolmogorov*, Studia Math., **44** (1972), 165–179.

[300] G. Pisier, *Riesz transforms: A simpler analytic proof of P.A. Meyer's inequality*, Lecture Notes in Math., 1321, Springer-Verlag, Berlin-New York, 1988, pp. 485–501.

[301] H. Poincaré, *Theorie des groupes Fuchsiens*, Acta Math., **1** (1882), 1–62.

[302] I. Prause, *A remark on quasiconformal dimension distortion on the line*, Ann. Acad. Sci. Fenn. Math., **32** (2007), 341–352.

[303] F. Przytycki and S. Rohde, *Rigidity of holomorphic Collet-Eckmann repellers*, Ark. Mat., **37** (1999), 357–371.

[304] C. Pucci, *Limitazioni per soluzioni di equazioni ellittiche*, Ann. Mat. Pura. Appl. (4), **74** (1966), 15–30.

[305] C. Pucci, *Operatori ellittici estremanti*, Ann. Mat. Pura. Appl. (4), **72** (1966), 141–170.

[306] T. Radó, *Aufgabe 41*, Jahresber. Deutsch. Math.-Verein., **35** (1926), 49.

[307] M.M. Rao and Z.D. Ren, *Theory of Orlicz spaces*, Pure and Applied Math., 146, John Wiley, New York, 1991.

[308] E. Reich, *Some estimates for the two dimensional Hilbert transform*, J. Anal. Math., **18** (1967), 279–293.

[309] M. Reimann, *Functions of bounded mean oscillation and quasiconformal mappings*, Comment. Math. Helv., **49** (1974), 260–276.

[310] M. Reimann, *Ordinary differential equations and quasiconformal mappings*, Invent. Math., **33** (1976), 247–270.

[311] Yu. G. Reshetnyak, *Space mappings with bounded distortion*, Translations of Mathematical Monographs, 73, American Mathematical Society, Providence, RI, 1989.

[312] S. Rickman, *Quasiregular mappings*, Springer-Verlag, Berlin-New York, 1993.

[313] T. Rivière and D. Ye, *Resolutions of the prescribedd volume form equation*, Nonlinear Differential Equations Appl., **3**, (1996), 323–369.

[314] R. Rochberg and G. Weiss, *Analytic families of Banach spaces and some of their uses*, Recent Progress in Fourier Analysis, North-Holland Math. Stud., 111, North-Holland, Amsterdam-New York, 1985, pp. 173–201.

[315] V. Ryazanov, U. Srebro and E. Yakubov, *BMO-quasiconformal mappings*, J. Anal. Math., **83** (2001), 1–20.

[316] S. Saks, *Theory of the integral*, 2nd ed., Dover, New York, 1964.

[317] D. Sarason, *Functions of vanishing mean oscillation*, Trans. Amer. Math. Soc., **207** (1975), 391–405.

[318] C. Sbordone, *Alcune questioni di convergenza per operatori differenziali del 2° ordine*, Boll. Unione Mat. Ital., **10** (1974), 672–682.

[319] C. Sbordone, *Rearrangement of functions and reverse Hölder inequalities*, Ernio De Giorgi Colloquium, Res. Notes in Math., 124, Pitman, Boston, 1985, pp. 139–148.

[320] J. Schauder, *Der Fixpunktsatz in Funktionalräumen*, Studia Math., **2** (1930), 171–180.

[321] J. Schauder, *Über lineare elliptische Differentialgleichungen zweiter ordung*, Math. Z., **38** (1934), 257–282.

[322] J. Schauder, *Numerische Abschätzungen in elliptischen linearen Differentialgleichungen*, Studia Math., **5** (1935), 34–42.

[323] R. Schoen and S.T. Yau, *On univalent harmonic maps between surfaces*, Invent. Math., **44** (1978), 265-278

[324] L. Schwartz, *Théorie des distributions*, Vols. I and II, Act. Sci. Ind., 1091, 1122, Hermann et Cie, Paris, 1951.

[325] J. Serrin, *Local behaviour of solutions of quasilinear equations*, Acta Math., **111** (1964), 247–302.

[326] M. Shishikura, *On the quasiconformal surgery of rational functions*, Ann. Sci. École Norm. Sup., **20** (1987), 1–29.

[327] M. Shishikura, *Complex dynamics and quasiconformal mappings*, additional chapter to L.V. Ahlfors' *Lectures on Quasiconformal Mappings*, 2nd ed., Amer. Math. Soc. University Lecture Series, American Mathematical Society, Providence, RI, 2006, pp. 119–143.

[328] S. Siltanen, J. Müller and D. Isaacson, *Reconstruction of high contrast 2-D conductivities by the algorithm of A. Nachman*, Radon Transforms and Tomography, Contemp. Math., **278** (2001), 241–254.

[329] Z. Słodkowski, *Holomorphic motions and polynomial hulls*, Proc. Amer. Math. Soc., **111** (1991), 347–355.

[330] Z. Słodkowski, *Holomorphic motions commuting with semigroups*, Studia Math., **119** (1996), 1–16.

[331] A.I. Šnirel'man, *The degree of a quasiruled mapping and the nonlinear Hilbert problem*, Mat. Sb. (N.S.), **89 (131)** (1972), 366–389; or USSR-Sb. **18** (1973), 373–396.

[332] S. Spagnolo, *Sulla convergenza di soluzioni di equazioni paraboliche ed ellittiche*, Ann. Scuola Norm. Sup. Pisa, **22** (1968), 571–597.

[333] U. Srebro and E. Yakubov, *Branched folded maps and alternating Beltrami equations*, J. Anal. Math., **70** (1996), 65–90.

[334] E.M. Stein, *Note on the class $L \log L$*, Studia Math., **32** (1969), 305–310.

[335] E.M Stein, *Singular integrals and differentiability properties of functions*, Princeton University Press, Princeton, NJ, 1970.

[336] E.M. Stein, *Harmonic Analysis*, Princeton University Press, Princeton, NJ, 1993.

[337] N. Steinmetz, *Rational iteration: Complex analytic dynamical systems*, de Gruyter Studies in Math., 16, Walter de Gruyter, Berlin, 1993.

[338] S. Stoilow, *Brief summary of my research work*, Analysis and topology, A volume dedicated to the memory of S. Stoilow, World Scientific Publishing, River Edge, NJ, 1998.

[339] K. Strebel, *Extremal quasiconformal mappings*, Results Math., **10** (1986), 168–210.

[340] D. Sullivan, *On the ergodic theory at infinity of an arbitrary discrete group of hyperbolic motions*, Riemann surfaces and related topics, Ann. of Math. Stud. 97, Princeton University Press, Princeton, NJ, 1981, pp. 465–496.

[341] D. Sullivan, *Quasiconformal homeomorphisms and dynamics I, II*, Ann. of Math., **122** (1985), 401–418; II. Acta Math., **155** (1985), 243–260.

[342] D. Sullivan and W. Thurston, *Extending holomorphic motions*, Acta Math., **157** (1986), 243–257.

[343] V. Šverák, *Examples of rank-one convex functions*, Proc. Roy. Soc. Edinburgh, Sect. A, **114A** (1990), 237–242.

[344] V. Šverák, *Rank-one convexity does not imply quasiconvexity*, Proc. Roy. Soc. Edinburgh Sect. A, **120** (1992), 185–189.

[345] V. Šverák, *New examples of quasiconvex functionals*, Arch. Ration. Mech. Anal., **119** (1992), 293–300.

BIBLIOGRAPHY

[346] J. Sylvester and G. Uhlmann, *A global uniqueness theorem for an inverse boundary value problem*, Ann. of Math., **125** (1987), 153–169.

[347] T. Tao and Ch. Thiele, *Nonlinear Fourier analysis*, Lecture Notes; LAS Park City Summer School, 2003.

[348] L. Tartar, *Compensated compactness and applications to partial differential equations*, Nonlinear Analysis and Mechanics, Res. Notes in Math., 39, Pitman, Boston, Mass.-London, 1979, pp. 136–212.

[349] O. Teichmüller, *Untersuchungen über konforme und quasiconforme Abbildung*, Deutsche Math. **3** (1938), 621–678; or *Collected works*, Springer-Verlag, Berlin-New York, 1982.

[350] O. Teichmüller, *Extremale quasiconforme Abbildungen und quadraticshe Differentiale*, Abh. Preuss. Akad. Wiss., Math.-Naturw. Kl, **22** (1939), 1–197; or *Gesammelte Abbildungen-Collected papers*, Springer-Verlag, Berlin-New York, 1982.

[351] G.O. Thorin, *Convexity theorems generalizing those of M. Riesz and Hadamard with some applications*, Comm. Sem. Math. Univ. Lund, **9** (1948), 1–58.

[352] W.P. Thurston, *Three-dimensional Geometry and Topology*, Princeton University Press, Princeton, NJ, 1997.

[353] X. Tolsa, *Painlevé's problem and the semiadditivity of analytic capacity*, Acta Math., **190** (2003), 105–149.

[354] H. Triebel, *Interpolation theory, function spaces, differential operators.* North-Holland Mathematical Library, 18, North-Holland, Amsterdam-New York, 1978.

[355] N.S. Trudinger, *On imbedding into Orlicz spaces and some applications*, J. Math. Mech., **17** (1967), 473–483.

[356] P. Tukia, *On Two Dimensional Quasiconformal Groups*, Ann. Acad. Sci. Fenn. Ser. A. I. Math., **5** (1980), 73–78.

[357] P. Tukia, *Quasiconformal extension of quasisymmetric mappings compatible with a Möbius group*, Acta Math., **154** (1985), 153–193.

[358] P. Tukia, *Compactness properties of μ-homeomorphisms*, Ann. Acad. Sci. Ser. A. I. Math., **16** (1991), 47–69.

[359] P. Tukia and J. Väisälä, *Quasisymmetric embeddings of metric spaces*, Ann. Acad. Sci. Fenn. Ser. A. I. Math., **5** (1980), 97–114.

[360] P. Tukia and J. Väisälä, *Lipschitz and quasiconformal approximation and extension*, Ann. Acad. Sci. Fenn. Ser. A. I. Math., **6** (1981), 303–334.

[361] A. Uchiyama, *On the compactness of operators of Hankel type*, Tohoku Math. J., **30** (1978), 163–171.

[362] K. Uhlenbeck, *Regularity for a class of non-linear elliptic systems*, Acta Math., **138** (1977), 219–250.

[363] I. Uriarte-Tuero, *Sharp examples for planar quasiconformal distortion of Hausdorff measures and removability*, Int. Math. Res. Notices IMRN, (2008), 43 pp.

[364] J.V. Vainio, *Conditions for the possibility of conformal sewing*, Ann. Acad. Sci. Fenn., Diss., **43** (1985).

[365] J. Väisälä, *Lectures on n-dimensional quasiconformal mappings*, Lecture Notes in Math., 229, Springer-Verlag, Berlin-New York, 1972.

[366] I.N. Vekua, *Generalized analytic functions*, Pergamon Press, Oxford, 1962.

[367] M. Vuorinen, *Conformal geometry and quasiregular mappings*, Lecture Notes in Math., 1319, Springer-Verlag, Berlin-New York, 1988.

[368] A. Weitsman, *Univalent harmonic mappings of annuli and a conjecture of J.C.C. Nitsche*, Israel J. Math., **124** (2001), 327–331.

[369] H. Weyl, *Zur Infinitesimalgeometrie; Einordnung der projektiven und conformalen Auffassung*, Göttingen Nachr, (1922), 99–112.

[370] G.T. Whyburn, *Analytic topology*, Amer. Math. Soc. Colloq. Publ., vol. XXVIII, American Mathematical Society, Providence, RI, 1963.

[371] G.T. Whyburn, *Monotoneity of limit mappings*, Duke Math. J., **29** (1962), 465–470.

[372] M. Wolf, *The Teichmller theory of harmonic maps*, J. Differential Geom., **29** (1989), 449–479.

[373] B. Yan, *On the weak limit of mappings with finite distortion*, Proc. Amer. Math. Soc., **128** (2000), 3335–3340.

[374] E. Zeidler, *Nonlinear functional analysis and its applications*, I. Fixed-point theorems, Springer-Verlag, Berlin-New York, 1986.

[375] K. Zhang, *Biting theorems for Jacobians and their applications*, Ann. Inst. H. Poincaré Anal. Non Linéaire, **7** (1990), 345–365.

[376] W.P. Ziemer, *Weakly differentiable functions*, Graduate Texts in Math., 120, Springer-Verlag, New York, 1989.

[377] M. Zinsmeister, *Thermodynamic formalism and holomorphic dynamical system*, SMF/AMS Texts and Monographs, 2, 2000.

Index

A-Harmonic operator, 405
A_p-weight, 130
absolute continuity, 631
ACL (absolute continuity on lines), 631
Ahlfors, 27, 37, 40, 50, 162, 189, 337
Ahlfors' three point property, 333
Alessandrini, 204
Alexandrov, 472
analytic dependence on parameters, 188
area distortion
    theorem, 325
    weighted, 320
area formula, 41
argument principle, 41
Arzela-Ascoli theorem, 39

$BMO(\Omega)$, 104
$BMO_2(\mathbb{C})$, 135
$BV(\Omega, \mathbb{C})$, 620
Baernstein, 446, 521
Ball's theorem, 520
Banach contraction principle, 240
Beals, 499
Becker, 335
Beltrami
    operator, 172, 363, 456
    reduced equation, 197
    system, 14, 275
Beltrami coefficient, 27
    composition formula, 182, 281
Bernoulli's theorem, 421
Bers, 28
Beurling transform, 94
    $\mathcal{S}_\Delta$, 155, 157
    Hölder estimates, 147
    in Ω, 151
    maximal, 97

    for Dirichlet problem, 155, 157
    in multiply connected domains, 159
    of $\chi_\mathbb{D}$, 96
    pointwise convergence, 97
Beurling-Ahlfors extension, 192
bilipschitz mapping, 32
Bishop, 194
Bloch's theorem, 40
Bojarski, 6, 162, 174, 231
Bojarski's theorem, 174
bounded variation, 620
Bourgain's theorem, 122
Brown, 491
Burkholder's theorem, 521

$C^{\ell,\alpha}$-regularity, 257, 390, 432, 446, 482
$C_0(\hat{\mathbb{C}})$, 99
Caccioppoli inequality, 174, 476
    second-order, 477
Calderón, 94, 490, 491
Calderón's problem, 490
Cantor
    function, 56
    set, 32
capacity
    BMO, 354
    analytic, 348
Carathéodory, 30, 70, 193
Carathéodory condition, 238
Cauchy principal value, 94
Cauchy problem, 289
Cauchy transform, 93
    $\mathcal{C}_\Omega$, 157
    formal adjoint, 94
    modified, 131
    for Dirichlet problem, 155, 157
    in $\mathbb{D}$, 152

672   INDEX

   in multiply connected domains, 159
Cauchy-Riemann
  equations, 4, 10
  inhomogeneous equation, 149
  operators, 95, 213
  polar coordinates, 611
Chirka, 298
Choquet-Radó-Kneser theorem, 587
circular distortion, 81
coercive, 442
Coifman, 123, 130, 499, 546
Coifman-Fefferman theorem, 130
Coifman-Rochberg-Weiss theorem, 143
commutators, 143
compact embedding, 637
compact operator, 116
complex dilatation, 27
complex geometric optics solutions, 493
complex gradient, 210
complex Riesz transforms, 125
composition formula, 182, 281
condition $\mathcal{N}$, 72
condition $\mathcal{N}^{-1}$, 72
conformal, 22
  deformation, 275
  equivalence, 14, 279
  mapping, 36
  structure, 14
conformal modulus, 583, 606
conformal welding, 193
conjugate
  A-harmonic, 408
  p-harmonic, 426
  equation, 409
content
  s-dimensional, 352
continuity equation, 420
convolution, 626
critical interval, 172, 364

$\mathcal{D}'(\Omega, \mathbb{V})$, 625
David, 554
deformation, 314
degree of a mapping, 32
dilatation
  antisymmetric, 337

complex, 27
dimension
  Hausdorff, 31
  of quasicircles, 336
Dirac, 624
  delta, 625
directional derivatives, 25
Dirichlet integral, 3, 419, 610
Dirichlet problem, 231, 441, 449, 486, 490
Dirichlet-to-Neumann map, 491
discrete mapping, 56, 178
distortion, 24
  finite, 528
  average, 606
  function, 379
  inequality, 24
  linear, 21, 587
  of Hausdorff dimension, 332
  of Hausdorff measure, 352
  tensor, 22
distribution, 624, 627
  positive, 627
  regular, 627
distributional derivatives, 628
div-curl couple, 415
Douady, 8, 161, 191
Douady-Earle extension, 191
Douglas, 595
Duren, 587

Earle, 191
elliptic
  degenerate, 222
  strongly, 212
  uniformly, 222
  uniformly, nonlinear case, 228
  uniformly, quasilinear case, 223
  uniformly, second-order, 233
ellipticity
  of linear differential operator, 212
embedding, 637
energy, A-harmonic, 406
equicontinuity, 39
Euler-Lagrange equations, 4, 5, 603, 604
exact packing, 567

INDEX

existence of solutions
    linear equations, 369, 486
    nonlinear equations, 237, 362, 442
extension theorem
    Beurling-Ahlfors, 192
    Douady-Earle, 191
    Jerison-Kenig, 189

Fabes-Stroock theorem, 206
factorization, 184
Faraco, 482
Fatou set, 297
Fefferman, 130, 546
finite distortion, 528
Fourier
    transform, 98
    multiplier, 101
    transform, nonlinear, 500
free Lagrangians, 612
Frostman's lemma, 354
Fuchsian group, 17

G-compact, 457
G-convergence, 456
G-limit, 457
$GL(2, \mathbb{R})$, 18
Gehring, 28, 316, 532
Gehring-Lehto theorem, 53
generalized analytic, 289
geodesic, 37
Goldstein, 530
good approximation lemma, 171
Grötzsch, 27
Grötzsch Problem, 588
gradient field, 445
gradient quasiregular mappings, 470
Green's formula, 35
Green's function, 156
Grunsky, 40

Haïssinsky, 9
Hadamard's inequality, 4
Hamilton, 194
Hardy space $H^1(\mathbb{C})$, 108
Hardy-Littlewood-Sobolev theorem,
    $1 < p < 2$, 110

Harmonic factorization, 419
Harnack inequality, 419
Hausdorff
    dimension, 30
    measure, 30
Hausdorff-Young inequality, 100
Hecke identities, 101
Heinonen, 48
Heinonen-Koskela theorem, 25
Hencl, 592
Hessian matrix, 473
Hilbert
    adjoint, 95
    transform, 495
Hilbert-Schmidt norm, 409
Hodge
    conjugate operator, 434
    star, 408
hodograph transform, 423, 430
Hölder
    regularity of solutions, 390
    weak reverse Hölder inequality, 477
    reverse inequalities, 207, 329, 640
    space, 115
holomorphic axiom of choice, 301
holomorphic motion, 294
homogeneous Sobolev space, 77, 406, 628
homotopic operators, 217
Hubbard, 161
Hurwitz's theorem, 40
hyperbolic
    domain, 37
    metric, 15

impedance tomography, 490
incompressible flow, 421
integrability
    borderline, 330
    exponential, 387
    higher, 328
    of the Jacobian, 524
interior regularity, 481
interpolation, 117
inverse conductivity problem, 491
isoperimetric inequality, 80

Jääskeläinen, 204
Jørgensen, 296
Jacobian, 21
    fundamental inequality for, 515
Jerison-Kenig extension, 189
John-Nirenberg inequality, 132
Jordan domain, 193, 259

Kaufman, 354
Kleinian group, 296
Kneser, 587
Koebe
    $\frac{1}{4}$-theorem, 44
    distortion theorem, 45
    transform, 45
Koskela, 48, 592
Kovalev, 86, 446, 560
Král-Kaufman theorem, 354

$L_\omega^p(\mathbb{C})$, 130
$L^{2\pm}(\mathbb{C})$, 249
$L_f^\varepsilon(z)$, 59
λ-lemma
    extended, 298
    Mañé-Sad-Sullivan, 294
Lacey, 352
Laplacian component, 231
Lavrentiev, 6, 591
Lehto, 28, 52
    condition, 581
Lehto's theorem, 584
Lewy, 587
lim inf theorem, 25
linear distortion, 21
linear partial differential operators
    homotopic, 217
Lions, 546
Liouville theorem, 176
Lipschitz mappings, 32
Loewner property, 583
logarithmic potential, 92
    for Dirichlet problem, 155, 156
Lusin
    condition $\mathcal{N}$, 72, 74
    condition $\mathcal{N}^{-1}$, 72, 75
    measurable, 224
    theorem, 72, 223
Lusin-Egoroff convergence, 244

Manfredi, 532
Marcellini, 461
Marcinkiewicz
    interpolation, 119
    multiplier theorem, 101
    space, 109, 124
Mateu, 97
Mattila, 30, 357
maximal derivative, 59
maximal function
    Fefferman-Stein, 122
    Hardy-Littlewood, 121
    sharp, 122
    spherical, 121
    mollified, 638
maximum principle
    Alexandrov, 473
    sharp, 478
mean distortion, 588
metric
    Riemannian, 13
Meyer, 546
Minkowski, 31
Minty-Browder theorem, 442
modulus
    method, 420, 583
    of an annulus, 583
mollifiers, 626
moment, 108
monodromy theorem, 167
monotone
    δ-, 84
    operators, 442
    strictly, 442
    topologically, 531
    weakly, 532
Montel's theorem, 40
Montgomery-Smith, 521
Mori's theorem, 81
Morrey, 6, 80, 161, 518
Morrey conjecture, 520
Muckenhoupt weight, 130
multiplier, 101

INDEX

$\mathcal{N}$, 72
$\mathcal{N}^{-1}$, 72
Nachman, 491
Nesi, 204
Neumann
    problem, 449
    series, 163
Newtonian
    potential, 104
Nirenberg, 6
Nitsche
    bound, 607
    conjecture, 607
    map, 608
normal family, 39
normalized solution, 170, 201, 238, 578
    uniqueness, 181, 201, 256
null Lagrangian, 469, 519

Onninen, 592
open mapping, 52
operator, 172, 363, 456
optimal L^p-regularity, 328, 369
orientation-preserving, 33
Orlicz spaces L^Φ and L^Ψ, 437
oscillation lemma, 65, 532

p-Laplacian, 423
p-harmonic equation, 210, 423
Painlevé's theorem, 349
Parseval identity, 99
Pavlović, 599
Petermichl-Volberg inequality, 130
Petrovsky elliptic, 226
Pfluger, 27
Poincaré
    disk, 15
    inequality, 633
Poincaré-Sobolev inequality, 633
polyconvex, 519
Pommerenke, 335
potential equation, 422
potential function, 476
primary solutions, 462
principal
    mapping, 318
    solution, 165, 235
    solution, degenerate equations, 529
    symbol, 212
pseudoanalytic, 289
Pucci
    conjecture, 474
    example, 488

Q-homeomorphism, 581
quasicircle, 309, 333
quasiconformal, 24
    analytic definition, 24, 48
    field, 415
    geometric definition, 23
    group, 285
    local compactness of families, 49
    optimal regularity, 328
    reflection, 192
quasiconvex, 519
quasilinear operator, 222
quasiregular
    gradient field, 445
    linear families, 464
    mapping, 25
    weakly, 178
quasisymmetric, 50
    extension theorem, 189
    weakly, 69, 85

$R(\Omega)$, 77
Radó, 587
radial stretchings, 28
rank-one convex, 520
reduced Beltrami equation, 197
reflection principle, 192
regular set, 465
Reich, 28, 316
Reimann's theorem, 343
Rellich-Kondrachov, 637
Riemann mapping theorem, 36, 259
Riemann-Hilbert problem, 451, 452
Riemannian
    metric, 13
    structure, 12, 279
Riesz
    complex transforms, 125

potential, 104
representation theorem, 626
transform, 103
Riesz-Thorin interpolation, 117
Rochberg, 123, 143
Rohde, 336
Royden Algebra, 77
Ruelle, 335

$\mathbf{S}(2)$, 12
$\mathscr{S}(\mathbb{C})$, 98
Šverák, 520
Słodkowski's, 298
Saksman, 362
Sarason, 639
Sawyer, 352
Sbordone, 204
Schauder estimates, 389
Schauder fixed-point theorem, 240
Schottky domain, 270
Schrödinger, 624
Schwartz class, 98
Schwarz-Pick lemma, 37
Semmes, 546
Serrin, 6, 411
sewing, 193
Shoen, 586
Sierpiński gasket, 31
singular set, 465
Smirnov, 336
Sobolev, 624
    mapping, 25, 55
    space, 628
    space, homogeneous, 77, 406, 628
solution
    normalized, 170
    principal, 165, 235
Spagnolo, 457
spherical metric, 39
Stein, 94, 637, 638
Stoilow factorization, 56, 179, 376, 565
    for elliptic systems, 196
stream function, 408, 422
Strebel, 28
strongly elliptic, 212
structural field, 433

δ-monotone, 435
sub-additive, 119
subexponential growth, 504
Sullivan, 8, 285, 293
Sylvester, 491
system, 14, 275
Székelyhidi, 482

Tao, 492
Teichmüller, 27
Thiele, 492
Tolsa, 351, 357
total variation of argument, 269
Trudinger inequality, 635
Tukia, 50, 189, 285

Uchiyama, 145
Uhlmann, 491
uniformization theorem, 36
uniformly, 222
upper gradient, 59
Uriarte-Tuero, 352, 361

$VMO(\mathbb{C})$, 105
$VMO(\hat{\mathbb{C}})$, 382
$VMO_*(\mathbb{C})$, 105
Väisälä, 50
Vainio, 194
Vamanamurthy, 312
Varg, 269
variable
    dependent, 222
    independent, 222
Vekua, 289
Verdera, 97
Vodop'yanov, 530
Volberg, 130
volume derivative, 57
Vuorinen, 312

$W_{\mathbf{n}}^{1,2}(\Omega)$, 450
$\mathbb{W}^{1,p}(\Omega)$, 77, 406
weak convergence, 636
weak-L^p, 61, 109
weakly
    quasiregular, 178

INDEX

quasisymmetric, 69
weight, 130, 320
weighted L^p-spaces, 130
Weiss, 143
welding, 193
Weyl lemma, 636
Whitney decomposition, 242
Whyburn, 531
winding number, 32

$X[G]$, 18
x-density, 52

Yau, 586
Young's inequality, 439

Zinsmeister, 335
Zygmund, 94